OF TRANSMISSION (1000–1500)

Florence, Italy outlaws Hindu-Arabic numerals (1299)
Chu Shih-Chieh — binomial coefficients (1303)
Ockham — formal logic (1320)
Burley — formal logic (1325)
Oresme — fractional exponents (1360)

EARLY MODERN PERIOD (1450 TO 1800)

Regiomontanus — trigonometry (1464)
Hindu-Arabic numerals take present form (1479)
Chuquet — notation for +, – of fractions (1484)
Widmann — addition (+), subtraction (–) signs (1489)
Calandri — long division process (1491)
Rudoff — square root symbol (√) (1525)
Apianus — notation for ×, ÷ of fractions (1527)
Cardano — cubic equations (1545)
Rheticus — right triangle trigonometry (1551)
Diez Feile — writes 1st math book in Mexico (1556)
Recorde — symbol for the equal sign (=) (1557)
Bombelli — imaginary numbers (1572)
Clavius — the dot (•) for multiplication (1583)
Stevin — decimal fractions (1585)
Napier — logarithms (1614)
Oughtred — multiplication sign (×), slide rule (1621)
Kepler and Briggs — tables of logarithms (1624)
Cavalieri — indivisibles in calculus (1635)
Descartes — analytic geometry (1637)
Desargues — projective goemetry (1640)
Fermat — probability, analytic geometry, number theory, calculus (1629–1654)
Pascal — projective geometry, calculating machine, probability, binomial coefficients (1642–1654)
Wallis — negative exponents, infinity symbol (∞) (1655)
Rahn — division sign (÷) (1659)
Isomura — magic circles (1660)
Barrow — calculus, differentiation (1663)
Leibniz — symbolic logic, calculating machine, calculus (1666–1684)
Seki Kōwa — determinants, magic squares (1685)
Newton — calculus (1687)
William Jones — symbol for pi (π) (1706)
Jakob Bernoulli — theory of probability (1713)
First math book printed in America (1719)
DeMoivre — actuary math, complex numbers (1720)
Greenwood — writes 1st American math book (1729)
Maria Agnesi — popular math text book (1748)
Achenwall — the word *statistik* (1749)
Gabriel Cramer — systems of equations (1750)
Euler — uses the symbols *i*, *e*, and Σ (1750)
Bayes — statistics, origin of polls (1763)
First Bank of the U.S. (1791)

MODERN PERIOD (1800 TO PRESENT)

Gauss — statistics, normal curve, the term *complex number* (1809–1832)
LaPlace — probability theory (1814)
Babbage — steam powered calculator (1823)
Lobachevsky — non-Euclidean geometry (1829)
Carl Jacobi — theory of determinants (1841)
Weierstrass — absolute value symbol (| |), fo[r] calculus (1841–1874)
Ada Byron — computer programming (1843)
George Boole — logic, Boolean algebra (1847)
Georg Riemann — non-Euclidean geometry, calculus (1850–1854)
Möbius — topology (1865)
John Venn — logic, Venn diagrams (1880)
Hollerith — electrical tabulating device (1880
Burroughs — practical adding machine (1894)
Markov — probability theory, Markov Chains
Whitehead, Russell — Principia Mathematica (1910)
Federal Reserve Act — U.S. banking system (
Lukasiewicz, Post, Wittgenstein — truth tables (1920)
Ronald Fisher — statistics sampling technique
Thomas Watson forms IBM (1924)
Mauchley and Eckert — ENIAC computer (19
Gertrude Cox — statistical design (1950)
Kilby, Texas Instruments — integrated circuit (1958)
Mandelbrot — fractal geometry (1967)
Gilbert Hyatt — computer microprocessor chip (1968)
Edward Roberts — first personal computer (1971)
Hewlett Packard — programmable calculator (
Seymour Cray — CRAY super computer (197
Chudnovsky — 480 million digits of π (1989)
227,832 digit prime found on Cray computer (
Slowinski) (1992)
Proposed proof of Fermat's Last Theorem by Andrew Wiles (1993)

# The Mathematical Palette

# The Mathematical Palette

*Second Edition*

Ronald Staszkow · Robert Bradshaw
Ohlone College

**Saunders College Publishing**
**Harcourt Brace and Company**
Fort Worth / Philadelphia / San Diego / New York / Orlando / Austin
San Antonio / Toronto / Montreal / London / Sydney / Tokyo

Text Typeface: Times Roman
Compositor: CRWaldman
Acquisitions Editor: Deirdre Lynch
Developmental Editor: Anita M. Fallon
Managing Editor: Carol Field
Project Editor: Laura Shur
Copy Editor: Sue Nelson
Manager of Art and Design: Carol Bleistine
Art Director: Robin Milicevic
Art and Design Coordinator: Sue Kinney
Cover Designer: Nicky Lindeman
Text Artwork: Grafacon
Director of EDP: Tim Frelick
Production Manager: Carol Florence
Marketing Manager: Monica Wilson

Cover Credit: Jim Goldsmith/Photonica

Printed in the United States of America

The Mathematical Palette, 2/e

ISBN: 0-03-000897-2

Library of Congress Catalog Card Number: 94-065398

90123 039 109876

This book was printed on acid-free recycled content paper, containing
**MORE THAN**
**10% POSTCONSUMER WASTE**

# Preface

*The Mathematical Palette 2/e* makes mathematics enjoyable, practical, understandable, and informative, stimulating the creativity of the liberal arts student. Just as an artist mixes paints on a palette, so *The Mathematical Palette* mixes the history of mathematics, its mathematicians, and its problems with a variety of real-life applications. The text presents this sampling of mathematics in a straightforward, interesting manner and is designed expressly for the liberal arts student, not for a mathematician.

*The Mathematical Palette* is intended for the liberal arts student who has a background in high school algebra or who has completed intermediate algebra in college. Students with a beginning algebra background should review the material covered in the Chapter Preliminaries sections found in most chapters. The text is designed not only to meet college general education requirements but also to help generate a positive attitude toward and an interest in mathematics. The text stresses learning mathematics rather than just learning about mathematical ideas. The intent is the development of problem-solving skills by actually solving problems. We feel that a greater appreciation of the beauty and power of mathematics is gained when students become active participants.

*The Mathematical Palette* provides additional material not ordinarily included in mathematics texts. The Brief Histories, Check Your Reading Questions, Research Questions, and Projects provide students with the opportunity to write about mathematics and mathematicians. The bibliography of sources used in preparing this text is included at the end of the book and gives students guidance in further investigating the historical aspects of mathematics. The exercises at the end of each section are divided into three categories, Explain, Apply, and Explore. The Explain questions ask the student to give written answers explaining the what, how, or why of topics covered in the section. The Apply questions ask the student to solve problems with a direct application of the methods presented in the section. The Explore problems require that the student go beyond the basic problems by exploring other applications or implications of the concepts of the section. In general, the Explore problems are of a higher degree of difficulty than the Apply problems and ask the student to look a little deeper into the material.

## Features

Instructors will find *The Mathematical Palette* a very teachable text, and students will find that it is a very readable text. Its organization, style, and format have been developed with both the student and the instructor in mind. Each chapter is a self-contained unit and may be taught in any order according to the instructor's or student's interests. The history section at the beginning of each chapter presents material and questions that are ideal for classroom discussions, research papers, projects, speeches, and presentations. The text's readable style, clear explanations, numerous solved examples, Explain-Apply-Explore problem sets, chapter summaries, and accurate answer section are valuable resources for the instructor and the student. In particular, the following features of each chapter are real assets to teaching and learning:

- Overviews describe and introduce the chapter.
- Interviews with successful people describe why the mathematics in the chapter is important in the real world.
- Short history sections present the development of the mathematics and introduce students to important dates, mathematicians, and events.
- Questions at the end of each history section review the content of the section and relate mathematical events to other historical milestones.
- Research questions and projects at the end of each history section go beyond the material and serve as an ideal source for projects, written papers, and are ideal for group cooperative learning experiences.
- Preliminary algebra concepts needed in a chapter are presented at the beginning of the chapter.
- Clear explanations along with worked examples and illustrations are found throughout the text.
- Explain-Apply-Explore problem sets include a variety of writing and skill-building exercises.
- Summaries at the end of each chapter present terminology, formulas, and objectives, and serve as a comprehensive review of the chapter.
- The full-color format highlights important concepts and formulas, and provides students with visual guideposts.
- Artwork and photographs that relate to the mathematics being discussed make the text visually stimulating.
- Consistent use of scientific calculators eliminates the time spent on paper-and-pencil computation.

## Changes in the Second Edition

Suggestions from the reviewers and users of the first edition of *The Mathematical Palette* have prompted the following changes in the second edition. These changes make the text even more alive to the liberal arts student by emphasizing not only problem solving but writing and critical thinking.

- Problems sets are organized into Explain-Apply-Explore categories.
- Each chapter begins with an overview.
- An interview with a successful person supports the importance of studying the mathematics contained in each chapter.
- More real-life problems and data are used throughout the text.
- Projects have been included in each chapter that can be used for group or individual reports and presentations.
- Linear programming has been added to Chapter 2.
- Geometric proofs have been removed from Chapter 4.
- The emphasis of Chapter 3 is now geometry and art.
- The analysis of paradoxes and flowcharts has been added to Chapter 4.
- Amortization tables are now included in Chapter 6.
- The study of probability in Chapter 7 now begins with more intuitive concepts and applications of probability.
- The emphasis of Chapter 8 is now statistics and voting.
- A new chapter on the history and uses of calculating devices has been added as Chapter 9.

## Organization

Each chapter is self-contained and can be treated as a separate unit. The ten chapters contain ample material for a one-semester three-credit course.

**Chapter 1, What's in a Number?** introduces the student to the basic building blocks of mathematics—numbers. It begins with ancient systems of numeration, progresses through the Hindu-Arabic system and the number systems of different bases, and ends with some creative investigation of numbers, including curiosities such as magic squares and circles.

**Chapter 2, Modeling with Algebra,** examines the use of algebraic functions as a model in various situations. Linear programming and linear, quadratic, exponential, and logarithmic functions are used to model real-life situations.

**Chapter 3, Geometry and Art,** begins with a review of some of the postulates, definitions, and theorems of Euclidean geometry and an introduction to non-Euclidean geometry. With this as a base, the student is introduced to an investigation of the same geometry in art such as the Golden Ratio, perspective, polygons, stars, tessellations, and fractals.

**Chapter 4, Sets, Logic, and Flowcharts,** investigates logic and some of its applications. It begins with sets, basic statements, and types of reasoning, then proceeds to the use of Venn diagrams, syllogisms, and truth tables to analyze arguments and paradoxes. The chapter concludes by looking at flow charts as a means of arriving at logical decisions.

**Chapter 5, Trigonometry: A Door to the Unmeasurable,** focuses on some of the practical applications found in the study of trigonometry. The trigonometry of right triangles and acute triangles is applied to a variety of situations.

**Chapter 6, Finance Matters,** continues to use mathematics as a practical tool in examining the world of finance through a realistic study of simple and compound interest, loans, and annuities.

**Chapter 7, Probability and Games People Play,** examines another very practical topic, probability. It investigates methods of counting, probability, odds, and expected values by emphasizing games of chance.

**Chapter 8, Statistics and Voting,** looks at basic statistics as a means of arranging and reporting data, especially those connected with voting. It begins with simple statistical graphs, examines measures of central tendency, and investigates the normal distribution. The chapter ends by looking at the statistics used in analyzing polls and apportioning representatives of a legislative body.

**Chapter 9, From Fingers to the Computer,** presents the history of calculating devices from fingers and the abacus to the hand-held electronic calculator and the personal computer. It explains how calculating devices work and how they are used in our society.

**Chapter 10, Calculus and the Infinitesimal,** completes the palette of mathematics by looking at calculus. The derivative and the integral are introduced along with applications of those two important concepts of mathematics.

## Ancillary Materials

The following ancillary materials are available to all adopters of *The Mathematical Palette 2/e*:

- *Student Study Guide* is available for purchase by students. It includes summaries that emphasize important concepts and techniques along with solutions to odd-numbered problems, crossword puzzles that will challenge students as well as assist them in reviewing the concepts discussed in each chapter, and suggestions for group projects.
- *Instructor's Resource Manual* contains answers and worked-out solutions to all Explain-Apply-Explore problems, 99 transparency masters created from text illustrations, a list of helpful teaching aids such as software and videos, two group projects per chapter (projects are also in the Student Study Guide), and a mini-

manual correlating sectional content with the Annenberg/ CPB video series ''For All Practical Purposes.''

- The printed *Test Bank* contains over 1,000 questions that can be used for testing or as a source of additional problems.
- *ExaMaster*™ computerized testing system allows instructors to quickly create, edit, and print tests from the set of test questions in the test bank. It is available for both MacIntosh and IBM environments.

## Custom Publishing

Liberal arts mathematics courses are structured in various ways that differ in length, content, and organization. To cater to these differences, Saunders College Publishing is offering *The Mathematical Palette* in a custom-published format. Instructors can rearrange, add, or delete chapters to produce a text that best meets their needs. Since *The Mathematical Palette* has no chapter dependencies, instructors using custom publishing are free to choose the topics they want to cover in the order they want to cover them, thereby creating a text that follows their course syllabi.

Saunders College Publishing is working hard to provide the highest quality service and textbooks for your courses. If you have any questions about custom publishing, please contact your local Saunders sales representative.

## Acknowledgments

We would like to thank those who were instrumental in developing this text:

Our wives, Dianne and Theresa, for continuing to put up with us.

David McLaughlin for his art work, art research, and contributions to Chapter 3.

Kurt Viegelmann for his photographs.

John O'Connor and Russ Lee for checking the accuracy of all examples, problems, and answers.

John Emert for writing all the questions and solutions in the *Test Bank* and Anna Cox for checking their accuracy.

Mina Kirby for developing and writing the twenty group projects included in the *Instructor's Resource Manual* and *Student Study Guide*.

Our Ohlone College students for their corrections and suggestions.

Anita Fallon, Deirdre Lynch, Laura Shur, and the staff at Saunders College Publishing for their concern for every detail in producing our book.

Deborah Blumberg, Michael Fallon, David Hopkins, George Landavazo, David McLaughlin, Frank Riolo, Audrey Talley, Blanche M. Lambert, Timothy White, and Perry Wong for their interviews that introduce each chapter.

We wish to thank the following instructors for their insightful reviews of the first edition:

Joanne W. Anderson, West Valley Joint Community College District
Gary Brown, College of St. Benedict
Warren J. Burch, Brevard Community College
John W. Emert, Ball State University
Mark Greenhalgh, Fullerton College
Suzanne Larson, Loyola Marymount University
Thomas J. Miles, Central Michigan University
Steven L. Thomassin, Ventura College

And of the second edition:

Lothar Doshe, University of North Carolina, Asheville
Marsha J. Driskill, Aims Community College
Theresa Geiger, St. Petersburg Junior College
Peter U. Georgakis, Santa Barbara Junior College
Jerome A. Goldstein, Louisiana State University, Baton Rouge
Rhonda L. Hatcher, Texas Christian University
Fred Hoffman, Florida Atlantic University
Robert W. Hunt, Humboldt State University
Mina Kirby, East Los Angeles College
Sandra Z. Keith, St. Cloud State University
Allan Robert Marshall, Cuesta College
Norm Martin, Northern Arizona University
Susan E. McLoughlin, Union County College
Duncan J. Melville, St. Lawrence University
ViAnn Olson, Rochester Community College
Margaret Rejto, Normandale Community College
George C. Trowbridge, University of New Mexico, Albuquerque
Carroll Wells, Western Kentucky College
Peggy White, Allan Hancock College
Robert J. Wisner, New Mexico State University, Las Cruces
Dana M. Updegrove, Bellevue Community College

**R. Staszkow**
**R. Bradshaw**
*August 1994*

# Contents

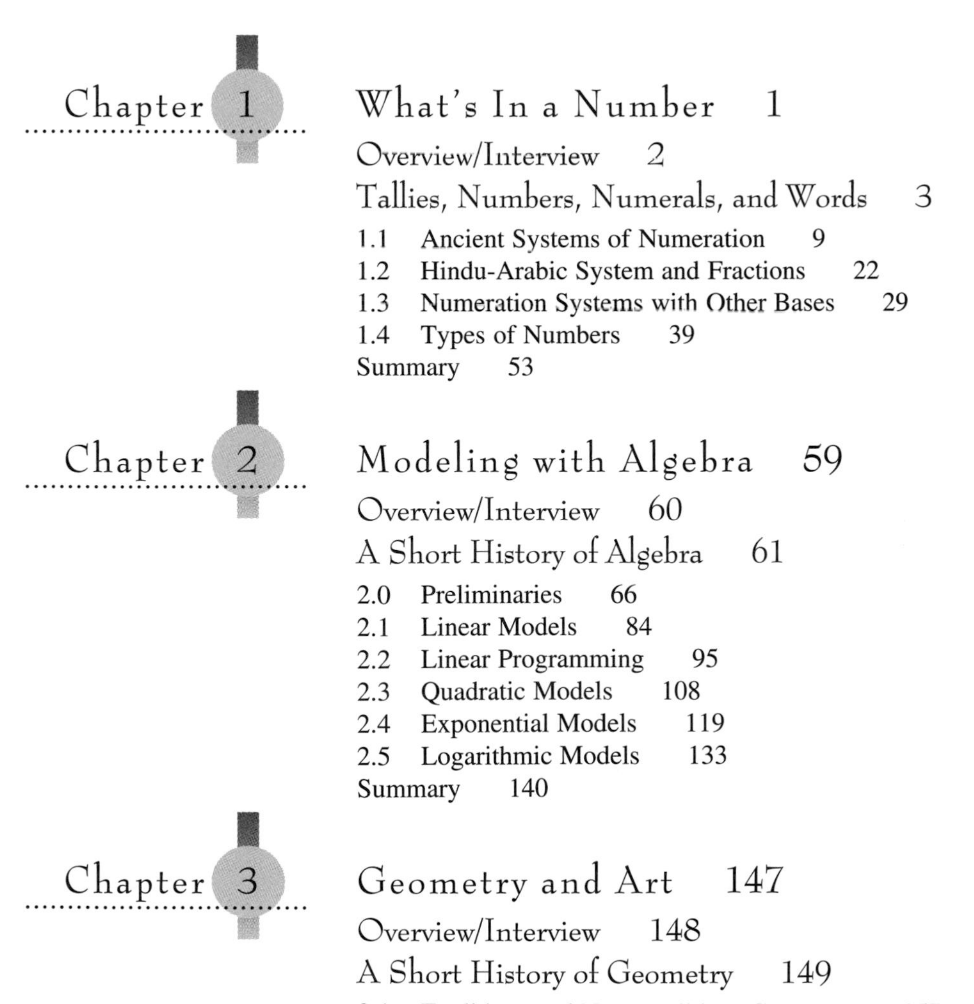

Chapter 4

Chapter 5

Chapter 6

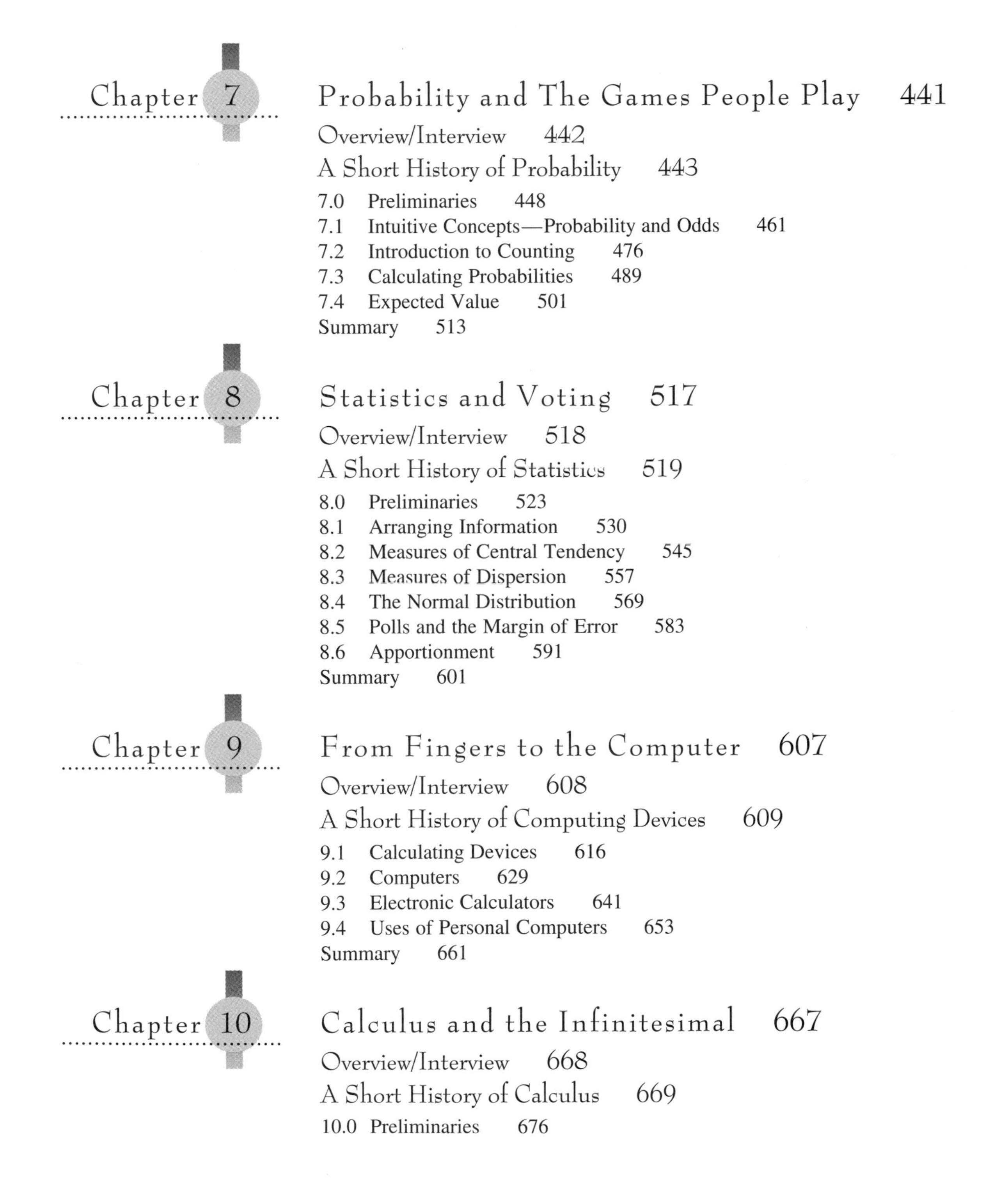

# The Mathematical Palette

# 1

# What's in a Number?

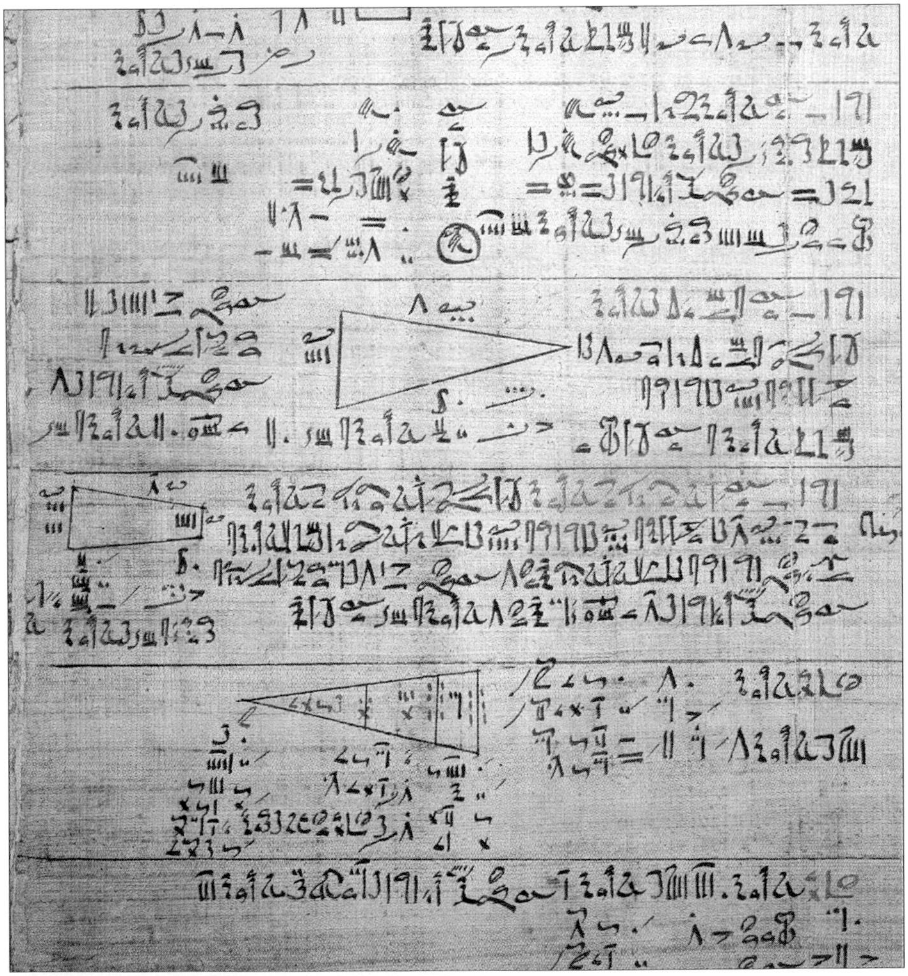

The formal study of mathematics has intrigued man for centuries. An example is shown on this mathematical treatise written on Rhind papyrus dating from the middle of the 16th century B.C. (Art Resources, Inc.)

# Overview

# Interview

The numerals 0 through 9 that we use every day did not merely appear out of thin air. Even though they are generally accepted throughout the world, they were not the first numerals invented. In fact, many cultures have had entirely different systems of numeration, based on symbols that may seem entirely foreign to us. In this chapter, you will learn about different systems of numeration from the past and about different systems that are in use today. You will also learn about special types of numbers that have names of their own such as perfect numbers, friendly numbers, and pentagonal numbers. You will investigate number phenomena such as pi ($\pi$), magic squares, and Egyptian fractions.

## Timothy White

TIMOTHY WHITE, PH.D., PROFESSOR OF ANTHROPOLOGY, UNIVERSITY OF CALIFORNIA, BERKELEY

TELL US ABOUT YOURSELF. I received my Bachelors of Science in Biology and Anthropology at the University of California, Riverside, and got my Masters of Arts and Ph.D. at the University of Michigan, Ann Arbor. Since 1977 I have been a full professor with research and teaching duties at University of California, Berkeley. My past and current research has included the codiscovery of the earliest human ancestor species, *Australopithecus afarensis* in 1978; excavation of the earliest human ancestor footprints, Laetoli, Tanzania, 1978; field research in Turkey, Jordan, Ethiopia, Kenya, Tanzania, and Malawi; and laboratory research in Europe, Africa, Asia, and North America. I have also authored *Human Osteology* (Academic Press, 1991), *Prehistoric Cannibalism* (Princeton University Press, 1992), and numerous scientific papers.

HOW IS MATHEMATICS USED IN ANTHROPOLOGY? Mathematics is of critical importance to social and biological scientists in anthropology. Anthropology, the study of human beings, encompasses the study of modern human culture (social anthropology or ethnology), past human culture (archeology), and human biology (physical anthropology). It includes such different research areas as folklore, linguistics, paleontology, anatomy, stratigraphy, and primatology. Mathematics is used by anthropologists working in these fields and many other subfields. Particularly important is the use of statistics in studies of human variation, population dynamics, and human behavior.

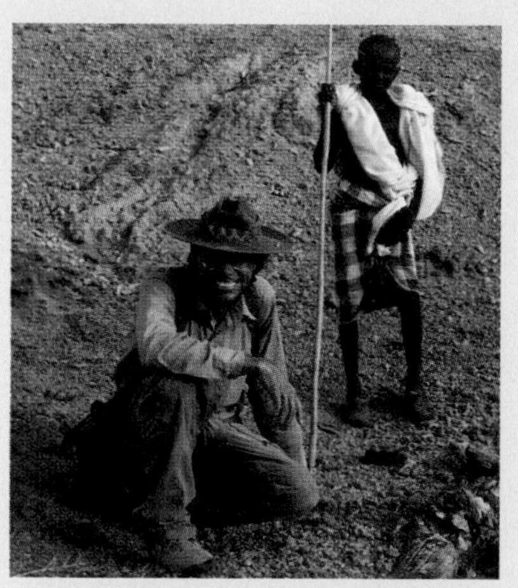

*Tim White with the fossilized skull of* Pelorovis, *a giant extinct buffalo found with human ancestor fossils at the site of Bouri, in the Afar depression of Ethiopia. (Photo by Gen Suwa).*

WHY IS THE STUDY OF MATHEMATICS IMPORTANT TO THE ANTHROPOLOGIST? Quantification of all the cultural and biological variables having to do with modern humans and their ancestors is important to all anthropologists. It is impossible to think of anthropologists operating without quantifying their observations, whether they are observations on human ancestors, prehistoric stone tools, modern primates, or diverse cultures. By the use of statistics, the anthropologist can test the significance of research findings and communicate the research results to other anthropologists and the rest of the world.

*I Saw the Figure 5 in Gold* by Charles Henry Demuth (The Metropolitan Museum of Art) and *Figure 7* by Jasper Johns (National Gallery of Art): modern artists' visual interpretations of the Hindu-Arabic digits five and seven.

## Tallies, Numbers, Numerals, and Words

It seems appropriate to embark on our journey through mathematics by studying math's basic building blocks, numbers. Ever since prehistoric times, human beings have concerned themselves with numbers. During the Paleolithic period (1,000,000 B.C.–10,000 B.C.), men and women were almost totally consumed with survival. However, it is believed that they possessed a basic sense of numbers and had the ability to distinguish between more, less, and equal. Their language may have been lacking in the words to represent numbers, but anthropologists agree that, at the very least, they possessed a visual number sense and an awareness of form.

As humans progressed through the Neolithic period (8000 B.C.), they became civilized. They grew crops, domesticated animals, wove cloth, made pottery, and lived in villages. Their number sense also grew. Though we have no written records, archaeological findings and opinions of anthropologists suggest that by the beginning of recorded history (3000 B.C.), humans had developed the ability to tally and count.

A **tally** is a mark that represents the object being counted. This process of tallying took the form of scratches on the ground or on cave walls, as knots on ropes or vines, as piles of pebbles or sticks, and as notches on pieces of bone or wood. For example, to count the number of days between full moons, you could make a mark each evening on a wall until the next full moon appeared. Such a tally might look like this:

///// ///// ///// ///// ///// ///

However, if the number of objects to be counted is very large, the tally method becomes very cumbersome. For example, if you wanted to record the population of the United States (252.7 million in 1991) using the tally system shown above, and the tallies filled both sides of standard $8\frac{1}{2}$-by-11-inch paper, you would end up with a pile of more than 19,000 pieces of paper.

As society became more involved in measurement, commerce, and taxes, more efficient means of representing numbers were needed, so different systems of numeration were developed. In this chapter, we will study systems of numeration.

## WHAT IS A SYSTEM OF NUMERATION?

A **number** is a quantity that answers the questions, "How much?" or "How many?" Numbers are given a name in words and are represented by symbols. The symbols that are used to represent numbers are called **numerals**. In common usage number and numeral are used interchangeably, but the number is really the abstract concept, the amount or value, and the numeral is a symbolic representation of that amount or value. For example, the quantity of trees shown below can be represented by the numeral 10 in our Hindu-Arabic system, $=$ in the Mayan system, $\cap$ in the Egyptian system, X in Roman numerals, $\triangle$ in the Attic system, 1010 in the binary system, $+$ in the traditional Chinese system, and so on. The word used for that amount of trees is *ten* in English, *zehn* in German, *diez* in Spanish, *i'wes* in Ohlone Indian of California, *decem* in Latin, *'umi* in Hawaiian, *dix* in French, *shyr* in Chinese, *tiz*, in Hungarian, *daca* in Sanskrit, *desiat* in Russian, and so on.

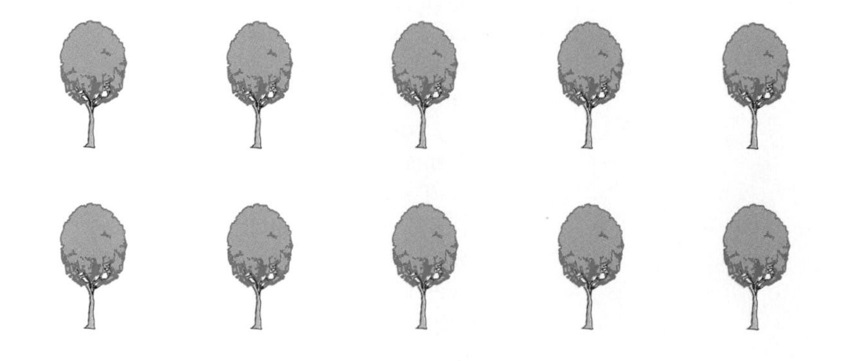

No matter what symbols or words are used, it is understood that there are ten trees.

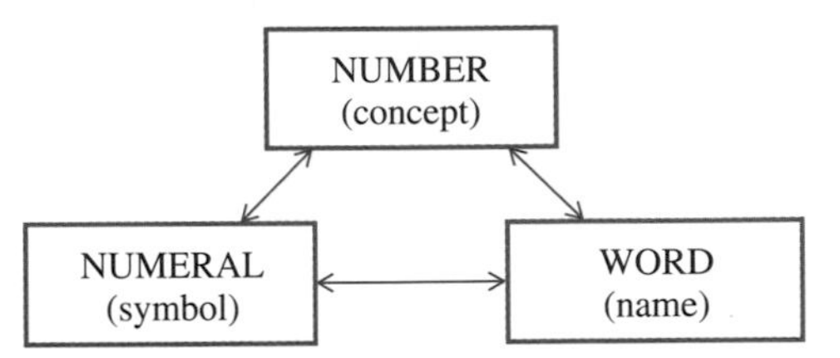


A system of numeration gives us a means of finding the numeral that is associated with a given number. A system of numeration consists of a set of symbols and a method for combining those symbols to represent numbers. As we study different systems of numeration, our goal is to develop a richer understanding of the system of numeration that we presently use, the Hindu-Arabic system.

Some of the important dates, people, and events associated with the development of that system and basic arithmetic are as follows:

c. 150 B.C. Brahmi numerals appeared in the Nana Ghat inscriptions. These numerals are the origins of the Hindu-Arabic numerals.

A.D. 825 Base 10 place-value numeration system (with a symbol for zero) was well established in India, as seen in the work of Persian mathematician Al-Khowârizmî.

1202 Leonardo Fibonacci encouraged the use of Hindu-Arabic numerals in his work *Liber Abaci*.

1299 City-state of Florence outlawed the Hindu-Arabic system of numeration.

1479 Hindu-Arabic digits first appeared in their present form.

1484 Nicolas Chuquet introduced notation for adding and subtracting fractions.

1489 Johann Widmann used the $+$ and $-$ signs to denote addition and subtraction.

1491 Filippo Calandri introduced the long-division process that is still used today.

c. 1500 The Hindu-Arabic system of numeration replaced Roman numerals in Europe.

1527 Petrino Apianus introduced notation for multiplying and dividing fractions.

1556 Spaniard Juan Diez Feile wrote the first math book in Mexico, *Sumario Compendioso*.

1585 Simon Stevin gave the first systematic account of decimal fractions and their use.

1617 John Napier introduced present-day notation by using the decimal point when writing numerals.

1631 William Oughtred used the $\times$ sign for multiplication.

1659 J. H. Rahn introduced the $\div$ sign for division.

1719 The first math book was printed in America: James Hoddler's *Arithmetick*, printed in Boston.

1729 Isaac Greenwood was the first American to write a mathematics book, *Arithmetick, Vulgar and Decimal.*

## Check Your Reading

1. Compare the terms ''tally,'' ''numeral,'' and ''number.'' How are they different? How are they the same? Use examples in your explanation.
2. Suppose a fisherman has caught twenty fish. Use our Hindu-Arabic numeration system to:
   (a) Explain how tallies are used in conjunction with the idea of ''twenty fish.''
   (b) Explain how numerals are used in conjunction with the idea of ''twenty fish.''
   (c) Explain how numbers are used in conjunction with the idea of ''twenty fish.''
   (d) Explain how ''twenty'' is used in conjunction with the idea of ''twenty fish.''
3. In the opinion of anthropologists, by what time in history did people learn to count?
4. By 146 B.C., when the Roman Empire was established, which numerals had made an appearance in India?
5. In the year 1484, the Portuguese navigator Diego Cam discovered the mouth of the Congo River. What discoveries were made in arithmetic during that year?
6. When did the present form of our numerals appear?
7. The Roman symbol representing 𝍸 𝍸 is X. Is X a numeral or a number? Explain
8. When the first court jesters appeared in European courts in 1202, what was Leonardo Fibonacci doing?
9. The year 1490 marked the births of King Henry VIII of England and Ignatius Loyola, founder of the Jesuit order. What was Filippo Calandri introducing to the world at about this time?
10. When Columbus landed in North America in 1492, what arithmetic process had just been introduced in Europe?
11. When the Vikings were discovering Greenland in A.D. 985, were the modern forms of fractions and decimals in use? About what time did they come into use?

12. When the new settlement was started in Jamestown (Virginia) in 1607, were our modern symbols for multiplication and division in common use?
13. In the year 1617, the Native American princess Pocahontas died. What was John Napier introducing in Scotland during the same year?
14. Write the numeral for the number of days between consecutive full moons.
15. Match each of the following names with the correct mathematical discovery or event.

| | |
|---|---|
| (a) Al Kowârizmi | Adding and subtracting fractions |
| (b) Calandri | Early user of base 10 place-value system |
| (c) Chuquet | Decimal fractions |
| (d) Fibonacci | Long-division process |
| (e) Oughtred | Wrote *Liber Abaci* |
| (f) Rahn | $+$ for addition and $-$ for subtraction |
| (g) Stevin | $\times$ for multiplication |
| (h) Widmann | $\div$ for division |

## • Research Questions

To answer the following questions, you will need to refer to material not contained in the text. Possible sources of information are listed in the bibliography at the end of the book.

1. In *The World of Mathematics*, edited by James Newman, and *Number: The Language of Science*, by Tobias Dantzig, studies are cited that indicate that certain birds and insects also have a sense of numbers. Explain these studies and their results.
2. Do some research on two of the people mentioned in this section. What other interests did they have? What else are they famous for?
3. The word used to represent a certain number varies from language to language. What are some ways in which the numerals from 0 to 9 are written in different languages?
4. What are Fibonacci numbers? What applications in the real world have been found for these numbers?
5. Trace the development of some of the basic symbols used in arithmetic for addition, subtraction, multiplication, and division.
6. What is a ''googol''? What is a ''googolplex''? Give a short description of the history of these terms.
7. Magic squares are discussed in Section 1.4 of this text. What cultures have studied magic squares? Expand on the history of magic squares that is given in Section 1.4.
8. Do some research on the largest known prime number. How many digits does it have?

9. Investigate and describe the way ancient Egyptians performed addition and multiplication with their system of numeration (described in Section 1.1).
10. What is cryptography? What is the RSA method invented by Rivest, Shamir, and Adelman?
11. The Chinese rod-numeral system using scientific Chinese numerals is an example of a place-value system that allowed for the representation of decimal fractions. Do some research and explain how this system works.
12. In the 1960s, the Duodecimal Society of America recommended the use of a base 12 system of numeration. Why did they propose this? What are the basic digits in the proposed system? How are numbers represented in this system? Give examples.

## • Projects

1. The Hindu-Arabic digits that we use today are 0, 1, 2, 3, 4, 5, 6, 7, 8, 9. Trace the development of these digits from their origins to their present forms. How did the shape of each digit change through time? What civilizations were involved with the origin and development of each digit?
2. There are methods for creating the magic squares introduced in Section 1.4. Discuss at least two of these methods. Who developed these methods? Show how to generate magic squares of various sizes using these methods.
3. Investigate some methods of doing computation using your ten fingers. In particular, investigate and describe the Chisanbop method of finger computation developed by Edwin Lieberthal. Give examples of how each finger computation system works.
4. What is numerology? How does the system of numerology work? Perform some numerological calculations.

A copy of a wall painting from the tomb of Menna, Thebes, XVIII Dynasty (c. 1420 B.C.), shows ancient Egyptians using tallying to record the amount of wheat harvested. (The Metropolitan Museum of Art)

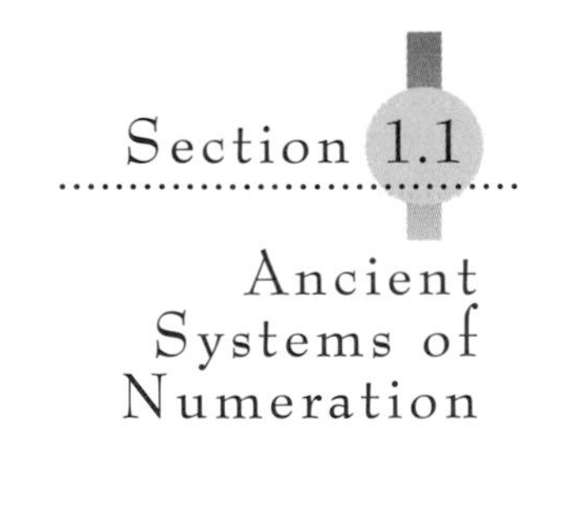

## Section 1.1 Ancient Systems of Numeration

In this section we examine the four basic types of ancient systems of numeration. Each system uses symbols to represent numbers or a set of tallies. We examine how different systems of numeration use symbols to represent numbers. We do not expect you to be an expert in ancient systems of numeration. We merely want to expose you to a very important area of mathematics. Furthermore, we do not expect you to memorize all the symbols in this section. As you read this section and solve the problems, you will need to refer frequently to the lists of symbols for a given system of numeration.

## Systems Using Addition

### *Egyptian Hieroglyphic System*

The oldest type of numeration system using addition is the Egyptian hieroglyphic system (c. 3400 B.C.). The numerals used in this system are shown in the table. The value of a number is simply obtained by finding the sum of the values of its numerals.

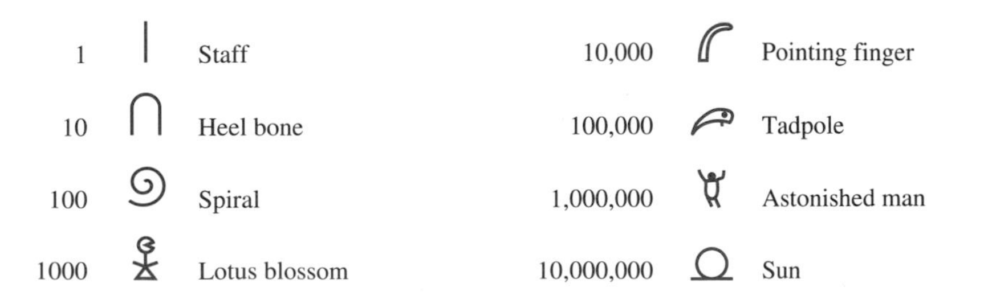

| Value | Name | Value | Name |
|---|---|---|---|
| 1 | Staff | 10,000 | Pointing finger |
| 10 | Heel bone | 100,000 | Tadpole |
| 100 | Spiral | 1,000,000 | Astonished man |
| 1000 | Lotus blossom | 10,000,000 | Sun |

**Example 1**

Find the number represented by

**Solution:** Using the table, add the value of each of the numerals:

1,000,000 + 10,000 + 10,000 + 10 + 10 + 10 + 1 + 1
= 1,020,032 ■

**Example 2**

Write the number 1753 as a numeral in the Egyptian hieroglyphic system.

**Solution:** 1753 = 1000 + 700 + 50 + 3, which in Egyptian hieroglyphics is

In such an additive system the symbols can be placed in any order.

123 = ■

### *Attic System*

Another example of a system of numeration that uses an additive grouping scheme was found in records in Athens, Greece, around 300 B.C. The Attic system uses the numerals

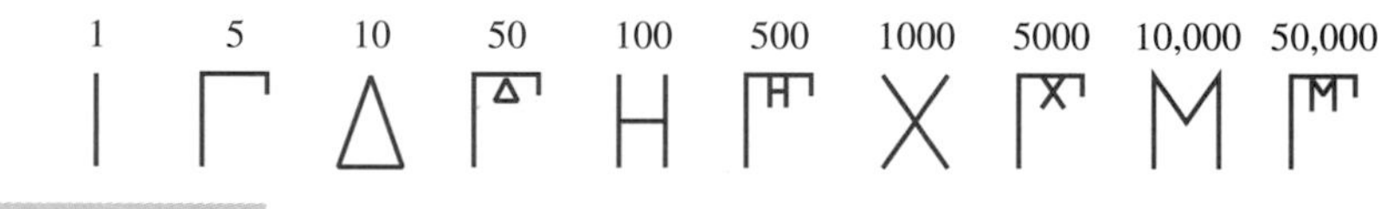


**Example 3**

Find the number represented by the Attic numeral

**Solution:** Using the chart above, add the value of each numeral.

10,000 + 10,000 + 500 + 100 + 100 + 100
+ 10 + 5 + 1 + 1 + 1 + 1 = 20,819 ■

**Example 4**

Write 6376 in the Attic system of numeration.

**Solution:** Using the chart above, we get

6376 = 5000 + 1000 + 100 + 100 + 100 + 50 + 10 + 10 + 5 + 1
= ■

## Systems Using Addition and Subtraction

### *Roman System*

Around 200 B.C., the Romans also developed a system of numeration that used grouping symbols in conjunction with addition. This system also utilized an abacus to perform computation. In the following table you will find the standardized numerals used in the Roman system.

| 1 | 5 | 10 | 50 | 100 | 500 | 1,000 |
|---|---|---|---|---|---|---|
| I | V | X | L | C | D | M |
| | 5000 | 10,000 | 50,000 | 100,000 | 500,000 | 1,000,000 |
| | $\overline{\text{V}}$ | $\overline{\text{X}}$ | $\overline{\text{L}}$ | $\overline{\text{C}}$ | $\overline{\text{D}}$ | $\overline{\text{M}}$ |

In this system the value of a numeral was originally obtained by adding the value of each symbol. Later, Roman numerals also included subtraction. Subtraction was used when the numerals were written with ''4's'' and ''9's''—that is, 4 (IV), 9 (IX), 40 (XL), 90 (XC), 400 (CD), 900 (CM), and so on. With these numbers, the numeral representing a smaller number is placed before the numeral that represents a larger number. This indicates that the smaller numeral is to be subtracted from the larger one.

**Example 5**

What number is represented by $\overline{\text{D}}\text{M}\overline{\text{V}}\text{CCLIV}$ in Roman numerals?

**Solution:** Using the table above, we get

$$\overline{\text{D}} = 500{,}000$$

$$\text{M}\overline{\text{V}} = 5000 - 1000 = 4000$$

$$\text{CC} = 100 + 100 = 200$$

$$\text{L} = 50$$

$$\text{IV} = 5 - 1 = 4$$

Thus,

$$\overline{\text{D}}\text{M}\overline{\text{V}}\text{CCLIV} = 500{,}000 + 4000 + 200 + 50 + 4 = 504{,}254$$ ■

**Example 6**

Write 1989 in the Roman numeration system.

**Solution:**

$$1989 = 1000 + 900 + 80 + 9$$
$$1000 = \text{M}$$
$$900 = \text{CM}$$
$$80 = \text{LXXX}$$
$$9 = \text{IX}$$

This gives 1989 = MCMLXXXIX ■

## Systems Using Addition and Multiplication

### *Traditional Chinese System*

The traditional Chinese system of numeration appearing in the Han dynasty around 200 B.C. also used grouping symbols and addition of numerals to represent numbers. However, instead of repeating a symbol when many of the same symbols are needed, multiplication factors are placed above the numeral. Numerals are written vertically with the symbols

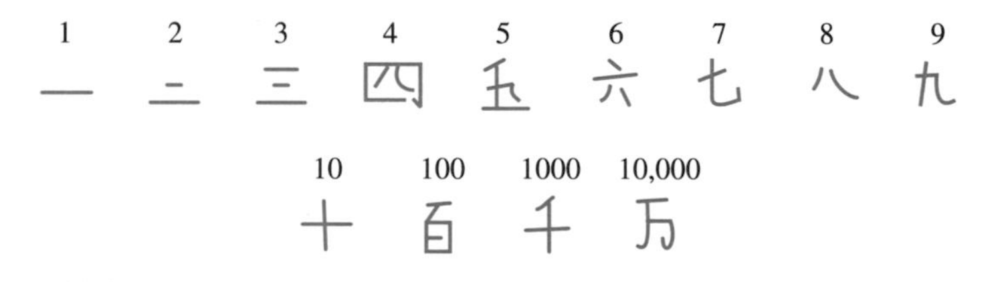


**Example 7**

Find the number represented by

三
千

六
百

一
十

八

**Solution:**

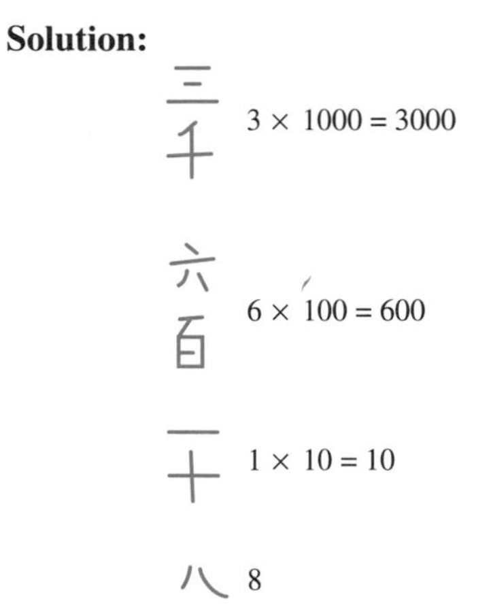


This gives a total of 3618.

Notice that each digit of the numeral 3618, except for the units digit, is represented by two characters in the Chinese system. ■

**Example 8**

Write 453 in the traditional Chinese system.

**Solution:**

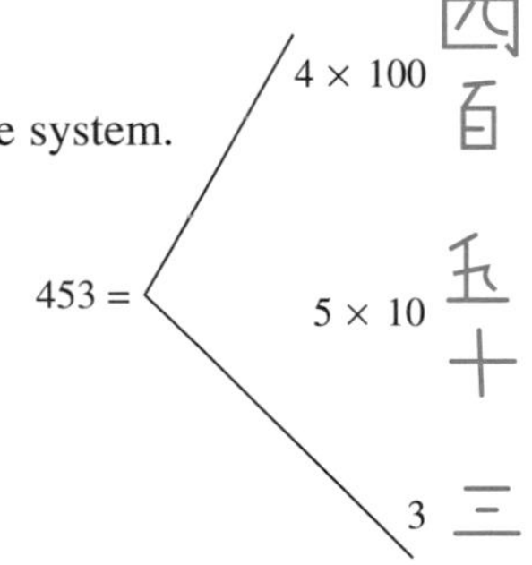


■

## *Ionic Greek System*

The Greeks (c. 450 B.C.) used the 24 letters of their alphabet along with 3 ancient Phoenician letters for 6, 90, and 900, to represent numbers. Originally the capital Greek letters were used, but they gave way to the lowercase letters. The symbols used to represent numbers in the Ionic Greek system are shown in the chart below.

| | | | | | |
|---|---|---|---|---|---|
| 1 | $\alpha$ | 10 | $\iota$ | 100 | $\rho$ |
| 2 | $\beta$ | 20 | $\kappa$ | 200 | $\sigma$ |
| 3 | $\gamma$ | 30 | $\lambda$ | 300 | $\tau$ |
| 4 | $\delta$ | 40 | $\mu$ | 400 | $\upsilon$ |
| 5 | $\epsilon$ | 50 | $\nu$ | 500 | $\phi$ |
| 6 | ς | 60 | $\xi$ | 600 | $\chi$ |
| 7 | $\zeta$ | 70 | $o$ | 700 | $\psi$ |
| 8 | $\eta$ | 80 | $\pi$ | 800 | $\omega$ |
| 9 | $\theta$ | 90 | Q | 900 | T |

To represent a number from 1 to 999, write the appropriate symbols next to each other; for example, $\pi\delta = 84$ and $\omega\kappa\gamma = 823$. To obtain numerals for the multiples of 1000, place a prime to the left of the symbols for 1 to 9 to signify that it is multiplied by 1000; for example,

$$'\alpha = 1000 \qquad '\zeta = 7000 \qquad '\theta = 9000$$

One way to obtain the numerals that represent the multiples of 10,000 is to place the symbols for 1 to 9 above the letter M; for example,

$$\overset{\alpha}{\mathrm{M}} = 10{,}000 \qquad \overset{\beta}{\mathrm{M}} = 20{,}000 \qquad \overset{\theta}{\mathrm{M}} = 90{,}000$$

A system in which many different symbols are used to represent the digits is called a ciphered system. This type of a system allows numbers to be written in a simple compact form but requires memorization of many different symbols.

### Example 9

What number is represented by $\overset{\eta}{\mathrm{M}}\ '\theta\phi\xi\epsilon$ ?

**Solution:**

$$\overset{\eta}{\mathrm{M}} = 80{,}000$$
$$'\theta = 9{,}000$$
$$\phi = 500$$
$$\xi = 60$$
$$\epsilon = 5$$

This gives $80{,}000 + 9000 + 500 + 60 + 5 = 89{,}565$. ■

### Example 10

Write 2734 in the Ionic Greek system.

**Solution:**

$$2734 = 2000 + 700 + 30 + 4 = {}'\beta\psi\lambda\delta$$ ■

## Systems Using Place Values

### *Babylonian System*

The most advanced numeration systems are those that not only use symbols, addition, and multiplication, but also give a certain value to the position a numeral occupies. The Babylonian system (c. 2300 B.C.) is an example of this type of system. The sexagesimal system, based on 60, uses only two symbols, formed by making marks on wet clay with a stick.

Groups of numerals separated from each other by a space signify that each group is associated with a different place value. Groups of symbols are given the place values from right to left. The place values are

$$60^0 = 1, \quad 60^1 = 60, \quad 60^2 = 3600, \quad 60^3 = 216{,}000, \ldots$$

For example, means

$$\begin{array}{ccccccc} 12 \times 60^2 & + & 31 \times 60^1 & + & 23 \times 60^0 & & \\ 12 \times 3600 & + & 31 \times 60 & + & 23 \times 1 & & \\ 43{,}200 & + & 1860 & + & 23 & = & 45{,}083 \end{array}$$

The Babylonian system, however, did not contain a symbol for zero to indicate the absence of a particular place value. Some Babylonian tablets have a larger gap between numerals or the insertion of the symbol , which indicates a missing place value.

## Example 11

Find the number represented by

**Solution:** The groupings represent 2, 11, 0, and 34, which gives

$$\begin{array}{ccccccccc} 2 \times 60^3 & + & 11 \times 60^2 & + & 0 \times 60^1 & + & 34 \times 60^0 & & \\ 2 \times 216{,}000 & + & 11 \times 3600 & + & 0 \times 60 & + & 34 \times 1 & & \\ 432{,}000 & + & 39{,}600 & + & 0 & + & 34 & = & 471{,}634 \end{array}$$

## Example 12

Represent 4507 in the Babylonian numeration system.

**Solution:** The place values of the Babylonian system that are less than 4507 are 3600, 60, and 1. To determine how many groups of each place value are contained in 4507, we can use the division scheme shown below.

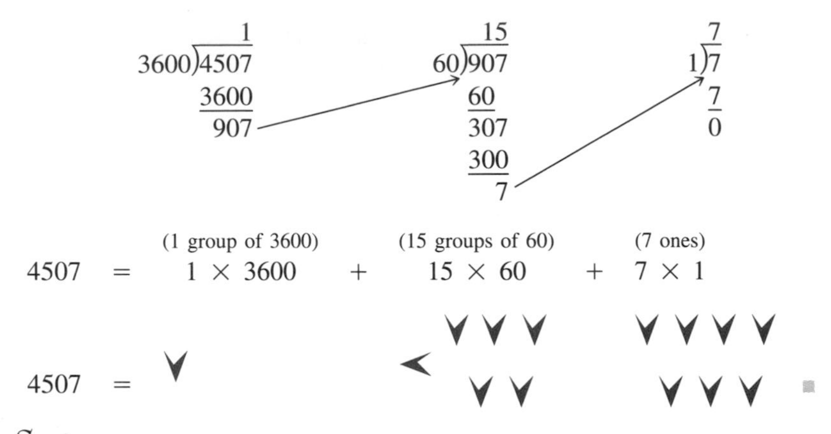


## *Mayan System*

In about 300 B.C., the Mayan priests of Central America also developed a place-value numeration system. Their system was an improvement on the Babylonian system because it was the first to have a symbol for zero. The Mayan system is based on 20 and 18. It uses the numerals shown below.

| | | | | | | | |
|---|---|---|---|---|---|---|---|
| 0 | (shell) | 5 | — | 10 | —<br>— | 15 | —<br>—<br>— |
| 1 | · | 6 | ·<br>— | 11 | ·<br>—<br>— | 16 | ·<br>—<br>—<br>— |
| 2 | ·· | 7 | ··<br>— | 12 | ··<br>—<br>— | 17 | ··<br>—<br>—<br>— |
| 3 | ··· | 8 | ···<br>— | 13 | ···<br>—<br>— | 18 | ···<br>—<br>—<br>— |
| 4 | ···· | 9 | ····<br>— | 14 | ····<br>—<br>— | 19 | ····<br>—<br>—<br>— |

The numerals are written vertically, with the place value assigned from the bottom of the numeral to the top of the numeral. The positional values are

$$20^0 = 1, \quad 20^1 = 20, \quad 18 \times 20^1 = 360,$$
$$18 \times 20^2 = 7200, \quad 18 \times 20^3 = 144{,}000, \ldots$$

Instead of the third position having a place value of $20^2$, the Mayans gave it a value of $18 \times 20^1$. This was probably done so that the approximate number of days in a year, 360 days, would be a basic part of the numeration system.

For example, in the Mayan system, 168,599 is written as

| | | | | | | |
|---|---|---|---|---|---|---|
| · | → | $1 \times (18 \times 20^3)$ | = | $1 \times 144{,}000$ | = | 144,000 |
| ··· | → | $3 \times (18 \times 20^2)$ | = | $3 \times 7200$ | = | 21,600 |
| ···<br>— | → | $8 \times (18 \times 20^1)$ | = | $8 \times 360$ | = | 2,880 |
| — | → | $5 \times 20^1$ | = | $5 \times 20$ | = | 100 |
| ····<br>—<br>—<br>— | → | $19 \times 20^0$ | = | $19 \times 1$ | = | 19 |
| | | | | | | 168,599 |

The Mayan system does have a zero, but, because of the use of 18 in the third position, its place-value feature is irregular.

## Example 13

Find the number represented by ····

**Solution:**

$$= 4 \times 7200 = 28{,}800$$

$$= 0 \times 360 = 0$$

$$= 7 \times 20 = 140$$

$$= 12 \times 1 = 12$$

This gives a total of 28,952. ■

## Example 14

Write 17,525 in the Mayan numeration system.

**Solution:** The place values less than 17,525 in the Mayan system are 7200, 360, 20, and 1. To determine how many groups of each place value are contained in 17,525, we use the division scheme below.

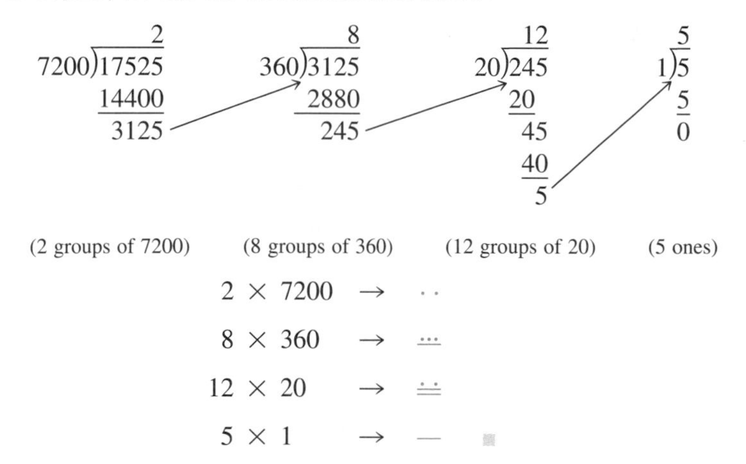


Our system of numeration originated with the Hindus in India in about 150 B.C., with its base 10 place-value feature and zero placeholder being introduced before 628. By 900, this Hindu-Arabic numeration system had reached Spain, and by 1210 it had been spread to the rest of Europe by traders on the Mediterranean Sea and scholars who attended universities in Spain. By 1479, the digits appeared in the form that is used today.

The Hindu-Arabic system of numeration made computation a reasonable task. However, those that used Roman numerals and calculated with their counting board abacus, the "abacists," were opposed to the "algorists," who used and computed with the Hindu-Arabic numerals. In fact, in 1299, the city-state of Florence in Italy outlawed the use of the Hindu-Arabic system. Many banks in Europe also forbade the use of these numerals because they were easily forged and/or altered on bank drafts. However, by 1500 the Hindu-Arabic system had won the battle and became the prevalent numeration system. In the next section, we look more closely at the Hindu-Arabic system of numeration.

## Section 1.1 PROBLEMS

## *Explain ➡ Apply ➡ Explore*

### *Explain*

1. What is meant by a system of numeration that uses addition and subtraction? Give some examples.
2. What is meant by a system of numeration that uses addition and multiplication? Give some examples.
3. Compare numeration systems that use addition and subtraction with those that use addition and multiplication. What are the advantages and disadvantages of each type of system?
4. What is meant by a system that uses place values? Give some examples.
5. Compare numeration systems that use addition and multiplication with those that use place values. What are the advantages and disadvantages of each type of system?
6. What are some advantages of a system that uses place values over a system that does not use place values?
7. If used today, which ancient systems of numeration would require that you invent new symbols for larger numbers? Which systems would not? Explain.
8. After studying seven ancient systems of numeration, you can better understand the advantages of the Hindu-Arabic system we presently use. What are some of these advantages?
9. Explain the statement "For any number, there are many numerals."
10. Which of the systems of numeration described in this section would you prefer to use? Give reasons for your choice.

*Apply*

What number is represented by each Egyptian hieroglyphic numeral?

**11.** 𓎆 𓎆 𓆐𓁨

**12.** 𓎆 𓍢𓍢𓍢

**13.** 𓆼 𓎆 𓍢𓏺

What number is represented by each Attic numeral?

**14.** Μ 𐅅 𐅄

**15.** Χ Η 𐅃 Ι

**16.** 𐅈 𐅈 Μ Η Η

What number is represented by each Roman numeral?

**17.** DCCXXXIV

**18.** $\overline{\text{C}}$MCM

**19.** $\overline{\text{XL}}$MMCDLVII

What number is represented by each traditional Chinese numeral?

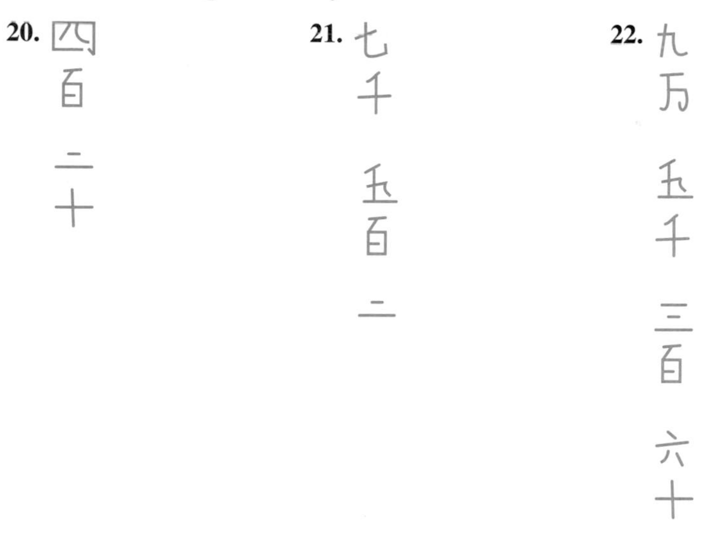

What number is represented by each Ionic Greek numeral?

**23.** $\zeta$

**24.** $\phi\mu\delta$

**25.** $\overset{\beta}{\text{M}}, \epsilon\xi\beta$

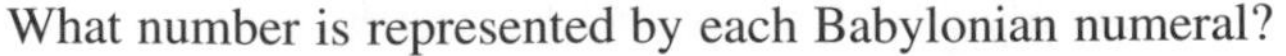

What number is represented by each Babylonian numeral?

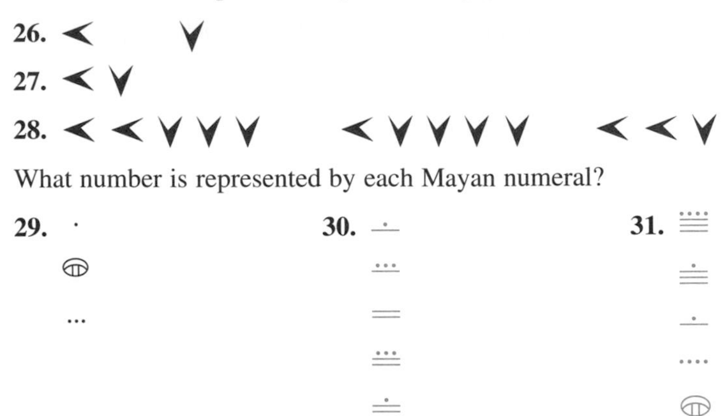

26.

27.

28.

What number is represented by each Mayan numeral?

29. 30. 31.

32. Represent the speed of sound (750 mph) in each of the following numeration systems:
    (a) Egyptian hieroglyphic
    (b) Roman
    (c) Ionic Greek
    (d) Mayan

33. Represent the number of minutes in a day (1440 min) in each of the following numeration systems:
    (a) Attic
    (b) Traditional Chinese
    (c) Babylonian

*Explore*

34. (a) Represent the number of chairs shown below in each of the systems of numeration of Problem 33.
    (b) Represent the number of legs on the chairs shown below in each of the systems of numeration of Problem 32.

**35.** Suppose that the Egyptian numeral 𓂭 𓆼 𓆼 𓍢 𓍢 𓍢 𓎆 | | | | | were converted by people from different cultures into a numeral in their own systems of numeration. How would the Egyptian numeral be represented in the following systems of numeration?

(a) Roman system
(b) Traditional Chinese system
(c) Ionic Greek system
(d) Babylonian system
(e) Mayan system

**36.** The birth date of one of the authors is 11-8-1944. Written in Roman numerals that would be XI-VIII-MCMXLIV.

(a) Write that birth date in Mayan, traditional Chinese, and Egyptian.
(b) Write your birth date in three of the systems of numeration described in this section.

**37.** Suppose ancient cultures had telephones. How would you write the phone number, 1-800-YEA-MATH, in three of the systems of numeration described in this section?

**38.** Using the symbols * for 1, / for 5, ¥ for 10, ∇ for 50, $ for 100, and □ for 1000, design a multiplicative system to represent numbers. Explain how the system works and represent 324 and 1995 in the system.

**39.** Using the symbols → for 1, ← for 9, ↓ for 81, and ↑ for 729, design an additive system of numeration. Use that system to represent the

(a) number of chairs in Problem 34.
(b) number of days in a leap year.
(c) number of feet in a mile.
(d) number of grams in a kilogram.

**40.** Using the existing symbols, modify the Roman system so that it is a multiplicative system. Explain your modifications and give three examples of numerals using your system.

**41.** Modify the existing Mayan system of numeration to create a base ten system that uses the same format and symbols. Explain your modifications and give three examples of numerals using your system.

**42.** Using the existing symbols, modify the Attic Greek system so that it is a multiplicative system. Explain your modifications and give three examples of numerals using your system.

**43.** When adding in the Hindu-Arabic system, we use a ''carrying'' procedure. For example, to find 36 + 48 we add the digits in the one's place (6 + 8) and get

14. We write the 4 in the one's place of the sum and carry the 1 to the ten's place as shown below.

$$\begin{array}{r} {}^{1}\phantom{6} \\ 36 \\ +\ 48 \\ \hline 84 \end{array}$$

For each ancient system of numeration listed below, explain how this carrying procedure would work and use the procedure to add 178 + 47.

(a) Egyptian
(b) Roman
(c) Traditional Chinese

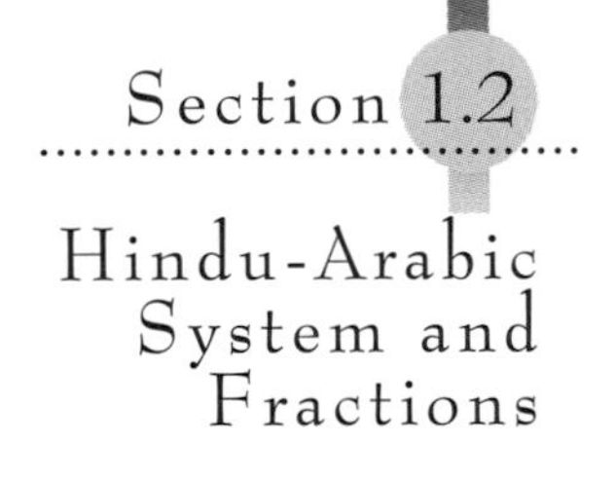

## Section 1.2 Hindu-Arabic System and Fractions

The Hindu-Arabic system is the numeration system used in the United States and many other parts of the world. It is also called the decimal system, from the Latin *deci*, meaning tenth. (Note that the Chinese system of numeration could also be called a decimal system.) The base of the Hindu-Arabic system is 10, and the symbols used in the system are the digits 0, 1, 2, 3, 4, 5, 6, 7, 8, and 9. The position a digit holds in a numeral gives it a certain value. The numeral 1,389,260,547 has place values as follows:

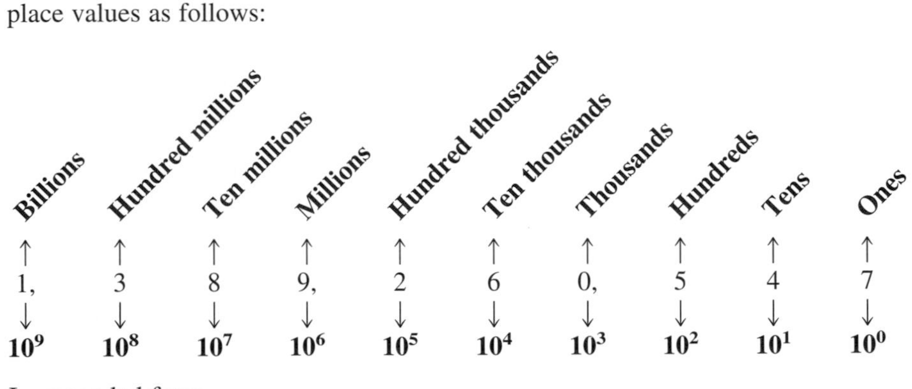


In expanded form,

$$\begin{aligned} 1{,}389{,}260{,}547 &= 1 \times 10^9 + 3 \times 10^8 + 8 \times 10^7 + 9 \times 10^6 \\ &\quad + 2 \times 10^5 + 6 \times 10^4 + 0 \times 10^3 + 5 \times 10^2 \\ &\quad + 4 \times 10^1 + 7 \times 10^0 \end{aligned}$$

The position of a digit tells us what it really represents. The 9 represents 9 millions

(9,000,000), the 6 represents 6 ten thousands (60,000), the 4 represents 4 tens (40), and so on.

**Example 1**

Write 4,175,280 in expanded form.

**Solution:**

$$\begin{aligned} 4{,}175{,}280 &= 4{,}000{,}000 + 100{,}000 + 70{,}000 + 5000 + 200 + 80 \\ &= 4 \times 10^6 + 1 \times 10^5 + 7 \times 10^4 + 5 \times 10^3 \\ &\quad + 2 \times 10^2 + 8 \times 10^1 \end{aligned}$$ ■

**Example 2**

In the numeral 576,239, what do the 5, 6, and 2 represent?

**Solution:**

5 represents 5 hundred thousands (500,000).
6 represents 6 thousands (6000).
2 represents 2 hundreds (200). ■

## Decimal Fractions

The decimal system also gives an efficient means of representing numbers that are less than a whole, numbers that fall between 0 and 1. Such numbers can be represented in two different forms, as a fraction or as a decimal (decimal fraction). For example, the shaded portion of the block shown below can be represented by the fraction 1/2 or the decimal 0.5.

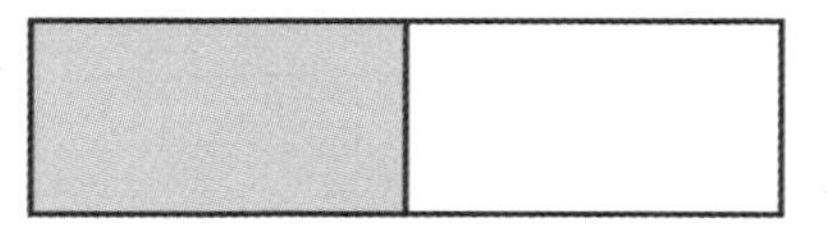

The decimal numbers simply extend the place-value system by using negative powers of 10. For example, 0.943271 means

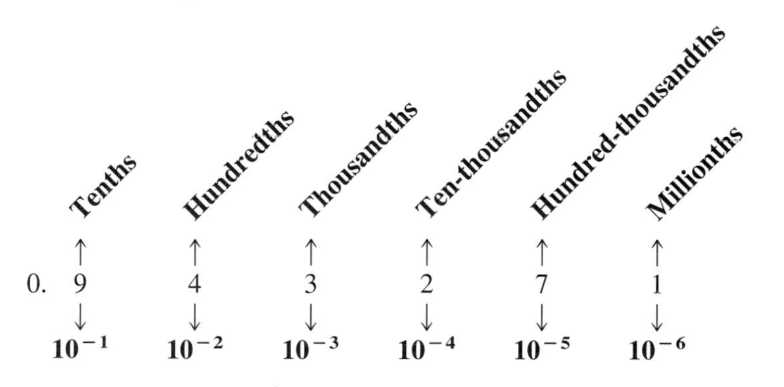

In expanded form,

$$\begin{aligned} 0.943271 &= 9 \times 10^{-1} + 4 \times 10^{-2} + 3 \times 10^{-3} + 2 \times 10^{-4} \\ &\quad + 7 \times 10^{-5} + 1 \times 10^{-6} \\ &= \frac{9}{10} + \frac{4}{100} + \frac{3}{1000} + \frac{2}{10{,}000} + \frac{7}{100{,}000} + \frac{1}{1{,}000{,}000} \end{aligned}$$

The position a digit holds to the right of the decimal point tells us what value it represents. The 9 represents 9 tenths (9/10), the 3 represents 3 thousandths (3/1000), the 1 represents 1 millionth (1/1,000,000), and so on.

### Example 3

In the numeral 1.7312, what do the digits 7, 3, and 2 represent?

**Solution:**

The 7 represents 7 tenths (7/10).
The 3 represents 3 hundredths (3/100).
The 2 represents 2 ten-thousandths (2/10,000). ■

### Example 4

Write 75.324 in expanded form.

**Solution:**

$$\begin{aligned} 75.324 &= 70 + 5 + 3/10 + 2/100 + 4/1000 \\ &= 7 \times 10^1 + 5 \times 10^0 + 3 \times 10^{-1} \\ &\quad + 2 \times 10^{-2} + 4 \times 10^{-3} \end{aligned}$$ ■

Some of the ancient systems of numeration discussed in the previous section also had methods of representing fractions. Although these systems were not as advanced as the decimal system, a look at them should prove interesting. You will have to refer to the numeration systems in the previous section to obtain the symbols used in ancient fractions.

## Egyptian Hieroglyphic Fractions

In this system the sign for a mouth, ◎, which in this context meant a part, was placed above a numeral to allow the Egyptians to represent fractions with a numerator of 1.

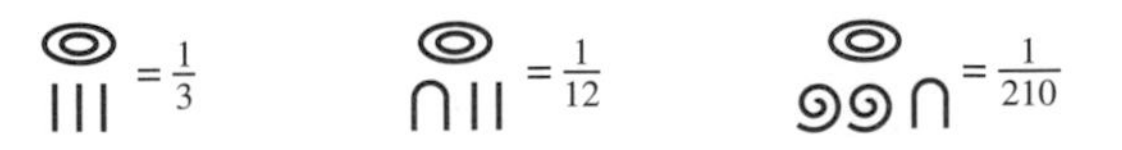

Any fraction whose numerator was not equal to 1 was represented as the sum of distinct fractions whose numerators were equal to 1.

◎ ◎ / ∩ || $= \frac{1}{10} + \frac{1}{2} = \frac{3}{5}$ ◎ ◎ ◎ / ||| ||||| ∩ $= \frac{1}{3} + \frac{1}{5} + \frac{1}{10} = \frac{19}{30}$

## Babylonian Fractions

The use of the symbols 𒀹𒀹 in the initial position of a numeral indicated that the number being represented was a fraction. A group of symbols following the 𒀹𒀹 became the numerator of a fraction, and successive powers of 60 (60, 3600, 216,000, etc.) were understood to be the denominator for each group of symbols. The Bablyonian fraction being represented was the sum of the individual fractions.

𒀹𒀹𒁹 $= \frac{1}{60}$ 𒀹𒀹 < 𒁹 𒁹 𒁹 < 𒁹 $= \frac{13}{60} + \frac{11}{3600} = \frac{791}{3600}$

## Roman Fractions

For the Romans fractions were mainly used in connection with their system of weights, 1 *as* (pound) = 12 *unciae* (ounces). Hence, their fractions were limited to parts of 12 (12ths). The table below gives the fractions of the Romans.

| | | | | | | | |
|---|---|---|---|---|---|---|---|
| . | or | – | → 1/12 | S . | or | S – | → 7/12 |
| . . | or | = | → 2/12 or 1/6 | S . . | or | S = | → 8/12 or 2/3 |
| . . . | or | = – | → 3/12 or 1/4 | S . . . | or | S = – | → 9/12 or 3/4 |
| . . . . | or | = = | → 4/12 or 1/3 | S . . . . | or | S = = | → 10/12 or 5/6 |
| . . . . . | or | = = – | → 5/12 | S . . . . . | or | S = = – | → 11/12 |
| | | S | → 6/12 or 1/2 | | | I | → 12/12 or 1 |

## Ionic Greek Fractions

The Greek scheme for representing fractions involved using a prime (′) on the right side of the numeral. Fractions with a numerator of 1 were written with a single prime to the right of the numeral.

$$\theta' = \frac{1}{9} \qquad \pi\epsilon' = \frac{1}{85} \qquad \rho' = \frac{1}{100}$$

Fractions with numerators greater than 1 were represented with the numerator written as a normal numeral and the denominator written twice with a prime to the right of each.

$$\epsilon\eta'\eta' = \frac{5}{8} \qquad \iota\alpha\lambda\epsilon'\lambda\epsilon' = \frac{11}{35}$$

### Example 5

For each fraction below, write its equivalent fraction in the Hindu-Arabic system.

(a) Egyptian: 𓂋𓏼 𓂋𓎆𓏤 (b) Babylonian: 𒐖 𒌍

(c) Roman: S . . (d) Greek: $\theta\pi\epsilon'\pi\epsilon'$

**Solution:**

(a) $1/3 + 1/11 = 11/33 + 3/33 = 14/33$

(b) $2/60 + 30/3600 = 4/120 + 1/120 = 5/120 = 1/24$

(c) $8/12 = 2/3$

(d) $\theta = 9$, $\pi\epsilon = 85$, so the fraction is $9/85$ ■

## Section 1.2 PROBLEMS

## *Explain → Apply → Explore*

### *Explain*

1. What is a decimal fraction? How is it different from a fraction?
2. How are exponents used in the Hindu-Arabic system of numeration?
3. What is the expanded form of a Hindu-Arabic numeral?
4. What is the major difficulty in writing a fraction such as 3/8 in Egyptian fractions?
5. How is the Babylonian system of fractions similar to the decimal system of fractions?
6. Compare Egyptian, Babylonian, Roman, and Ionic Greek fractions. Which system is easiest to use? Explain.

### *Apply*

In Problems 7–18, write each number in expanded form

| | |
|---|---|
| **7.** 139 | **13.** 543,867 |
| **8.** 0.53 | **14.** 0.03874 |
| **9.** 437.15 | **15.** 53.171 |
| **10.** 1,032,742 | **16.** 5083 |
| **11.** 0.314 | **17.** 0.62193 |
| **12.** 5.23 | **18.** 1.043 |

**19.** In Problems 7–18, what does the digit 3 represent?

In Problems 20–23, write the equivalent Hindu-Arabic fraction for each Egyptian fraction

**20.** ⊚ ⊚ ⊚
II IIII ∩

**21.** ⊚ ⊚
∩ IIII

**22.** ⊚ ⊚
୭∩ ୭

**23.** ⊚ ⊚ ⊚
𓀀 ୭୭୭ ∩∩

In Problems 24–27, write the equivalent Hindu-Arabic fraction for each Babylonian fraction given

**24.** ◂◂ ‹‹‹‹

**25.** ◂◂ ‹ ∨ ∨

**26.** ◂◂∨ ∨ ∨ ∨   ‹‹

**27.** ◂◂ ‹   ‹   ‹

In Problems 28–31, write the equivalent Hindu-Arabic fraction for each Roman fraction

**28.** . .

**29.** = =

**30.** S .

**31.** S = –

In Problems 32–35, write the equivalent Hindu-Arabic fraction for each Greek fraction

**32.** $\kappa\delta'$

**33.** $\xi\gamma'$

**34.** $\iota\lambda\beta'\lambda\beta'$

**35.** $\mu\theta\phi\alpha'\phi\alpha'$

### *Explore*

The sum of a whole number and a fraction is called a mixed number. In the Hindu-Arabic system 5 + 2/3 is written as 5 2/3. In Problems 36–39, find the mixed numbers represented by each ancient numeral.

**36.** Egyptian: 𓆼 𓍢 𓂋∩ 𓂋IIII

**37.** Babylonian: < ∨ ∨ < < ∨ ∨ ∨ 𒁹𒁹 < ∨ ∨

**38.** Roman: X X I V S =

**39.** Greek: $\rho\alpha\lambda\delta'$

Changing a Hindu-Arabic fraction into the corresponding Egyptian fraction requires that you represent the Hindu-Arabic fraction as the sum of distinct fractions that have a numerator of 1. In Problems 40–44, find the Egyptian fractions

**40.** 3/4

**41.** 5/12

**42.** 7/10

**43.** 8/15

**44.** 31/100

In Problems 45–48, write $1/3 + 1/2 = 5/6$ in the ancient system of numeration.

**45.** Egyptian

**46.** Babylonian

**47.** Roman

**48.** Greek

**49.** How would a Roman merchant write the following calculation?

$$12 \times 3\frac{3}{4} = 45$$

**50.** How would an Ancient Egyptian merchant write the following calculation?

$$20 \times 3\frac{3}{5} = 72$$

**51.** How would an Ionic Greek, merchant write the following calculation?

$$120 \times 3\frac{3}{4} = 450$$

**52.** How would a Babylonian merchant write the following calculation?

$$70 \times 3\frac{3}{5} = 252$$

**53.** Devise a system that would allow the Romans to write fractions other than 12ths. Give three examples and explain how your system works.

**54.** Devise a system that would allow the Mayans to write fractions. Your method should be consistent with the existing Mayan system. Give three examples and explain how your system works.

**55.** The Traditional Chinese system of numeration has a way to represent fractions. Do some research to find out how this is done. Give three examples and explain how the system works.

## Section 1.3 Numeration Systems with Other Bases

The decimal system uses the powers of 10 to determine the place value of each digit used in a numeral. The base of 10 is the result of human beings having ten fingers. However, as we saw in Section 1.1, other place-value systems did not use a base of 10. The Babylonians had a system based on 60, whereas the Mayan system was based on 20. Primitive tribes have been discovered that had a system of numeration based on 5, the number of fingers on one hand. The Duodecimal Society of America in the 1960s advocated the change to a base of 12. Computers, on the other hand, use a base of 2 since an electric pulse is in one of two states—on or off. If animals could develop a system of numeration, a horse might use a base 4 system, an octopus a base 8, an ant a base 6, and so on. In this section, we investigate how to write numbers in different bases and how to convert a numeral written in one base to the corresponding numeral in another base.

### The Place-Value System for Any Base

The first component of any place-value system is its base. The base of a numeration system is a whole number that is larger than 1. The integer powers of the base give each position in a numeral its place value. If we let a dot separate the whole number part and the fractional part of a number, for any base $b$ to the left of the dot the place values are the nonnegative integer powers of the base and to the right of the dot the place values are the negative powers of the base.

| Whole-number part | | | | Fractional part | | | |
|---|---|---|---|---|---|---|---|
| # | # | # | # | # | # | # | # |
| ↓ | ↓ | ↓ | ↓ | ↓ | ↓ | ↓ | ↓ |
| $\ldots b^3$ | $b^2$ | $b^1$ | $b^0$ | $b^{-1}$ | $b^{-2}$ | $b^{-3}$ | $b^{-4}\ldots$ |

The second component of a place-value system is the set of digits. The digits are symbols that represent the quantities from 0 to 1 less than the base. The base determines the number of symbols that are in the system. In the decimal system the base is 10, and it has 10 digits (0, 1, 2, 3, 4, 5, 6, 7, 8, and 9). The following is a summary of some numeration systems with various bases. It includes the base, the digits, and the place value for whole numbers from the right to the left of the numeral. Notice that base 12 and base 16 require the use of letters to represent some numbers greater than 9. For example, the letter A represents the number 10.

| System | Base | Digits | Place Values |
|---|---|---|---|
| Binary | 2 | 0, 1 | 1, 2, 4, 8, 16, . . . |
| Quintary | 5 | 0, 1, 2, 3, 4 | 1, 5, 25, 125, 625, . . . |
| Octal | 8 | 0, 1, 2, 3, 4, 5, 6, 7 | 1, 8, 64, 512, 4096, 32,768, 262,144, . . . |
| Duodecimal | 12 | 0, 1, 2, 3, 4, 5, 6, 7, 8, 9, A(10), B(11) | 1, 12, 144, 1728, 20,736, 248,832, . . . |
| Hexadecimal | 16 | 0, 1, 2, 3, 4, 5, 6, 7, 8, 9, A(10), B(11), C(12), D(13), E(14), F(15) | 1, 16, 256, 4096, 65,536, 1,048,576, . . . |

When you read a numeral in a base other than 10, read each digit separately. For example, the numeral $123_5$ is read "one two three, base five" and not "one hundred twenty-three, base five." The reason is that the concept of "hundreds" is part of the base 10 system and therefore should not be used in other bases.

## Converting to a Decimal Numeral

To understand what number or amount is being represented by a numeral in a base other than 10, we need to convert it to the system we are familiar with, the decimal system. For example, the numeral 23014 in base 5, written $23014_5$, represents an amount. To understand what amount that is, we will convert it to a decimal numeral.

### Example 1

Write $23014_5$ in base 10.

**Solution:** Use the powers of 5 for the place value for each digit in the numeral.

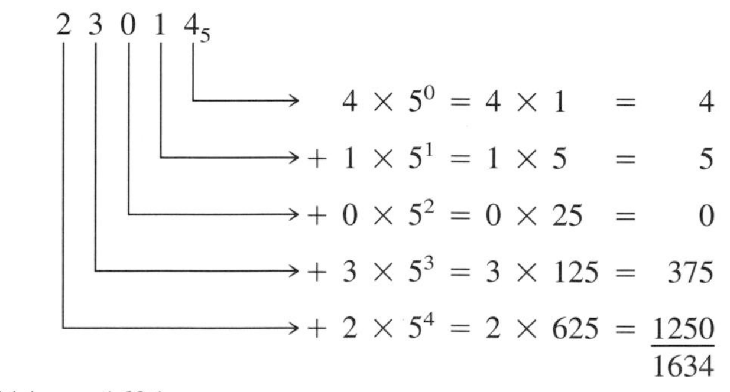

$2\ 3\ 0\ 1\ 4_5$
$4 \times 5^0 = 4 \times 1 = 4$
$+ 1 \times 5^1 = 1 \times 5 = 5$
$+ 0 \times 5^2 = 0 \times 25 = 0$
$+ 3 \times 5^3 = 3 \times 125 = 375$
$+ 2 \times 5^4 = 2 \times 625 = 1250$
1634


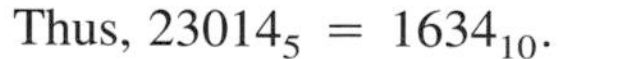

Thus, $23014_5 = 1634_{10}$. ■

### Example 2

Write $17A6_{12}$ in base 10.

**Solution:** Use the powers of 12 for the place value of each digit in the numeral.

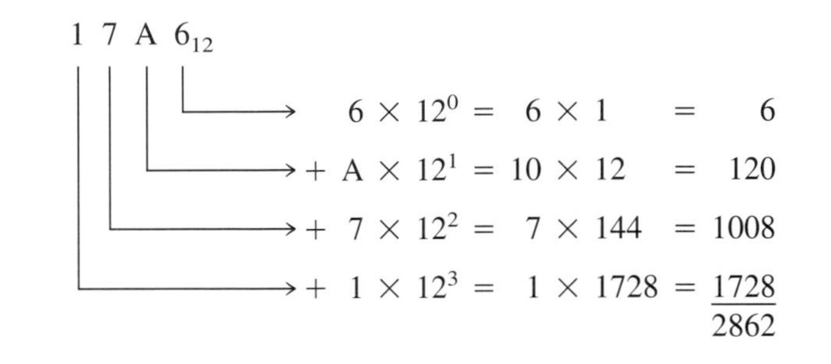


Thus, $17A6_{12} = 2862_{10}$.

### Example 3

Write $101\ 101_2$ in base 10.

**Solution:** Use the powers of 2 for the place values of each digit in the numeral.

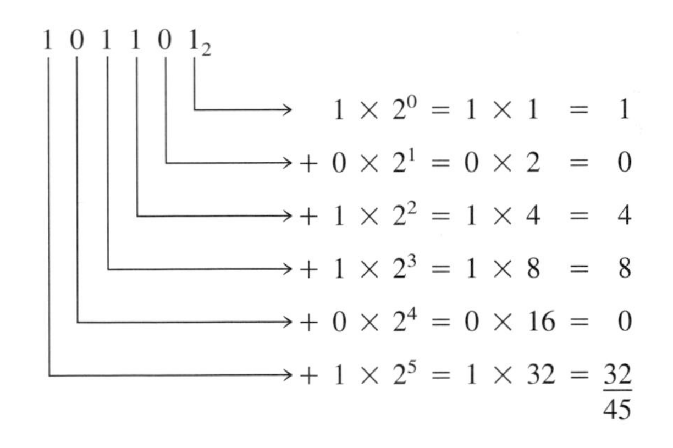


Thus, $101\ 101_2 = 45_{10}$.

### Example 4

Write $86DC0_{16}$ in base 10.

**Solution:** Use the powers of 16 for the place values of each digit in the number.

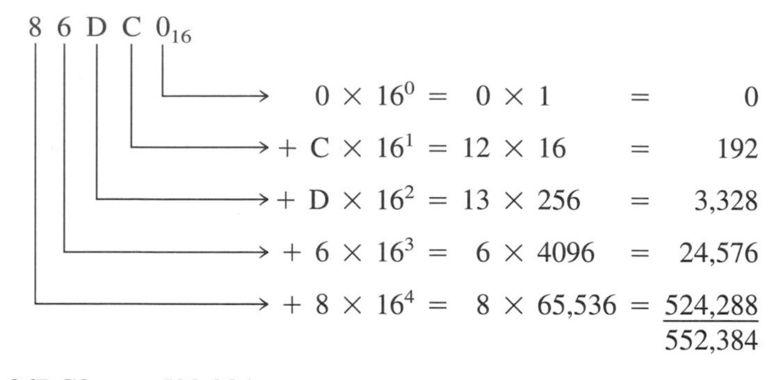

8 6 D C $0_{16}$
$0 \times 16^0 = 0 \times 1 = 0$
$+ C \times 16^1 = 12 \times 16 = 192$
$+ D \times 16^2 = 13 \times 256 = 3{,}328$
$+ 6 \times 16^3 = 6 \times 4096 = 24{,}576$
$+ 8 \times 16^4 = 8 \times 65{,}536 = 524{,}288$
552,384


Thus, $86DC0_{16} = 522{,}384_{10}$ ■

## Converting Decimal Numerals to Other Bases

To convert a decimal number to another base, we need to find how many groups of each appropriate place value are contained in the decimal number. The division scheme shown in the examples that follow will allow you to convert a base 10 numeral into another base.

### Example 5

Write 89 in base 5.

**Solution:** The powers of 5 that are less than 89 are 25, 5, 1. We need to find how many groups of each of those place values are contained in 89. We can do that by using the division scheme shown below.

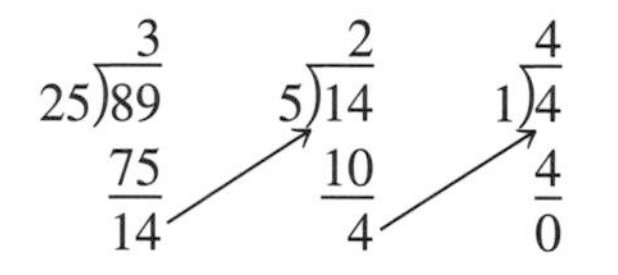


The number 89 contains 3 groups of 25, 2 groups of 5, and 4 groups of 1.

$$\begin{aligned} 89_{10} &= 3 \times 25 + 2 \times 5 + 4 \times 1 \\ &= 3 \times 5^2 + 2 \times 5^1 + 4 \times 5^0 \\ &= 324_5 \end{aligned}$$ ■

### Example 6

Write 204 in binary.

**Solution:** The powers of 2 that are less than 204 are 128, 64, 32, 16, 8, 4, 2, and 1. Use the division scheme of Example 5.

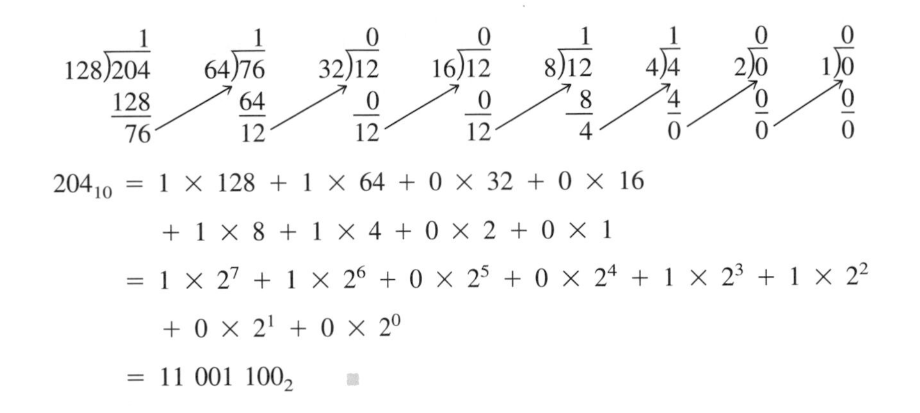

$$
\begin{aligned}
204_{10} &= 1 \times 128 + 1 \times 64 + 0 \times 32 + 0 \times 16 \\
&\quad + 1 \times 8 + 1 \times 4 + 0 \times 2 + 0 \times 1 \\
&= 1 \times 2^7 + 1 \times 2^6 + 0 \times 2^5 + 0 \times 2^4 + 1 \times 2^3 + 1 \times 2^2 \\
&\quad + 0 \times 2^1 + 0 \times 2^0 \\
&= 11\ 001\ 100_2
\end{aligned}
$$

## Example 7

Write the decimal number 10,000 in base 8.

**Solution:** The powers of 8 that are less than 10,000 are 4096, 512, 64, 8, and 1. Use the division scheme of previous examples.

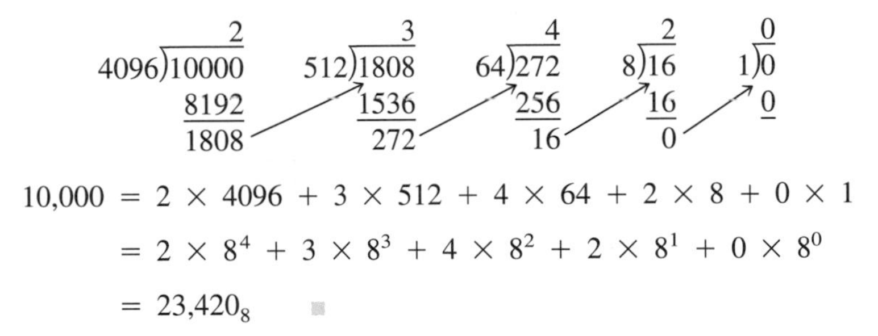

$$
\begin{aligned}
10{,}000 &= 2 \times 4096 + 3 \times 512 + 4 \times 64 + 2 \times 8 + 0 \times 1 \\
&= 2 \times 8^4 + 3 \times 8^3 + 4 \times 8^2 + 2 \times 8^1 + 0 \times 8^0 \\
&= 23{,}420_8
\end{aligned}
$$

## Example 8

Write the decimal number 40,600 in base 16.

**Solution:** The powers of 16 that are less than 40,600 are 4096, 256, 16, and 1. Use the division scheme again.

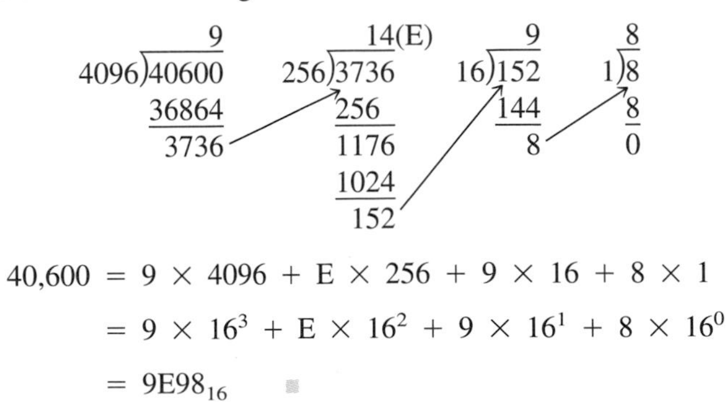

$$
\begin{aligned}
40{,}600 &= 9 \times 4096 + E \times 256 + 9 \times 16 + 8 \times 1 \\
&= 9 \times 16^3 + E \times 16^2 + 9 \times 16^1 + 8 \times 16^0 \\
&= 9E98_{16}
\end{aligned}
$$

## Fractions in Other Bases

A base 10 number less than 1 can be represented using a decimal point or negative integer powers of 10. For example,

$$0.2358 = 2 \times 10^{-1} + 3 \times 10^{-2} + 5 \times 10^{-3} + 8 \times 10^{-4}$$
$$= 2 \times \frac{1}{10} + 3 \times \frac{1}{100} + 5 \times \frac{1}{1000} + 8 \times \frac{1}{10{,}000}$$

In other bases, the dot used to separate the whole number part of a number and the fractional part has different names. It is called the quintary point in base 5, the binary point in base 2, the octal point in base 8, and so on. However, the generic term used for the dot in any base is the **basimal** point. Fractional numbers can be represented in any base by using methods similar to those used with decimal numbers.

### Example 9

Write $0.2314_5$ as a base 10 numeral.

**Solution:**

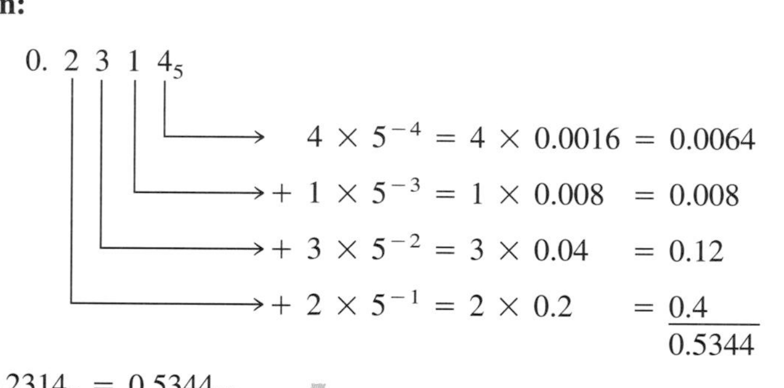

0. 2 3 1 $4_5$
$4 \times 5^{-4} = 4 \times 0.0016 = 0.0064$
$+ 1 \times 5^{-3} = 1 \times 0.008 = 0.008$
$+ 3 \times 5^{-2} = 3 \times 0.04 = 0.12$
$+ 2 \times 5^{-1} = 2 \times 0.2 = 0.4$
0.5344


Thus, $0.2314_5 = 0.5344_{10}$. ■

### Example 10

Write $53.72_8$ as a base 10 numeral.

**Solution:**

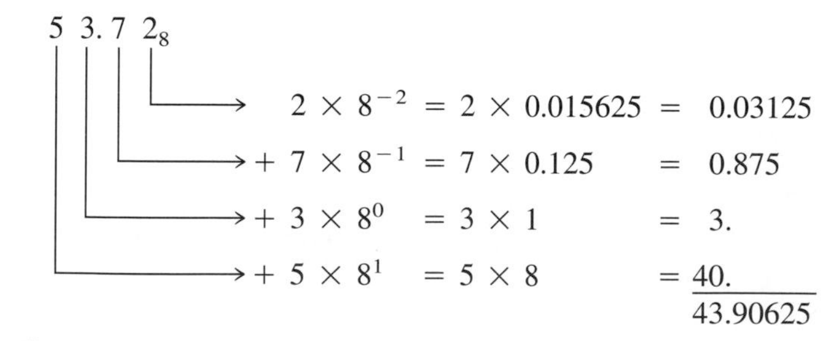

5 3. 7 $2_8$
$2 \times 8^{-2} = 2 \times 0.015625 = 0.03125$
$+ 7 \times 8^{-1} = 7 \times 0.125 = 0.875$
$+ 3 \times 8^{0} = 3 \times 1 = 3.$
$+ 5 \times 8^{1} = 5 \times 8 = 40.$
43.90625


Thus, $53.72_8 = 43.90625_{10}$. ■

Looking back at the survey of the different systems of numeration, you can appreciate the advantages of the Hindu-Arabic system. First, the Hindu-Arabic system uses a small number of uncomplicated symbols. After memorizing these ten digits, a person can write any numeral. This is an improvement over the large number of symbols used by the Ionic Greeks and the elaborate characters of the Egyptian system.

Using too few characters also has drawbacks. While the Babylonian system and base 2 use only two characters, both methods require using the symbols many times to express a relatively small number. For example,

$$59 = \begin{array}{c} <<<<< \\ \vee\vee\vee\vee\vee \\ \vee\vee\vee\vee \end{array} = 111\ 011_2$$

Our system strikes a balance between using a large number of symbols and requiring a large number of digits to write a relatively small number.

A second advantage of the Hindu-Arabic system is its use of place value. While initially more difficult to learn, a place value system is advantageous for writing most large numbers. The advantages of a place value system become clear when writing 888 in a system such as the Roman system:

$$888 = \text{DCCCLXXXVIII}$$

Finally, we leave it to your imagination to envision a long division problem such as 10,488 ÷ 23 in any system other than our own.

## Section 1.3 PROBLEMS

## *Explain ➡ Apply ➡ Explore*

### *Explain*

1. How many symbols are used to create numerals in base 10?
2. How many symbols would be needed to create numerals in base 17? What are they?
3. What is the common belief as to why modern cultures use a base ten system?
4. Why should the number $135_7$ be read "one three five base seven," rather than "one hundred thirty-five base 7"?
5. Describe in words the method for converting the numeral $12_8$ into a decimal numeral.
6. Describe in words the method for converting the Hindu-Arabic numeral 12 into a numeral in base 8.

### *Apply*

Write each of the following as a decimal numeral.

**7.** $302_5$

**8.** $140.32_5$

**9.** $101.101_2$

**10.** $100\ 111\ 001_2$

**11.** $5610_8$

**12.** $70{,}037.2_8$

**13.** $7A4.3_{12}$

**14.** $123ABC.D_{16}$

**15.** $253_7$

**16.** $1068_9$

**17.** $1111_2$

**18.** $1111_5$

Write the following decimal numerals in the specified base.

**19.** Write 401 as a numeral in base 5.

**20.** Write 186 as a numeral in base 4.

**21.** Write 990 as a numeral in base 8.

**22.** Write 88,888 as a numeral in base 8.

**23.** Write 777 as a numeral in base 7.

**24.** Write 7777 as a numeral in base 7.

**25.** Write 1860 as a numeral in base 12.

**26.** Write 4235 as a numeral in base 12.

**27.** Write 129 as a numeral in base 16.

**28.** Write 16,016 as a numeral in base 16.

**29.** Write 186,000 as a numeral in base 9.

**30.** Write 32 as a numeral in base 2.

**31.** Write 222 as a numeral in base 2.

**32.** Write 2222 as a numeral in base 2.

Represent the number of days in a leap year (366) in each of the following systems of numeration.

**33.** Quintary

**34.** Binary

**35.** Octal

Represent the number of pounds in a ton (2000) in each of the following systems of numeration.

**36.** Duodecimal

**37.** Hexadecimal

**38.** Base 6

Represent the number 1,000,000 in each of the following systems of numeration.

**39.** Quintary

**40.** Binary

**41.** Octal

**42.** Duodecimal

**43.** Hexadecimal

**44.** Base 6

*Explore*

In each pair of numerals, determine which one has a larger value.

**45.** $254_9$ or $12202_3$

**46.** $6C_{16}$ or $253_6$

**47.** $101\ 101_2$ or $3033_4$

**48.** $13.421_5$ or $1.421_8$

**49.** Solve the following $45_6 + 67_8 = \_\_\_\_\__7$

**50.** Solve the following $23_4 + 23_5 = \_\_\_\_\__6$

**51.** Suppose $200_B = 128$. Find B.

**52.** Suppose $301_B = 76$. Find B.

**53.** (a) Write the numbers 45, 100, 200 in base 3 and base 9.
(b) What do you notice about the number of digits in the numerals as the base increases?
(c) Write each of the base 3 numerals in groups of two digits, starting at the right most digit. Compare these with the base 9 numerals. Do you notice the pattern? Explain the pattern.

**54.** (a) Write the numbers 6, 18, and 45 in base 2, base 4 and base 8.
(b) What do you notice about the number of digits in the numerals as the base increases?
(c) Write each of the base 2 numerals in groups of two digits, starting at the right most digit. Compare these with the base 4 numerals. Do you notice the pattern? Explain the pattern.
(d) Write each of the base 2 numerals in groups of three digits, starting at the right most digit. Compare these with the base 8 numerals. Do you notice the pattern? Explain the pattern.

**55.** Create a system of numeration for base 32. What are its digits? Write two 4-digit numerals in base 32 and determine the equivalent Hindu-Arabic numerals.

**56.** Create a system of numeration for base 64. What are its digits? Write two 4-digit numerals in base 64 and determine the equivalent Hindu-Arabic numerals.

**57.** Consider your telephone number as a 3-digit number followed by a 4-digit number. Write your telephone number in at least two different bases greater than 10. What do you notice about the number of digits in the telephone number when written in these bases?

**58.** Consider your telephone number as a 3-digit number followed by a 4-digit number. Write your telephone number in at least two different bases less than 10. What do you notice about the number of digits in the telephone number when written in these bases?

*O*, *e*, and *i* by David McLaughlin (courtesy of the artist) and $\pi$ by Tom Marioni (courtesy of Crown Point Press): artists' representation of four historically significant numerical constants.

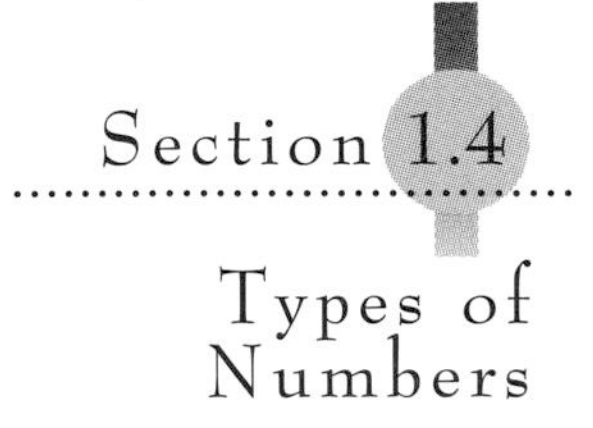

# Section 1.4 Types of Numbers

Numbers are a basic part of our daily lives. They are all around us. As seen in this chapter, they have been of interest to human beings from earliest times. Just as humans classified the animals, insects, plants, and objects around them, they also classified their numbers. So as we continue our study of mathematics, it would be appropriate to study the many types of numbers that have been classified.

## Real Numbers

The **natural numbers**, also called counting numbers, consist of the numbers {1, 2, 3, 4, . . .}. The **whole numbers** consist of the natural numbers and zero {0, 1, 2, 3, 4, . . .}. The **integers** consist of the whole numbers and the negatives of the whole numbers, {. . . −4, −3, −2, −1, 0, 1, 2, 3, 4, . . .}.

The **rational numbers** are numbers that can be represented by the quotient $a/b$, where $a$ and $b$ are integers and $b \neq 0$. It can be shown that when these quotients are converted into decimals, they are either terminating or repeating decimals. For example, the rational number 3/5 is the terminating decimal 0.6, −7/4 is −1.75, 53 is 53.0, 156 17/25 is 156.68, 2/3 is the repeating decimal 0.666 . . . , −36/11 is −3.2727 . . . , and 43 1/7 is 43.142857142857 . . . .

The **irrational numbers** are decimal numbers that do not terminate and do not repeat (as the rational numbers do). It can be shown that an irrational number cannot be written as a ratio of two integers. Included in this group of numbers are radical numbers, such as $\sqrt{2}$, $\sqrt{19}$, $\sqrt[4]{101}$, $\sqrt[3]{34}$, and mathematical constants, such as $\pi$ and $e$. The root spiral in Section 5.0 gives a visual representation of the square roots of the natural numbers.

The **real numbers** consist of all the rational and irrational numbers. They can be visualized by using a horizontal number line where a zero point (the origin) and a unit length are marked off. Each real number corresponds to exactly one point on the line, and each point on the line corresponds to exactly one real number. Numbers that are larger than zero, the positive numbers, are placed to the right of zero, and numbers that are less than zero, the negative numbers, are placed to the left of zero.

The Real Number Line

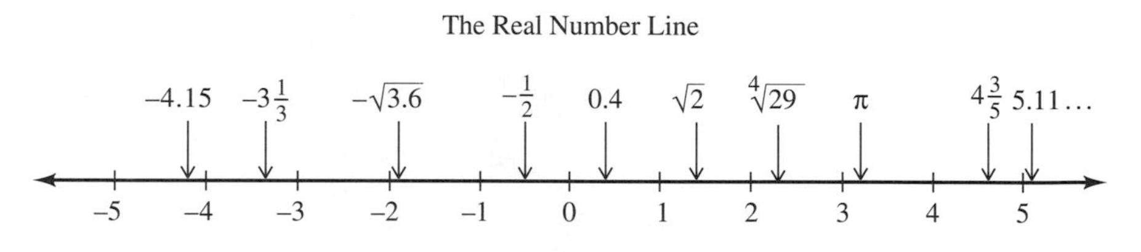


The following chart shows the classification of the real numbers.

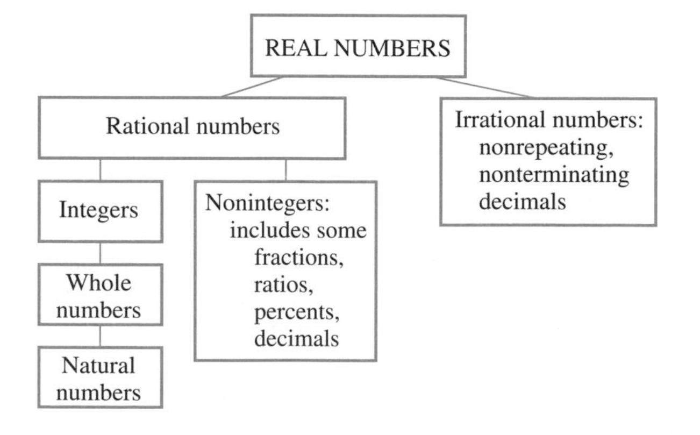

**Example 1**

$-10,\ -4.67,\ -\sqrt{17}\ -\ 3\frac{5}{8},\ -\sqrt[5]{2.9},\ -1,\ -\frac{2}{3},\ 0,\ 0.333\ldots,\ \sqrt[3]{8.41},\ \frac{5}{2},$
$\sqrt{25},\ \sqrt{27},\ 7.87,\ 27$

From the list above, determine which of the numbers are (a) integers, (b) irrational numbers, (c) natural numbers, (d) rational numbers, (e) real numbers, and (f) whole numbers.

**Solution:**

(a) Integers: $-10,\ -1,\ 0,\ \sqrt{25},\ 27$

(b) Irrational numbers: $-\sqrt{17},\ -\sqrt[5]{2.9},\ \sqrt[3]{8.41},\ \sqrt{27}$

(c) Natural numbers: $\sqrt{25},\ 27$

(d) Rational numbers: all except those in part (b)

(e) Real numbers: all numbers in the list

(f) Whole numbers: $0,\ \sqrt{25},\ 27$ ■

**Example 2**

Explain why $\sqrt{36}$ can be classified as a rational number, a natural number, a whole number, and an integer but not an irrational number.

**Solution:** Since $\sqrt{36} = 6$, it is a rational number (6/1), a natural number (1, 2, 3, 4, . . .), a whole number (0, 1, 2, 3, 4, . . .), and an integer (. . . −3, −2, −1, 0, 1, 2, 3, . . .). It is not an irrational number because it can be expressed as a quotient of integers (6/1) and because, as a decimal, it terminates (6.0). ■

## Zero and Negative Numbers

Before we continue our investigation of other types of numbers, we will look at the origins of zero and negative numbers. Though sometimes taken for granted, zero frequently finds its way into our daily activities. The symbol for zero, 0, is used quite often when we write numerals and perform computations. Negative numbers have also gained acceptance today. We use a negative number for an amount less than zero, such as the temperature at the North Pole or a checking account balance after writing a check for more money than is in the account. Surprisingly, this modern use and acceptance of zero and negative numbers did not occur overnight; it took many centuries for these mathematical concepts to be accepted by the scientific community.

Let's examine some facts about zero. The concept of zero, which indicates the absence of a quantity, has most likely been understood since prehistoric times. Each time a hunter came home without any game, the number zero was experienced. Though the early Egyptian, Greek, and Roman civilizations understood the concept of zero, they had no symbol for zero. Their systems of numeration did not need a symbol for zero. Zero was, however, represented on ancient Babylonian tablets and Chinese counting boards as a blank space used for a missing place value in a numeral. In the fourth century B.C., the symbol was used by the Babylonians, and the symbol was used by the Mayans of South America in place of the blank space. The symbol 0 is believed to have originated in India some 20 years after forms of the other nine numerals appeared. This symbol for zero developed before A.D. 870, since it was contained on an A.D. 870 inscription in Gwalior, India. However, some experts believe that this symbol may have come to India by way of Indochina, since inscriptions in Cambodia and Sumatra (A.D. 683) used this same symbol for zero.

In Western culture, the use of the symbol 0 is a fairly recent development. It became well established in Europe during the late 1400s when the Hindu-Arabic system replaced Roman numerals. Finally, the word "zero" comes from the Latin word *ziphrum*. *Ziphrum* is a translation from the Arabic word *sifr*, which came from the Hindu word *sunya*, meaning void or empty.

As with the number zero, negative numbers have an interesting history. There is no trace of negative numbers in ancient Egyptian, Babylonian, or Greek writings. In A.D. 270, when negative numbers occurred as solutions to equations, the Greek mathematician Diophantus dismissed them as being absurd. This had such an effect that it was not until the Renaissance (14th and 15th centuries) that mathematicians began to be more receptive to negative numbers.

Negative numbers gained acceptance in Europe in the 16th century through the work *Ars Magna* by Girolamo Cardano (1545), who used negative numbers as solutions to equations, and the work of Michael Stifel (1544), who described negative numbers as those numbers that are less than zero. The terms "positive" and "affirmative" were used to indicate positive numbers, and the terms "privative," "negative," and "minus" were used for negative numbers.

In non-Western cultures, such as China, India, and Arabia, however, negative numbers were readily accepted. In the second century B.C., Chinese counting boards used red or triangular rods to represent positive numbers and black or square rods to represent negative numbers. In India, c. A.D. 628, Brahmagupta mentioned negative numbers, and the Hindu and Arabian mathematicians that followed continued to use negative numbers in their arithmetic and algebra.

## Complex Numbers

Besides the real numbers described previously, there is another set of numbers based on $\sqrt{-1}$. These numbers are called **imaginary numbers** and use the letter $i$, where $i = \sqrt{-1}$. For example, $\sqrt{-16} = 4\sqrt{-1} = 4i$, $\sqrt{-27} = 3\sqrt{3}\sqrt{-1} = 3\sqrt{3}i$, and $\sqrt{-93} = \sqrt{93}i$. If a real number and an imaginary number are added together, the sum is called a **complex number** and is written in the form $a + bi$, where $a$ and $b$ are real numbers. Thus,

$$5 + \sqrt{-9} = 5 + 3i \qquad \text{and} \qquad -7.2 - \sqrt{-20} = -7.2 - 2i\sqrt{5}$$

Imaginary numbers are a recent mathematical development. Until the 1500s, square roots of negative numbers were considered an impossibility. The work of Girolamo Cardano (1545) and Rafael Bombelli (1572) introduced imaginary numbers as roots of equations. René Descartes (1637) called them ''imaginary,'' and Leonhard Euler (1748) used $i$ to represent $\sqrt{-1}$. Though imaginary numbers are not used in everyday transactions, they are used to solve problems in mathematics, electronic circuit design, vibration analysis, and other branches of science and engineering.

## Numbers Based on Geometric Shapes

Polygonal numbers are numbers that were devised to conform to basic geometric shapes. These geometric-based numbers were of interest especially to the Greeks because of their simplistic geometric beauty and the many different mathematical patterns found between the terms of each number and between different polygonal numbers.

**Triangular numbers** take the shape of triangles, as pictured below.

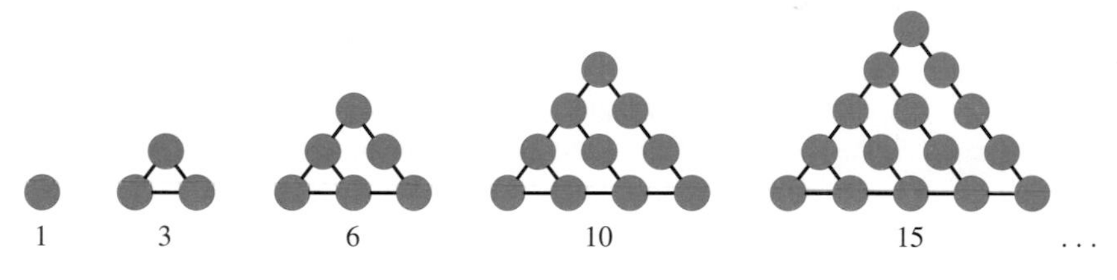


**Square numbers** take the shape of squares, as pictured below.

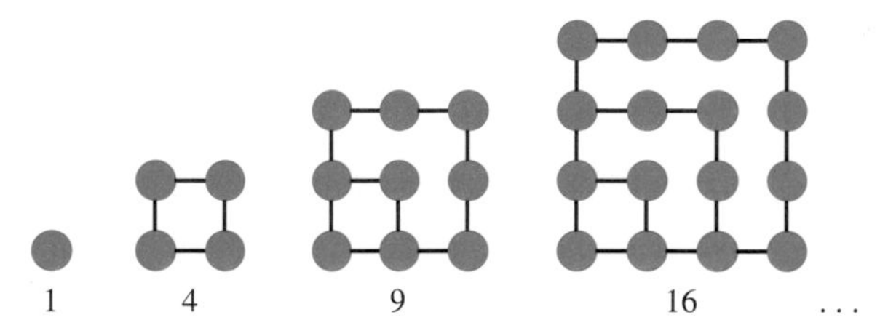


**Pentagonal numbers** take the shape of pentagons, as pictured below.

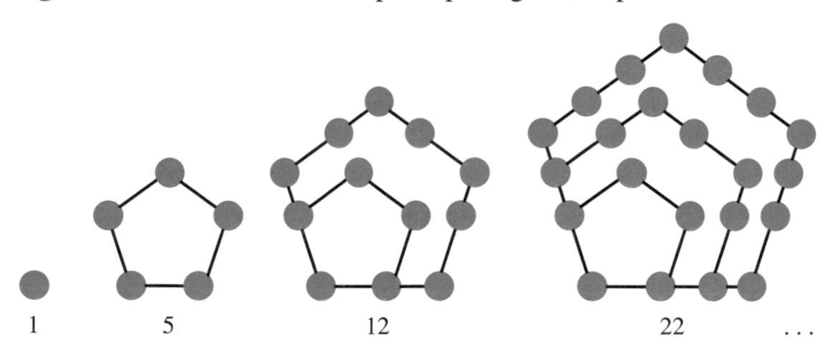

Other polygonal numbers can be formed by using other regular geometric shapes such as hexagons, octagons, and decagons.

### Example 3

Find the next two triangular numbers and describe the pattern that exists in going from one triangular number to the next.

**Solution:** The next two triangular numbers are 21 and 28. If you list the triangular numbers and find the difference between successive terms, the differences are the whole numbers 2, 3, 4, 5, 6, . . . . Thus, to get from one triangular number to the next, just continue adding consecutive integers. ■

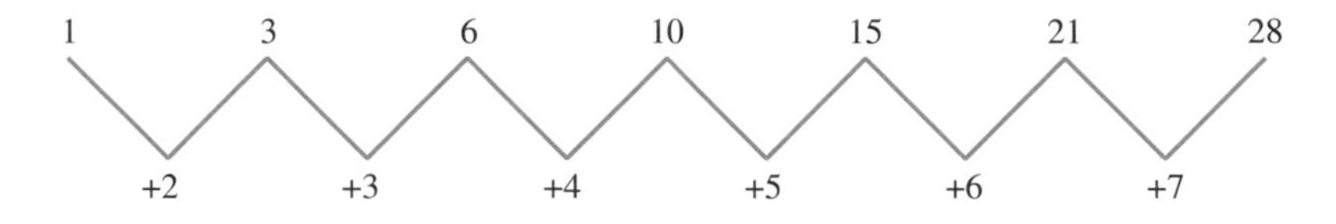


## Magic Squares and Cubes

Humankind's fascination with numbers also led to the creation of magic squares and cubes. The ancients believed that square arrays of numbers that had the same sum horizontally, vertically, and diagonally contained mystical powers while exhibiting the harmony of numbers, mathematical regularity, and symmetry. Probably the most ancient of these squares, the Lo Shu square, dates back to the Chinese emperor Yu the Great, who reigned from 2205 to 2198 B.C. In the square array of numbers represented by knots on strings, the gold knots represent even numbers and the green knots represent odd numbers. In the square, all rows, columns, and diagonals have a sum of 15.

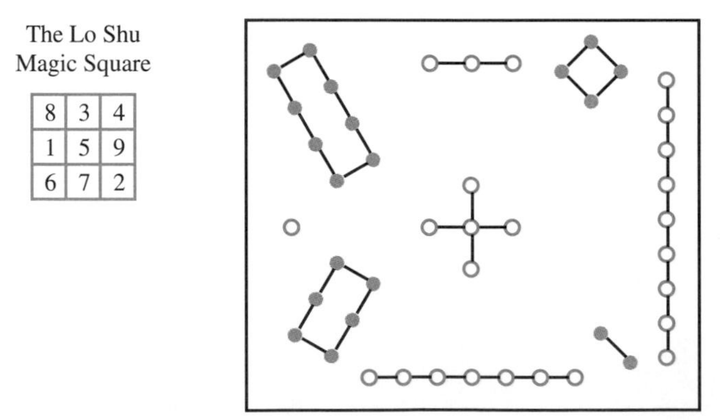

The Lo Shu Magic Square

| | | |
|---|---|---|
| 8 | 3 | 4 |
| 1 | 5 | 9 |
| 6 | 7 | 2 |

Any one of eight possible arrangements of those nine numbers can be used to give $3 \times 3$ magic squares. Magic squares have also been found of larger sizes in

**Figure 1.4.1**

| 14 | 7 | 11 | 2 |
|---|---|---|---|
| 1 | 12 | 8 | 13 |
| 4 | 9 | 5 | 16 |
| 15 | 6 | 10 | 3 |

| 17 | 24 | 1 | 8 | 15 |
|---|---|---|---|---|
| 23 | 5 | 7 | 14 | 16 |
| 4 | 6 | 13 | 20 | 22 |
| 10 | 12 | 19 | 21 | 3 |
| 11 | 18 | 25 | 2 | 9 |

various cultures throughout history. Japanese mathematicians of the 1600s were especially attracted to magic squares. Muramatsu (1663) determined magic squares up to 19 rows by 19 columns, and Seki Kōwa (1666) developed rules for creating magic squares of various large dimensions. Figure 1.4.1 displays $4 \times 4$ and $5 \times 5$ magic squares.

In a San Francisco Bay Area science fair project in 1982, Kevin Staszkow, a son of one of the authors, used an Apple II computer to generate magic cubes. In these cubes the whole numbers from 1 to 27 were arranged in three layers with nine numbers in each layer. Forty-two was the identical sum of the three numbers in a horizontal row, a vertical column, or one of the cube's four diagonals. Kevin did not know that magic cubes were studied in the past without the use of computers, especially by Japanese mathematicians Tanaka Kisshin (1662) and Kurushima Gita (1757). Kurushima Gita determined a $4 \times 4 \times 4$ magic cube that uses the whole numbers from 1 to 64 and has a magic sum of 130. Kevin's experience does, however, show that mathematics can be rediscovered. It can cause as much excitement and sense of accomplishment in the new discoverer as it did for the original discoverer. The three layers of one of Kevin Staszkow's Apple II–generated magic cubes are shown in Figure 1.4.2.

## Example 4

Verify that the diagonals of the magic cube shown in Figure 1.4.2 do have a sum of 42.

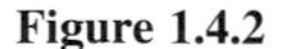

**Figure 1.4.2**

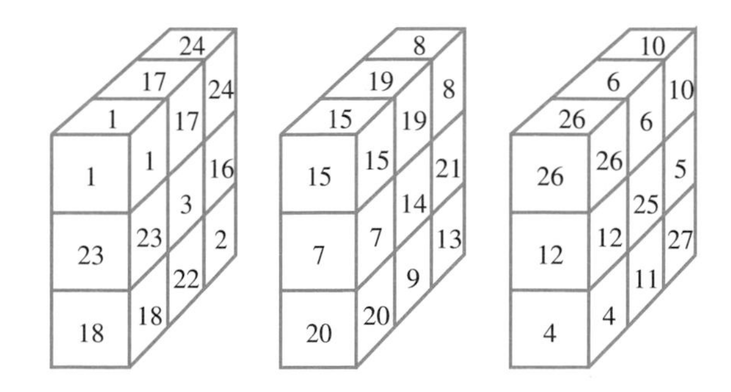

**Solution:** The diagonals of a cube go from one corner of the cube through the center of the cube to the opposite corner. In the cube shown the diagonals are

(a) $1 + 14 + 27 = 42$
(b) $18 + 14 + 10 = 42$
(c) $2 + 14 + 26 = 42$
(d) $24 + 14 + 4 = 42$ ■

## Numbers Based on Factors

A **proper factor** of a natural number is a natural number less than the number that divides evenly into the number. For example, the proper factors of 8 are 1, 2, and 4. A **prime number** is a natural number whose only proper factor is 1, and a **composite number** is a natural number that has proper factors greater than 1. According to these definitions, the number 1 is not a prime number. The first ten prime numbers are 2, 3, 5, 7, 11, 13, 17, 19, 23, and 29. It has been proved that there are an infinite number of primes and that every composite number can be represented as a product of prime numbers. The Greek scholar Eratosthenes (274–194 B.C.). invented an arithmetical sieve for finding prime numbers. For example, to find all the prime numbers less than 102, write down all the natural numbers from 2 to 101 (Fig. 1.4.3). The first number, 2, is a prime. Draw a box around it and cross out all other multiples of 2 (every second number: 4, 6, 8, 10, . . .). The next number that is not crossed, 3, is a prime. Draw a box around it and cross out all other multiples of 3 (every third number: 6, 9, 12, 15, 18, . . .). Continuing in this manner, you can see that there are 26 prime numbers less than 102.

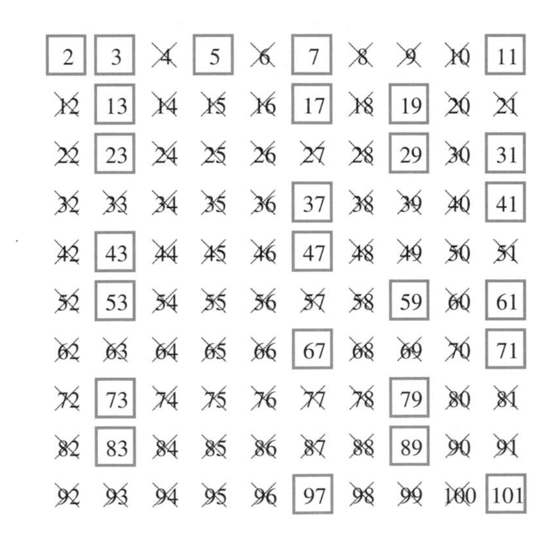


**Figure 1.4.3**

Since every composite number has proper factors, the properties of these factors have also been studied. The following types of numbers are a result of these investigations.

A **perfect number** is a natural number with the property that the sum of its proper factors equals the number. Six is the first perfect number because its proper factors, 1, 2, and 3, have a sum of 6 ($1 + 2 + 3 = 6$). Another perfect number is 496 because the sum of its proper factors, 1, 2, 4, 8, 16, 31, 62, 124, and 248 equals 496. In 1952, there were only 12 known perfect numbers. Since then, with the aid of computers, more perfect numbers have been found. Some of these have more than 100,000 digits.

An **abundant number** is a natural number with the property that the sum of its proper factors is greater than the number. Twelve is an abundant number because its proper factors 1, 2, 3, 4, and 6 have a sum of 16. One hundred is an abundant number because its proper factors 1, 2, 4, 5, 10, 20, 25, and 50 have a sum of 117.

A **deficient number** is a natural number with the property that the sum of its proper factors is less than the number. Twenty-seven is a deficient number because the sum of its proper factors 1, 3, and 9 equals 13. Forty-three is a deficient number because its only proper factor is 1.

**Amicable numbers** are two natural numbers such that the sum of the proper factors of one number equals the other number, and vice versa. The numbers 220 and 284 are amicable because the sum of the proper factors of 220 (1, 2, 4, 5, 10, 11, 20, 22, 44, 55, 110) is 284 and the sum of the proper factors of 284 (1, 2, 4, 71, 142) is 220. This pair of numbers has been ascribed to the Greek mathematician Pythagoras (c. 540 B.C.). The fact that each amicable number generates the other gives them an intimate relationship that played a role in mysticism and superstition through the ages. In 1636, the French mathematician Pierre Fermat discovered a second amicable pair, 17,296 and 18,416. In 1638, René Descartes discovered a third pair, and in 1750, Leonhard Euler found 60 other pairs. In 1866, a 16-year-old Italian, Nicolo Paganini, astounded the mathematical world when he discovered the small amicable pair of 1184 and 1210. There are more than 900 known amicable pairs of numbers, and they are still intriguing to mathematicians and computer enthusiasts.

## Example 5

Classify the numbers 18, 28, 31, and 45 as (a) composite or prime and (b) perfect, abundant, or deficient.

**Solution:**

18: (a) Composite number because 1 is not the only proper factor.
(b) Abundant number because the sum of its proper factors (1, 2, 3, 6, 9) equals 21, and $21 > 18$.

28: (a) Composite number because 1 is not the only proper factor.
(b) Perfect number because the sum of its proper factors (1, 2, 4, 7, 14) equals 28.

31: (a) Prime number because its only proper factor is 1.
(b) Deficient number because its only proper factor is 1, and $1 < 31$.

45: (a) Composite number because 1 is not its only proper factor.
(b) Deficient number because the sum of its proper factors (1, 3, 5, 9, 15) equals 33, and $33 < 45$. ■

### Example 6

Show that the numbers 17,296 and 18,416 found by Pierre Fermat are amicable numbers.

**Solution:** The proper factors of 17,296 are 1, 2, 4, 8, 16, 23, 46, 47, 92, 94, 184, 188, 368, 376, 752, 1081, 2162, 4324, and 8648. The sum of these factors is 18,416. The proper factors of 18,416 are 1, 2, 4, 8, 16, 1151, 2302, 4604, and 9208. The sum of these factors is 17,296. Thus, 17,296 and 18,416 are amicable numbers. ■

---

**Notes on $\pi$:** Another number that has been of interest since ancient times is the ratio of the circumference ($C$) of a circle to its diameter ($D$), given by $C/D$. No matter what size circle is considered, this ratio has the same value. By about 2000 B.C., the Babylonians used a value of $25/8$ for this ratio. Many brilliant minds have worked on obtaining approximate values for this ratio. Here are some of them:

Archimedes of Syracuse (200 B.C.): $211875/67441 \approx 3.141635$

Astronomer Ptolemy (160): $377/120 \approx 3.141667$

Liu Hui (263): $157/50 = 3.14$

Āryabhata (499): $626832/200{,}000 = 3.13416$

Valentin Otho (1573): $355/113 \approx 3.141593$

In 1706, William Jones used the symbol $\pi$ (the Greek letter pi) to represent the ratio. With usage of this symbol in 1736 by Leonhard Euler, $\pi$ became a standard. Between 1500 and 1800, others used trigonometry and calculus to approximate $\pi$ to more than 500 decimal places. In 1766, Johann Lambert proved that $\pi$ could not be represented by the ratio of two natural numbers and was therefore an irrational number. In the 20th century, calculators and computers have been used to determine $\pi$ to thousands of decimal places. In 1993 a supercomputer was used to calculate $\pi$ to 1,262,612 digits. However, without the use of computers evidence of accurate estimations of $\pi$ have been noted. For example, in the Great Pyramid of Gizeh in Egypt (2600 B.C.), the ratio of twice the width of the pyramid ($w = 230.364$ m) to the height ($h = 146.599$ m) of the pyramid gives $\pi$ accurate to the hundredths place ($2w/h \approx 3.14278$).

The first 501 digits of $\pi$ are as follows:

3.14159 26535 89793 23846 26433 83279 50288 41971 69399 37510
58209 74944 59230 78164 06286 20899 86280 34825 34211 70679
82148 08651 32823 06647 09384 46095 50582 23172 53594 08128
48111 74502 84102 70193 85211 05559 64462 29489 54930 38196
44288 10975 66593 34461 28475 64823 37867 83165 27102 19091
45648 56692 34603 48610 45432 66482 13393 60726 02491 41273
72458 70066 06315 58817 48815 20920 96282 92540 91715 36436
78925 90360 01133 05305 48820 46652 13841 46951 94151 16094
33057 27036 57595 91953 09218 61173 81932 61179 31051 18548
07446 23799 62749 56735 18857 52724 89122 79381 83011 94921
(never stops or repeats)

## Section 1.4 PROBLEMS

## *Explain → Apply → Explore*

### *Explain*

**1.** What are prime numbers?

**2.** What are composite numbers?

**3.** What are abundant numbers?

**4.** What are perfect numbers?

**5.** Is it possible to have a real number that is not rational? Explain.

**6.** What are magic squares?

**7.** Trace the history of negative numbers and zero.

**8.** Answer true or false for each statement. If a statement is false, explain why and give an example to show that it is false.

(a) All rational numbers are real numbers.
(b) All real numbers are rational numbers.
(c) All irrational numbers are real numbers.
(d) All real numbers are irrational numbers.
(e) All integers are whole numbers.
(f) All whole numbers are integers.
(g) All rational numbers are irrational numbers.
(h) All irrational numbers are rational numbers.
(i) All imaginary numbers are irrational numbers.
(j) All irrational numbers are imaginary numbers.
(k) All radicals are irrational.
(l) Complex numbers of the form $a + bi$ are real numbers when $b = 0$.

**9.** Even numbers are integers that end in 0, 2, 4, 6, or 8, and odd numbers end in 1, 3, 5, 7, or 9. How can even numbers be defined by using the concept of a factor? Explain.

**10.** Why is 2 the only even prime number?

**11.** Why is every prime number a deficient number?

## *Apply*

**12.** From the list below, choose the numbers belonging to each category.

$-11, \quad -9.4, \quad -8\frac{2}{9}, \quad -\sqrt{50}, \quad -4, \quad 1, \quad -\sqrt[3]{7.3}, \quad 0, \quad \frac{3}{4}, \quad \sqrt{-2},$
$\sqrt{2}, \quad 6.1212\ldots, \quad \sqrt{49}, \quad 9, \quad 10.12$

(a) Natural numbers
(b) Whole numbers
(c) Integers
(d) Rational numbers
(e) Irrational numbers
(f) Real numbers
(g) Imaginary numbers
(h) Noninteger rational numbers

**13.** From the list below, choose the numbers belonging to each category.

$-14.785, \quad -7, \quad -\sqrt{64}, \quad -\sqrt{-25}, \quad \frac{-5}{16}, \quad 0 \quad \sqrt[5]{19}, \quad \sqrt{-8}, \quad \sqrt{8},$
$\pi, \quad 5\frac{7}{8}, \quad 9.76555\ldots, \quad \sqrt{100}, \quad 19$

(a) Natural numbers
(b) Whole numbers
(c) Integers
(d) Rational numbers
(e) Irrational numbers
(f) Real numbers
(g) Imaginary numbers
(h) Nonradical irrational numbers

**14.** List five numbers that satisfy each description.

(a) Real numbers that are not rational
(b) Rational numbers that are not integers
(c) Irrational numbers that are not square roots
(d) Real numbers that are not irrational
(e) Integers that are not natural numbers
(f) Noninteger rational numbers

**15.** What are the fifth and sixth pentagonal numbers? Make a sketch of each number.

**16.** The first three hexagonal numbers are 1, 6, 15. Make a sketch of these three polygonal numbers.

**17.** Classify each of the following numbers as (a) prime or composite and (b) perfect, abundant, or deficient.

(a) 31
(b) 77

(c) 145
(d) 1988
(e) 8128
(f) 9000

*Explore*

**18.** Show that the pair of numbers, 1184 and 1210, found by 16-year-old Nicolo Paganini in 1866, is an amicable pair of numbers.

**19.** Determine whether the following are magic squares.

| | | |
|---|---|---|
| 7 | 5 | 3 |
| 2 | 9 | 4 |
| 6 | 1 | 8 |

| | | | |
|---|---|---|---|
| 14 | 7 | 11 | 2 |
| 1 | 12 | 8 | 13 |
| 4 | 9 | 5 | 16 |
| 15 | 6 | 10 | 3 |

| | | | | |
|---|---|---|---|---|
| 2 | 21 | 20 | 14 | 8 |
| 18 | 12 | 6 | 5 | 24 |
| 25 | 19 | 13 | 7 | 1 |
| 11 | 10 | 4 | 23 | 17 |
| 9 | 3 | 22 | 16 | 15 |

**20.** Example 3 showed that there is a pattern going from term to term in the triangular numbers. Find the patterns for the square and pentagonal numbers and use the patterns to find the first ten of each type of number.

**21.** The following "very" magic $4 \times 4$ square is found in an Albert Dürer engraving.

| | | | |
|---|---|---|---|
| 16 | 3 | 2 | 13 |
| 5 | 10 | 11 | 8 |
| 9 | 6 | 7 | 12 |
| 4 | 15 | 14 | 1 |

In parts (a)–(e), verify the following properties of this magic $4 \times 4$ square.

(a) The $2 \times 2$ squares in each corner and the center have the property that the sum of the four numbers in each square is 34.
(b) The sum of the squares of the numbers in the two top rows is the same as the sum of the squares of the numbers in the two bottom rows.
(c) The sum of the squares of the numbers in the first and third rows is the same as the sum of the squares of the numbers in the second and fourth rows.
(d) The sum of the numbers on the diagonals is the same as the sum of the numbers that are not on the diagonals.
(e) The sum of the squares of the numbers on the diagonals is the same as the sum of the squares of the numbers not on the diagonals.
(f) What other groups of four numbers besides the horizontal rows, vertical columns, diagonals, and the $2 \times 2$ squares mentioned in (a) also have a sum of 34?

**22.** The ancient Chinese considered even numbers to be female and odd numbers to be male. Using the Lo Shu magic square as a pattern, create the female magic square, using the even integers 2, 4, 6, 8, 10, 12, 14, 16, and 18, and the male magic square, using the odd integers 1, 3, 5, 7, 9, 11, 13, 15, and 17.

**23.** The three layers shown contain some of the numbers of a $3 \times 3 \times 3$ magic cube. Determine the missing numbers.

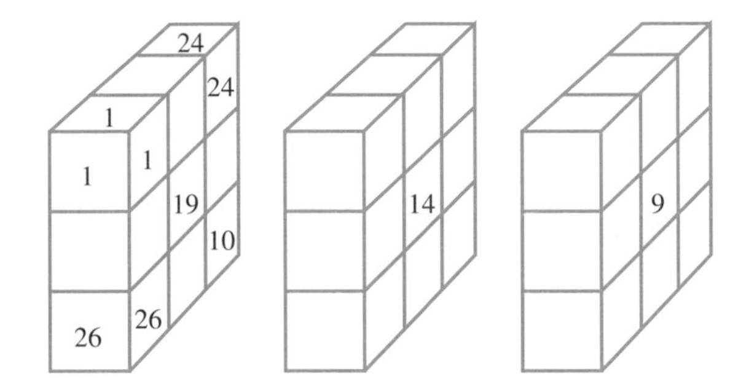


**24.** An old basic math text stated that $\pi = 22/7$. Why is this statement incorrect? What is the correct relationship between $\pi$ and $22/7$?

**25.** Explain why the number 1 is not considered a prime number.

**26.** The Japanese mathematician Kittoku Isomura (c. 1660) did a great deal of work on magic circles. In these circles, consecutive natural numbers, starting at 1, are placed on the diagram shown. If we add the numbers on any circle and the number in the center of the diagram, we get the same result as the sum of the numbers on each of the diagonals of the circle.

(a) Place the numbers 1 to 9 in the diagram so that you have created a magic circle.

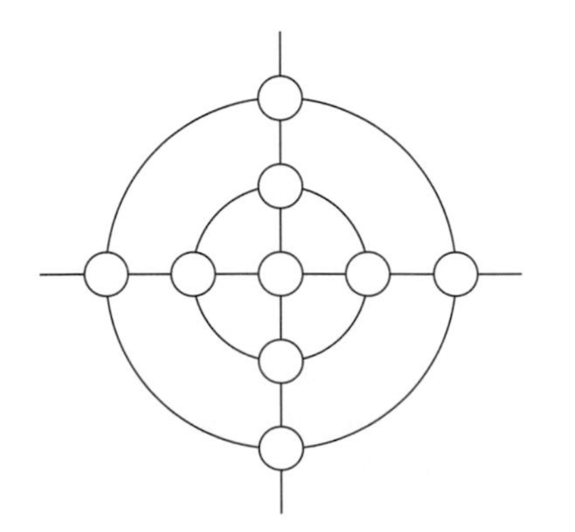

(b) Place the numbers 1 to 19 in the diagram so that you have created a magic circle.

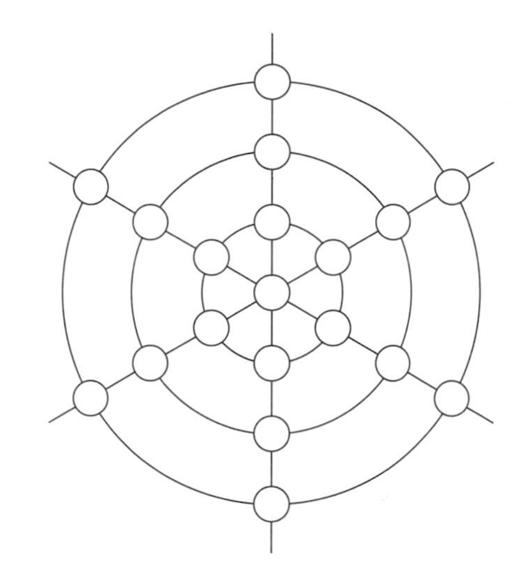

## Chapter 1 • Summary

### Key Terms, Concepts, and Formulas

The important terms in this chapter are

**Binary:** A number system with a base of 2 using the digits 0 and 1. p. 30

**Complex number:** The sum of a real number and an imaginary number. p. 42

**Composite number:** A whole number that has proper factors greater than 1. p. 46

**Duodecimal:** A number system with a base of 12 using the digits 0, 1, 2, 3, 4, 5, 6, 7, 8, 9, A, and B. p. 30

**Hexadecimal:** A number system with a base of 16 using the digits, 0, 1, 2, 3, 4, 5, 6, 7, 8, 9, A, B, C, D, E, and F. p. 30

**Imaginary number:** A number involving $\sqrt{-1}$ ($i$). p. 42

**Integer:** A number from the set $\{\ldots, -4, -3, -2, -1, 0, 1, 2, 3, 4, \ldots\}$. p. 40

**Irrational number:** A number that cannot be represented as the ratio of two integers; as a decimal, it does not terminate or repeat. p. 40

**Magic square:** A square array of numbers that has the same sum vertically, horizontally, and diagonally. p. 44

**Natural number:** A number from the set {1, 2, 3, 4, 5, . . .}. p. 40

**Number:** A measure of a quantity or amount. p. 4

**Numeral:** A symbol used to represent a number. p. 4

**Octal:** A number system with a base of 8 using the digits 0, 1, 2, 3, 4, 5, 6, and 7. p. 30

**Perfect number:** A natural number with the property that the sum of its proper factors equals itself. p. 47

**Pi ($\pi$):** The ratio of the circumference of a circle to its diameter ($\approx$ 3.14159). p. 48

**Place value:** The value given to the position a digit holds in a numeral. p. 14

**Polygonal numbers:** Numbers based on geometric shapes, such as triangular, square, pentagonal, and hexagonal numbers. p. 43

**Prime number:** A natural number whose only proper factor is 1. p. 46

**Proper factor of a natural number *N*:** A natural number less than *N* that divides evenly into *N*. p. 46

**Rational number:** A number that can be represented as the ratio of two integers; as a decimal it terminates or repeats. p. 40

**Real number:** A number that is either rational or irrational; each real number corresponds to a point on the number line. p. 40

**System of numeration:** A scheme for representing numbers by using a set of symbols. p. 4

**Tally:** A mark used to represent an object being counted. p. 3

**Whole number:** A number from the set {0, 1, 2, 3, 4, 5, . . .}. p. 40

After completing this chapter, you should be able to:

1. Explain the difference between a number, a tally, a numeral, and the word used to verbalize a quantity or amount. p. 3
2. Represent numbers in ancient systems of numeration that use grouping symbols along with:
   (a) Addition—Egyptian hieroglyphic system
   (b) Addition and subtraction—Roman numeral system
   (c) Addition and multiplication—traditional Chinese system, Ionic Greek system
   (d) Place values—Babylonian system, Mayan system p. 9
3. Represent numbers in the Hindu-Arabic system (decimal system), which is a place-value system using a base of 10. p. 22
4. Show how fractions are formed in the decimal system and in some of the ancient systems of numeration. p. 23

5. Represent numbers in a place-value system that has any base and be able to convert between decimal numerals and numerals in other bases. p. 29
6. Classify different types of numbers, such as real numbers, complex numbers, numbers based on geometric shapes, and numbers based on factors. p. 40

## • Summary Problems

1. Represent the tally on the right as a numeral in each of the following systems of numeration:
   (a) Egyptian hieroglyphic
   (b) Roman numeral
   (c) Traditional Chinese
   (d) Ionic Greek
   (e) Babylonian
   (f) Mayan
   (g) Hindu-Arabic
   (h) Binary
   (i) Base 5
   (j) Octal
   (k) Duodecimal
   (l) Hexadecimal

   √√√√ √√√√ √√√√ √√√√
   √√√√ √√√√ √√√√ √√√√
   √√√√ √√√√ √√√√ √√√√
   √√√√ √√√√ √√√√ √√√√
   √√√√ √√√√ √√√√ √√√√
   √√√√ √√√√ √√√√ √√√√
   √√√√ √√√√ √√√√ √√√√
   √√√√ √√√√ √√√

2. Using the symbols $\rightarrow$ for 1, $\leftarrow$ for 5, $\downarrow$ for 25, $\uparrow$ for 125, $\leftrightarrow$ for 625, and $\updownarrow$ for 3125, design an additive system of numeration. Use that system to represent the:
   (a) number of tallies in Problem 1
   (b) number of days in a leap year
   (c) number of feet in a mile
   (d) number of grams in a kilogram

3. Using the letters of the alphabet: a for 1, b for 2, c for 3, d for 4, e for 5, f for 6, g for 7, h for 8, i for 9, and dots placed above a letter to indicate that a certain digit is being multiplied by a power of 10. (**.** placed above a letter indicates it is being multiplied by 10, **..** placed above a letter indicates it is being multiplied by 100, **...** placed above a letter indicates it is multiplied by 1000, and so on), form a system of numeration that uses both addition and multiplication. Use that system of numeration to
   (a) find the value of the following numerals:
      (i) $\dot{e}$ d

(ii) $\ddot{\mathrm{f}}\ \dot{\mathrm{a}}\ \mathrm{d}$
(iii) $\dddot{\mathrm{h}}\ \ddot{\mathrm{i}}\ \dot{\mathrm{d}}\ \mathrm{e}$
(iv) $\dddot{\mathrm{c}}\ \dot{\mathrm{a}}\ \mathrm{b}$
(v) $\ddddot{\mathrm{g}}\ \dddot{\mathrm{a}}\ \ddot{\mathrm{g}}\ \dot{\mathrm{e}}$
(vi) $\ddddot{\mathrm{a}}\ \ddot{\mathrm{d}}$

(b) represent the quantities asked for in Problem 2.

**4.** Devise a method for representing fractions in the system of numeration in Problem 3. Give examples showing how various fractions can be written by using the system.

**5.** Using the symbols $*$ for 0, / for 1, $\times$ for 2, and ▼ for 3, design a base 4 place-value system of numeration. Represent the same quantities in this place-value system as asked for in Problem 2.

**6.** Explain the components of a base 7 system of numeration. Show how whole numbers and fractional numbers can be represented in this system. Represent the quantities asked for in Problem 2 as base 7 numerals.

**7.** The octal computer code for the letter S is 123. What is the decimal representation of that code?

**8.** The algebraic expression $n^2 - n + 41$ generates prime numbers for $0 \leq n \leq 40$. Find the first nine prime numbers generated by this algebraic expression. Show that this expression generates a composite number when $n = 41$. Find another value for $n$ where this expression generates a composite number.

**9.** Classify the four numbers 400, 461, 496, and 512 as (a) composite or prime and (b) as perfect, abundant, or deficient.

**10.** The following four layers contain some of the numbers that form a $4 \times 4 \times 4$ magic cube. Determine the missing numbers.

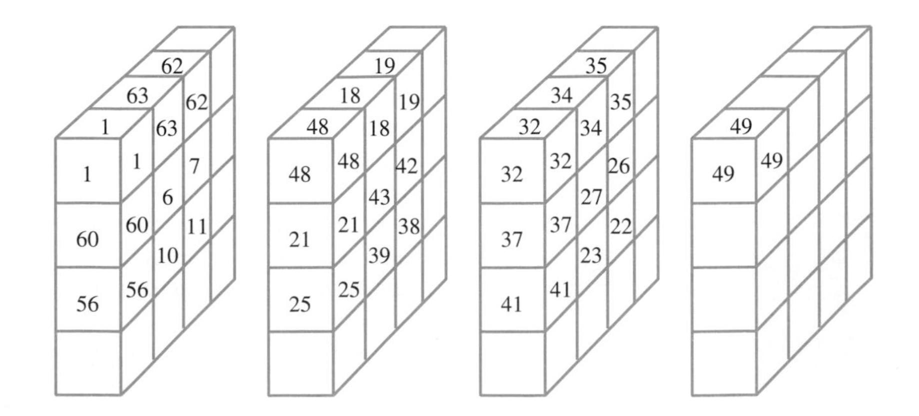

**11.** A figure is created by placing a number at each intersection point on the lines of a star. If the sum of the numbers along each line of the star is the same as every other such sum in that star, the star is called a magic star. Verify that star (a) is a magic star and determine the missing values in stars (b), (c), and (d) that will make them magic stars.

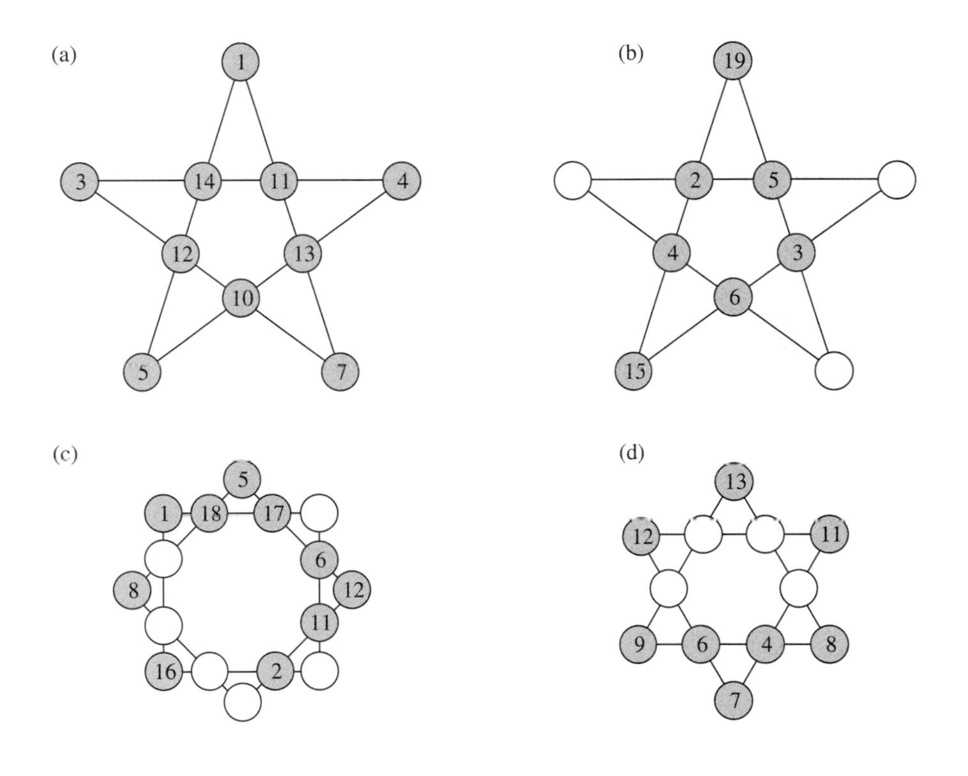


**12.** $-7.45,\ -\sqrt{40},\ -5,\ \sqrt[3]{-27},\ -\frac{5}{9},\ 0,\ \sqrt[4]{19},\ \pi,\ 3.1416,\ 4.6161\ldots,\ 7\frac{2}{3},\ 13,\ \sqrt{-400}$

From the list above determine which of the numbers are

(a) Imaginary
(b) Integers
(c) Irrational
(d) Natural
(e) Rational
(f) Real
(g) Whole

**13.** Design a system of numeration using your own symbols and your own scheme to represent numbers. How would you represent fractions in your system? Give examples and explanations of the components of your system.

# 2

# Modeling with Algebra

Parabolas, curves, and slopes, figures commonly graphed in algebra, form the image of Joseph Stella's *The Brooklyn Bridge: Variation on an Old Theme*. (Whitney Museum of Art, New York)

# Overview Interview

The primary focus of your math classes prior to this one has most likely been algebra. In this chapter, you will review some of the algebra from previous courses and then learn of some of the many applications of algebra. You will not investigate the typical mixture, work, interest, or motion problems of those previous classes. Instead, you will look at how algebraic functions can act as models for real-life situations. You will determine the profits at Carmen's Coffee Shop and study the path of a jumping frog. You will investigate how to determine the maximum profit for a company manufacturing compact disks if the company is faced with restrictions on the numbers of hours the assembly line can operate. You will also learn about applications of logarithms and exponents to population growth or how long it takes a cup of coffee to cool to a drinkable temperature.

## Perry Wong

PERRY WONG, ECONOMIST, THE WEFA GROUP, A MERGER OF CHASE ECONOMETRICS AND WHARTON ECONOMETRIC FORECASTING ASSOCIATES, PHILADELPHIA, PENNSYLVANIA

TELL US ABOUT YOURSELF. I was born in mainland China and moved to the United States to attend Temple University in Philadelphia, where I received my bachelor's degree in computer information science and did graduate work in economics. Since 1990, I have been a senior economic analyst and associate economist with The WEFA Group. My responsibilities include making forecasts about regional industrial output and employment for all the 50 states, along with other economic forecasts for the state of New Jersey and states in the midwest. I am also responsible for the maintenance and operation of the computers used in these forecasts.

*Perry Wong in front of the WEFA Group office. (Courtesy of Amy Creighton)*

HOW IS MATHEMATICAL MODELING USED IN ECONOMICS? An economist utilizes various mathematical and statistical models to present and demonstrate theories. For example, an economist may utilize linear and logarithmic models to anticipate changes in consumer purchasing patterns. An economist may use an exponential function to establish a model that predicts production by a factory. Such a production function might answer the question, ''If there is a one-unit increase in capital outlay, how much increase in output can be expected?''

WHY IS THE STUDY OF MATHEMATICS IMPORTANT TO THE ECONOMIST? Mathematics is the most important tool to the economist. Mathematics enables the economist to extract information from complex regional, national and world economies. Mathematics assists the economist in describing the status and the well-being of the economy in an objective and precise manner. Without a knowledge of mathematics, an economist cannot accurately describe the structure of an economic system, predict the course of the economy, or communicate with fellow economists.

*Melancholia I* by the 16th-century artist Albrecht Dürer shows the contemplation engendered by the study of mathematics. (The Metropolitan Museum of Art)

## A Short History of Algebra

As ancient men and women investigated the geometry of the world around them and used their number systems to count, measure, and calculate, they began to generalize the procedures of arithmetic and apply them to unknown quantities. Anthropologists believe that before 2000 B.C. the Chinese, Persians, Babylonians, and people of India may have begun this process and had some elementary knowledge of what we now call algebra. However, the first definite evidence of algebra is found in the Rhind papyrus (c. 1650 B.C.). In this work the Egyptian mathematician Ahmes included problems such as this:

*If a "heap" and a seventh of a "heap" are 19, what is the value of the "heap"?*

The ancient Greeks (650 B.C.–A.D. 200) contributed much to the development of mathematics, but their main concern was geometry, not algebra. This left algebra in a stage where its problems and solutions were stated only in words and used mainly in reference to geometric figures. Around A.D. 250 a major step in the development of algebra occurred with the work of the Greek mathematician Diophantus (210–290). He worked out a system of his own to solve problems by using symbols to replace numbers and operations. For example, he used ⋔ for subtraction, ⌊ for equals, and ς for an unknown quantity. Diophantus made significant contributions to mathematical notation and expanded the scope of algebra. For this he is considered by many to be the father of algebra.

The period after the disintegration of the Roman Empire in the fourth and fifth centuries is called the Dark Ages. These were years of very little progress in the development of algebra in Europe. The main advances came from India and Arabia. Hindu mathematicians like Brahmagupta (c. 625) followed the lead of Diophantus by continuing to use symbols in the solutions of mathematical problems. Around 825, Al-Khowârizmî, an Arabian teacher of mathematics in Baghdad, used the word we know as algebra in his work *Ilm al-jabr walmuqabalah*, meaning the science of transposition and cancellation. Through his writing, algebra became known as the study of solving equations. The interest in algebra also spread to Persia, where famed poet and mathematician Omar Khayyam (1050–1123) wrote a book on algebra. The Arabian and Indian influence did much to improve number notation and the symbolism of algebra.

As Europe emerged from the Dark Ages, contributions to the development of algebra by Europeans again appeared. Italian merchant Leonardo de Pisa (1202), commonly known as Fibonacci, summarized Arabian algebra and introduced the Hindu-Arabic number system to Europe in his work *Liber Abaci*. John of Holywood (1240) wrote the standard mathematics text that was used for centuries in European universities. In 1247, Ch'in Kiu-shao showed the high degree of sophistication of Chinese mathematics in his works on solving higher-degree equations by numerical methods, a discovery not made in Europe until 1819. In 1303, Chinese scholar Chu Shih-Chieh displayed the binomial coefficients more than 200 years before they were published in Europe. In 1360, Nicole Oresme introduced fractional exponents in his unpublished work *Algorismus Proportionum*.

The European Renaissance was a time of great progress and creativity in the development of algebra and its notation:

Johann Widmann (1489) used $+$ and $-$ for positive and negative numbers.

Christoff Rudolff (1525) introduced $\sqrt{\phantom{x}}$ for square roots.

In 1527, the coefficients for binomial expansions were published.

Scipione del Ferro, Nicolo Fontana Tartaglia, and Girolamo Cardano (1545) found general solutions to cubic equations.

Robert Recorde (1557) used the symbol $=$ to represent equality.

Rafael Bombelli (1572) published the first consistent treatment of imaginary numbers.

Christopher Clavius (1583) initiated the use of a dot for multiplication.

Francois Vièté (1591) systematically used letters to represent unknowns.

John Napier (1614) invented logarithms.

Johann Kepler and Henry Briggs (1624) published a table of logarithms.

Probably the most notable contributions to elementary algebra are credited to René Descartes (1596–1650). Even though Pierre de Fermat (1601–1655) worked on similar material before Descartes, Descartes was the first to publish his work. In *La Geometrie*, he introduced notation similar to what we find in present-day algebra. He used $x$, $y$, and $z$ for unknowns and the superscript ($x^3$) for cubes. But most significantly, he brought algebra and geometry together by creating the $(x, y)$ rectangular coordinate system. He made it possible for the equations of algebra to be represented graphically, laid the foundation for algebraic geometry (analytic geometry), and made the development of calculus possible.

Following Descartes and Fermat, Sir Isaac Newton (1642–1727) and Gottfried Leibniz (1646–1716) independently developed the calculus, and for the next 200 years mathematicians spent a great deal of effort on calculus and its applications. Along with these advancements, however, came progress in algebra. The major contributors and their contributions were:

Thomas Harriot (1631) introduced the inequality symbols $>$ and $<$.

John Wallis (1655) used algebraic notation very similar to what is currently used, including negative exponents and $\infty$ for infinity.

Seki Kōwa of Japan (c. 1683) introduced a system of determinants for solving equations some 10 years before Gottfried Leibniz suggested it.

Maria Agnesi (1748) published a widely used math text covering topics from algebra through calculus.

Leonhard Euler (c. 1749) defined algebraic functions and used $i$ to represent $\sqrt{-1}$, $e$ for the base of natural logarithms, and $\Sigma$ for summations.

Gabriel Cramer (1750) gave a general rule for solving systems of $n$ linear equations in $n$ unknowns.

David Rittenhouse (1799) made America's earliest contribution to mathematics research when he wrote on methods of computing with logarithms.

Carl Friedrich Gauss (1832) initiated the use of the term "complex number" for the sum of a real number and an imaginary number.

Carl Jacobi (1841) developed the theory of determinants.

Karl Weierstrass (1841) introduced the absolute value symbol $|\ |$.

Arthur Cayley (1857) formulated the algebra of matrices.

Equations of degree higher than 3 were investigated throughout the 1800s by many mathematicians, such as Niels Abel, L. Ferrari, Evariste Galois, and Peter Roth.

Some mathematicians also examined the underlying structure of algebra.

These discoveries opened the door to the study of modern, or abstract, algebra.

Because of the work of such dedicated individuals, we find elementary algebra to be a mathematical system that

(a) Has a rich history of development.
(b) Uses symbols to represent numbers and operations.
(c) Allows us to graphically represent and analyze mathematical concepts.

(d) Has a logical foundation.
(e) Gives us a powerful problem-solving tool.

This chapter contains problems and concepts from this system of elementary algebra. We will study the use of algebraic functions as models representing various real-life situations.

## Check Your Reading

1. Show how to solve the problem on the Rhind papyrus contained in the first paragraph of this section.
2. In 1624, the first English settlement was established in eastern India. What important event occurred in mathematics during this year?
3. The coefficients for the binomial expansion were published in the same year that Sebastian Cabot built the fortifications of Espiritu Santo in Paraguay. What year was this?
4. In 1557, Spain and France went bankrupt. What mathematical symbol was first used during this year?
5. While the Swedish army was occupying Cracow and Warsaw in Poland, what symbol was John Wallis introducing to mathematics?
6. In the late 1740s, Giacobbo Rodriguez Pereire invented a sign language for deaf people. Who wrote the popular mathematical textbook used during this period?
7. From 2500 to 2000 B.C. was a significant period in history. Not only did painted pottery appear in China, papyrus appear in Egypt, and chickens first become domesticated in Babylon, but algebra was being developed in which ancient civilizations?
8. In 1832, the New England Anti-Slavery Society was founded in Boston and Japanese artist Ando Hiroshige published his famous art series, ''Fifty-three Stages of the Tokaido.'' What term did Carl Gauss initiate in this year?
9. Match each of the following names with the appropriate symbol or word.

<table>
<tr><td>(a) Al-Khowârizmî</td><td>$|\ |$</td></tr>
<tr><td>(b) Euler</td><td>$+$ and $-$</td></tr>
<tr><td>(c) Gauss</td><td>Algebra</td></tr>
<tr><td>(d) Harriot</td><td>Complex number</td></tr>
<tr><td>(e) Napier</td><td>$=$</td></tr>
<tr><td>(f) Recorde</td><td>$<$ and $>$</td></tr>
<tr><td>(g) Rudolff</td><td>$i$ and $e$</td></tr>
<tr><td>(h) Wallis</td><td>$\infty$</td></tr>
<tr><td>(i) Weierstrass</td><td>Logarithms</td></tr>
<tr><td>(j) Widmann</td><td>$\sqrt{\ }$</td></tr>
</table>

## Research Questions

To answer the following questions, you will need to refer to material not contained in the text. Possible sources of information are listed in the bibliography at the end of the text.

1. Many of the mathematicians mentioned in the short history of the development of algebra had other interests besides mathematics. Explain some of these other interests. What anecdotes have been recorded about these mathematicians?
2. Throughout the history of mathematics, there has been controversy over the person given credit for a particular discovery. Discuss one of the following controversies:
   (a) René Descartes and Pierre de Fermat over algebraic (analytic) geometry.
   (b) Scipione del Ferro, Nicolo Fontana Tartaglia, and Girolamo Cardano over the solution of cubic equations.
3. What are Diophantine equations? Give some examples and solutions.
4. Descartes developed the $(x, y)$ rectangular coordinate system, but there is another coordinate system used in mathematics, the polar coordinate system. What is the polar coordinate system? How does it differ from the rectangular coordinate system?
5. What was the School Mathematics Study Group (SMSG)? Who were some of the people involved in the group, and what were the objectives of the group?
6. Investigate the work of Girolamo Cardano in solving cubic equations. What are some of his formulas?
7. What is Pascal's Triangle? How can it be used to raise the binomial $x + y$ to the fourth power?
8. Interview a science teacher at your school to determine how algebra is used in that area of science.
9. What is a matrix? How is it used to represent a system of equations? What are some uses of matrices?

## Projects

1. Compare the Dark Ages to the Renaissance. What factors in the Dark Ages contributed to the stagnation of intellectual pursuits during this period? What factors in the Renaissance contributed to the rebirth of intellectual pursuits during this period? What happened to mathematics during those two periods?
2. What is Fermat's Last Theorem? Why has it caused great interest over the centuries? Investigate the proofs of the theorem submitted in 1989 and 1993. What do these proofs suggest about Fermat's statement made in the margin of a book?

3. Investigate the simplex method used in linear programming. Who invented the method? When was it invented? What are the advantages of the simplex method over the graphical method given in the text?
4. What are the conic sections? Discuss how are they produced, both geometrically and algebraically. What applications do they have? In particular what are the uses of parabolas in the field of optics and satellite communication technology?

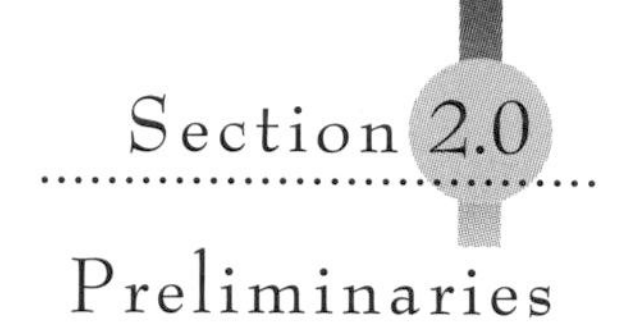


This section contains some of the algebraic facts, terminology, and techniques needed in Chapter 2.

## Functions

The concept of a function in mathematics flows naturally from everyday experiences. For example, the size of a projected image on a screen depends on (is a function of) the distance the projector is from the screen. The temperature of a pot of soup on the stove depends on (is a function of) the length of time it has been on the hot burner. In the first example, for each distance from the screen there is one image size. In the second example, for each length of time the soup has a certain temperature. This concept can be summarized into the mathematical definition of a function. A **function** is a collection of values arranged in pairs, usually written as $(x, y)$. In a function, for each value of $x$ there is exactly one value of $y$. The set of $x$ values is called the **domain** and the set of $y$ values is called the **range** of the function.

## Linear Functions

A **linear function** is one whose graph is a line or part of a line, has an equation of the form $y = mx + b$, and has a single value that measures the slope of its graph. The slope is a measure of the steepness of the graph of the line. In the equation, the slope is given by $m$, the coefficient of $x$. The point where the graph of this line crosses the $y$ axis is called the $y$ intercept. In the equation, the location of the $y$ intercept is given by the constant $b$.

To calculate the slope, divide the change in the $y$ values by the change in the $x$ values. For a line passing through the points $(x_1, y_1)$ and $(x_2, y_2)$, the slope is

**Slope**

$$m = \frac{\text{change in the } y \text{ values}}{\text{change in the } x \text{ values}} = \frac{y_2 - y_1}{x_2 - x_1}$$

To find the equation of a line and represent it as a function, we will use the slope-intercept form.

## Equation of a Line

$$y = mx + b \qquad \text{where} \begin{cases} m = \text{slope} \\ b = y \text{ intercept} \end{cases}$$

### Example 1

Graph the linear function $y = 2x - 2$.

**Solution:** Since there are no restrictions on the domain ($x$ values) of this function, $x$ can be any real number. Since two points determine a line, simply plot two points that satisfy the equation and draw a line connecting both points.

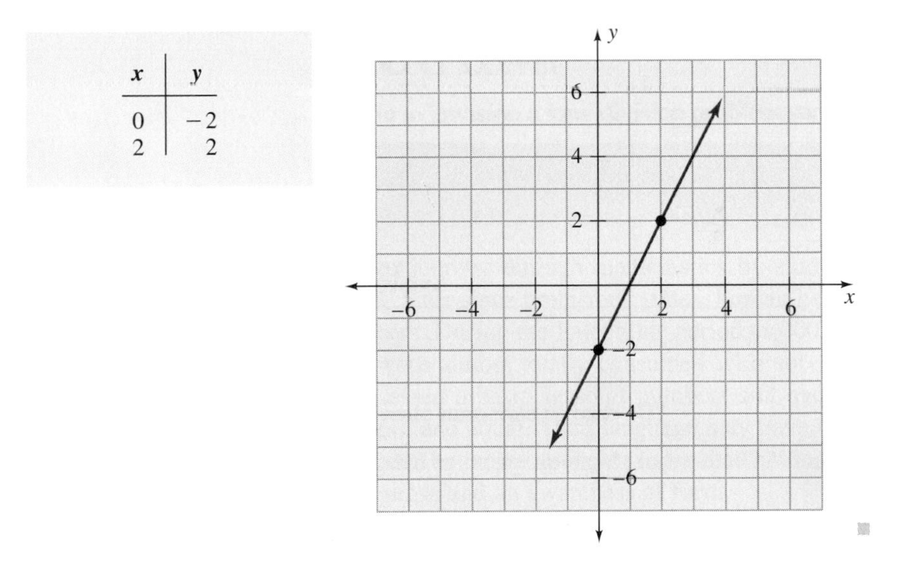

| $x$ | $y$ |
|---|---|
| 0 | $-2$ |
| 2 | 2 |



### Example 2

Graph $y = 2x - 2$ where $0 \leq x \leq 4$.

**Solution:** The restriction on the domain of the function allows us to use only the values from 0 through 4 for $x$. The graph is only part of a line, a line segment.

| $x$ | $y$ |
|---|---|
| 0 | −2 |
| 4 | 6 |

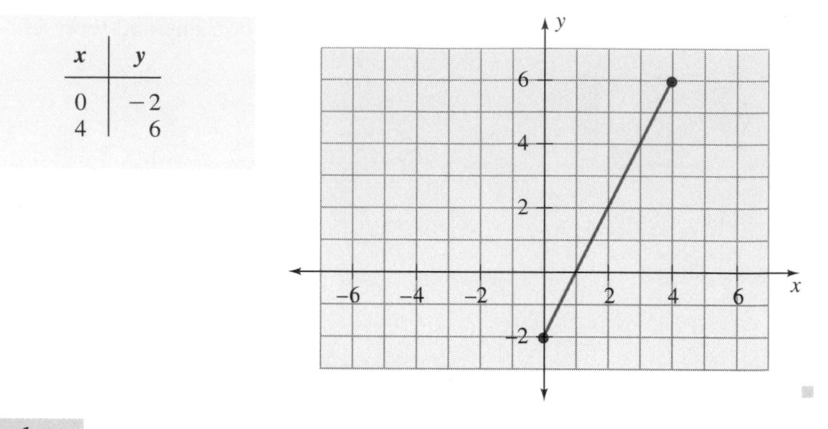


## Example 3

Graph $y = 2x - 2$ where $0 \leq x \leq 4$ and $x$ is a whole number.

**Solution:** The restriction on the domain of the function here forces us to use only the whole number values from 0 through 4 for $x$. The graph is a series of five points. It is a **discrete graph**, not a continuous one.

| $x$ | $y$ |
|---|---|
| 0 | −2 |
| 1 | 0 |
| 2 | 2 |
| 3 | 4 |
| 4 | 6 |

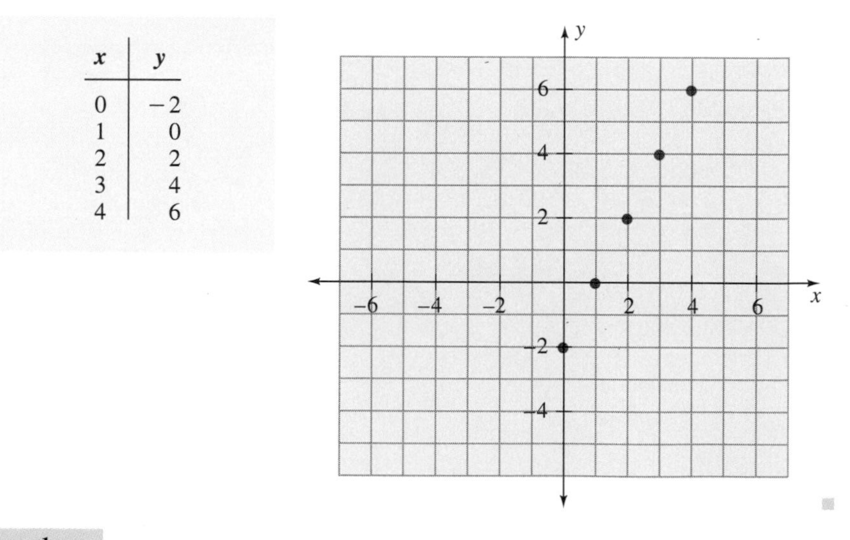


## Example 4

Find the equation of the linear function passing through the points $(2, -3)$ and $(4, 7)$.

**Solution:** The solution requires two steps. The first step is to find the slope:

$$m = \frac{y_2 - y_1}{x_2 - x_1} = \frac{7 - (-3)}{4 - 2} = \frac{10}{2} = 5$$

The second step is to use the equation for a linear function:

$$y = mx + b$$

$$y = 5x + b$$

Find $b$ by substituting known values for $(x, y)$, either $(2, -3)$ or $(4, 7)$. Using $(2, -3)$ we get

$$-3 = 5(2) + b$$

$$-3 = 10 + b$$

$$-13 = b$$

Thus, $y = 5x - 13$ is the desired linear function. ■

## Solving Systems or Linear Equations

A **system of linear equations** is a set of two or more linear equations. An example of a system with two linear equations and two unknowns is

$$2x + 3y = 8$$

$$5x - 2y = 1$$

Since the equations given in the system are linear, the graphs of the equations are lines. Thus, if there is a single solution to this system, the solution is graphically represented as the intersection of the two lines. Although there are many ways to solve such a system, we are primarily concerned with two of them, the *graphing method* and the *addition method.*

When using the **graphing method**, the solution to the system is found by carefully graphing the lines represented by the equations and then using the graph to determine the point of intersection of the lines. The coordinates of the intersection point simultaneously satisfy both equations and are called the solution of the system of equations.

### Example 5

Use the graphing method to find the solution to the system of equations

$$2x + 3y = 8$$

$$5x - 2y = 1$$

**Solution:** To graph the equation $2x + 3y = 8$, we can use any two points that satisfy the equation, such as $(4, 0)$ and $(-2, 4)$. Similarly, to graph the equation $5x - 2y = 1$, we can use the points $(3, 7)$ and $(-1, -3)$.

Examining the graph, we see the solution is the point (1, 2).

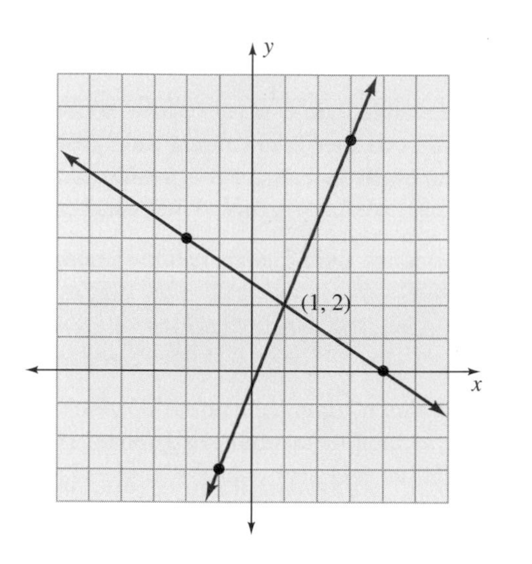


A second way to solve a system of equations, the **addition method**, involves adding multiples of the two equations so that one of the variables is eliminated. The objective here is to create a single equation with only one variable.

## Example 6

Use the addition method to find the solution to the system of equations

$$\begin{cases} 2x + 3y = 8 \\ 5x - 2y = 1 \end{cases}$$

**Solution:** To solve this system, we first select a variable that we wish to eliminate from the equations. If we choose $y$ to be eliminated, the two coefficients of $y$ must be the same number but with opposite signs. To do this, we determine a number into which both the coefficients of $y$ divide evenly. Since 3 and $-2$ both divide into 6 evenly, we multiply each equation by a constant so that one equation contains the term $6y$ and the other equation contains the term $-6y$. Thus, we must multiply the first equation by 2 and the second equation by 3.

$$(2x + 3y = 8) \xrightarrow{\times 2} 4x + 6y = 16$$

$$(5x - 2y = 1) \xrightarrow{\times 3} 15x - 6y = 3$$

Adding these equations eliminates the $y$ variable and leaves an equation containing only $x$.

$$\begin{aligned} 4x + 6y &= 16 \\ 15x - 6y &= 3 \\ \hline 19x \qquad &= 19 \end{aligned}$$

Dividing both sides of this last equation by 19 gives the result $x = 1$. Substituting $x = 1$ into $2x + 3y = 8$ provides the following:

$$\begin{aligned} 2(1) + 3y &= 8 \\ 2 + 3y &= 8 \\ 3y &= 6 \\ y &= 2 \end{aligned}$$

From this, we find the solution to be $x = 1$ and $y = 2$.

Notice that the same result may be obtained by substituting $x = 1$ into the other equation, $5x - 2y = 1$:

$$\begin{aligned} 5(1) - 2y &= 1 \\ 5 - 2y &= 1 \\ -2y &= -4 \\ y &= 2 \end{aligned}$$

■

## Linear Inequalities

**Linear inequalities** are linear relationships in which the equality sign has been replaced by an inequality sign. Two examples of linear inequalities are

$$3x + 5y \leq 30 \qquad \text{and} \qquad y > 5x - 7$$

To graph a linear inequality, the first step is to graph the relationship as if an equal sign had been used. If the problem uses a strict inequality, $<$ or $>$, draw a dotted line. If the relationship uses an inequality containing an equal sign, $\leq$ or $\geq$, draw a solid line.

The second step in graphing an inequality is to shade the region on the side of the line for which the relationship is true. This can be done by picking a point not on the line and checking to see if the point satisfies the inequality. If it does, the side of the line containing the point should be shaded. If the point does not satisfy the relationship, the other side of the line should be shaded.

### Example 7

Sketch the graph of the inequality $3x + 5y \leq 30$.

**Solution:** First, graph the line $3x + 5y = 30$. Two of the points that satisfy the equation are (10, 0) and (0, 6). Plot the points and, since the inequality uses a $\leq$ sign, draw the graph using a solid line.

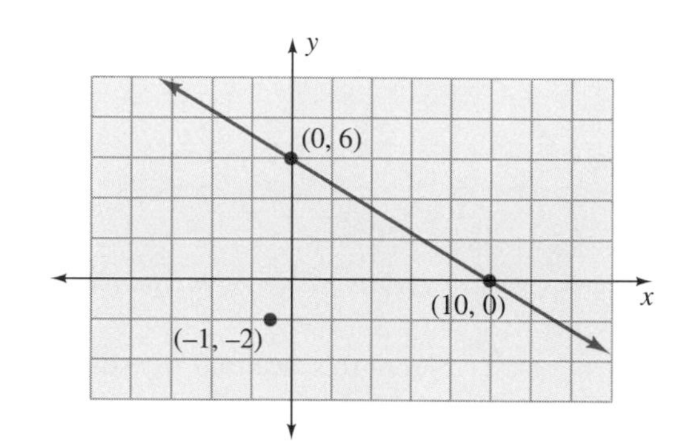


Now, pick any point not on the line, such as $(-1, -2)$, and substitute the point into the inequality.

$$3(-1) + 5(-2) \leq 30$$
$$-3 - 10 \leq 30$$
$$-13 \leq 30$$

Since this is a true statement, we know that the side of the line containing the point $(-1, -2)$ should be shaded. This gives the graph of the linear inequality as

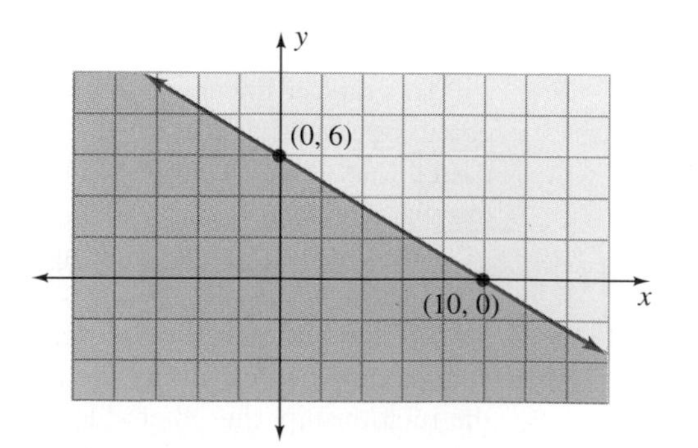

The shaded region indicates that any point in the region will satisfy the original inequality $3x + 5y \leq 30$. ■

## Example 8

Sketch the graph of the inequality $y > 5x - 7$.

**Solution:** Graph the line $y = 5x - 7$. Two of the points that satisfy the equation are $(0, -7)$ and $(2, 3)$. Since the inequality uses a $>$ sign, the graph is drawn using a dotted line.

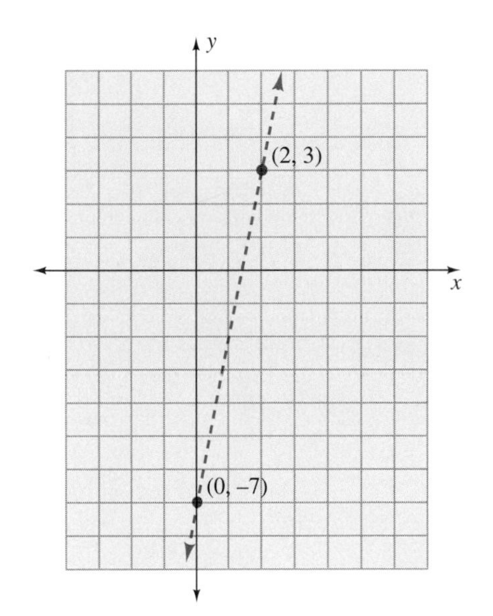


Pick a point not on the line, such as $(4, 0)$, and substitute the point into the inequality $y > 5x - 7$.

$0 > 5(4) - 7$

$0 > 13$ Note that this is false.

Since this is a false statement, we know that the side of the line that does *not* contain the point (4, 0) should be shaded. This gives the graph of the linear inequality as

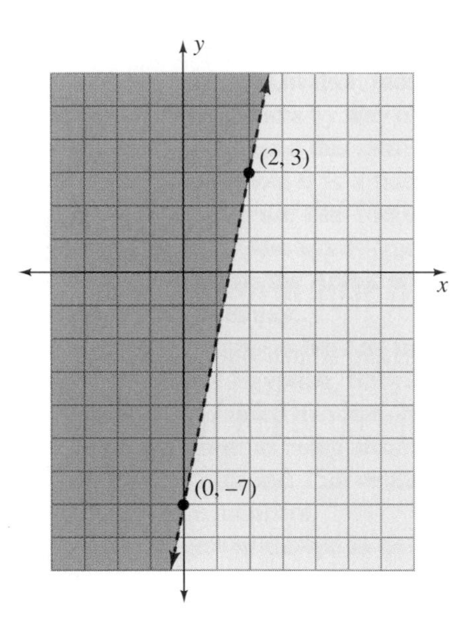


The shaded region indicates that any point in the region satisfies the inequality $y > 5x - 7$. ■

## Systems of Linear Inequalities

A system of linear inequalities is a set of two or more linear inequalities. An example of a system with four linear inequalities is

$$x + y \leq 18$$
$$x + 2y \leq 24$$
$$x \geq 0$$
$$y \geq 0$$

To solve a system of linear inequalities, we graph each of the inequalities and shade only the region common to each of the graphs. Any point in the resulting region will satisfy all of the inequalities in the system.

## Example 9

Sketch the solution to the following system of inequalities.

$$x + y \le 18$$

$$x + 2y \le 24$$

$$x \ge 0$$

$$y \ge 0$$

**Solution:** To solve the system, graph each of the inequalities in the system on the same coordinate axes. Because of the inequalities $x \ge 0$ and $y \ge 0$, we can restrict the solution to only those points that are in the first quadrant.

Next, graph the line $x + y = 18$. Two of the points that satisfy the equation are (18, 0) and (0, 18). Plot the points and, since the inequality uses a $\le$ sign, draw the graph using a solid line. Notice that we are restricting the graph to the first quadrant.

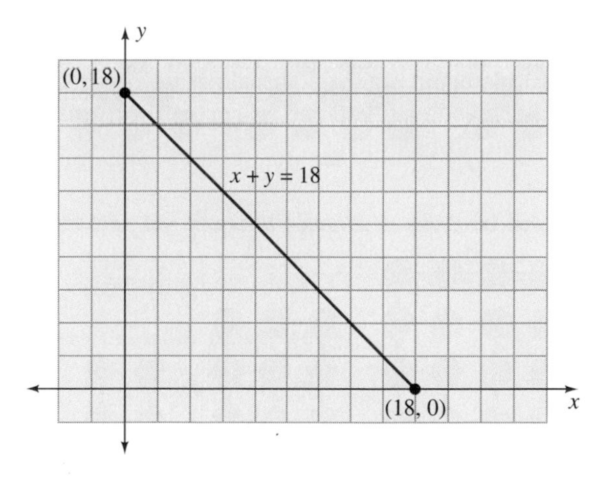


Pick a point not on the line, such as (0, 0), and substitute the point into the inequality $x + y \le 18$.

$$0 + 0 \le 18$$

Since this is a true statement, we know that the side of the line that contains the point (0, 0) should be shaded. This gives the graph of the linear inequality as

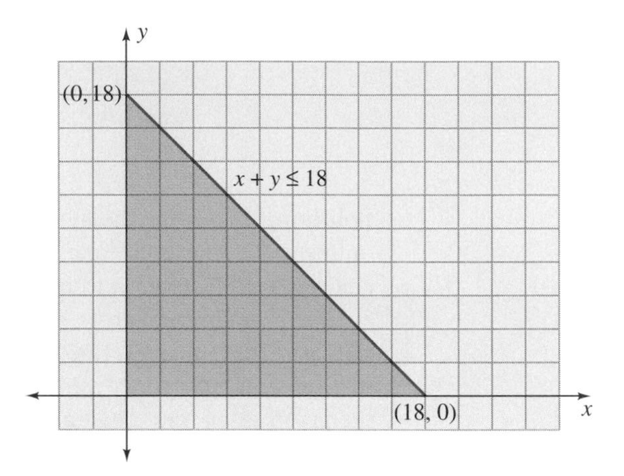


Now, using similar techniques, we must include the graph of the inequality $x + 2y \leq 24$ in the same picture. Two of the points that satisfy the equation $x + 2y = 24$ are (24, 0) and (0, 12). Plot the points and, since the inequality uses a $\leq$ sign, draw the graph using a solid line.

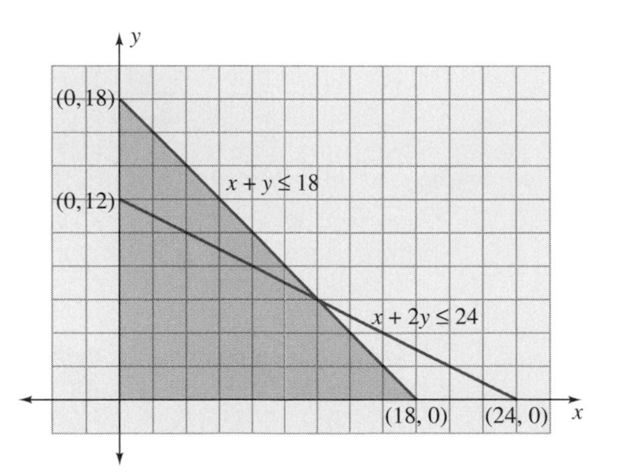


Again, pick a point not on the line, such as (0, 0), and substitute the point into the inequality $x + 2y \leq 24$.

$$0 + 2(0) \leq 24$$

Since this is a true statement, we know that the side of the line that contains the point (0, 0) should be shaded. This gives the graph of the linear inequality as

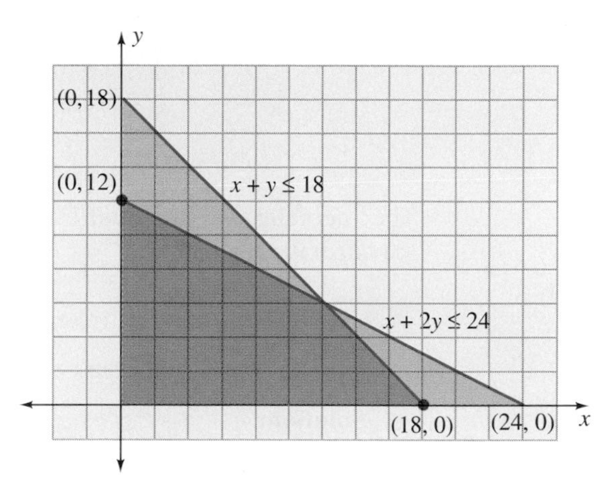


As the final result, we shade only the portion of the graph that was shaded by each of the inequalities. Every point in the shaded region satisfies all four of the inequalities in the original system. Thus, the shaded region is the solution to the system of inequalities.

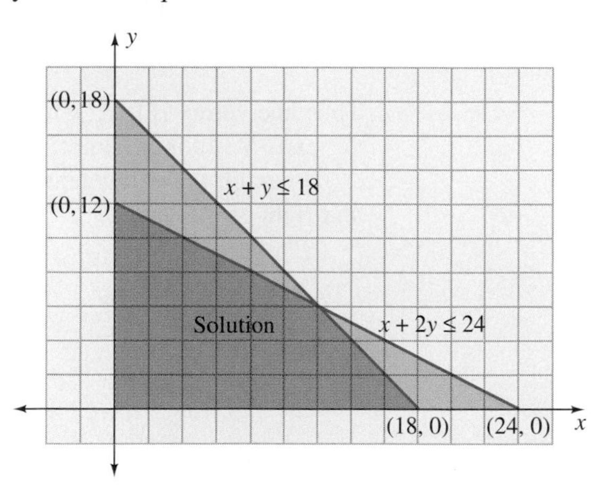


## Quadratic Functions

A **quadratic function** is a function whose graph is a parabola and has an equation of the form

$$y = ax^2 + bx + c, \qquad \text{where } a \neq 0$$

To graph a quadratic function it is best to find the vertex (the maximum or minimum point) of the parabola it represents. The $x$ coordinate of the vertex can be determined by the equation

$$x = \frac{-b}{2a} \quad \text{where} \begin{cases} \text{the parabola opens upward if } a > 0 \\ \text{the parabola opens downward if } a < 0 \end{cases}$$

The $y$ coordinate of the vertex can be found by substituting the value found for $x$ into the quadratic function.

### Example 10

Find the vertex and graph of $y = x^2 - 6x + 4$.

**Solution:**

The $x$ coordinate of the vertex is

$$x = \frac{-b}{2a} = \frac{-(-6)}{2(1)} = \frac{6}{2} = 3 \quad \xrightarrow{\text{substitute } x = 3}$$

The $y$ coordinate of the vertex is

$$y = x^2 - 6x + 4$$
$$y = (3)^2 - 6(3) + 4$$
$$y = 9 - 18 + 4$$
$$y = -5$$

Thus, the vertex is $(3, -5)$. Since $a > 0$, the parabola opens upward and the vertex is a minimum point.

To graph the parabola, calculate and plot points using $x$ values to the left and right of the vertex.

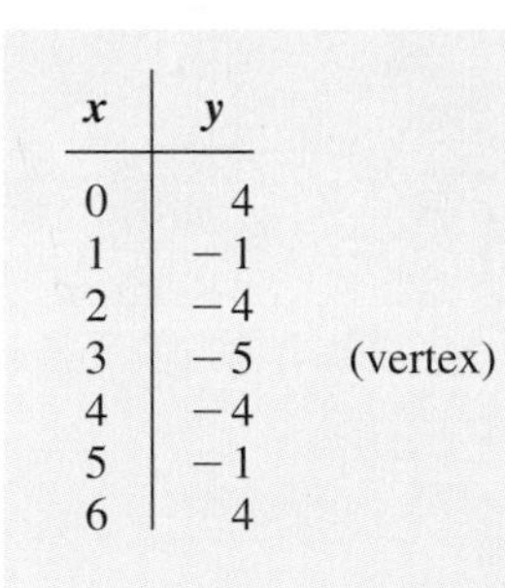

| $x$ | $y$ | |
|---|---|---|
| 0 | 4 | |
| 1 | −1 | |
| 2 | −4 | |
| 3 | −5 | (vertex) |
| 4 | −4 | |
| 5 | −1 | |
| 6 | 4 | |

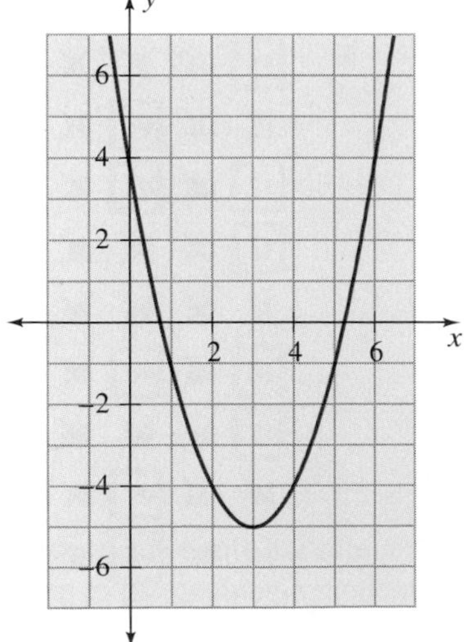

**Example 11**

Graph $y = -2x^2 + 5x$.

**Solution:** Find the vertex.

$$x = \frac{-b}{2a} = \frac{-5}{2(-2)} = 1.25$$

$$y = -2(1.25)^2 + 5(1.25) = 3.125$$

Since $a < 0$, the parabola opens downward and the vertex (1.25, 3.125) is the maximum point on the parabola.

Calculate and plot points using $x$ values to the left and right of the vertex:

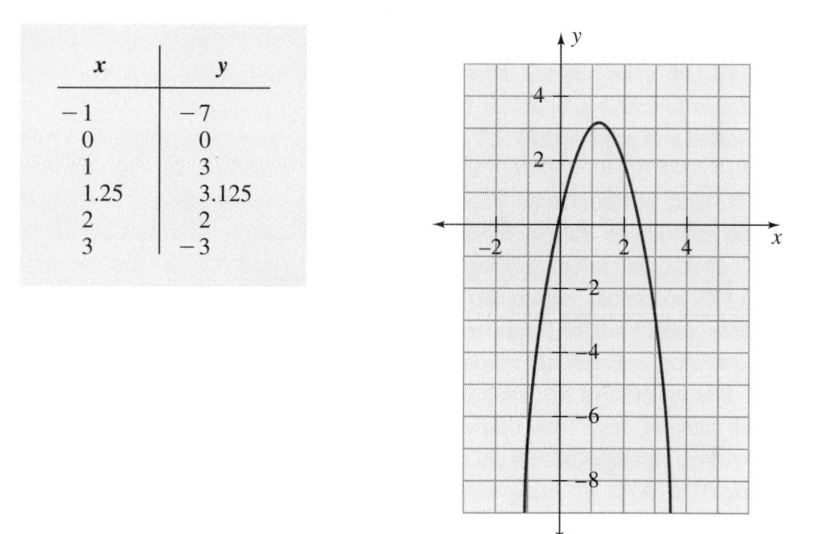

| $x$ | $y$ |
|---|---|
| −1 | −7 |
| 0 | 0 |
| 1 | 3 |
| 1.25 | 3.125 |
| 2 | 2 |
| 3 | −3 |



## Quadratic Formula

If the $y$ value of a quadratic function is zero, we have the quadratic equation

$$ax^2 + bx + c = 0$$

We can solve for $x$ and find the $x$ intercepts of the graph of the quadratic function by using a calculator and the quadratic formula.

Quadratic Formula

$$x = \frac{-b \pm \sqrt{b^2 - 4ac}}{2a}$$

where $a \neq 0$ and $a$, $b$, and $c$ are the constants in the quadratic equation.

If the quantity under the square root sign is negative, the solutions are not real numbers. Since most calculators use only real numbers, your calculator may give you an "error" message when the square root of a negative number is attempted. The solutions to the quadratic equations in this text are real numbers. Therefore, if the "error" message occurs, please check your computation.

### Example 12

Solve for $x$: $x^2 - 5x + 7 = 4$.

**Solution:** First, get zero on one side of the equation by adding $-4$ to both sides of the equation:

$$x^2 - 5x + 7 = 4$$

$$x^2 - 5x + 3 = 0$$

$$x = \frac{-(-5) \pm \sqrt{(-5)^2 - 4(1)(3)}}{2(1)}$$ Use the quadratic formula with $a = 1, b = -5$, and $c = 3$.

$$x = \frac{5 \pm \sqrt{25 - 12}}{2} = \frac{5 \pm \sqrt{13}}{2} \approx 4.30 \text{ or } 0.70.$$

## Section 2.0 PROBLEMS — *Explain* ➡ *Apply*

### *Explain*

1. In your own words, describe what is meant by the slope of a line.
2. In your own words, describe what is meant by the vertex of a parabola.
3. In the equation $y = mx + b$, what does $m$ represent? What does $b$ represent? What do $x$ and $y$ represent?
4. If you start with a quadratic equation, $y = ax^2 + bx + c$, and let $y = 0$, the resulting equation may be solved using the quadratic formula. What points on the graph of $y = ax^2 + bx + c$ do the solutions from the quadratic formula represent?
5. How can you determine which side of a line to shade when graphing an inequality?
6. In the graph of $3x + 5y \leq 30$, (see page 81) what is the significance of the shaded area? Give an example. What is the significance of the nonshaded area? Give an example.

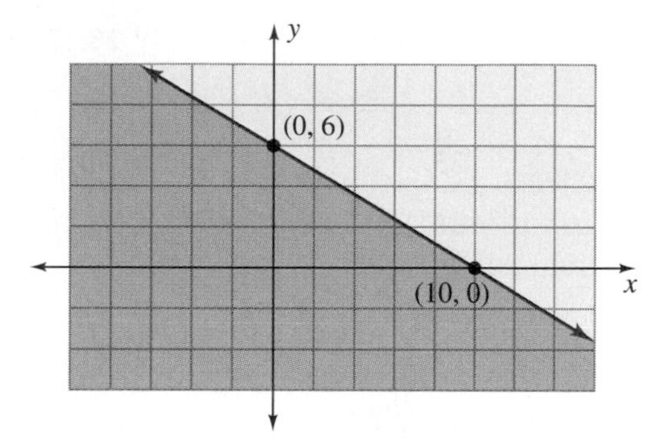


7. How does the graph of a linear equation differ from the graph of a linear inequality?

8. How do the graphs of linear inequalities containing $\leq$ or $\geq$ differ from those containing $<$ or $>$?

9. Why is the addition method for solving systems of equations sometimes called the elimination method?

## *Apply*

10. Sketch the graph of $y = -2x + 5$.

11. Sketch the graph of $y = -2x + 5$ where $x \geq -1$.

12. Sketch the graph of $y = -2x + 5$ where $x \leq 6$ and $x$ is a whole number.

13. Sketch the graph of $y = \frac{1}{4}x - 3.5$.

14. Sketch the graph of $y = \frac{1}{4}x - 3.5$ where $4 \leq x \leq 10$.

15. Sketch the graph of $y = \frac{1}{4}x - 3.5$ where $4 \leq x \leq 10$ and $x$ is a whole number.

16. Sketch the graph of $y = 5x$.

17. Sketch the graph of $y = -5x$.

In Problems 18–21, find an equation for the linear function that

18. has a slope of 4 and passes through the point $(3, -5)$.

19. passes through the points $(-4, -5)$ and $(8, -1)$.

20. passes through the points $(0, 525)$ and $(4, 800)$.

21. passes through the points $(1, 2300)$ and $(12, 5900)$.

In Problems 22–31, graph each linear inequality.

22. $y \geq 2x - 3$

23. $y < -3x + 8$

24. $y \le -x + 6$

25. $y > 2x + 1$

26. $2x + 3y < 12$

27. $5x - 6y > 60$

28. $7x - 2y \ge 28$

29. $3x + 4y \le -24$

30. $y > 3x$

31. $y \le 3x$

In Problems 32–35, use the graphing method to find the solution to the system of equations.

32. $y = 3x + 5$ and $y = 6x + 2$

33. $y = 6x$ and $y = -2x$

34. $2x + 6y = -4$ and $3x - 4y = 7$

35. $7x + 2y = -14$ and $6x - 4y = -12$

In Problems 36–39, use the addition method to find the solution to the system of equations.

36. $x + 6y = -4$ and $3x - 4y = 10$

37. $2x + 6y = 0$ and $4x - 5y = 0$

38. $3x + 6y = 15$ and $5x - 4y = -3$

39. $7x + 2y = -25$ and $6x - 4y = -10$

In Problems 40–43, sketch the solution to the system of inequalities.

40. $$\begin{aligned} 4x + y &\le 28 \\ 2x + 3y &\le 24 \\ x &\ge 0 \\ y &\ge 0 \end{aligned}$$

41. $$\begin{aligned} x + 3y &\le 12 \\ 3x + 4y &\le 21 \\ x &\ge 0 \\ y &\ge 0 \end{aligned}$$

42. $$\begin{aligned} 3x + 5y &\le 40 \\ 5x + 2y &\le 35 \\ x &\ge 0 \\ y &\ge 0 \end{aligned}$$

43. $$\begin{aligned} x + 6y &\le 48 \\ 7x + 2y &\le 56 \\ x &\ge 0 \\ y &\ge 0 \end{aligned}$$

In Problems 44–49, find the vertex and sketch the graph of each quadratic function.

44. $y = x^2 + 8x - 7$

45. $y = -x^2 + 6x - 5$

46. $y = -4x^2 - 7x$

47. $y = -3.2x^2 - 5.6$

48. $y = \frac{3}{4}x^2 + 12x + 8$

49. $y = \frac{1}{2}x^2 + 9x$

In Problems 50–57, solve for $x$ using the quadratic formula. Round off answers to the nearest hundredth.

**50.** $x^2 + 8x - 7 = 0$

**51.** $x^2 = 7x + 8$

**52.** $-x^2 + 7x - 5 = 0$

**53.** $2x^2 = 11x - 9$

**54.** $6x^2 - 5x - 11 = 0$

**55.** $3x^2 - 6x + 3 = 4$

**56.** $2.7x^2 - x - 25.3 = 20$

**57.** $-1.3x^2 - 2.5x - 2.3 = -20$

The Australian *Xanthorrhoea quadrangulata* plant can be algebraically described by a set of lines emanating from the same point. (Courtesy of Kurt Viegelmann)

## Section 2.1 Linear Models

In algebra, we study a system in which symbols (usually letters) are used to represent numbers. We can use this system to create mathematical models for various kinds of situations. The first model we will look at is one in which the situation can be graphically displayed by a line or part of a line. In this linear model we assume that the rate at which the quantities change (the slope) is constant. Also, we must realize that a model may not give an exact description of the situation and that it may have limitations depending on the situation. Let us look at some situations where a linear function would make an appropriate model.

In Section 2.0 we reviewed how to graph a line. We saw that for a line passing through the points $(x_1, y_1)$ and $(x_2, y_2)$

### Slope

$$m = \frac{\text{change in the } y \text{ values}}{\text{change in the } x \text{ values}} = \frac{y_2 - y_1}{x_2 - x_1}$$

## Equation of a Line

$$y = mx + b \qquad \text{where } \begin{cases} m = \text{slope} \\ b = y \text{ intercept} \end{cases}$$

We will use these equations to find linear functions that can act as models in various situations.

### Example 1

The equation to convert Celsius (°C) to Fahrenheit (°F) temperature is the linear function

$$F = \frac{9}{5}C + 32$$

(a) Graph this equation.
(b) Find its slope and explain what it tells us about the different temperature scales.
(c) Find the Fahrenheit equivalent of 40°C.

**Solution:**

(a) To graph a linear function, we find and plot two pairs of values that satisfy the equation. Since the value for $C$ (the domain) can be any real number, we draw a continuous line.

| °C | °F |
|---|---|
| 0 | 32 |
| 10 | 50 |

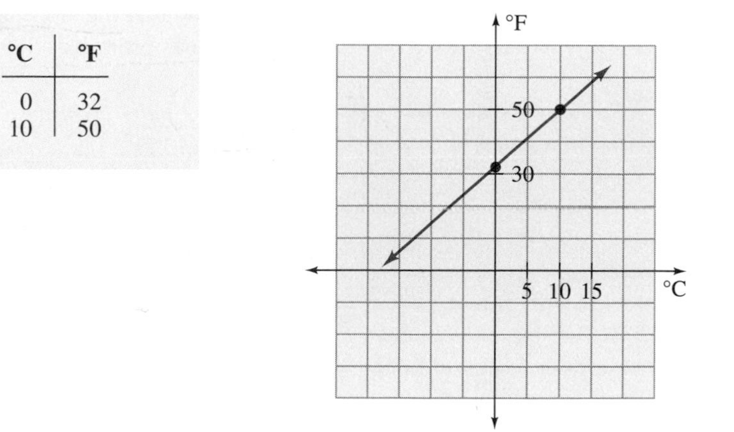

(b) From our equation $F = \frac{9}{5}C + 32$, the slope $m = 9/5$:

$$m = \frac{9}{5} = \frac{\text{change in Fahrenheit (°F)}}{\text{change in Centigrade (°C)}}$$

Thus a 9° change in $F$ corresponds to a 5° change in $C$.

(c) If $C = 40°$, $F = \frac{9}{5}(40) + 32 = 72 + 32 = 104°$. ■

## Example 2

Carmen's Coffee Shop had a net loss of $300 in its first month of operation, January. In April it had a net profit of $240. If business continues to grow at this rate, how much profit would the coffee shop make in December? or next April?

**Solution:** Carmen is looking for a way to predict her monthly profit, assuming that her profit will continue to increase at the present rate. A linear function would satisfy those conditions. Let

$$t = \text{time in months } (t \text{ is a whole number greater than } 0)$$

$$p = \text{net monthly profit}$$

$$(t, p) = \text{ordered pairs relating time and profit}$$

We are looking for the linear equation that gives us the profit $p$ based on the time $t$, using the facts:

$$\text{January, loss of \$300} \rightarrow (1, -300)$$

$$\text{April, profit of \$240} \rightarrow (4, 240)$$

The equation of a line is normally $y = mx + b$. The equation of the line, using the ordered pairs $(t, p)$ instead of $(x, y)$, becomes

$$p = mt + b$$

We must now find the slope $m$ and the $y$ intercept $b$ to get the linear function that determines profit based on time.

(a) Find $m$: $m = \dfrac{p_2 - p_1}{t_2 - t_1}$ (Note: $t$ and $p$ are used instead of $x$ and $y$)

$$m = \frac{240 - (-300)}{4 - 1} = \frac{540}{3} = 180$$

(b) Find $b$: $p = mt + b$. Since $m = 180$, $p = 180t + b$. To find $b$, substitute either known pair for $(t, p)$; $(4, 240)$ is used here.

$$240 = 180(4) + b$$

$$240 = 720 + b$$

$$-480 = b$$

Thus the linear function that determines profit based on time for Carmen's Coffee Shop for any month ($t$) is

$$p = 180t - 480$$

In December, $t = 12$ and $p = 180(12) - 480 = \$1680$; next April, $t = 16$ and $p = 180(16) - 480 = \$2400$. The graph of Carmen's profit equation is as follows:

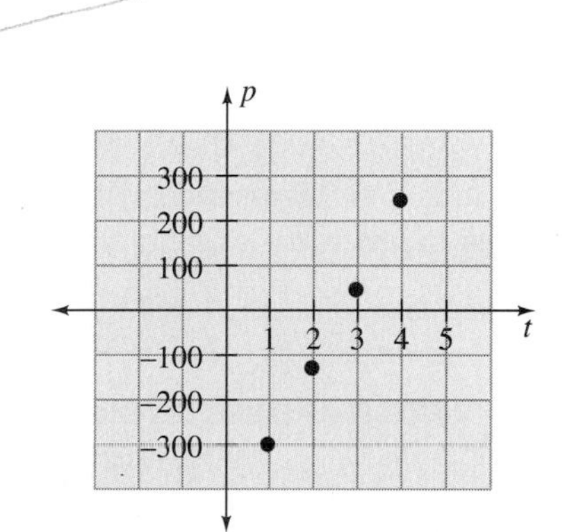


**Note** Since profit is determined at the end of the month, the graph consists of a dot for each month. The graph is a discrete one; it shows that there are no values at fractional parts of months.

This linear model for Carmen's Coffee Shop is based on the assumption that profit will continue to increase at the same rate. Factors such as competition, prices, salaries, weather, and advertising, can affect profit but have not been included in our model. In any linear model, the approximations become less accurate as you move away from the ordered pairs used to establish the model. Our model does, however, give us a means of making approximations based on present facts. ■

## Example 3

The speed of sound has been calculated to be approximately 1090 ft/s when the temperature is 32°F. However, as the temperature rises above 32°F, the speed at which sound travels increases at a constant rate. At 50°F, the speed of sound is about 1110 ft/s. Find the linear equation that relates the speed of sound to the Fahrenheit temperature and determine the speed of sound at 100°F.

**Solution:** Let

$$T = \text{temperature in Fahrenheit where } T \geq 32°\text{F}$$

$$s = \text{speed of sound}$$

$$(T, s) = \text{ordered pairs relating temperature and speed}$$

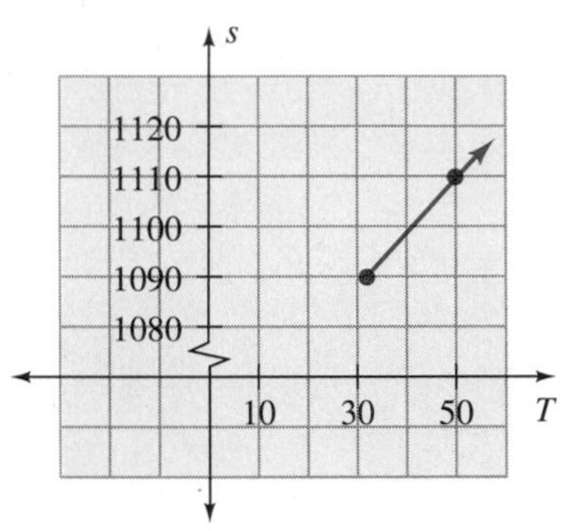


Equations using $T$ and $s$ instead of $x$ and $y$ are

$$s = mT + b \qquad \text{where } m = \frac{s_2 - s_1}{T_2 - T_1}$$

Known pairs for $(T, s)$ are (32, 1090) and (50, 1110).

(a) Find $m$:

$$m = \frac{s_2 - s_1}{T_2 - T_1} = \frac{1110 - 1090}{50 - 32} = \frac{20}{18} = \frac{10}{9}$$

(b) Find $b$: $s = mt + b$. Substitute $m = 10/9$ and either known pair for $(T, s)$.

$$1090 = \frac{10}{9}(32) + b \qquad (32, 1090) \text{ is used here.}$$

$$1090 = 35.6 + b$$

$$1054.4 = b$$

Thus the linear function that gives the approximate speed of sound based on the Fahrenheit temperature is

$$s = \frac{10}{9}T + 1054.4$$

When $T = 100°$F,

$$s = \frac{10}{9}(100) + 1054.4 = 1165.5 \text{ ft/s} \qquad ■$$

Section 2.1
PROBLEMS

## *Explain* ➡ *Apply* ➡ *Explore*

### *Explain*

**1.** What assumption is made when a linear equation is used to model a situation?

**2.** Given the equation $C = mx + b$, where $C$ is the cost of producing $x$ items, what does the value of $b$ represent? Explain.

**3.** Given the equation $C = mx + b$, where $C$ is the cost of producing $x$ items, what does the value of $m$ represent? Explain.

**4.** Given the equation $V = mx + b$, where $V$ is the value of a car after it has depreciated for $x$ years, what does the value of $m$ represent?

**5.** Given the equation $V = mx + b$, where $V$ is the value of a car after it has depreciated for $x$ years, what does the value of $b$ represent?

**6.** Give two real-life situations where the $x$ values in the equation $y = mx + b$ are restricted to being positive real numbers.

**7.** Give two real-life situations where the $x$ values in the equation $y = mx + b$ are restricted to being positive integers.

### *Apply*

In Problems 8–12, conversions from U.S. customary units to metric units are given by linear functions. Graph each equation if the number of U.S. customary units varies from 0 through 10.

**8.** $c = 2.54i$ $\quad$ $c$ = centimeters, $i$ = inches

**9.** $k = 2.2p$ $\quad$ $k$ = kilograms, $p$ = pounds

**10.** $g = 454p$ $\quad$ $g$ = grams, $p$ = pounds

**11.** $l = \dfrac{50}{53}q$ $\quad$ $l$ = liters, $q$ = quarts

**12.** $k = \dfrac{1000}{621}m$ $\quad$ $k$ = kilometers, $m$ = miles

In Problems 13–16, the profit ($P$) of a small business after time ($t$) in months is modeled by a linear function. Graph each equation for integer values of $t$ with $1 \le t \le 12$.

**13.** $P = 750t + 1000$ $\qquad$ **14.** $P = 750t - 1000$

**15.** $P = -750t + 1000$ **16.** $P = -750t - 1000$

**17.** For the four profit functions given above, give a description of the profit outlook for each small business.

## *Explore*

**18.** During the summer, as the temperature gets over 80°F, the chickens on a chicken farm drink more water. This behavior is modeled by the equation

$$w = 25t - 1250 \quad \text{where} \begin{cases} w = \text{number of gallons of water drunk per hour} \\ t = \text{Fahrenheit temperature } (t \geq 80°) \end{cases}$$

(a) Sketch a graph of this function.
(b) What is the slope of the function and what does it tell us about the situation?
(c) How many gallons of water are used in an hour when the temperature is 100°F?

**19.** Clover, a local department store, notices that there is a direct relationship between the gross revenue $R$ in dollars on a given day and the number $n$ of customers entering the store. It is determined that the equation $R = 2.3n$ approximates this revenue.

(a) Sketch a graph of this equation.
(b) What is its slope and what does it tell us about this situation?
(c) If 1500 people enter the store on a given day, what is the approximate revenue?

**20.** The equation $W = 1.7x + 0.3$ gives the weight $W$ in pounds of a $\frac{3}{4}$-in. galvanized pipe with protective thread caps. The variable $x$ represents the length of the pipe in feet.

(a) Graph the equation giving the linear relationship between the length and weight of the pipe.
(b) What is the slope of the equation? What is the physical significance of the slope in the problem?
(c) Find the weight of an 8-ft-long piece of pipe.
(d) If a piece of pipe weighs 19.85 lb, how long is the pipe?

**21.** The equation $C = 450x + 1200$ gives the cost $C$ in dollars when $x$ French wine barrels are produced.

(a) Graph the equation giving the linear relationship between the number of barrels produced and the cost of producing those barrels.
(b) While you may draw the graph as a line, what should be noted about the graph?
(c) What is the slope of the equation? What is the physical significance of the slope in the problem?

(d) Find the cost of producing 100 barrels.
(e) How many wine barrels may be completed if you are allowed to spend \$100,000?

**22.** You purchase a new Ford Explorer for \$26,500. A year later the car is only worth \$24,800. If the value of the car continues to depreciate at that rate,

(a) Find the linear equation that determines the value of the car based on the number of years you own it.
(b) When will the car be worth \$500?

**23.** As a weather balloon rises in altitude from sea level to 6 mi above sea level, the temperature decreases at a fairly constant rate. If the temperature is 59°F at sea level and 55.5°F at 1000 ft,

(a) Find the linear equation that relates the temperature $t$ to the altitude $a$.
(b) What is the temperature at an altitude of 24,000 ft?

**24.** When water turns to ice, its volume increases about 9%. For example, 100 ml of water has a volume of 109 ml after being frozen. Find the linear function that determines the volume of water after it has been frozen.

**25.** The foundation for a brick wall rises 10 in. above ground level. The bricks used for the wall are 8 in. high.

(a) Find a function that determines the height of the wall in inches based on the number of layers of brick that have been laid.
(b) Sketch a graph of the function.
(c) Find the equation that would give the height of the wall in feet.

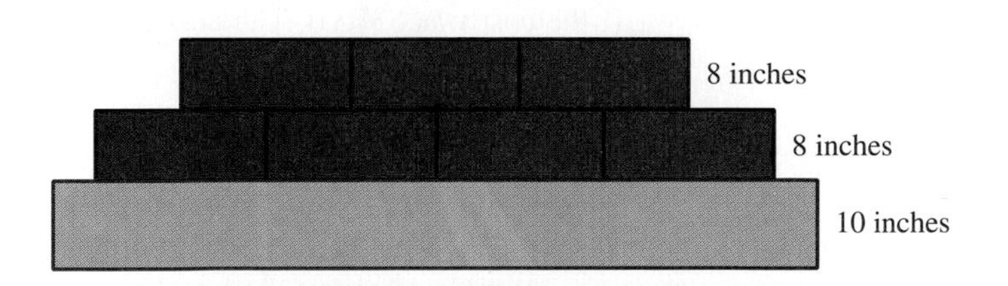


**26.** A fish tank is sitting on a stand 27 in. high. The tank is 36 in. tall and is being filled with water. The water is rising in the tank at a rate of 3 in./min.

(a) Find a function that would determine the distance the water level is above the floor at any given moment.
(b) What is the domain of the function?
(c) Sketch a graph of the function.

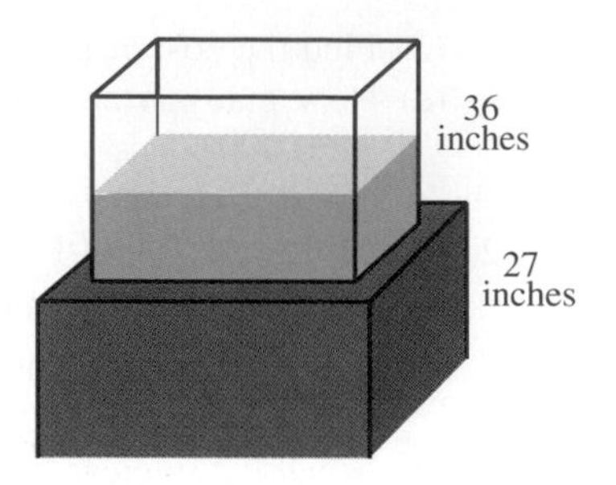


**27.** At higher altitudes, water boils at a lower temperature. At sea level, water boils at approximately 212°F. In Asheville, N.C., at 2000 ft, water boils at approximately 208°F.

(a) Assuming that this relationship is linear, find an equation that relates altitude to the boiling temperature of water.

(b) Find the boiling temperature of water at the top of Pike's Peak (Colorado), elevation 14,110 ft.

**28.** The Morgan Hill Car Rental Agency charges a one-time fixed charge of \$35 for a Ford Taurus and a mileage charge of \$0.20 per mile.

(a) Write a linear equation to determine the cost ($C$) of driving $m$ miles.

(b) Use the equation to find the cost of driving from San Francisco to Los Angeles and back (806 miles).

(c) How far can you drive for \$200?

**29.** For a typical four bedroom house in Pleasanton, California, during February 1994, Pacific Gas and Electric Company used a linear function to determine the cost of electricity usage. For the baseline quantity of 310.5 kwh (kilowatt-hours), the cost was \$37.10. The cost of each additional kilowatt-hour was \$0.13737.

(a) Write a linear equation to determine the cost ($C$) for electricity usage of ($k$) kwh over the baseline amount.

(b) If 262 kwh more than the baseline amount were used, what was the total cost of electricity?

(c) If the electricity bill was \$112.80, how many kilowatt-hours of electricity were used above the baseline amount?

**30.** According to the tax rate schedules given in the 1993 Federal Income Tax booklet, a married person filing separately, with a taxable income greater than or equal to \$125,000, will owe a tax of \$37,764.25 plus 39.6% of the amount of taxable income exceeding \$125,000.

(a) Determine the taxes paid on \$125,000 of taxable income and \$135,000 of taxable income.

(b) If $x$ represents your taxable income and $t$ represents the federal tax to be paid, use the results of part (*a*) to determine a linear equation expressing the tax ($t$) in terms of the amount of taxable income ($x$).

(c) Use the formula determined in part (*b*) to find the income tax on a taxable income of $1,000,000.

**31.** Is there a linear relationship between weight and height of men and women? The lists that follow show average weights of 20- to 24-year-old Americans by height.

y = mx + b

*Average Weight in Pounds of 20- to 24-Year-Old Americans*

| Men | | Women | |
|---|---|---|---|
| **Height** | **Weight** | **Height** | **Weight** |
| 5′7″ | 153 | 5′0″ | 112 |
| 5′9″ | 162 | 5′2″ | 120 |
| 5′11″ | 171 | 5′4″ | 128 |

(a) Why does a linear equation fit these lists?

(b) Find the equation that determines the weight for 20- to 24-year-old men and for 20- to 24-year-old women.

(c) Check the equations out on various people. Do the equations work? When? For whom?

**32.** Windchill factor is a combination of the actual temperature and wind speed. The wind makes it feel colder than it really is. Below are the windchill Fahrenheit temperatures when the wind speed is 10 mph.

| x<br>Actual Temperature | y<br>Windchill Temperature (At 10 mph) |
|---|---|
| 40°F | 28°F |
| 30°F | 16°F |
| 20°F | 3°F |
| 10°F | −9°F |
| 0°F | −22°F |
| −10°F | −34°F |
| −20°F | −46°F |
| −30°F | −58°F |

(a) Explain why a linear function does not exactly fit this chart.
(b) Find a linear equation that relates windchill to the actual temperature by using the first two pairs of information.
(c) How much error will that equation have in determining the windchill temperature when the actual temperature is $-30°F$?
(d) Plot the given windchill data and the graph of the equation from part (b) on the same coordinate system.
(e) How does the linear equation compare to the actual data?

**33.** The graph below shows the profit of a company based on the number of years the company has been in business.
(a) Why does a linear function seem to be an appropriate model for this graph?
(b) Use the data on the graph to find a linear function that estimates the profit $P$ of the company as a function of time $t$ where $t \geq 0$.
(c) Use the linear function determined in part (b) to estimate the company's profit after 10 yr.

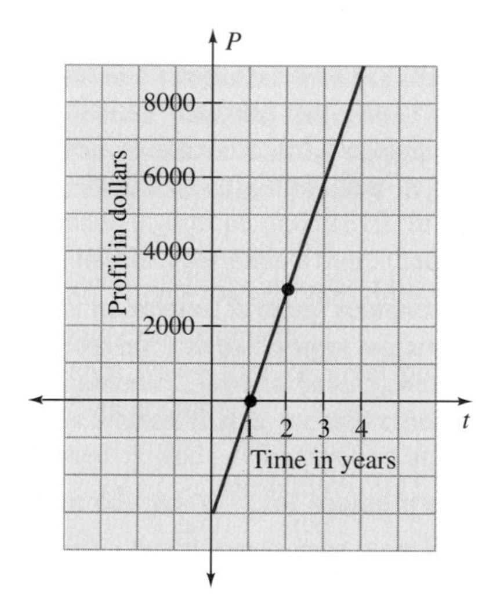


***34.** Is there a linear relationship between the length of a woman's foot and her shoe size? Do some measurements of the length of women's feet, compare this with their shoe size, formulate linear equations to match the results, test out the equations on other women, and so forth. What can you say about the relationship between the length of a woman's foot and her shoe size? Do the same problem for men and compare the results.

## Section 2.2 Linear Programming

Suppose you are running a business. A goal for your business is to keep profits to a maximum as you continue to manufacture, package, and distribute your product. A typical situation for a company producing CD players might be as follows.

Lovelace Electronics manufactures two types of CD players, a standard model and a deluxe model. The profit from selling a single standard model is \$25, whereas the profit from selling a deluxe model is \$28. Although the profit would increase if you sold more CD players, production is restricted by the size of your assembly and testing facilities. Each CD player must be assembled and tested prior to shipping. The assembly line is available for 24 worker-hours each shift. The testing line is available for 18 worker-hours each shift. The standard model requires one hour to assemble and one hour to test. The deluxe model requires two hours to assemble and one hour to test. How many of each type of CD player should be manufactured during each shift if the goal is to maximize the profit?

When tackling a problem containing so much information, it is important to determine the variables. Also, it is often helpful to organize the information using a table. Reading the last line of the problem, we find that the problem asks us to determine the number of each type of CD player to be manufactured so as to maximize the profit. Since there are two types of CD players being considered and we are also concerned about profit, it makes sense to have three variables. We will use the variables $x$, $y$, and $P$ to represent the following:

$$x = \text{the number of standard model CD players}$$

$$y = \text{the number of deluxe model CD players}$$

$$P = \text{the profit}$$

Organizing the information from the problem in the following table, we have

| | **Standard Model** | **Deluxe Model** | **Time Available** |
|---|---|---|---|
| Profit | \$25 | \$28 | |
| Assembly time | 1 hr | 2 hr | 24 hr |
| Testing time | 1 hr | 1 hr | 18 hr |

Since profit will increase for each additional CD player manufactured, we have the profit equation

$$P = 25x + 28y$$

If we keep producing more of each model of CD player, the profit will continue to increase. However, the assembly and testing facilities are resources with limitations. As we determine the solution to the problem, we cannot exceed the amount of time available. However, if it will increase the profit, we can use less than the available amount of time. Therefore, the relationships for the time required are as follows:

$$\text{Assembly time:} \quad 1x + 2y \leq 24$$

$$\text{Testing time:} \quad 1x + 1y \leq 18$$

Notice that we use a less than or equal sign to allow for less than total usage of a resource.

A final consideration in the setup of the problem is that since $x$ and $y$ represent the number of CD players produced, $x$ and $y$ should be non-negative values. Therefore, there are two more inequalities

$$x \geq 0 \quad \text{and} \quad y \geq 0$$

We have now completed the setup of our problem. We have changed the original word problem into a set of four inequalities and one equation. The inequalities are called **constraints** and the equation is called the **objective function**. The goal is to find the values of $x$ and $y$ that maximize the value of the objective function while staying within the boundaries defined by the constraints.

$$\text{Maximize } P = 25x + 28y\} \quad \textbf{objective function}$$

$$\left.\begin{aligned} \text{subject to } x + 2y &\leq 24 \\ x + y &\leq 18 \\ x &\geq 0 \\ y &\geq 0 \end{aligned}\right\} \quad \textbf{constraints}$$

**Linear Programming** is a mathematical technique used to solve problems in which there is a stated objective function that must be either maximized or minimized and a series of constraints that limit the values of the variables. In these problems, both the objective function and the constraints must be linear. That is, none of the variables may have exponents or be multiplied by any other variables. There are many methods used in linear programming. In this book, we will limit the discussion to one called the graphing method. The **graphing method** relies on graphing the constraints and then determining the maximum or minimum value of the objective function.

To solve our problem using the graphing method, we first graph the four inequalities and shade the appropriate region, using the methods discussed in Section 2.0. Notice that the graph is restricted to the first quadrant with its $x$ and $y$ axes since the problem uses only non-negative values for $x$ and $y$.

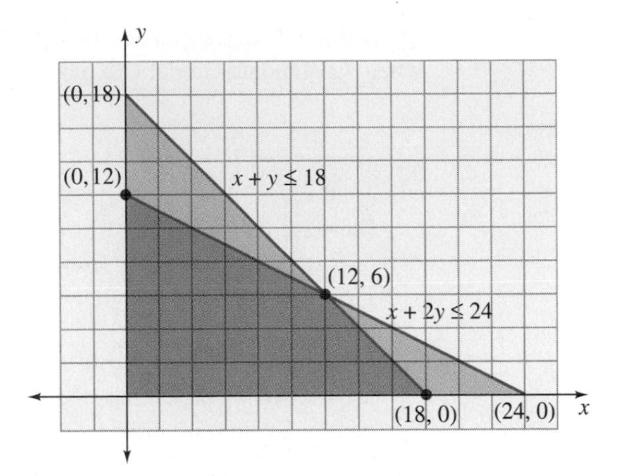


The region that is satisfied by all four constraints is shaded and is called the **feasible region**. The points within the region and on the boundary of the region all satisfy the constraints. Any point that is outside the region does not satisfy at least one of the constraints.

To determine the maximum value of the profit $P$, we use the Fundamental Theorem of Linear Programming.

## The Fundamental Theorem of Linear Programming

The maximum or minimum value of the objective function will occur at a corner point of the feasible region.

To determine the maximum value of $P$, we substitute the values of the corner points into the objective function $P = 25x + 28y$ and determine its maximum value. Substituting the corner points (0, 0), (1, 0), (12, 6), and (0, 12) into the profit function, we get

| $x$ | $y$ | $P$ |
|---|---|---|
| 0 | 0 | 0 |
| 18 | 0 | 450 |
| 12 | 6 | 468 |
| 0 | 12 | 336 |

From this table, we can see that the maximum profit will be \$468 and will occur when 12 standard model CD players and 6 deluxe models are produced per shift.

### Example 1

Suppose that the profit from the standard model CD players was reduced to \$15 each and the profit from the deluxe model was increased to \$32 each. Find the maximum profit.

**Solution:** The new profit function is $P = 15x + 32y$. Since the constraints are the same, we can use the feasible region and the corner points found above. Evaluating the new profit equation gives us

| $x$ | $y$ | $P$ |
|---|---|---|
| 0 | 0 | 0 |
| 18 | 0 | 270 |
| 12 | 6 | 372 |
| 0 | 12 | 384 |

Thus, the maximum profit is \$384 and occurs when 12 deluxe model CD players are produced per shift.

Notice that if the decision regarding production of the CD players is made solely on the basis of profit, the company would decide to produce none of the standard models. However, consumer demands will often influence the decision making process, causing the company to continue producing a limited number of the less profitable models. ■

### Example 2

Sketch the region bounded by the constraints

$$\begin{aligned} 2x + 3y &\geq 18 \\ 8x + 2y &\geq 32 \\ x &\geq 0 \\ y &\geq 0 \end{aligned}$$

Find the minimum value of the cost function $C = 20x + 16y$.

**Solution:** Sketching the graph, we have

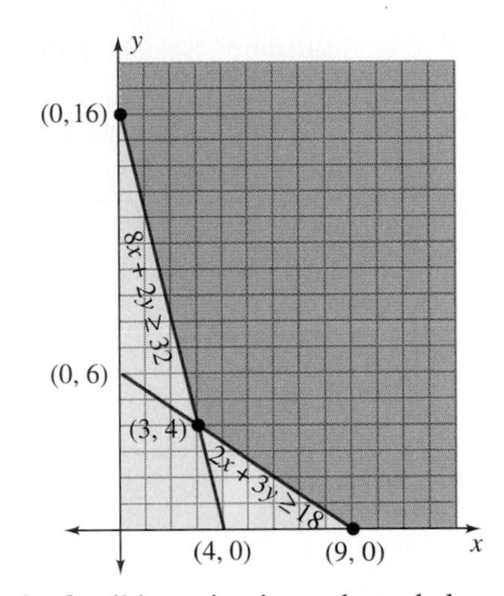


In this problem, the feasible region is not bounded on the top and right side. However, using the corner points (9, 0), (3, 4), and (0, 16), we still find the solution by substituting the values of the corner points into the cost function, $C = 20x + 16y$.

| $x$ | $y$ | $C$ |
|---|---|---|
| 9 | 0 | 180 |
| 3 | 4 | 124 |
| 0 | 16 | 256 |

Thus, we can see that the minimum cost is \$124 and occurs when $x = 3$ and $y = 4$. ■

## Example 3

Find both the maximum and minimum values of the function

$$R = 30x + 25y$$

given the constraints

$$\begin{aligned} 5x + y &\geq 14 \\ -5x + 6y &\leq 49 \\ 7x + 2y &\leq 77 \\ 4x + 9y &\geq 44 \\ x &\geq 0 \\ y &\geq 0 \end{aligned}$$

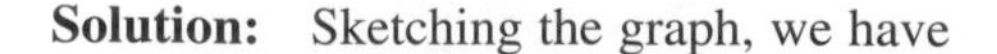

**Solution:** Sketching the graph, we have

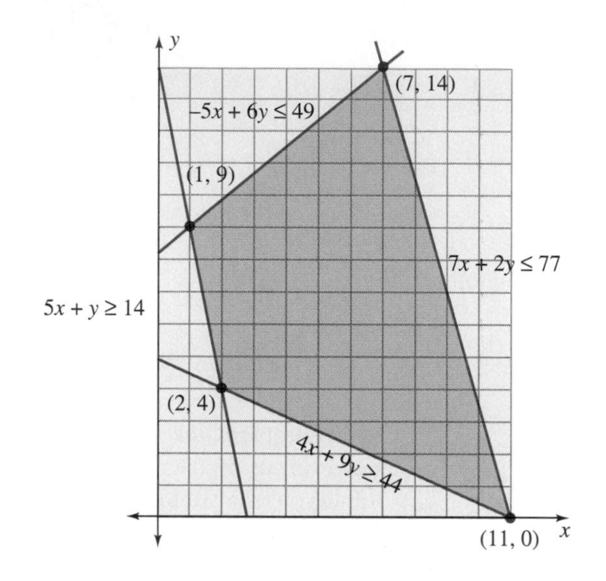


The corner points of the feasible region are (1, 9), (7, 14), (11, 0), and (2, 4). Evaluating $R = 30x + 25y$ at these points, we get

| $x$ | $y$ | $C$ |
|---|---|---|
| 1 | 9 | 255 |
| 7 | 14 | 560 |
| 11 | 0 | 330 |
| 2 | 4 | 160 |

Thus, the minimum value of $R$ is 160 and occurs when $x = 2$ and $y = 4$. The maximum value of $R$ is 560 and occurs when $x = 7$ and $y = 14$. ■

## Example 4

A veterinarian is creating a special diet for a sick dog. The diet will be a combination of two food mixes, A and B. Mix A contains 0.25 g of protein, 0.15 g of fiber, and 0.03 g of fat per gram of mix. Mix B contains 0.10 g of protein, 0.18 g of fiber, and 0.02 g of fat per gram of mix. The vet wants to create a diet that contains at least 25 g of protein and 27 g of fiber. How many grams of each type of food should be used to create a diet that contains the required amount of protein and fiber and at the same time contains the minimum amount of fat?

**Solution:** The first step in solving this problem is to arrange the information in a table.

| | Protein | Fiber | Fat |
|---|---|---|---|
| Mix A | 0.25 | 0.15 | 0.03 |
| Mix B | 0.10 | 0.18 | 0.02 |
| Goal | $\geq 25$ g | $\geq 27$ g | minimum amount |

Since the question asks for how many grams of mix A and mix B, we let the variables $x$ and $y$ represent the desired amounts, namely,

$$x = \text{number of grams of mix A}$$

$$y = \text{number of grams of mix B}$$

To begin, since we cannot use negative amounts of either mix A or mix B, we have

$$x \geq 0 \qquad \text{and} \qquad y \geq 0$$

Since there are 0.25 grams of protein in each gram of mix A, there must be $0.25x$ grams of protein in $x$ grams of mix A. Similarly, there must be $0.10y$ grams of protein in $y$ grams of mix B. Thus, the total amount of protein from $x$ grams of mix A and $y$ grams of mix B is given by

$$0.25x + 0.10y$$

Since the veterinarian wants a diet containing at least 25 g of protein, we obtain the constraint

$$0.25x + 0.10y \geq 25$$

In a similar way, for the required amount of fiber, we obtain the constraint

$$0.15x + 0.18y \geq 27$$

Since the vet wants to minimize the amount of fat, we have as an objective function

$$F = 0.03x + 0.02y$$

Combining this information, we can state the linear programming problem as

$$\begin{array}{ll} \text{Minimize} & F = 0.03x + 0.02y \\ \text{subject to the constraints} & 0.25x + 0.10y \geq 25 \\ & 0.15x + 0.18y \geq 27 \\ & x \geq 0 \\ & y \geq 0 \end{array}$$

We graph the constraint $0.25x + 0.10y \geq 25$ by plotting the points (100, 0) and (0, 250), drawing the line through the points, and shading the region above the line. Similarly we use the points (180, 0) and (0, 150) to graph the constraint $0.15x + 0.18y \geq 27$.

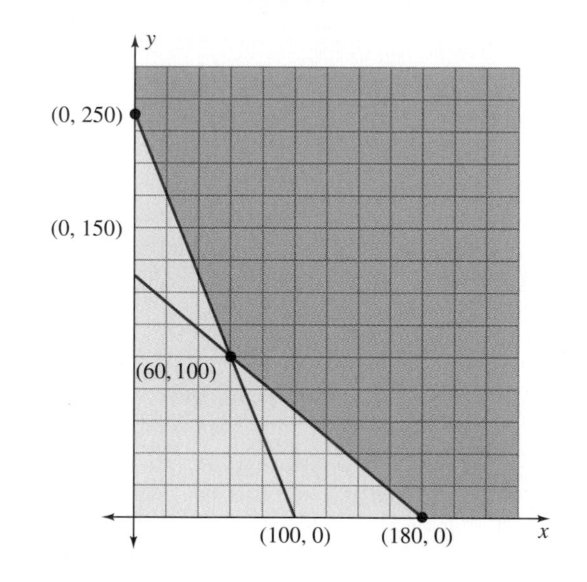


Evaluating the objective function $F = 0.03x + 0.02y$ at each of the corner points (0, 250), (60, 100), and (180, 0) gives

| $x$ | $y$ | $F$ |
|---|---|---|
| 0 | 250 | 5.0 |
| 60 | 100 | 3.8 |
| 180 | 0 | 5.4 |

Thus, we minimize the fat while still maintaining the required amounts of protein and fiber if we use 60 g of mix A and 100 g of mix B. ■

## Section 2.2 PROBLEMS

## *Explain* ➡ *Apply* ➡ *Explore*

### *Explain*

1. Explain what is meant by the feasible region.
2. Explain what is meant by the objective function.

3. Explain how the maximum value of a linear programming problem can be determined.

4. Explain how the minimum value of a linear programming problem can be determined.

5. Explain what is meant by a constraint.

6. Explain if each of the points $A$ through $E$ in the figure can be used as a possible maximum or minimum point for an objective function.

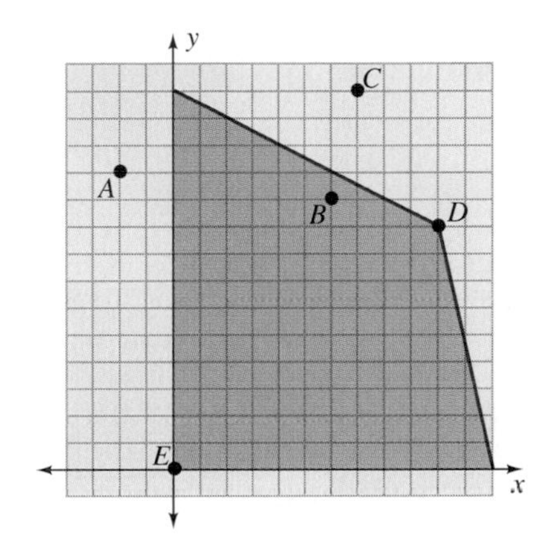


## *Apply*

In Problems 7–14, find the desired maximum and minimum values over the desired region.

7. Maximize $P = 3x + 5y$

8. Maximize $P = 11x + 12y$

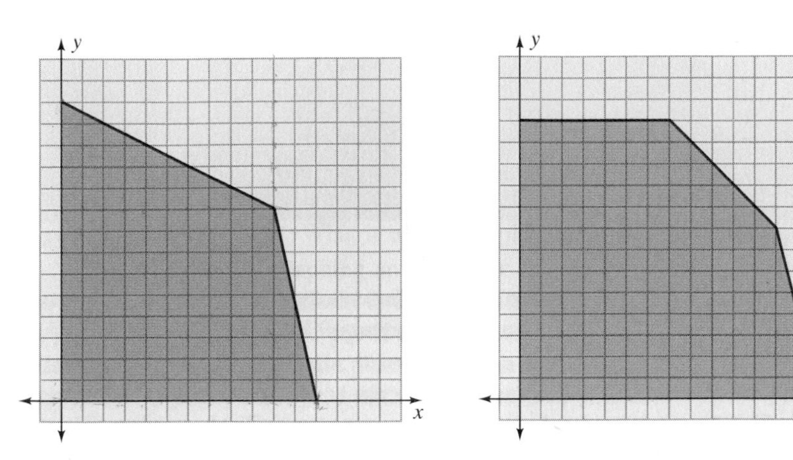

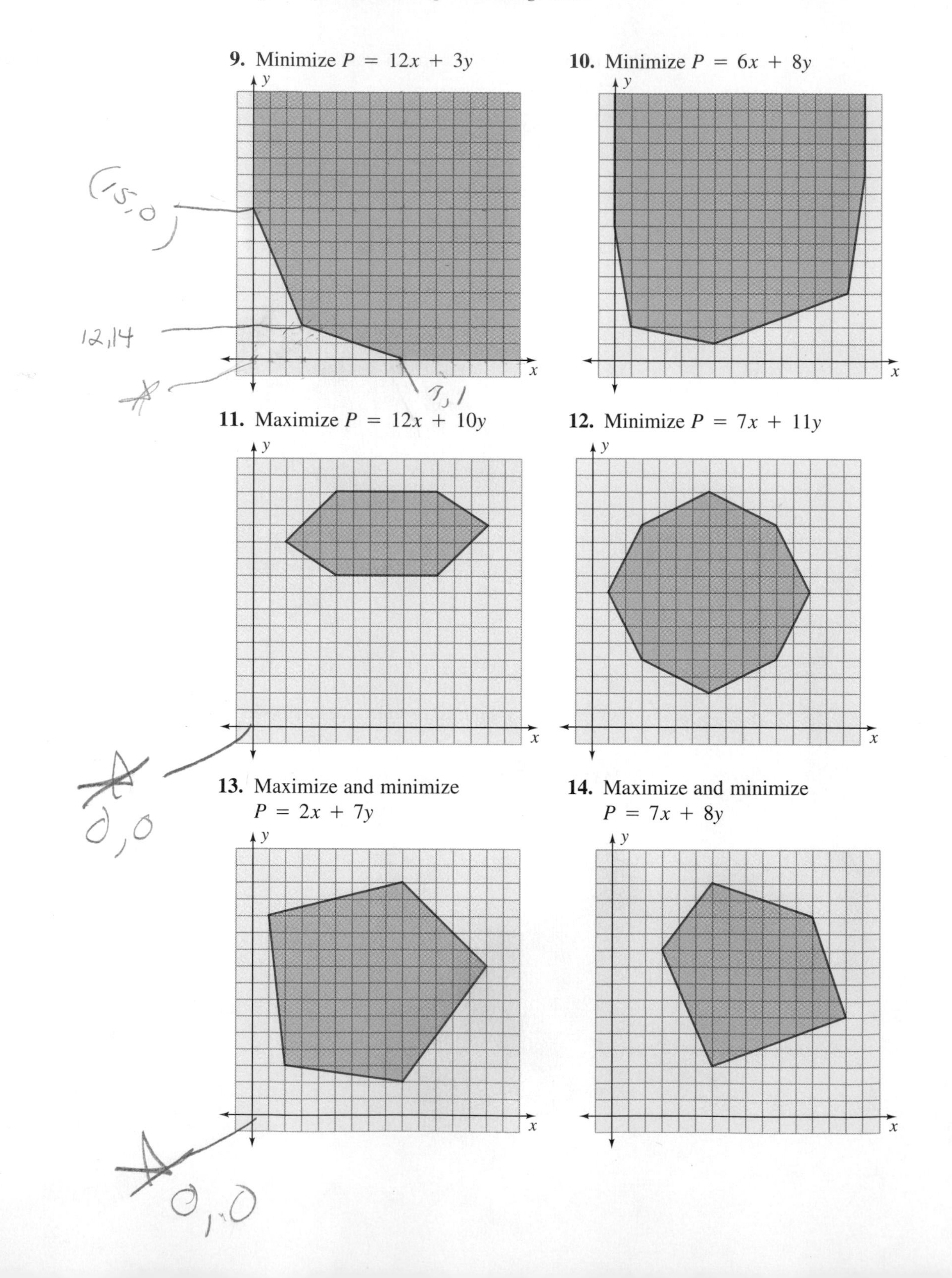
9. Minimize $P = 12x + 3y$
10. Minimize $P = 6x + 8y$
11. Maximize $P = 12x + 10y$
12. Minimize $P = 7x + 11y$
13. Maximize and minimize $P = 2x + 7y$
14. Maximize and minimize $P = 7x + 8y$
x
y

In Problems 15–24, solve the linear programming problem by sketching the region bounded by the constraints.

**15.** Maximize $P = 3x + 5y$

subject to the constraints $\begin{cases} 4x + y \leq 28 \\ 2x + 3y \leq 24 \\ x \geq 0 \\ y \geq 0 \end{cases}$

**16.** Maximize $P = 10x + 3y$

subject to the constraints $\begin{cases} x + 3y \leq 12 \\ 3x + 4y \leq 21 \\ x \geq 0 \\ y \geq 0 \end{cases}$

**17.** Maximize $P = 2x + 11y$

subject to the constraints $\begin{cases} 3x + 5y \leq 40 \\ 5x + 2y \leq 35 \\ x \geq 0 \\ y \geq 0 \end{cases}$

**18.** Maximize $P = 6x + 7y$

subject to the constraints $\begin{cases} x + 6y \leq 48 \\ 7x + 2y \leq 56 \\ x \geq 0 \\ y \geq 0 \end{cases}$

**19.** Minimize $C = 7x + 2y$

subject to the constraints $\begin{cases} 4x + y \geq 28 \\ 2x + 3y \geq 24 \\ x \geq 0 \\ y \geq 0 \end{cases}$

**20.** Minimize $C = x + 2y$

subject to the constraints $\begin{cases} x + 3y \geq 12 \\ 3x + 4y \geq 21 \\ x \geq 0 \\ y \geq 0 \end{cases}$

**21.** Minimize $C = 11x + 8y$

subject to the constraints $\begin{cases} 3x + 5y \geq 40 \\ 5x + 2y \geq 35 \\ x \geq 0 \\ y \geq 0 \end{cases}$

**22.** Minimize $C = 2x + 8y$

subject to the constraints $\begin{cases} x + 6y \geq 48 \\ 7x + 2y \geq 56 \\ x \geq 0 \\ y \geq 0 \end{cases}$

**23.** Find the maximum and the minimum of $P = 2x + 4y$

subject to the constraints $\begin{cases} x + y \leq 16 \\ 5x + y \leq 40 \\ -3x + 2y \leq 12 \\ 3x + 4y \geq 24 \end{cases}$

**24.** Find the maximum and the minimum of $P = 4x + 2y$

subject to the constraints $\begin{cases} x + y \leq 16 \\ 5x + y \leq 40 \\ -3x + 2y \leq 12 \\ 3x + 4y \geq 24 \end{cases}$

### *Explore*

**25.** A veterinarian is creating a special diet for a sick cat. The diet will be a combination of two mixes, A and B. Mix A contains 0.25 g of protein, 0.15 g of fiber, and 0.03 g of fat per gram of mix. Mix B contains 0.10 g of protein, 0.18 g of fiber, and 0.03 g of fat per gram of mix. The vet wants to create a diet that contains at least 25 g of protein and 27 g of fiber. How many grams of each type of food should be used to create a diet that contains the required amount of protein and fiber and at the same time contains the minimum amount of fat? What is the minimum amount of fat?

**26.** Use the information from Problem 25 except that mix A contains 0.04 g of fat per gram of mix and mix B contains 0.1 g of fat per gram of mix. How many grams of each type of food should be used to create a diet that contains the required amount of protein and fiber and at the same time contains the minimum amount of fat? What is the minimum amount of fat?

**27.** An orchid specialist wants to mix two types of fertilizers. One type of fertilizer contains 3% nitrogen, 2% phosphorus, and 1% potassium and costs $9 per pound. The other type of fertilizer contains 6% nitrogen, 1% phosphorus, and 1% potassium and costs $6 per pound. In order to properly feed all the orchids in the greenhouse, the specialist needs 27 lb of nitrogen, 12 lb of phosphorus, and 8 lb of potassium. Find the number of pounds of each fertilizer needed to minimize the total cost of the fertilizer while meeting the other requirements. What is the minimum cost?

**28.** Use the information in the preceding problem except that the first type of fertilizer costs $5/lb and the second type costs $7/lb. Find the number of pounds of each fertilizer needed to minimize the total cost of the fertilizer while meeting the other requirements. What is the minimum cost?

**29.** Suppose Huffy® produces two types of bicycles, mountain bikes and touring bikes. Suppose a mountain bike requires 2 hr to manufacture and 2 hr to assemble, while a touring bike requires 2 hr to manufacture and 3 hr to assemble. Suppose a mountain bike provides Huffy with $70 of profit while a touring bike provides the company with $100 of profit. If Huffy has 40 hr/wk of manufacturing time and 42 hr/wk of assembly time, how many bikes of each type should they produce in order to maximize the profit? What is the maximum profit?

**30.** In the preceding problem, how could the profit be maximized if all conditions were the same except that the profit was $95 for a mountain bike and $80 for a touring bike? What is the maximum profit?

The fireworks display can be algebraically described by a set of parabolas passing through the same point. (Courtesy of Kurt Viegelmann)

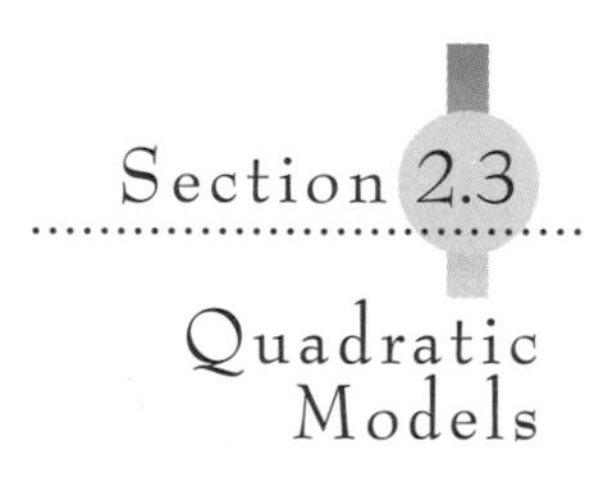

## Section 2.3 Quadratic Models

The linear function studied in Section 2.1 is only one of many functions that can be used as a mathematical model. There are situations in which the appropriate mathematical model is a quadratic function with its parabolic graph, as reviewed in Section 2.0.

As our first example, let us consider Dan'l Webster, the notorious jumping frog of Calaveras County, described in a short story by Mark Twain. That frog could outjump any frog in Calaveras County. One of the feats attributed to Dan'l Webster was his ability to jump from the floor to the top of a counter to catch a fly. If we assume that he took off from a point 2 ft from the counter, that the counter was 3 ft high, and the apex of his jump occurred at the edge of the counter, it is possible to find the equation of the parabola that approximates the flight of Dan'l Webster.

If we let the origin of a coordinate system (0, 0) be the frog's take-off point, where

$x$ = the length of the jump in feet

$y$ = the height of the jump in feet

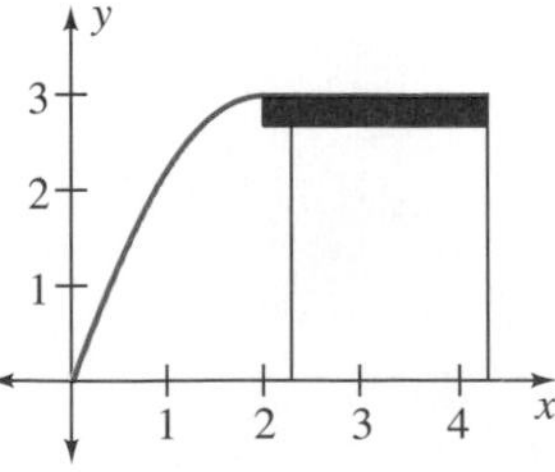

and the top of the counter (2, 3) be the vertex of the jump, we can find the equation of the parabola by using the general quadratic equation

$$y = ax^2 + bx + c$$

All we need to do is find the constants $a$, $b$, and $c$. This can be done as follows:

**1.** Since (0, 0) is on the graph,

$$y = ax^2 + bx + c$$
$$0 = a(0)^2 + b(0) + c$$
$$0 = c$$

**2.** Since (2, 3) is on the graph and $c = 0$,

$$y = ax^2 + bx$$
$$3 = a(2)^2 + b(2)$$
$$3 = 4a + 2b$$

**3.** Since the vertex is at (2, 3) and the $x$ coordinate of the vertex can be determined by $x = \frac{-b}{2a}$, we get

$$2 = \frac{-b}{2a}$$
$$4a = -b$$
$$-4a = b$$

**4.** Now using the results of parts (2) and (3), we can solve for $a$ and $b$. Starting with $4a + 2b = 3$ and substituting $b = -4a$ gives

$$4a + 2(-4a) = 3$$
$$4a - 8a = 3$$
$$-4a = 3$$
$$a = \frac{-3}{4}$$

Substituting this result into $b = -4a$ gives $b = 3$.

Thus, assuming the frog's jump follows a parabolic path, the equation for Dan'l Webster's jump to the top of the counter is

$$y = \frac{-3}{4}x^2 + 3x \qquad \text{where } \begin{cases} x = \text{horizontal distance in feet} \\ y = \text{vertical distance in feet} \end{cases}$$

**Note** To find the equation of a quadratic function, you can use the coordinates of the vertex and one other point.

With this quadratic model for the jump of Dan'l Webster, we can do some mathematical investigation. Although Mark Twain's narrative does not tell us if there was a stool located in front of the counter where Dan'l Webster leaped, we can determine the possibility of a stool being there by using the quadratic model discovered above.

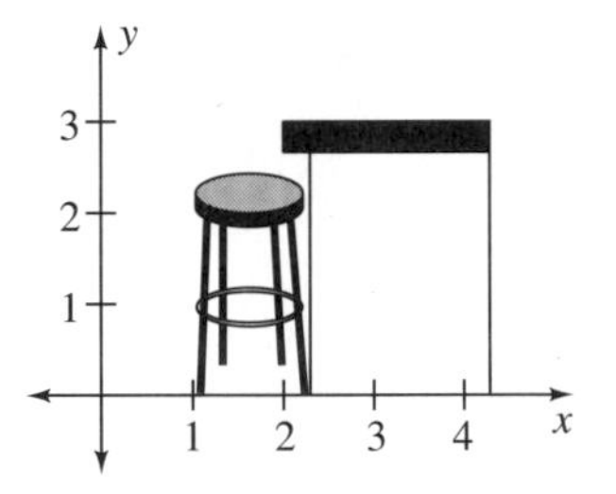


If we assume that a standard stool is 28 in. ($2\frac{1}{3}$ ft) tall and has a circular seat with a diameter of 12 in., could Dan'l Webster have cleared the stool and made it to the top of the counter? All we need to do is simply find the height of the frog when he is one foot from the take-off point. If $x = 1$,

$$y = \frac{-3}{4}(1)^2 + 3(1) = \frac{9}{4} = 2\frac{1}{4}\text{ft} = 27 \text{ in.}$$

Thus, when Dan'l Webster was 1 ft from his take-off point, he was 27 in. off the ground. Since we assumed the stool to be 28 in. tall, we can conclude that Dan'l would not have cleared such a stool. Since he did make it to the top of the counter, we can logically say that there was no such stool in front of Dan'l Webster in Mark Twain's story.

## Example 1

A recent competitor in the annual Calaveras County Frog Jumping Contest held in Angel City, CA, was Rosie the Ribiter. Her three consecutive jumps totaled 21 ft $5\frac{3}{4}$ in. If her first jump was 7 ft (84 in.) in length and reached a height of 18 in. at its apex, find the quadratic function that models the parabolic path taken by Rosie on that first of three jumps.

**Solution:** Let

$$x = \text{length of jump in inches}$$
$$y = \text{height of jump in inches}$$
$$(0, 0) = \text{starting point of jump}$$
$$(84, 0) = \text{landing point of jump}$$

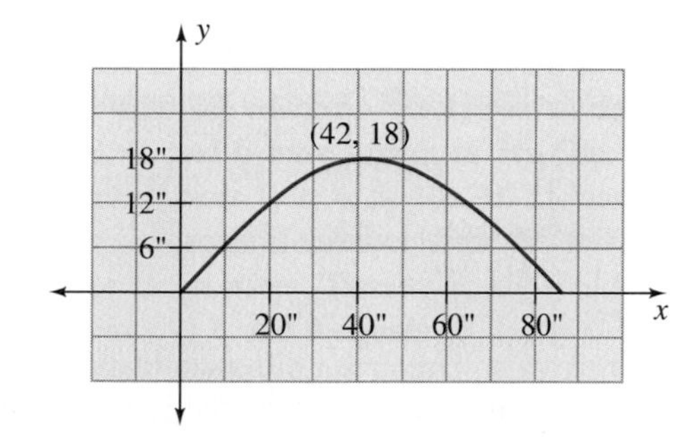


Then, the apex of the jump occurs half way between $x = 0$ and $x = 84$. So the vertex of the parabola is (42, 18). Now we will again use the general equation for a quadratic function.

(a) Since (0, 0) is on the graph,

$$y = ax^2 + bx + c$$
$$0 = a(0)^2 + b(0) + c$$
$$0 = c$$

(b) Since (84, 0) is on the graph and $c = 0$,

$$y = ax^2 + bx$$
$$0 = a(84)^2 + b(84)$$
$$0 = 7056a + 84b$$

(c) Since (42, 18) is also on the graph,

$$y = ax^2 + bx$$
$$18 = a(42)^2 + b(42)$$
$$18 = 1764a + 42b$$

(d) Solve the simultaneous system of equations obtained from parts (b) and (c).

$$\begin{aligned} 7056a + 84b &= 0 &\rightarrow\quad 7056a + 84b &= 0 \\ [1764a + 42b &= 18](-2) &\rightarrow\quad -3528a - 84b &= -36 \\ & & 3528a &= -36 \\ & & a &= \frac{-1}{98} \end{aligned}$$

Substituting that value for $a$, we get $b = \frac{6}{7}$.

Thus, the quadratic function that models the first jump of Rosie the Ribiter is

$$y = \frac{-1}{98}x^2 + \frac{6}{7}x \quad ■$$

As with the quadratic model for Dan'l Webster's jump, this equation can be used to further analyze the jump. For example, how far from the take-off was Rosie the Ribiter when she was 10 in. off the ground? By replacing $y$ with 10 in. we can solve for the distance from the take-off point.

$$\begin{aligned} y &= \frac{-1}{98}x^2 + \frac{6}{7}x \\ 10 &= \frac{-1}{98}x^2 + \frac{6}{7}x \\ 980 &= -x^2 + 84x && \text{multiplying both sides by 98} \\ x^2 - 84x + 980 &= 0 && \text{getting zero on one side} \\ x &= \frac{84 \pm \sqrt{7056 - 3920}}{2} && \text{using the quadratic formula} \\ x &= 14 \text{ or } 70 \end{aligned}$$

Thus, Rosie was 10 in. off the ground at two times in her jump, 14 in. and 70 in. from her take-off.

## Example 2

Quadratic functions can also be used to analyze the paths of objects that are moving under the force of gravity. (Air resistance is ignored.) For example, if you throw a baseball straight upward at 43.6 mph (64 ft/s) it slows down because of the force of gravity, reaches its highest point, and returns to hand level at approximately the same speed that it left your hand. If we chart the height $h$ that a vertically thrown ball reaches on one axis compared to the time $t$ the ball is in the air on the other axis, we get a graph that is a parabola. The equation for this motion on earth has been determined to be

$$h = -16t^2 + v_0t + s_0 \quad \text{where} \begin{cases} h = \text{height above the ground in feet} \\ t = \text{time in seconds} \\ v_0 = \text{initial upward velocity in ft/s} \\ s_0 = \text{initial height in feet} \end{cases}$$

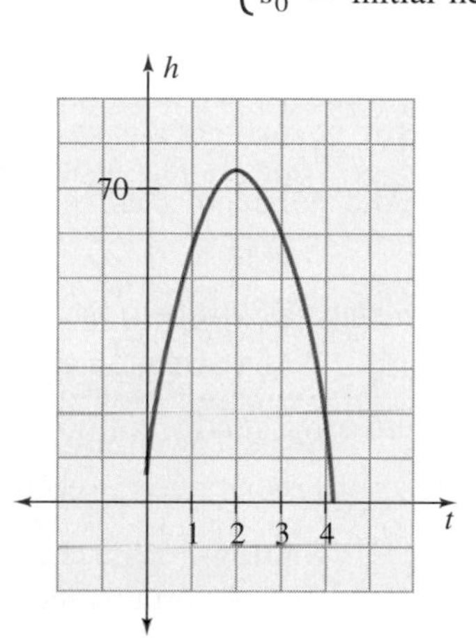


By initial height we mean the height above the ground at which the object began its motion and by initial velocity we mean the speed at which the object began its motion. Furthermore, as an object moves away from the earth its velocity is a positive quantity and as it moves toward the earth its velocity is a negative quantity. The speed of an object is actually the absolute value of the velocity and is measured only as a positive quantity.

Thus, if a vertically thrown ball leaves your hand 6 ft from the ground at 64 ft/s, $v_0$ is $+64$ ft/s, and $s_0$ is 6 ft. The equation of motion for the thrown baseball is

$$h = -16t^2 + 64t + 6$$

This equation can be used to analyze the path of the ball.

(a) What is the maximum height reached by the ball?
(b) How long does it take the ball to reach the ground?

**Solution:**

(a) The ball is at its highest point at the vertex of the parabola. The $t$ coordinate of that point can be found using $t = -b/(2a)$.

$$t = \frac{-64}{-32} = 2 \text{ s}$$

So the ball reaches its maximum height 2 s after leaving your hand and the height reached is

$$h = -16(2)^2 + 64(2) + 6 = 70 \text{ ft}$$

(b) To determine how long it takes the ball to return to the ground, let $h = 0$.

$$h = -16t^2 + 64t + 6$$

$$0 = -16t^2 + 64t + 6$$

$$t = \frac{-64 - \sqrt{4096 + 384}}{-32} \approx 4.1 \text{ s}$$

*Note:* In this instance only the negative root gives a positive result. ■

## Example 3

Suppose a baseball is thrown directly toward the ground from the top of the Sears Tower in Chicago, Illinois, at 60 mph. How far is the ball above the ground after 5 seconds?

**Solution:** Since the Sears Tower is 1454 ft tall, $s_0 = 1454$. In addition, our motion formula requires the initial velocity to be in ft/s, we must convert 60 mph to ft/s. The conversion fact for that change is

$$15 \text{ mph} = 22 \text{ ft/s}$$

We can set up a proportion relating mph to ft/s as follows. Let $v_0 =$ the number of ft/s that is equivalent to 60 mph. Then,

$$\frac{15}{22} = \frac{60}{v_0}$$

$$15v_0 = 1320$$

$$v_0 = 88$$

Since the object is traveling toward the earth, the initial velocity is negative ($v_0 = -88$). Thus, the equation of the height of the baseball thrown from the Sears Tower is

$$h = -16t^2 - 88t + 1454$$

To determine the height of the ball 5 s after it is thrown, let $t = 5$:

$$h = -16(5)^2 - 88(5) + 1454 = 614 \text{ ft}$$ ■

The quadratic function gives us another tool to use in analyzing problems. The exercises that follow will give you more examples in which the quadratic model can be used.

## Section 2.3 PROBLEMS

## *Explain* ➡ *Apply* ➡ *Explore*

### *Explain*

**1.** Describe how the vertex of a parabola can be found.

**2.** In the equation $h = at^2 + bt + c$, $h$ is the height of a thrown object and $t$ represents the time in seconds since the object was thrown. What is the physical meaning of $c$?

**3.** Describe how you can distinguish between the equation of a line and the equation of a parabola.

**4.** From the equation of a parabola, $y = ax^2 + bx + c$, explain how you can determine if the parabola has a maximum point or a minimum point.

### *Apply*

In Problems 5–12, find the vertex and graph each parabola.

**5.** $y = -x^2 + 5$

**6.** $y = x^2 + 5$

**7.** $y = 3x^2 - 6x + 8$

**8.** $y = 2x^2 + 8x + 17$

**9.** $y = \frac{1}{2}x^2 + x + 1$

**10.** $y = \frac{1}{4}x^2 - 2x + \frac{3}{4}$

**11.** $y = 0.4x^2 + 0.64x + 9.6$

**12.** $y = -0.3x^2 + 1.8x - 2.7$

### *Explore*

**13.** The path that the feet of a typical college pole vaulter take to the top of the crossbar in a successful attempt can be described by the quadratic function $h = -\frac{1}{8}d^2 + 3d$, where $h$ = the height of his feet above the ground measured in feet, $d$ = the horizontal distance from his take-off point to the bar measured in feet, and $0 \leq d \leq 12$.

(a) Graph the path followed by the vaulter's feet.
(b) How high did the vaulter's feet get on the vault?

**14.** At a local frog jumping contest, Rivet's jump can be approximated by the equation $y = -\frac{1}{6}x^2 + 2x$ and Croak's jump can be approximated by $y = -\frac{1}{2}x^2 + 4x$, where $x$ = the length of the jump in feet and $y$ = the height of the jump in feet.

(a) Which frog jumped the highest? How high did it jump?
(b) Which frog jumped the farthest? How far did it jump?

**15.** The equation for the cross section of a satellite dish with a diameter of 16 ft is given by

$$y = 0.002x^2$$

where $x$ and $y$ are measured in inches and $-96 \leq x \leq 96$.

(a) Graph the equation describing the cross section of the satellite dish.
(b) How far above the $x$ axis is the outer edge of the dish?

**16.** A farmer wants to construct a series of pig pens shown in the diagram. He has 120 ft of fencing available. The total area of the pens is given by $A = 600x - 2x^2$. Find the length $x$ that gives the maximum total area. (*Hint:* Find the vertex.)

**17.** A diver does a swan dive from a platform 20 ft above the water. The path followed by the feet of the diver can be approximated by a parabola as shown in the figure.

(a) Find the quadratic equation that gives a mathematical description of the dive.
(b) How far from the point directly below the take-off point will the diver's feet enter the water?

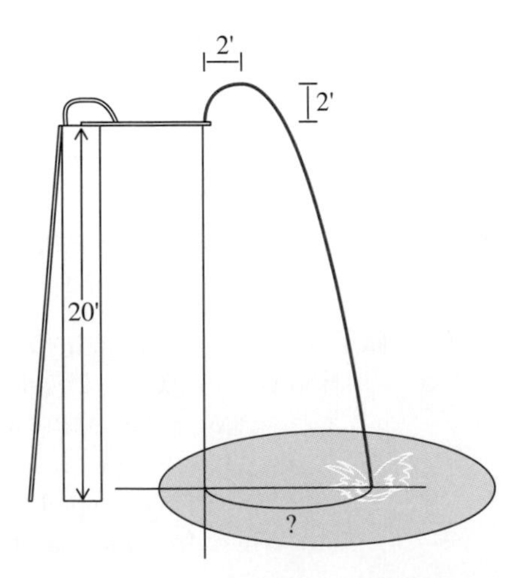

**18.** The numbers shown below, 1, 3, 6, 10, . . . , are called the triangular numbers. The first triangular number has a value of 1, the second has a value of 3, the third has a value of 6, and so forth. The relationship between the value of the triangular number $V$ and its position in the sequence $n$ is quadratic.

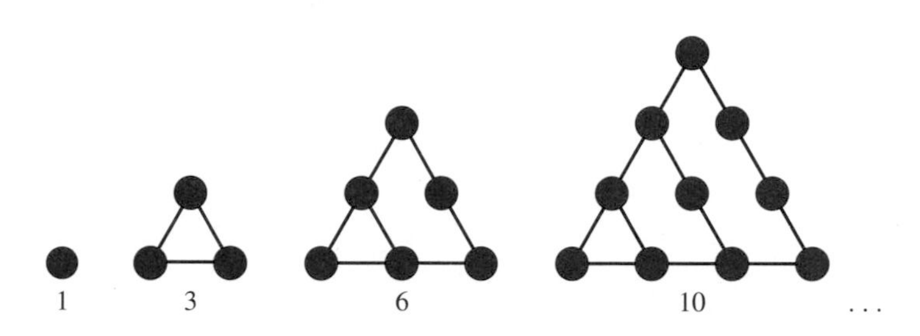


(a) Find the equation that expresses $V$ in terms of $n$.
(b) Find the 100th triangular number.

**19.** A professional baseball pitcher releases a 139.33-ft/s (95-mph) fastball from a point 9 ft above the ground, sending it on a vertical path upward into the air.

(a) Find the quadratic function that determines the height of the ball at any given instant.
(b) Find the maximum height reached by the ball.
(c) Determine how long the ball remains in the air.

**20.** A bullet is fired directly upward with a muzzle velocity of 860 ft/s from a height of 7 ft above the ground.

(a) Find the quadratic function that determines the height of the bullet at any given time.
(b) How long does it take the bullet to reach a height of 100 ft?
(c) How long is the bullet in the air?

**21.** The water from Bridal Veil Falls in Yosemite National Park falls 640 ft to a pool at the bottom of the falls. A rock falls from the top.

(a) Find the quadratic function that determines the height of the rock above the bottom of the falls at any given time. (*Hint:* $v_0 = 0$.)
(b) How far has the rock fallen in 3 s?
(c) How long does the rock take to reach the pool at the bottom of the falls?

**22.** Consider the quadratic function that approximates the leaps of our notorious frogs. They have all been of the form $y = -\frac{1}{d}x^2 + bx$. Suppose $1 \leq d \leq 10$ and $1 \leq b \leq 10$. What values for $d$ and $b$ would produce a jump that is both the longest and highest?

**23.** The approximate distance it takes to stop a car, based on the speed the car is traveling, is given in the chart below:

| Miles per Hour | Stopping Distance (ft) |
|---|---|
| 25 | 62 |
| 35 | 106 |
| 45 | 161 |
| 50 | 195 |
| 55 | 228 |
| 65 | 306 |

(a) Find a quadratic function based on the stopping distance for 25 mph and 50 mph, and the fact that at 0 mph the stopping distance is 0 feet.
(b) Use that equation to predict the stopping distances for 55 mph and 65 mph and compare them with the distances given in the chart.
(c) How accurate is the quadratic function?
(d) If the equation continues to be valid for higher speeds, how many feet would it take to stop a drag racer that reaches a speed of 230 mph?

**24.** A dog breeder has 260 ft of fencing material to make a kennel for the dogs along the side of a garage as shown in the figure. Let $x$ represent the width of the kennel and $y$ represent the length of the kennel.

(a) Find the equation that represents $y$ in terms of $x$.
(b) Find the values for $x$ and $y$ that give the maximum area for the kennel. (*Hint:* Find the vertex of a parabola.)

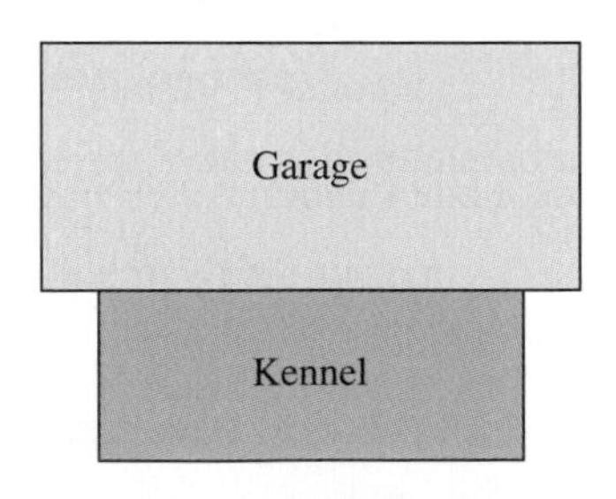


**25.** To get maximum distance of an arrow shot from a bow, the arrow should be aimed at about a 45° angle with the horizontal. The equation of motion for such an arrow can be approximated by

$$y = \frac{-32x^2}{v_0^2} + x \qquad \text{where} \begin{cases} x = \text{horizontal distance traveled in feet} \\ y = \text{height reached in feet} \\ v_0 = \text{initial velocity of arrow in ft/s} \end{cases}$$

Assume that an arrow leaves the bow with a speed of 192 ft/s.

(a) Find the equation of motion for the arrow.
(b) Find the height the arrow reaches.
(c) Find the distance the arrow reaches.
(d) How much higher would the arrow get if it is shot directly upward?

***26.** In this section we have looked at the parabola from a purely algebraic point of view. However, there are physical properties of a parabolic shaped object that make it very useful in many everyday objects. What objects have this parabolic shape, and what property of parabolas makes them useful?

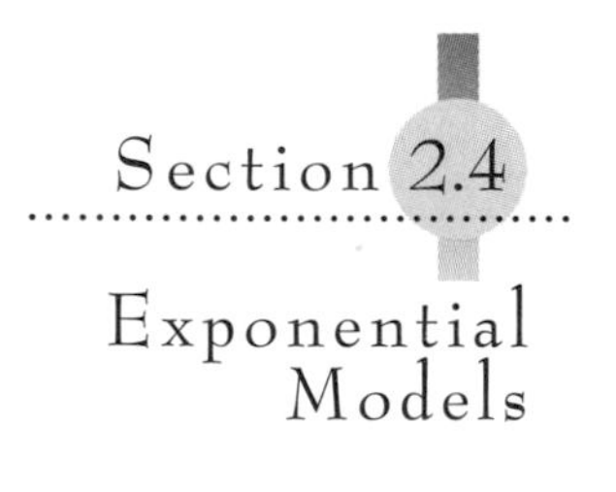


In the earlier sections of this chapter we studied two types of models, linear and quadratic. Linear equations are used to model situations that have a steady rate of increase or decrease, and quadratic equations are used to model situations displaying parabolic behavior. In this section, we will examine exponential functions.

## Exponential Functions

An **exponential function** is a function that contains an exponent that is a variable and a base that is a constant, such as $y = 2^x$, $y = 1.025^x$, $y = e^x$, and $y = 10^x$. The base of an exponential function is greater than zero and not equal to one. The equation

$$y = b^x \qquad \text{where } b > 0; \quad b \neq 1; \quad x \text{ is a real number}$$

defines the basic exponential function. You must understand that there is a significant difference between a quadratic and an exponential function. You can see this difference by examining a chart of values and the graphs of the quadratic function $y = x^2$ and the exponential function $y = 2^x$.

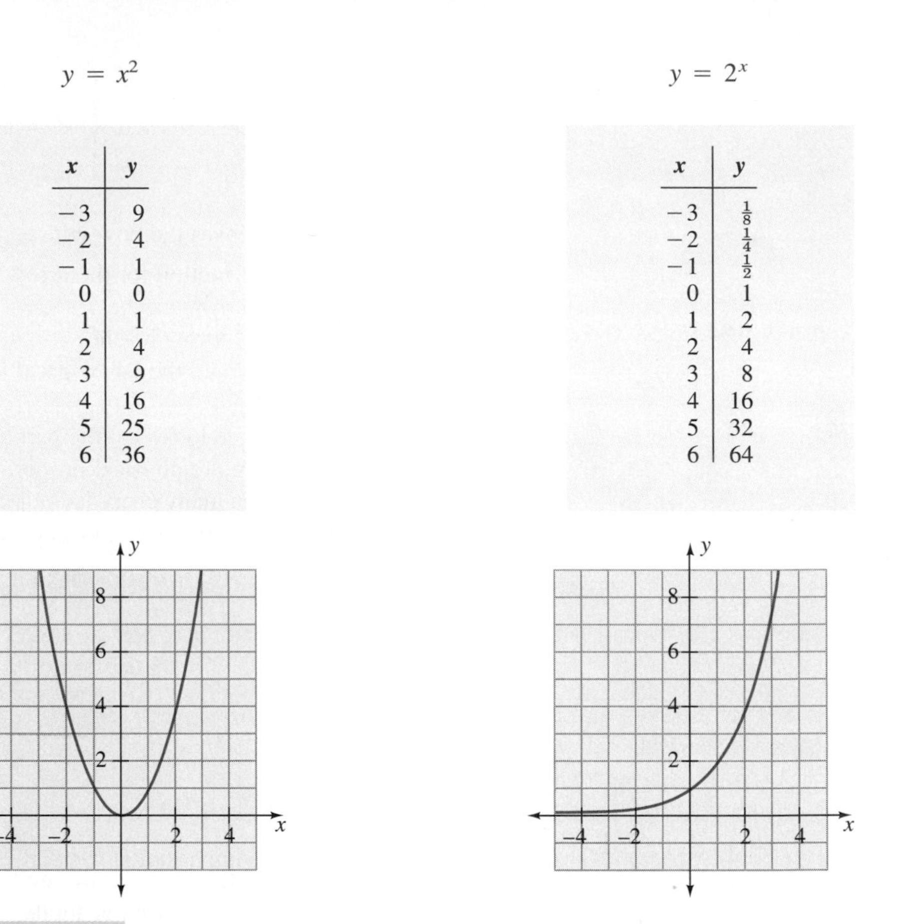

$y = x^2$

| $x$ | $y$ |
|---|---|
| −3 | 9 |
| −2 | 4 |
| −1 | 1 |
| 0 | 0 |
| 1 | 1 |
| 2 | 4 |
| 3 | 9 |
| 4 | 16 |
| 5 | 25 |
| 6 | 36 |

$y = 2^x$

| $x$ | $y$ |
|---|---|
| −3 | $\frac{1}{8}$ |
| −2 | $\frac{1}{4}$ |
| −1 | $\frac{1}{2}$ |
| 0 | 1 |
| 1 | 2 |
| 2 | 4 |
| 3 | 8 |
| 4 | 16 |
| 5 | 32 |
| 6 | 64 |

## Example 1

Graph the exponential function $y = 2^{-x} = (\frac{1}{2})^x$.

**Solution:** Calculate a chart of values for $(x, y)$ and plot the points.

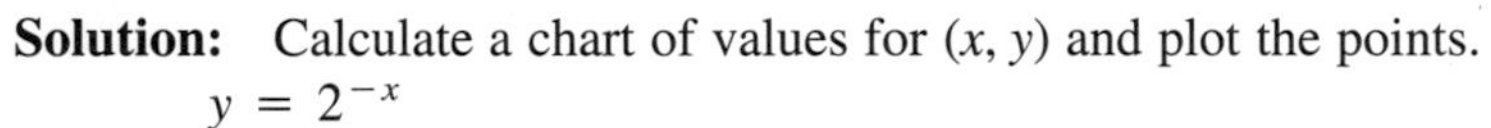

$y = 2^{-x}$

| $x$ | $y$ |
|---|---|
| −3 | 8 |
| −2 | 4 |
| −1 | 2 |
| 0 | 1 |
| 1 | $\frac{1}{2}$ |
| 2 | $\frac{1}{4}$ |
| 3 | $\frac{1}{8}$ |

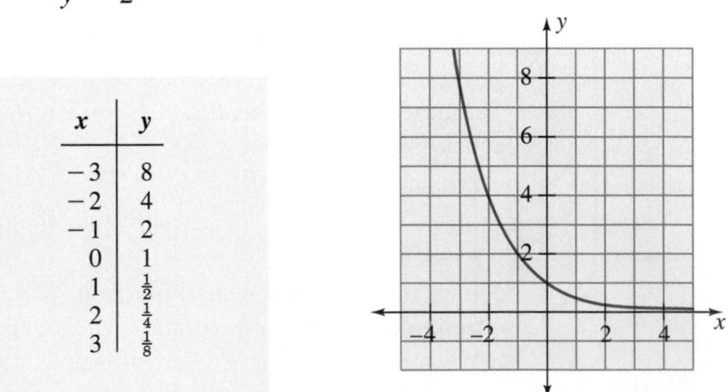

By examining exponential functions of the form $y = b^x$, where $b > 0$ and $b \neq 1$, we can summarize the following properties of an exponential function:

1. Its graph is a continuous curve that has the shape of a "banana" or "hockey stick."
2. Its graph passes through the point (0, 1).
3. Its graph approaches the $x$ axis but never touches it.
4. If $b > 1$, then $b^x$ increases as $x$ increases.
5. If $0 < b < 1$, then $b^x$ decreases as $x$ increases.

In this section, we will consider exponential functions that include other constants along with the basic exponential form of $b^x$. We will examine exponential functions of the form

$$y = a + c(b)^{kx} \qquad \text{where } \begin{cases} b > 0 \text{ and } b \neq 1 \\ a, c, \text{ and } k \text{ are constants} \end{cases}$$

The constants $a$, $c$, and $k$ simply move the basic exponential graph up and down or change the rate at which the $y$ values increase or decrease. No matter what constants are used, the graph of the exponential function retains the basic "banana" or "hockey stick" shape. The graph of an exponential function can be determined by calculating and plotting a sufficient number of points that satisfy the function. To perform the computations involved in exponential functions, we will use the following keys on a calculator:

If your calculator does not have each of those keys, please check with the manual for your calculator or with your instructor for a possible alternate method of performing calculations that involve those keys.

(Note: $e \approx 2.71828$ is an irrational number that can be approximated by the expression $(1 + \frac{1}{n})^n$ for large values of $n$.

To determine $3^7$:

| Press | Display |
|---|---|
| 3 $y^x$ 7 = | 2,187 |

To determine $8(6.5)^4$:

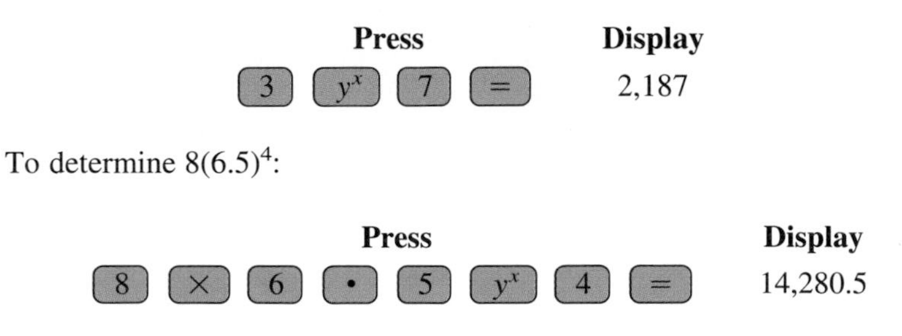

To determine $10^{-1.56}$:

| Press | Display |
|---|---|
| 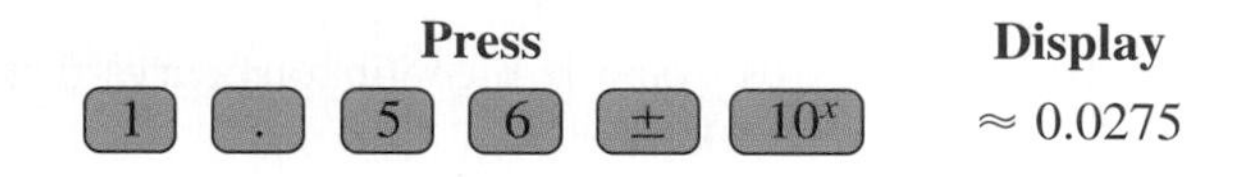 | $\approx 0.0275$ |

**Note** The $\pm$ key on the calculator changes the sign of the number in the window of the calculator.

To determine $9e^{-3.1}$:

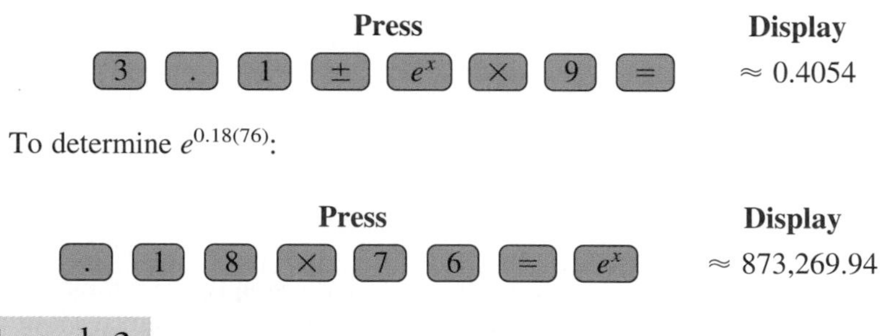

| Press | Display |
|---|---|
| 3 . 1 ± $e^x$ × 9 = | $\approx 0.4054$ |

To determine $e^{0.18(76)}$:

| Press | Display |
|---|---|
| . 1 8 × 7 6 = $e^x$ | $\approx 873{,}269.94$ |

## Example 2

Graph $y = 4 + 10^x$.

**Solution:** With the help of a calculator, determine and plot points that satisfy the exponential function. A sufficient number of points must be determined so that the exponential function can be graphed.

| $x$ | $y$ |
|---|---|
| −3 | 4.001 |
| −2 | 4.01 |
| −1 | 4.1 |
| 0 | 5 |
| 1 | 14 |
| 2 | 104 |

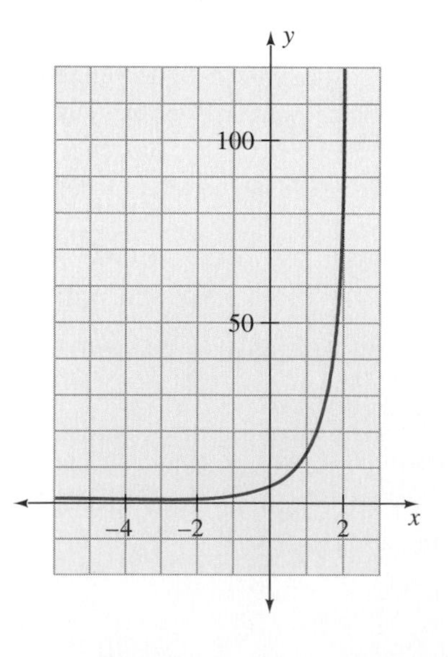

## Example 3

Graph $y = 50e^{0.3x}$.

**Solution:** With the help of a calculator, determine and plot points that satisfy the exponential function. A sufficient number of points must be determined so that the exponential function can be graphed.

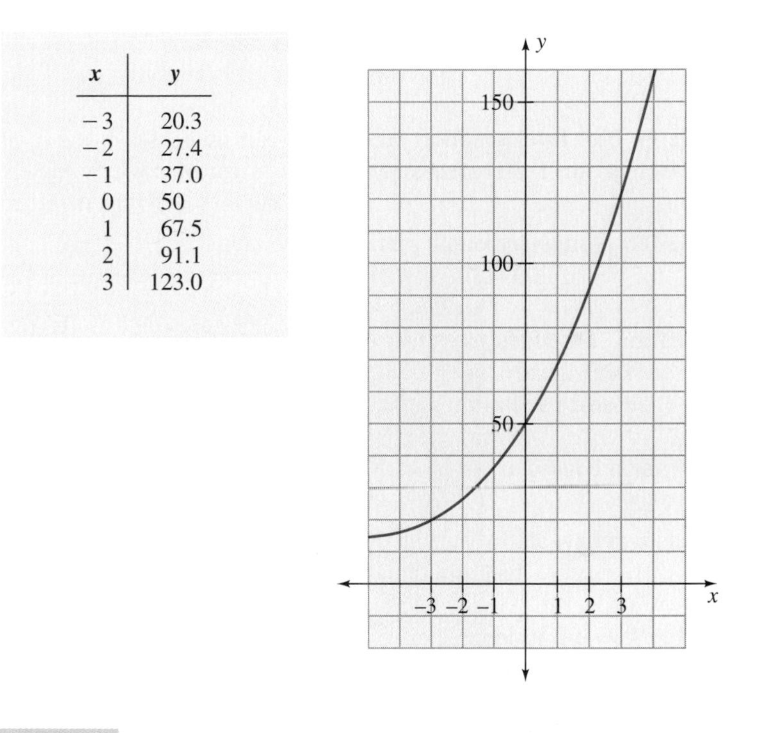

| $x$ | $y$ |
|---|---|
| −3 | 20.3 |
| −2 | 27.4 |
| −1 | 37.0 |
| 0 | 50 |
| 1 | 67.5 |
| 2 | 91.1 |
| 3 | 123.0 |



## Example 4

Graph $y = 100 + 250(4)^{-0.3x}$ where $x \geq 0$.

**Solution:** Using $x \geq 0$, calculate and plot points that satisfy the function.

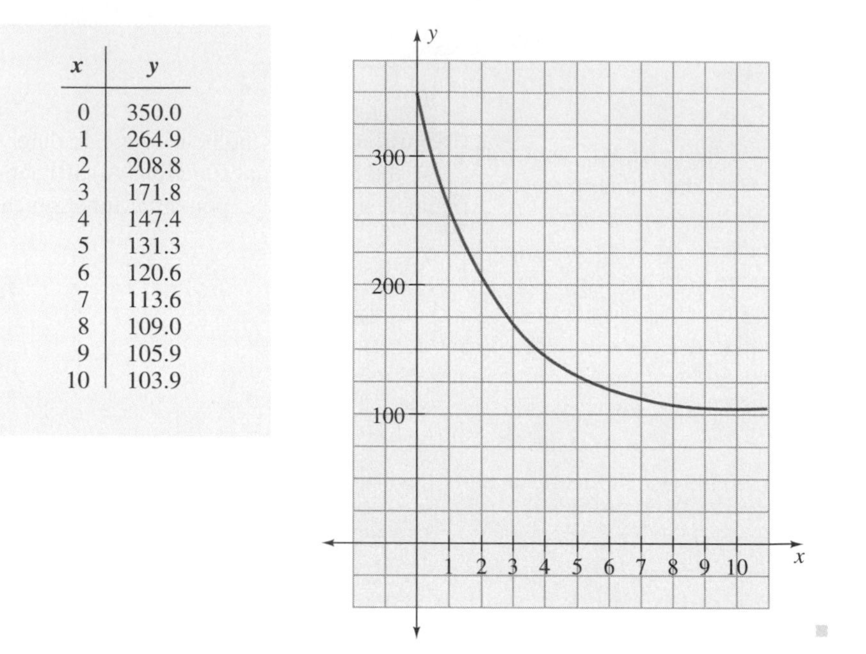

| $x$ | $y$ |
|---|---|
| 0 | 350.0 |
| 1 | 264.9 |
| 2 | 208.8 |
| 3 | 171.8 |
| 4 | 147.4 |
| 5 | 131.3 |
| 6 | 120.6 |
| 7 | 113.6 |
| 8 | 109.0 |
| 9 | 105.9 |
| 10 | 103.9 |



## Exponential Models

As we have seen in the previous sections, mathematical models can be used to answer questions and to make predictions about observed events. Exponential functions can also be used as mathematical models of real-life situations and scientific phenomena.

Exponential functions can be used to model situations in which as the values for one variable increase at a steady rate, the values for the other variable either

**(a)** decrease rapidly and then decrease slowly, approaching a limiting value, or
**(b)** increase slowly and then increase more and more rapidly.

An example of the latter can be seen in the common practice of determining the amount of money in an account after interest is compounded.

### Example 5 **Compound Interest**

Suppose you invest $5000 in an account that earns 1% on the amount in the account each month. If you make no withdrawals from the account and do not deposit any more money into the account, how much will you have in the account at the end of one year and at the end of ten years?

**Solution:** To determine the amount in the account after one year, we could calculate the interest each month using the formula $i = prt$ (interest = principal × rate × time). We could then add this interest to the amount in the account to get the new balance in the account at the end of each month as shown below.

| Month | Interest | Amount in Account |
|---|---|---|
| 1 | $I = (5000.00)(.01)(1) = \$50.00$ | $\$5000.00 + 50.00 = \$5050.00$ |
| 2 | $I = (5050.00)(.01)(1) = \$50.50$ | $\$5050.00 + 50.50 = \$5100.50$ |
| 3 | $I = (5100.50)(.01)(1) = \$51.01$ | $\$5100.50 + 51.01 \approx \$5151.51$ etc. |

Such a process could get us the amount in the account after 20 years, but it would require tedious computations. Luckily, there is an easier method to do this. Let us examine those first three months again.

| Month | Amount in Account |
|---|---|
| 1 | $5000 + (.01)(5000) = 5000(1 + .01) = 5000(1.01) = \$5050$ |
| 2 | $5050 + (.01)(5050) = 5050(1 + .01) = 5050(1.01) = \$5100.50$ |

Since $\$5050 = 5000(1.01)$, $5100.50 could be rewritten as

$$5000(1.01)(1.01) = 5000(1.01)^2$$

The pattern seen in the first two months suggests that the amount in the account at the end of the third month might possibly be determined by $5000(1.01)^3 = 5000(1.030301) \approx \$5151.51$. This result matches our previous total for the third month.

As this discussion suggests and as we shall see in Chapter 6, if $5000 is invested in an account that pays 1% interest each month with no withdrawals or other deposits, the amount $A$ in the account after $n$ months can be determined by the exponential function

$$A = 5000(1.01)^n$$

Thus, after one year (12 months),

$$A = 5000(1.01)^{12} = \$5634.13$$

After ten years (120 months),

$$A = 5000(1.01)^{120} = \$16,501.93$$ ■

To give you a better understanding of an exponential function, let's examine a chart of values and a graph of the exponential function $A = 5000(1.01)^n$.

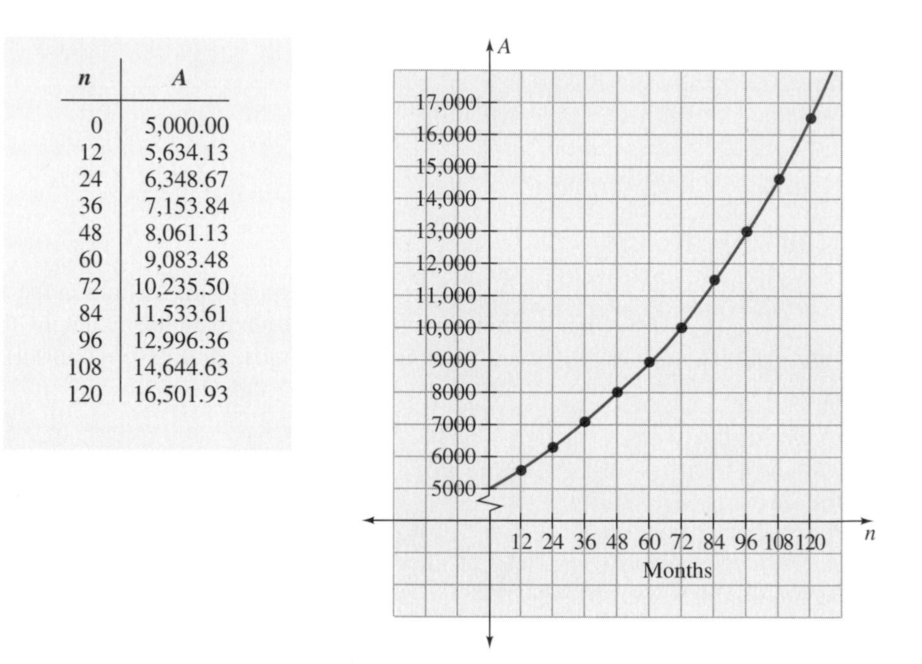

| $n$ | $A$ |
|---|---|
| 0 | 5,000.00 |
| 12 | 5,634.13 |
| 24 | 6,348.67 |
| 36 | 7,153.84 |
| 48 | 8,061.13 |
| 60 | 9,083.48 |
| 72 | 10,235.50 |
| 84 | 11,533.61 |
| 96 | 12,996.36 |
| 108 | 14,644.63 |
| 120 | 16,501.93 |



By observing the distances between successive amounts on the vertical scale, you will notice that the amount in the account increases slowly at the start and increases more rapidly as time goes on. For example, in the second year the account increased by \$714.54; in the tenth year the account increased by \$1857.30; and in the 30th year the account will increase by \$20,230.80. As we shall see in the next example, other situations in which a quantity increases by continual multiplication by the same amount lead to other exponential functions.

**Example 6** **Exponential Growth**

Suppose you decide to form a new club that meets every Friday. At the first meeting, just you and a friend show up. The next week, each of you brings a new member to the meeting. Now the club has four members. The third Friday, each of the four members brings another person to the meeting, making a club of eight people. If every Friday each present member brings a new member to join the club, how many members will the club have after six months?

**Solution:**

| Week | Members |
|---|---|
| 1 | 2 |
| 2 | 4 |
| 3 | 8 |
| 4 | 16 |

To answer the question we could continue the table until we reached the 26th week, but that would be a lot of work. Let's look at that problem a little more closely. We want to find a relationship between the week $w$ the meeting is held and the number of members $M$ at the meeting, assuming that every present member always brings a new member to the next meeting. By analyzing the data above, the number of members at any meeting is a power of 2 ($2 = 2^1, 4 = 2^2, 8 = 2^3, 16 = 2^4$, etc.) with the exponent being the week of the meeting. So the equation that determines $M$ is

$$M = 2^w$$

If $w = 26$, $M = 2^{26} = 67{,}108{,}864$.

As you can see, the growth in membership is astounding. By the 21st week the increase in the number of members per week is over a million. In fact, by the 28th meeting every person in the United States would be a member of your club. This example gives us a good indication of what is meant by a quantity growing exponentially.

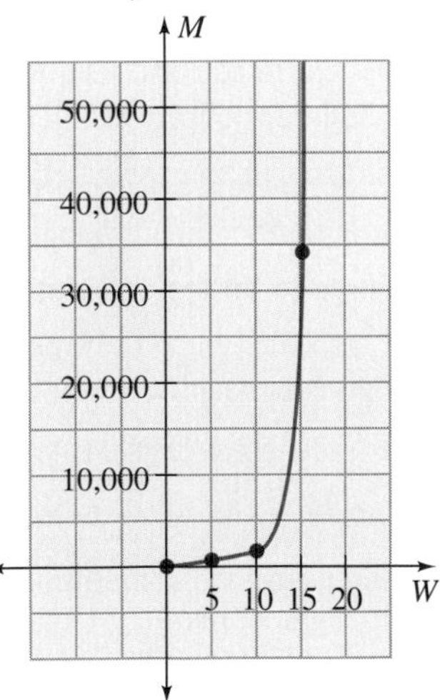


*Note:* The graph should be a discrete graph consisting of just the dots. The curve was added to make the exponential effect more evident.

**Example 7** **Exponential Growth of a Population**

Population increases when there are more births than deaths. Thomas Robert Malthus (1798) determined a model for predicting population based on the assumption that the rate of births $B$ and the rate of deaths $D$ remain constant and no other factors are considered. In this model the population $P$ is given by the following exponential function:

$$P = P_0 e^{kt} \qquad \text{where} \begin{cases} P = \text{population at any time} \\ P_0 = \text{initial population} \\ k = \text{annual growth rate } (B - D) \\ t = \text{time in years} \end{cases}$$

In 1970 the population of the United States was 205,052,000, the birth rate was 18.4 per 1000 population, and the death rate was 9.5 per 1000 population. Use this information to estimate the number of people in the United States in 1991.

**Solution:** Using $P_0 = 205{,}052{,}000$, $t = 21$ years, and $k = \frac{18.4}{1000} - \frac{9.5}{1000} = 0.0089$ with the formula $P = P_0 e^{kt}$ gives

$$P = 205{,}052{,}000e^{(0.0089)(21)}$$

$$P \approx 247{,}191{,}566$$

The actual population at the end of 1991 was 252,688,000. As you can see, the **Malthusian population model** did not give the exact 1991 population. One of the reasons for this is that both the birth rate and death rate have changed since 1970. The model did, however, give a reasonable approximation based on the facts that were available in 1970. ■

**Example 8** **Atmospheric Pressure**

Atmospheric pressure is produced by the weight of air from the top of the atmosphere as it presses down upon the layers of air below it. At sea level, air pressure is about 14.7 pounds per square inch. As the distance from the earth's surface increases, the air pressure decreases. This phenomenon can be observed when a sealed bag of potato chips becomes puffed out like a balloon when taken into the mountains. The following exponential function relating air pressure ($P$) and altitude ($a$) can approximate the atmospheric pressure at altitudes up to 50,000 ft.

$$P = 14.7(10)^{-0.000018a} \qquad \text{where} \begin{cases} P = \text{pressure measured in lb/in.}^2 \\ a = \text{altitude measured in feet} \end{cases}$$

The graph of this function is

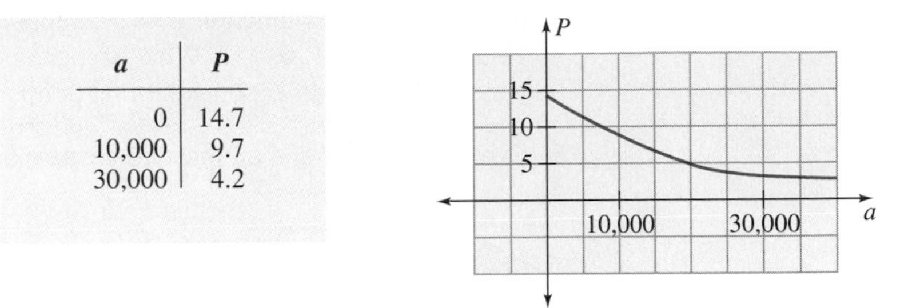

| $a$ | $P$ |
|---|---|
| 0 | 14.7 |
| 10,000 | 9.7 |
| 30,000 | 4.2 |



This function can also be used to estimate the air pressure at any altitude up to 50,000 ft. What is the air pressure on the top of the world's tallest mountain, Mount Everest in Nepal-Tibet, which has an altitude of 29,028 ft?

**Solution:**

$$P = 14.7(10)^{-0.000018a}$$

$$P = 14.7(10)^{-0.000018(29{,}028)}$$

$$P \approx 4.4 \text{ lb/in.}^2$$

This section has attempted to show you that exponential functions can be used as mathematical models for actual occurrences in the world. There are many other places where these functions are used, but many of them are beyond the scope of this book and require a greater knowledge of the areas in which they are used. The problems that follow, however, will let you experiment with other exponential functions used as mathematical models. ■

## Section 2.4 PROBLEMS

## *Explain* ➡ *Apply* ➡ *Explore*

### *Explain*

**1.** Compare the graphs of an exponential function and a quadratic function. Which function seems more appropriate to describe the growth of a population. Explain.

**2.** In the exponential function $y = b^x$, where $b > 1$, what happens to the graph as

(a) $x$ gets larger?
(b) $x$ gets smaller?

**3.** In the exponential function $y = b^x$, where $0 < b < 1$, what happens to the graph as

(a) $x$ gets larger?
(b) $x$ gets smaller?

**4.** In the exponential function $y = b^x$, why does $b \neq 1$?

**5.** In the exponential function $y = b^x$, suppose $b = -4$. Determine the value of $y$ if $x = -2, -1, 0, 1, 2$. What happens if $x = \frac{1}{2}$? Use these results to explain why the base of an exponential function is positive.

**6.** Explain why $10^x$ and $e^x$ must be greater than zero for all real values of $x$.

## *Apply*

In Problems 7–24, use your calculator to determine the answers. Round answers off to the nearest thousandth.

**7.** $5.2^4$

**8.** $7.3^{-2.56}$

**9.** $10^{-2.56}$

**10.** $10^{4.45}$

**11.** $e^{-2.56}$

**12.** $e^{4.45}$

**13.** $1.08 + 10^{-2.56}$

**14.** $7.34 - 10^{1.45}$

**15.** $9.76 + e^{-1.34}$

**16.** $7.24 - e^{3.4}$

**17.** $7.6e^{1.3454}$

**18.** $8.34e^{-2.56}$

**19.** $12.4(10^{2.34})$

**20.** $8.36(10^{-3.45})$

**21.** $9(10)^{(47.89-45.9)}$

**22.** $-8.2(10)^{(19.6-21.2)}$

**23.** $-7.1(e^{(52.3-51.7)})$

**24.** $5.67(e^{(14.2-15.7)})$

In Problems 25–30, sketch a graph of each exponential function.

**25.** $y = 0.25e^x$

**26.** $y = 50e^{0.3x}$

**27.** $y = 500 + 10^x$

**28.** $y = 200 - 10^{-x}$

**29.** $y = 100(1.06)^{12x}$

**30.** $y = 200(1.015)^{4x}$

## *Explore*

**31.** If the \$5000 of Example 5 were deposited in a savings account that paid interest daily, the interest rate would be 0.0328767% (12 ÷ 365) each day and the amount $A$ in the account after $n$ days would be given by the exponential function, $A = 5000(1.000328767)^n$.

(a) Sketch a graph of that exponential function.
(b) Find the amount in the account after one year.
(c) Find the amount in the account after 20 years.

**32.** Suppose Parker Brothers determines that the profit $P$ for a board game that is on the market for $t$ years is given by the equation:

$$P = 6000 + 20{,}000(3)^{-0.2t}$$

(a) Sketch the graph of this profit function.
(b) What is the profit after 25 years?
(c) What does the graph and the answer tell us about the profit for the board game?
(d) Is this exponential function a logical model for the profit from a board game? Give reasons for your answer.

**33.** The manufacturing company in Problem 32 is trying to stimulate sales of the board game through 30 days of television advertising in an area that has 250,000 viewers. The number of viewers $V$ in thousands who are made aware of the board game after $t$ days of advertising is expected to be $V = 250 - 250e^{-0.04t}$

(a) Sketch a graph of this exponential function for the 30 days.
(b) How many viewers were made aware of the board game after one day?
(c) How many viewers were made aware of the board game after two weeks?
(d) How many viewers were made aware of the board game after 30 days?

**34.** If an amount of principal $P$ is invested at an annual rate $r$ expressed as a decimal and is compounded continuously, the amount $A$ in the account at the end of $t$ years is given by the exponential function, $A = Pe^{rt}$. If \$5000 is invested at an annual rate of 9%,

(a) Write the exponential function that determines the amount in the account at the end of any year.
(b) Sketch a graph of the function.
(c) Find the amount in the account after 20 years.

**35.** In an attempt to promote world peace you decide to start a chain letter. You send a peace message to five friends asking each of them to send copies of the message to five of their friends by the end of the week. Suppose this process continues and every person sends the message on time to five new people. A chart of the number of people receiving messages each week would look like this:

| Week | Number of People Receiving Messages |
|---|---|
| 1 | 5 |
| 2 | 25 |
| 3 | 125 |
| 4 | 625 |

(a) Graph the data using the number of weeks on the horizontal axis and the number of people on the vertical axis.
(b) Find an exponential function relating the number of people $p$ that receive the peace message to the number of weeks $w$.
(c) How many people would receive the message by the end of 12 weeks?

**36.** Suppose you are gainfully employed and earn $200 a day. You are, however, offered a temporary job doing similar work for three weeks where you will be paid $0.01 the first day, $0.02 the second day, $0.04 the third day, and so on. Your daily wage will continue to double for each of the 21 days. Would it be more profitable for you to take the temporary job or keep your regular job for the three weeks? Justify your answer mathematically.

**37.** The population of the Soviet Union as of January 1, 1987 was 282,000,000. It had a birth rate of 20.1 per 1000 population and a death rate of 9.8 per 1000 population. Using the Malthusian population model, determine an estimate of the population in the Soviet Union in 1995.

**38.** Answer the same questions as you did in the above problem for Ethiopia. In 1989, Ethiopia had a population of 48,898,000, a birth rate of 49.9 per 1000, and a death rate of 25.4 per 1000.

**39.** Using the exponential function that gives a model for atmospheric pressure, find the atmospheric pressure on the top of Mount McKinley in Alaska, altitude 20,320 ft.

**40.** Suppose you take up the game of golf. You keep a record of your average score (the average number of strokes it takes to complete a round of golf) for each month of playing golf. Explain why an exponential function might make a good mathematical model for your scores in this endeavor.

***41.** A queen, wishing to reward a faithful maid, agreed to grant her one wish. The maid replied that she was a very humble woman and only wanted some corn as her reward. The maid requested that the corn be given to her in the following manner: Upon a chess board, place two kernels of corn on the first square, 4 on the second, 8 on the third, 16 on the fourth, and so on until the last (64th) square. At first the queen refused saying that this was not a just reward for such a faithful maid. However, after the maid insisted, the queen ordered a servant to bring in a bag of corn and give the maid her desired reward. How much corn did the maid receive? If one pound of corn contains about 3500 kernels, how many pounds of corn did the maid receive?

***42.** Exponential functions are also used in other areas, such as in the decay of radioactive material (half-life) and bacterial growth. Do some research on one of these areas and on any other area which uses exponential functions as a mathematical model. Explain how the function is used and give some examples.

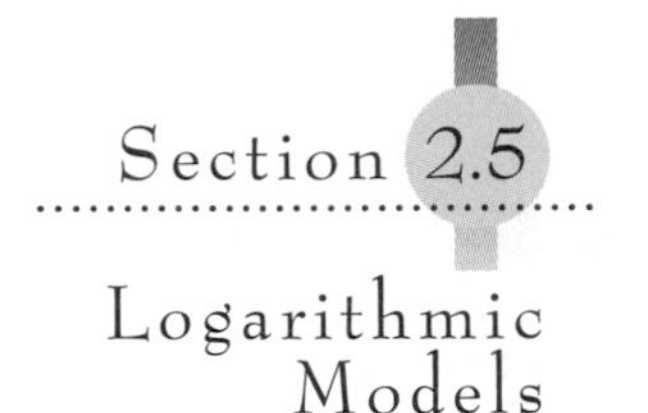

# Section 2.5 Logarithmic Models

## Logarithmic Functions

You may recall from algebra that inverse functions are functions that have the opposite effect. In arithmetic, multiplication and division are inverses of each other. For example, if you choose a number, say 7, and multiply it by 5, you get an answer of 35. If you then divide that answer by 5, you get back the 7 you started with. In algebra, equations such as $y = 2x$ and $y = \frac{1}{2}x$ are inverse functions. If you choose a number, say $x = 8$, and substitute $x = 8$ into the equation $y = 2x$, you get 16. Substituting $x = 16$ into the second equation, $y = \frac{1}{2}x$, gives you the answer 8, the number you started with.

Similarly, the inverse of the exponential function is the logarithmic function. For example, the inverse of the exponential function $y = 10^x$ is the function $y = \log x$. To see this inverse relationship, choose a number, say $x = 3$. Substituting $x = 3$ into the equation $y = 10^x$ gives us the answer $10^3 = 1000$. Substituting this result for $x$ in the function $y = \log x$ gives us log 1000. To determine this value, we can use a calculator.

To determine log 1000:

| Press | Display |
|---|---|
|  1 0 0 0 log | 3 |

Log 1000 = 3 is equivalent to the statement $10^3 = 1000$. The exponential function gives the result of raising 10 to the third power as its answer, while the logarithmic function gives the exponent of 10 as its answer. A logarithm is simply an exponent. Although logarithms can be evaluated with different bases, in this chapter we will use only common logarithms and natural logarithms. Common logarithms use the base 10 and are denoted by the LOG (log) button on a calculator. Natural logarithms use base $e$ ($e \approx 2.71828$) and are denoted by LN (ln) on a calculator. The logarithmic functions used in the text are

$$y = \log x, \text{ which is equivalent to } x = 10^y$$

$$y = \ln x, \text{ which is equivalent to } x = e^y$$

Since 10 raised to any power or $e$ raised to any power is always a positive quantity, $x$ is always positive. We can, therefore, only take the log or ln of positive quantities.

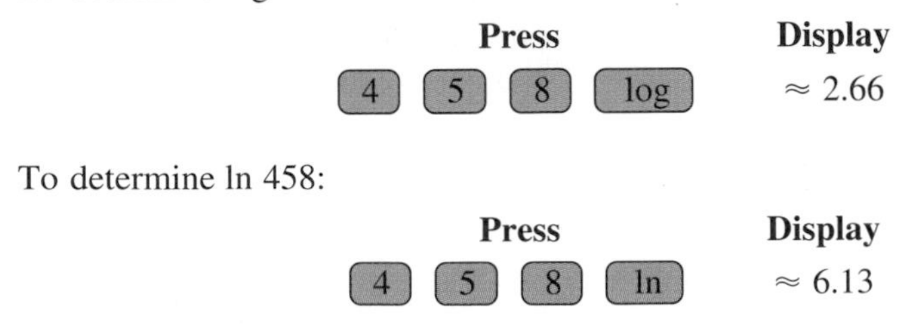

To determine log 458:

| Press | Display |
|---|---|
| 4 5 8 log | $\approx 2.66$ |

To determine ln 458:

| Press | Display |
|---|---|
| 4 5 8 ln | $\approx 6.13$ |

To determine 5 ln(9 − 3.6):

**Press** **Display**

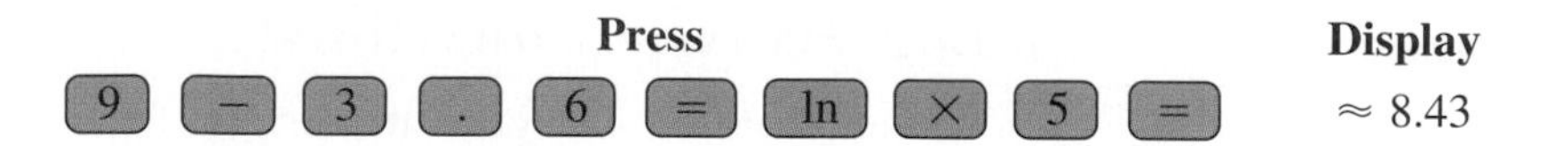


≈ 8.43

As we did with exponential functions, logarithmic functions can be graphed by calculating and plotting a sufficient number of points.

### Example 1

Graph $y = \ln x$.

**Solution:** With the help of a calculator, determine and plot points that satisfy the equation. Remember to choose $x$ values greater than 0, since it is not possible to compute the logarithm of a negative number.

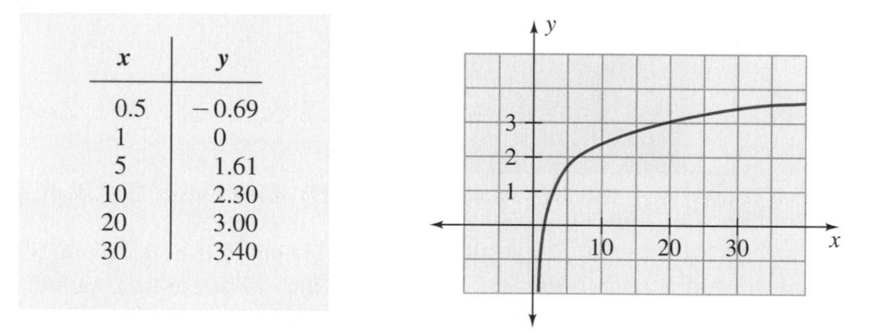

| $x$ | $y$ |
|---|---|
| 0.5 | −0.69 |
| 1 | 0 |
| 5 | 1.61 |
| 10 | 2.30 |
| 20 | 3.00 |
| 30 | 3.40 |



If we include other constants in a basic common or natural logarithmic function, the graph of the logarithmic function can be moved up or down or the rate at which it increases or decreases can be changed. In this chapter we will consider logarithmic functions of the form

$$y = a + b \log(x + c) \qquad \text{and} \qquad y = a + b \ln(x + c)$$

where $a$, $b$, and $c$ are constants.

The graphs of these functions have the same basic shape as an exponential function. They can be graphed by determining and plotting points with the help of a calculator. In this process, you must remember that you cannot take the logarithm of a negative number.

### Example 2

Graph $y = 1.2 \log x$.

**Solution:** With the help of a calculator, determine and plot points that satisfy the logarithmic function. Since it is impossible to take logarithms of negative numbers, use $x > 0$.

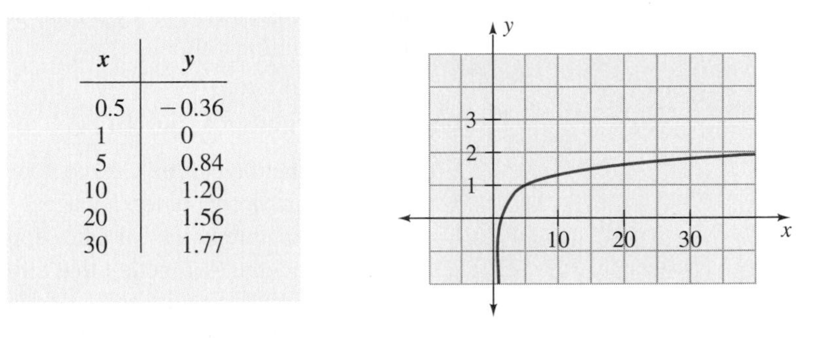

| $x$ | $y$ |
|---|---|
| 0.5 | −0.36 |
| 1 | 0 |
| 5 | 0.84 |
| 10 | 1.20 |
| 20 | 1.56 |
| 30 | 1.77 |



## Example 3

Graph $y = 4.7 + \ln(x - 2)$ where $x \geq 4$.

**Solution:** With the help of a calculator, determine and plot points that satisfy the logarithmic function using $x \geq 4$.

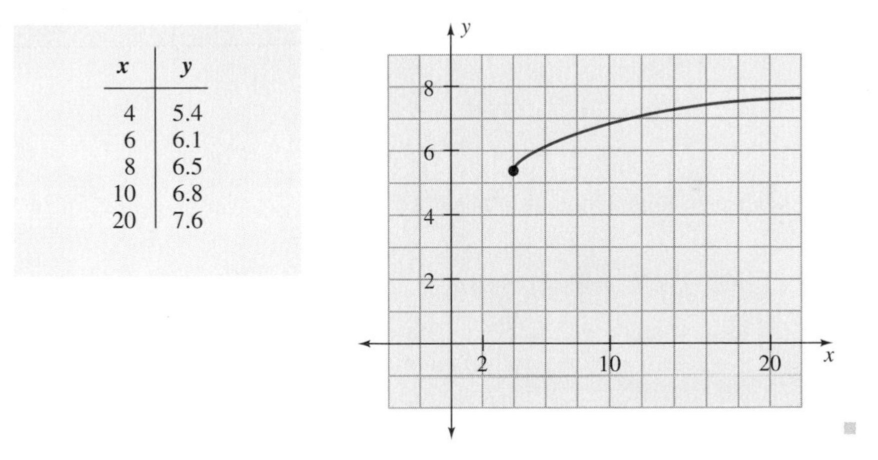

| $x$ | $y$ |
|---|---|
| 4 | 5.4 |
| 6 | 6.1 |
| 8 | 6.5 |
| 10 | 6.8 |
| 20 | 7.6 |



## Logarithmic Models

As we saw in the previous section, exponential functions can be used as mathematical models of situations in which a quantity experiences a period of gradual increase followed by a period of rapid increase. Logarithmic functions can also be used as mathematical models of real-life situations and scientific phenomena. Since logarithmic functions are inverses of exponential functions, they react in a manner that is opposite to the exponential function. The logarithmic functions we will be studying exhibit

**(a)** rapid initial increase followed by a long period of gradual increase, or
**(b)** rapid initial decrease followed by a long period of gradual decrease, but never approaching a limiting value.

## Example 4 Height of Children

A logarithmic function can be used to approximate the change in the height of a child as the child grows older. By age 2, most children have reached 50% of their adult or mature height. It takes approximately 16 years for the child to attain full adult height. A function that allows for a large initial change and then a gradual increase is a logarithmic function. In fact, the mature height of boys aged 0–16 can be approximated by the following function:

$$P = 29 + 48.8 \log(A + 1) \qquad \text{where } \begin{cases} P = \text{percentage of adult height} \\ A = \text{age in years} \end{cases}$$

The graph of this function is

| A | P |
|---|---|
| 0 | 29 |
| 5 | 67 |
| 10 | 80 |
| 15 | 88 |

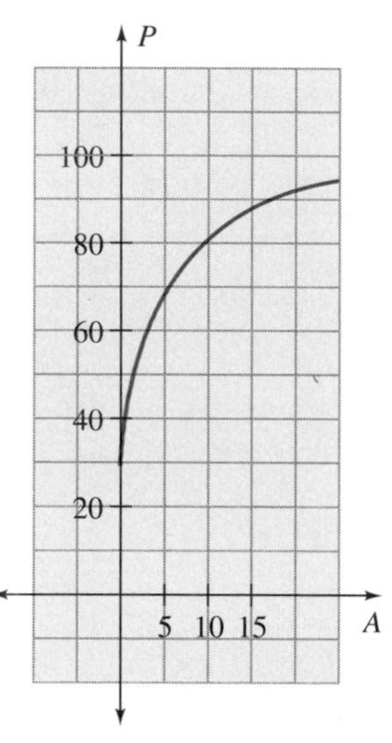


We can use that logarithmic model to answer questions about the growth of boys such as

(a) Approximately what percentage of his adult height is a boy at the age of 7?
(b) If that seven-year-old boy is 52 inches tall, how tall can we expect him to be when he is an adult?

**Solution:**

(a) Substituting $A = 7$ into the formula gives

$$P = 29 + 48.8 \log(A + 1)$$

$$P = 29 + 48.8 \log(7 + 1) \approx 73$$

So, a seven-year-old is 73% of his mature height.

(b) From part (a), the 52-inch-tall boy is 73% of his mature height. So, if $M =$ the mature height,

$$52 = 0.73M$$

$$M = 52 \div 0.73 \approx 71.2 \text{ inches}$$

Thus, we can expect the boy to become a little over 5′11″ tall. ■

## Example 5 Newton's Law of Cooling

From experiments on cooling bodies, Isaac Newton concluded that over moderate temperature ranges, the rate at which an object changes temperature is proportional to the difference between the temperature of the object and the temperature of the surrounding air.

If a hot cup of 200°F coffee is taken outdoors where the temperature is 35°F, it will begin to cool. If after 1 minute, the temperature of the coffee is 170°F, according to Newton's Law of Cooling, the time $t$ it takes for the coffee to reach a temperature $x$ is given by the formula

$$t = 25.5 - 5 \ln(x - 35) \qquad \text{where } \begin{cases} t = \text{time to reach temperature } x \\ x = \text{temperature of coffee} \\ \quad (35° < x \leq 170°) \end{cases}$$

Use this function to determine the following:

(a) When is the temperature of the coffee 98.6°F?
(b) When is the temperature of the coffee 40°F?

**Solution:**

(a) Substituting $x = 98.6°\text{F}$ into the equation gives

$$t = 25.5 - 5 \ln(98.6 - 35) \approx 4.7 \text{ min}$$

(b) Substituting $x = 40°\text{F}$ into the equation gives

$$t = 25.5 - 5 \ln(40 - 35) \approx 17.5 \text{ min}$$ ■

This section has attempted to show you that logarithmic functions can be used as mathematical models for actual occurrences in the world. There are many other places where these functions are used, but many of them are beyond the scope of this book and require knowledge of the areas in which they are used. The problems that follow will let you experiment with other logarithmic functions used as mathematical models.

Section 2.5
PROBLEMS

## Explain ➡ Apply ➡ Explore

### Explain

1. What are the basic characteristics of the function $y = \log x$?
2. Describe the general shape of the logarithmic models discussed in this section.
3. Suppose you take up the sport of weight lifting. You keep a record of the maximum number of pounds you can lift at the end of each week. Explain why a logarithmic function might make a good mathematical model for predicting the amount of weight you can lift from week to week.
4. You are advertising a new computer game. Consumer research is keeping track of how many people have heard of your new game (vertical axis of a graph) as the number of weeks of advertising continues (horizontal axis). Explain why a logarithmic function might make a good mathematical model for estimating the number of people who have heard of your game as the number of weeks of advertising increases.
5. What does your calculator do when you try to take the common logarithm of a negative number?
6. What does your calculator do when you try to take the natural logarithm of a negative number?

### Apply

In Problems 7–14, use your calculator to determine the answers. Round answers off to the nearest thousandth.

7. $12.7 \log(56.91)$
8. $-2.34 \log(17.95)$
9. $12.7 + \ln(56.91)$
10. $-4.56 - \ln(28.73)$
11. $3000 + 5600 \ln(56.8 - 4)$
12. $2000 - 4800 \ln(17.6 - 5.2)$
13. $4.5 - \log(3.45 - 3.1)$
14. $-5.6 + \log(5.68 + 5.29)$

In Problems 15–22, sketch a graph of each logarithmic function.

**15.** $y = \ln 2x$

**16.** $y = \log 5x$

**17.** $y = 4.2 \ln(x - 2.9)$

**18.** $y = 5.3 \log(x + 1)$

**19.** $y = 32 + 48.5 \log(x + 2)$

**20.** $y = 5 + 14.3 \ln(2x + 1.2)$

**21.** $y = \ln(3x - 1) - 15$

**22.** $y = -20 + \log(3x + 1)$

## *Explore*

**23.** At age 5, a girl's height is approximately 62% of her full adult height. At age 15, she has reached about 98% of her adult height. The logarithmic function below gives an approximate percentage $P$ of adult height a girl has reached at any age $A$ from 5 to 15 years.

$$P = 62 + 35 \log(A - 4)$$

(a) At age 10, what percentage of her height has a girl reached?
(b) If the girl is 4′6″ at age 10, how tall can she expect to be as an adult?

**24.** A logarithmic model to approximate the percentage $P$ of adult height a male has reached at any age $A$ from 13 to 18 is

$$P = 16.7 \log(A - 12) + 87$$

(a) Sketch a graph this function.
(b) What does the graph tell us about males that have reached the age of 18?

**25.** A roast, cooking for 2 hours, is taken out of the oven when the meat thermometer reads 140°F and is placed in a kitchen that is 68°F. After 6 minutes the thermometer reads 132°F. Newton's Law of Cooling states that the time $t$ for the roast to get to a temperature $x$ is given by

$$t = 217.9 - 50.94 \ln(x - 68) \qquad \text{where } 68° < x \leq 140°$$

(a) In how many minutes will the internal temperature of the roast be 110°F?
(b) In how many minutes will the internal temperature of the roast be 88°F?
(c) Sketch a graph of this function.

**26.** If \$1000 is invested in an account that earns 1% interest each month, the number of months $n$ for the account to grow to an amount $A$ is given by the formula

$$n = -694.2 + 231.4 \log A$$

where $A \geq \$1000$ and no withdrawals or other deposits are made to the account.

(a) Sketch a graph of this function.
(b) How many years would it take for the account to grow to $1 million?

**27.** If 1000 e. coli bacteria are placed in a culture, the time $t$ in hours it takes for the bacteria to grow to an amount $A$ can be approximated by the formula

$$t = -172.7 + 25 \ln A \qquad \text{where } A \geq 1000$$

(a) Sketch a graph of this function.
(b) How many hours would it take for the culture to contain a million bacteria?

**28.** If $500 is deposited into a savings account that pays 3.65% annual interest, compounded daily, the number of years ($y$) that it takes for $500 to grow to an amount $x$ is given by the equation $y = 27.4 \ln x - 170.3$.

(a) How long would it take for your money to double?
(b) How long would it take for you to have $5000 in the account?

**29.** Radioactive elements such as Uranium and Plutonium actually decrease in mass over a period of time. The half-life of an element is the length of time required for the mass of an element to decay to one-half the original amount. Suppose a nuclear storage site has been contaminated by 100 kilograms of radioactive Plutonium. The half-life of Plutonium is 24,360 years. The equation relating the mass ($m$) of the Plutonium to the elapsed time ($t$) is

$$t = \frac{24{,}360}{\ln 0.5} \ln \frac{m}{100}.$$

(a) Find the time required for the 100 kilograms of Plutonium to decay to 50 kilograms.
(b) Find the time required for the 100 kilograms of Plutonium to decay to 1 kilogram.

***30.** Logarithmic functions are also used in other areas, such as in the pH (acidity) of solutions, intensity of sound (decibels), and the intensity of earthquakes (Richter scale). Do some research on one of these areas or any other area that uses a logarithmic function as a mathematical model. Explain how the function is used and give some examples.

## Chapter 2 • Summary

### Key Terms, Concepts, and Formulas

The important terms in this chapter are:

**Addition method:** A method used to solve a system of linear equations. p. 70

**Common logarithm:** Logarithmic function that uses a base of 10, denoted by **log**. p. 133

**Constraints:** The inequalities that form the feasible region in a linear programming problem. These constraints restrict the values of the solution to certain values. p. 96

**Discrete graph:** A graph consisting of distinct and separate points. p. 68

**Domain:** The set of first coordinates of a function. p. 66

**Exponential model:** A representation of a situation by an exponential function of the form $y = a + c(b)^{kx}$. p. 124

**Feasible region:** The set of points that satisfy all the constraints in a linear programming problem. p. 97

**Function:** A set of ordered pairs such that for each value of the first coordinate there is exactly one value for the second coordinate. p. 66

**Graphing method:** A method used to solve a linear programming problems. p. 96

**Linear model:** A representation of a situation by a linear function of the form $y = mx + b$. p. 84

**Linear programming:** A mathematical technique in which a stated objective function must be either maximized or minimized while satisfying a set of constraints. p. 96

**Logarithmic model:** A representation of a situation by a logarithmic function of the form $y = a + b \log(x + c)$ or $y = a + b \ln(x + c)$. p. 135

**Malthusian population model:** A method of predicting population growth based on constant birth and death rates. p. 128

**Natural logarithm:** Logarithmic function that uses a base of $e$, denoted by **ln**. p. 133

**Objective function:** The function to be maximized or minimized in a linear programming problem. p. 96

**Quadratic model:** A representation of a situation by a quadratic function of the form $y = ax^2 + bx + c$. p. 108

**Range:** The set of second coordinates of a function. p. 66

**System of linear equations:** A set of two or more linear equations. p. 69

After completing this chapter, you should be able to:

1. Find and graph linear functions that serve as mathematical models for situations in which the rate at which quantities change is constant. p. 84
2. Use the graphing method to find the solutions to linear programming problems. p. 96
3. Find and graph quadratic functions whose parabolic shapes serve as models for given situations. p. 108
4. Graph and use exponential functions as models to analyze various situations. p. 124
5. Graph and use logarithmic functions as models to analyze various situations. p. 135

• Summary

## Problems

1. Find the vertex of the parabola $y = -2x^2 - 6$.
2. Find the equation of the line through the points (2, 5) and (6, −7).
3. Sketch the solution to the system of linear inequalities
$$3x + y \leq 6$$
$$5x + 2y \geq 9$$
4. Plot each of the following sets of points. Determine which type of algebraic model (linear, quadratic, exponential, logarithmic) best describes the points on the graph.
   (a) (0, 3), (3, 24), (6, 192), (−1, 1.5), (−3, 0.375)
   (b) (0, 3), (3, −6), (6, −15), (9, −24), (12, −33)
   (c) (0, 3), (3, 7), (6, 8), (9, 9), (12, 9.5)
   (d) (0, 3), (3, −6), (6, 3), (9, 30), (−3, 30)
5. Solve the following linear programming problem.

   Minimize $C = 6x + 4y$

   subject to the constraints $\begin{cases} x + y \geq 7 \\ x + 5y \geq 15 \\ x \geq 0 \\ y \geq 0 \end{cases}$
6. Solve the following linear programming problem.

   Maximize $P = 5x + 3y$

   subject to the constraints $\begin{cases} 3x + 4y \leq 24 \\ x + 4y \leq 16 \\ x \geq 0 \\ y \geq 0 \end{cases}$
7. Scientists use the Kelvin temperature scale where the lowest possible temperature (absolute zero) is zero degrees Kelvin (0°K). The linear function that relates the Kelvin scale to the Celsius scale is
$$K = C + 273 \text{ where } \begin{cases} K = \text{temperature on Kelvin scale} \\ C = \text{temperature on Celsius scale} \end{cases}$$
   (a) Sketch a graph of this function.
   (b) At what Celsius temperature is absolute zero?
   (c) Since water boils at 100°C, at what Kelvin temperature does it boil?

**8.** If a person walks at 5 mph for an hour, the approximate number of calories burned per hour, based on the person's weight, is given in the chart below:

| Weight in Pounds | Calories Burned per Hour |
|---|---|
| 110 | 440 |
| 132 | 500 |
| 154 | 560 |
| 176 | 620 |
| 198 | 680 |

(a) Explain why a linear function would be an appropriate model for this data.
(b) Find the linear function that determines the calories $c$ burned per hour based on the weight $w$ of the person walking at 5 mph.
(c) If you weigh 160 lb and walk at 5 mph, how many calories does your body burn per hour?

**9.** The first (bottom) row of a huge stack of logs in a lumber yard has 247 logs. The second row has 245 logs, the third row has 243 logs, and so on. Assume that each successive row continues to contain exactly two less logs than the previous row.

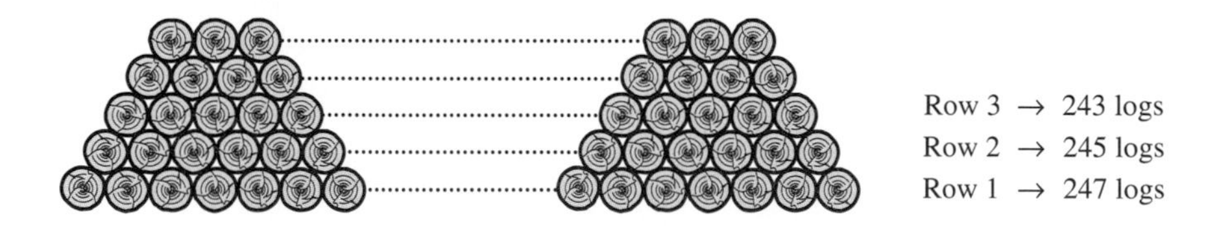


(a) Explain why a linear function could be used to predict the number of logs $L$ in any row $r$.
(b) Find the linear function that determines the number of logs in any given row.
(c) Sketch a graph of the function found in part (b).
(d) Find how many logs are in the 50th row.
(e) Find how many rows of logs the stack contains.

**10.** At age 9, Krista's stamp collection contained 102 stamps. When she entered high school at the age of 14, her collection had grown to 1567 stamps. Assume that Krista continues to collect stamps at that rate.

(a) Find a linear function that, based on her age, predicts the number of stamps she owns.

(b) Predict how many stamps she will have when she graduates from high school at age 18.
(c) Determine at what age her stamp collection would contain more than 10,000 stamps.

**11.** A rock is thrown vertically upward at 88 ft/s (60 mph) from a sheer cliff in the Grand Canyon, 5000 ft above the Colorado River. A quadratic function that approximates the height of the rock at any given time is

$$h = -16t^2 + 88t + 5000$$

(a) Sketch a graph of this quadratic function for $t \geq 0$.
(b) What is the maximum height the rock reaches?
(c) How long does it take for the rock to reach the Colorado River?

**12.** In the 1968 Olympics in Mexico City, Bob Beamon of the United States electrified the track and field world with his 29′2.5″ leap in the long jump. Assume that the path of the jump can be approximated by a parabola where the highest point occurred at the middle of the jump and at that point his feet were 4′6″ off the ground.

(a) Find the quadratic function that determines the height of this jump as a function of the length of the jump where distances are measured in inches.
(b) How far past the take-off board was Bob Beamon when his feet were 30 in. off the ground.

**13.** The ancient Greeks studied the pentagonal numbers as shown. The first pentagonal number is 1, the second one is 5, the third one is 12, the fourth one is 22, and so on. The algebraic function that relates the value of a pentagonal number to what term it is in the sequence of numbers is a quadratic function.

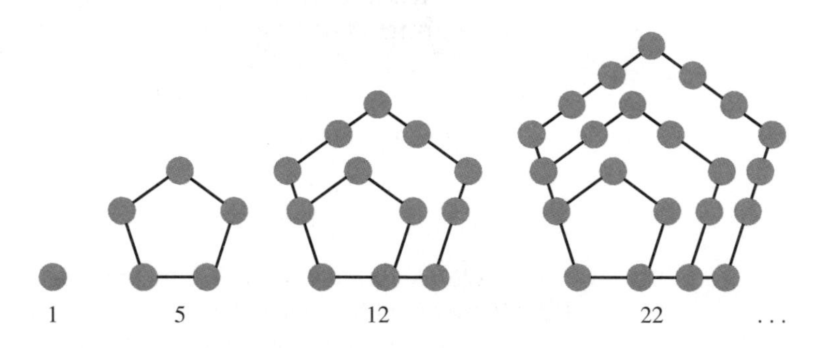


(a) Find the quadratic function that determines the value of a pentagonal number $P$ based on what term $t$ it is in the sequence.
(b) Sketch a graph of that function.
(c) Determine the 100th pentagonal number and describe its geometric shape.

**14.** If the guaranteed rate of return on an investment of \$6000 is compounded annually at 10% per year, the exponential function that determines the amount $A$ that investment is worth at time $t$ is given by

$$A = 6000(1.1)^t \qquad \text{where } \begin{cases} A = \text{amount investment is worth} \\ t = \text{time in years} \end{cases}$$

(a) Sketch a graph of this exponential function for $t \geq 0$.
(b) Determine the value of the investment after 5 years, 10 years, and 30 years.

**15.** Suppose the total cost for manufacturing a certain toy truck is given by the equation

$$C = 500 + 400 \ln x \quad \text{where } \begin{cases} C = \text{total cost} \\ x = \text{number of trucks and } 1 \leq x \leq 3000 \end{cases}$$

(a) Sketch a graph of this cost function.
(b) Determine the cost for manufacturing 1000 trucks and 2000 trucks.
(c) Determine the cost for manufacturing each truck when 1000 trucks and 2000 trucks are produced. What happens to the cost per truck as the number of trucks manufactured increases?

**16.** The Parker Brothers game of Monopoly™ has a game board that has 40 spaces where your game token can land and a bank that has \$15,140 in play money. Suppose you place \$3 on the start (Go), \$9 on the next space (Mediterranean Avenue), \$27 on the next space (Community Chest), and so on. You keep tripling the amount placed on each space as you go around the board.

(a) Find an exponential function that determines the amount $A$ placed on each space $S$ of the Monopoly board.
(b) How much must you place on the 9th space (Vermont Avenue)? Will you have enough money to do that?
(c) The Parker Brothers Company prints about \$40 billion in play money each year. With a year's worth of play money would you be able to put the required amount on the 24th space (Illinois Avenue)?
(d) How much play money would you need for the 40th space (Boardwalk)? How many years of play money production by Parker Brothers would you need to put the required amount on Boardwalk?

# Geometry and Art

3

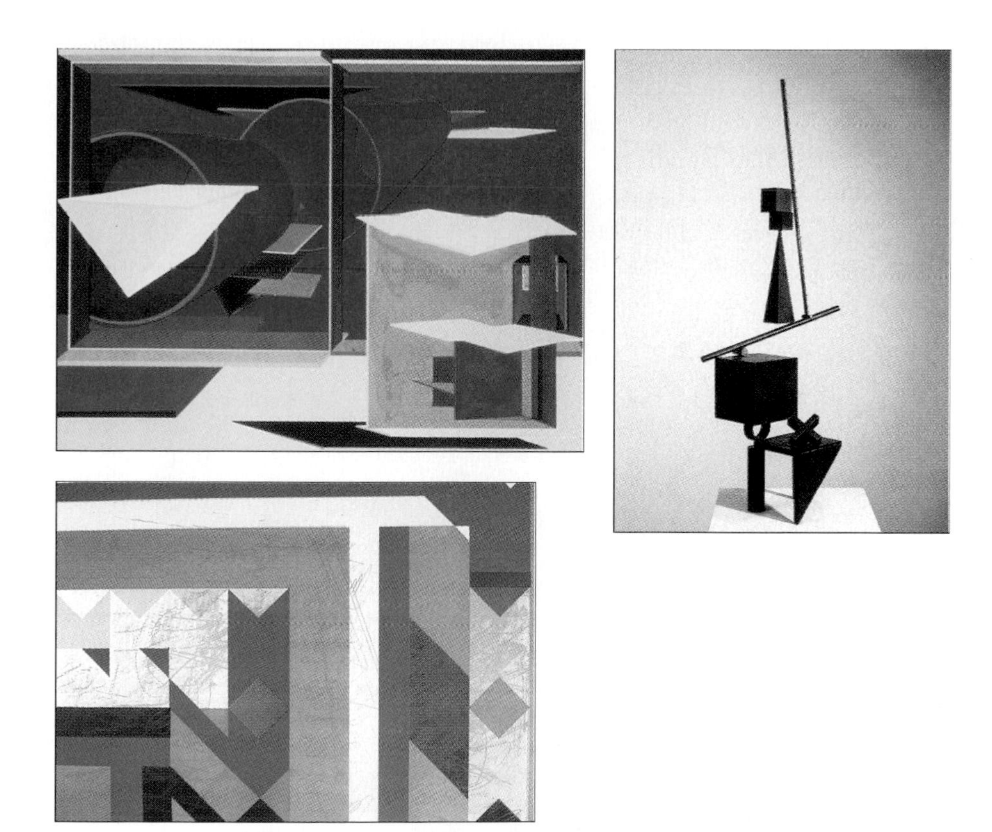

Artists look to geometry as a means of creating visual excitement. (*Upper left*) *Pachinko* by Al Held. (Courtesy of Crown Point Press) (*Upper right*) *Balanced-Unbalanced Table* by Fletcher Benton. (Courtesy of the John Berrgruen Gallery) (*Left*) *Kastle Keep* by Hassell Smith. (Courtesy of Magnolia Editions)

# Overview
# Interview

In this chapter you will learn that there is more to geometry than the definitions, postulates, theorems, proofs, and measurement formulas you may have studied in previous geometry courses. The chapter starts with a quick look at Euclidean geometry (the geometry taught from grade school through high school) and discusses the existence of non-Euclidean geometries. After that, the chapter leads you on a short journey through aspects of geometry in art, such as perspective, the Golden Ratio, polygons, stars, and tessellations. The chapter ends with a look at fractal geometry, a geometry that combines the use of algebraic equations and computers to produce some interesting visual effects. In this chapter, you will experience some of the artistic effects that can be produced with geometry.

## David McLaughlin

DAVID MCLAUGHLIN, PAINTER AND PROFESSOR OF ART AT OHLONE COLLEGE, FREMONT, CALIFORNIA

TELL US ABOUT YOURSELF. I received my bachelor's degree from the California College of Arts and Crafts and my master's degree in art from San Francisco State University. I have taught drawing, painting, and art history at Ohlone College since 1967. My deep interest in symbolism and classical mythology is revealed in my still life and landscape paintings. My works have been shown in numerous art galleries and museums, and I am currently represented by the Dorothe Barlett Gallery in Pleasanton, California.

HOW IS MATHEMATICS USED IN ART AND WHY IS IT IMPORTANT TO THE ARTIST OF TODAY? In art, elements such as line, shape, form, color, perspective, and space are combined and organized through intuitive and intellectual applications of balance, rhythm, proportion, unity, and variety. Often, the use of mathematics aids in this process. In modern painting and contemporary art, mathematics is frequently the foundation of thought-provoking combinations of space, lines, and forms. In realistic and surrealistic art, the mathematical tool of perspective creates the illusion of space and distance in a picture. An artist who understands perspective can draw or paint any subject, observed or imagined.

*David McLaughlin in his studio putting the finishing touches on a recent work of art.*

Present-day artists have also become involved in computer art, animation, and graphic design. The art produced in these areas reflects our technological society through the use of mathematics. Geometry, ratios, sequences, patterns, and proportions have become invaluable tools to work out the details for producing the creative images we see in animations, videos, paintings, sculptures, and designs.

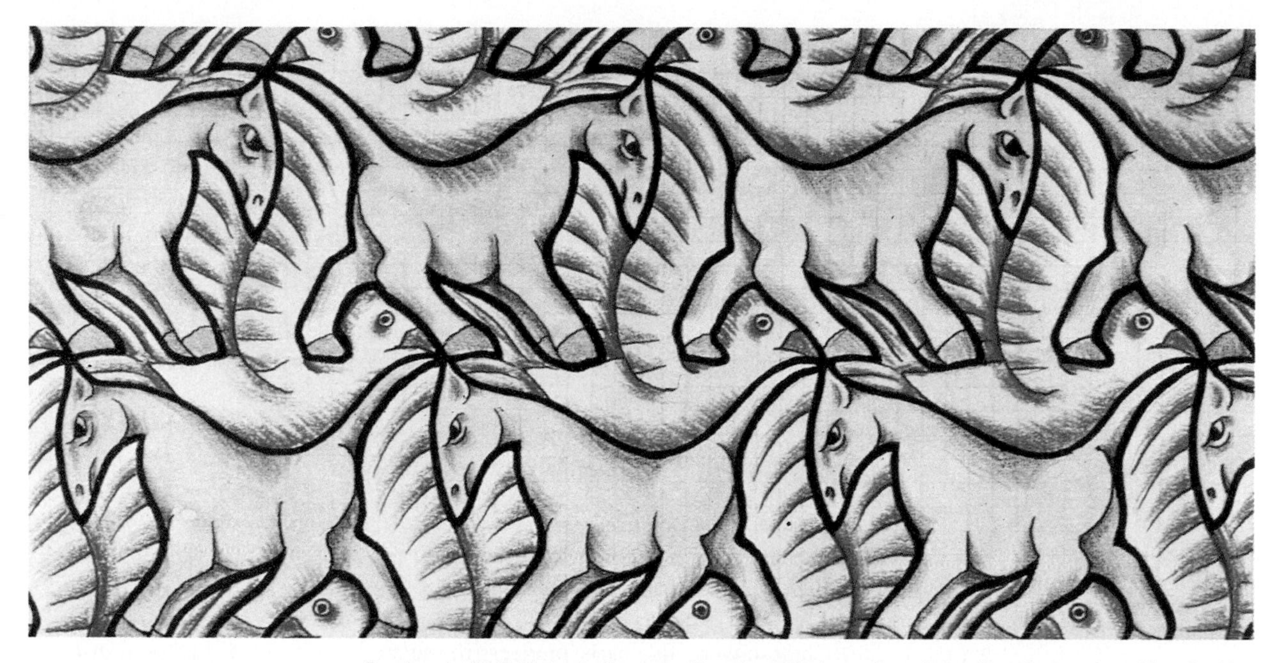

Dutch artist M.C. Escher's *Symmetry Drawing E76* shows the artistic effect that can be achieved with simple geometric tessellations. (M.C. Escher Heirs/Cordon Art-Baarn)

## A Short History of Geometry

The exact origin of geometry is not known, but its roots are believed to date back before recorded history. There is evidence that the intuitive concepts of geometry are universal. For example, prehistoric men and women probably realized that the shortest distance between two points is the straight line joining the two points and that the angle at which an arrow is shot affects the distance above the ground the arrow will reach. A basic understanding of geometric shapes is evident in the tools, weapons, and shelters designed by prehistoric men and women. Further, the drawings and handicraft of prehistoric men and women show a concern for spatial relationships. Their pottery, baskets, and weaving display examples of symmetry and sequences of designs. This concern for spatial relationships probably originated in the wonder of the world around them. Perhaps they marveled at the concentric circles formed when a rock is thrown into a pool of water, or the network of hexagons found in a beehive, or the intricate patterns found in snowflakes. Although we can only speculate on their knowledge of geometry, prehistoric interest in geometry seems to have originated from basic intuition, practical needs, and the aesthetics of order and design.

The first recorded evidence of geometry can be found with the Babylonians and Egyptians (3000 B.C.–300 B.C.). Cuneiform tablets, hieroglyphic papyri, inscriptions on temples and tombs, and construction feats show a variety of practical uses of geometric concepts. Since both civilizations were largely agricultural, much of their geometry was developed to parcel out land, determine areas and perimeters, and calculate volumes of their granaries. In Egypt, the periodic overflows and flooding of the Nile River made surveying the land for reestablishment of boundary lines a necessity. With the use of ropes, Egyptian surveyors, called ''rope-stretchers,'' accurately redetermined agricultural plots in the Nile Valley after the annual flooding had subsided. The Egyptians also showed their skill at measurement in the construction of their pyramids. For example, the Great Pyramid of Gizeh has a square base with sides that are about 756 ft long. Amazingly, the difference in the lengths of the these sides is less than two-thirds of an inch. Babylonian irrigation canals and their beautifully constructed temples, such as the Hanging Gardens of Babylon, also showed practical uses of geometric concepts. The Egyptians and Babylonians are credited with the formation of a geometry that consisted of practical uses of measurement techniques. Their knowledge of geometry was based on intuition, experimentation, and approximation. Tablets and papyri, however, contain only specific, concrete problems in geometry and show no evidence of general formulations, logical proofs, or mathematical abstraction. Thus, their geometry is considered as an empirical or experiential geometry.

The Chinese were also early pioneers in the study of geometry as seen in *K'iu-ch'ang Suan-shu* or *Arithmetic in Nine Sections*, which was probably prepared by Chóu-kung around 1100 B.C. However, since in 213 B.C. Emperor Shï Huang-ti of the Chin dynasty ordered all books to be burned, it is believed that Ch'ang Ts'ang collected writings of the ancients and wrote the version of *K'iu-ch'ang Suan-shu* that has been passed down through history. In this book, the Chinese showed an understanding of determining areas and volumes of geometric objects, finding the lengths of sides of figures, and using what is now called the Pythagorean theorem some 500 years before the time of Pythagoras.

Around 700 B.C. the Greeks took the empirical geometry of the Babylonians and Egyptians and began to show that geometric truths could be abstracted from the practical situations in which they arise. In fact, the word ''geometry'' comes from the Greek words *gē̄*, ''earth,'' and *met'ron*, ''measure.'' However, the Greeks took the study of geometry far beyond measurement. They developed a geometric system based on logic in which geometric facts follow from generally accepted statements called axioms or postulates. The position of the Greeks on the method and importance of geometry is summarized in quotes from two Greek scholars. Anaxagoras (499–427 B.C.) stated, ''Reason rules the world,'' and Plato (430–347 B.C.) stated, ''God eternally geometrizes.''

Major contributors and their contributions to the development of Greek geometry include:

- Thales (624–547 B.C.) gave logical proofs for geometric relationships and is considered one of the founders of mathematical science.
- Pythagoras (572–501 B.C.), a pupil of Thales, formed a society devoted to the

study of arithmetic, music, astronomy, and geometry. He and his followers proved many new theorems about triangles, circles, solids, and proportions. He is credited with discovering a geometric proof of what is now called the Pythagorean theorem (see Section 5.0).

- Philosophers Socrates (468–399 B.C.), Plato (430–347 B.C.), and Aristotle (385–322 B.C.) emphasized the need for making clear assumptions, formulating accurate definitions, and using sound logic in studying geometry.
- Euclid of Alexandria (c. 300 B.C.) wrote *Elements*, a collection of 13 books of which the first six and the last three are devoted to geometry. In *Elements*, Euclid logically developed and summarized the geometry known up to that time. This "Euclidean" model dominated the study of geometry until about 1700 and is still the basis of many geometry courses taught at the secondary level.
- Archimedes (287–212 B.C.) made contributions to finding the areas and volumes of geometric figures.
- Apollonius of Perga (262–190 B.C.) wrote on conic sections and named them ellipse, parabola, and hyperbola.

The Greeks also spent much time on three famous construction problems. Using a straight edge and a compass, they attempted to

1. Divide any angle into three equal angles (trisect an angle).
2. Draw a square equal in area to that of a given circle (square a circle).
3. Draw a cube the volume of which is twice that of a given cube (double a cube).

Greek mathematicians such as Anaxagoras (c. 440 B.C.), Antiphon of Athens (c. 430 B.C.), Hippocrates of Chios (c. 460 B.C.), Hippas of Elis (c. 425 B.C.), Archytas of Tarentum (c. 400 B.C.), Eudoxus (c. 370 B.C.), and Eratosthenes (c. 230 B.C.) worked on these construction problems. Although it was proved in the 19th century that these classical constructions are impossible, the attempts to solve them led to the investigation of many other important mathematical topics.

After Apollonius of Perga, the development in geometry began to stagnate. With the decline of the Greek city-states and the rise of the Roman Empire during the next 600 years, the centers of mathematical thought in Alexandria and Athens did not experience the advancements that were found in previous centuries. The only significant work of this period was the mathematical treatise *Collection* by Pappus of Alexandria (c. 320). In it, Pappus introduced some of his original work in geometry, but, more significantly, he provided a historical record of parts of Greek mathematics that would otherwise be unknown to us. The death of Hypatia (c. 415) in Alexandria, the execution of Boethius (c. 524) in Pavia, and the fall of the Roman Empire (c. 476) brought an end to the Greek period in mathematics. Because of their extensive contributions to the study of geometry, the Greeks can be considered the founders of geometry.

From the disintegration of the Roman Empire until about 1000, Western Europe experienced a period where intellectual pursuits were at a low. These five centuries are called the Dark Ages by historians. The study of mathematics during this period moved to the East, to India and Arabia. The Hindus and Arabs made advancements in systems of numeration, algebra, and trigonometry, but their contributions to geometry were limited to a few theorems by Āryabhata (c. 510), Brahmagupta (c. 628), Ṯâbit ibn Qorra (c. 870), and Bhāskara (c. 1150), among others. However, although the Arabs and Hindus did not contribute much to the development of geometry, they did preserve and keep alive the mathematics of the Greeks.

As Europe emerged from the Dark Ages (c. 1000), interest in geometry began to reappear. Mathematical works written in Arabic were translated into Latin. Among these were Adelard of Bath's translation of Euclid's *Elements* (c. 1120), Robert Chester's translation of al-Khowârizmî's algebra (c. 1140), and Johannes Campanus's widely published translation of Euclid's *Elements* (c. 1260). Creative work in geometry was also given impetus by the establishment of universities in Europe such as the University of Paris in 1200, Oxford in 1214, and Cambridge in 1231. During the Renaissance, the geometry of Euclid was expanded by Leonardo Fibonacci (c. 1220) in *Practica geometriae* and Jordanus Nemorarius (c. 1225) and Regiomontanus (c. 1464) in *De triangulis*. Henry Billingsley (c. 1570) also translated Euclid's *Elements* from Latin into English, and in 1607, Jesuit missionary Matteo Ricci and Chinese scholar Hsü translated the first six books of *Elements* into Chinese. Following Euclidean geometry, the major advancements in geometry settled into four distinct categories: analytic/algebraic geometry, projective/descriptive geometry, non-Euclidean geometry, and differential geometry.

## ANALYTIC/ALGEBRAIC GEOMETRY

In about 1629, Pierre de Fermat began applying methods of algebra to geometric objects. His work included the determination of equations for lines, circles, ellipses, parabolas, and hyperbolas. However, his work was not published until after his death in 1679. In 1637, *La Geometrie*, a discourse by René Descartes on analytic geometry, was published. In this discourse, Descartes introduced the *xy* coordinate system and allowed for the graphic representation of geometric curves given by equations. Because the work of Descartes was published first, he is credited with founding analytic (algebraic) geometry. In the Descartes-Fermat scheme, points became pairs of numbers and curves became sets of points generated by algebraic equations. Geometry became ''arithmetized.''

## PROJECTIVE/DESCRIPTIVE GEOMETRY

In the 15th and 16th centuries, a focus of geometry was to obtain the correct perspective in representing what one sees. Architects Filippo Brunelleschi (c. 1400) and Leon Alberti (c. 1435), along with artists Pietro Franceschi (c. 1490), Leonardo da Vinci (c. 1500), and Albrecht Dürer (c. 1525) worked on representing three-dimensional objects on a two-dimensional surface. This interest in projecting a space figure onto a surface led to the development of projective and descriptive geometry. The mathematics of these geometries was initiated by Girard Desargues (c. 1640) and Blaise Pascal (c. 1650) and developed by Gaspard Monge (c. 1795), Victor Poncelet (c. 1822), Jacob Steiner (c. 1832), K. G. C. von Staudt (c. 1847), and Felix Klein (c. 1871). The work of these architects, artists, and mathematicians led to an understanding of perspective, became the foundation of architectural and mechanical drawing, and initiated investigations of topology.

## NON-EUCLIDEAN GEOMETRY

Euclid's fifth postulate has been the source of much thought and mathematical investigation. In Euclid's *Elements* we find the fifth postulate.

> *That, if a straight line falling on two straight lines makes the interior angles on the same side less than two right angles, the two straight lines, if produced indefinitely, meet on that side on which the angles are less than two right angles.*

A simpler and more intuitive equivalent of the postulate was formulated by John Playfair in 1795.

> *Through a given point only one parallel can be drawn to a given straight line.*

Mathematicians felt that this postulate could be shown to be true as a result of logical deductions from Euclid's other postulates. In an attempt to do this, the efforts of Girolamo Saccheri (c. 1733), Johann Lambert (c. 1788), Adrien Legendre (c. 1794), János Bolyai (c. 1794), Carl Friedrich Gauss (c. 1816), Nicolai Lobachevsky (c. 1829), and Georg Riemann (c. 1854) led to what is known today as Lobachevskian geometry and Riemannian geometry. Lobachevskian geometry is based on a postulate that more than one line can be drawn parallel to a given line through an external point, while Riemannian geometry employs a postulate that no parallel line can be drawn through the point. Some of the details of these geometries will be discussed in Section 3.1.

The logical conclusions reached by assuming the Lobachevskian or Riemannian postulate contradicted Euclid's results. To the scholars of the 19th century, this was very perturbing because, for 2000 years, Euclidean geometry had been accepted as giving unquestionable truth about the real world. The study of non-Euclidean geometry did, however, lead to three conclusions:

1. Euclid's fifth postulate cannot be logically deduced from his other postulates.
2. Non-Euclidean geometries could be used to describe space just as Euclidean geometry did.
3. Mathematics does not give absolute truths about the physical world.

## DIFFERENTIAL GEOMETRY

Differential geometry is essentially the technique of applying calculus to the study of curves and surfaces. The development of this branch of geometry can be credited to the study of ''infinitesimals'' by Johann Kepler (c. 1604) and Bonaventura Cavalieri (c. 1635); to the application of the ''derivative'' and the ''integral'' to curves by Isaac Newton (c. 1680) and Gottfried Leibniz (c. 1682); to the investigation of ''neighborhoods'' of a point on a curve or surface by Carl Friedrich Gauss (c. 1827); and to the ''methods of analysis'' of Georg Riemann (c. 1850) and Jean Gaston Darboux (c. 1890).

## • Check Your Reading

1. What are the four categories in geometry that followed Euclidean geometry? Give a brief description of each category.
2. What were the major differences between the geometry of the Greeks and that of the Babylonians and the Egyptians?
3. In about 1200 B.C., Chinese mathematicians were working on magic squares. What were the Egyptian ''rope-stretchers'' doing at that time?
4. While French traders were settling at St. Louis in about 1637, what was French mathematician René Descartes doing?
5. What contributions did the Hindus and the Arabs make to geometry during the Dark Ages?
6. Why did the study of the non-Euclidean geometries cause controversy?
7. While Abraham Ortelius was producing the first modern atlas with 53 maps in 1570, what was Henry Billingsley doing?
8. While commonplace things like the use of black-lead pencils and pocket handkerchiefs came in to use in the early 1500s, what were artists like Franceschi and da Vinci doing in geometry?
9. What role did artists play in the development of geometry?
10. What were the three famous construction problems of geometry?
11. In 1829, Andrew Jackson was inaugurated as the seventh president of the United States. What was the Russian professor Nikolai Lobachevsky doing at that time?
12. In 1854, the Republican party was formed in the United States. What geometric proposition was being formulated by Georg Riemann?
13. Match each of the following names with the correct mathematical discovery, event, or concept.

| | |
|---|---|
| (a) Appollonius | $a^2 + b^2 = c^2$ named after him |
| (b) Archimedes | Artist involved in perspective |
| (c) Billingsley | Edited *Arithmetic in Nine Sections* |
| (d) Ch'ang Ts'ang | Famous geometrician who wrote the book *Elements* |

| | |
|---|---|
| (e) Descartes | Found area and volume formulas |
| (f) Dürer | A founder of mathematical science |
| (g) Euclid | Named the conic sections |
| (h) Lobachevsky | Devised non-Euclidean geometry with more than one parallel through a point not on a given line |
| (i) Plato | Devised non-Euclidean geometry with no parallels |
| (j) Pythagoras | Stated, "God eternally geometrizes." |
| (k) Riemann | Translated *Elements* into English |
| (l) Thales | Devised the *xy* coordinate system |

## Research Questions

To answer the following questions, you will need to refer to material not contained in the text. Possible sources of information are listed in the Bibliography at the end of the text.

1. Many of the mathematicians mentioned in this short history of geometry are known for other mathematical endeavors or had other interests besides mathematics. Do some research on two of the mathematicians mentioned in this section. Tell something about their lives, their achievements, and their interests.
2. One of the reasons for the creation of geometry was its use in the measurement of lengths, perimeters, areas, and volumes. Give at least five examples of how these practical applications of geometry are still used today.
3. The pyramids of Egypt are early examples of practical geometry at work. Discuss the geometry involved in the construction of the pyramids.
4. Do some research on the geometry of the ancient Chinese. What discoveries did they make before their European counterparts?
5. What were Euclid's first four postulates as stated in his book *Elements*? Make sketches of each and explain them in your own words. What were Euclid's five common notions?
6. Who were the Pythagoreans? What did this society do and what were some of its beliefs?
7. Who was Hypatia of Alexandria? What is she famous for?
8. What does symmetry mean to the mathematician? How does the artist consider symmetry? Compare the two interpretations and uses of the term symmetry.
9. Discuss the controversy over the development of analytic geometry by Pierre de Fermat and René Descartes. Why was such a discovery so important in the development of mathematics?
10. Discuss the controversy over the founding of non-Euclidean geometry by Carl Gauss, János Bolyai, and Nicolai Lobachevsky.

11. Constructions with a straight edge and a compass were a major concern of early geometric studies. What are four geometric operations that can be performed with such constructions?
12. What is topology? What are some of the objects studied in topology and what are some of the practical applications of topology?
13. Who was M. C. Escher? What were his contributions to geometry and art?
14. Tessellations and stars are described in Sections 3.4 and 3.5. How are these used in the mosques such as the Alhambra in Spain?

## Projects

1. What are optical illusions? Find examples of them. Artists like M. C. Escher have used them to produce interesting results. Investigate optical illusions in art.
2. Some credit for the advances of the Greeks in mathematics is given to the Greek schools. The most noted schools of mathematics were the Alexandrian School of Mathematics and the Platonic Academy in Athens. What were these schools like? How was mathematics taught in these schools? Who were some of the teachers in these schools? Compare and contrast those schools to the schools of today.
3. Contrast and compare some of the theorems of Euclidean geometry, Lobachevskian geometry, and Riemannian geometry. In particular, what do they say about parallel lines, triangles, quadrilaterals, and areas? Go beyond what is discussed in Section 3.1 of this chapter.
4. This chapter introduces you to some of the ways geometry can be used to produce "artistic" effects. What are some other ways in which geometry is used in art? Go beyond what is presented in this chapter.

Bruce Cohen's *Blue Table with Many Tulips* uses familiar geometric shapes from Euclidean geometry. (Courtesy of the artist)

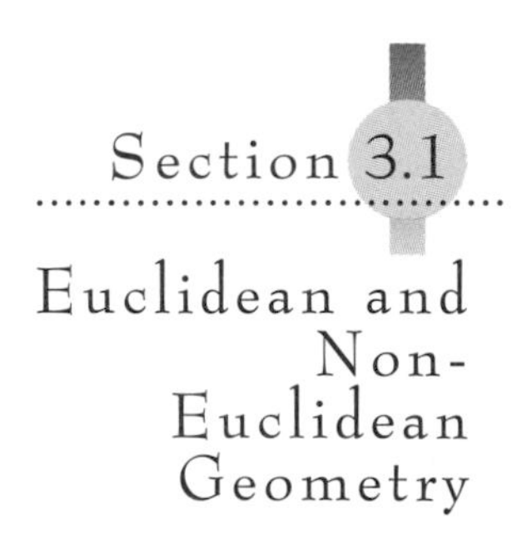


## Euclidean Geometry

The geometry studied in elementary and secondary schools is based on the system of geometry contained in *Elements*, a book written by the Greek mathematician Euclid in about 300 B.C. This work summarized much of the geometry known up to that time and presented the first systematic treatment of geometry. Starting with definitions, postulates, and common notions, Euclid proceeded to prove numerous propositions of a geometry now called Euclidean geometry. The intent of this section is not to cover all of Euclidean geometry but to make you aware of the development of this mathematical system and to summarize some of the basic components of a system of geometry.

## *Undefined Terms*

One of the most important features of a mathematical system is that the terms used in the system must have clear definitions. When making a definition, you should use words that are simpler than the term being defined. However, if you continually try to define terms by using simpler words, the process could conceivably go on forever. There comes a time when a definition must use a term whose meaning is assumed to be clear. Such terms are called **undefined terms**. They are used to begin the process of defining new terms. We will begin our study of Euclidean geometry by considering its undefined terms—point, line, and plane.

A **point** can be described as a location or position. It can be represented by a dot and is usually named by an uppercase letter. A **line** may be described as the set of points arranged along a straight path. It extends indefinitely in two opposite directions and can be named by either a single lowercase letter or by two points on the line with a double-headed arrow placed above the letters. A **plane** can be described as the set of all points that form a flat surface extending indefinitely in all directions. It can be represented by means of a parallelogram and named by an uppercase letter placed in one corner of the parallelogram.

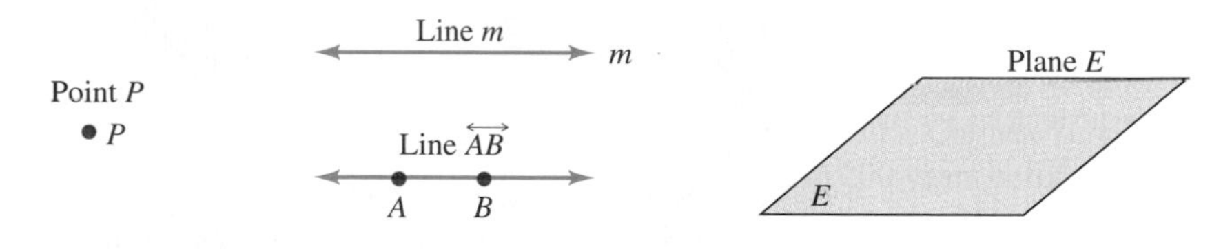


The preceding paragraph merely gives an intuitive description of a point, a line, and a plane. These are not considered definitions. Point, line, and plane are not defined in the system. However, as we proceed in the development of the Euclidean geometric system, we will become more aware of the properties of these three basic geometric objects.

## *Definitions*

With the use of the undefined terms, we can now give meanings to other terms and expressions of geometry. Here are some basic definitions:

A **line segment** is the set of points on a line consisting of two points, called end points, and all the points in between those two points.

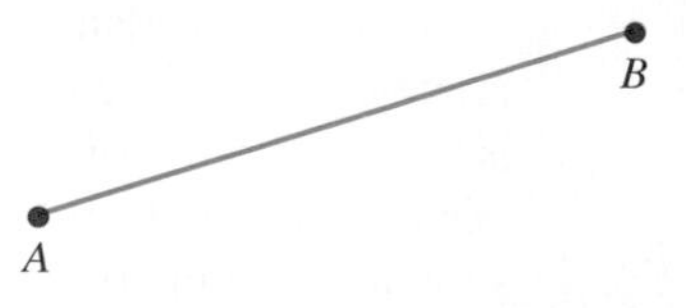


The line segment joining the points $A$ and $B$ is written $\overline{AB}$.

A **ray** is the figure formed by extending a line segment in only one direction.

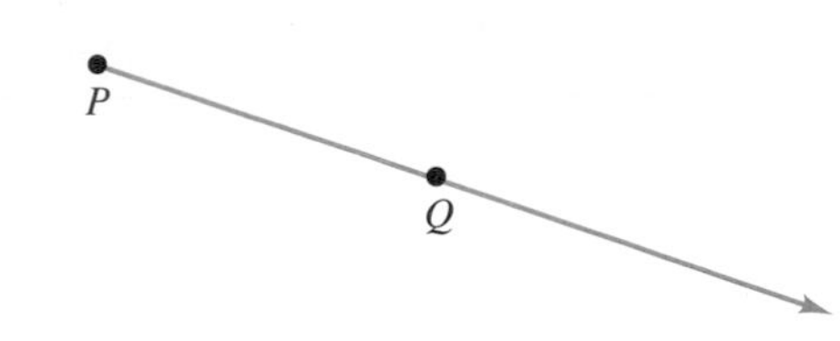


A ray with endpoint at point $P$ extending through point $Q$ is written $\overrightarrow{PQ}$.

An **angle** is the figure formed by two rays or line segments with a common end point.

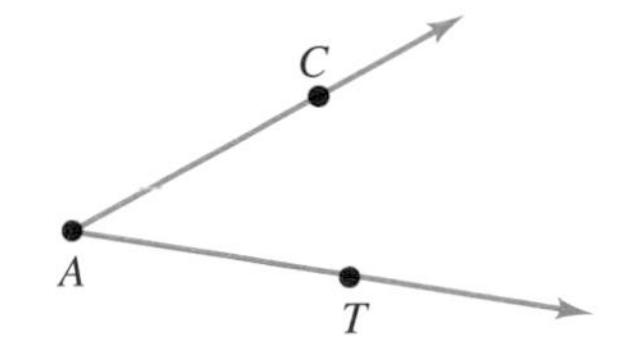


$\angle CAT$ has a vertex at point A which are the endpoints of $\overrightarrow{AC}$ and $\overrightarrow{AT}$.

Types of angles: A **straight angle** is an angle that forms a line and measures 180°; a **right angle** is an angle that has a measure of 90°; an **acute angle** has a measure of between 0° and 90°; an **obtuse angle** has a measure of between 90° and 180°.

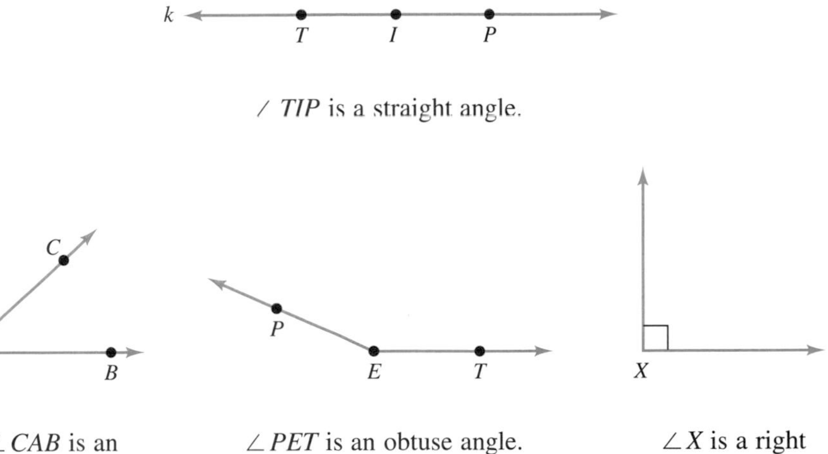


$\angle TIP$ is a straight angle.

$\angle CAB$ is an acute angle.

$\angle PET$ is an obtuse angle.

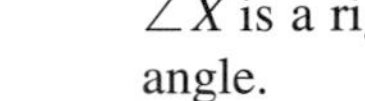

$\angle X$ is a right angle.

D5: A **triangle** is a figure consisting of three line segments determined by three points that are not on the same line.

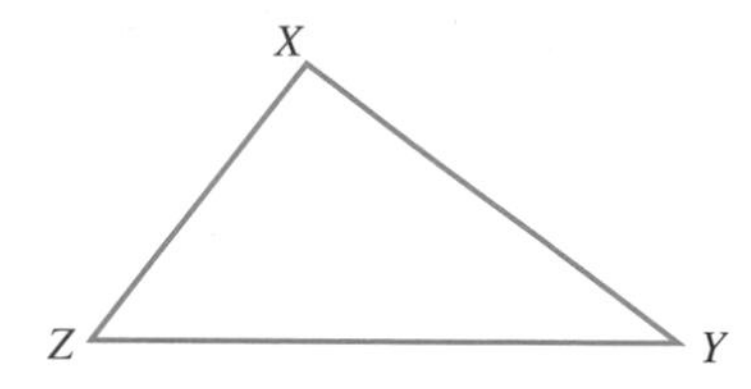


$\triangle XYZ$ has the vertices, the points at $X$, $Y$, and $Z$. It has three sides, $\overline{XY}$, $\overline{YZ}$, and $\overline{ZX}$. It has three angles, $LX$, $LY$, and $LZ$.

D6: An **exterior angle** is an angle formed outside a triangle by one side of the triangle and the extension of another side of the triangle.

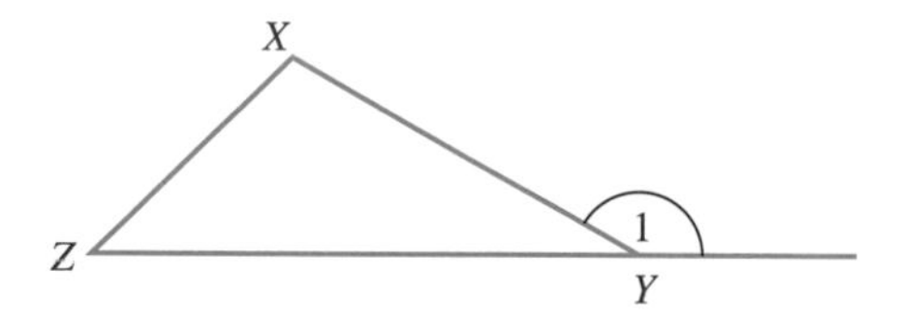


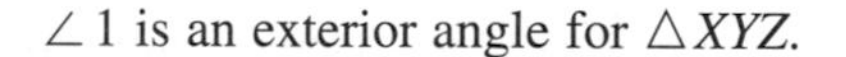

$\angle 1$ is an exterior angle for $\triangle XYZ$.

D7: Two lines, rays, or line segments are **perpendicular** ($\perp$) if they intersect and form a right angle. Two lines, rays, or line segments are **parallel** ($\|$) if they lie in the same plane and the lines that contain them do not intersect.

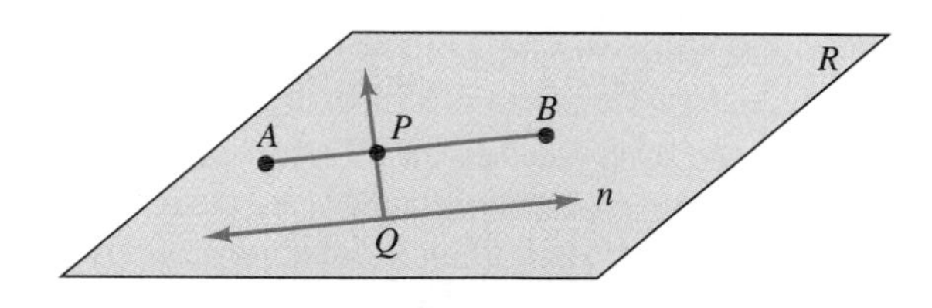


**Example:**

In plane $R$, segment $\overline{AB}$ is parallel to line $n$, written $\overline{AB} \| n$, and ray $\overrightarrow{QP}$ is perpendicular to line $n$, written $\overrightarrow{QP} \perp n$. ■

There are many more terms that we could define, but these are all we will need in our brief excursion into geometry.

## *Axioms and Postulates*

If we are to develop a system of geometry, each proposition must logically follow from other propositions. A system of geometry must start with some propositions that are accepted as true. We simply cannot deduce all statements. These assumptions are called **axioms** and **postulates**. Though these terms are used interchangeably, axioms refer to assumptions from arithmetic or algebra, whereas postulates refer to assumptions from geometry. Axioms and postulates are sometimes called self-evident truths. However, we shall see in our study of non-Euclidean geometry that this is not necessarily the case. We will start our study with axioms and postulates based on those of Euclid.

### Axioms

**A1:** A quantity may be substituted for its equal in any expression.

**Example:**

If $\angle A + \angle B = \angle C$ and $\angle A = 90°$, then $90° + \angle B = \angle C$. ■

**A2:** If quantities are equal to the same quantity, then they are equal to each other.

**Example:**

If $A = C$ and $G = C$, then $A = G$. ■

**A3:** If equal quantities are added to or subtracted from equal quantities, the results are equal.

**Example:**

If $\angle A = \angle R$ and $\angle B = \angle K$, then

$$\angle A + \angle B = \angle R + \angle K \quad \text{and} \quad \angle A - \angle B = \angle R - \angle K.$$ ■

**A4:** If equal quantities are multiplied by the same quantity or divided by the same nonzero quantity, the results are equal.

**Example:**

If $x = y$, then $5x = 5y$ and $x \div 7 = y \div 7$. ■

**A5:** A whole quantity is equal to the sum of its parts and is greater than any one of them.

**Example:**

$AD = AB + BC + CD$ and $AD > AB$, $AD > BC$,
$AD > CD$, $AD > AC$, $AD > BD$.

**Postulates**

**P1:** Through two given points, one and only one line can be drawn.

**Example:**

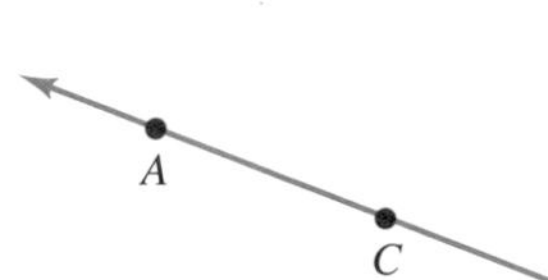


Through the points $A$ and $C$ only one line can be drawn.

**P2:** A line segment can be extended indefinitely in both directions.

**Example:**

The line segment $\overline{PQ}$ can be extended into line $\overleftrightarrow{PQ}$.

**P3:** If two points of a line lie in a plane, then the line through the two points lies in the plane.

**Example:**

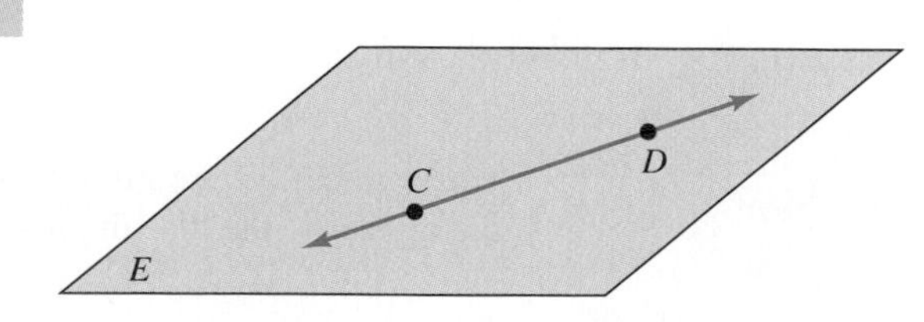


Points $C$ and $D$ lie in plane $E$. Therefore, the entire line passing through $C$ and $D$ lies in plane $E$.

To every pair of points there corresponds a unique positive number called its distance.

Example:

The line segment $\overline{PQ}$ has a unique distance; $PQ = 4$ cm. ■

**Parallel Postulate**
Through a given point, only one parallel can be drawn to a given line.

Example:

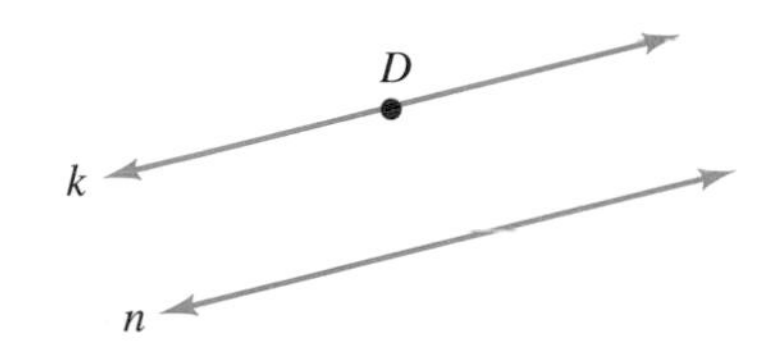


Through point $D$, line $k$ is the only line that can be drawn parallel to line $n$. ■

To every angle there corresponds a unique number between 0° and 180° called the measure of the angle.

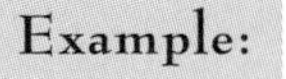


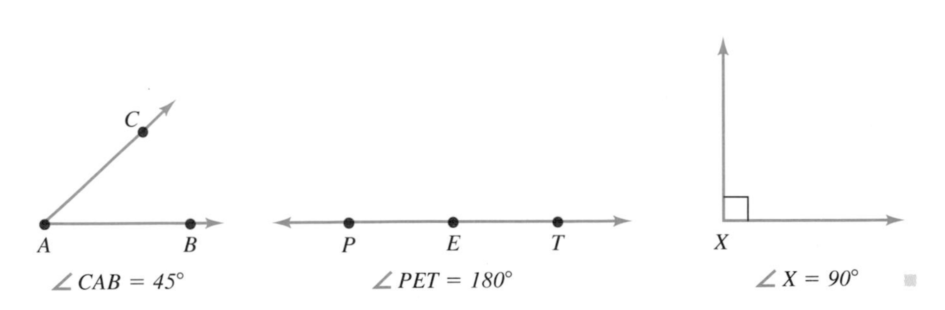


With the establishment of axioms, postulates, and definitions in this section, we can demonstrate the logical development of and prove some propositions of Euclidean geometry. Such proven propositions are called **theorems**. The **proof** of a theorem consists of a series of statements that logically show that the theorem is true based on axioms, postulates, definitions, and other proven theorems. The following theorems are examples of propositions from Euclidean geometry that can be logically deduced from the definitions, axioms, postulates, and theorems.

### *Theorems*

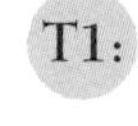
T1: If a line segment lies in a plane, then the line containing the segment lies in the plane.

T2: If two distinct lines intersect, they intersect in at most one point.

T3: If two lines in the same plane are parallel to the same line, then they are parallel to each other.

T4: An exterior angle of a triangle is greater than either nonadjacent interior angle.

T5: **The Triangle-Sum Theorem:**
The sum of the measures of the angles of a triangle is 180°.

These propositions and many others constitute the propositions of Euclidean geometry. There are, however, other geometries.

## Non-Euclidean Geometry

Any system of geometry that uses a consistent set of postulates, with at least one postulate that is not logically equivalent to one of Euclid's postulates, is a non-Euclidean geometry. In particular, mathematicians wondered if Euclid's parallel postulate was really a postulate. The statement of the postulate by Euclid seemed to lack the clarity and the ''self-evident'' character of his other postulates. In fact, Euclid did not use the parallel postulate until the proof of his 29th proposition. Mathematicians, therefore, attempted to derive the parallel postulate from Euclid's other postulates. These attempts, however, met with little success. It was not until the late 19th and early 20th centuries that mathematicians Eugenio Beltrami (1835–1900), Felix Klein (1849–1929), and Henri Poincaré (1854–1912) finally established that the parallel postulate could not be deduced from Euclid's other postulates. The attempts to prove the parallel postulate did, however, generate some interesting results. The work of the Jesuit priest Girolamo Saccheri (1667–1733) laid the groundwork for the creation of the two principal non-Euclidean geometries that bear the names of the mathematicians who spent their lifetimes studying them—Lobachevskian geometry and Riemannian geometry.

Russian mathematics professor Nicolai Lobachevsky (1793–1856), Hungarian army officer Janós Bolyai (1802–1860), and renowned German mathematician Carl Gauss (1777–1855) accepted the postulates of Euclid but replaced the parallel postulate with the following postulate:

**Lobachevskian Parallel Postulate**

Through a point that is not on a line, there is more than one line parallel to the given line.

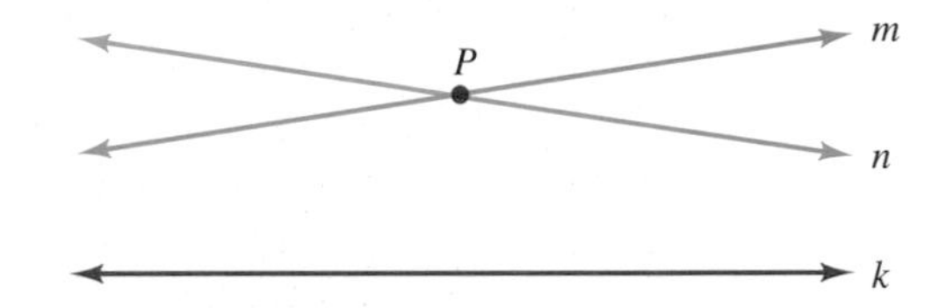


Through point $P$, lines $m$ and $n$ are parallel to line $k$ ($m \| k$ and $n \| k$).

The German mathematician and student of Carl Gauss, Georg Riemann (1826–1866), replaced Euclid's parallel postulate with this:

**Riemannian Parallel Postulate**

Through a point that is not on a line, there is no line parallel to the given line.

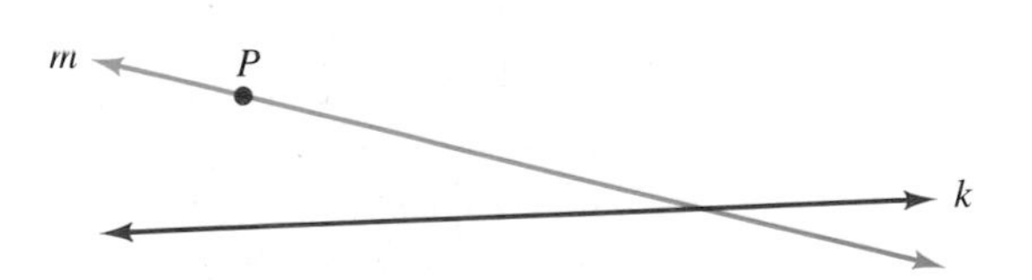


Any line $m$ through point $P$ will intersect line $k$.

These assumptions led to the creation of geometries that were logically consistent and contained no contradictions within themselves. However, the theorems deduced from these postulates contradicted some of the well-established theorems of Euclidean geometry. So controversial was the idea that one could deny a postulate of Euclid and arrive at conclusions that contradicted Euclid that even the renowned mathematician Carl Gauss was reluctant to publish his findings. Eventually, these non-Euclidean geometries were accepted by the mathematical community. However, the question of which geometry actually gives an accurate description of physical space remained. Euclidean geometry was so widely accepted as accurately describing physical space that these non-Euclidean geometries were considered as mere mental exercises. Mathematicians believed that the geometry of the physical world must be Euclidean. Although it is true that Euclidean geometry corresponds with our intuitive ideas about our surroundings, we would like to present some models and observations to suggest the existence of non-Euclidean geometries.

## *A Riemannian Model*

Besides changing Euclid's parallel postulate, Georg Riemann questioned if lines extended infinitely. He proposed that lines trace back on themselves, like circles. You could traverse a circle endlessly, but its length is still finite. Euclid's second postulate said that a line segment can be extended indefinitely in both directions. Riemann modified Euclid's second postulate to state that a line is endless but not necessarily infinite. Furthermore, do parallel lines never meet? Never? If one looks down a set of railroad tracks, it seems that the tracks get closer together. Maybe parallel lines in space eventually do meet.

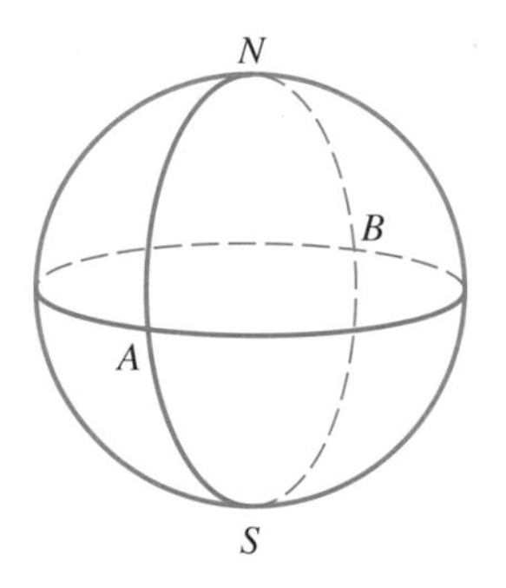


If we accept the two assumptions that lines are endless but not infinite and that parallel lines do not exist, we form a different geometry, the geometry of Georg Riemann. Its postulates and theorems cannot be easily displayed on the flat plane of Euclidean geometry. They can, however, be visualized on the surface of a sphere. In Riemannian geometry, the Euclidean plane becomes a sphere, the Euclidean line becomes a great circle on the sphere, and the Euclidean point becomes a point on the sphere along with its antipodal point (point on the sphere farthest from the first point). For example, although we consider the North and South Poles to be different points, in Riemannian geometry the North and South Poles are considered together as one point. Such a model allows us to visualize the propositions of Riemannian geometry. The great circles of a sphere have a finite length. Since any two great circles intersect, there are no parallel lines in such a model.

**Consequences of the Riemannian Parallel Postulate** In Riemannian geometry, all theorems that are consequences of the parallel postulate are different from those in Euclidean geometry. One of the most significant results of the Riemannian parallel postulate is that the sum of the measures of the angles of a triangle is greater than 180° rather than equal to 180°, as found in Euclidean geometry. If we look at the triangle drawn on the spherical model for Riemannian geometry in the figure, we see that $\triangle ANB$ has two 90° angles, so the sum of the angles of $\triangle ANB$ is greater than 180°.

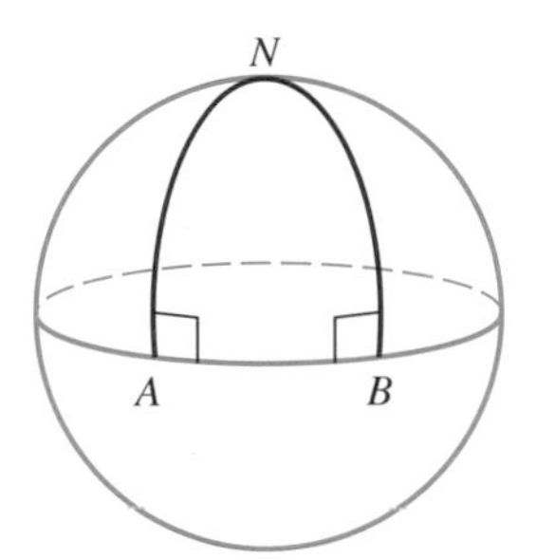


Riemannian Triangle-Sum Theorem

The sum of the measures of the angles of a triangle is greater than 180°.

## *A Lobachevskian Model*

With Riemannian geometry, we introduced the possibility that lines are not necessarily straight. The same holds true in Lobachevskian geometry. In fact, scientists believe that space is curved. A ray of light traveling through space does not take a straight path. According to Einstein's theory of relativity, the path of a ray of light is affected by the gravitational field of large objects in space. The gravitational field causes the ray of light to bend. The lines in Lobachevskian geometry also bend, as can be seen in the model in the figure.

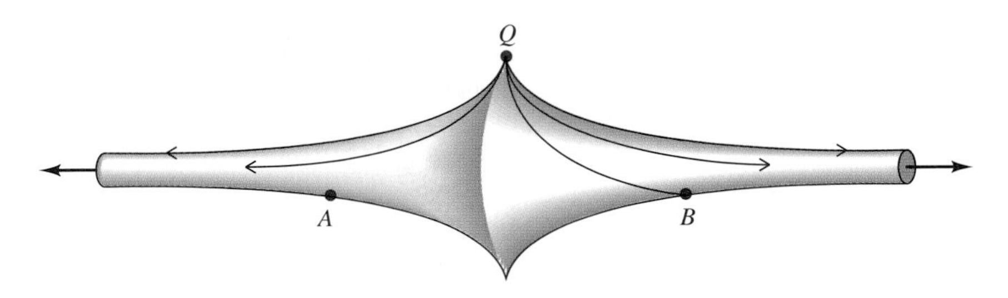

The model resembles two attached trumpets with the small ends extending indefinitely, a shape called a pseudosphere. The pseudosphere corresponds to the plane in Euclidean geometry and to the sphere in Riemannian geometry. Lobachevskian lines consist of two symmetric curves meeting at a point at the widest part of the pseudosphere (the bold line in the drawing passing through points $A$ and $B$). Through point $Q$ there is more than one "line" parallel to the "line" through $A$ and $B$. Through point $Q$ on the pseudosphere, you will also notice that "line" $QB$ intersects "line" $AB$ at point $B$.

**Consequences of the Lobachevskian Parallel Postulate** In Lobachevskian geometry, all theorems that are consequences of the parallel postulate are different from those in Euclidean geometry. A notable result of the Lobachevskian parallel postulate is that the sum of the measures of the angles of a triangle is less than 180° rather than equal to 180° (as found in Euclidean geometry) or greater than 180° (as found in Riemannian geometry). If we look at an equiangular triangle drawn on the pseudosphere model, each angle is less than 60° because the sides of the triangle curve inward. The sum of the angles of the triangle is visibly less than 180°.

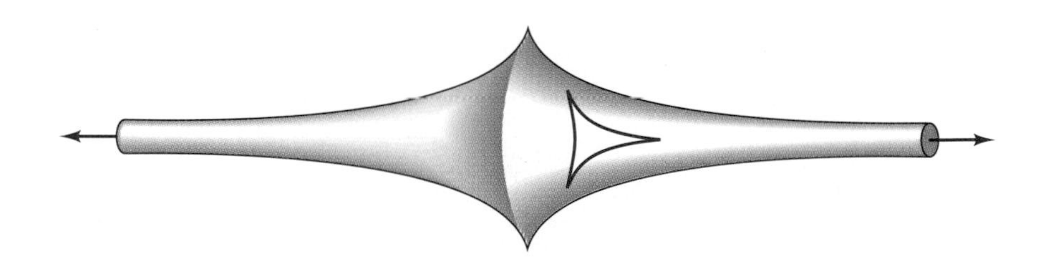

**Lobachevskian Triangle-Sum Theorem**

The sum of the measures of the angles of a triangle is less than 180°.

### *Is the World Euclidean, Riemannian, or Lobachevskian?*

The conclusions of each geometry are true based on the initial assumptions of each geometry, but is the world Euclidean, Riemannian, or Lobachevskian? We may never decide if Euclidean, Riemannian, or Lobachevskian geometry adequately describes our world since mathematics does not establish truths about the physical world. Geometry gives us a set of logical conclusions based on possible perceptions and assumptions about the physical universe. The mathematics of each geometry has an intrinsic beauty and logic. Instead of asking, ''Is the world Euclidean, Riemannian, or Lobachevskian?'' one might wonder, ''Which geometry is the best to apply in a given situation?'' Mankind's experience over thousands of years suggests that when working with measurement, travel, construction, and design, Euclidean geometry seems most useful. On the other hand, advancements in science over the last century suggest that, when investigating outer space or analyzing the inner space of atoms, properties of non-Euclidean geometries may be appropriate models.

## Section 3.1 PROBLEMS

## *Explain* ➡ *Apply* ➡ *Explore*

### *Explain*

1. Explain why it is necessary to have undefined terms and postulates in the development of a deductive system of geometry.
2. What is the difference between a ray and a line segment?
3. What are the differences between straight, right, acute, and obtuse angles?
4. Given lines $a$ and $b$, explain the difference between $a \parallel b$ and $a \perp b$. Make a sketch of each.
5. Even though $\overline{RS}$ does not intersect line $m$, explain why $\overline{RS}$ is not parallel to line $m$.

6. Explain why the following definition of parallel segments is faulty: ''Two segments are parallel if they do not intersect.'' Give a good definition for parallel segments.
7. What are axioms and postulates? What are the differences and similarities?
8. Explain the difference between a postulate and a theorem in geometry.
9. Explain why the figure is not possible for lines $m$ and $n$ in Euclidean geometry.

**10.** Explain why the figure is not possible in Euclidean geometry for points $A$ and $B$ and the line through $A$ and $B$ in plane $E$.

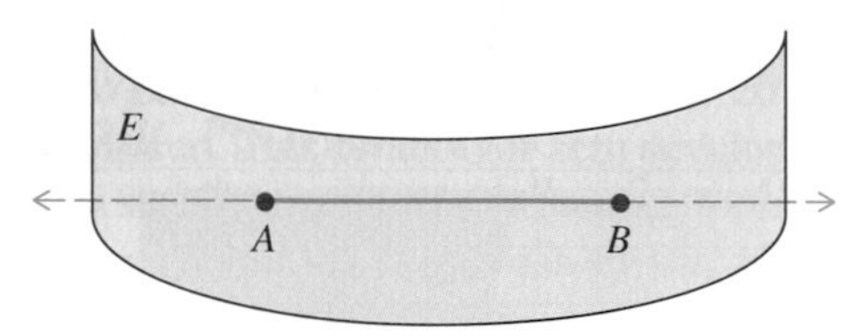


**11.** The transitive property of equality states that if $a = b$ and $b = c$, then $a = c$. Explain why this is equivalent to Axiom A2.

**12.** Describe the parallel postulates of Euclidean, Riemannian, and Lobachevskian geometries. Explain how they are the same and how they are different.

**13.** Contrast and compare the triangle-sum theorems of Euclidean, Riemannian, and Lobachevskian geometries.

## *Apply*

In Problems 14–21, in the Euclidean plane make a sketch that satisfies the conditions.

**14.** Obtuse $\angle FUN$, where $\overline{UF}$ and $\overline{UN}$ are line segments.

**15.** Acute $\angle RUN$, where points $R$, $U$, and $N$ lie in plane $M$.

**16.** Line $j$ and line $k$, where line $r$ makes a right angle with line $j$ and intersects line $k$.

**17.** Line $j$ and line $k$, where line $r$ is perpendicular to both line $j$ and line $k$.

**18.** $\overline{FX} \perp \overrightarrow{FG}$, line $y \| \overline{FX}$.

**19.** Right angle $\angle HER$ with $\overline{HB} \perp \overline{ER}$.

**20.** Right angle $\angle HER$ with $\overline{BE}$ making an acute angle with $\overline{HE}$.

**21.** $\triangle RAD$ with an exterior angle at each vertex.

In Problems 22–26, use specific points, lines, planes, and angles, to make a sketch of each theorem and state the conclusion of the theorem.

**22.** Theorem T1.

**23.** Theorem T2.

**24.** Theorem T3.

**25.** Theorem T4.

**26.** Theorem T5.

In Problems 27–30, in a Riemannian model make a sketch that satisfies the conditions.

**27.** $\triangle ABC$ in which $\angle A$, $\angle B$, and $\angle C$ are right angles.

**28.** $\triangle REI$ with $RE = EI = RI$ and $\angle E$ is an obtuse angle.

**29.** Acute $\angle FAD$ and $\overline{HE} \perp \overline{AD}$.

**30.** $\overleftrightarrow{XY}$ and $\overline{QV} \perp \overline{XY}$.

In Problems 31–34, in a Lobachevskian model make a sketch that satisfies the conditions.

**31.** Right $\triangle ABC$ with line $m$ intersecting $\overline{AB}$.

**32.** Acute $\angle FAD$ and $\overrightarrow{HE} \perp \overline{AD}$.

**33.** Lines $h$, $k$, and $n$ with $h \| k$, $n \| k$, and $h \| n$.

**34.** Lines $h$, $k$, and $n$ with $h \| k$, $n \| k$, and $h$ not parallel to $n$.

## *Explore*

**35.** Euclid defined a point as "that which has no part." Do you think this is a good definition? Why or why not?

**36.** Euclid defined a line as "breadthless length." Do you think this is a good definition? Why or why not?

**37.** Using a standard dictionary, show how giving the definition of the word "dimension" can lead you in a circular path; that is, continuing to define the words of previous definitions could lead you back to the word you began with.

**38.** Show how Axiom A3 can be used to solve for $x$ in the equation $5x - 3 = 4x + 9$.

**39.** Show how Axioms A3 and A4 can be used to solve for $x$ in the equation $5 + 6x = 4x - 11$.

**40.** Draw three pairs of parallel lines. Draw a line that intersects each pair of parallel lines. Use a protractor to measure each set of corresponding angles as shown in the figure. Corresponding angles, $\angle 1$ and $\angle 2$, are marked on the figure.

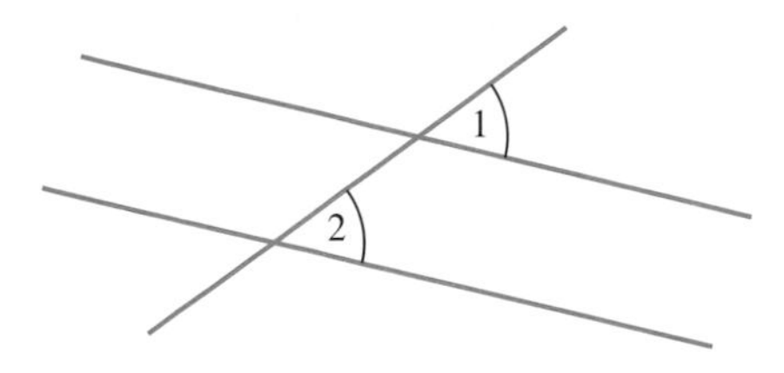


What seems to be true about these angles? Formulate a precise statement of your conjecture.

**41.** Draw three pairs of parallel lines. Draw a line that intersects each pair of parallel lines. Use a protractor to measure each set of alternate interior angles. Alternate interior angles, $\angle 1$ and $\angle 2$, are marked on the figure.

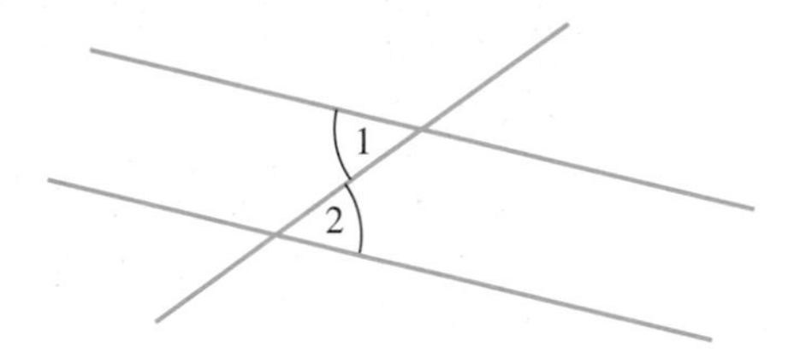


What seems to be true about these angles? Formulate a precise statement of your conjecture.

**42.** Draw three large triangles and an exterior angle for each triangle. Use a protractor to measure the exterior angle and the interior angles of each triangle. Using your results, make some conjectures about exterior angles and interior angles of a triangle.

**43.** What does the existence of non-Euclidean geometry tell us about mathematics and the truth about the physical world?

**44.** Make a sketch of a ray in the Riemannian plane. Explain why the ray is actually a line.

**45.** One of the theorems from Lobachevskian geometry states that parallel lines are not always the same distance apart. Use the Lobachevskian model to demonstrate this theorem.

In Problems 46–49, use the fact that a rectangle is a four-sided plane figure that has four right angles.

**46.** Explain why rectangles do not exist in Riemannian geometry.

**47.** Explain why rectangles do not exist in Lobachevskian geometry.

**48.** Draw a four-sided plane figure with three right angles in the Lobachevskian plane. What type of angle is the fourth angle?

**49.** Draw a four-sided plane figure with three right angles in the Riemannian plane. What type of angle is the fourth angle?

In Problems 50–52, use the Lobachevskian triangle-sum theorem to explain why each is possible.

**50.** In a right triangle, the sum of the two non-right angles is less than 90°.

**51.** Each angle of an equiangular triangle is less than 60°.

**52.** A triangle can have only one right angle or one obtuse angle.

In Problems 53–55, use the Riemannian triangle-sum theorem to explain why each is possible.

**53.** Each angle of an equiangular triangle is greater than 60°.

**54.** In a right triangle $\triangle ABC$, if $\angle A$ is a right angle, then $\angle B + \angle C > 90°$.

**55.** A triangle can have two right angles.

***56.** The five theorems of geometry stated in this section constitute a very small selection of the theorems of Euclidean geometry. Find five other theorems of Euclidean geometry. Make a labeled sketch of each theorem and state the conclusion of each theorem.

## Section 3.2 Perspective

Have you ever looked at a beautiful scene and wondered how an artist can draw it on a piece of paper? The paper is flat (two dimensional) and the visual scene is in real space (three dimensional). How can the artist place lines and shapes on the paper to display this real space? This art of drawing objects in such a way so as to give them depth and show their distance from the observer is called **perspective**. With a few simple techniques and geometric operations the artist is able to give a drawing the illusion of three-dimensional space. When perspective is used, the depth and distance of the actual three-dimensional scene are conveyed on the two-dimensional surface. With the use of the techniques presented in this section, you too can draw the three-dimensional world on a piece of paper.

## Overlapping Shapes

One of the easiest techniques for creating a sense of depth is using **overlapping shapes**. When shapes overlap each other in a drawing, an illusion of shallow space is created. When you see one shape placed in front of another, you are given the impression that the scene is not flat. You are led to conclude that the scene is three dimensional. Here are some examples of this technique.

Mary Cassatt was among many late 19th century artists that adopted this style of shallow perspective. In her etching, *The Letter* (1891), there are many subtle uses of overlapping shapes to create the illusion of dcpth. Thc dcsk that projccts into thc picture overlaps the woman's legs. The hands with the envelope overlaps the body and face. The letter overlaps the top of the desk. The overlapping causes us to conclude that a three-dimensional scene appears on the two-dimensional aquatint etching.

*The Letter* by Mary Cassatt. (National Gallery of Art)

## Diminishing Sizes

Since objects near to us appear larger than objects that are farther away, a second method of conveying a sense of depth is by systematically making objects smaller. This technique of **diminishing sizes** also creates the illusion of depth. Like the technique of overlapping shapes, it causes us to conclude that a three-dimensional scene appears on a two-dimensional surface. Here are some examples of that technique.

The technique of overlapping shapes and diminishing the size can be combined to create an even greater illusion of depth. In the example that follows, the trees decrease in size and overlap each other.

In André Derain's *The Old Bridge* (1910), the illusion of space is created by the positioning of smaller and smaller dwellings, progressing up the hill away from the bridge.

*The Old Bridge* by André Derain. (National Gallery of Art)

## Atmospheric Perspective

A third technique that can serve to create the illusion of depth on a flat surface is a gradual lessening of clarity and visual strength. When we look into the distance, objects at a further distance are often characterized by diminishing color and values of shading. The hills close by seem clear and colorful with defined details and shadows while the hills farther away appear softer and less detailed. As objects recede into the background, they become less distinct. Individual trees become a forest in the mist. People with faces become anonymous forms in a crowd. It seems that the effect of the earth's atmosphere makes the objects less identifiable as they get farther away. Thus, the technique used to produce this effect is called **atmospheric perspective**.

We can demonstrate this technique by shading the trees of the previous example. We can decrease the intensity of the shading and the crispness of the edges of the trees as we progress from the first to the last tree. A deeper, more illusionist effect can be seen in the figure that follows.

Chinese and Japanese ink and wash landscapes are famous for the use of this method of conveying space and distance. Leonardo Da Vinci developed an oil painting technique that creates this effect by scrubbing across previously painted forms with an almost dry brush. In his *Mona Lisa* (1505), he created softness of edges, a veil of mystery, and a subtle depth in the background using this ''sfumato'' or ''smokey effect'' method.

*Cattleya Orchid and Three Brazilian Hummingbirds* by Martin Johnson Heade. (National Gallery of Art)

In Martin Johnson Heade's *Cattleya Orchid and Three Brazilian Hummingbirds* (p.177) (1871), you will notice that the flower, birds, and trees at the front of the painting are very bright and clear. As you visually progress beyond the foreground, the objects become less defined and engulfed in an eerie mist.

## One-Point Perspective

These methods of creating depth were not enough for the artists of the Renaissance who desired to draw and paint more realistically. They wanted their paintings to be a window through which one looks at the real world. They wanted to frame reality. To accomplish this, they needed to understand how to describe the environment of man-made structures, such as streets, buildings, bridges, fences, and walls. Their study of the classical past failed to reveal any logical solutions. However, in Florence, Italy (c. 1400), Filippo Brunelleschi discovered a ''perspective'' technique that solved the problem for the artists. Brunelleschi was an architect but was also a lover of classical thought, philosophy, art, and mathematics. While his accomplishments in architecture were many, his greatest contribution was the discovery of linear, one-point perspective. This technique enabled artists to realistically represent three-dimensional space on a two-dimensional canvas by using the simple geometry of converging lines. When a drawing is created with a one-point perspective, objects seem to converge to a single fixed point. Objects are systematically shortened as they recede into the distance.

### *Creating One-Point Perspectives*

To create a one-point perspective drawing the artist must first focus on what is to be drawn, often by using a small rectangle as a frame or view finder. The artist establishes the window through which the scene will be drawn (see Figure 3.1).

**Figure 3.1**

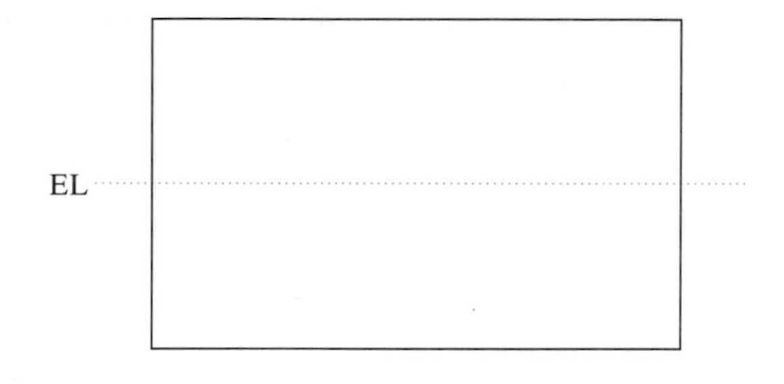


Figure 3.2

Once the limits of the scene have been determined, the following steps can be used to draw the scene in a one-point perspective.

1. Looking straight ahead, a line is placed across the frame of the picture to establish where everything should be located in terms of being above or below eye level. Such a line is called the **eye level line**. In Figure 3.2, the eye level line (EL) is placed near the center of the frame.
2. To create the illusion of space, a point is established on the eye level line to which all receding lines converge. This point is called the **vanishing point** (VP). For example, to draw a three-dimensional box, the front of the box can be given the illusion of receding to the vanishing point by connecting the corners of the box to the vanishing point as shown in Figure 3.3
3. The box can be given a sense of depth by drawing the back of the box using segments that are parallel to the front of the box and have endpoints on the lines to the vanishing point as shown in Figure 3.4
4. By darkening some of the lines and erasing all nonessential lines, the box can be seen as a solid object receding into space toward a fixed point as shown in Figure 3.5.
5. If the artist wants to add other objects to the drawing, the eye level line and the vanishing point remain the same. In Figure 3.6, the box becomes a part of a more complex scene.

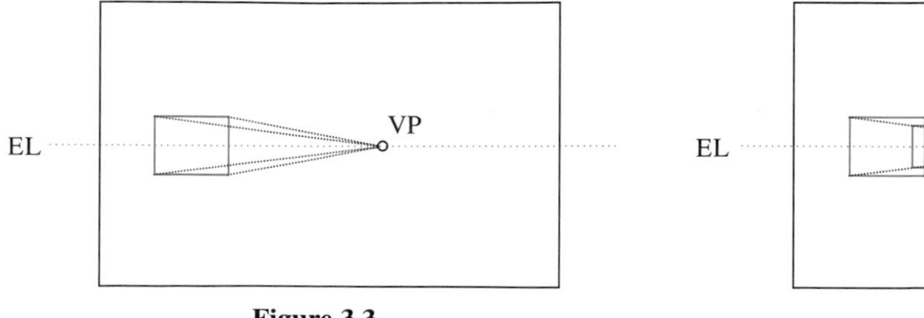


Figure 3.3

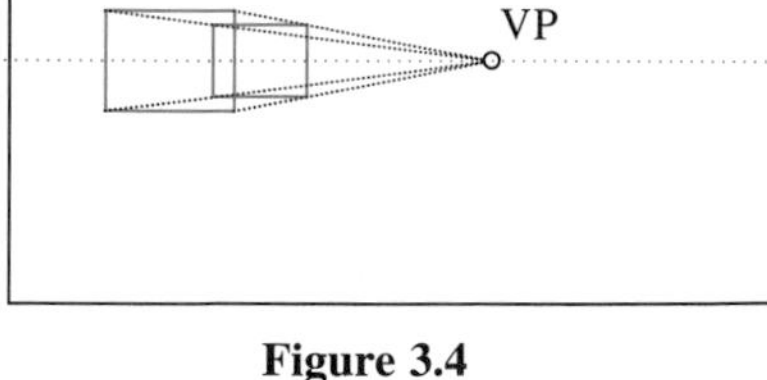


Figure 3.4

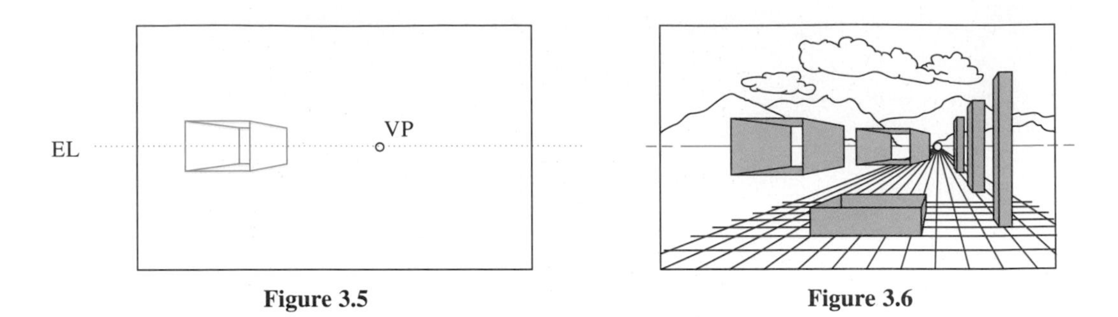


**Figure 3.5** **Figure 3.6**

## *Varying Positions of Eye Level Line and Vanishing Point*

When creating the illusion of space on a two-dimensional surface, there are an unlimited number of ways you can use the basic procedure for one-point perspectives. Changing the eye level line and vanishing point can create very different effects. The figures that follow show variations when the subjects being drawn are in different relative positions to the eye level line. The changes caused by moving the vanishing point are left as exercises at the end of the section.

**1.** Most of the subject is above the eye level line.

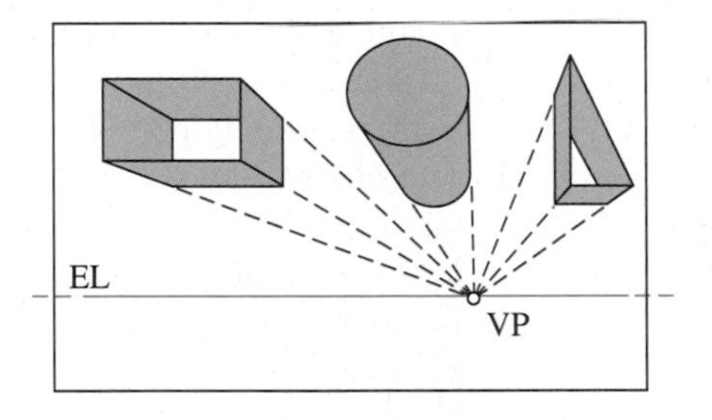


Notice that the eye level line is drawn low across the frame and the objects connected to the vanishing point appear above you.

**2.** Most of the subject is below the eye level line.

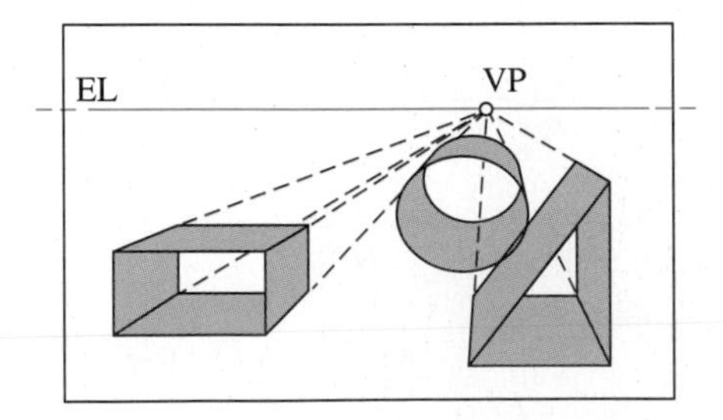


Notice that the eye level line is drawn high across the frame and the objects connected to the vanishing point appear below you.

**3.** The subject is distributed above, below, and on the eye level line.

V.P.

E.L.

Notice that the eye level line is drawn horizontally near the center of the frame and the objects connected to the vanishing point appear above, below, and at eye level.

In Vincent van Gogh's *Flower Beds in Holland* (1883), you can see that he used a one-point perspective with an eye level line below the center and a vanishing point at the center of the picture to give us a beautiful view of his home land.

*Flower Beds in Holland* by Vincent van Gogh. (National Gallery of Art)

## *Equally Spaced Objects*

If equally spaced vertical objects such as poles, posts, or trees are drawn in a one-point perspective, they not only get shorter, they also get closer together. A very efficient geometric method is used to determine the spacing between such upright objects in a one-point perceptive. Here is how it is done.

1. The first upright ($\overline{AB}$) is drawn, and the end points are connected to the vanishing point on the eye level line (see Figure 3.7).
2. A second upright ($\overline{CD}$) is drawn at an arbitrary distance from the first upright.
3. The third upright is mathematically placed at $E$. Point $E$ is positioned by finding where the line drawn from the top of the first upright ($A$) through the midpoint of the second upright ($M$) intersects the line to the vanishing point.
4. Other poles are systematically located and drawn between the lines to the vanishing point by using the third step.

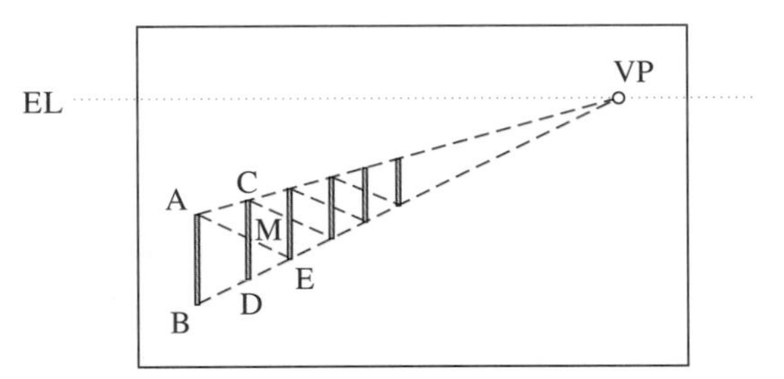


## *Proportions in a One-Point Perspective*

In a one-point perspective, objects of the same size are drawn proportionally smaller as they get closer to the vanishing point. This proportion is determined once the first object has been drawn and its distance from the vanishing point has been established. If another object of the same size is positioned closer to the vanishing point, the ratio of its height to its distance from the vanishing point is the same as the ratio of the height to the distance of the first object. If we let $h_1$ be the height of the first object, $d_1$ be its distance from the vanishing point, $h_2$ be the height of the added object, and $d_2$ its distance from the vanishing point, we get the following proportion for the objects in the drawing.

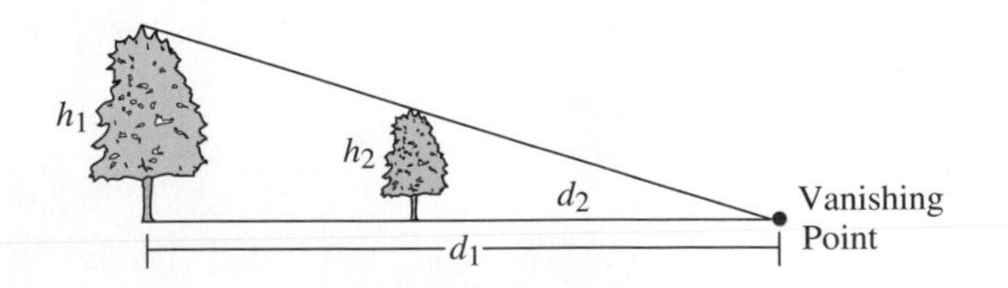


This proportion enables us to determine the heights or distances of other congruent objects that are positioned in a line of objects.

## Example 1

Suppose the first tree in the drawing that follows is 5 in. tall and 18 in. from the vanishing point.

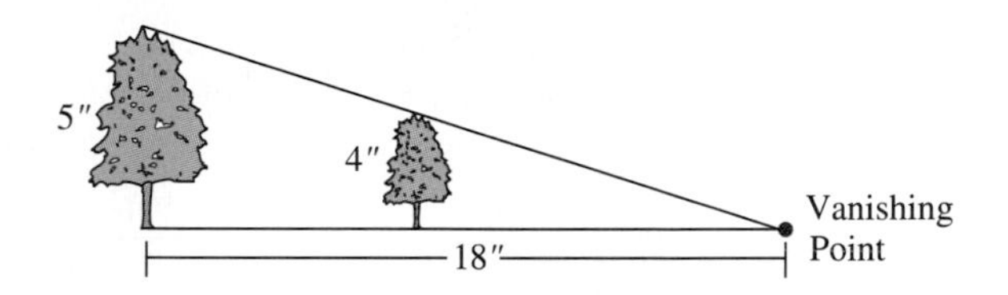


(a) How far from the vanishing point should you place a second tree that is to be 4 in. tall in the drawing?
(b) What is the distance between the first tree and the second tree?

**Solution:**

(a) Using the proportion for a line of objects of the same size in a one-point perspective, we get the following solution. Let $h_1 = 5$, $d_1 = 18$, $h_2 = 4$, and $d_2$ be unknown.

$$\frac{h_1}{d_1} = \frac{h_2}{d_2}$$

$$\frac{5}{18} = \frac{4}{d_2}$$

$$5d_2 = 18(4)$$

$$5d_2 = 72$$

$$d_2 = 14.4$$

Thus, the second tree is placed 14.4 in. from the vanishing point.

(b) Since the first tree is positioned 18 in. from the vanishing point and the second tree is positioned 14.4 in. from the vanishing point, the distance between the trees on the drawing is $18 - 14.4 = 3.6$ in. ■

## Example 2

Suppose in the drawing of the previous example you want to draw another tree 3 in. from the vanishing point. How tall should you make the tree in the drawing?

**Solution:** We again use the proportion for a line of objects of the same size this time with $h_1 = 5$, $d_1 = 18$, $d_2 = 3$, and $h_2$ being unknown.

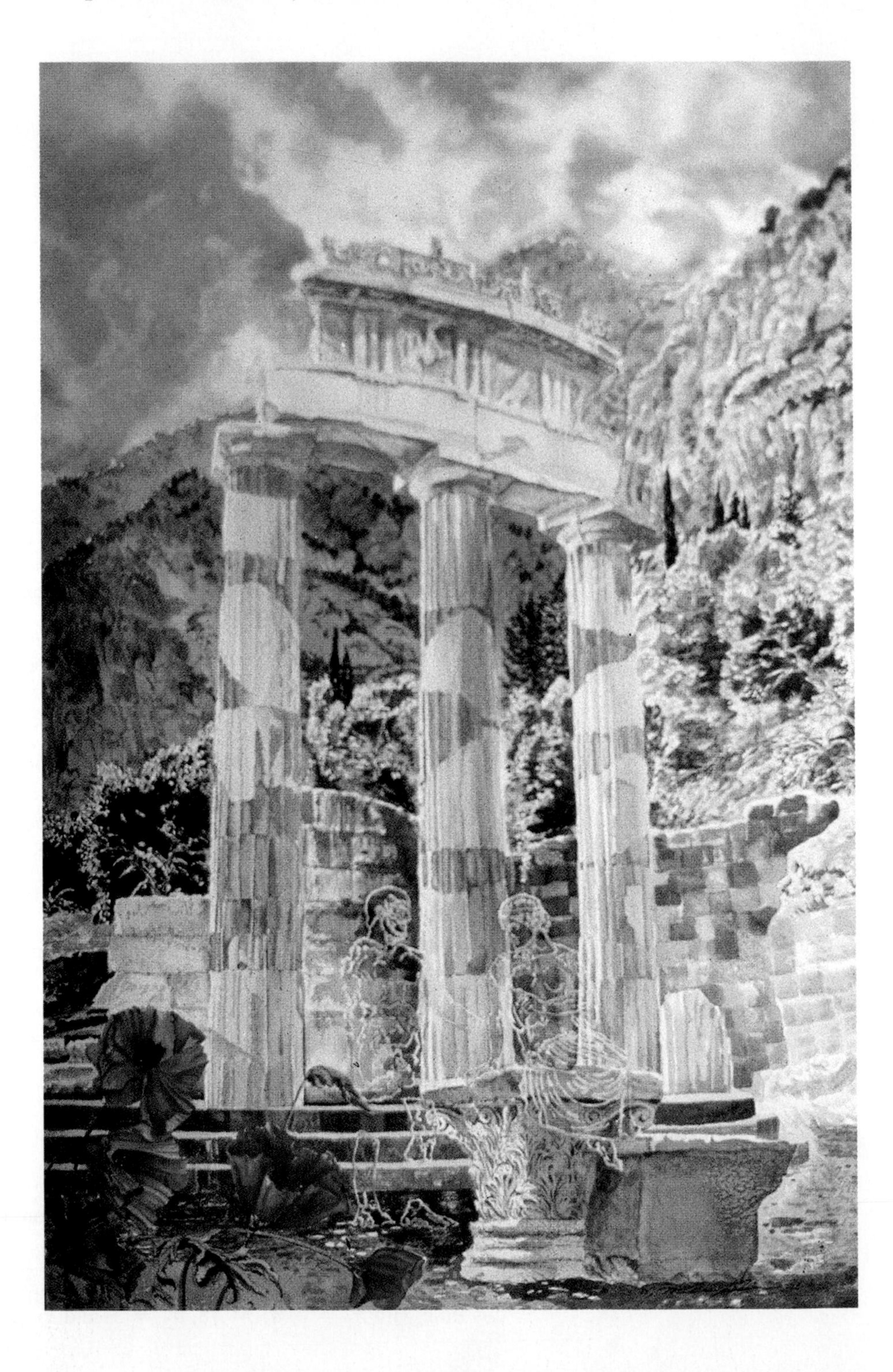

*Delphic Vision* by David McLaughlin. (Courtesy of the artist)

$$\frac{h_1}{d_1} = \frac{h_2}{d_2}$$

$$\frac{5}{18} = \frac{h_2}{3}$$

$$18h_2 = 5(3)$$

$$18h_2 = 15$$

$$d_2 = \frac{15}{18} = \frac{5}{6}$$

Thus, the tree should be $\frac{5}{6}$ in. tall. ∎

## More Complex Perspectives

Artists have gone beyond the one-point perspective discovered by Brunelleschi in 1400. They have perfected two-point and three-point perspectives and have integrated different techniques into a single painting. It is beyond the scope of this book to show you how these more complex perspectives are created. Our objective is to demonstrate how simple constructions from geometry can be used to create artistic effects. However, we will give you some examples of these more complex techniques.

### *Two-Point Perspective*

In *The Square at St. Mark's* by Canaletto (1735) you will notice that if you follow the receding lines to the right and left there are two vanishing points on the eye level line. You will also notice that both of these vanishing points are off the canvas.

### *Three-Point Perspective*

In David McLaughlin's *Delphic Vision* (1993), in addition to vanishing points to the right and left, the view sweeps upward, high above the ground, to a third point. This three-point perspective along with the overlapping of shapes, diminishing sizes, and atmospheric perspective give this watercolor a stunning three-dimensional effect on McLaughlin's two-dimensional canvas.

If you experiment with the techniques described in this section, you can draw objects on a piece of paper that give an illusion of being three dimensional. The exercises that follow will give you practice is doing just that.

Section 3.2 PROBLEMS

## *Explain → Apply → Explore*

### *Explain*

**1.** What is perspective?

**2.** What is the perspective technique of overlapping shapes? Sketch an example.

*The Square of St. Mark's* by Canaletto. (National Gallery of Art)

**3.** What is the perspective technique of diminishing sizes? Sketch an example.

**4.** What is the technique of atmospheric perspective? Sketch an example.

**5.** What is a one-point perspective? Sketch an example.

**6.** What are the eye level lines and vanishing points? Sketch an example.

**7.** What is the basic procedure for creating a one-point perspective?

**8.** In a one-point perspective, what procedure is used to position equally spaced objects of the same size?

### *Apply*

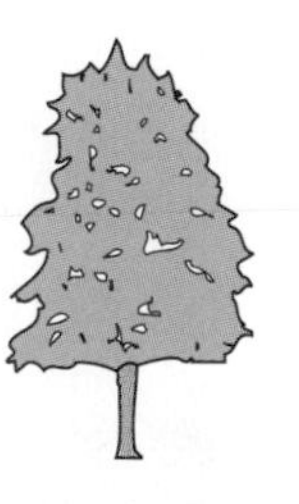

**9.** Use the following shape and make a sketch that gives a sense of depth with

(a) overlapping shapes.
(b) diminishing sizes.
(c) atmospheric perspective.

**10.** Use the following shape and make a sketch that gives a sense of depth with

(a) overlapping shapes.
(b) diminishing sizes.
(c) atmospheric perspective.

**11.** Create solid objects in a one-point perspective using the shapes given as the front of the object and the eye level line and vanishing point as shown.

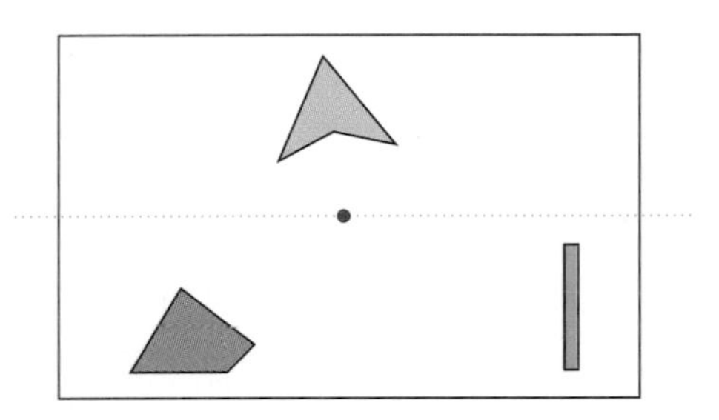

**12.** Create solid objects in a one-point perspective using the shapes given as the front of the object and the eye level line and vanishing point as shown.

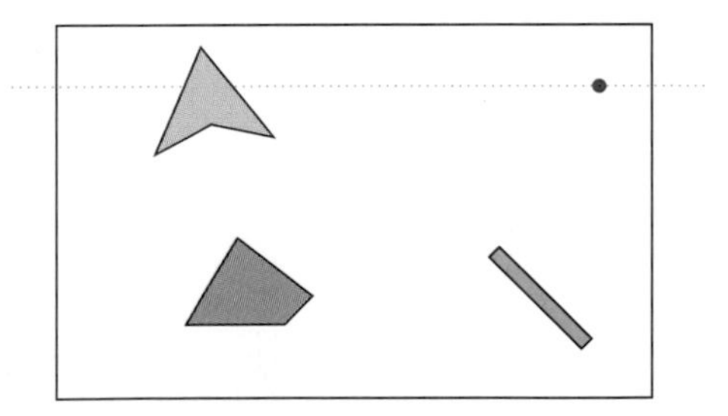

**13.** Create solid objects in a one-point perspective using the shapes given as the front of the object and the eye level line and vanishing point as shown.

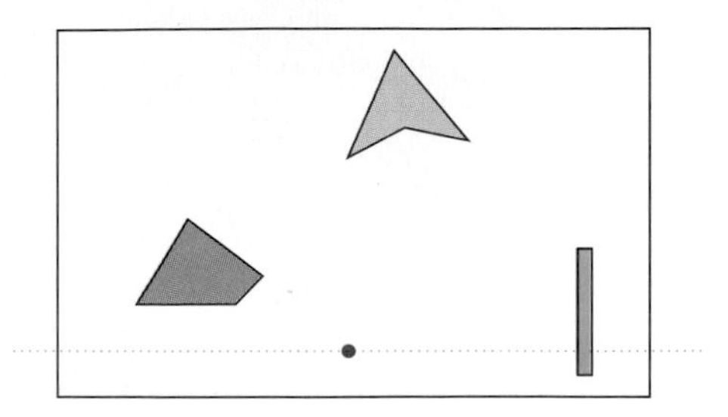

**14.** Draw a one-point perspective showing three billboards along a four-lane road that has a word on each billboard. The words on the billboards give the message, "EAT AT AL'S." The start of the road, the first billboard, the eye level line, and the vanishing point are shown in Figure 3.8

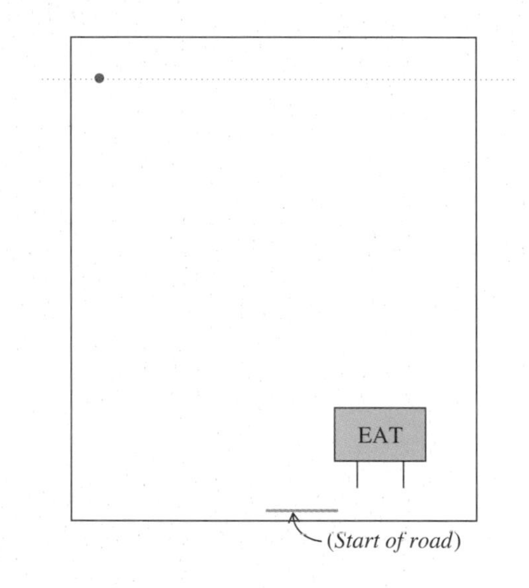


**Figure 3.8**

**15.** A scene showing a road through the desert is shown in Figure 3.9. There are a series of equally spaced telephone poles along the road. However, only one pole is shown in the picture. Using a one-point perspective draw at least five more poles.

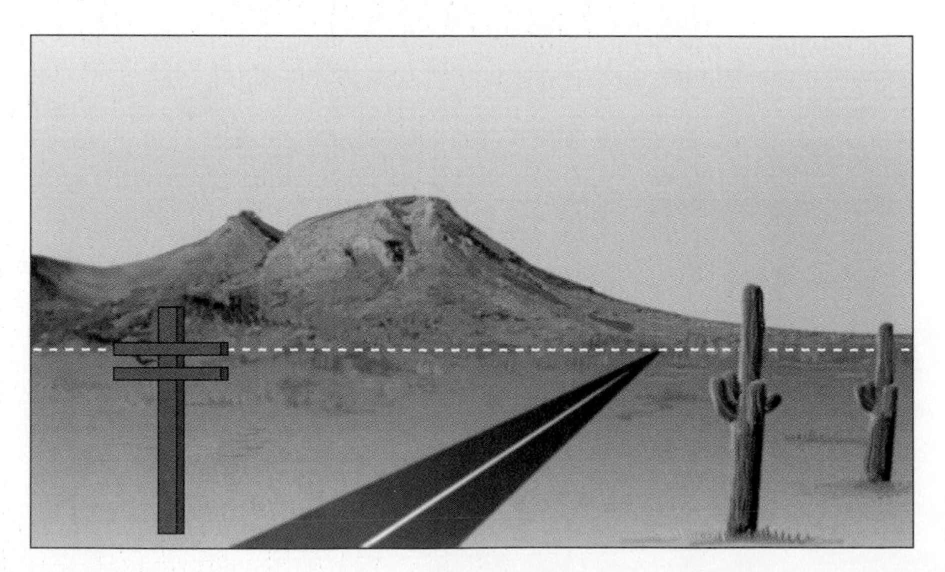

**Figure 3.9**

**16.** Put ''LIFE'' in perspective by using the picture frame, eye level line, and the vanishing point as shown. Use a one-point perspective to show the word ''LIFE'' systematically receding toward the vanishing point.

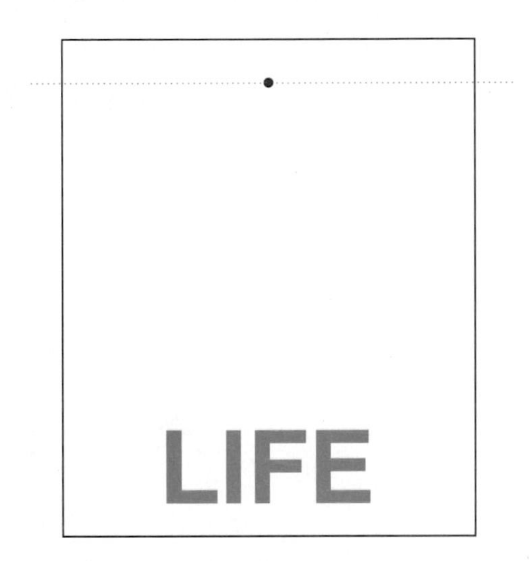


**Figure 3.10**

**17.** Put ''FIT'' in perspective by using the picture frame, eye level line, and the vanishing point as shown. Use a one-point perspective to show the word ''FIT'' systematically receding toward the vanishing point.

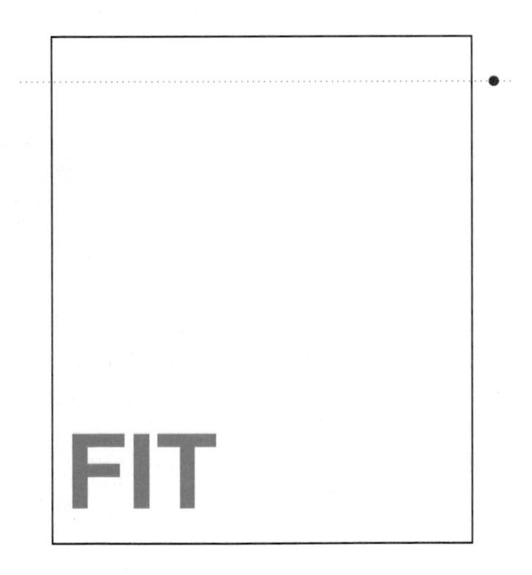


**Figure 3.11**

In Problems 18–23, a series of utility poles of the same size are a part of a one-point perspective drawing. The letters $a$, $b$, $c$, $d$, and $e$ represent heights or distances in the drawing as shown in Figure 3.12. Use the proportion for a line of congruent objects to determine the desired heights and distances in each problem.

**18.** $a = 3$ in., $e = 12$ in., $b = 2$ in., find $d$ and $c$.

**19.** $a = 4$ ft, $e = 20$ ft, $b = 2.5$ ft, find $d$ and $c$.

**20.** $a = 4$ m, $e = 10$ m, $d = 2$ m, find $b$ and $c$.

**21.** $a = 12.2$ cm, $e = 24.5$ cm, $d = 4.5$ cm, find $b$ and $c$.

**22.** $b = 3.6$ cm, $d = 1.2$ cm, $c = 6$ cm, find $e$ and $a$.

**23.** $a = 8$ in., $c = 5$ in., $e = 20$ in., find $d$ and $b$.

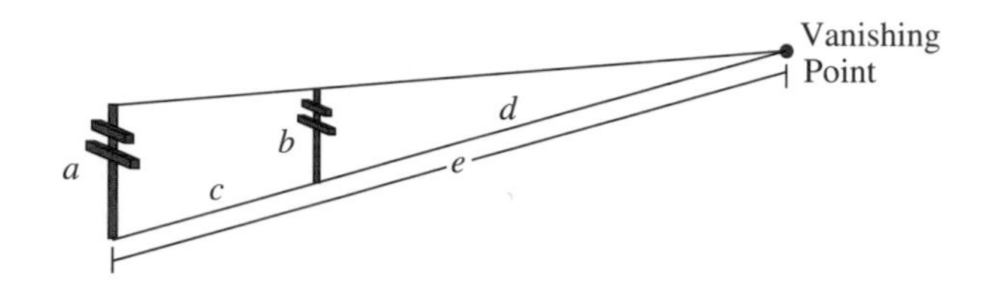


**Figure 3.12**

*Explore*

**24.** The wall along a stretch of a four-lane freeway has the pattern as shown in the figure that follows. Use a one-point perspective to give the wall an illusion of running along the freeway. Choose your own eye level line, vanishing point, and picture frame.

**25.** A wooden fence runs along a horse pasture. Use a one-point perspective to draw the fence, giving it an illusion of being three dimensional. Choose your own eye level line, vanishing point, and picture frame.

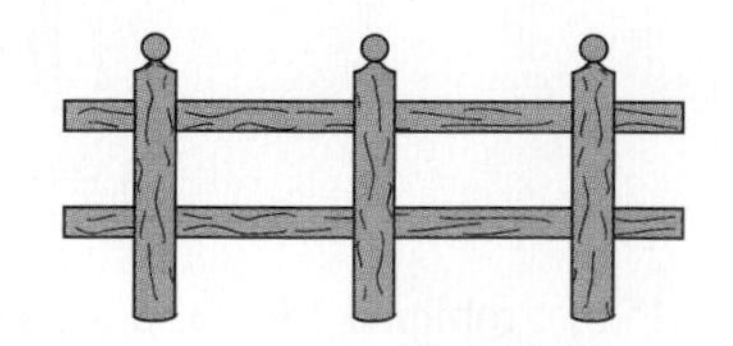

**26.** In the painting that follows, *The New Road* by Grant Wood (1939), what method(s) of perspective are evident? Explain.

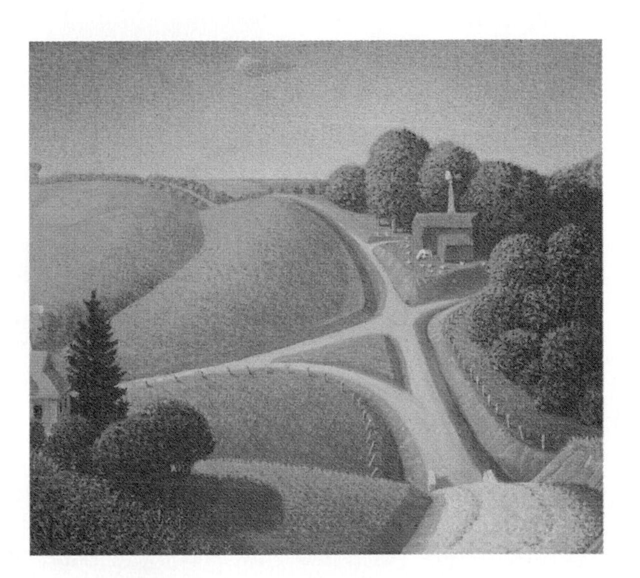

*The New Road* by Grant Wood. (National Gallery of Art)

**27.** In the painting that follows, *Quadrille at the Moulin Rouge* by Henri de Toulouse-Lautrec (1892), what method(s) of perspective are evident? Explain.

*Quadrille at the Moulin Rouge* by Henri de Toulouse-Lautrec. (National Gallery of Art)

**28.** In the painting that follows, *Cappriccio of Roman Ruins* by Marco Ricci (1720), what method(s) of perspective are evident? Explain.

*Capriccio of Roman Ruins* by Marco Ricci. (National Gallery of Art)

**29.** In the painting that follows, *Death and the Miser* by Hieronymus Bosch (1490), what method(s) of perspective are evident? Explain.

*Death and the Miser* by Hieronymus Bosch. (National Gallery of Art)

**30.** Examine some art of the ancient Egyptians. Do they make use of perspective? If so, which of the techniques discussed in the chapter are used?

**31.** The painting that follows is *Egyptian Mall* by Darice Rodrigues (1993). Are any method(s) of perspective used in this painting? Explain.

*Egyptian Mall* by Darice Rodrigues. (Courtesy of the artist)

**32.** An artist is using a one-point perspective to draw a line of Boeing 767 airliners parked along a runway. The first Boeing 767 in the line is to be 10 cm tall, and the last Boeing 767 is to be 1.5 cm tall. If the last Boeing 767 is drawn 8 cm from the vanishing point, how far from the last Boeing 767 should he draw the first Boeing 767 airliner?

**33.** An artist wants to have a fence running along a field in a one-point perspective drawing. She decides to use a 3-in. vertical post at the start of the fence and a vanishing point that is 16 in. away from the vertical post. If she wants the last vertical post positioned in the perspective to be $\frac{1}{2}$ in. tall, how far from the first pole should she place the last pole?

**34.** Examine some art of the ancient Chinese or Japanese. Do they make use of perspective? If so, which of the techniques discussed in the chapter are used?

**35.** Look down a set of railroad tracks or double center lines along a straight road. What do you notice? What perspective technique is suggested by this visual experience?

**36.** Examine some photographs of outdoor scenes such as landscapes, mountains, valleys, forests, or lakes. What perspective techniques are evident in such photographs?

37. Examine art shown in other sections of this text. Find an example of these perspective techniques:
    (a) overlapping shapes.
    (b) diminishing sizes.
    (c) atmospheric perspective.
    (d) one-point perspective.

38. Look at paintings and drawings in an art book. Find examples of the perspective techniques discussed in this chapter. Make photocopies of the art and clearly indicate the perspective techniques used.

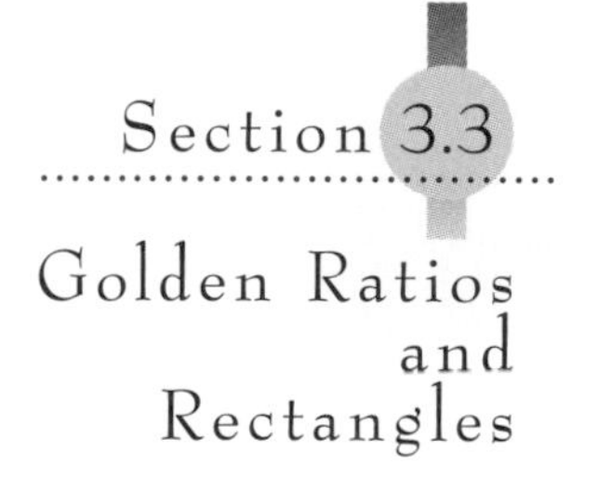

## Section 3.3 Golden Ratios and Rectangles

We continue our look at geometry and art by studying various geometric forms from a more aesthetic point of view. We look at the visual beauty contained in geometric objects and investigate some of the mathematics behind the creation of geometric designs.

Ancient architects, sculptors, and artists used a ratio of distances in their work that they deemed pleasing to the eye. Called the **Golden Ratio**, it is based on the division of a line segment into two parts such that the ratio of the longer piece to the shorter piece is the same as the ratio of the entire line segment to the longer piece. The Greek letter **phi**, $\phi$ (pronounced fī), was adopted in the early 20th century to represent this ratio because it is the first letter in the name of the Greek sculptor Phidias, who made extensive use of the Golden Ratio in his work.

### The Golden Ratio

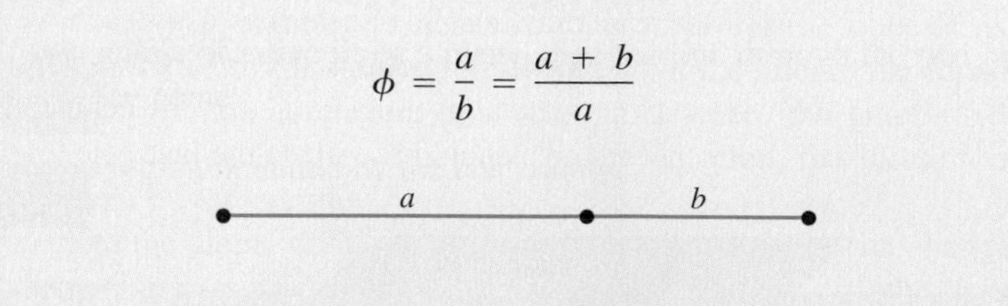


$$\phi = \frac{a}{b} = \frac{a + b}{a}$$

The exact value of the Golden Ratio can be determined by performing some algebraic manipulations on the ratio $a/b = (a + b)/a$. In the derivation that follows, we use the quadratic formula. If you need a review of the quadratic formula, see Section 2.0.

$$\frac{a}{b} = \frac{a+b}{a} \qquad \text{Golden Ratio}$$

$$a^2 = b(a+b) \qquad \text{Cross multiplying}$$

$$a^2 - ba - b^2 = 0 \qquad \text{Subtracting } b(a+b) \text{ from both sides}$$

$$a = \frac{b \pm \sqrt{b^2 + 4b^2}}{2} \qquad \text{Solving for } a \text{ by the quadratic formula}$$

Therefore,

$$a = \frac{b + b\sqrt{5}}{2} \qquad \text{The length } a \text{ must be positive.}$$

Thus,

$$\frac{a}{b} = \frac{\frac{b + b\sqrt{5}}{2}}{b} = \frac{b + b\sqrt{5}}{2}\left(\frac{1}{b}\right)$$

$$\phi = \frac{1 + \sqrt{5}}{2} = 1.618033988\ldots$$

The Golden Ratio is an irrational number because it contains $\sqrt{5}$. Its decimal representation is a nonterminating, nonrepeating decimal. However, because of errors inherent in measurement and for ease in computation, we will use 1.62 as the value of the Golden Ratio in this chapter.

A **Golden Rectangle** is a rectangle whose sides form the Golden Ratio. The Greeks believed that a rectangle having this ratio was more pleasing to the eye than was any other rectangle. In 1876, the psychologist Gustav Fechner analyzed the responses of many people about which rectangles they found most pleasing. From his research he concluded that people prefer rectangular shapes that are approximately Golden Rectangles. Experiments carried out by Witmer (1894), Lalo (1905), and Thorndike (1917) arrived at similar conclusions.

**The Golden Rectangle**

$$\frac{l}{w} = \frac{l + w}{l} = \phi$$

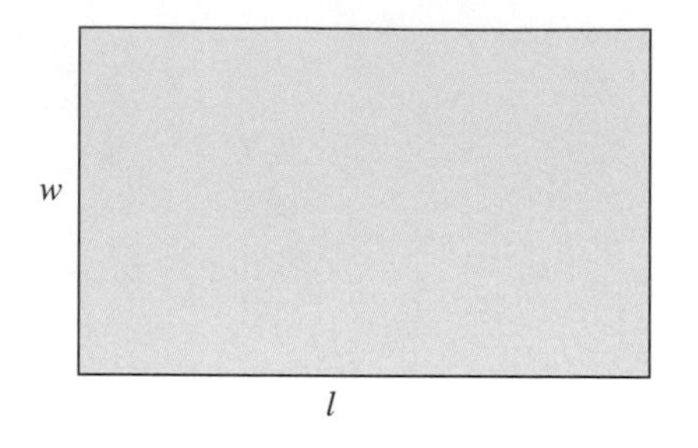


## Example 1

By measuring the rectangles below, determine if they approximate Golden Rectangles.

A

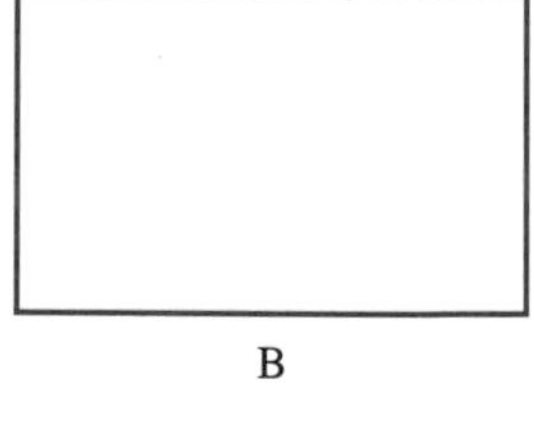

B

**Solution:** Measuring the rectangles, we find the dimensions of rectangle A to be 3 cm and 7 cm, whereas rectangle B has dimensions of 2.1 cm and 3.4 cm. If we take the ratio of the *longer* side to the *shorter* side in each rectangle, we get

***A.*** $7/3 \approx 2.33$, and that is not close to the Golden Ratio of approximately 1.62.

***B.*** $3.4/2.1 \approx 1.62$, which is approximately the same as the Golden Ratio.

Thus, rectangle *A* is not a Golden Rectangle, but rectangle *B* is very close to a Golden Rectangle. ■

## Example 2

Find a point on $\overline{AB}$ such that $\overline{AB}$ is divided into segments that form the Golden Ratio.

*A* *B*

**Solution:** Let *X* be a point that divides $\overline{AB}$ into the Golden Ratio as shown.

According to the Golden Ratio, $a/b = (a + b)/a \approx 1.62$. Since the length of $\overline{AB}$ is 7 cm, $a + b = 7$ and we get

$$\frac{7}{a} \approx 1.62$$

$$1.62a \approx 7$$

$$a \approx 4.3$$

Thus, the point $X$ is 4.3 cm from point $A$. ■

The Golden Ratio has an interesting history. Objects that contain the Golden Ratio have been studied and admired through the ages. In the rest of this section, we investigate some situations in which the Golden Ratio occurs.

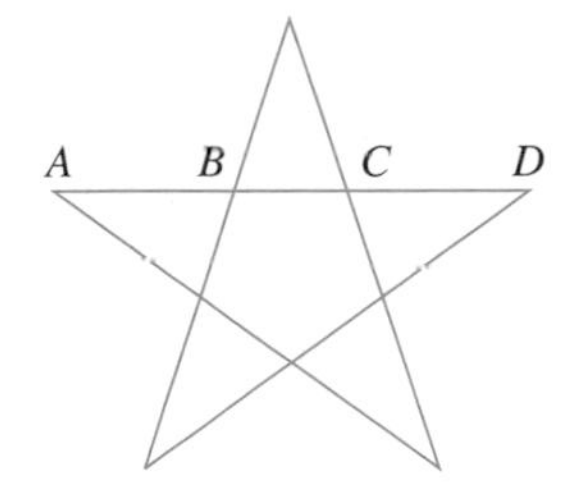

Figure 3.13

### Example 3

The pentagram, which dates back to ancient Babylonian and Greek cultures, was the mystic symbol and badge of the Society of Pythagoras. The pentagram contains the Golden Ratio many times. By actual measurement, find two occurrences of the Golden Ratio in the pentagram in Figure 3.13.

**Solution:** The ratios of $AC$ to $CD$ and $BD$ to $AB$ are approximately 1.62. ■

The Golden Ratio can also be found in the measurements of buildings of antiquity such as the Parthenon on the Acropolis in Athens. You can verify that the ratio of the length to the width of the overall dimensions of the Parthenon is approximately the Golden Ratio by measuring the scale drawing of the Parthenon in Figure 3.14.

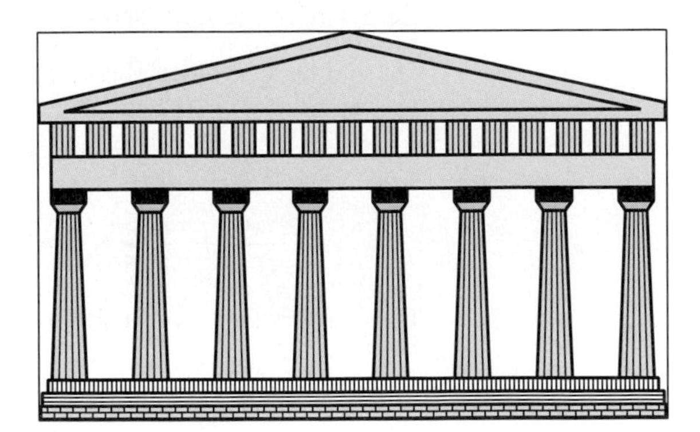

**Figure 3.14**

The Parthenon on the Acropolis in Athens

In *Geometry of Art and Life*, the author, Matila Ghyka, shows that the "perfect" human face can be viewed as sequences of Golden Ratios.

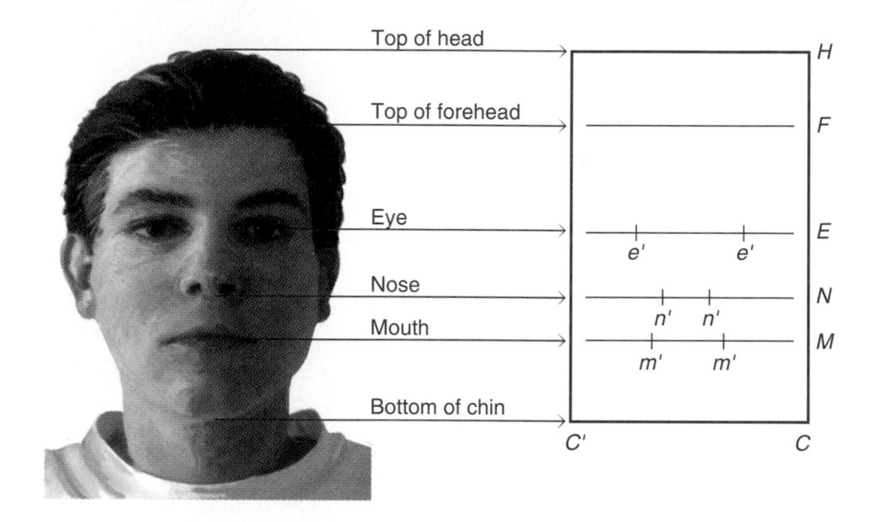


$C'C$ = Width of the face
$e'e$ = Distance between outsides of eyes
$n'n$ = Distance between outsides of nostrils
$m'm$ = Distance between ends of mouth

$$\frac{HC}{C'C} = \frac{HE}{FE} = \frac{EC}{NC} = \phi \quad \frac{FE}{EN} = \frac{EM}{EN} = \frac{NC}{MC} = \phi \quad \frac{C'C}{e'e} = \frac{e'e}{m'm} = \frac{m'm}{n'n} = \phi$$

## Example 4

Consider a "perfect" face that measures 2.5 in. between the outside edges of the eyes. How wide and how long is such a face?

**Solution:** In the figure, $C'C$ represents the width of the face, $HC$ represents the height of the face, and $e'e$ represents the distance between the eyes.

The following Golden Ratios can be used:

$$\frac{C'C}{e'e} = \frac{C'C}{2.5} \approx 1.62 \qquad C'C \approx 2.5(1.62) \approx 4.05$$

$$\frac{HC}{C'C} = \frac{HC}{4.05} \approx 1.62 \qquad HC \approx 4.05(1.62) \approx 6.56$$

Thus, such a face has a width of about 4 in. and a height of about $6\frac{1}{2}$ in. ■

Leonardo Fibonacci (c. 1200) studied sequences of numbers in which each successive number is the sum of the two previous numbers: 1, 1, 2, 3, 5, 8, 13, 21, 34,

55 .... In 1877, Edward Lucas named that sequence of numbers the **Fibonacci numbers**. This sequence has the unique property that the ratio of successive terms gets close to the Golden Ratio.

$$\frac{1}{1} = 1 \qquad \frac{2}{1} = 2 \qquad \frac{3}{2} = 1.5$$

$$\frac{5}{3} \approx 1.6666667 \qquad \frac{8}{5} = 1.6 \qquad \frac{13}{8} = 1.625$$

$$\frac{21}{13} \approx 1.6153846 \qquad \frac{34}{21} \approx 1.6190476 \qquad \frac{55}{34} \approx 1.6176470 \ldots$$

### Example 5

Find the ratio of the 15th and 16th Fibonacci numbers. How does this ratio compare to the Golden Ratio?

**Solution:** The 15th Fibonacci number is 610, and the 16th is 987; 987/610 ≈ 1.6180328. The first six digits of that ratio match those of the Golden Ratio. ■

The Golden Ratio was referred to by various names over the centuries. Luca Pacioli called it *divina proportione* (divine proportion) in 1509. Johann Kepler called it *sectio divina* (divine section) in 1610. The term "Golden Ratio" or "Golden Section" came into use around 1840. The Golden Ratio has stimulated mathematical interest for centuries and is still of interest to designers, botanists, sculptors, composers, and artists. The work of artists Georges Seurat, Piet Mondrian, and Juan Gris and some musical compositions by Bela Bartok utilize the Golden Ratio.

We end this section with a beautiful mathematical curve. It is known as the logarithmic spiral and is created by curves formed within Golden Rectangles. The spiral has no ending point. It grows outward or inward indefinitely, but its shape remains unchanged. In nature this spiral can be seen in the successive chambers of the nautilus sea shell.

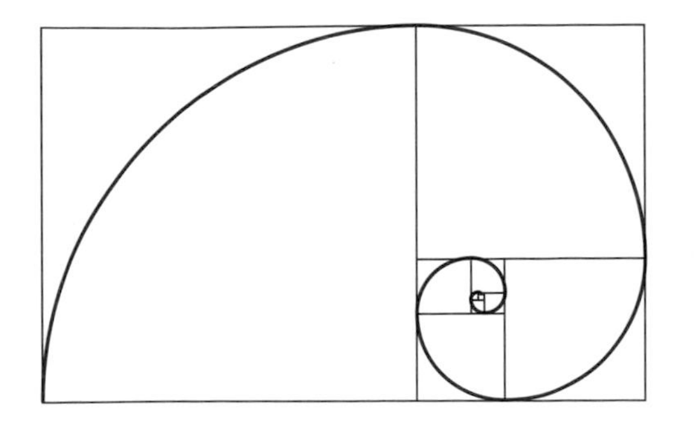

Logarithmic spiral

Nautilus seashell

Section 3.3
PROBLEMS

## *Explain ➡ Apply ➡ Explore*

### *Explain*

1. What is the Golden Ratio?

2. Why were the Greeks fascinated with the Golden Ratio?

3. What is the importance of the Golden Ratio in art?

4. How do you find the point that divides a line segment into the Golden Ratio?

5. What is a Golden Rectangle?

6. If you knew the measurement of the longer side of a Golden Rectangle, how would you find the length of the shorter side?

7. If you knew the measurement of the shorter side of a Golden Rectangle, how would you find the length of the longer side?

8. What are the Fibonacci numbers and what relationship do they have to the Golden Ratio?

### *Apply*

In Problems 9–12, find a point that approximately divides each segment into the Golden Ratio.

9. ______________________

10. ____________________________

11. __________________________________

12. ________________________________________

In Problems 13–16, determine which of the following rectangles are approximately Golden Rectangles.

13.

**14.**

**15.**

**16.**

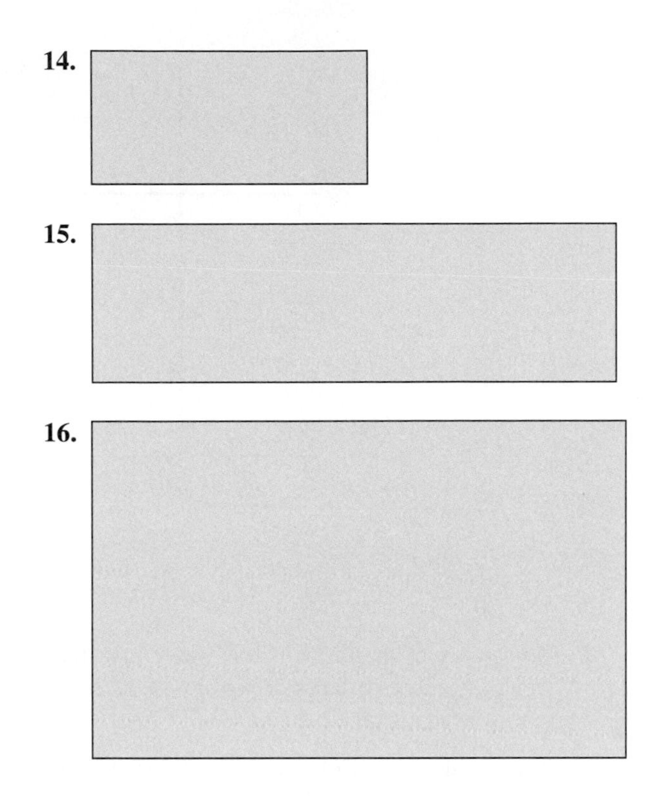

In Problems 17–23, one dimension of a Golden Rectangle is given. Find the two possible values for the other dimension of the Golden Rectangle.

**17.** 23 ft

**18.** 4.5 mi

**19.** 56.8 m

**20.** 45.5 cm

**21.** 9 cm

**22.** $10\frac{3}{4}$ in.

**23.** 36 in.

## *Explore*

If you were to construct a "Golden Cross," you could make the ratio of the height ($h$) to the width ($w$) of the cross equal to the Golden Ratio and the ratio of the bottom section of the cross ($b$) and the top section of the cross ($t$) equal to the Golden Ratio. Find the width, top section, and bottom section of the cross if the height of the cross is as given in Problems 24–26.

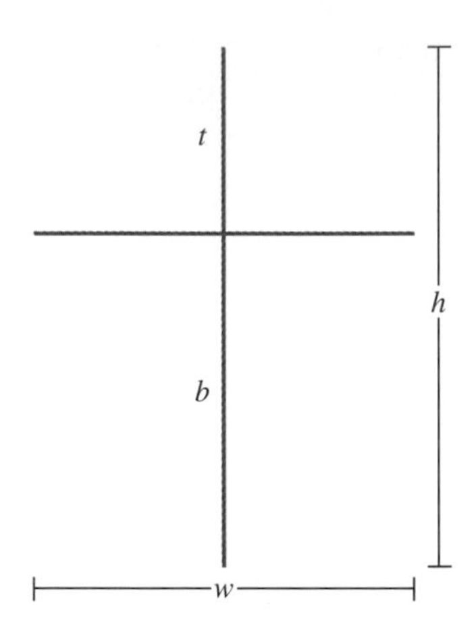


**24.** $h = 72$ in.

**25.** $h = 16$ cm

**26.** $h = 20$ ft

**27.** Do a study of the ratios within the human face shown in Example 4. By analyzing at least two faces or pictures of faces, what can you conclude about the ratios or the average of the ratios?

**28.** The architect Le Corbusier (c. 1940) developed the modular system of harmonious proportions. In this system, he established that in the visually ''perfect'' human form:

(a) The ratio of distance from the bottom of the neck to the navel to the distance from the top of the head to the bottom of the neck is the Golden Ratio.
(b) The ratio of the distance from the navel to the knee to the distance from the knee to the bottom of the foot is the Golden Ratio.

Measure people or pictures of people to determine if and when these ratios are golden. Do the average of the ratios you found approximate the Golden Ratio?

We can define a ''Golden Box'' as one whose height, width, and length satisfy the Golden Ratio. We are assuming here that the height is the shortest dimension and the length is the longest dimension of the box.

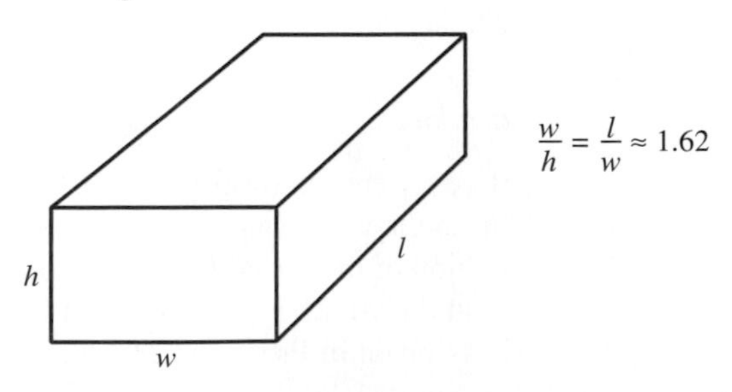

In Problems 29–34, one dimension of a "Golden Box" is given. Find the dimensions that are not given.

**29.** $h = 5$ in.

**30.** $l = 5$ in.

**31.** $w = 9$ cm

**32.** $w = 5$ in.

**33.** $h = 9$ cm

**34.** $l = 9$ cm

**35.** Construct a "Golden Box" that has a height of 3 in.

**36.** Measure various boxes. Do they form "Golden Boxes," as defined? Find the average of the heights, widths, and lengths of the boxes. Is the ratio of these averages approximately the Golden Ratio?

**37.** Explain why three consecutive Fibonacci numbers can be used to give approximate dimensions of "Golden Boxes."

**38.** By finding the ratio of the length to the width of ten common rectangular-shaped objects, such as boxes, cards, cushions, doors, appliances, and windows, determine which objects approximate Golden Rectangles and which objects do not approximate Golden Rectangles. Find the average of the ten ratios. How does your result compare to that of psychologist Gustav Fechner, who concluded that the average of the ratios of the sides of common rectangles was approximately the Golden Ratio? Since Golden Rectangles are supposed to be the most pleasing to the eye, why aren't all manufactured rectangular objects Golden Rectangles?

**39.** A "Golden Can" could be defined as a can that has its height greater than its diameter, where the ratio of its height to its diameter is equal to the Golden Ratio. Measure ten cans of different sizes. Which ones approximate the Golden Ratio, and which do not? Find the average of the ten ratios. How does this average compare to the Golden Ratio? Since the Golden Ratio is supposed to be the most pleasing to the eye, why aren't all cans "Golden Cans"?

**40.** Are Golden Rectangles really more pleasing to the eye? To help answer this question, do the following:

(a) On separate cards cut out rectangles that are 1 cm by 5 cm, 2 cm by 5 cm, 3 cm by 5 cm, 4 cm by 5 cm, and 5 cm by 5 cm.

(b) Place the cards in a random order and ask ten people to select the rectangle that they find the most pleasing or attractive.

(c) Tabulate your results and calculate the percentage of the people choosing each rectangle.

(d) Use your results to answer the question posed at the beginning of the problem.

(*Note:* The 3-by-5-cm rectangle is approximately a Golden Rectangle.)

**41.** Consider a parallelogram whose sides form the Golden Ratio (i.e., 3 cm by 5 cm). The shape of the parallelogram is determined by the angle between two adjacent sides as shown below.

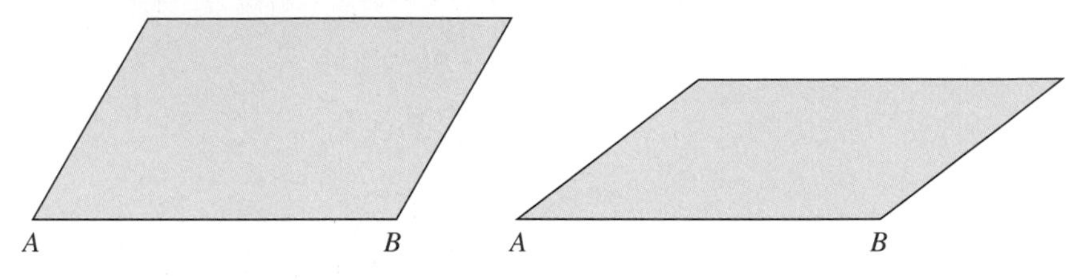


To make the parallelogram even more ''golden,'' let us also make the ratio of the measures of $\angle A$ to $\angle B$ approximately the Golden Ratio. Such a parallelogram would have $\angle A \approx 69°$ and $\angle B \approx 111°$. Is such a ''Golden Parallelogram'' more pleasing to the eye than other parallelograms? Construct an experiment to answer that question. (See Problem 40.)

**42.** In 1876, psychologist Gustav Fechner also did experiments with ellipses that had the ratio of the length of the major axis to the length of the minor axis approximately equal to the Golden Ratio. Again, he discovered that people found that ellipses close to ''Golden Ellipses'' were more pleasing to the eye. Use the ellipses shown to test Fechner's findings in present society. (*Note:* The first ellipse is the ''Golden Ellipse.'')

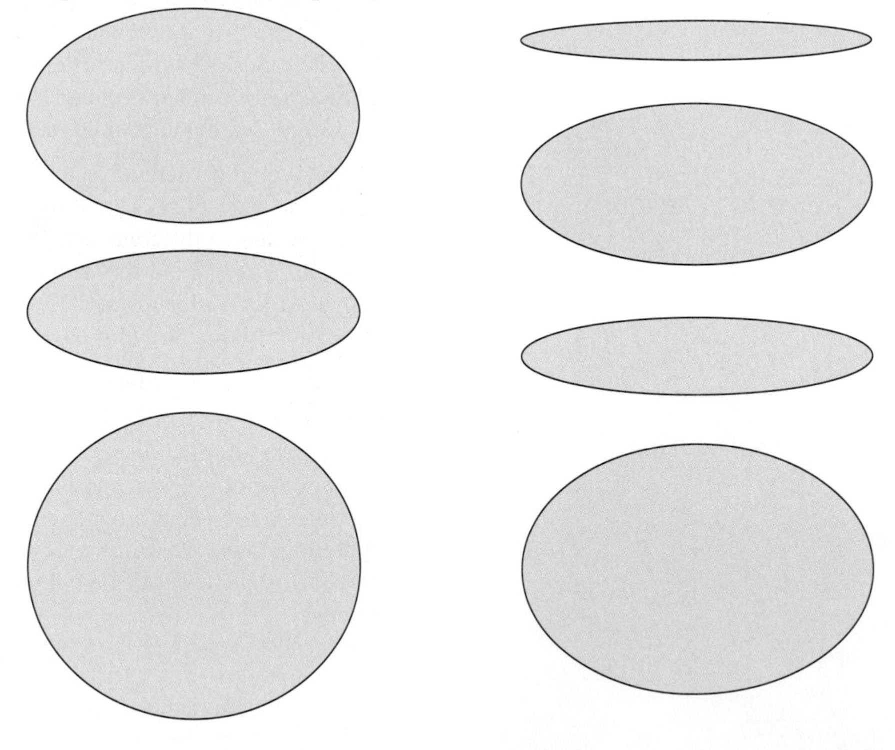

**43.** In Problem 41, find the exact value for $\angle A$ so that the ratio of $\angle A$ to $\angle B$ is the Golden Ratio. (*Hint:* Use the fact that $\angle A + \angle B = 180°$, and the ratio of $\angle B$ to $\angle A$ is $(1 + \sqrt{5})/2$.

**44.** The Golden Ratio $\phi$ is the only positive number that if decreased by 1 equals its reciprocal. Show algebraically that $\phi - 1 = \frac{1}{\phi}$.

**45.** In Example 5, we showed that the ratio of terms in the Fibonacci sequence approaches the Golden Ratio. Consider any sequence in which we choose any two starting numbers and generate the terms of the sequence by adding two consecutive numbers as in the Fibonacci sequence. For example, if we chose 5 and 2 as the starting numbers, the sequence would be 5, 2, 7, 9, 16, 25, 41, 66, 107, . . . . Examine the ratio of successive terms of any such sequence, as we did for the Fibonacci sequence in Example 5. What do you notice about successive ratios of terms?

Works of modern artisans employing polygons and stars. (*Left*) Nautical Stars pieced quilt by Judy Mathieson. (*Right*) Ukrainian Easter eggs (Pysanky) by Nancy Eddinger. (Courtesy of the artists)

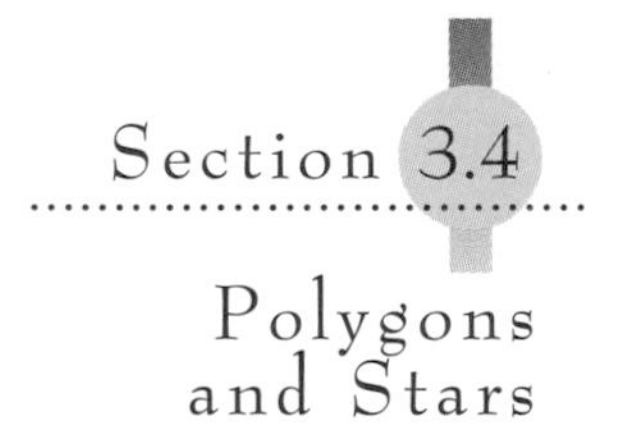

## Section 3.4 Polygons and Stars

The Golden Rectangle is just one type of a class of geometric objects called polygons. Polygons and their properties are carefully examined in most treatments of geometry. In this section, we will spend most of our efforts investigating how polygons can be used to create some interesting geometric designs. A **polygon** is a closed figure in a plane formed by line segments that intersect each other only at their end points. The following are examples of polygons.

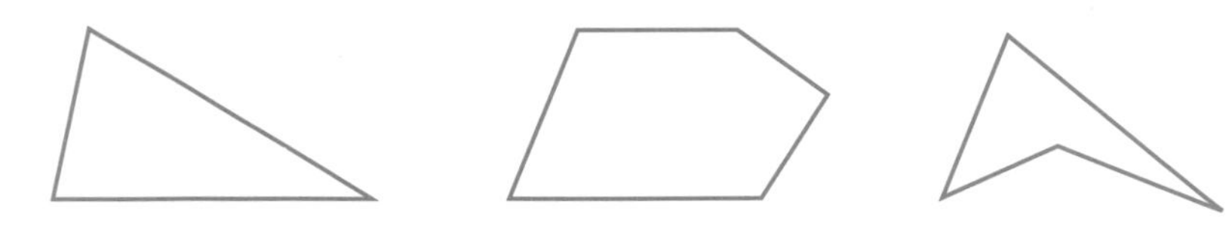

The following figures are not considered polygons.

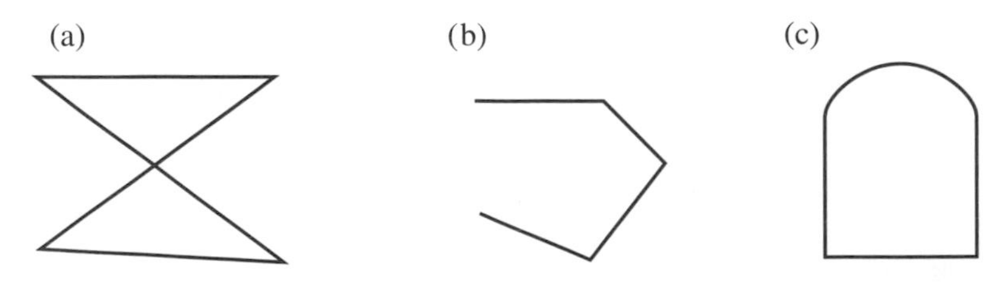

### Example 1

Explain why each of the figures shown in (a), (b), and (c) is not a polygon.

**Solution:** Each figure has a property that contradicts the definition of a polygon.

(a) The sides intersect at a point other than at the end points.
(b) It is not a closed figure.
(c) The top is a curve, not a line segment. ■

The line segments that form a polygon are called its **sides**, and the points at which the sides meet are called **vertices** (plural of ''**vertex**''). Polygons are named by the number of sides they contain. Specific names are given to polygons with from 3 to 12 sides. Other polygons are usually referred to as $n$-gons, where $n$ represents the number of sides of the polygon. For example, a polygon with 13 sides is called a 13-gon.

Polygons are classified as concave or convex. A polygon is **concave** if an extension of one of its sides enters the interior of the polygon; a polygon is **convex** if the extensions of its sides do not enter the interior of the polygon. Some polygons are classified as **regular polygons**. These polygons have sides of equal length and angles of equal measure. Some common objects with shapes that are regular polygons are pizza boxes, stop signs, and honeycombs.

The following is a list of the first ten polygons and their names. In column I, you will find regular polygons; in column II, convex polygons; and in column III, concave polygons.

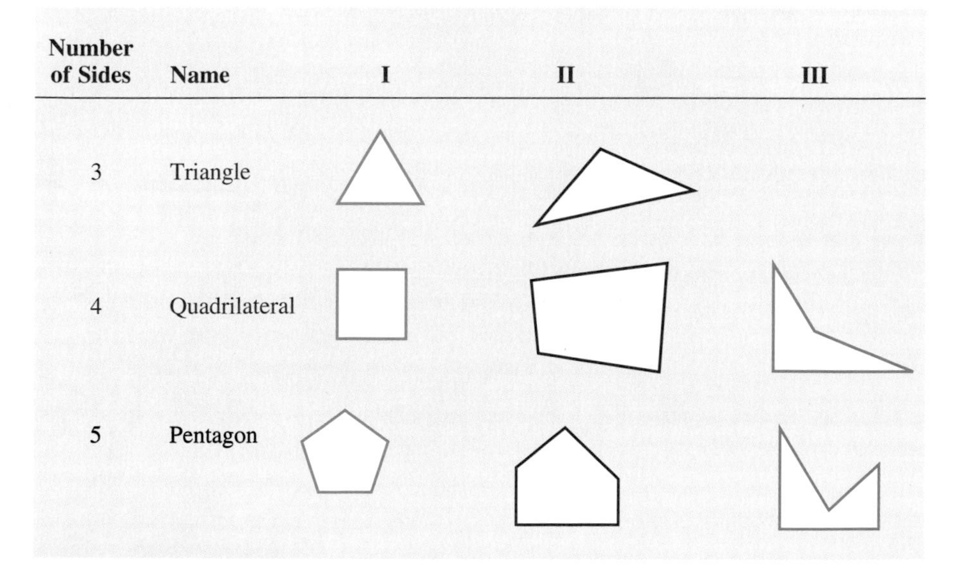

| Number of Sides | Name | I | II | III |
|---|---|---|---|---|
| 3 | Triangle | | | |
| 4 | Quadrilateral | | | |
| 5 | Pentagon | | | |

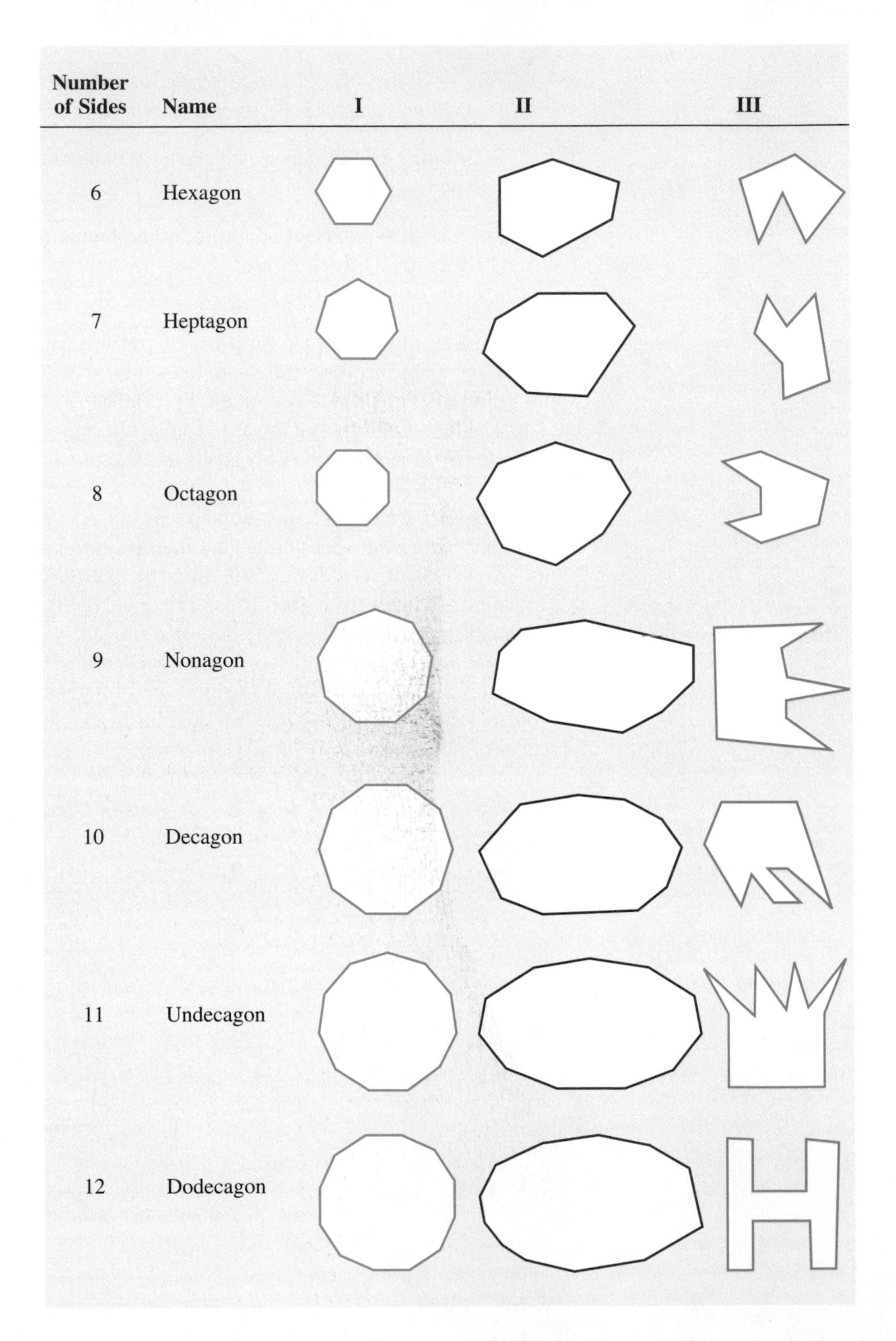

| Number of Sides | Name | I | II | III |
|---|---|---|---|---|
| 6 | Hexagon | | | |
| 7 | Heptagon | | | |
| 8 | Octagon | | | |
| 9 | Nonagon | | | |
| 10 | Decagon | | | |
| 11 | Undecagon | | | |
| 12 | Dodecagon | | | |

## Example 2

In each figure, (a) give its name and classify it as concave or convex and (b) determine if it is a regular polygon.

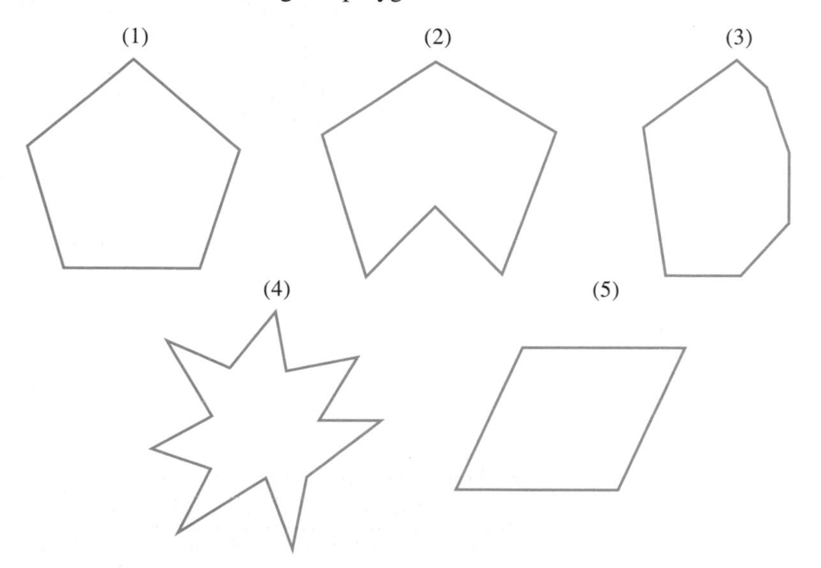


**Solution:**

(a) (1) Pentagon, convex
(2) Hexagon, concave
(3) Heptagon, convex
(4) 14-gon, concave
(5) Quadrilateral, convex

For example, Figure 2 is concave because an extension of the side $\overline{AB}$ enters the interior of the hexagon.

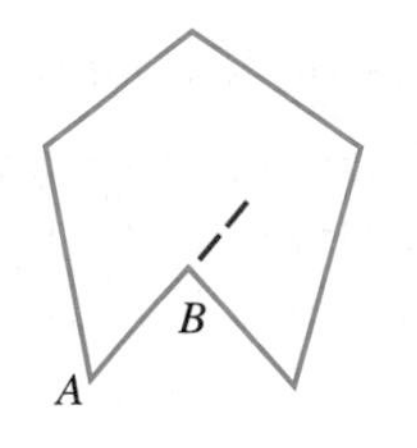


(b) The only regular polygon is Figure 1. Figure 5 is not a regular polygon because its angles are not of equal measure. ■

## *Polygons and Their Angles*

The angles of a polygon have a unique property. The sum of the angles for each type of polygon is always the same. For example, in Section 3.1, we learned in the triangle-sum theorem (T5) that the sum of the angles of a triangle is 180°. This means that in any triangle the sum of its three angles equals 180°. A proof of this theorem can be found in the original work, *Elements* by Euclid, and in most geometry textbooks. We will not present a proof for the triangle-sum theorem here, but we will give you a visual justification for the theorem. If you fold $\triangle ABC$ along the dotted lines as shown in the figure, you will obtain the following figure.

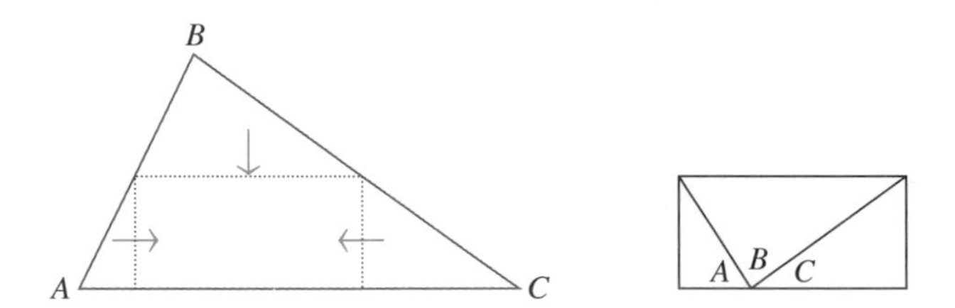


In the figure, $\angle A$, $\angle B$, and $\angle C$ form a straight angle. Since straight angles measure 180°, $\angle A + \angle B + \angle C = 180°$. The fact that the sum of the angles of a triangle is 180° will allow us to find the sum of the angles for any polygon. To determine this sum, we will cover a given convex polygon with nonoverlapping triangles originating at one vertex and use these triangles to determine the sum of the angles of the polygon.

| Sides | Triangles | Sum of Angles |
|---|---|---|
| **4** | 2 | $180°(2) = 360°$ |
| **5** | 3 | $180°(3) = 540°$ |
| **6** | 4 | $180°(4) = 720°$ |

These examples show a relationship between the number of nonoverlapping triangles that cover a convex polygon and the number of sides. The number of triangles is 2 less than the number of sides. Thus, if the polygon has $n$ sides, there are $n - 2$ triangles. Since the sum of the angles of each triangle is 180°, the sum of the angles ($S$) of the convex polygon is as follows.

**Sum of the Angles of a Polygon**

$$S = 180(n - 2)$$

(*Note:* this formula also works for concave polygons.)

If the polygon is a regular polygon, each of its angles has the same measure. Thus, to find the measure of each angle ($A$) of a regular polygon, we take the sum of its angles and divide by the number of angles or sides ($n$) in the polygon.

**Each Angle of a Regular Polygon**

$$A = \frac{180(n - 2)}{n}$$

**Example 3**

Find the sum of the angles in a regular decagon and find the measure of each of its angles.

**Solution:** A decagon has 10 sides. Thus, we use $n = 10$ in the formulas for the sum of the angles of a polygon and each angle of a regular polygon.

$$\begin{aligned} S &= 180(n - 2) \\ &= 180(10 - 2) = 180(8) = 1440° \\ A &= \frac{180(n - 2)}{n} \\ &= \frac{180(10 - 2)}{10} = \frac{1440}{10} = 144° \end{aligned}$$

To complete our discussion of polygons we will investigate methods for actually drawing them and creating designs based on them. Polygons have probably been admired and studied since prehistoric times. These shapes are present in almost every object made by man. The simple beauty and completeness of polygons, especially regular polygons, have stimulated many creative designs over the centuries. Designs using polygons can be found in religious symbols such as the Star of David; in logos such as the pentagon emblem of the Chrysler Corporation; in modern quilts, flooring, and wallpaper; and on wheel covers of cars and trucks. We are confident that this excursion into geometry will enable even the ''unartistic'' to create some beautiful results.

## Drawing Regular Polygons

If you study the regular polygons shown earlier in this section, you will notice that the vertices are equally spaced and that, as the number of sides increases, the regular polygon appears to be more and more like a circle. This suggests a method for actually creating regular polygons.

For example, to draw a regular pentagon, find five equally spaced points on a circle and use those points as vertices of the regular pentagon. We will use the fact that if an arrow with its end point at the center of a circle rotates completely around the circle like the second hand of a clock, it passes through 360°. Therefore, to create five equally spaced points on the circle, we simply divide 5 into 360. If we take the answer, 72, and mark off five angles at the center of the circle (central angles) of 72°, we find the desired points on the circle to use as vertices of the regular pentagon.

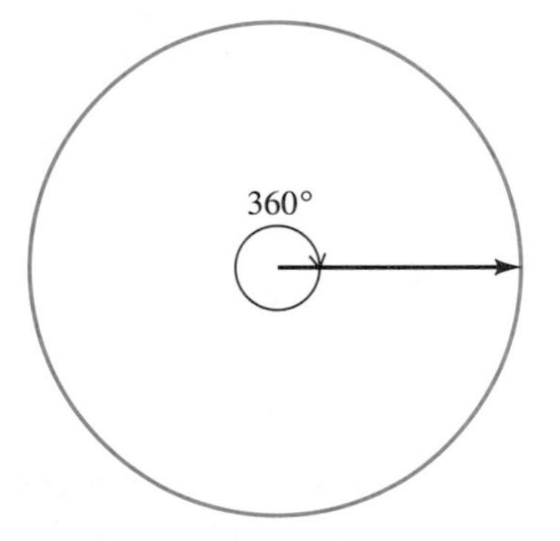


Let's actually draw a regular pentagon. With a compass, draw a large circle. Using a protractor, measure five 72° angles at the center of the circle. Extend the angles until they intersect the circle. Connect the five points marked on the circle to form the pentagon.

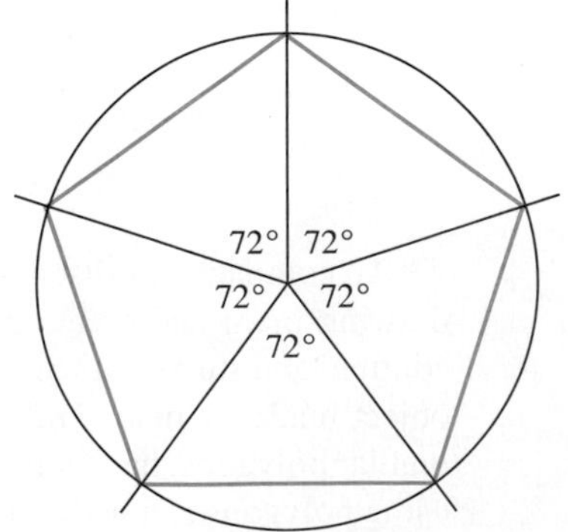

### Example 4

Draw a regular octagon.

**Solution:** Using the method described in the creation of the pentagon, we should mark off eight central angles of 45° each, since $360 \div 8 = 45$.

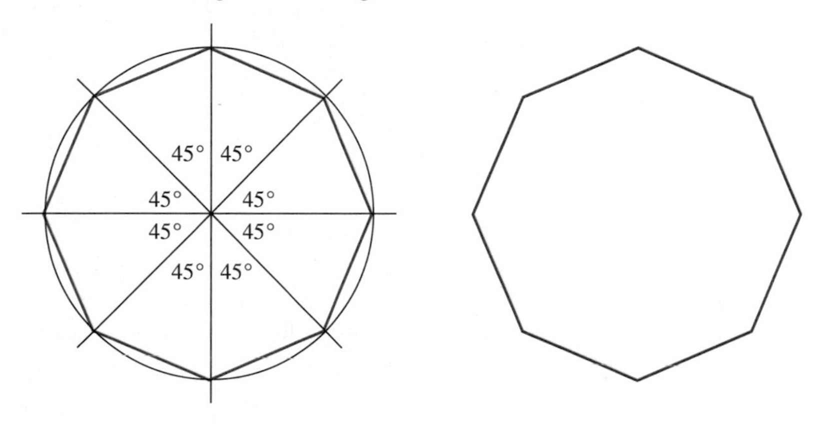


## Drawing Regular Stars

Finding equally spaced points on a circle can also be used to create some interesting geometric "stars." Consider seven equally spaced points on a circle obtained by marking off seven central angles of approximately 51.4° ($360 \div 7 \approx 51.429$). Instead of connecting each point in order to create a regular heptagon, connect every second point or every third point.

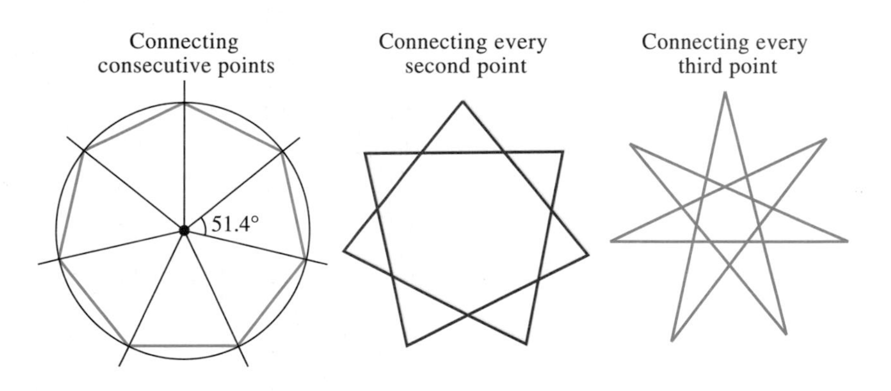


### Example 5

Draw a 12-point star by connecting every fifth point of a regular dodecagon.

**Solution:** Find 12 equally spaced points on a circle by marking off 12 central angles of 30° ($360 \div 12 = 30$).

Mark off each 30° central angle. Connect every fifth point.

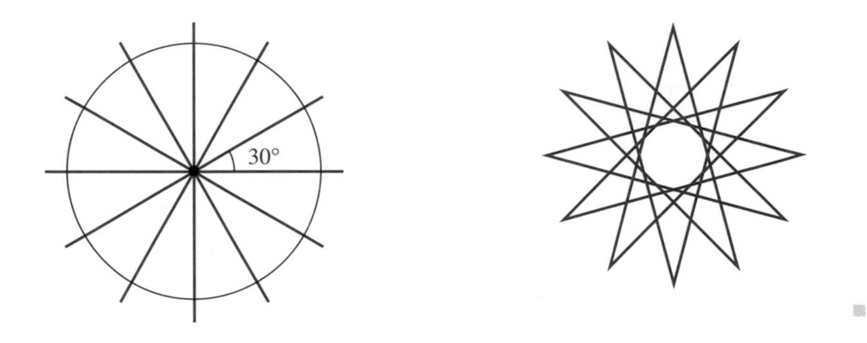


The process of locating equally spaced points on a circle can be used to create regular polygons and a variety of symmetrical stars. That technique combined with others such as those listed below can be used to create some interesting geometric designs.

1. Construct stars using points on a circle that are not equally spaced.
2. Create the illusion of curved lines by marking off equally spaced points on both sides of an angle and systematically connecting points on one side of the angle to points on the opposite side.

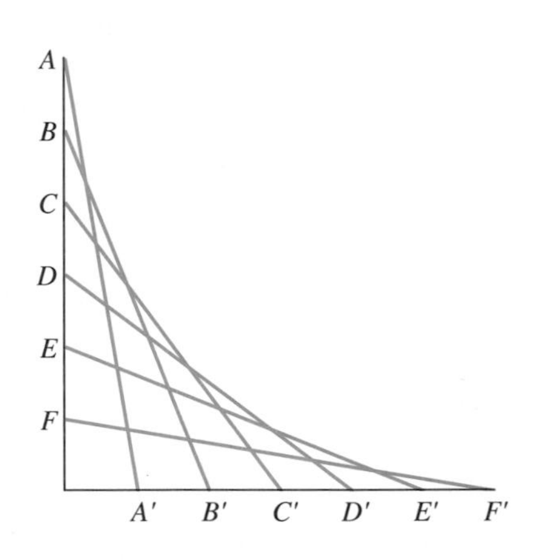


3. Use different colors to systematically shade various regions of the star.
4. Combine different polygons and stars along with various lines in the same design.

The results of experimenting with these and other techniques can be surprising. Don't be afraid to let your creative juices flow. Geometry can be beautiful. What follows is an example of a geometric design using some of those techniques.

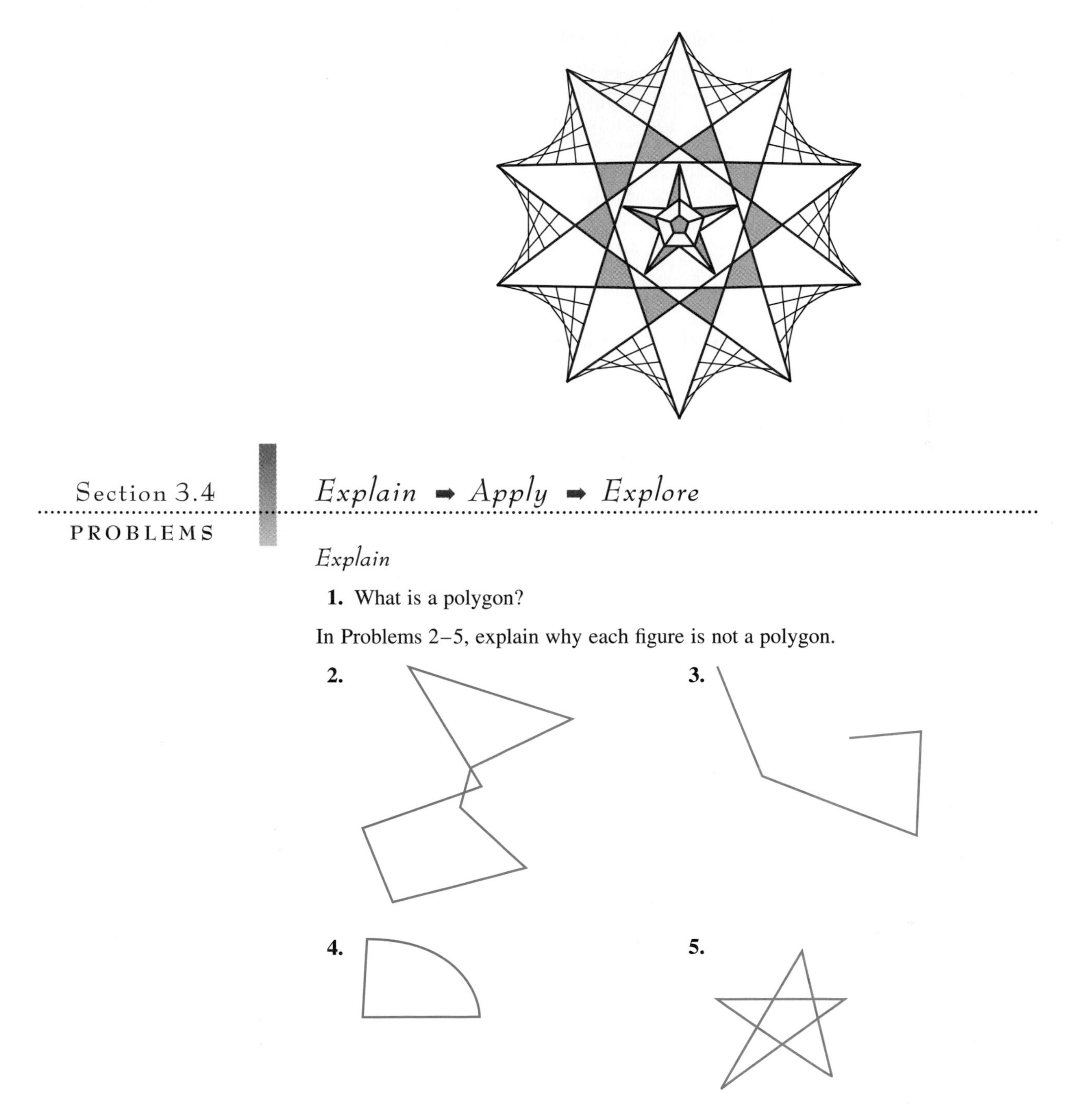

Section 3.4
PROBLEMS

## *Explain* ➡ *Apply* ➡ *Explore*

### *Explain*

**1.** What is a polygon?

In Problems 2–5, explain why each figure is not a polygon.

**2.**

**3.**

**4.**

**5.**

**6.** What is a regular polygon?

In Problems 7–10, explain why each figure is not a regular polygon.

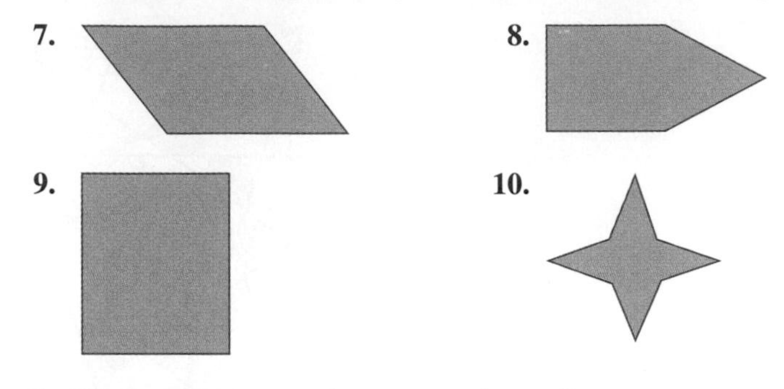


**11.** What are convex and concave polygons?

In Problems 12–15, give the name of each polygon and explain why it is concave or convex.

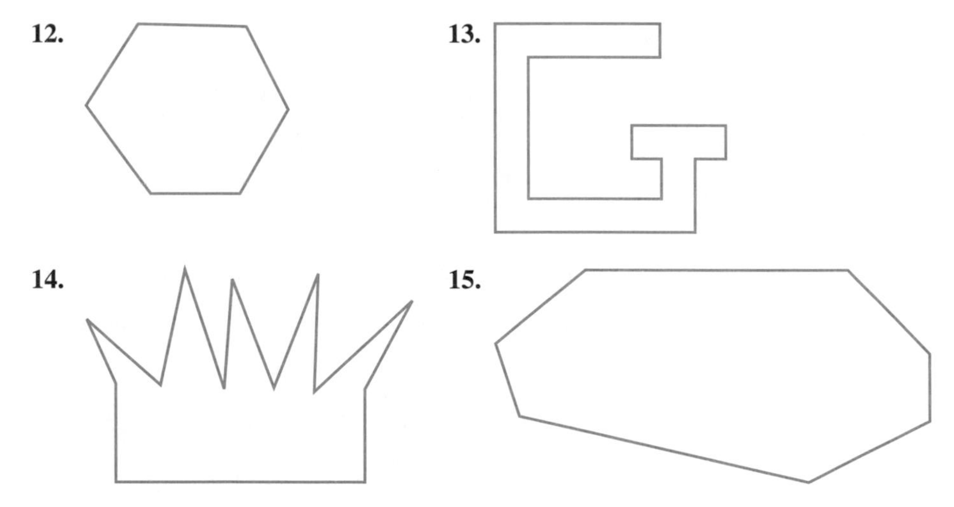


**16.** What steps should you follow to draw a regular polygon of $n$ sides?

**17.** What steps should you follow to draw a regular star of $n$ points?

**18.** How do you find the sum of the angles of a polygon?

**19.** How do you find the measure of each angle of a regular polygon?

## *Apply*

In Problems 20–26, make a sketch of each polygon and find the sum of the angles of each polygon.

**20.** (a) Concave quadrilateral
(b) Convex quadrilateral

**21.** (a) Concave pentagon
(b) Convex pentagon

**22.** (a) Concave hexagon
(b) Convex hexagon

**23.** (a) Concave 13-gon
(b) Convex 13-gon

**24.** (a) Concave 14-gon
(b) Convex 14-gon

**25.** (a) Concave decagon
(b) Convex decagon

**26.** (a) Concave dodecagon
(b) Convex dodecagon

**27.** Complete the following chart to determine the central angle needed to find equally spaced points of a circle.

| Number of Points | Central Angle | Number of Points | Central Angle |
|---|---|---|---|
| 5 | ________ | 13 | ________ |
| 6 | ________ | 14 | ________ |
| 7 | ________ | 15 | ________ |
| 8 | ________ | 16 | ________ |
| 9 | ________ | 18 | ________ |
| 10 | ________ | 20 | ________ |
| 11 | ________ | 30 | ________ |
| 12 | ________ | 36 | ________ |

In Problems 28–35, draw the regular polygon indicated and find the measure of each angle of the polygon.

**28.** Regular hexagon

**29.** Regular octagon

**30.** Regular nonagon

**31.** Regular undecagon

**32.** Regular 18-gon

**33.** Regular 20-gon

**34.** Regular 15-gon

**35.** Regular 14-gon

### *Explore*

In Problems 36–42, draw the regular star indicated. Color different regions in the star to give it an ''artistic'' effect.

**36.** A 10-point star connecting every second point

**37.** A 10-point star connecting every third point

**38.** A 12-point star connecting every fifth point

**39.** A 15-point star connecting every sixth point

**40.** A 20-point star connecting every fifth point

**41.** A 20-point star connecting every ninth point

**42.** A 30-point star connecting every tenth point

**43.** The designs of many wheel covers on cars and trucks use the techniques discussed in this section. Examine and sketch the designs of three different wheel covers, noting the make and model of the vehicle.

**44.** Designs for wheel covers on cars and trucks can be created by using the techniques discussed in this section. Use these techniques to design your own wheel cover.

**45.** Create an original geometric design using a combination of any techniques presented in this section.

***46.** Find a proof of the triangle-sum theorem in Euclid's *Elements* or in a geometry text book. What theorems are used in the proof?

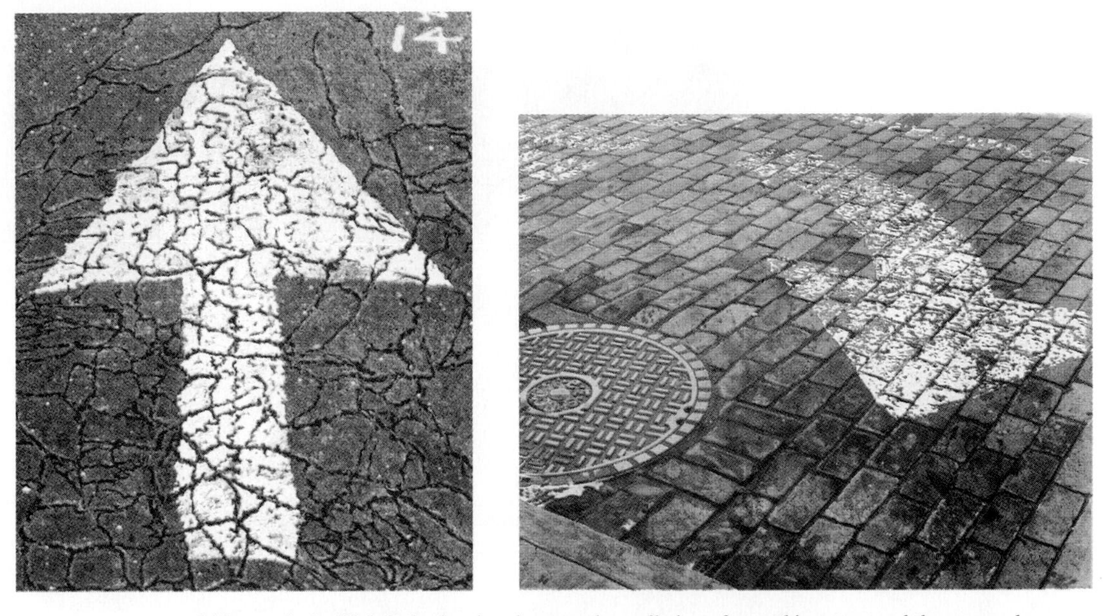


Lithographs by Rick Dula showing the natural tessellation of a cracking street and the manmade tessellation of cobblestones. (Courtesy of Magnolia Editions)

## Section 3.5 Tessellations

The study of polygons in the previous section enabled us to create various geometric designs based on placing equally spaced points on a circle. In this section we investigate combining polygons into geometric patterns called tessellations. The word "tessellation" comes from the Latin word *tessella*, which is a small square tile used in Roman mosaics. A **tessellation** is a pattern of one or more congruent shapes that covers an area in a plane without overlapping or leaving any gaps. A tessellation actually covers the entire plane if its basic pattern is continued in all directions. However, we will only be concerned with using tessellations to cover a small region in the plane. Simple tessellations or tilings are very common. They can be seen, for example, on walls, on clothing, in beehives, or in works of art (Fig. 3.15).

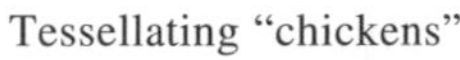
Tessellating "chickens"

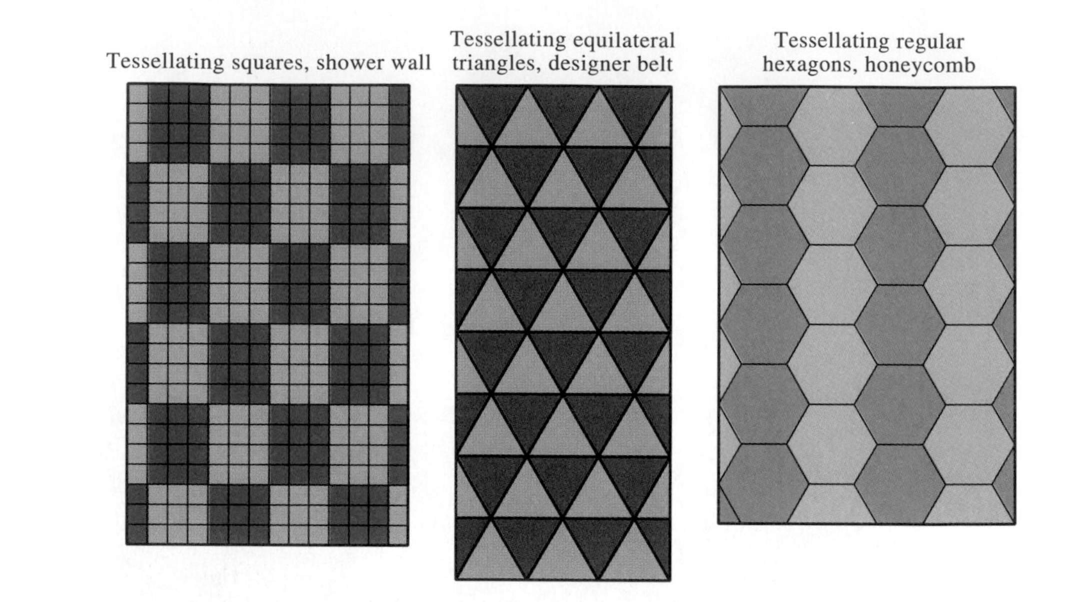

Tessellating squares, shower wall
Tessellating equilateral triangles, designer belt
Tessellating regular hexagons, honeycomb


Figure 3.15

Tessellations were used on window lattices in China; on painted ceilings in Egypt; on the mosques in Granada, Spain; on textile and basket patterns of the native peoples of Peru, India, Ghana, Zaire, America, Mexico, and other countries; and in the decorative arts of the ancient Greeks, Romans, Arabs, Japanese, Persians, and Celts. Tessellations can be observed in nature in the cells of an onion skin, in the design of a spider's web, and in the arrangement of seeds in a sunflower. Tessellations can be seen in our modern society as designs on wallpaper, linoleum, parquet flooring, ceramic tiles, patchwork quilts, crocheted placemats, and lace tablecloths. Tessellations have been made popular by the work of Dutch artist M. C. Escher. Inspired by the ornamental art of the Moors, he created intriguing designs that truly integrate mathematics and art. Tessellations can also be seen in the work of modern kinetic and optical artists such as Bridget Riley and Victor Vasarely and in the intricate tilings of Heinz Voderberg and Roger Penrose. In this section we introduce some of the principles and methods of creating basic tessellations.

## Tessellating with a Regular Polygon

The only regular polygons that tessellate the plane are the equilateral triangle, the square, and the regular hexagon, as shown in Figure 3.15. Any of the other regular polygons would overlap at a vertex and therefore not tessellate the plane.

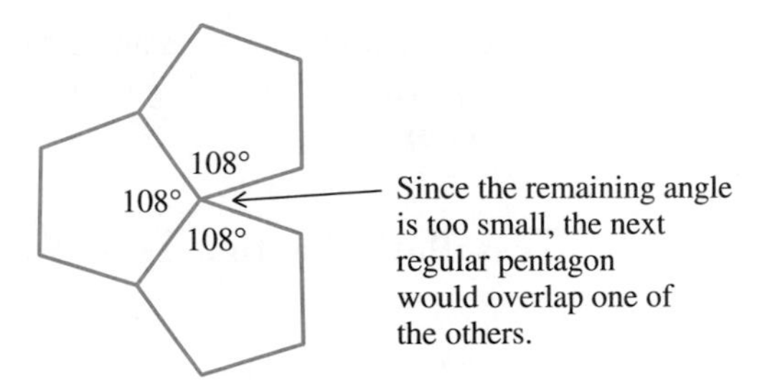


Since the remaining angle is too small, the next regular pentagon would overlap one of the others.

For example, if regular pentagons were placed around one vertex, as shown in the figure, overlapping would occur by the fourth regular pentagon. A similar situation occurs for any other regular polygon with the number of sides other than 3, 4, or 6. Table 3.5.1 will be helpful in verifying that fact. Its proof is a starred exercise in this section's problem set.

**Table 3.5.1** *Regular Polygons*

| Number of Sides | Measure of Each Angle | Number of Sides | Measure of Each Angle |
|---|---|---|---|
| 3 | 60° | 11 | $147\frac{3}{11}°$ |
| 4 | 90° | 12 | 150° |
| 5 | 108° | 15 | 156° |
| 6 | 120° | 18 | 160° |
| 7 | $128\frac{4}{7}°$ | 20 | 162° |
| 8 | 135° | 24 | 165° |
| 9 | 140° | 36 | 170° |
| 10 | 144° | 42 | $171\frac{3}{7}°$ |

## Example 1

Show that it is not possible to tessellate the plane with regular octagons.

**Solution:**

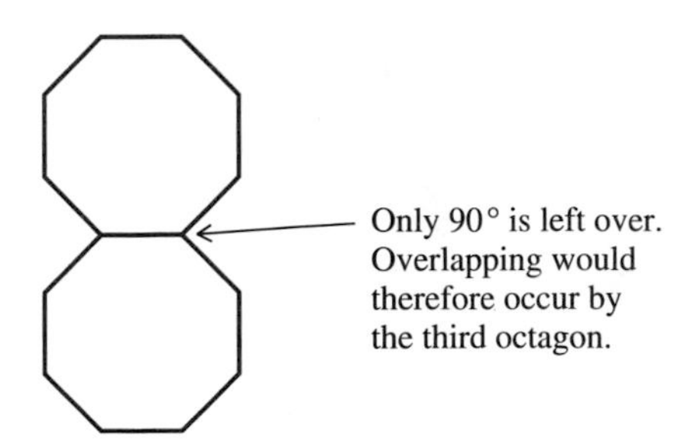

Only 90° is left over. Overlapping would therefore occur by the third octagon.


Since each angle of an octagon is 135°, two octagons placed at the same vertex would have an angle sum of 270°. That would leave 90° at the vertex. Thus, overlapping would occur if another octagon is placed at that vertex. ■

## Tessellating with a Triangle or Quadrilateral

We have seen that only three regular polygons tile the plane. If we allow the use of nonregular polygons, any triangle or quadrilateral can tessellate the plane. The basis for this is the fact that in order to tessellate the plane, there must be no overlapping of polygons and no gaps left by the polygons. Overlapping occurs when the sum of the angles placed around a vertex is greater than 360°, and gaps occur when the sum is less than 360°. Since the sum of the angles of a triangle is 180° and the sum of the angles of a quadrilateral is 360°, triangles and quadrilaterals can be placed around a vertex so that the sum of the angles is exactly 360°. In the case of the triangle, we need to use each angle of the triangle twice. In the case of the quadrilateral, we need to use each angle only once around a vertex.

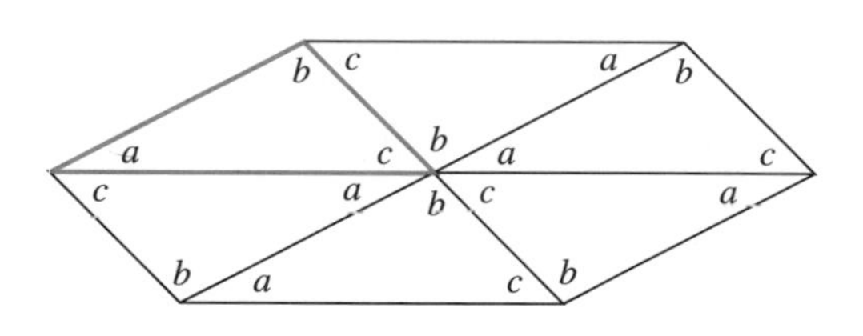

**Around the vertex at the center of the figure the three angles of the triangle appear twice. Since $a + b + c = 180°$, the sum of the six angles is exactly 360° and no overlapping occurs.**

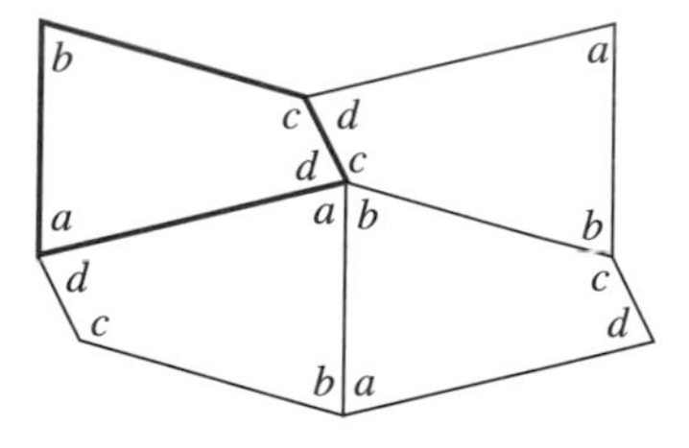

**Around the vertex at the center of the figure the four angles of the quadrilateral appear. Since $a + b + c + d = 360°$, no overlapping occurs.**

By actually tessellating the plane with different triangles and quadrilaterals, you will be convinced that any triangle or quadrilateral tessellates the plane.

**Example 2**

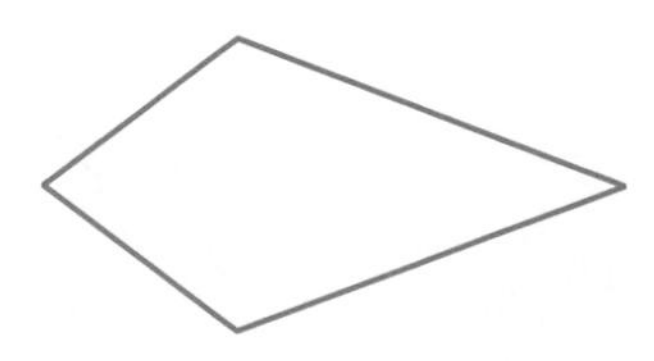

Arrange the angles of the kite-shaped quadrilateral in the figure so that four of them can be placed around a vertex without overlapping each other.

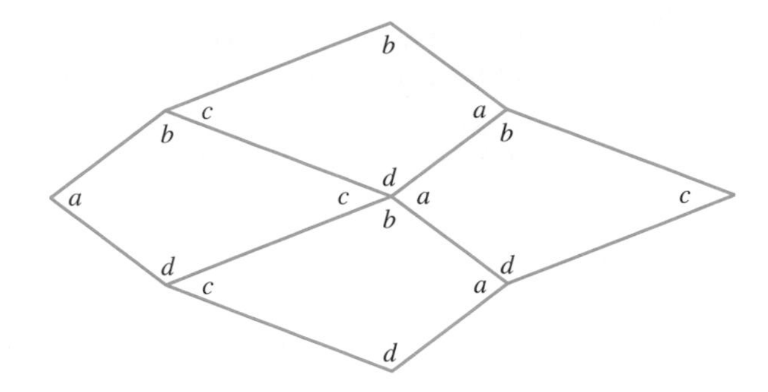


**Solution:**
By sliding and turning over the quadrilateral, we can arrange four of them around one vertex as shown. ■

You are probably wondering how one actually tessellates an area with a triangle or quadrilateral. This process, as you may have guessed, requires some mathematical operations that move geometric figures: translations and reflections.

**Translations**
**The triangle is translated a distance that is equal to each of its sides in the direction of each side.**

**Reflections**
**The triangle is reflected across one of its sides, across a horizontal line, and across a vertical line.**

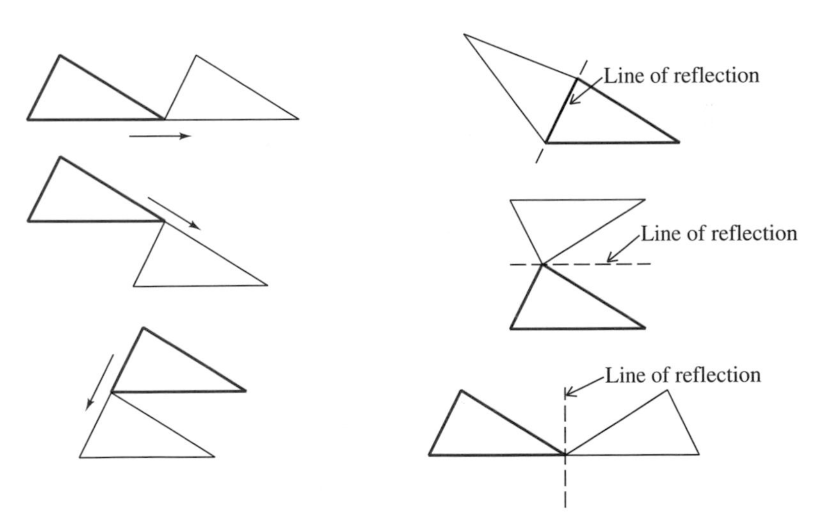


A **translation** slides an object a certain distance in a certain direction. A **reflection** gives the mirror image of an object across a certain line called the line of reflection. A reflection has the effect of turning an object over and placing it on the opposite side of its line of reflection.

Tessellations with a triangle can be formed by using translations or a combination of reflections and translations. Let us examine both methods.

### *Tessellating with a Triangle Using Translations*

To tessellate an area with a triangle, we can translate the triangle in the direction of two sides of the triangle a distance that is equal to the sides of the triangle. The tessellation can be created by repeatedly using two translations.

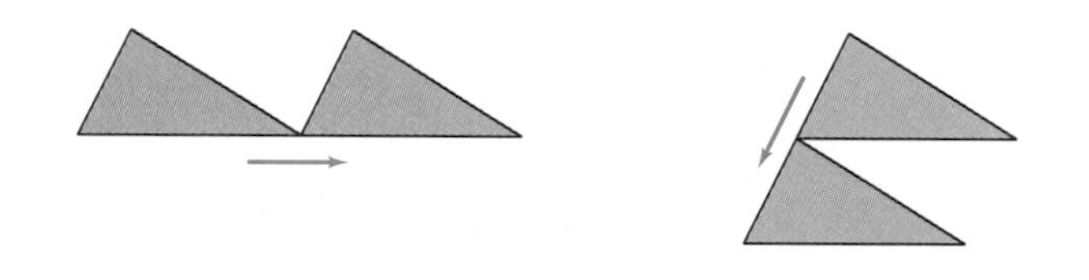

The details of actually creating the tessellation are as follows:

**1.** Create a primary line of triangles by repeatedly translating the original triangle in the direction of one of the sides of the triangle.

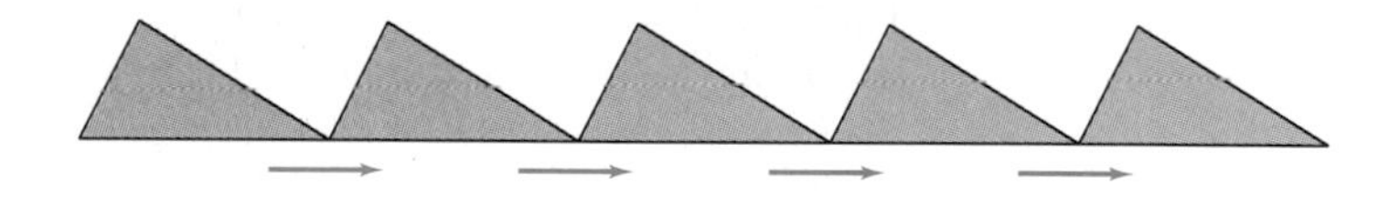

**2.** Using each triangle in that primary line repeatedly translate in the direction of another side of the triangle.

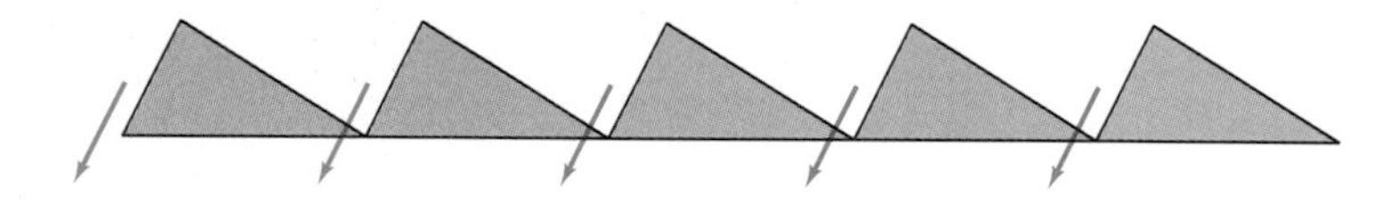

**3.** Connect any missing lines and the tessellation will be complete. You will notice that in the process blank spaces created half of the triangles.

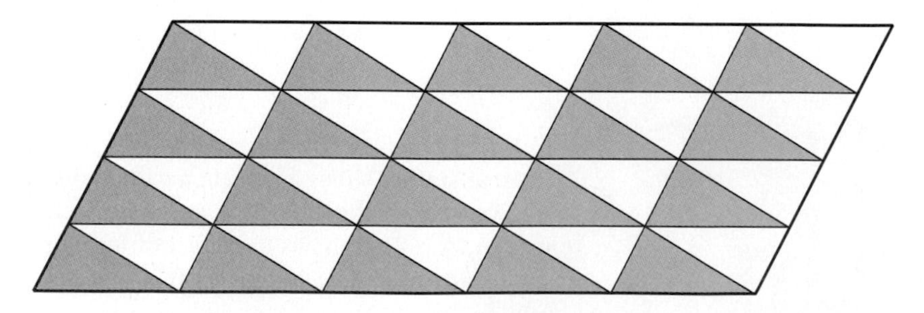

## Example 3

Using the method shown in the previous example, create a tessellation with the triangle in the figure.

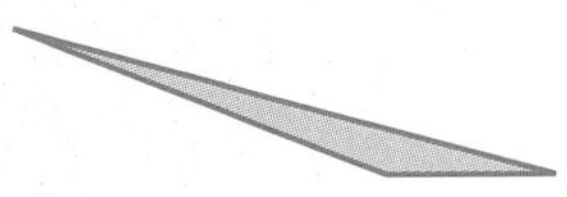

**Solution:** We can tessellate with the triangle by translating the triangle in the two directions shown.

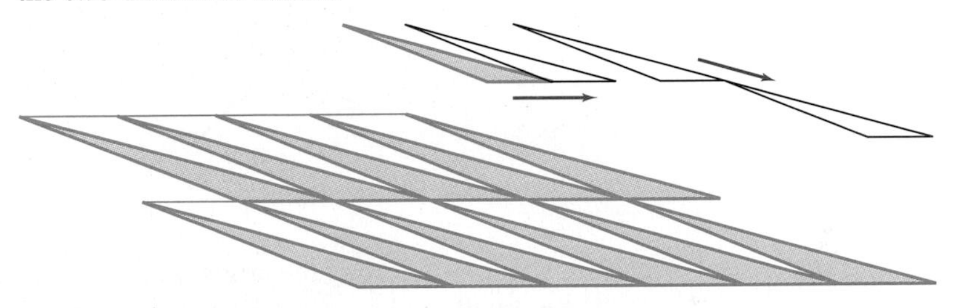

### *Tessellating with a Triangle Using Translations and a Reflection*

A second method of tessellating an area using a triangle involves translations and a reflection. The steps for creating such a tessellation from $\triangle ABC$ are as follows:

**1.** Reflect the triangle $\triangle ABC$ across line $AC$, forming quadrilateral $ABCB'$.

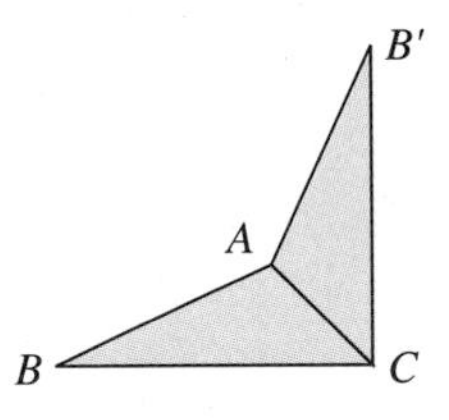


**2.** Use translations of quadrilateral $ABCB'$ in the direction of its diagonals $\overline{AC}$ and $\overline{BB'}$ as shown.

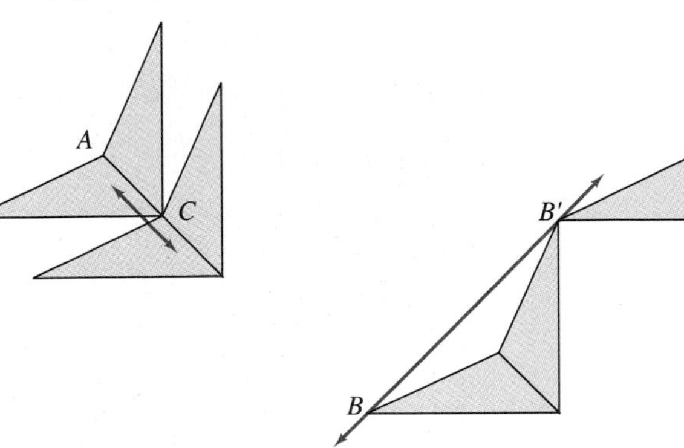

3. Create a primary line of quadrilaterals using the first translation, and then translate each of those quadrilaterals using the second translation. The blank spaces will form quadrilaterals tessellating in a direction opposite to that of ray $AC$.

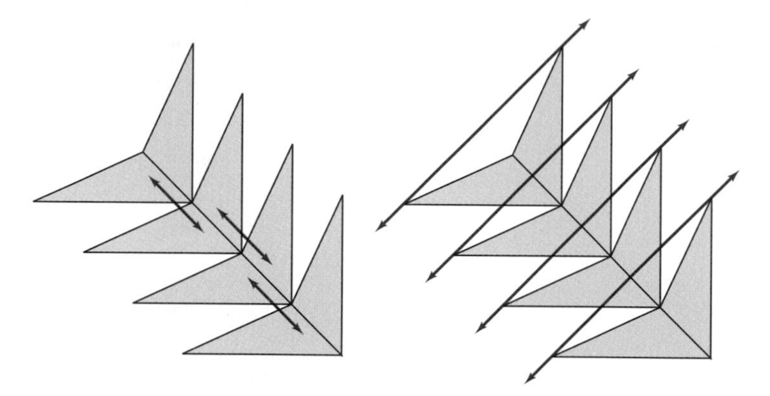

The result of reflecting the original triangle and then translating the resulting quadrilateral will produce a different tessellation than was previously created by using only translations.

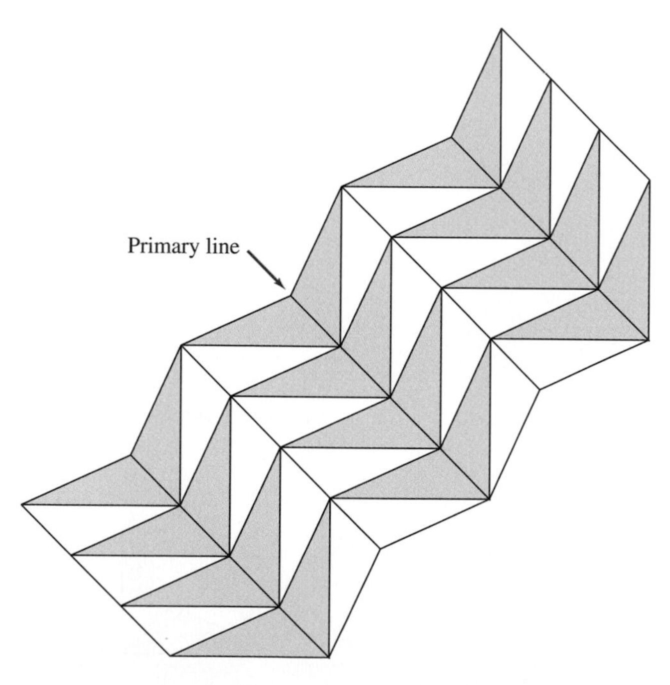


This method will also make it possible to tessellate the plane with any quadrilateral. All you need to do is translate the quadrilateral in the direction of its diagonals.

**Example 4**

Tessellate an area with the quadrilateral shown.

**Solution:** To tessellate the plane with the quadrilateral, repeatedly translate the quadrilateral in the direction of both diagonals as shown. ■

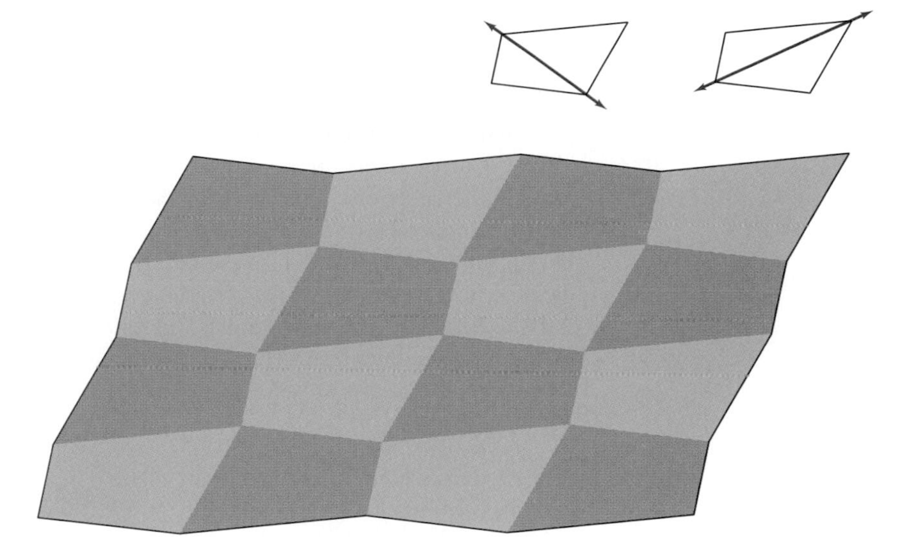

## Tessellating with Other Polygons

It is possible to tessellate the plane with certain pentagons and hexagons, but it is impossible to tessellate the plane with single polygons that have more than six sides. Some examples of tessellating pentagons and hexagons follow.

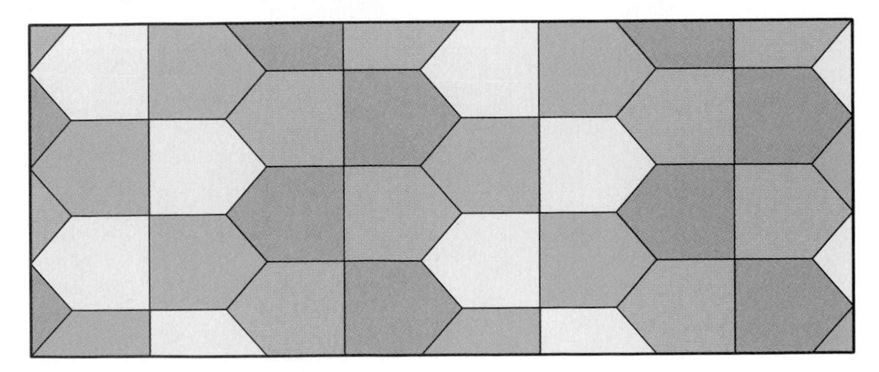

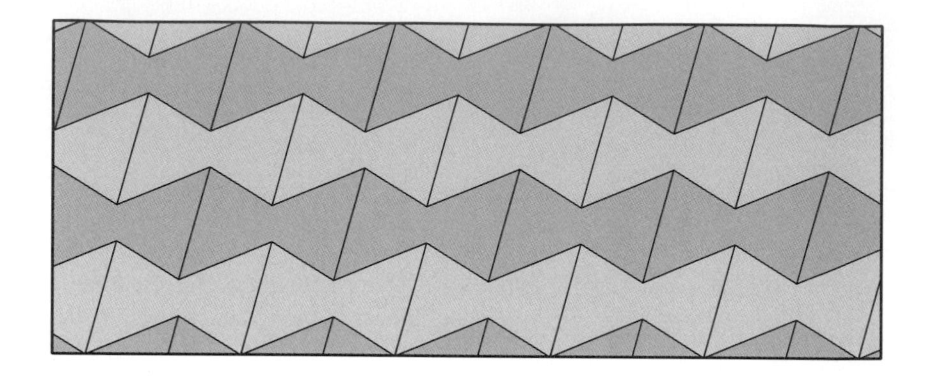

## Tessellating with More than One Polygon

The plane can also be tessellated with more than one polygon. There are a multitude of ways to tessellate an area by using nonregular polygons, but there are only 21 possible tessellations using combinations of regular polygons. Remember that for polygons to tessellate the plane, the sum of the angles at any vertex must be exactly 360°. Table 3.5.1, which lists the size of each angle of a regular polygon, will be helpful in determining which combination of regular polygons can be used to tessellate the plane. The tessellations that follow are examples of using more than one polygon to tile the plane.

### Example 5

Explain why it is possible to tessellate a plane by using two regular octagons and one square at each vertex.

**Solution:** The measure of an angle of a regular octagon is 135°, and the measure of an angle of a square is 90°. The sum of the angles at one vertex of two octagons and a square is 360° ($135 + 135 + 90 = 360$). With such a sum there will be no overlapping of polygons. ■

The next tessellation has two regular octagons and one square at each vertex.

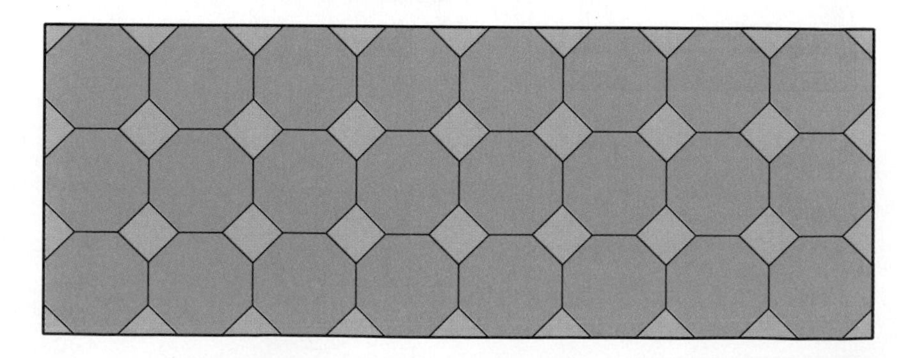

The list of ways to tessellate the plane with polygons is endless. This section has presented some basic tessellation principles. There is much more to be learned about techniques used in generating other types of tessellations and using tessellations to create Escher-like designs. The purpose of this section was simply to give you a taste of the many possibilities in this area of mathematics. We end this section with two tessellations. The first incorporates stars into a tessellation, and the second shows how simple modifications of a basic tessellation (regular hexagons) can produce some interesting results.

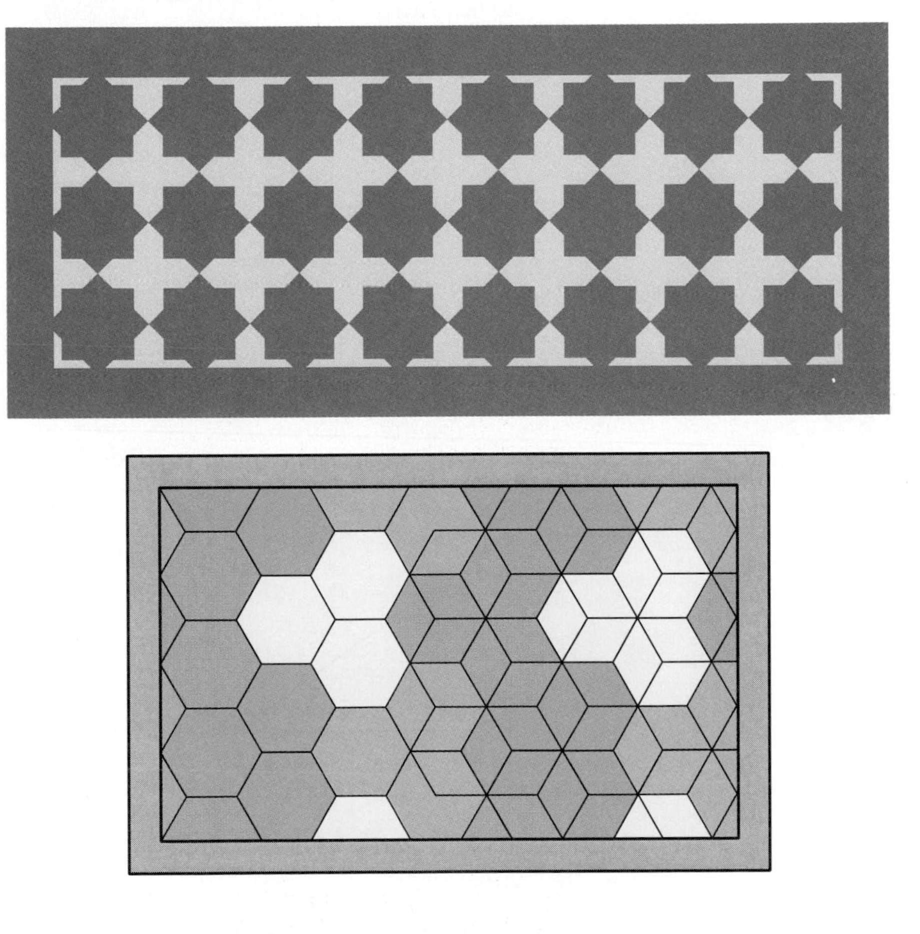

Section 3.5 PROBLEMS

## *Explain ➡ Apply ➡ Explore*

### *Explain*

**1.** What is a tessellation?

**2.** What is the origin of the word ''tessellation''?

3. Which three regular polygons can tessellate the plane?

4. Explain what must happen at each vertex for polygons to tessellate an area.

5. What is a translation?

6. What is a reflection?

7. If a given triangle is to tessellate an area with only translations, what steps need to be taken?

8. If a given triangle is to tessellate an area with only reflections and translations, what steps need to be taken?

9. How can a region be tessellated with a quadrilateral?

10. Give at least three examples of tessellations in your home.

## *Apply*

11. Show why it is impossible to tessellate the plane with regular heptagons.

12. Show why it is impossible to tessellate the plane with regular nonagons.

13. Show why it is impossible to tessellate the plane with regular decagons.

In Problems 14–17, tessellate a region with the triangle shown, using only translations.

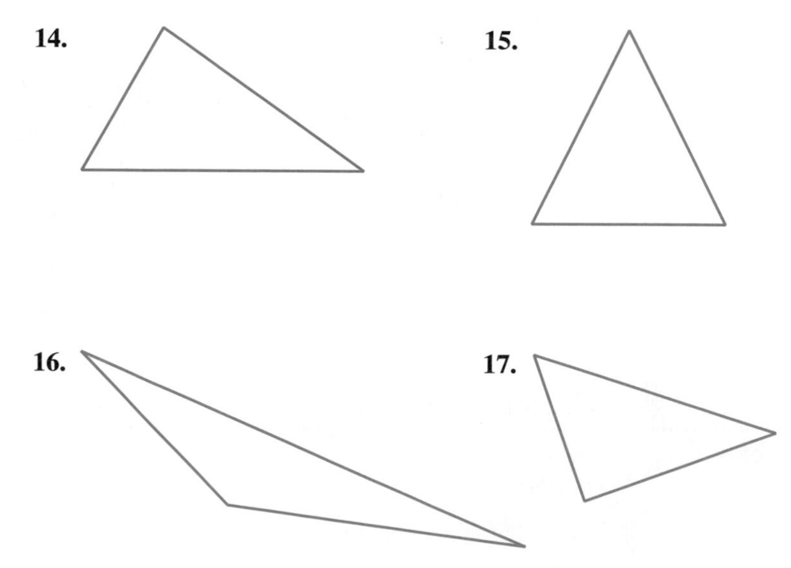

In Problems 18–21, tessellate a region with the triangle shown, using a reflection and translations.

**18.** **19.**

**20.** **21.**

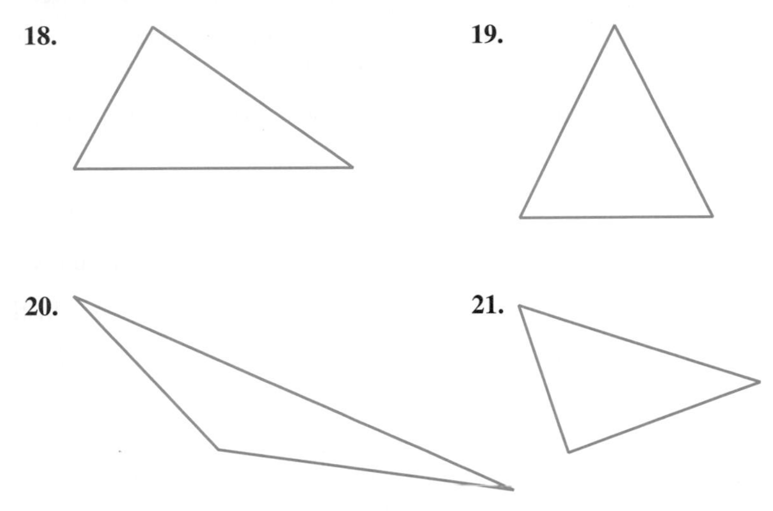

In Problems 22–25, tessellate a region with the quadrilateral shown.

**22.** **23.**

**24.** **25.**

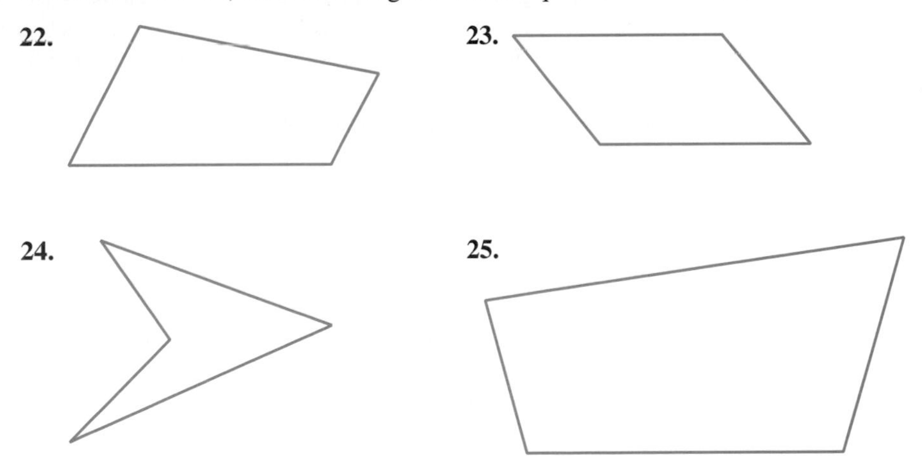

*Explore*

**26.** Explain why it is impossible to tessellate a plane with the following regular polygons located at one vertex: one regular hexagon, one regular pentagon, and one square.

**27.** Explain why it is impossible to tessellate a plane with the following regular polygons located at one vertex: two dodecagons and one octagon.

**28.** Tessellate a region with two regular hexagons and two equilateral triangles at each vertex.

**29.** Tessellate a region with one regular hexagon, two squares, and one equilateral triangle at each vertex.

**30.** Tessellate a region with one regular dodecagon, one regular hexagon, and one square at each vertex.

**31.** There are 21 possible combinations of regular polygons that can be placed around a vertex to tessellate the plane. Find at least ten of these combinations. (*Hint:* The sum of the angles around a vertex must be 360°.)

**32.** Draw sketches of the polygons for the combinations found in Problem 31.

**33.** Using the principles and techniques of the last two sections, create your own tessellation.

**34.** If you examine things you see each day, you will see many examples of natural and man-made tessellations. Find at least five examples of tessellations and make sketches of them.

**35.** If a box is packed with objects of the same size and shape so as to leave no gaps or empty spaces, a three-dimensional tessellation is formed. Sometimes it is not possible to pack a box so that a perfect tessellation is created. In order to save space and packaging costs, the objective for a manufacturer is to minimize the area of the gaps or empty spaces. Suppose a company that manufactures wooden miter boxes for the arts and crafts market is presently packing 12 boxes in a sturdy container. The miter box, the dimensions of its front face, and the top of a packed container are shown in the figure. This, however, is not the most efficient way to pack the container. More miter boxes can be put in the container if it is packed more efficiently.

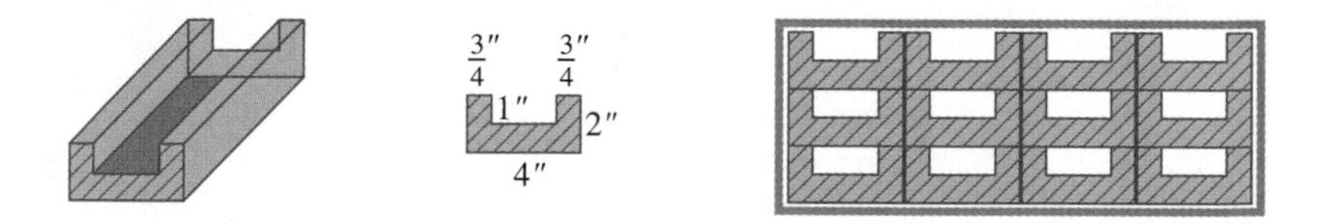


(a) Find a way to pack 14 miter boxes in the container.
(b) Make a sketch of the top surface of the container.

**36.** Suppose the manufacturing company in Problem 35 also made L-shaped wooden shelves that had the same length as the miter boxes and were packed in the same size container as the miter boxes. The inside dimensions of the top of the packing container are 16.1 in. by 6.1 in. The L-shaped shelf and the dimensions of its front face are shown in the figure.

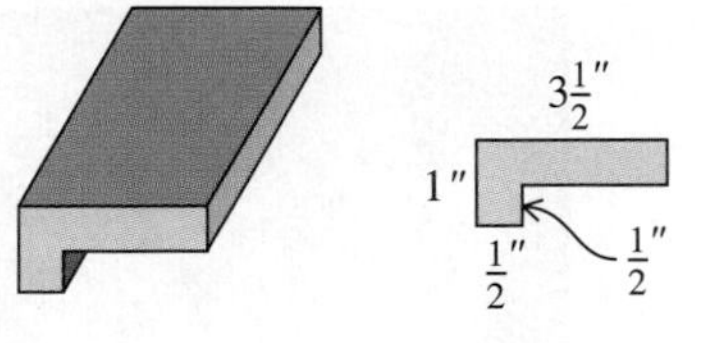


(a) Find the maximum number of L-shaped shelves that can be packed in the container.
(b) Make a sketch of the top surface of the container.
(c) How much empty space is on the top surface of the packed carton?

***37.** Using the formula for each angle ($A$) of a regular polygon with $n$ sides, show that the only regular polygons that tessellate the plane are the equilateral triangle, square, and regular hexagon. (*Note:* The expression $\frac{360}{A}$ must be an integer.)

(*Left*) A magnified photograph of a resin cast of the airways to the lung. (Courtesy of C. Quesada) (*Right*) A computer-generated fractal created by Yoichiro Kawaguchi. (Courtesy of the artist)

## Section 3.6 Fractals

Since the time of the ancient civilizations, geometry has been used to measure and describe the world in which we live. We have come to believe that much of the world can be described by the basic shapes of geometry. The planets and stars are in the shape of spheres; curved paths such as roads can be measured with straight line segments; a parabola can be used to describe the path of a thrown object; the orbits of the planets are elliptical.

As science has progressed, new discoveries have demonstrated that many of our early ideas about using simple geometric shapes to describe natural phenomena are only partially true. The earth is not really a sphere. It bulges along the equator. In addition, although the surface of a sphere is smooth, the surface of the earth is dimpled with craters, canyons, and oceans and has rolling hills and towering mountains jutting from its surface. The simple geometrical models of our predecessors do not accurately describe a world that has since been more carefully observed.

In addition to reexamining old models, scientists are taking a look at phenomena that were previously considered too complex to be described by mathematics. Why are ferns constructed the way they are? Is there a pattern to the branching designs of oak trees? In an oil spill, what determines the depth to which the oil penetrates the surface? How can the static of a radio transmission be described mathematically? Is there a way to describe the fluctuations of cotton prices over the last 100 years? These phenomena are too complex for the simple geometric models of earlier times. As with the imperfect surface of the earth, a new geometry is needed to accurately describe these details.

Through the use of computers, in the 1970s and 1980s we have seen the arrival of a new area of mathematics called **fractal geometry**. This geometry can model the complex situations mentioned above. The word ''fractal,'' coined by the mathematician Benoit Mandelbrot, comes from the same Latin root as does the word "fraction." Mandelbrot uses this term because simple one-, two-, and three-dimensional figures like lines, squares, and spheres cannot describe complex phenomena. To do so requires objects that have **fractional dimension**. In this section, we give a short history of fractals, describe what is meant by a fractional dimension, describe how fractals can be drawn, and show some examples of one of the most famous fractal sets, the Mandelbrot set.

## History

The 19th century was a time when new ideas extended mathematics outside the realm of the observed physical world. Mathematicians such as Georg Cantor, Giuseppe Peano, and David Hilbert used ideas that were at the frontiers of mathematical thought. One such idea was to have a "curve" (a path with length but no thickness) fill a two-dimensional area. One such curve, called Hilbert's curve, has its first four generations shown here.

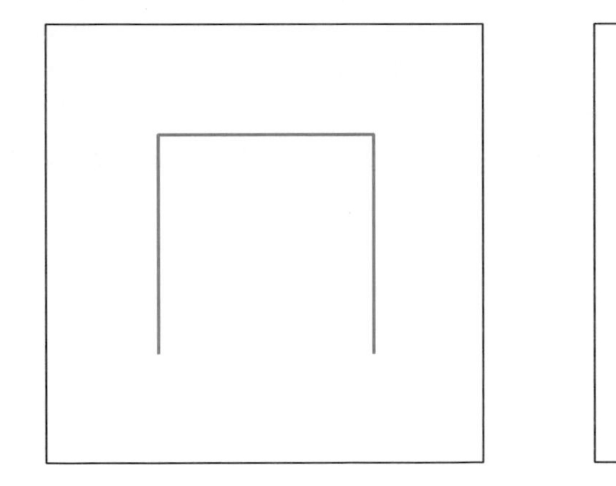

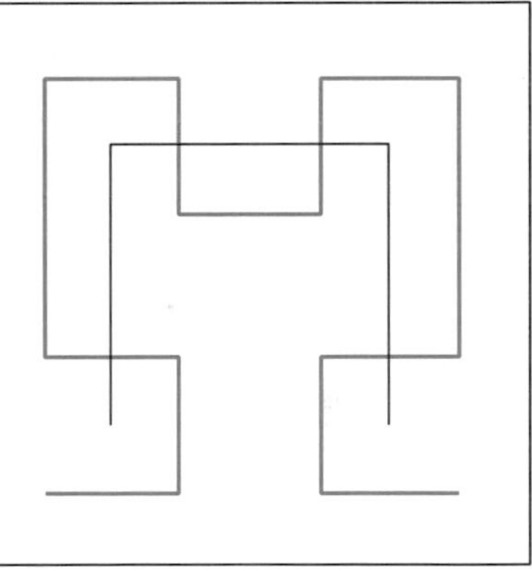

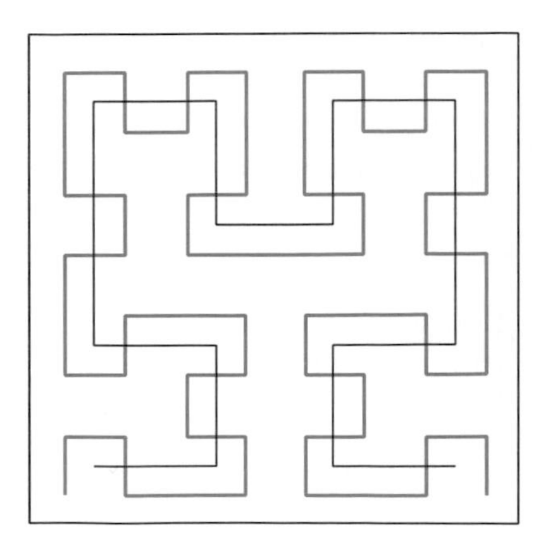

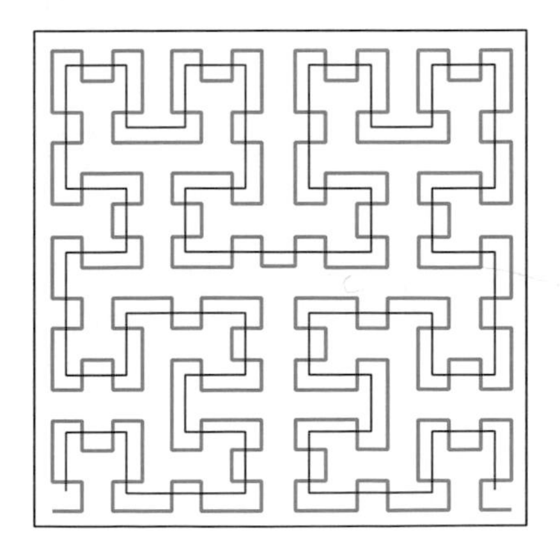

In each drawing, the boldface curve wraps around the previous generation of the curve, shown in the thinner line. There are two things to notice. First, the basic shape remains the same in each drawing. As the pictures become more detailed, the basic shape is simply drawn with a finer scale. Second, the amount of white space in each drawing is being reduced. Hilbert's remarkable curve eventually passes through all the points in the space, "filling" the entire space.

A question arising from the concept of space-filling curves led to a discovery that became a critical part of Mandelbrot's fractional dimensions. If a curve is a one-dimensional object having length but no thickness, and a square is a two-dimensional object having both length and width, what is the dimension of a space-filling curve? Since it is a curve, it should be a one-dimensional object. However, it fills the two-dimensional space of a square. In 1919, two mathematicians, Felix Hausdorff and A. S. Besicovitch, published a work that answered this question. A space-filling curve has a dimension that is between 1 and 2, with the dimension approaching 2 as the complexity of the curve increases.

A second major factor that influenced Mandelbrot in the development of fractals was the work on iteration of functions* and complex numbers† by the French mathematicians Pierre Fatou and Gaston Julia. Their work, along with the availability of the computer, enabled Mandelbrot to make many of his discoveries.

## Constructing Snowflakes, Carpets, and Other Fractals

To construct a fractal, we start with a geometric figure and divide it into smaller versions of itself. We then replace some of the smaller versions, as specified by the rule that generates the fractal. First let's look at a few interesting figures from fractal geometry.

### *The von Koch Snowflake*

The von Koch snowflake is constructed by starting with the line segments forming an equilateral triangle, dividing each of the line segments into three equal sections, and replacing the middle section of each segment by two additional sections. Shown here are the first three iterations of this process. If we start with a line segment, the second figure is constructed from four line segments whose lengths are one-third the length of the original line segment. As we repeat this process, notice that all four of the segments of the second diagram have been used to form the third diagram.

---

*When a function is iterated, it is used repeatedly. For example, suppose we are using the function $f(x) = \sqrt{x}$. If we pick a number, let's say 10, we want to know what happens when the number is put into the function, and that result is in turn put into the function and the process is continued for some time. With our example, $\sqrt{10} \approx 3.1623$, $\sqrt{3.1623} \approx 1.7783$, $\sqrt{1.7783} \approx 1.3335$, etc.

†Complex numbers are of the form $a + bi$, where $a$ and $b$ are real numbers and $i = \sqrt{-1}$.

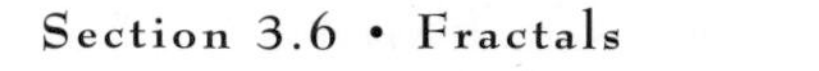

Since the initial step in the construction of the von Koch snowflake is to start with an equilateral triangle, we must perform this process on each of the original sides. Doing this on each side of the triangle creates a six-pointed star.

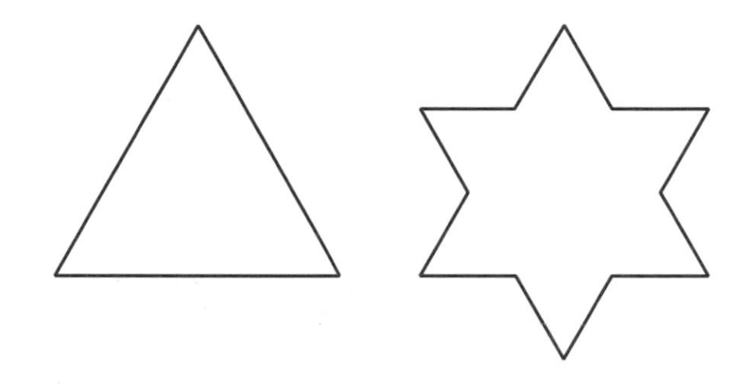

If we repeat this process for each of the 12 segments in the six-pointed star, we obtain a star with 18 points. This, in turn, gives us a 66-point star.

---

**Note:** Be sure to notice that the new points are constructed on each segment of the previous diagram.

---

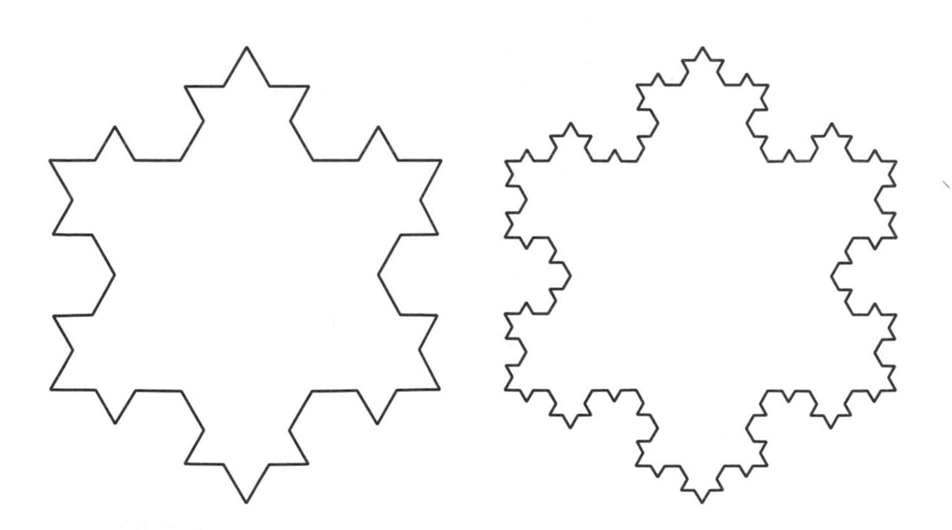

If we continue this process, we arrive at the next two iterations of the von Koch snowflake.

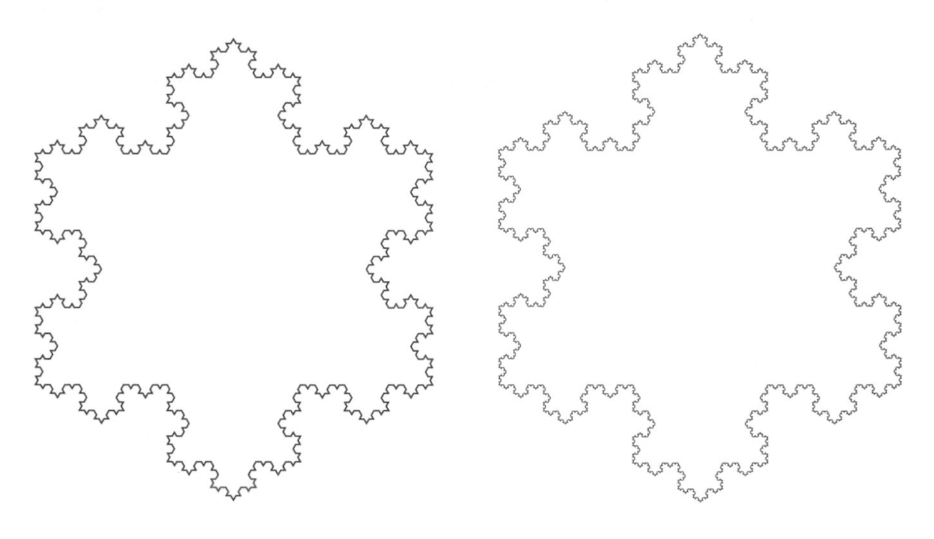

## Example 1

A fractal whose first two iterations are shown in Figures 3.16 and 3.17 is constructed by dividing each line segment into five equal segments and replacing the middle segment by three new segments, one-fifth the length of the original segment. Draw the next iteration of the fractal.

**Solution:** In order to construct the next iteration of the fractal, we need to understand the basic construction process. This consists of dividing each line segment into five equal-length sections and constructing a new three-sided figure to replace the middle of the five sections. This must be done for each line segment in the figure. Therefore, starting with one side of Figure 3.17, we have the next iteration given in Figure 3.18.

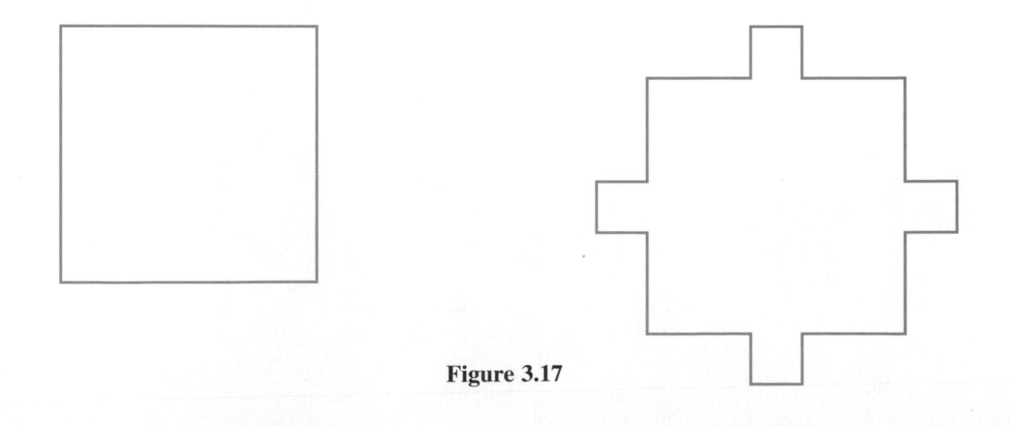

**Figure 3.16**

**Figure 3.17**

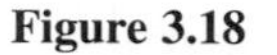

Figure 3.18

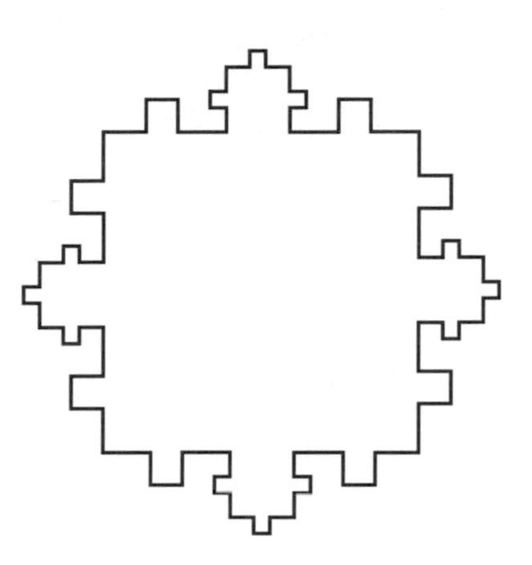

Figure 3.19

By applying this process to each side of Figure 3.18, we have the third generation of the fractal shown in Figure 3.19. ■

## *The Sierpinski Carpet*

With the von Koch snowflake and the fractal shown in Example 1, the construction of the fractal was done by working with line segments. The following fractal is created by working with a two-dimensional object, a square. As shown in Figure 3.20, the basic step of the construction is to start with a square, divide it into nine smaller squares of equal size, and remove the center square.

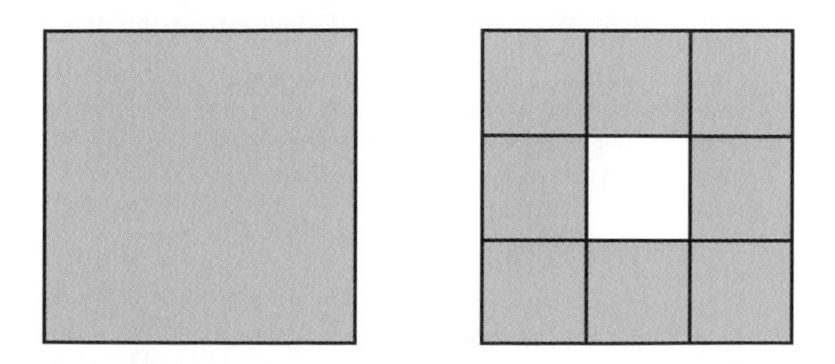

Figure 3.20

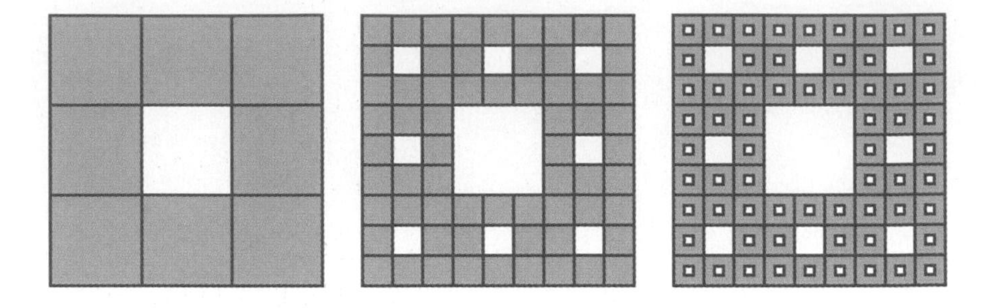

Figure 3.21

The Sierpinski carpet is shown in Figure 3.21. Starting with a square, it uses the process just described to create a plane region with progressively smaller holes.

## Dimension of a Fractal

The formula for the dimension of a fractal is

**The Dimension of a Fractal**

$$d = \frac{\log N}{\log (1/r)}$$

where

$r$ = ratio of the length of the new object to the length of the original object

$N$ = the number of new objects

Using this formula, we can determine the dimension of the fractals given in the foregoing examples.

### Example 2

Find the dimension of the von Koch snowflake.

**Solution:** To determine the dimension of the von Koch snowflake, we examine the basic construction process, the change that occurs in one step of the construction.

Since each line segment is divided into three equal sections, $r = \frac{1}{3}$. Because the construction step involves replacing the middle of the three sections by

two new sections, the new figure consists of four segments (each of which has a length that is one-third the length of the original segment); therefore, we have $N = 4$. Using the dimension formula gives the dimension of the von Koch snowflake as

$$d = \frac{\log N}{\log (1/r)} = \frac{\log 4}{\log \left(\frac{1}{1/3}\right)} = \frac{\log 4}{\log 3} \approx 1.26$$

This result can be found on a calculator by using the following keys:

| Press | Display |
|---|---|
| 4 log ÷ 3 log = | ≈ 1.26 |

## Example 3

Determine the dimension of the fractal generated in Example 1.

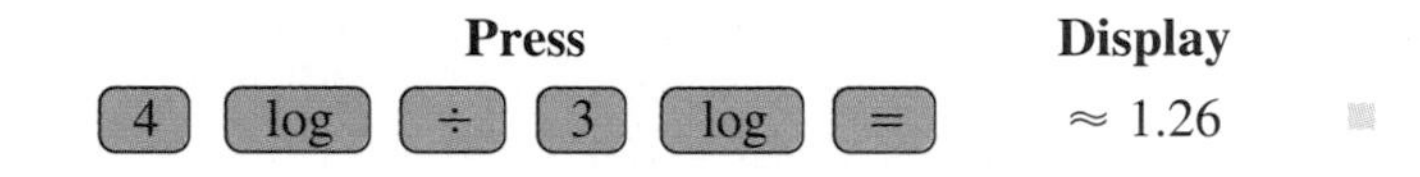

Figure 3.22

**Solution:** To determine the dimension of the fractal, we again examine the basic construction process. Referring to Figure 3.22 and to the directions for creating the fractal, we see that the fractal is created by dividing each line segment into five equal-length subsections. Each of the original line segments is replaced by the new shape. This shape is created from seven of the small subsections. Therefore, using the dimension formula with $N = 7$ and $r = \frac{1}{5}$, we have

$$d = \frac{\log N}{\log (1/r)} = \frac{\log 7}{\log \left(\frac{1}{1/5}\right)} = \frac{\log 7}{\log 5} \approx 1.21$$

## Example 4

Find the dimension of the Sierpinski carpet.

**Solution:** Since we are replacing one square by eight smaller ones, $N = 8$. Since the length of the side of a square is one-third the length of the side of the previous square, $r = \frac{1}{3}$. Using the dimension formula gives

$$d = \frac{\log N}{\log (1/r)} = \frac{\log 8}{\log \left(\frac{1}{1/3}\right)} = \frac{\log 8}{\log 3} \approx 1.89$$

## What Do We Mean by the *Dimension* of an Object?

To understand the concept of a fractional dimension, we first need to examine what we mean when we say that a figure is two dimensional. Let's look at a square with a side of length 1. The area of the square is given by $A = 1^2 = 1$. Dividing each side of the square by 5 gives 25 miniature versions of the original square.

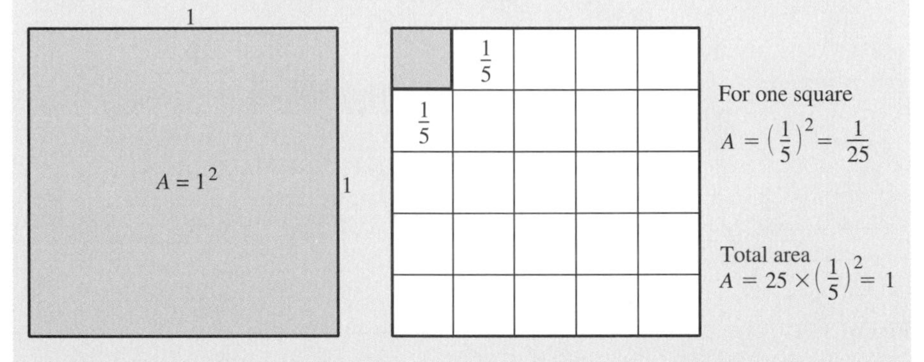


Each of the 25 new squares has sides of length $\frac{1}{5}$ and an area of $(\frac{1}{5})^2$. This gives a total area for the 25 squares of $A = 25 \times (\frac{1}{5})^2 = 1$.

**In general, if we let $r$ be the ratio of the length of the new object to the length of the original object, $N$ be the number of new objects, and $d$ be the dimension of the figure, we can say $Nr^d = 1$.** For the two-dimensional squares, we have used $N = 25$, $r = \frac{1}{5}$, and $d = 2$.

To determine a formula for the dimension of an object, we solve for $d$ in this formula.

$$Nr^d = 1$$

We start by dividing both sides by $r^d$ and using the rules of exponents.

$$N = \frac{1}{r^d}$$

$$N = \left(\frac{1}{r}\right)^d$$

Next, we take the common log of both sides and apply the rules of logarithms.

$$\log N = \log \left(\frac{1}{r}\right)^d$$

$$\log N = d \log \left(\frac{1}{r}\right)$$

Finally, dividing by $\log \frac{1}{r}$ gives the desired equation.

$$\frac{\log N}{\log (1/r)} = d$$

## *The Mandelbrot Set*

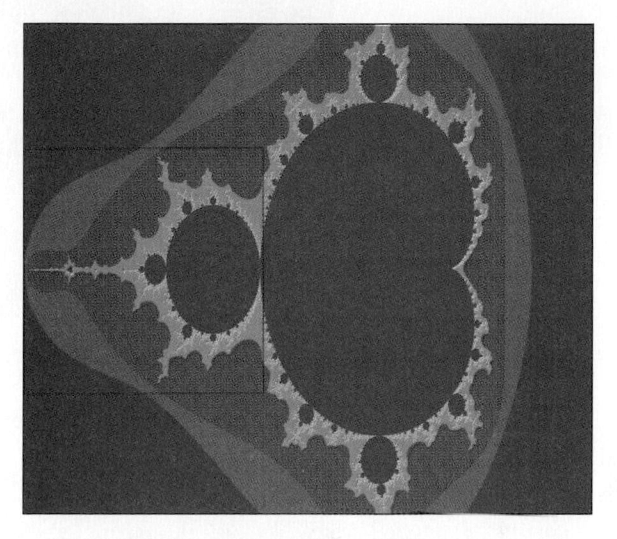

This diagram is called the Mandelbrot set. It is more complex than the fractals discussed earlier, but it still demonstrates the same repetitive patterns exhibited by the other fractals. Notice that the general shape of a circle with a little nob sticking off one side is repeated, at smaller scales, throughout the diagram. If we take the portion of the diagram located in the box and magnify it, we obtain the same structure. This process can be continued indefinitely, with each new iteration showing similar characteristics.

Examining the small nob at the bottom center of this diagram again gives a new, smaller version of the same diagram.

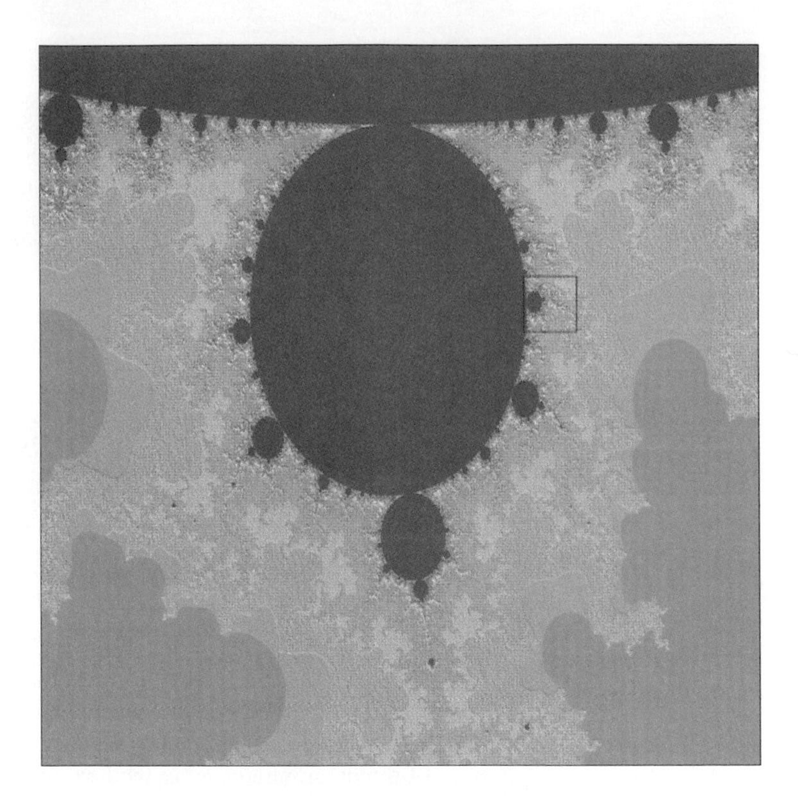

If we again zoom in on another nob that was only a dot on the original diagram, we get the following:

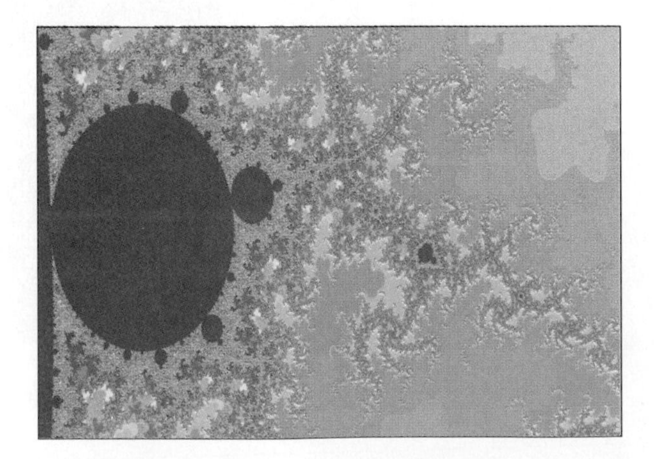

This set is generated by a very short set of instructions (see page 246), but it provides insights into many physical phenomena. The idea of a portion of a set being similar to itself (called **self-similarity**) can be seen in the mapping of coastlines. For example, the coastline of California has many small bays and peninsulas. These are usually measured on a scale of thousands of feet. If we examine a small portion of the coastline more closely, the smaller section is also seen to have inlets and peninsulas, this time measured on a scale of hundreds of feet. If this process is continued, we can examine a section of coastline only a fraction of an inch long. In this very small section, we can again see inlets in the form of grains of sand. As we examine the coastline on increasingly more detailed levels, we continue to see the same degree of intricacy.

This phenomenon of self-similarity can be seen in many other situations. Cotton prices since 1900 have fluctuated due to the vagaries of supply and demand. Yet, when Mandelbrot and the Harvard economist Hendrik Houthakker closely examined the fluctuations, they found that a decade-by-decade pattern of prices showed the same type of fluctuation as yearly prices or monthly prices. Noise in telephone lines has certain, apparently random, patterns when the signals are examined on an hourly basis. When the same signal is examined at the level of seconds or minutes, the same patterns, at smaller scales, reappear. Like the von Koch snowflake and the Sierpinski carpet, the same patterns reappear as the situation is examined in finer detail.

The intricate fractals that follow were created with a computer and a fractal-generating equation.

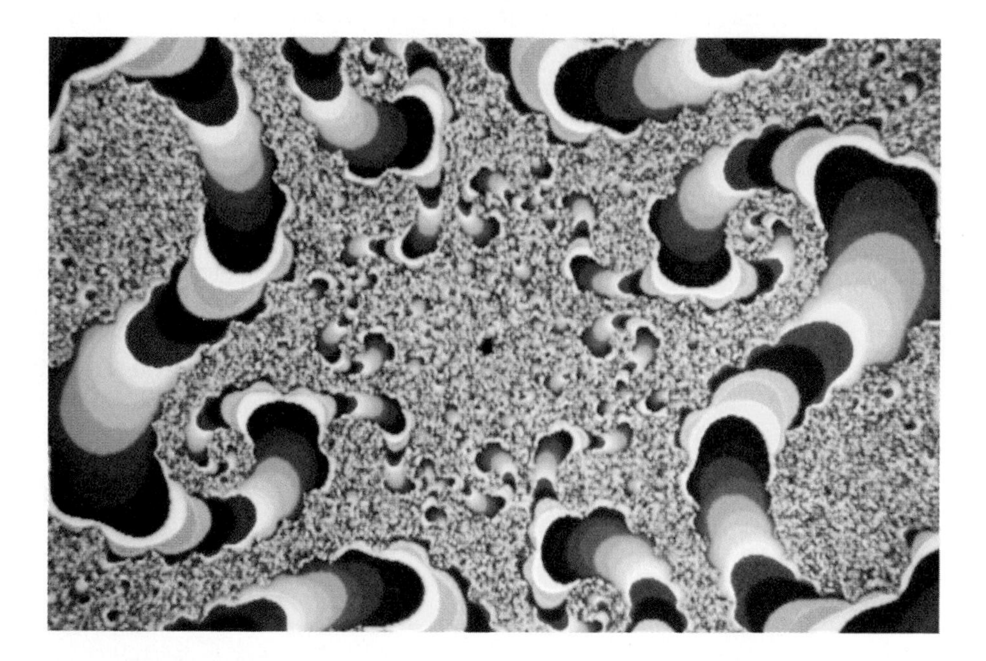

*SSS's:* These spiraling fractal tubes, based on the Mandelbrot set barely visible at the center, were created with the computer program FracTools™ and then photographed off a computer monitor using a 35-mm camera. (Courtesy of Bourbaki, Inc. and Lloyd W. Black)

*BENFORKS:* These fractal forks, based on the Mandelbrot set at the center, were created with the computer program FracTools™ and then photographed off a computer monitor using a 35-mm camera. (Courtesy of Bourbaki, Inc. and Lloyd W. Black)

## The Mathematics Behind the Mandelbrot Set

In algebra, you may have learned that complex numbers are numbers of the form $a + bi$, where $i = \sqrt{-1}$. You may also have learned that the magnitude of a complex number $z = a + bi$ is given by $\sqrt{a^2 + b^2}$.

The Mandelbrot set is generated by using the equation $z_n = (z_{n-1})^2 + C$, where $z_n$ and $C$ are complex numbers. A complex number $z_0$ is chosen and substituted into the equation for the $z$ value on the right side of the equation. For each complex number chosen, the equation is used repeatedly until the magnitude of $z$ exceeds 2 or until it is determined that the magnitude of $z$ will probably never exceed 2. If the final result exceeds 2, the starting point $z_0$ is not in the Mandelbrot set. If the final result does not exceed 2, the starting point is in the Mandelbrot set.

For example, let's look at $z = 0.75 + 0i$ and assume that the constant $C = 0$. Since the first $z$ value is 0.75, we write $z_1 = 0.75$. This gives the following sequence of values:

$$z_2 = (0.75)^2 = 0.5625$$

$$z_3 = (0.5625)^2 = 0.3164$$

$$z_4 = (0.3164)^2 = 0.1001$$

$$z_5 = (0.1001)^2 = 0.0100, \ldots$$

It is apparent that this sequence of values will approach zero and, hence, is less than 2. Therefore, the starting point $z_1 = 0.75 + 0i$ is in the Mandelbrot set.

If we start with a point such as $z_1 = 0.75 + 1.3i$, we arrive at the following sequence of values:

$$z_2 = (0.75 + 1.3i)^2 = -1.1275 + 1.95i$$

$$z_3 = (-1.1275 + 1.95i)^2 = -2.5312 - 4.3973i$$

$$z_4 = (-2.5312 - 4.3973i)^2 = -12.9293 + 22.2609i$$

Since these numbers are growing quickly, we can see that the magnitude will exceed the limit of 2. Hence, the point $z = 0.75 + 1.3i$ is not in the Mandelbrot set. Although these calculations are tedious, the availability of computers has made the computations and the drawing of the Mandelbrot set and other fractals a reality.

Section 3.6
PROBLEMS

## *Explain ➡ Apply ➡ Explore*

### *Explain*

1. What is a fractal and what properties does a fractal possess?
2. What is the dimension of a fractal? What formula is used to determine the dimension of a fractal? Explain what each variable in the formula represents.
3. What does it mean for a fractal to have a dimension of 1.6?
4. What does it mean for a fractal to have a dimension of 2.6?
5. What does self-similarity refer to when discussing fractals.
6. What is meant by a space-filling curve?
7. Examine a leaf of a fern. Explain why the parts of a leaf have properties of fractals.
8. Explain why cirrus clouds can be more accurately depicted by fractals than by standard geometric shapes.

### *Apply*

9. Create a fractal by starting with a square, dividing each line segment into three equal lengths, and replacing the middle third of each side with three line segments whose lengths are one-third the length of the original segment. The first

**19.** Find the dimension of the fractal in Problem 11.

**20.** Find the dimension of the fractal in Problem 12.

**21.** Find the dimension of the fractal in Problem 13.

**22.** Find the dimension of the fractal in Problem 14.

**23.** A fractal is created by the following process. A line segment of length 1 is drawn. It is divided into five equal sections. The second and fourth of these sections are then removed. This process is continued indefinitely. What is the dimension of this fractal?

**24.** A fractal is created by the following process. Start with a regular pentagon. Cut each side into five equal-length sections and replace the middle section with another regular pentagon. This process is continued indefinitely. What is the dimension of this fractal?

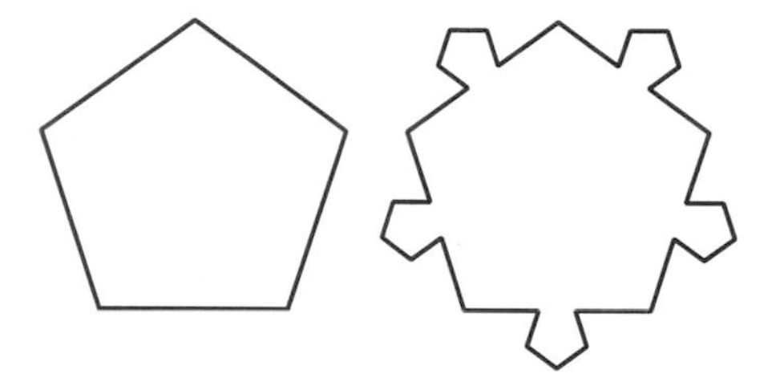

***25.** Investigate the resources in the bibliography at the end of the text and determine some of the other areas of study to which fractals are being applied.

***26.** The repetitive fractal patterns generated by computers can have a stunning visual effect. Find examples that show this ''artistic'' effect of fractals.

***27.** What is a Sierpinski gasket? Draw the first three iterations of the gasket.

***28.** What is a Cantor dust? In what area of study is it being applied?

## Chapter 3 • Summary

### Key Terms, Concepts, and Formulas

The important terms in this chapter are:

**Acute angle:** An angle that has a measure between 0° and 90°. p. 159

**Angle:** The figure formed by two rays or line segments with a common end point. p. 159

**Atmospheric perspective:** A technique of creating an illusion of depth on a two-dimensional surface by a gradual lessening of clarity and visual strength. p. 176

**Axiom:** A proposition that is assumed to be true. Historically, axioms referred to propositions of algebra or arithmetic. p. 161

**Concave polygon:** A polygon in which an extension of at least one of its sides enters the interior of the polygon. p. 207

**Convex polygon:** A polygon in which extensions of its sides do not enter the interior of the polygon. p. 207

**Diminishing sizes:** A technique of creating an illusion of depth on a two-dimensional surface by systematically making objects smaller. p. 175

**Euclidean geometry:** A phrase used to describe a system of geometry based on *Elements* written by the Greek mathematician Euclid. p. 157

**Exterior angle:** An angle formed outside a triangle by one side of the triangle and an extension of another side of the triangle. p. 159

**Eye level line:** A horizontal line across a picture in a one-point perspective that shows the observer's focus when looking straight ahead. p. 179

**Fibonacci numbers:** The sequence of numbers 1, 1, 2, 3, 5, 8, 13, 21, 34, 55, . . . , where each successive number is the sum of the two previous numbers. p. 199

**Fractal:** An object with a fractional dimension. p. 234

**Golden Ratio ($\phi$):** The ratio between distances $a$ and $b$ such that p. 194

$$\phi = \frac{a}{b} = \frac{a + b}{a} \approx 1.62$$

**Golden Rectangle:** A rectangle whose sides form the Golden Ratio. p. 195

**Line segment:** The set of points on a line consisting of two end points and all the points on the line between those two points. p. 158

**Non-Euclidean geometry:** A phrase used to describe a system of geometry that assumes a consistent set of postulates with at least one that is not logically equivalent to Euclid's postulates. p. 164

**Obtuse angle:** An angle that has a measure between 90° and 180°. p. 159

**One-point perspective:** A technique of creating an illusion of depth on a two-dimensional surface by using the geometry of converging lines. p. 178

**Overlapping shapes:** A technique of creating an illusion of depth on a two-dimensional surface by placing shapes in front of one another. p. 174

**Parallel:** Two lines, rays, or line segments are parallel ($\|$) if they lie in the same plane and the lines that contain them do not intersect. p. 163

**Parallel postulate:** A postulate of geometry that establishes the number of lines that are parallel to a given line through an external point. p. 163

**Perpendicular:** Two lines, rays, or line segments are perpendicular ($\perp$) if they intersect and form a right angle. p. 160

**Perspective:** The art of drawing objects on a two-dimensional surface so as to give an illusion of depth and show distance from the observer. p. 173

**Polygon:** A closed figure in a plane formed by the line segments that intersect each other only at their end points. p. 206

**Postulate:** A proposition that is assumed to be true. Historically, postulates referred to propositions of geometry. p. 160

**Proof:** A series of statements that logically show a conclusion follows from the hypothesis. p. 164

**Ray:** The figure formed by extending a line segment in only one direction. p. 159

**Reflection:** Movement of a shape in a plane by finding its mirror image across a specified line. p. 223

**Regular polygon:** A polygon that has sides of equal length and angles of equal measure. p. 207

**Right angle:** An angle that has a measure of 90°. p. 159

**Sides:** Line segments that form a polygon. p. 207

**Straight angle:** An angle that forms a line and has a measure of 180°. p. 159

**Tessellation:** A pattern of one or more congruent shapes that covers an area in a plane without overlapping or leaving any gaps. p. 219

**Theorem:** A proven proposition. p. 164

**Translation:** Movement of a shape in a plane by sliding it in a certain direction for a specified distance. p. 223

**Triangle-sum theorem:** A theorem of geometry that states the sum of the measures of the angles of a triangle. p. 164

**Undefined terms:** Terms that have meanings assumed to be clear. Point, line, and plane are undefined terms in geometry. p. 158

**Vanishing point:** A fixed point on the eye level line in a one-point perspective to which receding lines converge. p. 179

**Vertex** (pl. **vertices):** The intersection point of sides of a polygon. p. 207

After completing this chapter, you should be able to:

1. Explain and use undefined terms, definitions, axioms, postulates, and theorems in a system of Euclidean geometry. p. 158
2. State the differences between Euclidean, Lobachevskian, and Riemannian geometries, including the parallel postulate, model, and triangle-sum theorem of each system. p. 164
3. Show perspective on a two-dimensional surface using overlapping shapes, diminishing sizes, atmospheric perspective, or a one-point perspective. p. 173

4. Find a point on a line segment that divides it into the Golden Ratio. p. 197
5. Determine whether a rectangle is approximately a Golden Rectangle and explain the significance of Golden Rectangles in art, architecture, and other areas. p. 196
6. Design and construct regular polygons and stars. p. 212
7. Determine the sum of the angles of a polygon and each angle of a regular polygon. p. 211
8. Create tessellations with a triangle or quadrilateral using translations and reflections. p. 222
9. Recognize basic properties of fractals. p. 234
10. Given a geometric figure and a method of construction, draw the next iteration of a fractal and determine the dimension of the fractal. p. 236

• Summary

## Problems

1. Explain the difference between
   (a) Undefined terms and definitions
   (b) Postulates, axioms, and theorems
2. Explain why it is necessary to have undefined terms and postulates in a geometric system.
3. Draw $\triangle TRY$ with $\angle T$ an obtuse angle on the model for the plane in Euclidean, Lobachevskian, and Riemannian geometries.
4. Draw lines, $h$, $j$, and $k$ intersecting at point $P$ on models for the Euclidean plane, the Riemannian plane, and the Lobachevskian plane.
5. In light of the non-Euclidean geometries, explain the statement ''Mathematics does not give truth about the real world.''
6. Using the given shape and each of the following techniques, make a sketch that gives a sense of depth.
   (a) overlapping shapes
   (b) diminishing sizes
   (c) atmospheric perspective

**7.** Use a one-point perspective with the given shapes, line at eye level, and vanishing point to make solid objects of varying lengths.

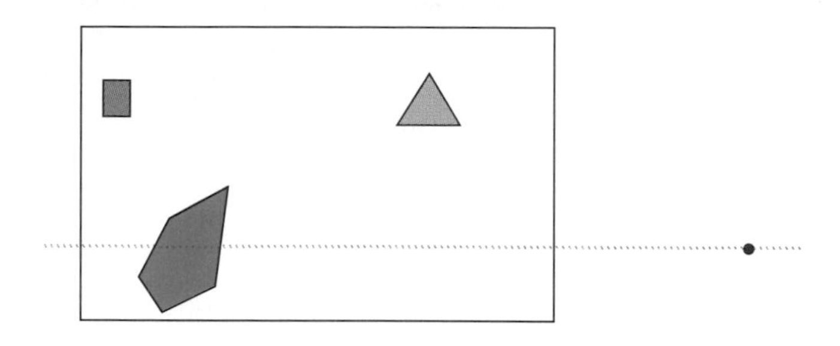

**8.** What methods of perspective did the artist Giovanni Paolo Pannini use in his painting, *The Interior of the Pantheon*, found on page 395 in Chapter 6 of this text? Explain.

**9.** Find a point that approximately divides the segment into a Golden Ratio.

**10.** If the width of a rectangle is to be 8.5 in., how long should its length be to make it a Golden Rectangle?

**11.** A kite is a quadrilateral with two distinct pairs of equal adjacent sides as shown below. Create a definition for a “Golden Kite.” Draw some of these “Golden Kites.” (*Hint:* There are two different ratios of segments that could be used to make the “Golden Kite.”)

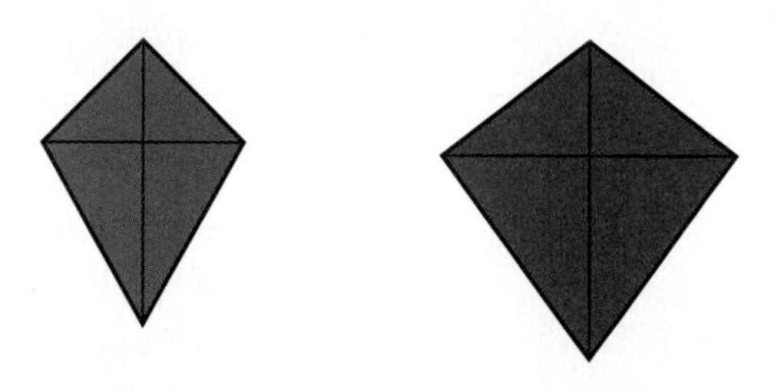

**12.** Construct the following regular polygons:

(a) dodecagon (b) 18-gon

**13.** Construct stars from the regular polygons in Problem 12 by connecting the following points:

(a) Every third point of the polygon (b) Every fifth point of the polygon

**14.** Tessellate a region with the following triangles:

(a) Using only translations (b) Using reflections and translations.

**15.** Tessellate a region with the following figures:

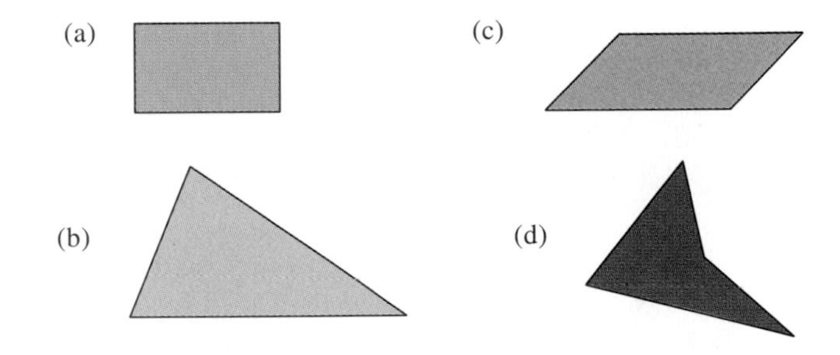


**16.** Explain why it is possible to tessellate a plane with a square, a regular hexagon, and a regular dodecagon placed at each vertex, but it is impossible to tessellate a plane with an equilateral triangle, a regular pentagon, and a regular heptagon placed at each vertex.

**17.** Tessellate a region with a regular hexagon, squares, and equilateral triangles.

**18.** Use the following procedure to construct a fractal. Start with a rectangle whose length is twice its width. Divide each line segment into four sections of equal length. Replace the center two sections with three sides of a rectangle, as shown in the diagram. The length of the longer of the three added segments should equal one-half the length of the original segment. The length of the other two added segments should be one-fourth the length of the original segment. The first iteration is shown. Repeat this process and draw the next iteration of the fractal.

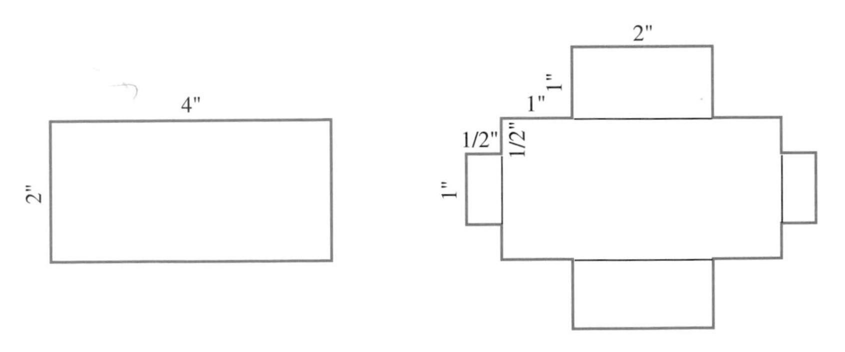

**19.** Determine the dimension of the fractal discussed in Problem 18. Remember that if the length of the original line segment is 4 in., the length of the next iteration of that segment is formed by six sections, for a total of 6 in.

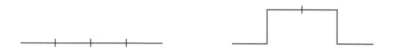

# Sets, Logic, and Flowcharts

4

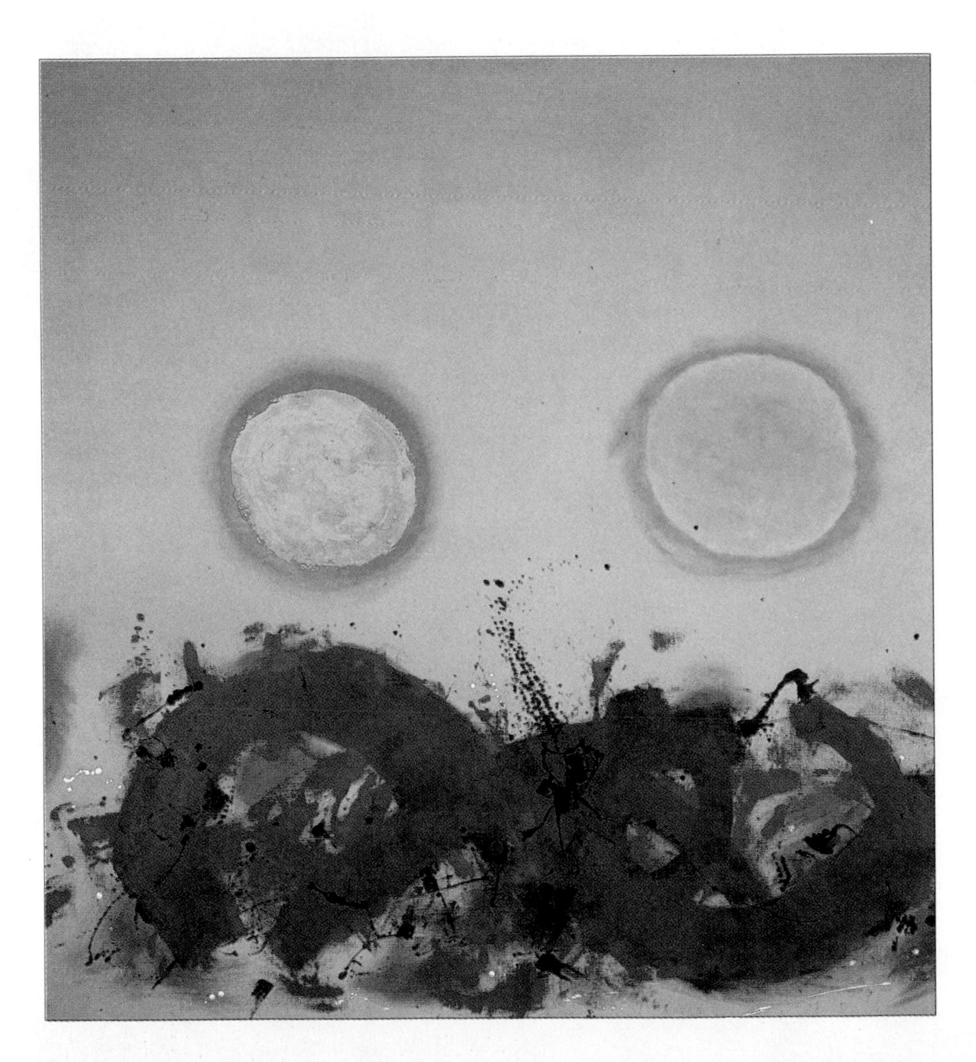

*Duet*, an artistic interpretation of two empty sets by Adolph Gottlieb. (High Museum of Art)

# Overview Interview

Logical thinking is a key to making sound decisions and solving complex problems. You use logic in everyday events such as determining why your car won't start, planning the route you are going to take when shopping in the city, or filling out your 1040 Tax Form in April. In this chapter, we will examine a few of the many facets of logic in the hope of helping you become an even better thinker and problem solver. We will begin this journey by discussing some basic ideas about sets. We then examine premises and conclusions; proceed to methods of logical argument; analyze truth tables, Venn diagrams, and paradoxes; and end with the use of flow charts as a means of making decisions.

## Audrey C. Talley

AUDREY C. TALLEY, ESQUIRE, PARTNER WITH THE LAW FIRM OF STRADLEY, RONON, STEVENS & YOUNG, PHILADELPHIA, PA

TELL US ABOUT YOURSELF. I received my Bachelor of Arts degree in Philosophy from Vanderbilt University, my Master of Arts degree in Drama from the University of Southern California, and my Juris Doctorate from Boston University Law School. My law practice is focused on securities and corporate law, primarily for the financial services industry. I am concerned with the development of investment products such as stocks and the compliance with laws that govern these investments.

HOW IS LOGIC USED BY A LAWYER? WHY IS MATHEMATICS IMPORTANT IN YOUR POSITION? Lawyers solve problems and prevent problems. We do this by reviewing facts and circumstances that are presented to use and the laws that apply to that particular situation. Correct reasoning is the road the lawyer travels from understanding the problem to the solution that is found at the end of the road. For a lawyer, being able to determine the relationships among the facts is much like working through the steps of a math problem to find the answer. When I analyze a legal problem, I begin with certain steps and then proceed to other steps that must follow, one after the other, in a logical sequence. In that respect, education in mathematics is useful in developing the logical thinking skills a lawyer needs.

*Audrey Talley at work.*

In my particular area of law, being able to do mathematics is also very important. If someone selling stocks characterizes a stock as a good investment, that generally means that mathematical analysis has been used to conclude that the stock will increase in value over a period of time. I review the formulas used to arrive at those conclusions and the graphs used to display them. The ability to comprehend such mathematical calculations is important because numbers are always used in describing the value of investments like stocks and bonds.

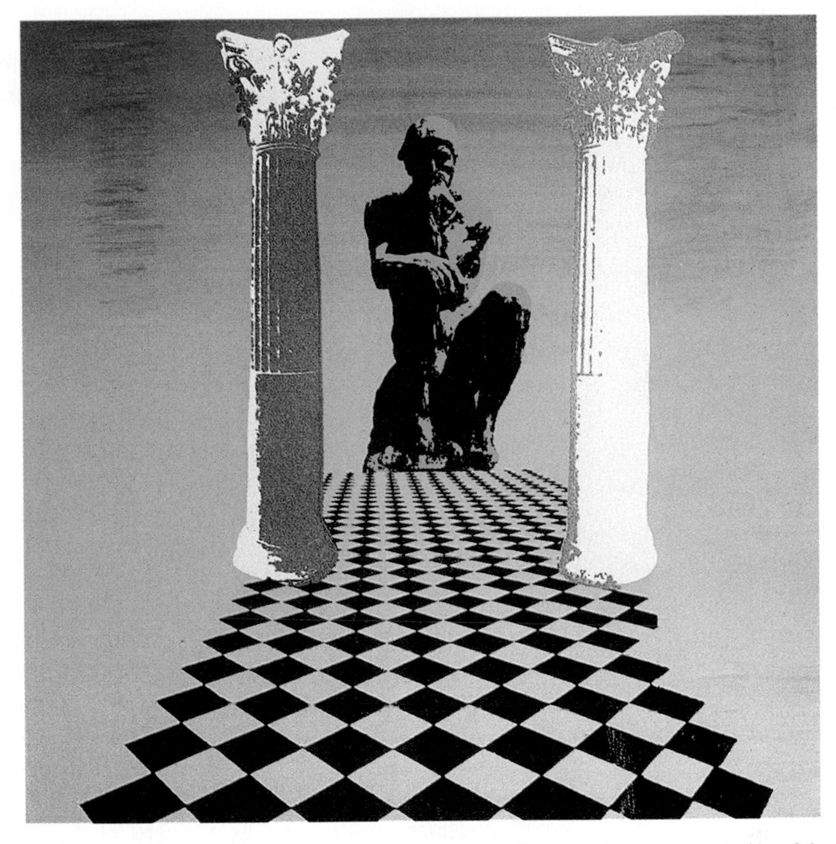

*Le Penseur* by Tami Whitt-Zenoble: a modern interpretation of a classic representation of the human thinking process. (Courtesy of the artist)

## A Short History of Logic

The ability to think logically is an innate human ability. From earliest times, humans have used this power. However, the first systematic study of logic is credited to the Greek philosopher Aristotle (384–322 B.C.). In his work *Organon*, Aristotle systematized principles of reasoning and laws of logic. He is known for his work on a type of argument called a syllogism that consists of two statements and a conclusion. The Stoic and Megarian schools in Greece (c. 300 B.C.) continued his study of logic. They formed a logic of propositions and valid inference schemes and were very interested in paradoxes. The most productive of the Stoic-Megarian logicians was Chrysippus.

Logic in the Middle Ages was marked by several different schools of thought. The Roman philosopher Boëthius (c. 480–520) was instrumental in passing on the logical traditions of the Greeks to European monks. These monks sought to preserve the logic found in classical Greek texts. Meanwhile, the Nyaya schools in India also made contributions to the field of logic during this period. As the Middle Ages came to an end, scholastic logic made its entrance. Scholastic logic was characterized by its use of Latin and the influence of Christian theology. It was developed for discussions of theological questions. Noted contributors in the development of logic during this period were Peter Abelard (c. 1130), Robert Grosseteste (c. 1240), St. Thomas Aquinas (c. 1250), and Petrus Hispanus (c. 1260).

With the Renaissance period in Europe, advances in formal logic again appeared. William of Ockham (c. 1320), in *Summa Logicae*, and Walter Burley (c. 1325), in *De puritate artis logicae*, improved upon Aristotle's logic. Gottfried Leibniz (1666) initiated the study of symbolic and mathematical logic in his essay *De arte combinatoria*. In this work and others over the next 25 years, Leibniz implied that mathematics can be derived from the principles of logic. His work began what is considered the period of mathematical logic. Leonhard Euler (c. 1770) adopted a method of visually checking syllogisms by means of circle diagrams. John Stuart Mill (c. 1843) made contributions to the development of inductive logic. Augustus De Morgan (c. 1850), in *Formal Logic*, and George Boole, in his works *The Mathematical Analysis of Logic* (1847) and *An Investigation of the Laws of Thought* (1854), applied algebraic operations to logic and placed logic on a mathematical basis. John Venn (1880) introduced the Venn diagram, which is a modification of the circle diagrams of Euler. From 1910–1913 Alfred North Whitehead and Bertrand Russell published the three volumes of *Principia Mathematica*, which attempted to develop mathematics from only undefined concepts and principles of logic. Jan Lukasiewicz, Emil Post, and Ludwig Wittgenstein between 1920 and 1921 independently introduced truth tables as a means of reaching logical conclusions. Developments in logic continue to the present day. In this chapter we will not look at the details of formal logic, but we will investigate some of the principles and applications of logic.

## Check Your Reading

1. What was studied during the Greek period of logic?
2. What is scholastic logic?
3. In 1912, the first successful parachute jump occurred. What notable attempt was being made by Bertrand Russell and Alfred North Whitehead in that year?
4. Who initiated the study of symbolic and mathematical logic?
5. What three mathematicians introduced truth tables?
6. When Marco Polo was making his journeys c. 1260, what was Petrus Hispanus doing?

7. What were truth tables used for?
8. Match each of the following names with the correct mathematical discovery, event, or concept.

| | |
|---|---|
| (a) Aquinas | Applied algebraic operations to logic |
| (b) Aristotle | Devised circle diagrams to check syllogisms |
| (c) Boole | Modified Euler diagrams |
| (d) Chrysippus | Co-authored *Principia Mathematica* |
| (e) Euler | Was a scholastic logician |
| (f) Lukasiewicz | Was a Stoic-Megarian logician |
| (g) Venn | Worked on syllogisms |
| (h) Whitehead | Devised truth tables |

## Research Questions

To answer the following questions, you will need to refer to material not contained in the text. Possible sources of information are listed in the bibliography at the end of the book.

1. What is the Socratic method? What was it used for? On what principles is it based?
2. Who is Bertrand Russell? What is Russell's paradox? Explain the reasoning behind Russell's paradox.
3. What is Boolean algebra? How is it related to logic?
4. What are Euler diagrams? How are they related to Venn diagrams? Give examples of each and show how they are used.
5. What were some of the theological questions the scholastic logicians were trying to answer?
6. Many of the mathematicians mentioned in this short history of logic are known for other mathematical endeavors or had other interests besides mathematics. Do some research on two of the mathematicians mentioned in this section. Tell something about their lives, their achievements, and their interests.
7. What are optical illusions? Give examples and show how they defy logic.
8. In the study of geometry, theorems are shown to be true using proofs. What are the basic components of a proof in geometry? Find some examples of proofs of geometrical theorems.
9. Magicians perform feats that defy logic. Examine some famous magic tricks and explain why the feat supposedly being accomplished in the trick is not logical.
10. Advertisements in magazines make use of logic to persuade us to buy certain products. Examine some ads and determine the ''logic'' used in the ad.

## • Projects

1. In 1972, Kenneth Arrow shared the Nobel Prize for economics for his studies involving voting. Part of his studies involve a voting paradox called Arrow's paradox. Discuss the background and results of this paradox.
2. The electric circuits that are used in the design of electronic equipment such as stereos and computers are sometimes called logic circuits. Investigate the connections between truth tables (studied later in this chapter) and logic circuits.
3. Who was Zeno? What are Zeno's paradoxes? Explain the reasoning behind each of Zeno's paradoxes.
4. Interview an attorney and determine how logic is used in court cases. Is there a difference between the types of logical strategies used in a civil case and those used in criminal cases? Explain.

*Golden Venn* by David McLaughlin makes use of Golden Ratios and colorfully represents the Venn diagram of three intersecting sets. (Courtesy of the artist)

## Section 4.1 Sets, Venn Diagrams, and Paradoxes

A **set** is simply a collection of items. Although we have not previously defined sets and their use, we have used and will continue to use the idea of a set throughout the book. For example, our system of numeration uses a set of ten symbols to represent numbers. When we graph an equation, the set of points on the graph represents the solutions of the equation. In statistics, a collection of data is called a data set. In this section, we want to examine the rules for working with sets and some of their uses.

### Sets and Their Symbols

#### *Set Notation and Members of a Set*

To begin our study of sets, we will discuss the notation used with sets. Since a set can contain numbers, words, equations, or even other sets, we need a very general term to describe an item inside a set. In mathematics, we use the word **element**. Each item in a set is an element of the set.

Suppose we want to have a set consisting of all the even integers between 1 and 9. Using the symbols { and } (called braces) to form a set, we represent the set in two different ways. Since the even integers between 1 and 9 are 2, 4, 6, and 8, we can represent the set by enclosing these numbers inside the braces:

$$\{2, 4, 6, 8\}$$

This method is called the **listing method**. Although it is easy to use, it is not convenient if a set contains a large number of elements.

A second method of representing a set is called the **descriptive method**. When using the descriptive method, simply describe the contents of the set rather than listing every element in the set. Instead of {2, 4, 6, 8}, we can write

$$\{x \mid x \text{ is an even integer with } 1 < x < 9\}$$

The vertical bar between the $x$'s means "such that." Therefore, the notation is read, "the set of $x$ such that $x$ is an even integer with $1 < x < 9$." Using the descriptive method allows us to write sets with a large number of elements without resorting to the tedium involved in writing out each element.

To indicate that a certain item is an element of a set or is not an element of a set, we use the symbols:

$\in$ for "is an element of"

$\notin$ for "is not an element of"

Thus, $6 \in \{0, 3, 6, 9, 12, 15, 18\}$, but $8 \notin \{0, 3, 6, 9, 12, 15, 18\}$.

**Example 1**

Use set notation to write all the whole numbers less than 13 that are divisible by 3.

**Solution:** We can do this in two ways. If we use the descriptive method, the set is given by

$$\{x \mid x \in \text{whole numbers}, x < 13, \text{ and } x \text{ is divisible by } 3\}$$

If we use the listing method, we have {0, 3, 6, 9, 12}. ■

## *The Empty Set*

The **empty set** is the set that does not contain any elements. It can be expressed in two ways. The first way uses the standard set notation of braces. Since the empty set does not contain any elements, the empty set is written as a pair of braces that do not enclose any elements:

$$\text{empty set} = \{\ \}$$

The second way to indicate the empty set is to use the symbol $\varnothing$. Thus,

$$\text{empty set} = \varnothing$$

The empty set can describe impossible events, such as a square circle or a natural number that is both even and odd.

### *Notation Indicating a Set*

To differentiate between sets and other mathematical objects, whenever we are discussing a set we use an uppercase script variable, such as $\mathcal{X}$. This will provide a visual clue that the symbol being used represents a set.

### *Subsets*

Set $\mathcal{B}$ is called a subset of a set $\mathcal{A}$ if every element in the set $\mathcal{B}$ is also an element of the set $\mathcal{A}$. The symbol used to indicate that $\mathcal{B}$ is a subset of $\mathcal{A}$ is $\subset$ ($\mathcal{B} \subset \mathcal{A}$). For example, if we let $\mathcal{X}$ = {ingredients in pizza} and $\mathcal{Y}$ = {ingredients in Italian cooking}, then we can say $\mathcal{X} \subset \mathcal{Y}$ because every element in the set $\mathcal{X}$ is a member of the set $\mathcal{Y}$. To indicate that a set is not a subset of another set, we use the symbol $\not\subset$. Thus for the sets $\mathcal{X}$ and $\mathcal{Y}$, since there are some ingredients in Italian cooking that are not used in pizza, $\mathcal{Y}$ is not a subset of $\mathcal{X}$. This may be represented as $\mathcal{Y} \not\subset \mathcal{X}$. An important technical note about the notation for sets is that the subset and element symbols serve different purposes. The subset symbol can only occur between two sets, and the element symbol can only occur between an element and a set. The following example gives correct and incorrect usage for each symbol.

**Example 2**

Let

$$\mathcal{A} = \{\text{all the planets in our solar system}\}$$
$$\mathcal{B} = \{\text{all the celestial objects}\}$$

Which of the following statements make correct use of the subset and element symbols?

(a) $\mathcal{A} \in \mathcal{B}$  (b) The earth $\in \mathcal{B}$  (c) $\mathcal{B} \subset \mathcal{A}$  (d) The earth $\subset \mathcal{A}$

**Solution:**

(a) $\mathcal{A} \in \mathcal{B}$ is an incorrect usage of the element symbol because $\mathcal{A}$ is a subset of $\mathcal{B}$, not an element of $\mathcal{B}$. The correct statement is $\mathcal{A} \subset \mathcal{B}$.
(b) The earth $\in \mathcal{B}$ is a correct usage of the element symbol since the earth is one element in the set consisting of all celestial objects.
(c) $\mathcal{B} \subset \mathcal{A}$ is a correct usage of the subset symbol since both $\mathcal{B}$ and $\mathcal{A}$ are sets. However, the statement is not true since not every celestial object is a planet. A correct statement is $\mathcal{B} \not\subset \mathcal{A}$.
(d) The earth $\subset \mathcal{A}$ is not a correct usage of the subset symbol since the earth is an element of $\mathcal{A}$, not a subset of it. A correct statement is "The earth $\in \mathcal{A}$." ■

## *Operations with Sets*

There are two set operations that will be used in this text, the operations of union and intersection. The **union** of two sets creates a new set containing all the elements of the two sets. The symbol used to indicate the union of two sets is $\cup$.

### Example 3

Given the sets $\mathcal{A} = \{0, 3, 6, 9\}$ and $\mathcal{B} = \{1, 3, 5, 7, 9\}$, find the union of $\mathcal{A}$ and $\mathcal{B}$.

**Solution:** The union of $\mathcal{A}$ and $\mathcal{B}$ is given by $\mathcal{A} \cup \mathcal{B}$. Since the union must contain any element that is in either of the original sets, we have

$$\mathcal{A} \cup \mathcal{B} = \{0, 1, 3, 5, 6, 7, 9\}$$

Notice that the elements in the union can be in either of the original sets or in both of the original sets. If there is an element that is contained in both of the original sets, that element is written only once in the union of the two sets. For example, the number 3 was in both of the original sets but was written only once in $\mathcal{A} \cup \mathcal{B} = \{0, 1, 3, 5, 6, 7, 9\}$. ■

The other set operation is the intersection operation. The **intersection** of two sets is a set consisting of all the elements that are in both sets. The symbol representing the intersection operation is $\cap$.

### Example 4

Find the intersection of the sets $\mathcal{A} = \{0, 3, 6, 9\}$ and $\mathcal{B} = \{1, 3, 5, 7, 9\}$.

**Solution:** The intersection of $\mathcal{A}$ and $\mathcal{B}$ is given by $\mathcal{A} \cap \mathcal{B}$. Since the intersection contains any element that is in both of the original sets, we have

$$\mathcal{A} \cap \mathcal{B} = \{3, 9\}$$ ■

In summary, we will be using the following symbols when we discuss sets.

## Symbols Used with Sets

| English | Symbol |
|---|---|
| Is an element of | $\in$ |
| Is not an element of | $\notin$ |
| Is a subset of | $\subset$ |
| Union | $\cup$ |
| Intersection | $\cap$ |
| Empty set | $\varnothing$ or $\{\ \}$ |

### *Translating Words into Set Operations*

In algebra, certain key words are translated into mathematical symbols or operations. For example, the word "**is**" often is represented by the symbol "=" when a word problem is translated into mathematical symbols. The same is true for set operations. When the word "**or**" is used, it may be represented in set operations by the union symbol.

#### Example 5

Find the set of whole numbers less than 13 that are divisible by 3 or 2.

**Solution:** The whole numbers less than 13 that are divisible by 3 can be written as the set $\mathcal{A} = \{0, 3, 6, 9, 12\}$. Similarly, the set of whole numbers less than 13 that are divisible by 2 is the set $\mathcal{B} = \{0, 2, 4, 6, 8, 10, 12\}$. Since the problem asks for those numbers that are divisible by 3 *or* 2, we want the union of $\mathcal{A}$ and $\mathcal{B}$.

$$\mathcal{A} \cup \mathcal{B} = \{0, 2, 3, 4, 6, 8, 9, 10, 12\}$$ ■

In a similar fashion, the word "**and**" is used to indicate the intersection of two sets.

#### Example 6

Find the set of whole numbers less than 13 that are divisible by 3 and 2.

**Solution:** As before, we let $\mathcal{A} = \{0, 3, 6, 9, 12\}$ and $\mathcal{B} = \{0, 2, 4, 6, 8, 10, 12\}$. Since the problem asks for those numbers that are divisible by 3 *and* 2, we want the intersection of $\mathcal{A}$ and $\mathcal{B}$.

$$\mathcal{A} \cap \mathcal{B} = \{0, 6, 12\}$$ ■

**Note:** It is important to remember that "and" and "or" have very specific meanings when used in connection with sets. In particular, "or" indicates that an element can be in either of the original sets or in both of the original sets. This contradicts the common usage meaning one set or the other but not both.

## Venn Diagrams

**Venn diagrams** (named after John Venn, a 19th-century English mathematician) are pictures depicting sets. Typically, a Venn diagram consists of a rectangular region containing several circles. Shown here is a Venn diagram consisting of two intersecting sets, $\mathcal{A}$ and $\mathcal{B}$, depicted by circles. The boundary rectangle is labeled $\mathcal{U}$, for universal set. The **universal set** is a set that contains all elements of the type under consideration. As we shall see in the following examples, Venn diagrams can be used to help us interpret mathematical problems.

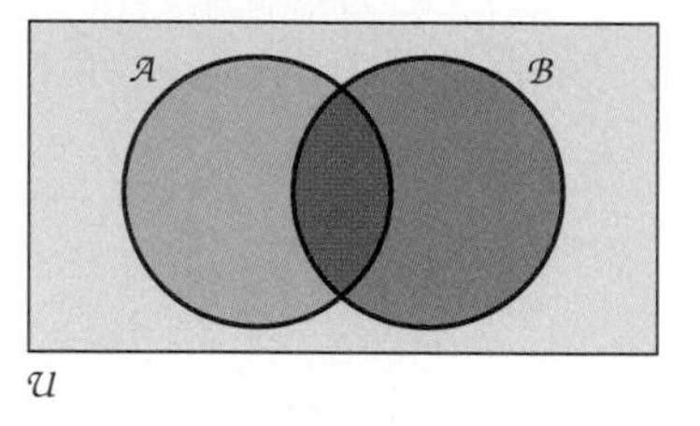


## Example 7

Examine the Venn diagram shown, where $\mathcal{U}$ is the set of all music, the circle $\mathcal{C}$ represents music available on cassettes, and the circle $\mathcal{D}$ represents music available on compact discs (CD's). Determine what is represented by each of the regions *a, b, c*, and *d*.

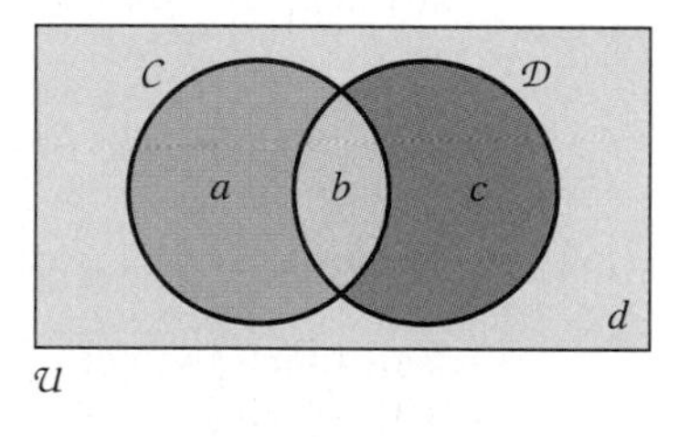


**Solution:** Region *a* is inside the circle $\mathcal{C}$ but outside the circle $\mathcal{D}$. Therefore, region *a* represents the music available on cassettes but not on CD's. Similarly, region *c* is inside the circle $\mathcal{D}$ but outside the circle $\mathcal{C}$. Therefore, region *c* represents the music available on CD's but not on cassettes.

Region *b* is inside both the circle $\mathcal{C}$ and the circle $\mathcal{D}$. Therefore, region *b* represents the music that is available on both cassettes and CD's.

Region *d* is outside of both circles. Therefore, region *d* represents the music that is not available on either cassettes or CD's. ■

## Example 8

Let

$\mathcal{Q}$ = the set of irrational numbers

$\mathcal{N}$ = the set of natural numbers = $\{1, 2, 3, 4, 5, \ldots\}$

$I$ = the set of integers = $\{\ldots -3, -2, -1, 0, 1, 2, 3, \ldots\}$

Draw the Venn diagram for these sets and label the regions as was done in Example 7. Use $\mathcal{U}$ = the set of real numbers.

**Solution:** In this example, the set $\mathcal{N}$ is a subset of $I$. Therefore, when we draw the Venn diagram, the set $\mathcal{N}$ should be completely enclosed inside the set $I$. In addition, $\mathcal{N}$ and $\mathcal{Q}$ have no common element, that is, $\mathcal{N} \cap \mathcal{Q} = \emptyset$. Thus they are shown as nonoverlapping circles.

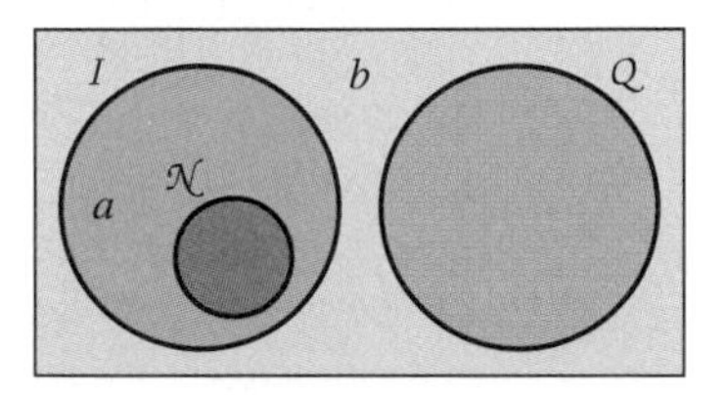


The region inside the circle $\mathcal{N}$ is the set of natural numbers. The region labeled $a$ is the set consisting of 0 and negative integers. The region $b$ contains all noninteger rational numbers. ■

## *Using Venn Diagrams*

Venn diagrams can answer specific questions about a situation. For each example, each region of a Venn diagram can be used to represent one of the possible outcomes of a problem. As each part of a problem is read, some of the regions are eliminated, leaving only the solution(s) to the problem. Because this method involves reading and drawing pictures, it is easy to use. It is limited, however, by the complexity of the pictures that can be drawn.

Let's examine a case in which we have a Venn diagram containing three circles, as shown in the figure. We will use this diagram to help understand the situation involving the intersection of 3 sets. The shaded section below shows $a \sim b \sim c$.

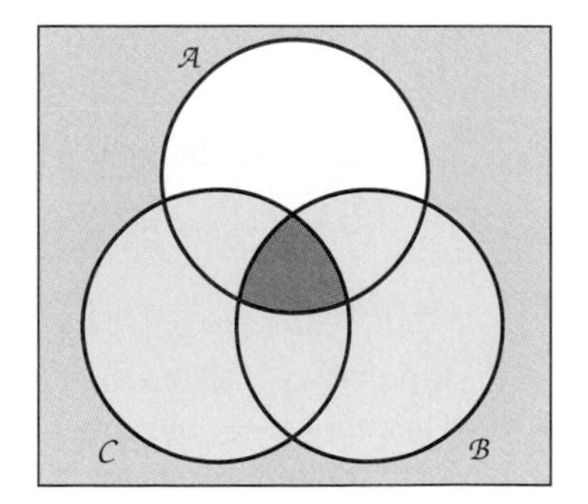

### Example 9

A survey of 200 employees determined that 22 of the employees do not have any medical insurance, while 43 of the employees are covered by the company's policy and an additional policy. If 160 employees are covered by the company's insurance policy, how many employees are covered only by some other policy?

**Solution:** To determine the solution of this problem, we will use three sets: $\mathcal{U}$, $C$, and $O$. The set $\mathcal{U}$ is the 200 employees in the survey. Let $C$ be the set of employees covered by the company's policy, and let $O$ be the set of employees covered by some other policy. Since 43 employees were covered by two or more policies, there are 43 people in the set $C \cap O$. Since there is a total of 160 employees covered by the company's policy, there must be $160 - 43 = 117$ employees covered only by the company's policy. Also, we know that there are 22 people who do not have any medical insurance. This information can be summarized in the Venn diagram shown here.

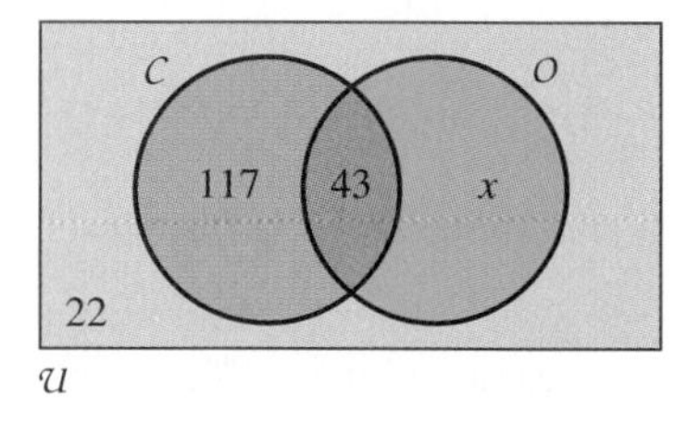


The intersection contains 43 people, the remainder of set $C$ contains 117 people, and there are 22 people not contained in either circle. The number of people who are insured only by some policy other than the company's are represented by $x$. Since the total number of people surveyed is 200, we have the following equation and solution:

$$22 + 117 + 43 + x = 200$$
$$182 + x = 200$$
$$x = 18$$

Therefore, there are 18 people whose only insurance is not through the company policy. ■

## *Paradoxes*

A **paradox** is a statement (or group of statements) that is in contradiction with itself. This very short introduction to paradoxes is intended to demonstrate how sets and Venn diagrams can be used to explain why a particular situation is considered a paradox.

## Example 10

After a meal at a Chinese restaurant in Lake Tahoe, California, one of the authors, Ron, received the following fortune in his fortune cookie:

*Don't take advice from a fortune cookie.*

Explain why this is a paradox.

**Solution:** This situation can be represented by a Venn diagram. There are two possibilities for Ron after reading this fortune. He can either take the advice of the fortune or he can ignore it. We will let $\mathcal{A}$ be the set of those who take the advice of the fortune cookie and let $I$ be the set of those who ignore the advice of the fortune cookie. Since he cannot both take the advice and ignore the advice, the intersection of the two sets is empty, $\mathcal{A} \cap I = \varnothing$.

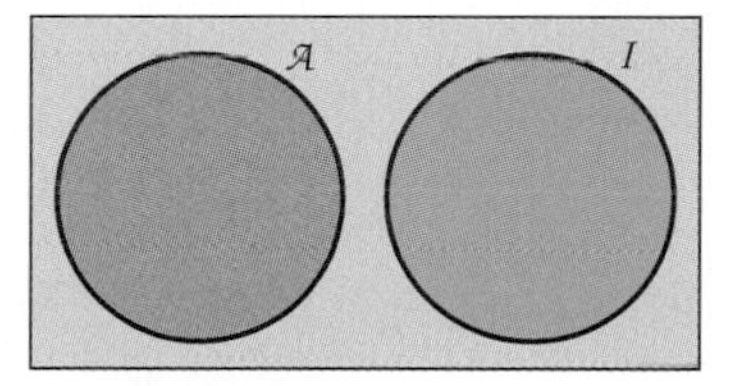


Suppose that when Ron reads the fortune, he decides to take its advice. That decision makes Ron an element of the set $\mathcal{A}$, Ron $\in \mathcal{A}$. However, by making that decision, Ron agrees to not take advice from a fortune cookie. This decision makes Ron an element of set $I$, Ron $\in I$. Since $\mathcal{A} \cap I = \varnothing$, his decision cannot be an element of both sets, as shown in the diagram by the disjoint sets.

If Ron's decision is to ignore the advice of the fortune, his decision is an element of $I$, Ron $\in I$. However, by ignoring the fortune, Ron is taking the advice of the fortune. Thus, Ron's decision is an element of $\mathcal{A}$, Ron $\in \mathcal{A}$. Again, since there is no intersection of $\mathcal{A}$ and $I$, it is not possible for Ron's decision to be an element of both $\mathcal{A}$ and $I$.

Since Ron cannot consistently take the advice of the fortune or ignore it, the fortune is considered a paradox. ■

As our second example of a paradox, we give the following puzzle.

## Example 11

Show that the following statement is a paradox.

*This sentence contaaans eggzactly three errors.*

**Solution:** There are two obvious errors in the sentence, the misspellings of the words "contains" and "exactly." Let's consider whether the word "three" is correct or not.

First, let $C$ be the set containing correct sentences and let $s$ be the given sentence. Then we have two choices, Either $s \in C$, or $s \notin C$. Now let's consider the word "three" in the given sentence.

Suppose that the sentence is correct, $s \in C$. Then the word "three" is correct. However, since, aside from the two misspellings, there are no other errors, there are only two errors. This means that the word "three" is incorrect and the sentence is not correct, $s \notin C$. This gives us a contradiction.

Now, suppose that the sentence is incorrect, $s \notin C$. Then the word "three" must be incorrect; so there are not three errors. But if the sentence is incorrect, and there are two spelling errors, there are three errors so the sentence is correct, $s \in C$. Since this again gives a contradiction, the sentence is a paradox. ■

## Section 4.1 PROBLEMS

## *Explain* ➡ *Apply* ➡ *Explore*

### *Explain*

1. What is a set?
2. What is the listing method of showing the items in a set?
3. What is the descriptive method of showing the items in a set?
4. What is meant by the intersection of two sets?
5. What is meant by the union of two sets?
6. What does it mean if a set $\mathcal{A}$ is a subset of set $\mathcal{B}$?
7. What does it mean for $\mathcal{A} \cup \mathcal{B} = \mathcal{B}$? Draw a picture to illustrate your reasoning.
8. What does it mean for $\mathcal{A} \cap \mathcal{B} = \varnothing$? Draw a picture to illustrate your reasoning.
9. If $\mathcal{A} \subset \mathcal{B}$ and $\mathcal{B} \subset \mathcal{A}$, what can you say about sets $\mathcal{A}$ and $\mathcal{B}$? Give an example.
10. What is a paradox?

### *Apply*

11. Write the set of even whole numbers less than or equal to 10, using the
    (a) Descriptive method
    (b) Listing method

12. Write the set of odd whole numbers less than or equal to 12, using the
    (a) Descriptive method
    (b) Listing method

13. Let $\mathcal{M}$ = {apples, bananas, peaches, tomatoes} and $\mathcal{T}$ = {beans, peas, sprouts, tomatoes}. Determine the following.
    (a) $\mathcal{M} \cup \mathcal{T}$
    (b) $\mathcal{M} \cap \mathcal{T}$

14. Let $\mathcal{F}$ = {oranges, apples, apricots, peaches} and $\mathcal{N}$ = {coconuts, filberts, almonds}. Determine the following.
    (a) $\mathcal{F} \cup \mathcal{N}$
    (b) $\mathcal{F} \cap \mathcal{N}$

In Problems 15–26, use the sets $\mathcal{T} = \{3, 6, 9, 12\}$, $\mathcal{F} = \{5, 10, 15, 20, 25\}$, $I$ = {all the integers} to determine if the subset and element symbols are used correctly. If the symbol is being used correctly, is the statement true? If not, write a true statement.
If the symbol is being used incorrectly, write a true statement using the correct symbol.

15. $\mathcal{F} \subset I$

16. $I \subset \mathcal{F}$

17. $\mathcal{T} \subset \mathcal{F}$

18. $\mathcal{T} \subset I$

19. $5 \subset I$

20. $5 \subset \mathcal{F}$

21. $5 \notin I$

22. $5 \notin \mathcal{F}$

23. $5 \in I$

24. $5 \notin \mathcal{T}$

25. $\mathcal{F} \in I$

26. $\mathcal{T} \notin I$

## *Explore*

27. Let $\mathcal{U}$ = the set of all people, $\mathcal{R}$ = the set of all people with red hair, and $\mathcal{W}$ = the set of all women. Draw a Venn diagram for these sets and describe the type of people contained in each region of the diagram.

**28.** Let $\mathcal{U}$ = the set of all books, $\mathcal{N}$ = the set of all novels, and $\mathcal{P}$ = the set of all books of poetry. Draw a Venn diagram for these sets and describe the type of books contained in each region of the diagram.

**29.** Let $\mathcal{U}$ = the set of all items in the library, $\mathcal{N}$ = the set of novels in the library, $\mathcal{B}$ = the set of books in the library, and $\mathcal{V}$ = the set of all videos in the library. Draw a Venn diagram for these sets and describe the type of item contained in each region of the diagram.

**30.** In order to construct the budget for English and English as a Second Language (ESL) teachers, the principal at Elizabeth Haddon Elementary School needs to figure out how many new students are not fluent in English. Let $\mathcal{U}$ = the set of all new students at the school, $\mathcal{E}$ = the set of new students whose first language is English, $\mathcal{S}$ = the set of new students whose first language is Spanish, and $\mathcal{V}$ = the set of new students whose first language is Vietnamese. Draw a Venn diagram for these sets and describe the type of people contained in each region of the diagram.

**31.** For the set $\mathcal{Q} = \{*\}$, there are two subsets, $\{\ \}$ and $\{*\}$.

(a) How many subsets can you find for the set $\mathcal{A} = \{*, \times\}$?
(b) How many subsets can you find for the set $\mathcal{B} = \{*, \times, \dagger\}$?
(c) Suppose a set $\mathcal{C}$ contains four elements. How many possible subsets are there?
(d) Suppose a set $\mathcal{D}$ contains 20 elements. Use the results from the previous parts of this problem to conjecture the total number of possible subsets.

**32.** The Venn diagrams presented have been constructed from circles and rectangles, but any shape may be used to construct Venn diagrams. Construct a Venn diagram, containing four sets $\mathcal{A}$, $\mathcal{B}$, $\mathcal{C}$, and $\mathcal{D}$, that has 16 distinct regions representing all possible intersections.

In Problems 33–40, use Venn diagrams to determine the solution to the problem.

**33.** A marketing firm is interested in what people are drinking with breakfast. The Venn diagram shown gives the results of a survey in which 300 people were asked if they drink coffee ($\mathcal{C}$) or orange juice ($\mathcal{O}$) for breakfast. Determine the number of people who drink both coffee and orange juice for breakfast.

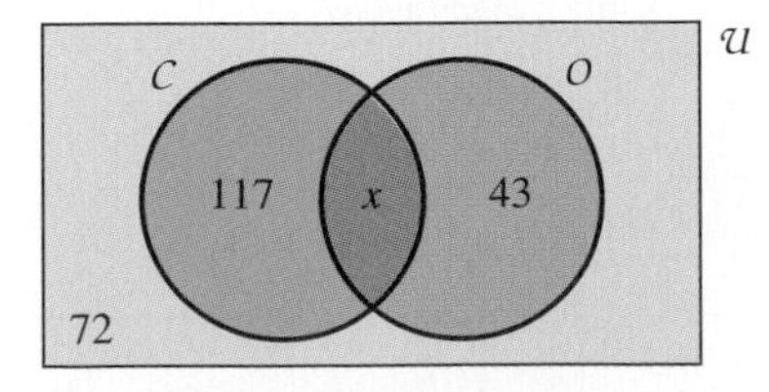


**34.** A marketing firm is interested in what people are drinking with breakfast. The Venn diagram shown gives the results of a survey in which 500 people were asked if they drink coffee ($\mathcal{C}$) or orange juice ($\mathcal{O}$) for breakfast. Determine the number of people who drink neither coffee nor orange juice for breakfast.

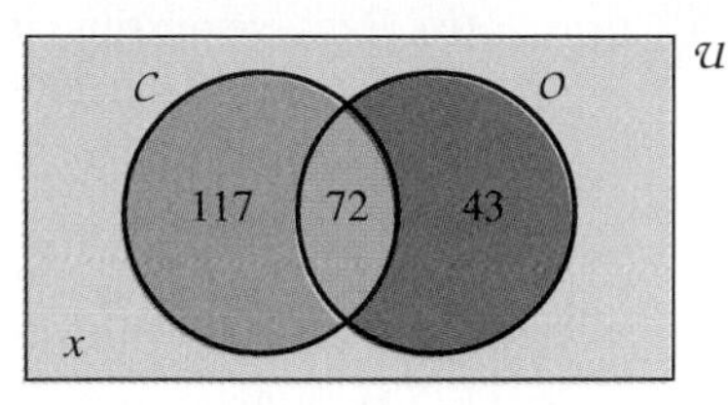


**35.** Given the Venn diagram shown, what can be said about $\mathcal{A} \cap \mathcal{B}$?

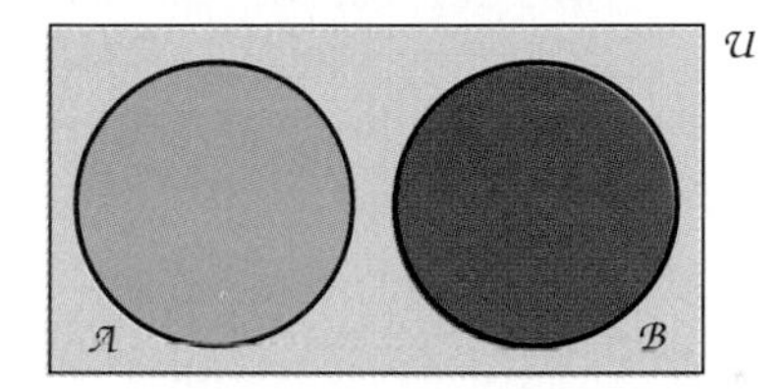


**36.** Draw a Venn diagram in which $\mathcal{A}$ contains a total of 30 elements, $\mathcal{B}$ contains a total of 50 elements, and $\mathcal{U}$ contains a total of 80 elements. (There are many possible answers.)

**37.** For marketing purposes, suppose the makers of CoffeeMate™ want to find out how many people put milk in their coffee or tea. A survey of 1000 people who drink coffee or tea has found that 625 people drink coffee with milk, 370 drink tea with milk, and 130 do not drink coffee or tea with milk. Draw a Venn diagram to represent this information and determine the number of people who drink both coffee and tea with milk.

**38.** Five hundred apples were examined for traces of pesticides. Forty of the apples showed traces of only malathion, 35 showed traces of both malathion and diazinon, and 420 showed traces of neither malathion nor diazinon. How many showed traces of only diazinon?

**39.** A survey of 1000 people determined the following results about their radio listening habits:

In the morning, between 5:00 and 7:00 A.M., 620 people listen to the radio.

During the evening, between 4:00 and 7:00 P.M., 640 people listen to the radio.

During the day, between 7:00 A.M. and 4:00 P.M., 450 people listen to the radio.

During all three periods, 210 people listen.

During the morning and the evening but not the day, 220 people listen.

During the morning and the day but not the evening, 70 people listen.

During the day and the evening but not the morning, 130 people listen.

Suppose the American Red Cross wants to broadcast a public service message.

(a) Because air time is expensive during the morning commute, the Red Cross is considering not broadcasting the message during the morning. How many people listen only in the morning and will therefore miss hearing the message?
(b) How many people listen only during the day?
(c) How many people listen only during the evening?
(d) How many people will not hear the message because they do not listen to the radio?

**40.** Suppose the American Heart Association is conducting a survey of 500 adults to determine their exercise habits. The results are as follows:

Two hundred forty-two adults participate in aerobics.

Two hundred seventy-eight adults participate in weight lifting.

Two hundred ninety-eight adults participate in running.

Forty-three adults participate only in aerobics and weight lifting.

Fifty adults participate only in aerobics and running.

Ninety-two adults participate only in weight lifting and running.

Seventy-five participate in all thrcc activities.

(a) How many participate only in aerobics?
(b) How many participate only in weight lifting?
(c) How many participate only in running?
(d) How many participate in none of these three activities?

***41.** Find an example of a paradox not contained in the book. Explain why the situation is a paradox.

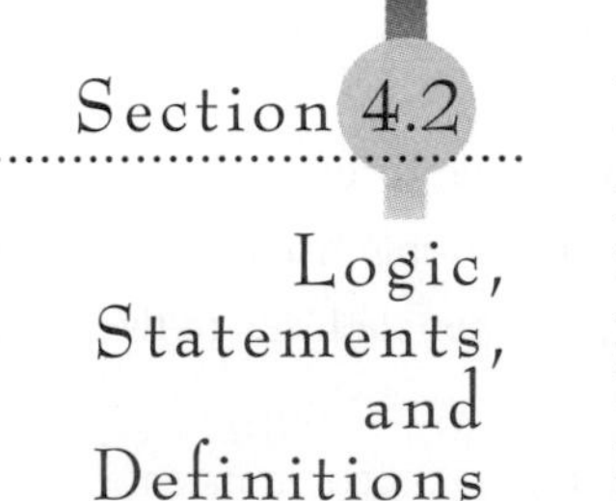


## Logic

Even though humans have the ability to think, they do not always reason correctly. The science of correct reasoning is called **logic**. An understanding of logic will help you correctly arrange supporting evidence that leads to a conclusion. It will also help you understand the process of proving mathematical facts. This process of proving mathematical propositions is probably one of the most challenging aspects of mathematics. The mathematician wants to be sure that a certain proposition actually follows logically from what has already been accepted or, just as importantly, that a proposition does not follow logically from what has been accepted. To a mathematician, creating an original proof is as gratifying as finishing a painting is to an artist, as setting a record is to an athlete, or as finding a cure to a disease is to a scientist.

## Statements

The basic components of logic are its statements. Statements in logic must have a clear meaning and be either true or false. Statements cannot be both true and false at the same time. They must have only one truth value—that is, either true or false. In general, questions, commands, or vague sentences cannot be used as statements in logic because they cannot be judged to be true or false. The following list gives examples of some sentences that are statements and some that are not.

| Statements | Nonstatements |
|---|---|
| Today is a holiday. | Are we having fun yet? |
| Pigeons fly. | Do your homework! |
| The square of 7 is 49. | Don't worry, be happy! |
| This book contains history sections. | It smells like whatchamacallit. |
| I did my homework. | This statement is true. |

In the study of logic, letters are used to represent statements just as letters are used to represent numbers in algebra. For example, the letter $Q$ can be used to represent the statement, ''All the players on this year's team are over 6 feet tall.''

### Example 1

For each of the following, classify $S$ as a statement or nonstatement.

(a) $S$: All men are mortal.
(b) $S$: Yea, team!
(c) $S$: Euclid did not study geometry.
(d) $S$: Finish your dinner.

**Solution:**

(a) $S$ is a statement.
(b) $S$ is a nonstatement.
(c) $S$ is a statement.
(d) $S$ is a nonstatement. ■

## *Negation of Statements*

If you change a statement to one that has the opposite meaning, you form the negation of the statement. The negation of a statement has a truth value that is the opposite of the truth value of the given statement. That is, if a statement is true, then its negation is false. Similarly, if a statement is false, its negation is true. If $S$ represents a statement, then $\sim S$ represents the negation of the statement and is read "not S." For example, if $S$ represents the statement "A triangle has three sides," then $\sim S$ represents the statement "A triangle does not have three sides." Notice that $S$ is true, but its negation $\sim S$ is false. In most statements, forming the negation is simply a matter of changing the action in the statement by adding or deleting the word "not." For example:

| Statement | Negation |
|---|---|
| Kai is running. | Kai is not running. |
| Kristi does not smile. | Kristi does smile. |
| The two amounts are equal. | The two amounts are not equal. |
| Logic is not a five-letter word. | Logic is a five-letter word. |
| He is a guitar player. | He is not a guitar player. |

In statements that involve the words "all," "every," "some," "none," or "no," forming the negation is not as easy as in the previous examples. For example:

| Statement | Negation |
|---|---|
| All men are mortal. | Some men are not mortal. |
| Some of the numbers are not positive. | All of the numbers are positive. |
| No birds are fish. | Some birds are fish. |
| Some women can swim. | No women can swim. |
| None of the flashlights worked. | Some of the flashlights worked. |

The basic forms for negating statements that involve "all," "every," "some," "none," or "no" can be summarized as follows:

| Statement | Negation |
|---|---|
| All/every | Some . . . not |
| Some . . . not | All/every |
| None/no | Some |
| Some | None/no |

**Example 2**

Write the negation of each of the following statements.

(a) My car did not start.
(b) Some of the cars did not start.
(c) None of the cars started.
(d) Every car started.
(e) Some of the cars started.

**Solution:**

(a) My car did start.
(b) All of the cars started.
(c) Some of the cars started.
(d) Some cars did not start.
(e) No car started. ■

## *Conditional Statements*

A very important type of statement used in logic is the **conditional statement**. It is a statement formed by two individual statements joined by the words "if . . . then . . . ." In the conditional statement "If $A$, then $B$," the letter $A$ represents the "if" clause or the **hypothesis** or **antecedent**, and the letter $B$ represents the "then" clause or the **conclusion** or **consequent**. For example, in the statement "If you are a student, then you should study," "you are a student" is the hypothesis and "you should study" is the conclusion. We will see in the next section that conditional statements are used extensively in formulating logical arguments.

The same conditional statement can be written in different ways.

| | |
|---|---|
| If $A$, then $B$ | If you are a student, then you should study. |
| $A$ implies $B$ | Being a student implies that you should study. |
| $A \to B$ | Being a student → one should study. |
| All $A$ are $B$ | All students should study. |

Related to the conditional statement $A \to B$ are three other basic types of statements.

| | |
|---|---|
| Converse: | $B \to A$ |
| Inverse: | $\sim A \to \sim B$ |
| Contrapositive: | $\sim B \to \sim A$ |

These statements are important because they are used in creating valid arguments. It can be shown that if a statement is true, then its contrapositive is always true, and if a statement is false, then its contrapositive is also false. That is, a conditional statement and its contrapositive are logically equivalent. However, if a statement is true, its inverse and converse may be either true or false. A conditional statement and its converse or its inverse are not logically equivalent.

---

**Note:** A conditional statement is logically equivalent to its contrapositive. A conditional statement can be replaced with its contrapositive and keep its same truth value. ($A \to B$ is logically equivalent to $\sim B \to \sim A$.)

---

This interrelationship between these statements can be seen by studying the following examples:

### Example 3

Write the converse, inverse, and contrapositive of the true conditional statement below. Determine if each of the statements is true or false.

If it is an IBM PC, then it is a computer.

**Solution:** Let

$A$: It is an IBM PC.

$B$: It is a computer.

$A \to B$: the conditional statement

Converse: $B \to A$: If it is a computer, then it is an IBM PC. (false)

Inverse: $\sim A \to \sim B$: If it is not an IBM PC, then it is not a computer. (false)

Contrapositive: $\sim B \to \sim A$: If it is not a computer, then it is not an IBM PC. (true) ■

### Example 4

Write the converse, inverse, and contrapositive of the false conditional statement below. Determine if each of the statements is true or false.

If $x$ is an even number, then the last digit of $x$ is 2.

**Solution:** Let

$A$: $x$ is an even number.

$B$: The last digit of $x$ is 2.

$A \to B$: the conditional statement

| | | |
|---|---|---|
| Converse: | $B \rightarrow A$: | If the last digit of $x$ is 2, then $x$ is an even number. (true) |
| Inverse: | $\sim A \rightarrow \sim B$: | If $x$ is not an even number, then the last digit of $x$ is not 2. (true) |
| Contrapositive: | $\sim B \rightarrow \sim A$: | If the last digit of $x$ is not 2, then $x$ is not an even number. (false) ■ |

### Example 5

Write the converse, inverse, and contrapositive of the true conditional statement below. Determine if each of the statements is true or false.

If two lines are perpendicular, then the two lines form a right angle.

**Solution:** Let

$A$: Two lines are perpendicular.

$B$: Two lines form a right angle.

$A \rightarrow B$: the conditional statement

| | | |
|---|---|---|
| Converse: | $B \rightarrow A$: | If two lines form a right angle, then the two lines are perpendicular. (true) |
| Inverse: | $\sim A \rightarrow \sim B$: | If two lines are not perpendicular, then the two lines do not form a right angle. (true) |
| Contrapositive: | $\sim B \rightarrow \sim A$: | If two lines do not form a right angle, then the two lines are not perpendicular. (true) ■ |

In Example 5, notice that the statement and its converse are both true statements; that is, $A \rightarrow B$ and $B \rightarrow A$ are both true. In such a situation, the two statements are combined into a **biconditional statement**, which is written in the following ways:

$A$ if and only if $B$

$A$ iff $B$

$A \leftrightarrow B$

Thus, the results of Example 5 could be written as follows:

Two lines are perpendicular if and only if the two lines form a right angle.

Two lines are perpendicular iff the two lines form a right angle.

Two lines are perpendicular $\leftrightarrow$ the two lines form a right angle.

## Definitions

Besides conditional statements, definitions of terms are used as basic building blocks of a mathematical system. A definition states properties of the term being defined, gives us a way of recognizing what is defined, and provides a way of distinguishing what is being defined from other objects. A definition must:

1. Name the term being defined.
2. Use words that have already been defined or already understood.
3. Be biconditional. The statement of the definition and its converse must both be true.

Besides those three necessary properties, a good definition should also:

1. Place the term in the smallest or nearest group to which it belongs.
2. Use the minimum information needed to distinguish the object from other objects.

### Example 6

Explain why the following statements are not examples of definitions.

(a) It is a place where tennis matches are played.
(b) Charity is the act of being eleemosynary.
(c) A mother is a parent of a child.
(d) A social insect is a bee.

**Solution:**

(a) The term that is being defined is not included.
(b) Charity is being defined by a word that is more difficult to understand and is probably not previously understood.
(c) The statement is not biconditional. The converse statement ''A parent of a child is a mother'' is false.
(d) The definition is not biconditional. The statement is false because a social insect could also be an ant. ■

### Example 7

Even though the following statements satisfy the three properties of a definition, they are not good definitions. Explain why.

(a) A treasurer is in charge of finances of an organization.
(b) A triangle is a plane figure with three sides, three angles, and three vertices.
(c) A Chevy is a Chevrolet automobile.

**Solution:**

(a) The group to which a treasurer belongs is not included. A treasurer is a *person* in charge of finances of an organization and not a machine, report, or computer program.

(b) More information is given than is needed to distinguish the object from all other plane figures.

(c) The definition does not place the term into the group to which it belongs. ''Chevy'' is a *slang* term for a Chevrolet automobile. ■

In this section, we have looked at the various kinds of statements and two ways of combining statements. We should pay particular attention to these concepts:

1. If a statement is true, its negation is false, and vice versa.
2. If a conditional statement is true, then its contrapositive is also true.
3. A definition must be biconditional.

## Section 4.2 PROBLEMS

## Explain ➡ Apply ➡ Explore

### Explain

1. Explain the difference between a statement and a nonstatement.
2. Explain how to form the converse of a statement.
3. Explain how to form the inverse of a statement.
4. Explain how to form the contrapositive of a statement.
5. Explain how to form the negation of a statement.
6. What is meant by a ''biconditional'' statement?

In Problems 7–14, explain why each statement is not an example of a definition.

7. A skean is a falchion.
8. To gasconade is to vaunt.
9. It is used to remove the skin of a potato.
10. We call it the period from noon to sunset.
11. An integer is a positive or negative number.
12. An obtuse angle is not a 90° angle.
13. A Toyota is an imported Japanese automobile.
14. A pencil is a writing implement.

In Problems 15–18, show how each statement satisfies the three necessary conditions of a definition.

15. A puppy is a young dog.

**16.** A crook is a person who steals or cheats.

**17.** A quadratic equation is an equation of the form $ax^2 + bx + c = 0$, where $a$, $b$, and $c$ are real numbers and $a$ is not equal to zero.

**18.** A rational number is a number that can be represented as the ratio of two integers $a/b$, where $b$ is not equal to zero.

### *Apply*

In Problems 19–26, determine which of the following are considered statements and which are not considered statements. Be sure to give your reasoning.

**19.** Don't eat the daisies!

**20.** My dog is a dalmatian.

**21.** Do you enjoy reading novels?

**22.** The jokes are great.

**23.** Mozart composed classical music.

**24.** The camera is not a Kodak™.

**25.** This statement is false.

**26.** Use the quadratic formula on that one.

In Problems 27–40, write the negation of the statements.

**27.** My car is in the shop.

**28.** Fred did not do his research paper.

**29.** I hate sitting around doing nothing.

**30.** The two lines are parallel.

**31.** That is an example of an exponential equation.

**32.** No rational number is irrational.

**33.** All fish can live under water.

**34.** Every chef knows how to boil water.

**35.** Some numbers are not prime numbers.

**36.** Some dogs do not have long tails.

**37.** Some trees are always green.

**38.** Some TV shows are boring.

**39.** None of the numbers are positive.

**40.** My uncle did not like what you did to his lawn.

In Problems 41–48 write the converse, inverse, and contrapositive of each conditional statement.

**41.** If you get a busy signal, the phone is in use.

**42.** If there is a leak in the tube, it will become flat.

**43.** If it is a point on the circle, then it will be 16 in. from the center of the circle.

**44.** If a whole number ends in 3, then it is an odd number.

**45.** When G. H. Mutton speaks, I listen.

**46.** When I am asleep, nothing bothers me.

**47.** If the figure has five sides, it is not a hexagon.

**48.** If it is an ellipse, then its graph is not a circle.

## *Explore*

In Problems 49–60, if possible, find an example of a conditional statement ($P$) that satisfies the stated condition.

**49.** $P$ is true and its inverse is false.

**50.** $P$ is true and its inverse is true.

**51.** $P$ is true and its converse is true.

**52.** $P$ is true and its converse is false.

**53.** $P$ is true and its contrapositive is false.

**54.** $P$ is true and its contrapositive is true.

**55.** $P$ is false and its inverse is false.

**56.** $P$ is false and its inverse is true.

**57.** $P$ is false and its converse is true.

**58.** $P$ is false and its converse is false.

**59.** $P$ is false and its contrapositive is false.

**60.** $P$ is false and its contrapositive is true.

For the terms in Problems 61–65, write definitions that satisfy the three necessary conditions of a definition.

**61.** Logic

**62.** Induction

**63.** Deduction

**64.** Syllogism

**65.** Paradox

**66.** Using a standard dictionary, find three words which have definitions that satisfy the three necessary conditons of a definition.

**67.** Using a standard dictionary, find three words which have definitions that do not satisfy the three necessary conditions of a definition.

**68.** Construct the contrapositive of the inverse of $A \rightarrow B$. What is this equivalent to?

**69.** Construct the inverse of the contrapositive of $A \rightarrow B$. What is this equivalent to?

The works of modern artists Donald Farnsworth (*left*, oil) and David McLaughlin (*right*, watercolor) show the continuing influence of the Greeks. (*Left*, courtesy of Magnolia Editions; *right*, courtesy of the artist)

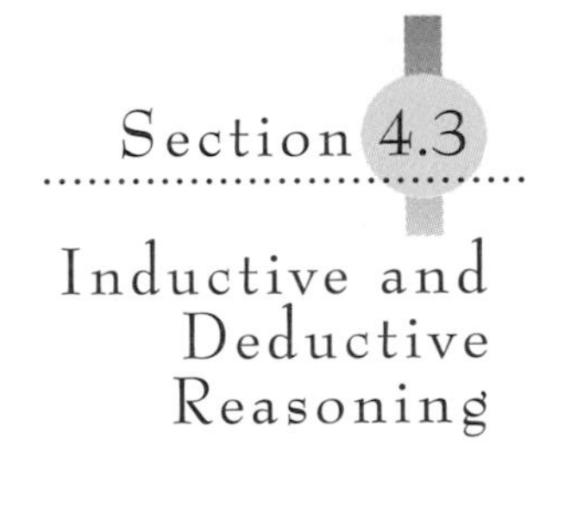

# Section 4.3 Inductive and Deductive Reasoning

We can now use the statements discussed in Section 4.2 to formulate arguments. An argument consists of statements of supporting evidence organized to show that a conclusion is true. Two of the reasoning processes used in creating an argument are induction and deduction.

## Induction

**Induction** is the process of reasoning in which conclusions are based on experimentation or experience. When using induction, we make a conclusion about a situation after observing results, analyzing experiences, citing authorities, or presenting statistics. We predict future experiences by extending patterns seen in present experiences.

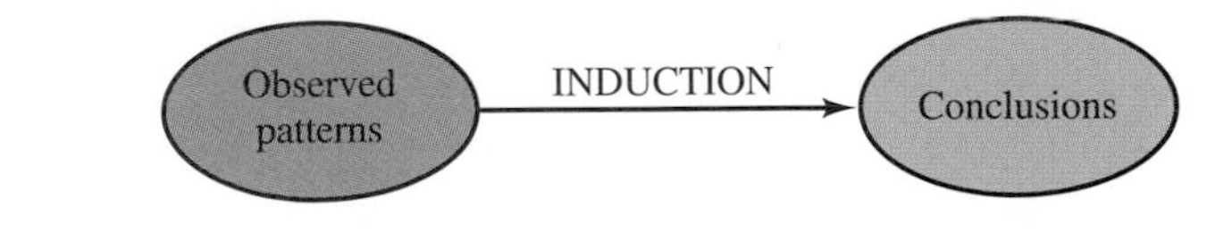


### Example 1

On a cold winter night there is a fire burning in the fireplace. A baby crawls up to the fireplace and touches the screen covering the fireplace. The baby burns his little hand and cries. A few weeks later, the baby does the same thing and burns his hand again. Because of these experiences, the baby stays away from the fireplace when he sees a fire burning. He has used the process of induction. Based on his experience, the baby has made the conclusion that touching an object heated by a fire will cause his hand to hurt. ■

## Example 2

In the triangles, use a ruler to measure each side and use a protractor to measure each angle. Is there a relationship between the length of a side and the size of the angle that is opposite that side, that is, between $\angle A$ and side $a$, $\angle B$ and side $b$, and $\angle C$ and side $c$?

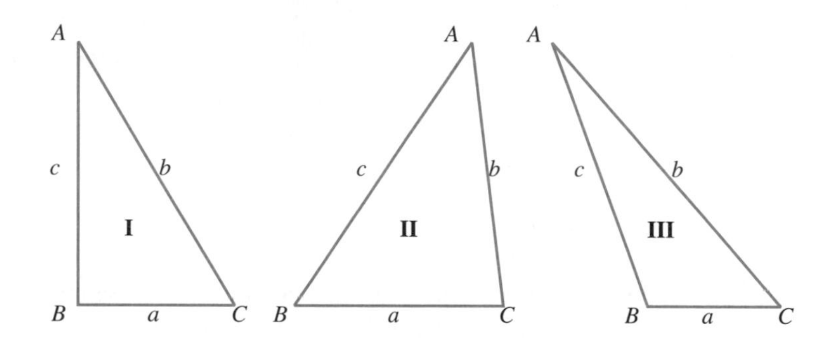


**Solution:**

| Triangle I | | Triangle II | | Triangle III | |
|---|---|---|---|---|---|
| $a = 2$ cm | $\angle A = 30°$ | $a = 2.7$ cm | $\angle A = 42°$ | $a = 1.7$ cm | $\angle A = 20°$ |
| $b = 3.9$ cm | $\angle B = 90°$ | $b = 3.4$ cm | $\angle B = 56°$ | $b = 4.4$ cm | $\angle B = 111°$ |
| $c = 3.3$ cm | $\angle C = 60°$ | $c = 4.1$ cm | $\angle C = 82°$ | $c = 3.6$ cm | $\angle C = 49°$ |

From studying the results we can conclude that in each triangle, opposite the longest side is the largest angle and opposite the shortest side is the smallest angle. ■

## Example 3

Consider the expression $n^2 - n + 11$. It seems to generate prime numbers when whole number values are substituted for $n$.

$$n = 0 \quad n^2 - n + 11 = 0 - 0 + 11 = 11$$
$$n = 1 \quad n^2 - n + 11 = 1 - 1 + 11 = 11$$
$$n = 2 \quad n^2 - n + 11 = 4 - 2 + 11 = 13$$
$$n = 3 \quad n^2 - n + 11 = 9 - 3 + 11 = 17$$
$$n = 4 \quad n^2 - n + 11 = 16 - 4 + 11 = 23$$
$$\vdots$$

Using inductive reasoning, you might conclude that the expression $n^2 - n + 11$ always generates prime numbers for whole numbers, $n$. This, however, is not true. When $n = 11$, the value of the expression is 121, and 121 is not a prime number. By not investigating a sufficient number of cases, we could have made a false conclusion. ■

Inductive reasoning is the process of determining a general conclusion by examining individual cases or particular facts. It can show us that there is a good chance that our conclusion is true, but we will not be absolutely certain. For example, if the first ten people you meet at a new school are very helpful and friendly, you may generalize that the people at the school are really nice. However, the next person you meet may be extremely hostile. Based on the first ten people, your conclusion seemed true. However, the 11th person disproved your conjecture. When reasoning inductively, one has to make sure that there is a sufficient number of facts or specific cases to warrant a conclusion. Scientists, for example, repeat experiments many times before making conclusions. Again, a good inductive argument only gives a high probability that a statement is true or an action should be performed.

### Example 4

Give an inductive argument to persuade your friend Maria to vacation in Hawaii.

**Solution:** The following list of premises gives an example of an inductive argument.

1. The weather is great in Hawaii, and the beaches are fantastic.
2. My friend has a condo you can rent for only $150 a week.
3. The airlines are having a special on fares to Hawaii this month.
4. I went there last year and had a wonderful time.
5. The people there were friendly and treated me kindly.
6. There are lots of nice guys vacationing there. You'll have a great time and are bound to meet that special man you have been looking for.

7. In a recent travel magazine, 95% of the vacationers polled said they enjoyed their vacation in Hawaii.
8. Anna Holiday, worldwide traveler and economist, in her book *Travels to Paradise*, states that a vacation to Hawaii is the best bet for your travel dollar.

Such an argument shows that vacationing in Hawaii makes good sense. It implies that there is a good chance that Maria would enjoy a vacation in Hawaii based on past experience, statistics, and the opinions of experts. However, even if all the premises of this argument were true. Maria could still have a miserable time in Hawaii. ■

Even though inductive reasoning does not ensure certainty, it is the basis of many everyday decisions and is used to extend our knowledge by making suppositions based on experimentation. If certainty is desired, we can in some situations use a second reasoning process, deduction.

## Deduction

**Deduction** is the process of reasoning in which conclusions are based on accepted premises. These premises are usually articles of faith, laws, rules, definitions, assumptions, or commonly accepted facts. The conclusions we reach are either explicitly or implicitly contained in the premises.

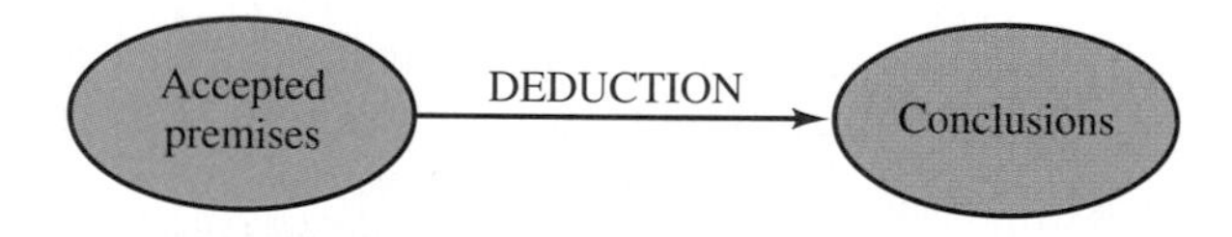


A deductive argument is a series of statements consisting of premises and a conclusion. The premises are the statements of evidence from which the conclusion is drawn. In deductive arguments, the premises are usually written as conditional statements. Arguments may take many different forms. One of the common forms is the **syllogism**. The basic syllogism consists of two statements or premises, and a logical conclusion drawn from them. According to Aristotle, ''A syllogism is a discourse in which, certain things being posited, something else follows from them by necessity.'' In this chapter, we discuss three types of syllogisms: hypothetical syllogisms, affirming the antecedent, and denying the consequent.

### *Hypothetical Syllogism*

If $A$, $B$, and $C$ represent statements, a **hypothetical syllogism** is constructed from the statements, the first two lines being the premises and the third being the conclusion. The hypothetical syllogism can be written in three different ways:

| | | |
|---|---|---|
| $A \rightarrow B$ | *A* implies *B*. | If *A*, then *B*. |
| $B \rightarrow C$ | *B* implies *C*. | If *B*, then *C*. |
| $\therefore A \rightarrow C$ | Therefore, *A* implies *C*. | ∴ If *A*, then *C*. |

In the hypothetical syllogism, the argument is correct even when one or both of the premises are false. The truth or falsehood of the premises does not affect the logic of the argument. Logic deals with the relation between premises and conclusion, not the truth of the premises. To say that a deductive argument is correct means that the premises are related to the conclusion in such a way that, if the premises are true, the conclusion must be true. A conclusion cannot be false if the logical form is correct and the premises are true.

**Note:** To reason deductively toward a true conclusion using a syllogism, you must have the correct form and true premises.

Thc following are examples of correct deductive arguments that use hypothetical syllogisms and lead to true conclusions.

If you live in Palolo, then you live on Oahu.
If you live on Oahu, then you live in Hawaii.
Therefore, if you live in Palolo, then you live in Hawaii.

If a triangle is isosceles, then it has two equal sides.
If a triangle has two equal sides, then it has two equal angles.
Therefore, an isosceles triangle has two equal angles.

### Example 5

Is the following argument a hypothetical syllogism? Why or why not?

If you have a party, you should invite your friends.

If you are graduating from college, you should invite your friends.

Therefore, if you are having a party, you are graduating from college.

**Solution:** The argument is not a hypothetical syllogism. The premises do not link properly. The conclusion of the first premise should be the hypothesis of the second premise, and no logical rearrangement can accomplish the proper linking of the statements. ■

### Example 6

Even though the conclusion of this argument is true, explain why the following argument is a poor one.

If you are over 18 years old, then you can read.

If you can read, you can vote.

Therefore, if you are over 18 years old, then you can vote.

**Solution:** The argument has the form of a hypothetical syllogism so it is a correct argument. However, it is a poor argument, since neither of the premises are true. The argument does not actually prove its conclusion.

## *Affirming the Antecedent*

If $A$ and $B$ represent statements, an argument that affirms the antecedent has the form

| | |
|---|---|
| Major premise: | $A \rightarrow B$ |
| Minor premise: | $A$ |
| Conclusion: | $\therefore B$ |

The major premise is a conditional statement. The minor premise states that the hypothesis of the major premise is true or has occurred. This is called **affirming the antecedent**. An example of an argument of this form is as follows.

If I study for 6 hours, I will pass the exam.

I studied for 6 hours.

Therefore, I will pass the exam.

This classical argument is another example of affirming the antecedent.

All men are mortal.

Socrates is a man.

Therefore, Socrates is mortal.

This can be rewritten so that the correct form is apparent.

If one is a man, then one is mortal.

Socrates is a man.

Therefore, Socrates is mortal.

If an argument has the correct form, it is a logical argument. However, if it is to be a convincing argument with a true conclusion, its premises must also be true. You can affirm the antecedent to reason deductively if the argument has the correct form and true premises.

**Example 7**

Is the following argument a good one? Explain.

> If you want to run a marathon, then you should train for the race.
>
> Kerry wants to run a marathon.
>
> Therefore, Kerry should train for the race.

**Solution:** The argument has the correct form for affirming the antecedent. If we take its first premise as true because of commonly accepted notions about the physical stamina needed to run a marathon (26.2 mi), the argument is a good one. ■

**Example 8**

What is wrong with the following argument?

> All good chess players wear glasses.
>
> Sylvia is a good chess player.
>
> Therefore, Sylvia wears glasses.

**Solution:** Rewriting the argument into conditional statements, we get the following:

> If one is a good chess player, then one wears glasses.
>
> Sylvia is a good chess player.
>
> Therefore, Sylvia wears glasses.

Even though the argument has the correct form of an argument using the technique of affirming the antecedent, the major premise is not true. Thus, it is a correct argument but it does not arrive at a true conclusion. You need both the correct form and true premises to ensure true conclusions. ■

### *Denying the Consequent*

If $A$ and $B$ represent statements, an argument that denies the consequent has the form

| | |
|---|---|
| Major premise: | $A \rightarrow B$ |
| Minor premise: | $\sim B$ |
| Conclusion: | $\therefore \sim A$ |

Examples of arguments of this form are as follows:

If John is at the beach, then he wears sun screen on his nose.
John does not have sun screen on his nose
Therefore, John is not at the beach.

If you pay the bill on time, then you are not charged a penalty.
You are charged a penalty.
Therefore, you did not pay the bill on time.

The major premise is a conditional statement. The minor premise is a denial (negation) of the consequent (conclusion) of the conditional statement. For this reason this argument is called **denying the consequent**. This form of argument is based on the contrapositive principle in which the statement $A \rightarrow B$ is logically equivalent to $\sim B \rightarrow \sim A$. We can see that this form of argument is correct by observing that it is really an application of affirming the antecedent.

| | | | |
|---|---|---|---|
| Major premise: | $A \rightarrow B$ | is equivalent to | $\sim B \rightarrow \sim A$ |
| Minor premise: | $\sim B$ | | $\sim B$ |
| Conclusion: | $\therefore \sim A$ | | $\therefore \sim A$ |

## Example 9

Is the following argument a good one? Explain.

If a number is not positive, then the number is negative.
Zero is not negative.
Therefore, zero is positive.

**Solution:** The argument has the form of an argument using denying the consequent, so it is logically correct. However, its first premise is not true, since if a number is not positive it could be either negative or zero. Thus, the argument is faulty. ■

We can also make correct arguments from premises that do not at first glance seem to be one of our standard logical forms, as in the next example.

## Example 10

Construct a logically correct argument from the following premises.

If $P$, then $\sim Q$.
If $\sim R$, then $Q$.
If $R$, then $\sim S$.

**Solution:** To have the correct form of the hypothetical syllogism, the conclusion of one statement must be the hypothesis of the next statement. Since we know that if a statement is true its contrapositive is true, we can use that principle on the second premise, that is "If $\sim R$, then $Q$" implies "If $\sim Q$, then $R$."

| | | |
|---|---|---|
| If $P$, then $\sim Q$. | | If $P$, then $\sim Q$. |
| If $\sim R$, then $Q$. | $\rightarrow$ | If $\sim Q$, then $R$. |
| If $R$, then $\sim S$. | | If $R$, then $\sim S$. |
| | | $\therefore$ If $P$, then $\sim S$. |

Thus, the conclusion of this argument is: If $P$, then $\sim S$. ■

While you may use the contrapositive statement in a logical argument, do not use the inverse or converse.

---

**Watch out for the following:**

**Converse:** $A \rightarrow B$ does *not* necessarily imply $B \rightarrow A$.
**Inverse:** $A \rightarrow B$ does *not* necessarily imply $\sim A \rightarrow \sim B$.

---

Although there are other forms of syllogisms and methods of reasoning deductively, the three forms of syllogisms explained in this section will be adequate for our brief excursion into logic.

## The Roles of Induction and Deduction

You have been introduced to two methods of reaching reasonable conclusions, induction and deduction. Both processes play a role in the formulation of a mathematical system of geometry. Induction is used to conjecture facts about geometry. Early Egyptians, Babylonians, and Greeks used experience and experimentation to hypothesize relationships about geometric objects. So too, modern mathematicians discover mathematical relationships by looking at specific cases and examples and generalizing their findings. The deductive process is used to determine if the findings follow logically from what has been previously accepted as true in the mathematical system. This process is sometimes very difficult. For example, it took nearly 2000 years to prove that, in general, it is impossible to trisect an angle with only a straight edge and a compass.

Section 4.3
PROBLEMS

## *Explain ➡ Apply ➡ Explore*

### *Explain*

**1.** Explain what is meant by ''inductive reasoning.''

**2.** Explain what is meant by ''deductive reasoning.''

**3.** What is a syllogism?

**4.** Explain what is meant by ''affirming the antecedent.''

**5.** Explain what is meant by ''denying the consequent.''

**6.** Give an example where a statement is true but the converse of the statement is not true.

**7.** Give an example where a statement is true but the inverse of the statement is not true.

**8.** Explain why it is possible to have a correct argument even if the conclusion is false.

### *Apply*

Determine whether or not the following arguments are correct. For those that are not correct, (a) explain what is wrong with the argument; (b) change the minor premise and make a correct argument.

**9.** When it is midnight, I am asleep.
I was asleep.
Therefore, it was midnight.

**10.** All NBA basketball players are over 5 ft tall.
Russell is 6 ft tall.
Therefore, Russell plays in the NBA.

**11.** If you are a farmer in Polt County, then you grow corn.
Farmer Ron does not farm in Polt County.
Therefore, Farmer Ron does not grow corn.

**12.** All Rhode Island Red hens lay brown eggs.
My hen, Marguerite, is a Rhode Island Red.
Therefore, Marguerite lays brown eggs.

**13.** If *ABCD* is a square, it has four sides.
If it has four sides, then it is a quadrilateral.
Therefore, if *ABCD* is a square, it is a quadrilateral.

**14.** If a triangle is equilateral, then it has three equal sides.
$\triangle ABC$ does not have three equal sides.
Therefore, $\triangle ABC$ is not equilateral.

Create valid deductive arguments for the following statements.

**15.** All even numbers greater than 2 are not prime numbers.

**16.** If you are a shoplifter, you are dishonest.

**17.** My teacher, Mrs. Santos, is a nice person.

The deductive arguments based on the hypothetical syllogism have two premises and a conclusion. An argument may, however, have many premises that logically lead to a conclusion. What statement is proven in each of Problems 18–20?

**18.**
$$3(x + 4) = 18 \rightarrow 3x + 12 = 18$$
$$3x + 12 = 18 \rightarrow 3x = 6$$
$$3x = 6 \rightarrow x = 2$$

**19.** If you are serious about school, you should be a good student.
If you are a good student, you should study regularly.
If you study regularly, you will have less time to watch TV.

**20.** If $A$, then $B$.
If $B$, then $C$.
If $C$, then $\sim D$.
If $\sim D$, then $\sim E$.

**21.** In Problem 20, find statements for $A$, $B$, $C$, $D$, and $E$ that will make the argument correct and the conclusion true.

## *Explore*

**22.** Example 2 dealt with triangles with sides of unequal lengths. The triangles here have at least two sides with the same length. What conclusions can you induce from analyzing the lengths of their sides and the measures of their angles?

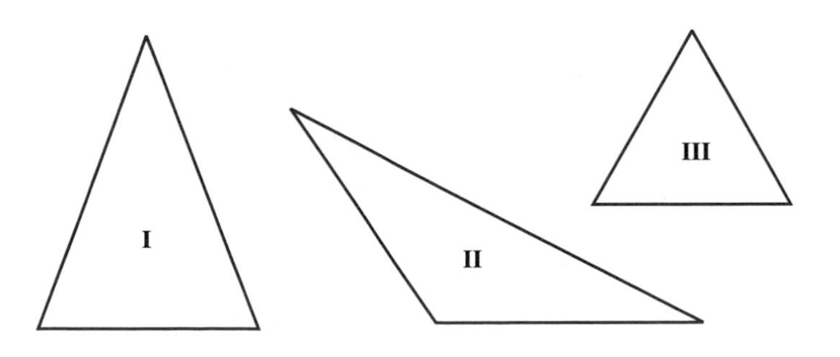

**23.** Combine the results of Example 2 and Problem 22 to formulate an accurate statement about the relationship between the lengths of the sides of triangles and the angles opposite those sides. Draw a few triangles that verify your statement.

**24.** Draw any triangle. Find the midpoints of two sides of the triangle and the length of the third side of the triangle. Draw the line segment connecting the two midpoints. Find the length of that segment. State the relationship that seems to exist between the length of the third side and the length of the segment joining the two midpoints. Test out your findings on a few other triangles.

**25.** Draw any quadrilateral (four-sided figure). Find the midpoint of each of the four sides. Use line segments to join the midpoints (in order) so that another quadrilateral is formed. Repeat this for several other quadrilaterals. What seems to be true about the quadrilaterals formed by connecting the midpoints?

**26.** Explain how Problems 22–25 are examples of inductive reasoning. Explain what role deduction would play in further analyzing the geometric principles involved in each problem.

**27.** Use a compass and straight edge to draw a large semicircle and its diameter. Mark any point on the semicircle and draw line segments joining each end point of the diameter to the point marked on the semicircle. Use a protractor to measure the angle formed by the two line segments. Repeat this for several other points marked off on the semicircle. What seems to be true about the angle formed by the line segments connecting the end points of the diameter to a point on a semicircle?

**28.** Show that the expression $n^2 - n + 17$ seems to generate prime numbers for whole number values of $n$. Explain how using induction might cause you to make a false conclusion in this case. What is the first value of $n$ that causes the expression to produce a composite number?

**29.** Give an inductive argument to convince a friend that he should not vacation in Hawaii.

**30.** Give both an inductive argument and a deductive argument to convince a friend to quit smoking.

**31.** Explain what type of argument is used to show that a frying chicken purchased at a local grocery store weighs less than 4 lb.

(a) Every frying chicken sold over the last 25 years has weighed less than 4 lb. Thus, the next chicken you purchase at a local grocery store will surely weigh less than 4 lb.

(b) If a frying chicken weighs 4 lb or more, it is not sold in grocery stores.
I purchased this frying chicken in a local grocery store.
Therefore, it must weigh less than 4 lb.

The following premises are not in the exact syllogistic form shown in this section. However, they can be rearranged or logically rewritten so that they correspond to one of the types of syllogisms. What conclusion follows from these premises?

**32.** All mice are not birds.
Some pets are mice.

**33.** Every integer is a rational number.
Some numbers are integers.

**34.** If you can hear, you are not hearing impaired.
If you can't hear, I'll use sign language.

**35.** If the car is full of gasoline, we can drive 300 mi.
If we haven't reached Yosemite Park, we haven't driven 300 mi.
If we can't see Vernal Falls, we haven't reached Yosemite Park.

**36.** $P \leftrightarrow Q$
$P \rightarrow R$
$S \rightarrow \sim R$

**37.** $P \leftrightarrow Q$
$Q \rightarrow \sim R$
$S \rightarrow R$

**38.** Using your solution to Problem 36, find statements for $P$, $Q$, $R$, and $S$ that make the conclusion true.

**39.** Using your solution to Problem 37, find statements for $P$, $Q$, $R$, and $S$ that make the conclusion true.

## Section 4.4 Symbolic Logic and Truth Tables

One of the common beliefs about mathematics is that everything can be turned into abstract symbols that must follow certain sets of rules. It is also believed that these rules always lead the mathematician to a correct solution of whatever problem is at hand and that there is no other correct answer. As we have begun to see, this is not always the case. Previous chapters have discussed developments in number systems and geometry systems that have led to different, and sometimes conflicting, systems of mathematics. Which system is best is not easy to say, particularly in the case of geometry. Still, mathematicians often try to systematize their work so that, as often as possible, a certain set of information always leads to one and only one solution.

In this chapter, we have seen several methods of logical thinking. This section covers what is, perhaps, the most mathematical method of logical thinking, symbolic logic. We have seen a small portion of symbolic logic in earlier sections. For example, when sets were discussed in Section 4.1, intersections were represented by the symbol ∩. In the section covering statements, the conditional statement ''If $A$, then $B$'', was abbreviated by using the symbols $A \rightarrow B$. In this section, we provide a list of symbols and their meanings, explain how to translate English into these symbols, and show how proofs may be developed using these symbols.

One important note before we start concerns our philosophy about this material. You should treat this material as you did the material in Chapter 1 on ancient numeration systems. We do not expect you to be an expert in symbolic logic. We merely want to expose you to another area of mathematics, to give you a view of one more of the many facets of mathematics.

## Translating English to Symbols

The following is a partial list of the symbols used in symbolic logic. While looking through the chart, you may have recognized the symbols ∴ and ~. They are found in earlier sections of this chapter. The remaining symbols have not been used, but some of the English terms for the symbols have been used. For example, when discussing Venn diagrams and sets, we used the symbol ∩ to indicate an intersection of two sets. It represents the common English word ''and.'' In logic, the symbol for ''and'' is ∧ and it is called a conjunction.

Symbols Used in Logic

| English | Symbol | Name |
|---|---|---|
| Not | $\sim$ | Negation |
| Therefore | $\therefore$ | |
| Implies: If . . . then . . . . | $\rightarrow$ | Conditional |
| Equivalent statements | $\equiv$ | Logical equivalency |
| And | $\wedge$ | Conjunction |
| Or | $\vee$ | Inclusive disjunction |

Our next step is to see how these symbols can be used to abbreviate English sentences.

### Example 1

Translate the following sentences into symbols.

(a) If it is raining, the skies are cloudy.

(b) If it is a dog and it can run quickly, then it has four legs.
(c) If my grade is not a C and my grade was better than average, then my grade must be an A or a B.

**Solution:**

(a) This is a conditional statement. If we let $R$ = it is raining and $C$ = the skies are cloudy, we can write

$$R \rightarrow C$$

(b) The premise of this statement involves two conditions, the dog and running quickly. To handle this, we need to use a conjunction as well as the conditional statement. Letting $D$ = dog, $R$ = run quickly, and $F$ = has four legs, we can write

$$(D \wedge R) \rightarrow F$$

We use the parentheses to enclose the portion of the statement that must be considered first.
(c) Like the earlier statements, this is a conditional. It contains a conjunction in the antecedent and a disjunction in the consequent. Letting $G$ = better than average grade and $A$, $B$, and $C$ represent those letter grades, we have

$$(\sim C \wedge G) \rightarrow (A \vee B)$$ ■

In our next example, we want to see how an entire syllogism can be translated into symbolic logic.

### Example 2

Translate the following syllogism into symbols.

If I own chickens, I can have fresh eggs for breakfast.

If I have fresh eggs for breakfast, I am in a good mood until lunchtime.

Therefore, if I own chickens, I am in a good mood until lunchtime.

**Solution:** We will use $C$ = owning chickens, $E$ = having fresh eggs for breakfast, and $G$ = being in a good mood until lunchtime. Using these letters and two of the symbols from the chart, we can write the syllogism as three statements:

$$C \rightarrow E$$
$$E \rightarrow G$$
$$\therefore C \rightarrow G$$ ■

**Example 3**

Translate the following paragraph into a set of symbolic logic statements.

> If the ground is rocky and the horse needs new shoes, its feet will be sore. However, if the horse does not need new shoes, its feet will not be sore. Therefore, either the ground is not rocky and the horse's feet are not sore, or the horse needs new shoes.

**Solution:** First, let $R$ = the ground is rocky, $N$ = the horse needs new shoes, and $S$ = the horse's feet are sore. Using these letters, we can abbreviate the paragraph as

If $R$ and $N$, then $S$.
If $\sim N$, then $\sim S$.
Therefore, $\sim R$ and $\sim S$, or $N$.

The problem now looks more approachable. The only difficult part remaining is that the first statement contains both a conjunction and a conditional statement whereas the third statement contains both a conjunction and a disjunction. To enable us to handle complex statements of this type, we use parentheses.

If ($R$ and $N$), then $S$.
If $\sim N$, then $\sim S$.
Therefore, ($\sim R$ and $\sim S$), or $N$.

Now, using our symbols for the conjunction and conditional, we have

$$(R \wedge N) \to S.$$
$$\sim N \to \sim S.$$
$$\therefore (\sim R \wedge \sim S) \vee N.$$ ■

With the completion of this last example, we have seen how a complicated paragraph in English can be reduced to a few short sentences in logic. To complete this section, we show how logical arguments can be used to answer questions about complicated situations. We do this by using a mathematical structure called a truth table.

## Truth Tables

A **truth table** is a chart consisting of all the possible true and false combinations of the clauses in a statement. Certain sequences of logical statements, such as syllogisms, have been discussed in earlier sections, but we have not been able to analyze the situation presented in Example 3. With care and patience, a truth table can provide this analysis.

The following three truth tables are the basis for all truth tables. They handle the possible situations that can arise in the disjunction, conjunction, and conditional statements.

## The Disjunction Statement

A **disjunction** is a logic statement used in connection with the English word "or." As a result, a disjunction is considered true whenever one or both of its clauses are true. Thus, for the disjunction $A \vee B$, we have the following truth table:

### Disjunction Truth Table

| $A$ | $B$ | $A \vee B$ |
|---|---|---|
| T | T | T |
| T | F | T |
| F | T | T |
| F | F | F |

Notice that the value of $A \vee B$ is true unless both $A$ and $B$ are false. This corresponds to our intuitive idea of the meaning of an *or* statement.

## The Conjunction Statement

A **conjunction** is a logic statement used in connection with the English word "and." A conjunction is true only if *both* of its clauses are true. As a result, we have the following truth table for conjunctions:

### Conjunction Truth Table

| $A$ | $B$ | $A \wedge B$ |
|---|---|---|
| T | T | T |
| T | F | F |
| F | T | F |
| F | F | F |

## *The Conditional Statement*

A **conditional** is a logic statement used when the statement is in the form ''if . . . then . . . .'' A conditional statement is logically true in three of the four possible cases, as shown here.

### *Conditional Truth Table*

| $A$ | $B$ | $A \rightarrow B$ |
|---|---|---|
| T | T | T |
| T | F | F |
| F | T | T |
| F | F | T |

At first, it may seem peculiar that the conditional statement is true even when the first clause (the antecedent) is false. To understand this, we need to realize what is meant by ''logically true.'' Recall from our work on deductive reasoning that if the premise of a syllogism is false, we can prove any statement based on the false assumption. Similarly, if the antecedent of a conditional statement is false, any conclusion is considered logically true. Therefore, the only situation that causes the conditional statement to be false is when the first clause (the antecedent) is true and the second clause (the consequent) is false.

Now that we have these three truth tables, we have the basic building blocks for all truth tables. We can use these tables to determine when a statement is logically true.

### Example 4

Use a truth table to determine the truth values of $A$ and $B$ that make the statement $\sim A \wedge \sim B$ logically true.

**Solution:** As in the three previous truth tables, we have rows that give all the possible truth values for $A$ and $B$.

| $A$ | $B$ | $\sim A$ | $\sim B$ | $\sim A \wedge \sim B$ |
|---|---|---|---|---|
| T | T | F | F | F |
| T | F | F | T | F |
| F | T | T | F | F |
| F | F | T | T | T |

The first two columns of the truth table give all possible combinations of the truth values of $A$ and $B$, and the second two columns of the truth table are merely negations of the first two columns. The statement $\sim A \wedge \sim B$ is the conjunction of the two statements $\sim A$ and $\sim B$. Therefore, the only case in which $\sim A \wedge \sim B$ is true is when both $A$ and $B$ are false. ■

## Example 5

Determine the truth values of $A$ and $B$ that make the statement $\sim(A \vee B)$ logically true.

**Solution:** As before, the truth table consists of four rows. The third column of the truth table gives the truth values for the disjunction $A \vee B$. The fourth column gives the truth values for the negation of the disjunction. Therefore, $\sim(A \vee B)$ is true only when both $A$ and $B$ are false.

| $A$ | $B$ | $A \vee B$ | $\sim(A \vee B)$ |
|---|---|---|---|
| T | T | T | F |
| T | F | T | F |
| F | T | T | F |
| F | F | F | T |

■

### *Equivalent Statements*

Notice that the truth values of $\sim(A \vee B)$ in Example 5 and $\sim A \wedge \sim B$ in Example 4 are identical. Whenever two statements have the same truth values, the statements are said to be **logically equivalent**. Referring to our chart used to convert English to symbols (found at the beginning of the section), we find that the symbol for equivalent statements is $\equiv$. Thus we can write

$$\sim(A \vee B) \equiv \sim A \wedge \sim B$$

## Example 6

Use a truth table to show that the conditional $A \rightarrow B$ and its contrapositive $\sim B \rightarrow \sim A$ are logically equivalent.

**Solution:** To do this, we construct a truth table containing both $A \rightarrow B$ and $\sim B \rightarrow \sim A$ and show that the two statements always have the same truth values. Being careful about the order of $\sim B$ and $\sim A$ in the fourth and fifth columns of the truth table shows us that the truth values of $A \rightarrow B$ and $\sim B \rightarrow \sim A$ are the same and, therefore, that $A \rightarrow B \equiv \sim B \rightarrow \sim A$.

| $A$ | $B$ | $A \rightarrow B$ | $\sim B$ | $\sim A$ | $\sim B \rightarrow \sim A$ |
|---|---|---|---|---|---|
| T | T | **T** | F | F | **T** |
| T | F | **F** | T | F | **F** |
| F | T | **T** | F | T | **T** |
| F | F | **T** | T | T | **T** |

## *Verifying Syllogisms*

A second use of truth tables is to verify the conclusion of a logical argument. To do this, we form a conditional statement. The hypothesis of the conditional will consist of a conjunction of all the premises. The conclusion of the conditional will be the conclusion of the syllogism. A syllogism has a true conclusion if the conditional formed in this manner is always true.

### Example 7

Use a truth table to verify that the following syllogism has a true conclusion:

$$P \rightarrow Q$$

$$P$$

$$\therefore Q$$

**Solution:** To verify the syllogism, we form a conditional statement whose hypothesis is the conjunction of the premises and whose conclusion is the conclusion of the syllogism. Doing so gives us the statement $[(P \rightarrow Q) \wedge P] \rightarrow Q$. Next we construct a truth table with this statement as its last column. If the last column always has a "true" value, the syllogism has a true conclusion.

| $P$ | $Q$ | $P \rightarrow Q$ | $(P \rightarrow Q) \wedge P$ | $[(P \rightarrow Q) \wedge P] \rightarrow Q$ |
|---|---|---|---|---|
| T | T | T | T | T |
| T | F | F | F | T |
| F | T | T | F | T |
| F | F | T | F | T |

Since the last column of this truth table is always true, the syllogism has a true conclusion. ■

## Example 8

Determine whether the following logical argument is correct:

$$P \rightarrow Q$$

$$\sim Q$$

$$\therefore P$$

**Solution:** To determine whether the argument is correct, we form a conditional from the premises and the conclusion of the argument. Doing so gives us the statement $[(P \rightarrow Q) \wedge \sim Q] \rightarrow P$. Next, we construct a truth table with this statement as its last column. If the last column always has a ''true'' value, the argument is correct.

| $P$ | $Q$ | $P \rightarrow Q$ | $\sim Q$ | $(P \rightarrow Q) \wedge \sim Q$ | $[(P \rightarrow Q) \wedge \sim Q] \rightarrow P$ |
|---|---|---|---|---|---|
| T | T | T | F | F | T |
| T | F | F | T | F | T |
| F | T | T | F | F | T |
| F | F | T | T | T | F |

Since the last column of this truth table is not always true, the argument is not correct. ■

## Example 9

Use a truth table to determine whether the following syllogism has a true conclusion:

$$S \rightarrow R$$

$$N \rightarrow S$$

$$\therefore N \rightarrow R$$

**Solution:** If $N \rightarrow R$ is the conclusion of $S \rightarrow R$ and $N \rightarrow S$, then the statement $[(S \rightarrow R) \wedge (N \rightarrow S)] \rightarrow (N \rightarrow R)$ must always be true.

Setting up a truth table for this statement requires eight lines because there are three variables: $R$, $N$, and $S$.

| $S$ | $R$ | $N$ | $S \to R$ | $N \to S$ | $(S \to R) \wedge (N \to S)$ | $N \to R$ | $[(S \to R) \wedge (N \to S)] \to (N \to R)$ |
|---|---|---|---|---|---|---|---|
| T | T | T | T | T | T | T | T |
| T | T | F | T | T | T | T | T |
| T | F | T | F | T | F | F | T |
| T | F | F | F | T | F | T | T |
| F | T | T | T | F | F | T | T |
| F | T | F | T | T | T | T | T |
| F | F | T | T | F | F | F | T |
| F | F | F | T | T | T | T | T |

Since the last column always has a true value, the conclusion must be true. ■

In summary, this section has presented a few topics from symbolic logic. The important topics are

**1.** The English to symbol dictionary

| **English** | **Symbol** | **Name** |
|---|---|---|
| Not | $\sim$ | Negation |
| Therefore | $\therefore$ | |
| Implies, if . . . then . . . | $\to$ | Conditional |
| Equivalent statements | $\equiv$ | Logical equivalency |
| And | $\wedge$ | Conjunction |
| Or | $\vee$ | Inclusive disjunction |

**2.** The three basic truth tables

| Disjunction Truth Table | | | Conjunction Truth Table | | | Conditional Truth Table | | |
|---|---|---|---|---|---|---|---|---|
| $A$ | $B$ | $A \vee B$ | $A$ | $B$ | $A \wedge B$ | $A$ | $B$ | $A \to B$ |
| T | T | T | T | T | T | T | T | T |
| T | F | T | T | F | F | T | F | F |
| F | T | T | F | T | F | F | T | T |
| F | F | F | F | F | F | F | F | T |

**3.** Truth tables can be used to determine whether two statements are logically equivalent and to determine the correctness of a logical argument.

## Section 4.4 PROBLEMS

## *Explain* ➡ *Apply* ➡ *Explore*

### *Explain*

**1.** What is a disjunction? What is the logic symbol used to represent a disjunction?

**2.** What is a conjunction? What is the logic symbol used to represent a conjunction?

**3.** What is a conditional? What is the logic symbol used to represent a conditional?

**4.** What does it mean for two statements to be logically equivalent? What is the logic symbol used for logical equivalence?

**5.** What is a truth table? What is one of the uses of a truth table?

**6.** What are the three basic truth tables? When is each used?

### *Apply*

In Problems 7–16, translate the given sentence(s) into symbols. State what each symbol represents.

**7.** If it was midnight, I was asleep.
I was asleep.
Therefore, it was midnight.

**8.** If a person plays basketball in the NBA, he is over 5 ft tall.
Russell (a person) plays NBA basketball.
Therefore, Russell is over 5 ft tall.

**9.** If you are a farmer in Polt County, then you grow corn.
Farmer Ron does not grow corn.
Therefore, Farmer Ron does not farm in Polt County.

**10.** If *ABCD* is a square, it has four sides.
*ABCD* does not have four sides.
Therefore, *ABCD* is not a square.

**11.** If $\triangle ABC$ is not scalene, then it has either two or three equal sides.
$\triangle ABC$ does not have three equal sides.
Therefore, $\triangle ABC$ has two equal sides.

**12.** If the correct answer is not yes, then the correct answer is either no or maybe.
The correct answer is not no.
Therefore, the correct answer is maybe.

**13.** If you are a winning professional golfer, then you have good hand-eye coordination and a positive attitude. Therefore, if you have good hand-eye coordination and a positive attitude, then you are a winning professional golfer.

**14.** Bill will study hard and pass the test, or Bill will not study hard and not pass the test.

**15.** If you voted for Bush or Perot, then you did not vote for Clinton.

**16.** If Joan plays sports or works out, then Joan will not be fat and will not be lazy.

In Problems 17–20, use a truth table to determine the truth values of $A$ and $C$ that make each statement true.

**17.** $(A \to C) \lor \sim C$

**18.** $A \to (C \lor \sim C)$

**19.** $(\sim A \to C) \to \sim C$

**20.** $(A \to C) \to \sim C$

In Problems 21–24, use a truth table to determine if the two statements are equivalent.

**21.** $\sim(A \land \sim B)$ and $\sim A \lor B$

**22.** $\sim A \to \sim B$ and $B \to A$

**23.** $(A \lor B) \land (A \lor \sim B)$ and $A$

**24.** $(A \land B) \lor (A \land \sim B)$ and $A$

In Problems 25–28, use truth tables to determine if the arguments are logically correct.

**25.** $P \rightarrow Q$
$P$
$\therefore Q$

**26.** $\sim Q \rightarrow \sim P$
$P$
$\therefore Q$

**27.** $\sim A$
$A \rightarrow Q$
$\therefore A$

**28.** $B$
$B \rightarrow \sim C$
$\therefore \sim C$

## *Explore*

In Problems 29–32, translate each argument into symbolic form and determine whether each is logically correct by constructing a truth table.

**29.** If elections become TV popularity contests, then good looking, smooth-talking candidates will get elected. Therefore, if elections do not become TV popularity contests, then good looking, smooth-talking candidates will not get elected.

**30.** If affirmative action policies are adopted, then minorities will be hired. If minorities get hired, then discrimination will be addressed. Therefore, if affirmative action policies are adopted, then discrimination will be addressed.

**31.** If high school graduates have poor math skills, then they will be less able to get a job in the computer industry. If high school graduates have poor writing skills, then they will be less able to get a job in the computer industry. Therefore, if high school graduates have poor math skills, then they have poor writing skills.

**32.** Money causes all the world's troubles, or money helps the poor. If money helps the poor, it is not the cause of all the world's troubles. Money is the cause of all the world's troubles. Therefore, money does not help the poor.

**33.** Determine if the argument in Example 3 is correct.

**34.** Lewis Carroll, author of *Alice in Wonderland*, was also an accomplished logician. One of his logic problems was the following.

No ducks waltz.

No officers ever decline to waltz.

All my poultry are ducks.

Therefore, officers are not my poultry.

If we let $D$ = ducks, $P$ = my poultry, $O$ = officers, and $W$ = willing to waltz, the problem becomes

$$D \rightarrow \sim W$$
$$O \rightarrow W$$
$$P \rightarrow D$$
$$\therefore O \rightarrow \sim P$$

Construct a 16-row truth table to give all the possible true-false combinations of $D$, $W$, $O$, and $P$. Use the truth table to verify if the conclusion of the argument is true.

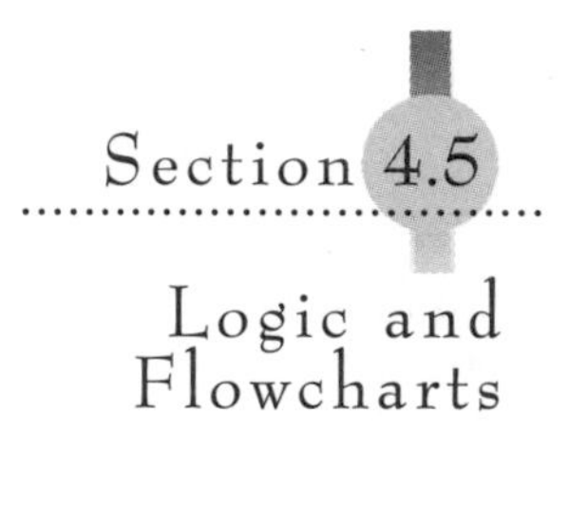

## Section 4.5 Logic and Flowcharts

Logic is used in many areas. It is particularly evident in a diagram called a flowchart. A **flowchart** describes the logical path that is followed when a decision is made. Formal flowcharts are used in computer programming to outline the decisions and steps that are followed by a computer as it performs its operations. Industry uses flowcharts to indicate how a process will occur as it progresses through different levels. Corporations, schools, and government bureaus also use flowcharts to display their organizational or management structures. In this section, we will examine how flowcharts are used to outline the logical solution to a problem or the structure of an organization.

### Symbols in Flowcharts

We will use four basic symbols: the start/stop symbol, the statement symbol, the decision symbol, and the flow line in constructing flowcharts. While a more thorough treatment of flowcharts will include more than these four symbols, they will be sufficient for our needs.

The **start/stop symbol** is an oval and indicates the beginning or end of a line of logic.

The **statement symbol** is a rectangle and is used to indicate either an action, person or result.

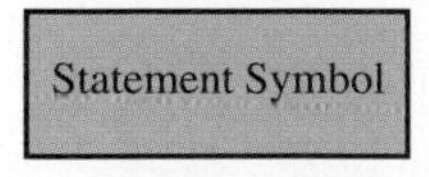

The **decision symbol** is a diamond-shaped object known as a rhombus. The decision symbol indicates a question in the flow chart.

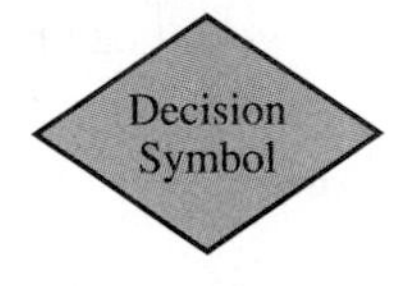


The final symbol, the **flow line**, is a directed line segment (a ray) that connects the other symbols and describes the path that will be followed as you progress through the flowchart.

Flow Line

## Constructing a Flowchart

To construct a flowchart, it is first necessary to have a clear idea of what situation the flowchart is to describe and what information is necessary to completely describe the situation. The following two examples show how a flowchart can be used to describe increasingly complex situations.

### Example 1

A company wants to construct a flowchart that represents the method it will use in answering a customer's telephone call. The initial plan is to have the telephone receptionist route the incoming calls to either the business department or the technical staff, depending on the nature of the call.

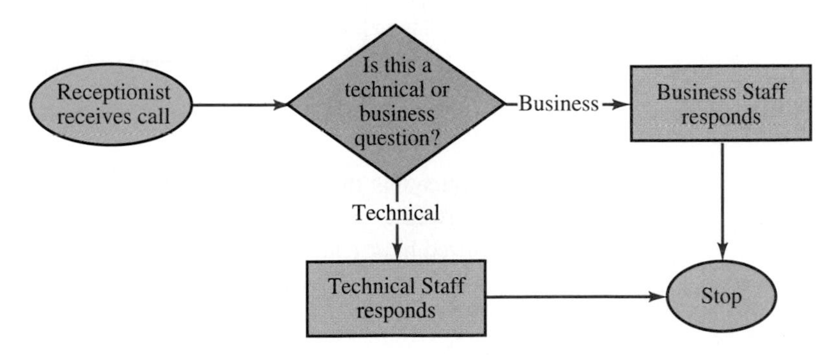


**Solution:** The flowchart drawn up by the company is shown in the figure. In a fairly straightforward process, the receptionist is expected to answer each call and forward it to the appropriate department. A member of that department will then respond to the call and the process ends. ■

The process described in the preceding example may seem straightforward and logical. However, because of the simplicity of the flowchart, it may not describe the process actually followed by the people working in the company. In such an instance, the flowchart can be expanded to handle a more complicated situation.

### Example 2

Occasionally, a customer will call the company and have a question that must be responded to by both the technical and business staffs. The customer may also have more than one question. How can the company add the capability to handle these situations into the flowchart?

**Solution:** After discussions among the staff in both departments and the telephone receptionist, it is decided that the best approach is to allow either department to transfer a call to the other department. The resulting flowchart is shown in this figure.

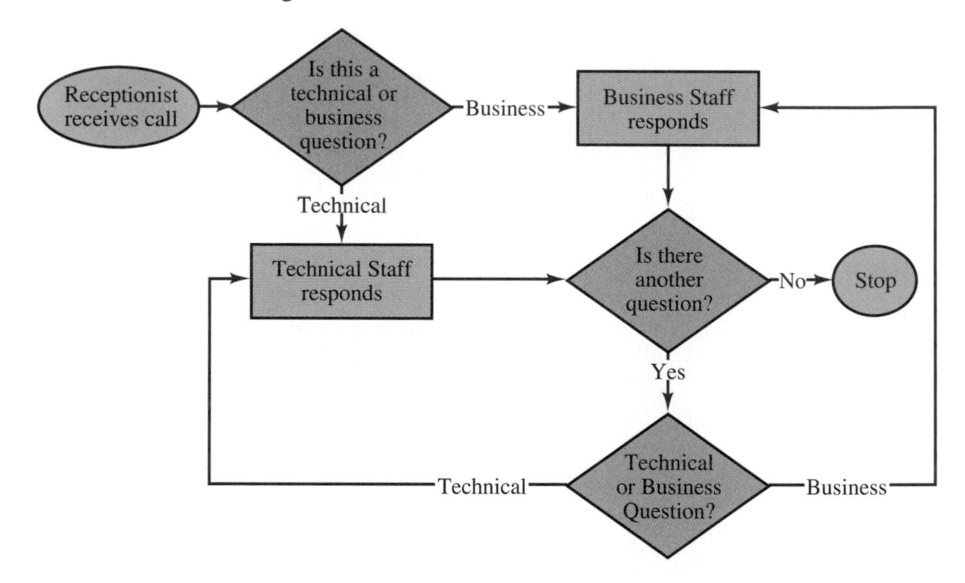


Notice that this new flowchart allows two new possibilities. The customer is allowed to ask more than one question in any department and also to be transferred back and forth between the two departments more than once. Only when the customer has no more questions does the process reach an end. ■

Throughout this chapter, we have shown how logic can be used to solve problems. The process of solving problems is one of analyzing both the available information and the desired goal and then determining a logical method of reaching that goal. For a simple problem, this can be accomplished without the use of flowcharts. However, as the problem becomes more involved, a flowchart provides a visual map to guide you to a solution. In the following example, we show how to construct a flowchart to answer a particular question.

## Example 3

Early one morning, your car won't start. Before calling for a tow truck, you want to make sure that the cause is not a simple problem that you can fix yourself. What steps should be used in reaching the decision to call the tow truck?

**Solution:** Since the goal is to call the tow truck only after all the simple causes for the car to not start are eliminated, the plan is as follows:

1. List the possible simple causes.
2. Determine the appropriate action if one of these causes is at fault.
3. Call the tow truck only as a last resort.

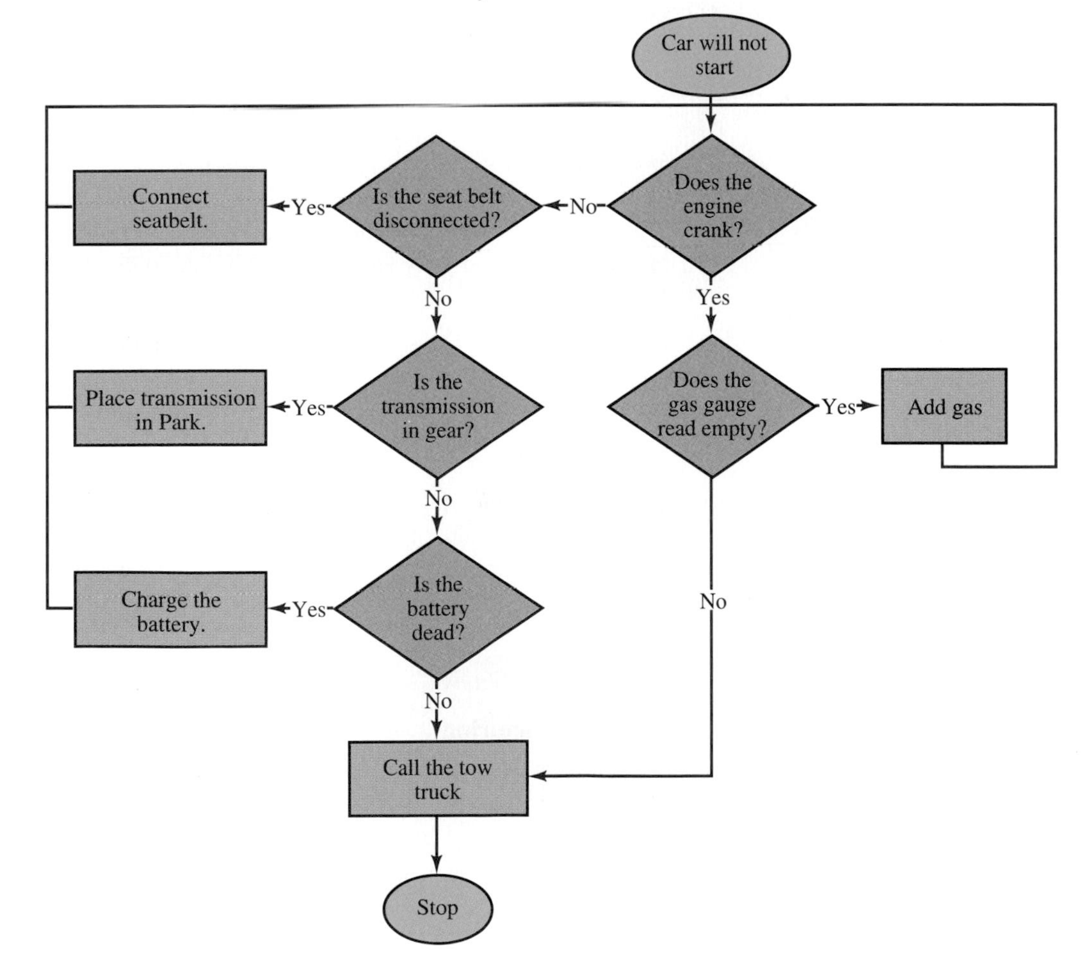

Although you are not a mechanic, you understand that if the engine is not receiving the correct amount of gasoline, the engine will crank but will not start. Alternatively, if the problem is electrical in nature, the engine will not crank. Thus, determining if the engine will crank is the first decision.

If the engine cranks but does not start, the car has an empty gas tank or there is some other problem with the fuel system. If the tank is empty, it can be partially filled with gas from a spare can in the tool shed. If there is some other problem with the fuel system, you will call the tow truck.

If the engine will not crank, you know of three possible causes: the seat belt is not connected, the automatic transmission is in gear, or the battery is dead. If none of these is the cause, you decide to call the tow truck. ■

## Organizational Charts

Our final example of how flowcharts are used is to describe the organizational structure of a company. While the flowchart itself does not contain any decision blocks, it does provide a diagram of how decisions are made within the company.

### Example 4

You have been given the following list of names and positions for the Vallejo Manufacturing Company and have been asked to create an organizational chart.

| | |
|---|---|
| **President** | P. Vallejo |
| **V.P. Finance** | P. Dang |
| Accounts Payable/Receivable | T. Tran |
| Controller | M. Yotter |
| **V.P. Engineering** | T. Hopkins |
| Engineering | T. Lee |
| Documentation | C. Wilson |
| **V.P. Marketing** | B. DeSousa |
| Sales | M. Yoshinobi |
| Customer Service | M. Sanchez |
| **V.P. Manufacturing** | M. Agredano |
| Purchasing | F. White |
| Assembly | J. O'Connor |
| Shipping | T. Root |

**Solution:** The organizational chart for Vallejo Manufacturing is shown in the figure.

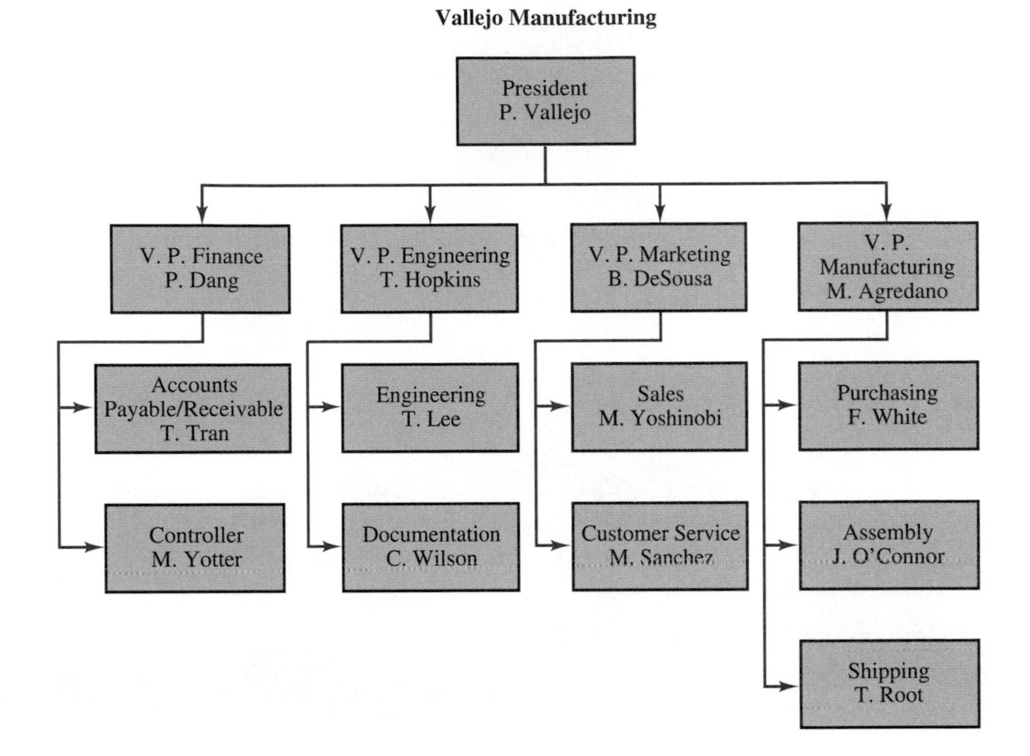


Based on the organizational chart, the chain of command in the company becomes clear. If a major policy change in the shipping department is being discussed, it must be approved first by T. Root, then by the V.P. of Manufacturing, M. Agredano, and finally by the President, P. Vallejo. ■

The three situations described in this section show how a flowchart can be an effective way to organize the logical path to the solution of a problem. The construction of a flow chart forces you to think clearly and organize an efficient solution to a problem. In essence, a flowchart is a picture of logic in use.

## Section 4.5 PROBLEMS

## *Explain ➡ Apply ➡ Explore*

### *Explain*

1. What is the purpose of a flowchart?
2. What is the purpose of a flow line in a flowchart? What symbol is used?
3. What is the decision symbol and how is it used in a flowchart?
4. What is the statement symbol and how is it used in a flowchart?

**5.** What is the purpose of an organizational chart?

**6.** Are decision statements used in organizational charts? Explain why or why not.

**7.** Explain the statement, ''A flowchart is a picture of logic in use.''

### *Apply*

In Problems 8 and 9, give a written description of the flowchart.

**8.**

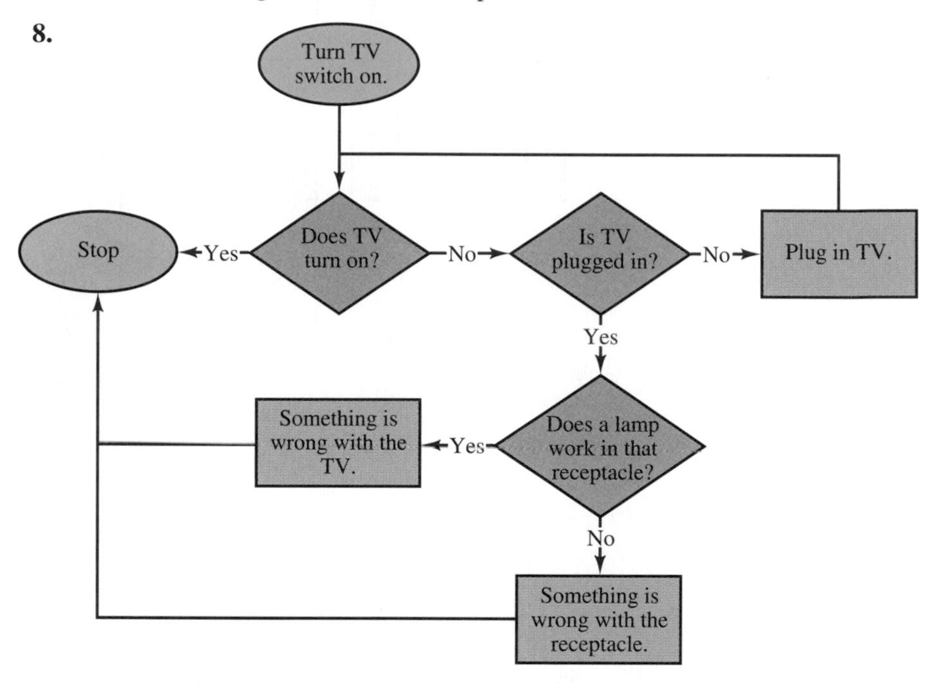

**9.**

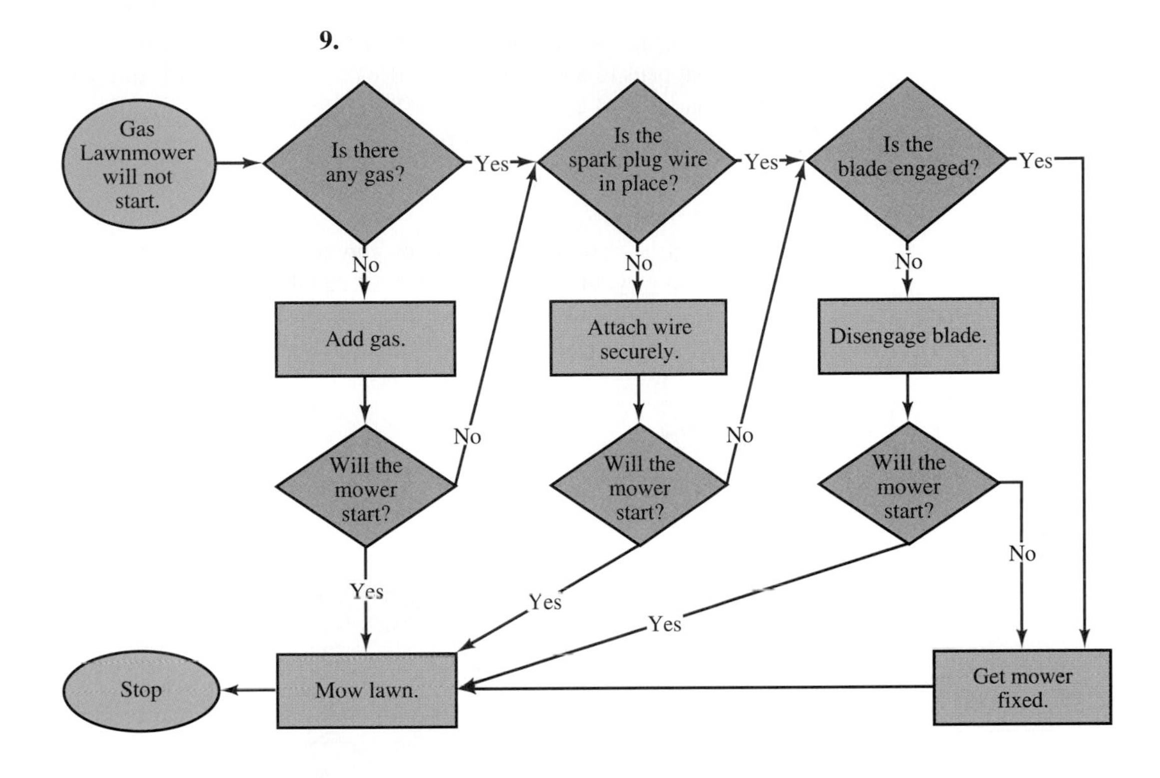


**10.** Given the following flowchart,

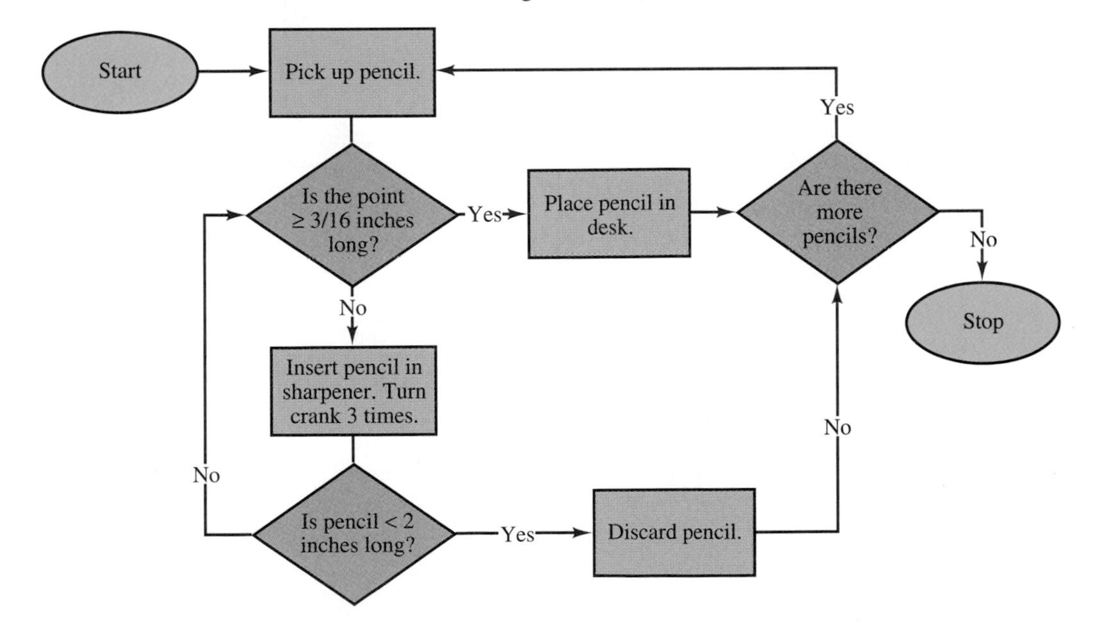

(a) What happens to a 4-in.-long pencil with a point that is $\frac{1}{4}$ in. long?
(b) What happens to a 6-in.-long pencil with a point that is $\frac{1}{8}$ in. long?
(c) What happens to a 1-in-long pencil?
(d) What happens to a 5-in.-long pencil?
(e) What happens to a pencil that has a point that keeps breaking as you sharpen it?

**11.** The following is a possible strategy for playing Blackjack. You win the game when the total of your cards is less than or equal to 21 points and greater than the total held by the dealer. An ace may count as either 1 or 11 points, and jacks, queens, and kings count as 10 points. All other cards are counted at face value.

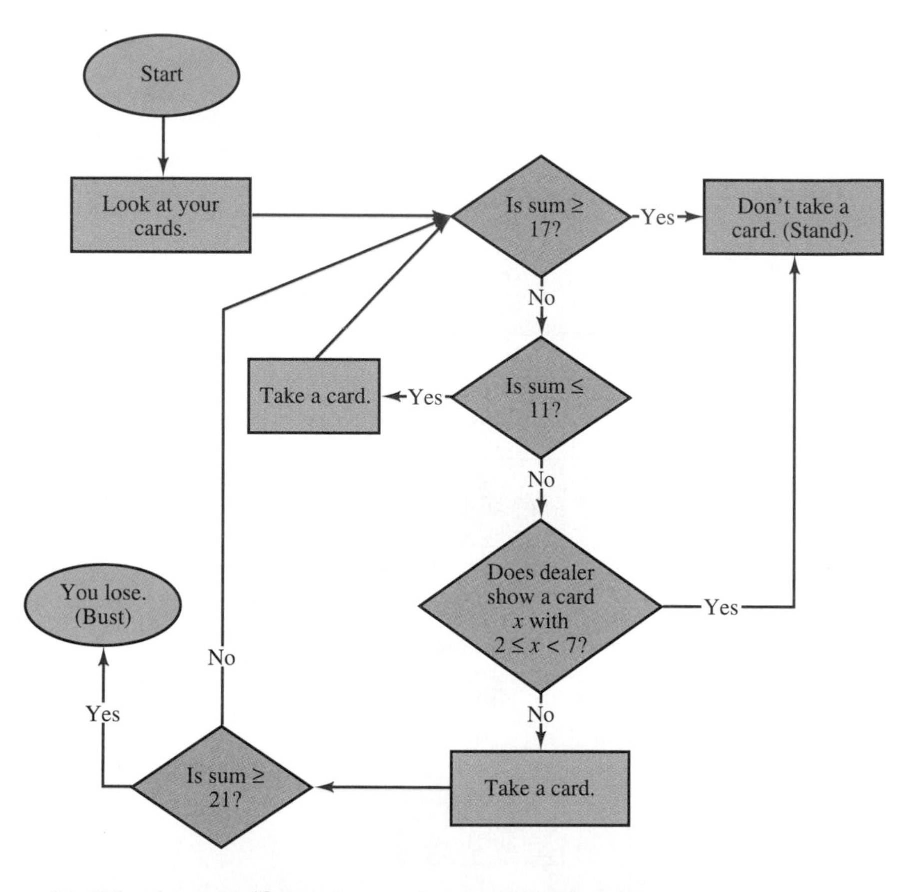


(a) What happens if your two cards are a 10 and an 8?
(b) What happens if your two cards are a 4 and an 8?
(c) What happens if your two cards are a 4 and a 2?
(d) What happens if your two cards are a 2 and a king when the dealer shows a queen?

(e) What happens if your two cards are a 2 and a king when the dealer shows a 6?

(f) Does this flowchart show how you can win the game? If so, explain. If not, modify the flowchart to show how the game can be won.

### *Explore*

In Problems 12–13, suppose that the company used in Example 2 decides to add an additional step to its existing flowchart. In each problem, modify the flowchart in Example 2 to accommodate the additional step.

**12.** Sometimes when a customer calls the technical staff, all the telephones are busy. Rather than requiring the customer to remain on hold, the company has instituted the following policy: If all lines are busy, the telephone receptionist is to ask customers if they wish to remain on hold or if they want to have their call returned by the next available representative of the technical staff. Change the flowchart in Example 2 in order to accommodate this situation.

**13.** Sometimes when a customer calls the business staff, the question is too complex for the staff to answer immediately. Rather than requiring the customer to remain on hold, the company has instituted the following policy: In situations in which an immediate answer cannot be given, the business staff have been instructed to ask customers if they wish to be placed on hold or if they want to have their call returned after the answer has been determined. Change the flowchart in Example 2 in order to accommodate this situation.

**14.** A member of the quality assurance staff of a company that manufactures and sells prefabricated, unassembled cabinets is responsible for determining if each carton contains the correct parts. If the carton contains the correct parts, it is sent on to shipping. If not, the carton is removed from the assembly line, has the necessary parts added to the carton, and then is sent to shipping. Write a flowchart for this process.

**15.** Write a flowchart that shows how to determine whether a positive integer is odd or even.

**16.** Write a flowchart that shows how to determine whether a positive integer has an integer square root.

**17.** A light on your desk no longer works. Describe in writing and construct a flowchart describing the process of determining whether it is the light bulb, the circuit breaker, or some other factor is the cause of the light not working.

**18.** A mechanical coin sorter puts coins in different bins depending on the size of the coins. The sorter is capable of handling quarters, dimes, nickels, and pennies. Write a flowchart that describes the process followed by the coin sorter. Assume that another machine has previously removed all foreign coins, U.S. coins other than those listed above, and all other objects.

**19.** Write a flowchart to tally the points scored by a basketball team during a game.

**20.** Write a flowchart to tally the points scored by a professional football team during a game.

**21.** Write a flowchart that describes a process to determine which new car you should buy.

**22.** Write a flowchart that a good student would use to determine if she has time to go to the beach.

**23.** Write a flowchart for the process of signing up for classes at your school.

**24.** Write an organizational chart describing the administrative structure at your school.

**25.** Write an organizational chart describing the separation of powers of the three major branches of the federal government.

**26.** Write an organizational chart describing the federal judiciary system.

## Chapter 4 • Summary

### Key Terms, Concepts, and Formulas

The important terms in this chapter are:

**Affirming the antecedent:** A deductive argument of the form

If *A*, then *B*.
*A*.
Therefore, *B*. (p. 292)

**Biconditional statement:** If *A* and *B* represent statements, the biconditional statement combines both the statement ''If *A*, then *B*'' and the statement ''If *B*, then *A*'' into one statement, ''*A* if and only if *B*.'' (p. 281)

**Conclusion:** The ''then'' clause of a conditional statement, also called the consequent. (p. 279)

**Conditional statement:** A statement formed by two individual statements joined by the words ''If . . . , then . . . .'' (p. 279)

**Conjunction:** A statement in which clauses are connected with the word ''and.'' A conjunction is true only if both of its clauses are true. (p. 303)

**Contrapositive:** If *A* and *B* represent statements, the contrapositive of the conditional statement ''If *A*, then *B*'' is the statement ''If not *B*, then not *A*.'' (p. 279)

**Converse:** If $A$ and $B$ represent statements, the converse of the conditional statement "If $A$, then $B$" is the statement "If $B$, then $A$." (p. 279)

**Decision symbol:** A diamond-shaped object that indicates a question in a flowchart. (p. 313)

**Deduction:** The process of reasoning in which conclusions are based on accepted premises. (p. 303)

**Denying the consequent:** A deductive argument of the form

If $A$, then $B$.
Not $B$.
Therefore, not $A$. (p. 293)

**Descriptive method:** The technique of representing a set by giving a rule for its elements. (p. 264)

**Disjunction:** A statement in which clauses are connected with the word "or." A disjunction is true whenever one or both of its clauses are true. (p. 300)

**Element:** An item in a set. (p. 263)

**Empty set (∅ or { }):** A set that does not contain any elements. (p. 264)

**Flowchart:** A diagram that shows the logical path that is followed when a decision is made. (p. 313)

**Flow line:** A directed line segment (or ray) that connects the symbols in a flowchart. (p. 312)

**Hypothesis:** The "if" clause of a conditional statement; also called the antecedent. (p. 279)

**Hypothetical syllogism:** A deductive argument of the form

If $A$, then $B$.
If $B$, then $C$.
Therefore, if $A$, then $C$. (p. 290)

**Induction:** The process of reasoning in which conclusions are based on experience or experimentation. (p. 287)

**Intersection:** An operation on two sets that creates a set containing elements common to both sets. (p. 266)

**Inverse:** If $A$ and $B$ represent statements, the inverse of the conditional statement "if $A$, then $B$" is the statement "if not $A$, then not $B$." (p. 279)

**Listing method:** The technique of representing a set by writing all its elements in braces. (p. 264)

**Logic:** The science of correct reasoning. (p. 276)

**Paradox:** A statement (or group of statements) that is in contradiction with itself. (p. 270)

**Set:** A collection of items. (p. 263)

**Start/Stop symbol:** An oval that is used to indicate the beginning or end of a flowchart. (p. 312)

**Statement symbol:** A rectangle used in a flowchart to indicate an action, person, or result. (p. 312)

**Subset:** A set whose elements are contained in another set. (p. 265)

**Symbolic logic:** A system of logic that uses symbols to represent statements. (p. 299)

The symbols used in this chapter are:

| Symbol | English | Name |
|---|---|---|
| $\sim$ | Not | Negation |
| $\therefore$ | Therefore | |
| $\rightarrow$ | Implies, if . . . then . . . | Conditional |
| $\equiv$ | Equivalent statements | Logical equivalency |
| $\wedge$ | And | Conjunction |
| $\vee$ | Or | Inclusive disjunction |

(p. 300)

**Truth table:** A chart consisting of all the true and false combinations of the clauses in a statement. (p. 302)

**Union:** An operation on sets that creates a set containing all the elements of both sets. (p. 266)

**Universal set:** A set that contains all elements of a certain category. (p. 268)

**Venn diagrams:** Pictures used to depict sets. (p. 267)

After completing this chapter, you should be able to:

1. Use sets and Venn diagrams to analyze a situation. (p. 270)
2. Distinguish between different types of conditional statements and use the contrapositive as the logical equivalent of a given statement. (p. 279)
3. Recognize and write good definitions. (p. 282)
4. Distinguish between inductive and deductive reasoning. (p. 287)
5. Make deductive arguments, using hypothetical syllogisms, affirming the antecedent and denying the consequent. (p. 290)
6. Use truth tables to determine the correctness of an argument or determine the equivalency of statements. (p. 305)
7. Construct flowcharts to show a decision making process or organizational structure. (p. 313)

• Summary

## Problems

**1.** For the conditional statement "I am in bed if it is after midnight,"
(a) State its hypothesis and conclusion.
(b) Write its converse.
(c) Write its inverse.
(d) Write its contrapositive.

**2.** What are the three necessary components of a definition? What additional properties are needed for a good definition? For each of the following, state what is wrong with the "definition" and change it to make it a good definition.
(a) Perpendicular line segments are segments that intersect each other.
(b) Microscopic: that which can only be seen through a microscope.
(c) Natural numbers are numbers that are greater than zero.

**3.** Write a statement that is true but whose converse and inverse are false.

**4.** Explain the difference between an inductive argument and a deductive argument. Give both an inductive argument and a deductive argument to persuade a friend to help you build a fence.

**5.** Rewrite the following in correct syllogistic form and construct a true conclusion from the premises.
(a) $X \rightarrow Y$
$Z \rightarrow \sim Y$
$\sim Z \rightarrow P$
(b) If the geometry is Riemannian, then there are no parallel lines.
If the geometry is Euclidean, then parallel lines exist.
If the geometry is non-Euclidean, then at least one of Euclid's postulates is changed.

**6.** Of 400 students, 360 students are taking a math course or an English course. Eighty-five of the students in math classes are also taking an English class. If a total of 190 students are taking a math course, how many are taking an English course but not a math course?

**7.** Translate the following statements into symbolic logic.
(a) If John has a headache, then John is either grumpy or silent.
(b) If the weather is not good, then we will not play baseball and we will not have a picnic.
(c) If you do not study math and you want a good job, it will be harder to advance in business.

**8.** Use truth tables to show that $(A \vee B) \wedge (A \vee \sim B)$ is equivalent to $A$.

**9.** Use a truth table to determine if the argument is logically correct.

Parrots can crack walnuts, and if it is my animal, it cannot crack walnuts.

Therefore, if it is my animal, it is not a parrot.

**10.** Write a flowchart that describes the process of locating a book in the library. Include directions on what to do if the book is not in the catalog, if the book is not on the shelves, and if the book is being held in the reserve section.

# 5

# Trigonometry: A Door to the Unmeasurable

Trigonometry is used in many aspects of sailing: establishing location by the stars, plotting courses, and positioning sails for maximum speed. Even the size of the triangular sails, which can be seen in Raoul Dufy's *Regatta at Cowes*, is a factor in maximizing speed. (National Gallery of Art)

# Overview Interview

In this chapter you will learn how to find the unknown sides and angles of triangles by using some of the basic components of trigonometry. With the use of trigonometric formulas and a scientific calculator, you will be able to determine distances that you may not be able to measure directly. For example, you will be able to find the height of a building, the width of a canyon, the altitude of a hot air balloon, or the distance to a forest fire. You will see that mathematics can be used to solve some very practical problems. You will see that a knowledge of trigonometry opens a door to determining unmeasurable distances.

## David Hopkins

DAVID HOPKINS, SENIOR TRANSPORTATION ENGINEER, CALIFORNIA DEPARTMENT OF TRANSPORTATION (RETIRED)

TELL US ABOUT YOURSELF. I graduated as a Civil Engineer from Santa Clara University in California and did graduate studies in engineering at the Massachusetts Institute of Technology. My career as an engineer has spanned 37 years, during which I supervised the design, construction, and environmental studies for various projects on the highway system in the State of California.

HOW IS TRIGONOMETRY USED IN ENGINEERING? Highways follow a path made up of a series of straight lines and arcs of circles. Trigonometry is used during the design stage of a highway project to plot the path of the highway on paper and to calculate the coordinates of points along the highway. Before construction can begin, that pathway must be staked on the ground by surveyors, who rely on their ability to use geometry and trigonometry to recreate in three dimensions the highway path described in the plans.

*Dave Hopkins, designer of many California bridges, at home.*

WHY IS THE STUDY OF MATHEMATICS IMPORTANT TO THE ENGINEER? Mathematics is the engineer's most important tool. Whether it is for the design of an electric circuit, the measurement of stress on a beam under the weight of a train, or the hundreds of other physical phenomena the engineer must deal with, mathematics is the key to understanding what is going on and how the desired results are achieved. Even though computers are now used to do much of the mathematics that was previously done by hand, a knowledge of mathematics is still necessary for a complete understanding of what is happening in a given situation. Finally, the studying and learning of mathematics is important because it helps develop the thought processes and precision that is needed by the engineer.

*The Lighthouse at Honfleur* by Georges Seurat. For the navigator at sea, a lighthouse serves as a permanent reference point for charting a course. (National Gallery of Art)

## A Short History of Trigonometry

As the understanding of numbers, algebra, and geometry developed, ancient scholars turned their attention to a quantitative study of the sun, the moon, the planets, and the stars. This interest in astronomy led to the formation of what is known today as trigonometry. The word "trigonometry" comes from the Greek words *tri'gonon* (triangle) and *met'ron* (measure). However, trigonometry began as the study of the relationship between arcs and chords of a circle as in a bow and a bowstring, progressed to the study of triangles on the surface of a sphere, and is now commonly known as the study of the relationship between the angles and sides of a triangle. The words used to describe these relationships have evolved into the present terms: sine, cosine, tangent, cotangent, secant, and cosecant.

The exact origins of trigonometry are not known. There is, however, evidence that astronomy (requiring trigonometry) was studied by the ancient Babylonians and Chinese. Ancient Egyptians investigated the ratios of sides of a triangle as recorded in the Ahmes papyrus (1550 B.C.). However, the first significant contributions to trigonometry were made by the Greeks. Motivated by their interest in astronomy, they systematically studied the relationship between arcs and chords of a circle. Hipparchus of Nicea (c. 140 B.C.), a mathematician and astronomer, wrote the first systematic study of trigonometry. He developed methods of studying spherical tri-

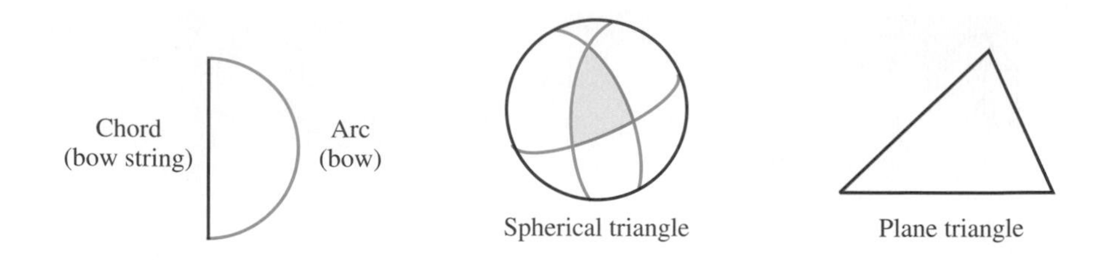


angles, is given credit for dividing the circle into 360 degrees, and is considered the founder of trigonometry. Menelaus of Alexandria (c. 100 B.C.) contributed much to the study of spherical triangles as related to the study of astronomy. Claudius Ptolemy of Alexandria (A.D. 125), in his work *Syntaxis Mathematica*, now known as *Almagest* (the greatest), displayed a table of chords that is equivalent to a table of sines for angles from 0° to 90° at 15-minute intervals. The *Almagest* exerted a profound influence on astronomy and trigonometry for the next century.

During the Dark Ages in Europe from 400 to 1000, the development of trigonometry moved to the East. Hindu astronomers displayed their knowledge of trigonometry in the book entitled *Sûrya Siddhānta* (c. 400) and in the works of Varâhamihira (c. 505) and Āryabhata (c. 510). The Arabs took what the Hindus had done in trigonometry and built upon that. Al-Battânî (c. 920) and Abû'l-Wefâ (c. 980) used tangents and cotangents as well as the Laws of Sines and Cosines to solve spherical triangles, and computed tables of sines, cosines, tangents, and cotangents. Kûshyâr ibn Lebbân (c. 1000) wrote a book on astronomy and trigonometry and Nasîr ed-dîn al-Tûsî (1250) wrote the first work on plane and spherical trigonometry as a branch of mathematics separate from astronomy. The Arabs' work in trigonometry during the ninth and the tenth centuries is one of the reasons this period is called the golden age of Arabian mathematics.

As Europe emerged from the Dark Ages and moved into the Renaissance, contributions to the development of trigonometry again appeared. The major contributions are outlined below.

Leonardo Fibonacci (1220), in *Practica Geometriae*, summarized Greek trigonometry.

Johann Müller (1464), known as Regiomontanus, in his work *De triangulis omnimodis*, wrote Europe's first treatment of plane and spherical trigonometry in which trigonometry is treated separately from astronomy.

Reiner Gemma Frisius (1533) modernized surveying by introducing trigonometric techniques of triangulation.

Peter Apian (1534) published a table of sine values for every minute of arc.

Georg Joachim von Lauchen (1551), known as Rhaeticus, defined the trigonometric functions as ratios of sides of a right triangle in his work *Opus palatinum de triangulis.*

Francois Viète (1579–1591) constructed a trigonometry table for every second of arc and systematically applied algebra to trigonometry, obtaining algebraic forms for many identities.

Paul Wittich and Tycho Brahe (1580) introduced prosthaphaeresis, a method of calculating products and quotients by using trigonometric identities involving addition and subtraction.

Christopher Clavius (1593) used trigonometric identities to introduce new methods of computation.

Thomas Blundeville (1594) wrote England's first complete treatment of trigonometry.

Bartholomaus Pitiscus (1600), professor of mathematics in Heidelberg, wrote the first text to have the title *Trigonometry*.

By 1658, the names of the trigonometric functions had taken their present forms. The sine function was called *jyâ* or *jîva*, meaning chord, by the Hindu mathematician Āryabhata (c. 510). The Arabs phonetically created the meaningless Arabic word *jiba* for the Hindu word *jîva*. Later the Arabs rearranged the letters of *jiba* into an actual Arabian word, *jaib*, meaning bay. Although *jaib* had nothing to do with the mathematical concept of a chord of an arc, it was widely used by the Arabs.

In about 1150 Gherardo of Cremona, when making translations from Arabic to Latin, used the Latin word for bay, *sinus*, as a translation of the Arabic *jaib*. Thus, our present word ''sine'' has no relation to the mathematical concept it represents.

The present term ''cosine'' also originated with Āryabhata, who called it *kotijyâ* in 510. Over the centuries the term ''cosine'' was referred to by different names. In 1620, Edmund Gunter used the terminology *co.sinus*, which, after being modified by John Newton in 1658, became the standard term, *cosinus*. The present word ''tangent'' originated in a work by Danish mathematician Thomas Fincke (1583), when he used the term *tangens*. After the more renowned writer Bartholomaus Pitiscus also used *tangens* in his writing, the name became permanent. The terms ''cotangent,'' ''secant,'' and ''cosecant'' also have different origins. Edmund Gunter (1620) introduced the word *cotangens*, Thomas Fincke (1583) used the word *secant*, and Rhaeticus (1596) adopted the word *cosecant*.

By 1750, trigonometry had become more than merely a tool for the astronomer and the surveyor. It was used by mathematicians John Bernoulli, Abraham de Moivre, Leonhard Euler, Isaac Newton, and Gottfried Leibniz in analyzing mathematical concepts, studying complex numbers, modeling periodic phenomena, and studying mathematical physics. In the 20th century, trigonometric functions are still important. They are used in the study of electricity, light, sound, radio, television, and other areas where values occur periodically. They can also be utilized in finding distances that cannot be measured directly and, consequently, are used by surveyors, astronomers, navigators, engineers, architects, and others. In this chapter you will learn more about the trigonometric functions and see how they can actually be used to solve problems in some of the areas mentioned in this section.

## Check Your Reading

1. Interest in what area led to the development of trigonometry?
2. What is the difference between a spherical triangle and a plane triangle?
3. When a Chinese encyclopedia of 1000 volumes was being written in A.D. 980, what was Abû'l-Wefâ doing in India?
4. In 140 B.C., Hu Shin compiled a dictionary of 10,000 characters, and the statue of Venus de Milo was sculpted. At that time, what was Hipparchus of Nicea doing?
5. What is a major reason for the period from 400 to 1000 being called the golden age of Arabian mathematics?
6. Trace the development of the word *jîva* to the word *sinus* for the trigonometric function sine.
7. As the Babylonian King Hammurabi established the first legal system in 1550 B.C., what was being written on the Ahmes papyrus?
8. In the year 1533, Francisco Pizarro executed Inca Indians in Peru. During that year, what was Reiner Frisius doing?
9. While tomatoes were being introduced into England in 1596, what was Rhaeticus doing?
10. While Italian cuisine became the culinary rage in 1580 in Europe, what were Paul Wittich and Tycho Brahe working on?
11. Match each of the following names with the correct mathematical event, discovery, or concept.

| | |
|---|---|
| (a) Apian | Called the sine function *jîva* |
| (b) Āryabhata | Coined word *tangens* |
| (c) Blundeville | Wrote first text entitled *Trigonometry* |
| (d) Fincke | Devised spherical trigonometry as related to astronomy |
| (e) Gherardo | Devised table of sines for every minute of arc |
| (f) Menelaus | Translated Arabic to Latin |
| (g) Pitiscus | Wrote *Almagest* |
| (h) Ptolemy | Wrote England's first complete treatment of trigonometry |

## Research Questions

To answer the following questions, you will need to refer to material not contained in the text. Possible sources of information are listed in the Bibliography at the end of the text.

1. Many of the mathematicians and astronomers mentioned in the short history of the development of trigonometry are also known for other mathematical endeavors or had other interests besides mathematics. Do some research on at least two

of the mathematicians mentioned in this section. Tell something about their lives, their achievements, and their interests.
2. What are sextants and transits? How and in what areas are they used?
3. What were some of the contributions of the Arabs to mathematics during the ninth and tenth centuries, the golden age of Arabian mathematics?
4. Explain how trigonometry is actually used in astronomy. How did the ancients use trigonometry in their astronomical endeavors?
5. Discuss the development of trigonometry in the Far East.
6. Before the scientific calculator, the values of trig functions were found in a "trig table." Find an example of such a table and show how it was used.
7. The three trigonometric functions sine, cosine, and tangent are defined in Section 5.1 of this chapter. There are, however, three other trigonometric functions: secant, cosecant, and cotangent. How are these defined and used in reference to a right triangle?
8. What is longitude and latitude? What is geodesy or geodetic surveying? How is geodetic surveying different from plane surveying?
9. Interview a surveyor to determine how trigonometry is used. What laws of trigonometry are used? What equipment is used?

## Projects

1. Actually use the trigonometric methods described in this chapter to find the height of a building or mountain in your area. Explain in detail how you did this and compare your results with the actual height of the building or mountain as published in the building specifications or on a map. If there is a difference in the results, explain the likely causes of the difference.
2. What does the graph of the sine function $y = \sin x$ look like in a rectangular coordinate system? What are the amplitude and period of that sine curve? Examine some phenomena that exhibit characteristics of sine curves, such as sound and electrical waves, biorhythm charts, and seismograph readings.
3. This chapter deals with the study of triangles on a plane surface. This is not the only trigonometry that has been studied. The study of triangles on a sphere is called spherical trigonometry. What are spherical triangles and what are some of their properties? What are some of the formulas used in spherical trigonometry? How do these formulas differ from those of plane trigonometry? Why is spherical trigonometry of considerable importance in navigation and astronomy?
4. The study of the motion of planets and stars led to the development of trigonometry. However, the theories of the astronomers themselves are quite interesting. Compare and contrast the astronomy of Ptolemy (A.D. 85–165), Nicolaus Copernicus (1473–1543), Tycho Brahe (1546–1601), and Johannes Kepler (1571–1630). Make diagrams of the movements of the earth, sun, moon, and planets according to each astronomer.

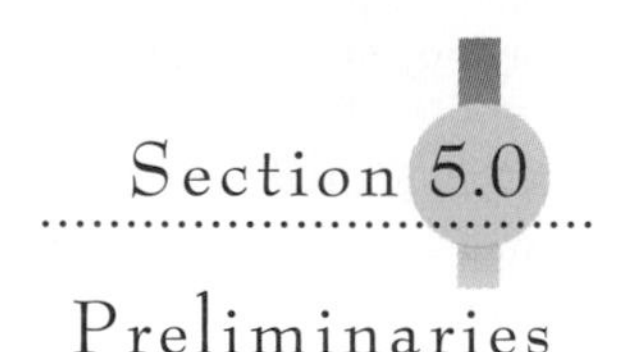


## Right Triangles

Before we start discussing trigonometry, we will look at some terms and concepts needed in the chapter. A major point is that all discussions in this chapter refer to Euclidean space. None of the material in this chapter is related to the non-Euclidean geometries you may have studied in Chapter 3. A **right triangle** is a triangle with a 90° angle. The two other angles are acute angles (less than 90°). The sides of a right triangle that form the 90° angle are the **legs** of the right triangle, and the other side, the one opposite the 90° angle, is the **hypotenuse** of the right triangle. The hypotenuse is always the longest side of a right triangle. In $\triangle ABC$,

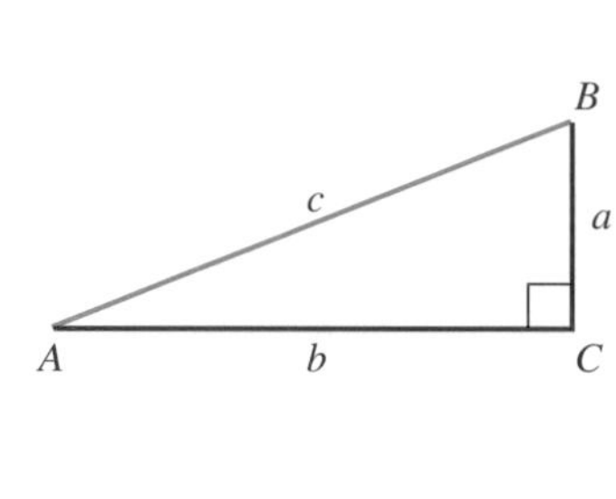


$\angle C =$ the right angle

$\angle A =$ an acute angle

$\angle B =$ an acute angle

$c =$ hypotenuse

$a =$ leg

$b =$ leg

### Example 1

In the right triangles, determine which sides are legs and which side is the hypotenuse.

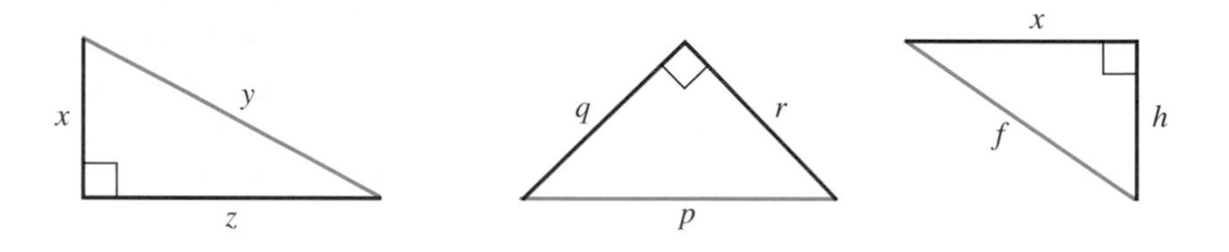


**Solution:**

First triangle: $y =$ hypotenuse, $x =$ leg, $z =$ leg

Second triangle: $p =$ hypotenuse, $q =$ leg, $r =$ leg

Third triangle: $f =$ hypotenuse, $x =$ leg, $h =$ leg ■

In a right triangle, the hypotenuse is always opposite the 90° angle, but the legs of the triangle can be referred to by their orientation with respect to one of the acute angles of the triangle. The leg across the triangle from an acute angle is the opposite side, and the leg that forms the angle is the adjacent side for that acute angle. For example, in $\triangle ABC$, side $c$ is the hypotenuse. With reference to $\angle A$, $b$ = adjacent side and $a$ = opposite side. With reference to $\angle B$, $a$ = adjacent side and $b$ = opposite side.

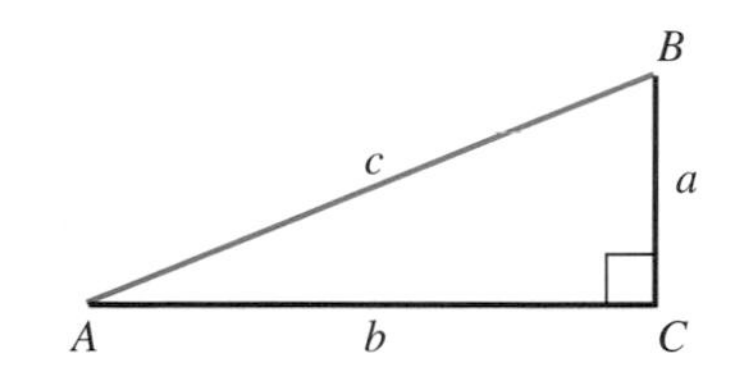


## Example 2

In the triangles, find sides opposite from and adjacent to $\angle K$.

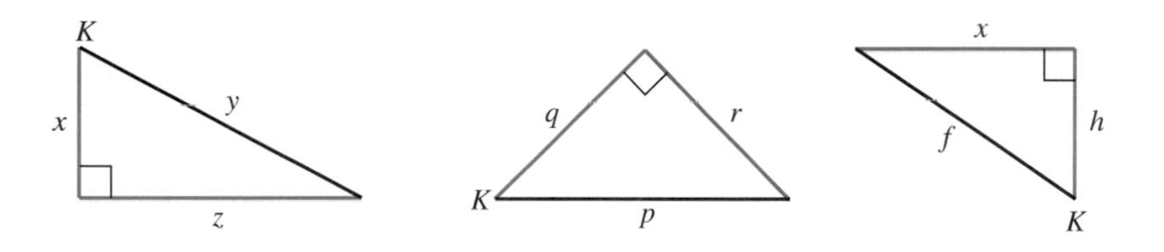

**Solution:** In reference to $\angle K$,

| | Adjacent Side | Opposite Side |
|---|---|---|
| First triangle | $x$ | $z$ |
| Second triangle | $q$ | $r$ |
| Third triangle | $h$ | $x$ |

## *The Pythagorean Theorem*

The Greek mathematician Pythagoras (c. 582–c. 501 B.C.) is given credit for the basic theorem concerning the sides of a right triangle, the **Pythagorean Theorem**. Geometrically, it states that the sum of the squares on the legs of a right triangle is equal to the square on the hypotenuse. In algebraic terms, if the lengths of the legs of a right triangle are $a$ and $b$ and the length of the hypotenuse is $c$, then $a^2 + b^2 = c^2$.

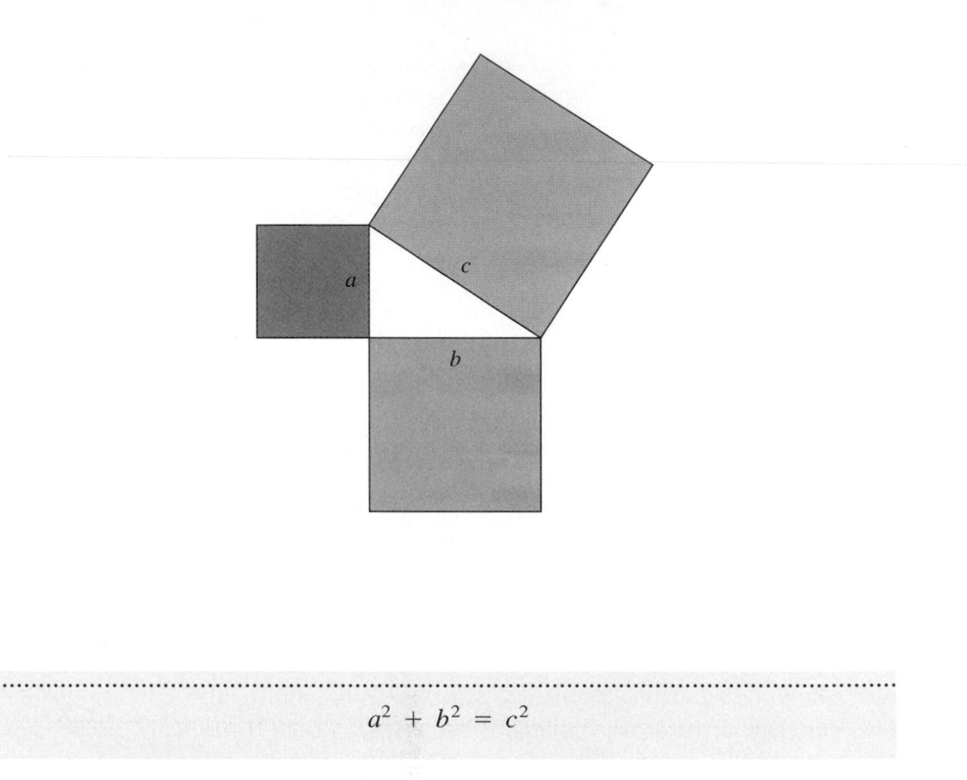


## Pythagorean Theorem

$$a^2 + b^2 = c^2$$

The Pythagorean Theorem is one of the most proved theorems in mathematics. In 1940, Elisha Scott Loomis compiled a book containing 370 proofs of the theorem. One of these proofs is credited to the 20th president of the United States, James A. Garfield. While a Congressman in 1876, he discovered what is now known as *Garfield's solution*. His proof is based on the formula for the area of a triangle, $A = 1/2 \times$ length of the base $\times$ the length of the height.

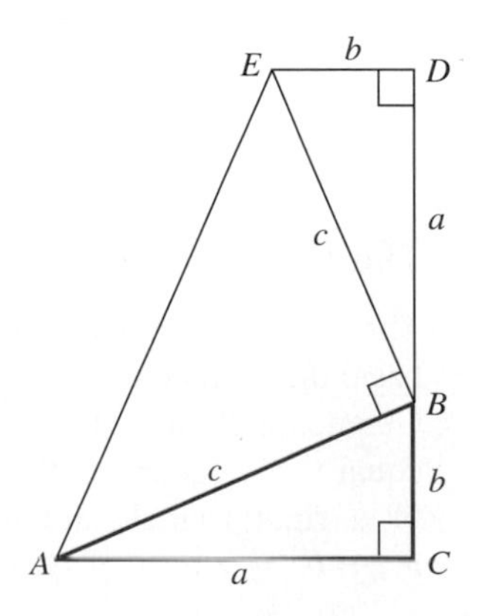

Starting with right $\triangle ABC$:

1. Extend side $\overline{CB}$ to $D$ so that $BD = a$.
2. Draw $\overline{DE}$ so that $\angle D = 90°$ and $DE = b$.
3. Draw $\overline{BE}$ and $\overline{AE}$.

The area of $ACDE$ can be determined by adding the areas of the right triangles $\triangle ABC$, $\triangle EDB$, and $\triangle ABE$.

$$\text{Area of } \triangle ABC = \frac{1}{2}ab$$

$$\text{Area of } \triangle EDB = \frac{1}{2}ab$$

$$\text{Area of } \triangle ABE = \frac{1}{2}c^2 \qquad \left[\text{Area} = \frac{1}{2}(\text{base})(\text{height}) = \frac{1}{2}c \cdot c\right]$$

$$\therefore \text{Area of } ACDE = ab + \frac{1}{2}c^2$$

The region $ACDE$ can also be divided into two triangles by drawing $\overline{EC}$. The area of $ACDE$ can be determined by adding the areas of $\triangle ACE$ and $\triangle EDC$. Noticing that the height in both triangles is $a + b$, we get

$$\text{Area of } \triangle ACE = \frac{1}{2}a(a + b) = \frac{1}{2}ab + \frac{1}{2}a^2$$

$$\text{Area of } \triangle EDC = \frac{1}{2}b(a + b) = \frac{1}{2}ab + \frac{1}{2}b^2$$

$$\therefore \text{Area of } ACDE = ab + \frac{1}{2}a^2 + \frac{1}{2}b^2$$

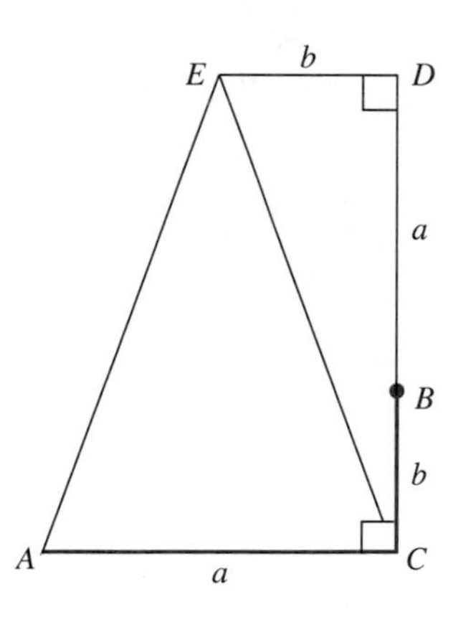

Since we have found the area of the region *ACDE* in two different ways, the two results must be equal.

$$ab + \frac{1}{2}a^2 + \frac{1}{2}b^2 = ab + \frac{1}{2}c^2$$

$$\frac{1}{2}a^2 + \frac{1}{2}b^2 = \frac{1}{2}c^2$$

$$\therefore a^2 + b^2 = c^2$$

## Example 3

If the legs of a right triangle are 5 in. and 12 in., find the length of the hypotenuse.

**Solution:** According to the Pythagorean Theorem,

$$c^2 = a^2 + b^2$$

$$c^2 = 12^2 + 5^2$$

$$c^2 = 144 + 25$$

$$c^2 = 169$$

$$c = \pm\sqrt{169}$$

$$c = 13$$

The hypotenuse is 13 in. (*Note:* The positive square root is used since a distance cannot be negative.) ■

## Example 4

If the hypotenuse of a right triangle is 6.5 cm and one leg is 3.3 cm, find the length of the other leg.

**Solution:**

$$a^2 + b^2 = c^2$$

$$3.3^2 + b^2 = 6.5^2$$

$$10.89 + b^2 = 42.25$$

$$b^2 = 31.36$$

$$b = \pm\sqrt{31.36}$$

$$b = 5.6$$

The other leg is 5.6. cm. ■

## Example 5

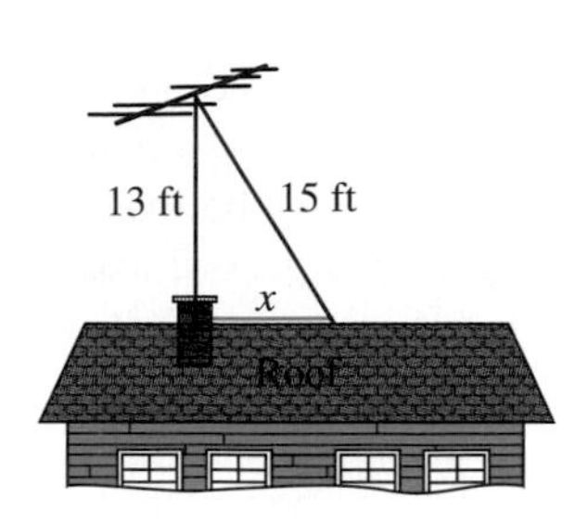


The 13-ft pole of a TV antenna stands vertically on the roof of a building. A 15-ft support wire is attached to the top of the pole and is tied to the roof. If the wire is taut, how far from the bottom of the antenna should the wire be attached?

**Solution:**
Let $x$ = the distance from the bottom of the antenna pole to the point where the wire is attached to the roof. By the Pythagorean Theorem:

$$x^2 + 13^2 = 15^2$$

$$x^2 + 169 = 225$$

$$x^2 = 56$$

$$x = \pm\sqrt{56}$$

$$x \approx 7.5$$

The wire is attached approximately 7.5 ft from the bottom of the antenna. ■

## Pythagorean Triples

Natural numbers that satisfy the Pythagorean Theorem are called **Pythagorean triples**. The first triple is 3, 4, and 5, since those numbers are the smallest natural numbers that satisfy the Pythagorean Theorem, $3^2 + 4^2 = 5^2$. Twenty-seven of these triples are listed below.

| | | | | | | | | |
|---|---|---|---|---|---|---|---|---|
| 3 | 4 | 5 | 12 | 35 | 37 | 24 | 32 | 40 |
| 5 | 12 | 13 | 14 | 48 | 50 | 24 | 45 | 51 |
| 6 | 8 | 10 | 15 | 20 | 25 | 27 | 36 | 45 |
| 7 | 24 | 25 | 15 | 36 | 39 | 28 | 45 | 53 |
| 8 | 15 | 17 | 16 | 30 | 34 | 30 | 40 | 50 |
| 9 | 12 | 15 | 18 | 24 | 30 | 33 | 44 | 55 |
| 9 | 40 | 41 | 20 | 21 | 29 | 33 | 56 | 65 |
| 10 | 24 | 26 | 20 | 48 | 52 | 36 | 48 | 60 |
| 12 | 16 | 20 | 21 | 28 | 35 | 39 | 52 | 65 |

## The Root Spiral

Right triangles with sides that are not natural numbers have also been of keen interest over the centuries. The study of these triangles enabled mathematicians to find approximations for the square roots of natural numbers that turn out to be irrational numbers, such as $\sqrt{2}$, $\sqrt{3}$, and $\sqrt{5}$. Today, to find $\sqrt{2}$ we simply press 2 and the square root button on a calculator and the answer 1.414213562 appears. But how could the ancients find $\sqrt{2}$? They did not have electronic calculators. Furthermore, can there really be a length $\sqrt{2}$ inches long? The answer to the question is yes. The Pythagorean Theorem allows us to find an approximation for the square roots of natural numbers and see line segments whose lengths are irrational numbers.

For example, to find $\sqrt{2}$, all we need to do is find the hypotenuse of a right triangle that has legs that are both one unit long. Using the Pythagorean Theorem, we get

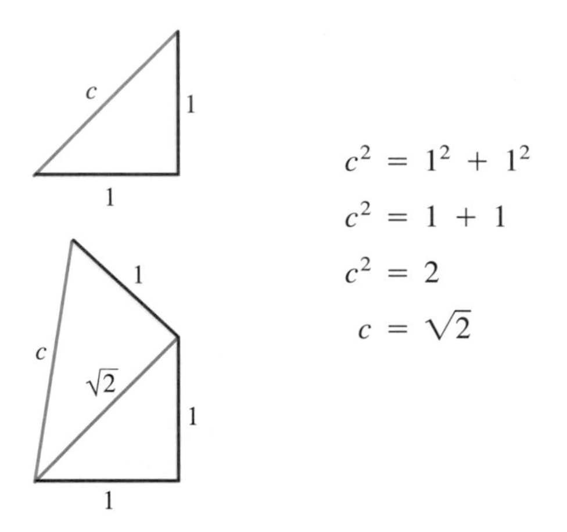


Similarly, we can find $\sqrt{3}$ by finding the hypotenuse of a right triangle that has one leg with a length of $\sqrt{2}$ and the other leg with a length of 1.

$$c^2 = 1^2 + (\sqrt{2})^2$$
$$c^2 = 1 + 2$$
$$c^2 = 3$$
$$c = \sqrt{3}$$

Repeating this process gives us a root spiral. Such a spiral has simple geometric beauty and allows us to see that square roots of the natural numbers actually exist as lengths of hypotenuses of right triangles.

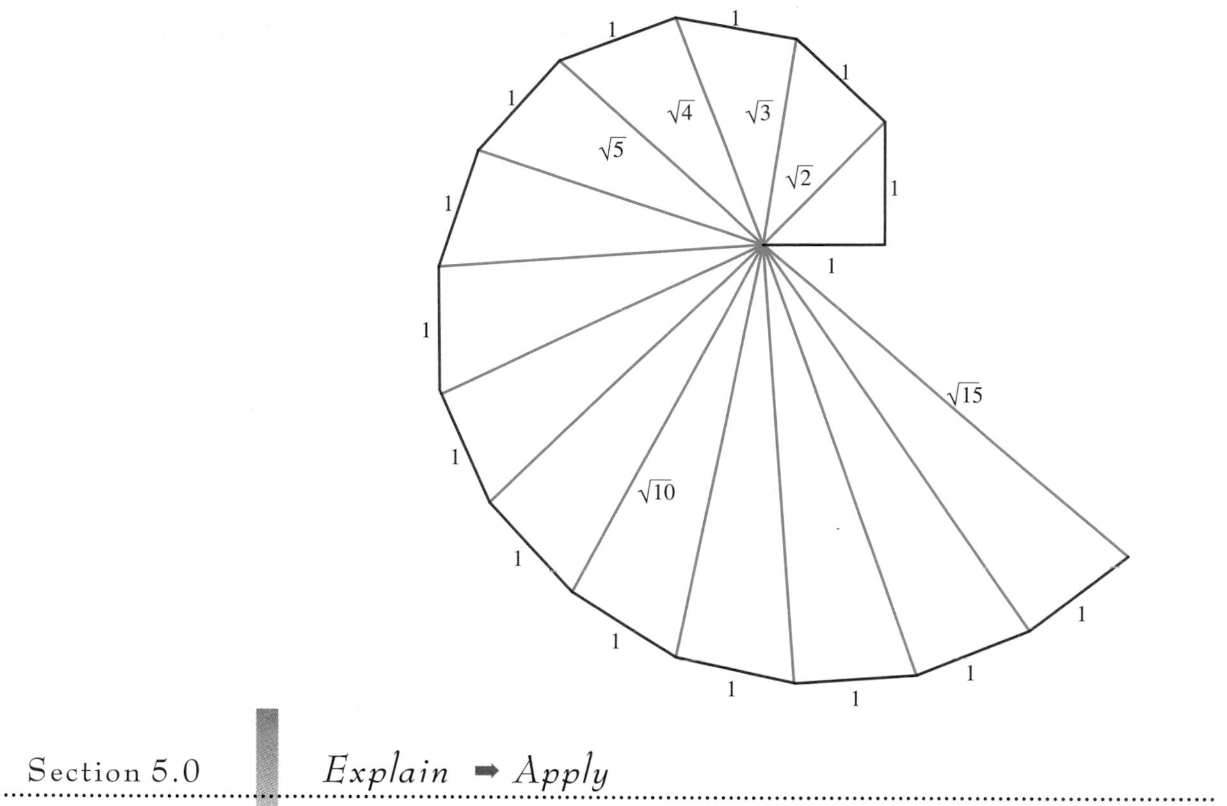


Section 5.0
PROBLEMS

## *Explain* ➡ *Apply*

### *Explain*

1. What is a right triangle?
2. What are the legs of a right triangle? What is the hypotenuse of a right triangle?
3. What is the Pythagorean Theorem?
4. What are Pythagorean triples?
5. Explain why you could not use the Pythagorean Theorem to find side $c$ in the triangle below.

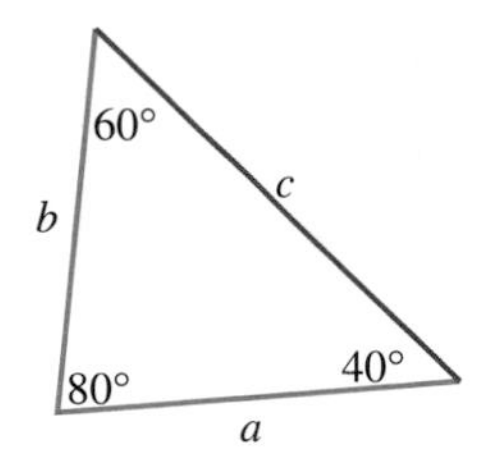

6. Explain how could you find a segment that is $\sqrt{2}$ inches long using a right triangle.

7. Explain why zero and negative integers cannot be used as the lengths of the sides of right triangles.

## *Apply*

In the right triangles, determine which sides are legs and which side is the hypotenuse.

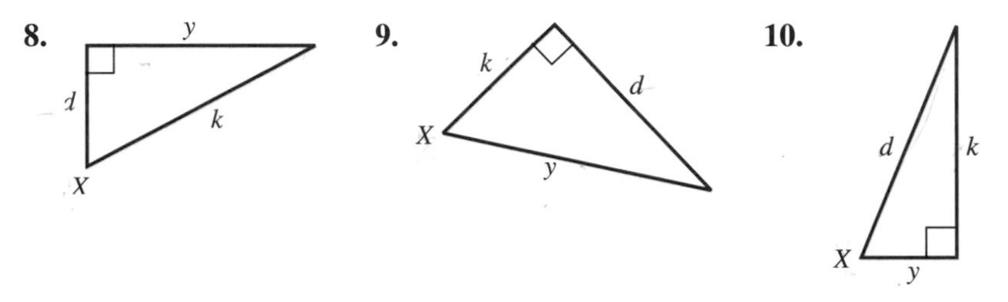


11. For each of the right triangles given in Problems 8–10, find the sides that are opposite from and adjacent to $\angle X$.

In Problems 12–20, find the side of the right triangle that is not given.

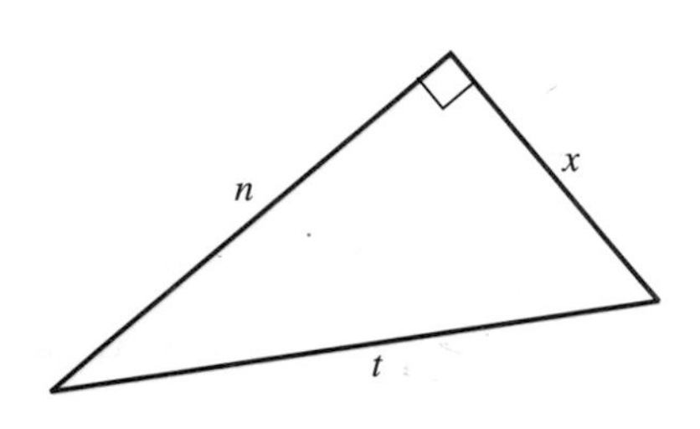


12. $x = 7,\quad n = 24$

13. $x = 10,\quad n = 31$

14. $x = 4.3,\quad n = 9.7$

15. $n = 9,\quad t = 41$

16. $n = 3.24,\quad t = 67.24$

17. $n = 54.1,\quad t = 76$

18. $x = 20\quad t = 25$

19. $x = 5,\quad t = 8.4$

20. $x = 1.5,\quad t = 1.7$

In Problems 21–28, determine if each set of numbers is a Pythagorean triple.

21. 36, 48, 60

22. 18, 25, 33

23. 16, 63, 82

24. 55, 132, 143

25. 28, 96, 110

26. 21, 72, 75

27. 35, 120, 125

28. 24, 70, 74

29. A rectangular gate is 5 ft wide and 6 ft high. Find the length of the support wire running diagonally across the gate.

**30.** A standard baseball diamond is a square 90 ft on a side. Find the distance of a throw from home plate to second base.

**31.** In softball the bases are 60 ft apart. Find the distance of a throw from first base to third base.

**32.** If a TV screen is a rectangle that has a 27-in. diagonal and a width of 21 in., what is the height of the TV screen?

**33.** If the diagonal of a square picture frame is 32 in., how long are its sides?

**34.** If you walk along the diagonal of a rectangular field that is 650 yd by 800 yd, how much shorter is that than walking along the length and width of the field?

**35.** In this section, the root spiral gave us $\sqrt{15}$. Continue drawing the root spiral to find $\sqrt{19}$.

**36.** Can two of the numbers of a Pythagorean triple be 16 and 21?

**37.** Can two of the numbers of a Pythagorean triple be 104 and 130?

***38.** Show that the Pythagorean triple 3, 4, 5 is the only triple with consecutive natural numbers.

***39.** Find all Pythagorean triples that consist of consecutive even integers.

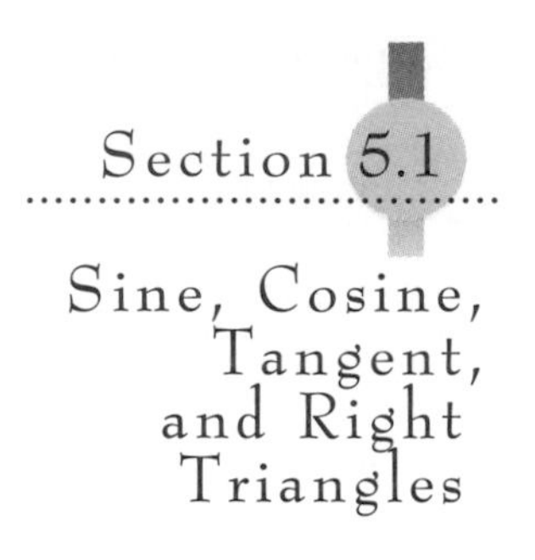

Section 5.1
Sine, Cosine, Tangent, and Right Triangles


The right triangle and the Pythagorean Theorem examined in Section 5.0 are the basic building blocks of trigonometry. We begin by designating the vertices of the acute angles of a right triangle with the uppercase letters $A$ and $B$ and the sides opposite those angles with the lowercase letters $a$ and $b$. The 90° angle of the right triangle is designated by the uppercase letter $C$ and the hypotenuse with the lowercase letter $c$. Later, we will use other uppercase letters for the angles at each vertex and the corresponding lowercase letter for the side opposite each vertex. The three fundamental trigonometric functions, **sine** (abbreviated **sin**), **cosine** (**cos**), and **tangent** (**tan**), are defined as ratios of the lengths of the sides of a right triangle. Whenever a side of a triangle is mentioned, we are referring to the *length* of that side.

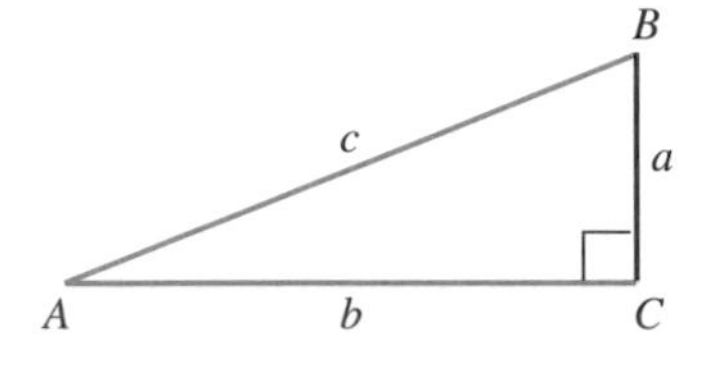


$$\frac{\text{side opposite } A}{\text{hypotenuse}} = \frac{a}{c} = \sin A \leftarrow \boxed{\frac{\text{opp}}{\text{hyp}}} \rightarrow \sin B = \frac{b}{c} = \frac{\text{side opposite } B}{\text{hypotenuse}}$$

$$\frac{\text{side adjacent } A}{\text{hypotenuse}} = \frac{b}{c} = \cos A \leftarrow \boxed{\frac{\text{adj}}{\text{hyp}}} \rightarrow \cos B = \frac{a}{c} = \frac{\text{side adjacent } B}{\text{hypotenuse}}$$

$$\frac{\text{side opposite } A}{\text{side adjacent } A} = \frac{a}{b} = \tan A \leftarrow \boxed{\frac{\text{opp}}{\text{adj}}} \rightarrow \tan B = \frac{b}{a} = \frac{\text{side opposite } B}{\text{side adjacent } B}$$

The three other trig functions, cosecant, secant, and cotangent, are the reciprocals of the sine, cosine, and tangent. They will not be used in this text.

### Example 1

In the right triangles, find $\sin A$, $\cos A$, $\tan A$, $\sin B$, $\cos B$, and $\tan B$.

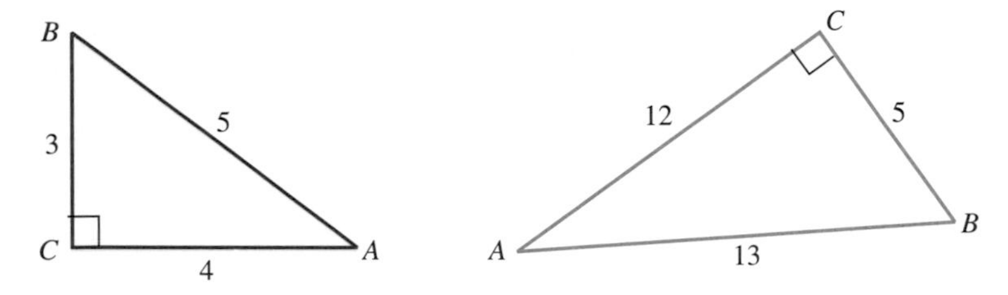


**Solution:**

First triangle:

$$\sin A = \left(\frac{\text{opp}}{\text{hyp}}\right) = \frac{3}{5} \qquad \sin B = \left(\frac{\text{opp}}{\text{hyp}}\right) = \frac{4}{5}$$

$$\cos A = \left(\frac{\text{adj}}{\text{hyp}}\right) = \frac{4}{5} \qquad \cos B = \left(\frac{\text{adj}}{\text{hyp}}\right) = \frac{3}{5}$$

$$\tan A = \left(\frac{\text{opp}}{\text{adj}}\right) = \frac{3}{4} \qquad \tan B = \left(\frac{\text{opp}}{\text{adj}}\right) = \frac{4}{3}$$

Second triangle:

$$\sin A = \left(\frac{\text{opp}}{\text{hyp}}\right) = \frac{5}{13} \qquad \sin B = \left(\frac{\text{opp}}{\text{hyp}}\right) = \frac{12}{13}$$

$$\cos A = \left(\frac{\text{adj}}{\text{hyp}}\right) = \frac{12}{13} \qquad \cos B = \left(\frac{\text{adj}}{\text{hyp}}\right) = \frac{5}{13}$$

$$\tan A = \left(\frac{\text{opp}}{\text{adj}}\right) = \frac{5}{12} \qquad \tan B = \left(\frac{\text{opp}}{\text{adj}}\right) = \frac{12}{5}$$

## Using a Calculator to Find Trigonometric Functions

A scientific calculator can be used to find the ratio of sides for any given angle. First, you must put your calculator in degree mode. Check your calculator instruction booklet or ask your instructor if you do not know how to do this. Now, to find the value of a trig function, for example sin 30°, press the following keys:

| **Press** | **Display** |
|---|---|
| 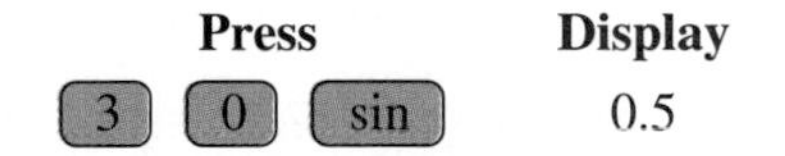 | 0.5 |

Thus, sin 30° = 0.5. This means that in a right triangle, the ratio of the side opposite a 30° angle to the hypotenuse of the triangle is 0.5 or 1/2. Each of the following right triangles has a 30° angle and its sine of 1/2.

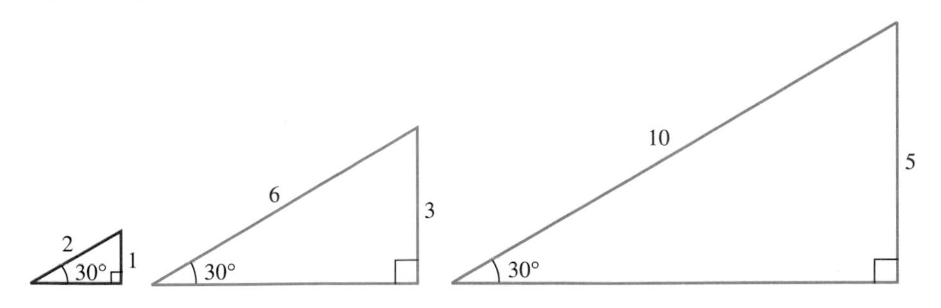


### Example 2

Use a calculator to find the following rounded off to four decimal places:

(a) sin 42°
(b) cos 67°
(c) tan 54.6°
(d) sin 5.79°
(e) tan 3.14°

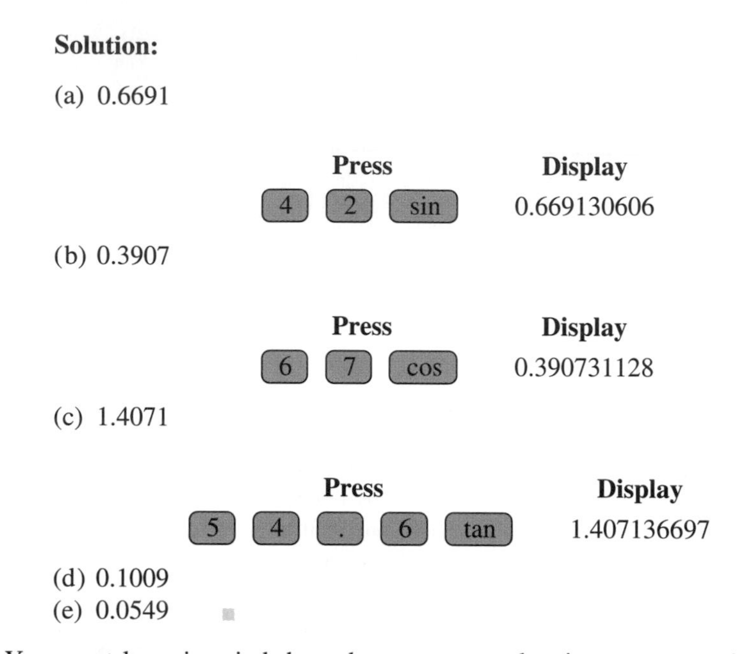

**Solution:**

(a) 0.6691

| Press | Display |
|---|---|
| 4 2 sin | 0.669130606 |

(b) 0.3907

| Press | Display |
|---|---|
| 6 7 cos | 0.390731128 |

(c) 1.4071

| Press | Display |
|---|---|
| 5 4 . 6 tan | 1.407136697 |

(d) 0.1009
(e) 0.0549 ■

You must keep in mind that when you press the sin, cos, or tan key for a given angle, the calculator gives you a ratio of sides of a right triangle in decimal form. The calculator can also be used to determine an angle of a right triangle when the ratio of two sides is known. For example, if $\sin A = 0.6$, the ratio of the length of opposite side to the length of the hypotenuse is $6/10$. We can determine the measure of $\angle A$ by pressing the following keys on a calculator:

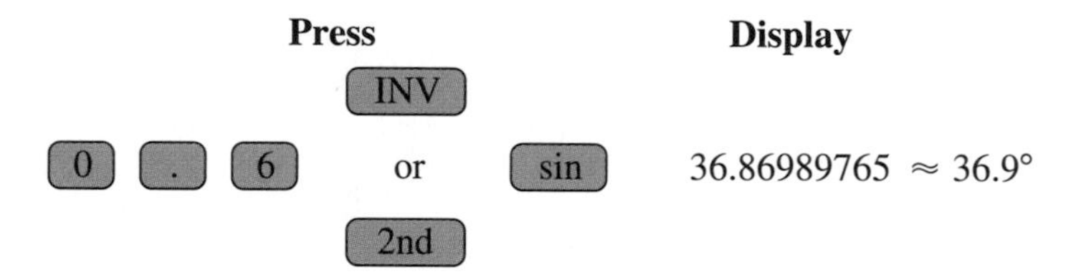

| Press | Display |
|---|---|
| 0 . 6 INV or 2nd sin | 36.86989765 $\approx$ 36.9° |

Your calculator may not do it in exactly the same manner. Check with the instruction booklet if this procedure does not work.

## Example 3

Find an angle that satisfies each of the following trig functions. Round off answers to the nearest tenth of a degree.

(a) $\cos B = 0.2588$
(b) $\tan A = 3.29$
(c) $\sin X = 2/9$

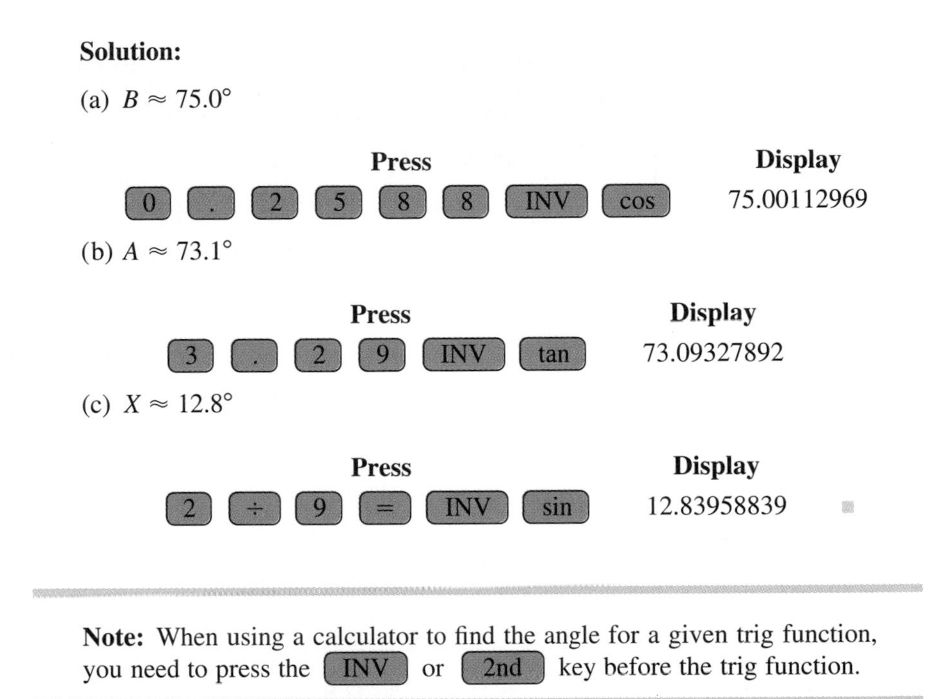

**Solution:**

(a) $B \approx 75.0°$

| Press | Display |
|---|---|
| 0 . 2 5 8 8 INV cos | 75.00112969 |

(b) $A \approx 73.1°$

| Press | Display |
|---|---|
| 3 . 2 9 INV tan | 73.09327892 |

(c) $X \approx 12.8°$

| Press | Display |
|---|---|
| 2 ÷ 9 = INV sin | 12.83958839 |

**Note:** When using a calculator to find the angle for a given trig function, you need to press the INV or 2nd key before the trig function.

## Solving Right Triangles

The trigonometric functions can be used to **solve triangles**—that is, to find unknown angles and sides of triangles. A triangle is solved when the lengths of all three sides and the measures of all three angles are known. Typically, some combination of three sides or angles will be given in a problem and we will find the values of the other three parts. To do this we use the three trig functions, the Pythagorean Theorem, the fact that the sum of the angles of a triangle equals 180°, and a calculator. In order to establish some uniformity in the answers we obtain in solving triangles, we use the following round-off rules in the rest of this chapter.

**Round-Off Rules**

The **final answers** to angles and sides of triangles that have more than one decimal digit will be rounded off to the **nearest tenth**. The final answers will be computed using intermediate results that have been rounded off to four decimal places (to the right of the decimal point).

## Example 4

In the right triangle, find $\angle B$, $a$, and $b$.

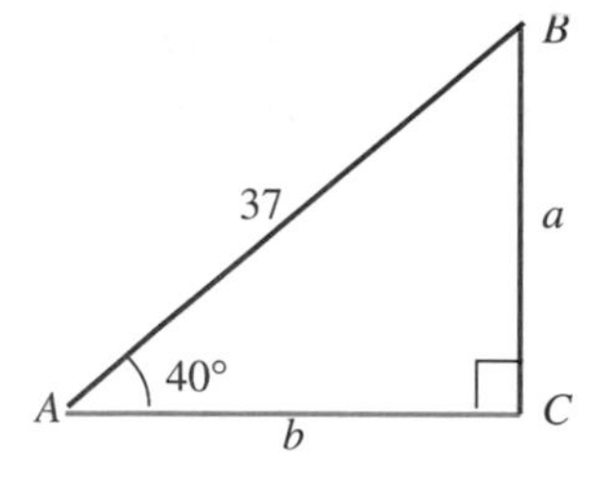


**Solution:** To find $\angle B$:

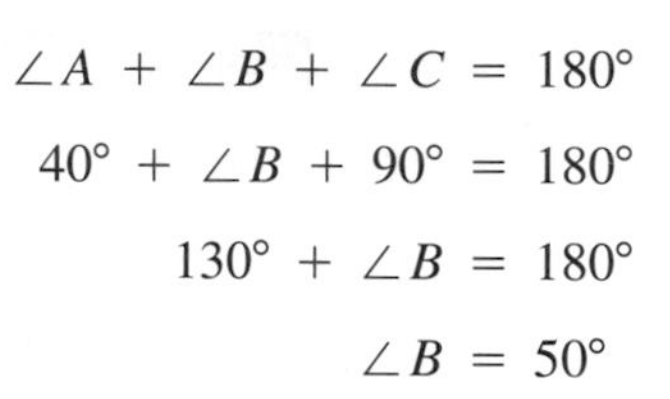

$$\angle A + \angle B + \angle C = 180°$$
$$40° + \angle B + 90° = 180°$$
$$130° + \angle B = 180°$$
$$\angle B = 50°$$

To find side $a$, we use sin 40°, since $a$ is the side opposite the 40° angle and we know the hypotenuse, 37.

$$\sin 40° = \frac{a}{37} \quad \text{(using a trig ratio where only one of the quantities is an unknown)}$$
$$0.6428 = \frac{a}{37}$$
$$37(0.6428) = a$$
$$23.8 = a$$

To find side $b$, we use cos 40°, since $b$ is the side adjacent to the 40° angle and we know the hypotenuse, 37.

$$\cos 40° = \frac{b}{37}$$
$$0.7660 = \frac{b}{37}$$
$$37(0.7660) = b$$
$$28.3 = b$$

■

## Example 5

In the right triangle, find $\angle X$, $\angle Y$, and $z$.

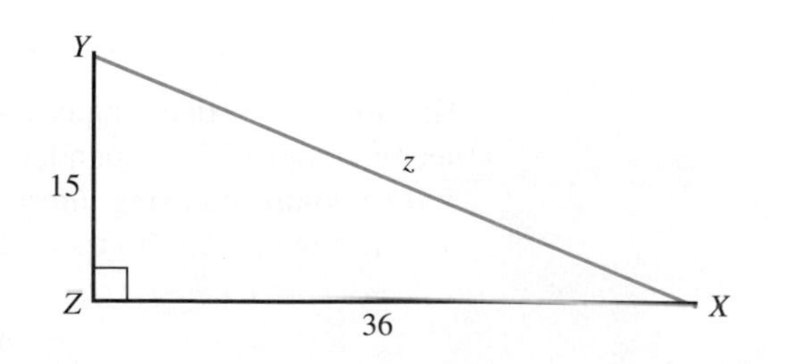

**Solution:** To find $X$, we can use the tangent function since we know the sides opposite and adjacent to angle $X$.

**Press**

$$\tan X = \frac{15}{36}$$

$$\tan X = 0.4167$$

$$X = 22.6°$$

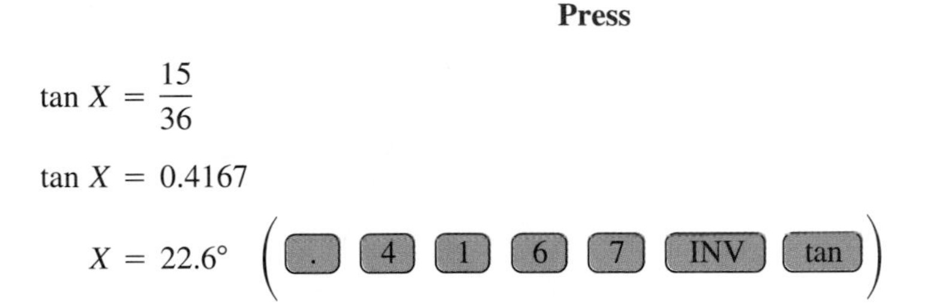


Since we know that the sum of the measures of the angles of a triangle is 180°, we can find the measure of angle $Y$.

$$\angle X + \angle Y + \angle Z = 180°$$

$$22.6° + \angle Y + 90° = 180$$

$$\angle Y = 67.4°$$

Finally, using the Pythagorean Theorem, we can find the value of $z$.

$$x^2 + y^2 = z^2$$

$$15^2 + 36^2 = z^2$$

$$225 + 1296 = z^2$$

$$1521 = z^2$$

$$39 = z$$

## Example 6

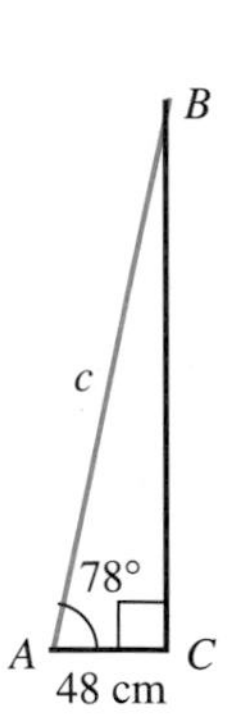


If an acute angle of a right triangle is 78° and the side adjacent to that angle is 48 cm, how long is the hypotenuse of the right triangle?

**Solution:** Let $c$ = the length of the hypotenuse.

$$\cos 78° = \frac{48}{c}$$

$$0.2079 = \frac{48}{c}$$

$$0.2079c = 48 \quad \text{(multiplying both sides by } c\text{)}$$

$$c = 230.9 \text{ cm} \quad \text{(dividing both sides by 0.2079)}$$

### Example 7

In the figure, find the length of side $h$.

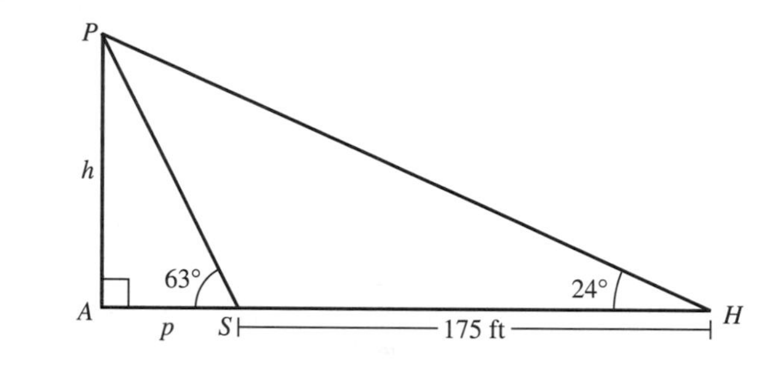


**Solution:** Since there is no right triangle with only one unknown side, we start by separating the figure into two right triangles, $\triangle SAP$ and $\triangle HAP$.

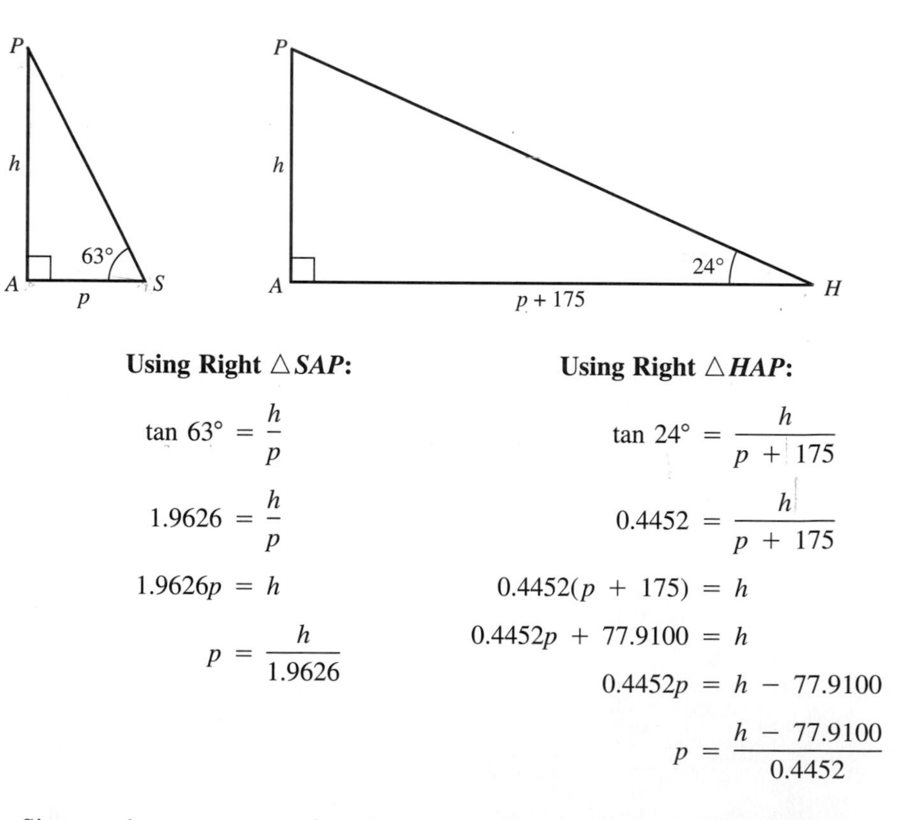


**Using Right $\triangle SAP$:**

$$\tan 63° = \frac{h}{p}$$

$$1.9626 = \frac{h}{p}$$

$$1.9626p = h$$

$$p = \frac{h}{1.9626}$$

**Using Right $\triangle HAP$:**

$$\tan 24° = \frac{h}{p + 175}$$

$$0.4452 = \frac{h}{p + 175}$$

$$0.4452(p + 175) = h$$

$$0.4452p + 77.9100 = h$$

$$0.4452p = h - 77.9100$$

$$p = \frac{h - 77.9100}{0.4452}$$

Since we have two expressions for $p$, we can set them equal to each other and solve for $h$.

$$\frac{h}{1.9626} = \frac{h - 77.9100}{0.4452}$$

$$0.4452h = 1.9626(h - 77.9100)$$

$$0.4452h = 1.9626h - 152.9062$$

$$-1.5174h = -152.9062$$

$$h = 100.8$$

This gives the height of the figure as 100.8 ft.

## Section 5.1 PROBLEMS

## *Explain → Apply → Explore*

### *Explain*

1. Give the right triangle definitions of the sine, cosine, and tangent functions.

2. In a right triangle, if $\sin R = \frac{p}{t}$, what do $R$, $p$, and $t$ represent?

3. In a right triangle, if $\cos M = \frac{k}{f}$, what do $M$, $k$, and $f$ represent?

4. In a right triangle, if $\tan H = \frac{r}{x}$, what do $H$, $r$, and $x$ represent?

5. In this chapter, what is the round-off rule for calculating angles of triangles?

6. In this chapter, what is the round-off rule for calculating sides of triangles?

In Problems 7–9, use the figure in the margin:

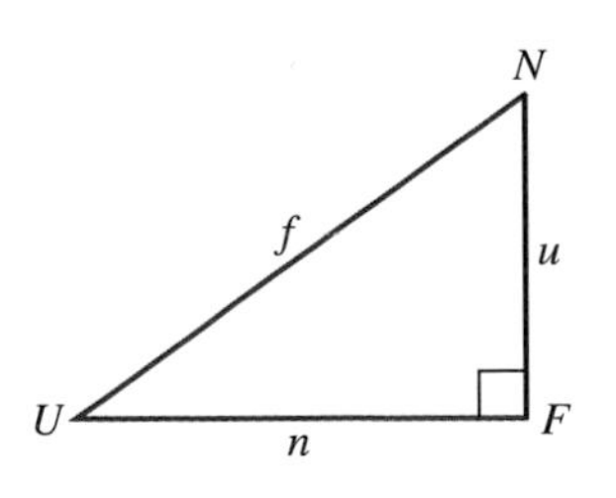


7. If you know that $\angle U = 38°$, what calculator keys should you press to obtain the ratio of side $u$ to side $f$?

8. If you know that $\angle U = 38°$, what calculator keys should you press to obtain the ratio of side $n$ to side $f$?

9. If you know that $\angle U = 38°$, what calculator keys should you press to obtain the ratio of side $u$ to side $n$?

10. If you know that the $\sin G = 0.25$, what calculator keys should you press to find the measure of $\angle G$?

11. If you know that the $\cos D = 0.25$, what calculator keys should you press to find the measure of $\angle D$?

12. If you know that the $\tan E = 0.25$, what calculator keys should you press to find the measure of $\angle E$?

## *Apply*

In Problems 13–15, find $\sin A$, $\cos A$, $\tan A$, $\sin B$, $\cos B$, and $\tan B$.

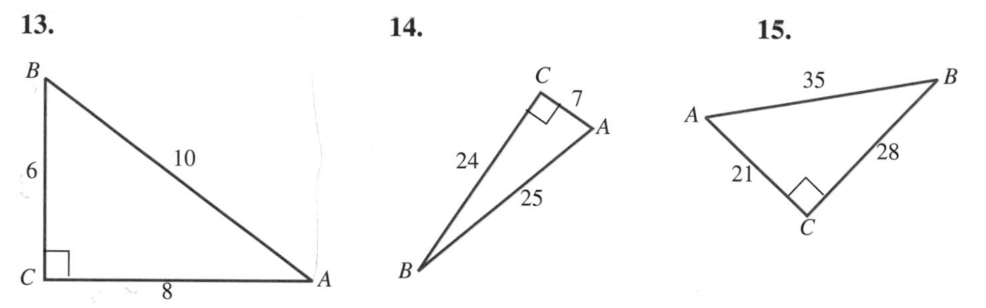


In Problems 16–25, use a calculator to find the following:

**16.** $\sin 59°$

**17.** $\sin 12.6°$

**18.** $\sin 8.25°$

**19.** $\cos 82°$

**20.** $\cos 5.9°$

**21.** $\cos 43.88°$

**22.** $\tan 33°$

**23.** $\tan 76.3°$

**24.** $\tan 17.95°$

**25.** $\cos 0.59°$

In Problems 26–34, use a calculator to find the angle for each trigonometric function.

**26.** $\sin A = 0.8974$

**27.** $\cos Y = 0.1230$

**28.** $\tan B = 6.548$

**29.** $\sin G = 0.67$

**30.** $\cos W = 0.5555$

**31.** $\tan D = 19.035$

**32.** $\sin F = 5/12$

**33.** $\cos K = 4/11$

**34.** $\tan J = 24/7$

## *Explore*

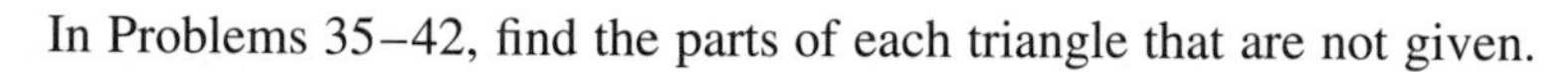

In Problems 35–42, find the parts of each triangle that are not given.

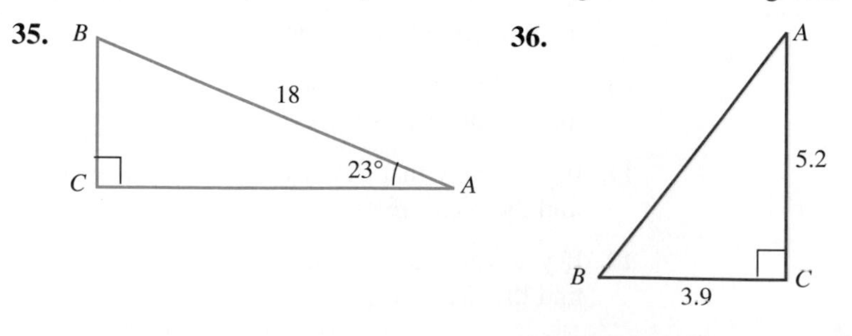

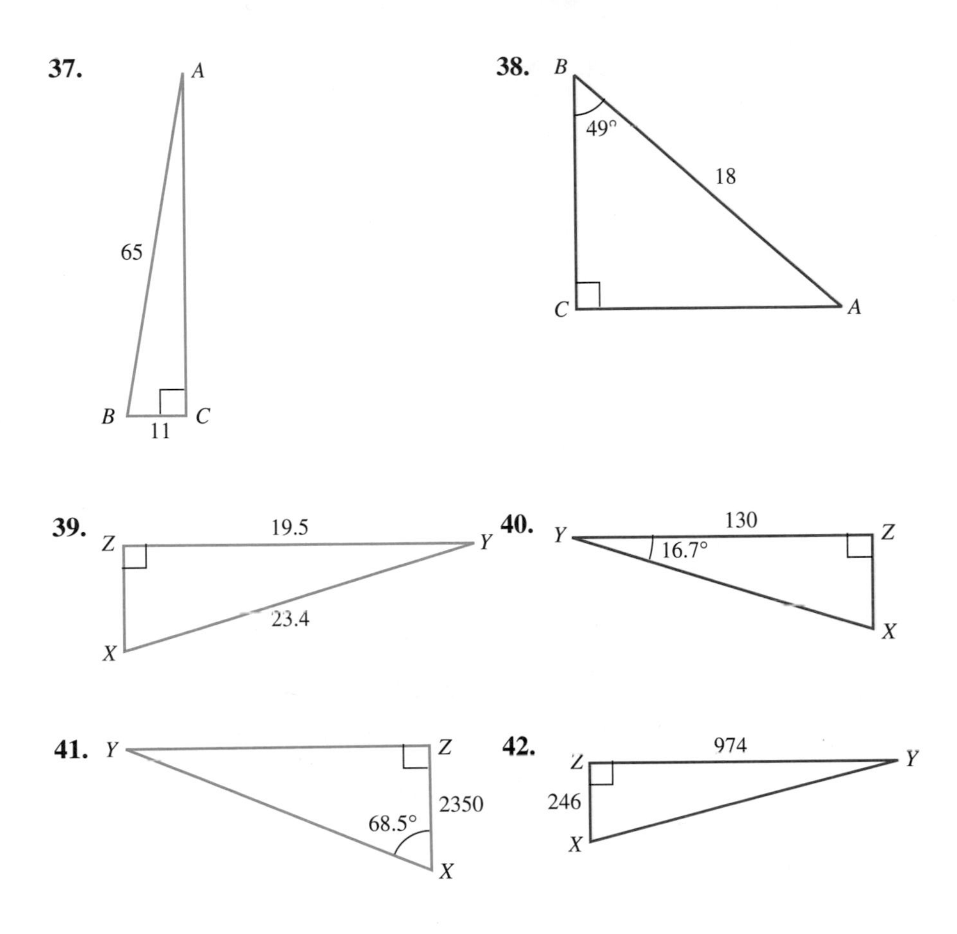


In Problems 43–48 find the angles of the right triangles whose sides are the Pythagorean triples listed.

**43.** 8, 15, 17 **45.** 18, 24, 30 **47.** 33, 44, 55

**44.** 9, 40, 41 **46.** 27, 36, 45 **48.** 102, 2600, 2602

In Problems 49–52 determine the value of $h$ in the following diagrams.

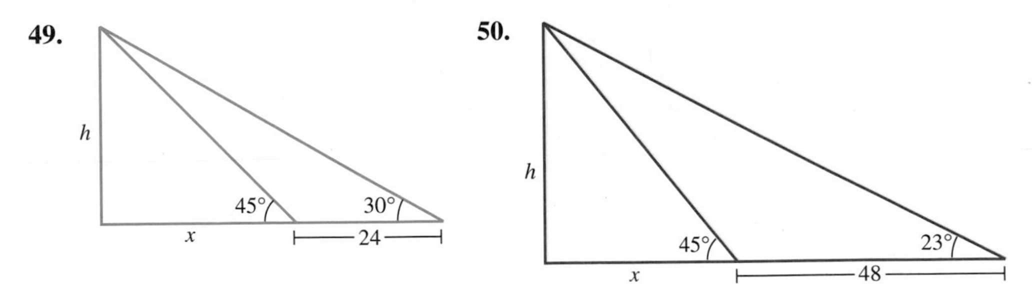

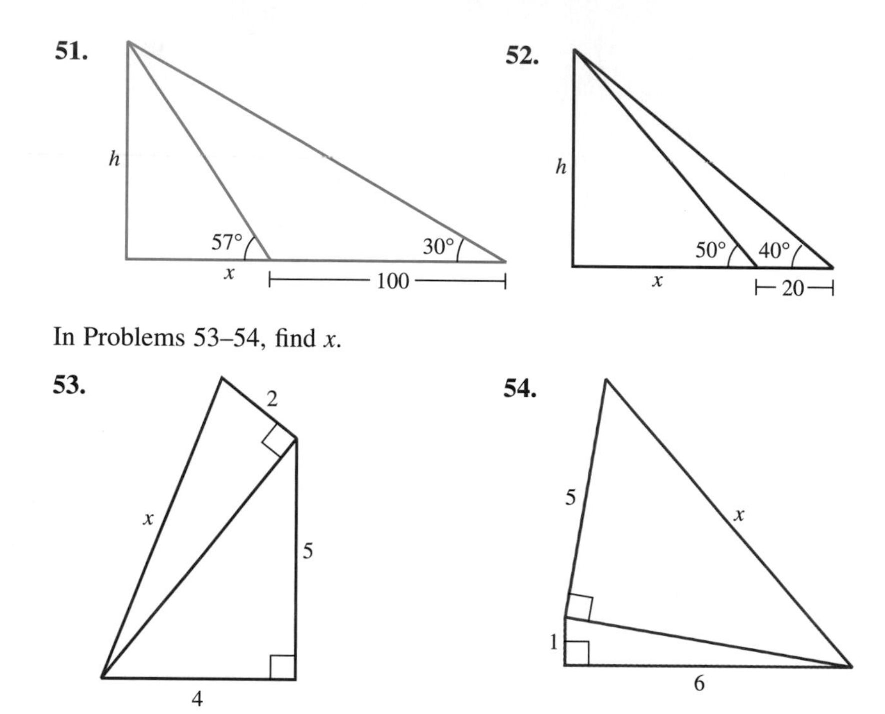


In Problems 53–54, find $x$.

*Eight Bells* by Winslow Homer shows one of the devices used by navigators in determining angles. (National Gallery of Art)

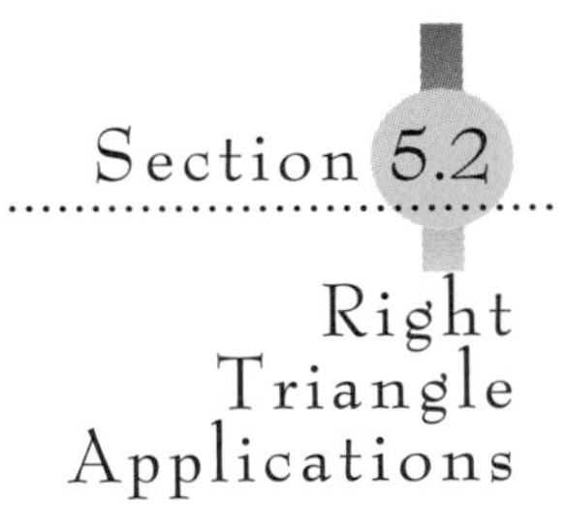


The trigonometric functions studied in the previous section can be used to solve many different kinds of practical problems. In solving these problems, you will find it helpful to do the following:

1. Sketch the situation and the right triangle involved.
2. Write the known angles or sides on the triangle.
3. Use letters to represent the unknown angles or sides.
4. Use trig functions to help solve the triangle, remembering to have only one unknown quantity in a trigonometric equation.

### Example 1

The instruction booklet for a 50-ft ladder states that, for safety reasons, when the ladder is leaning against a vertical surface, the angle the ladder makes with the horizontal ground must be from 60° to 75°. Under those constraints, what are the minimum and maximum distances that the ladder will reach on the wall?

**Solution:** The minimum angle and maximum angle are shown in the right triangles. If we let $h$ = the height the ladder reaches up the wall, using the sine function, we can solve for $h$ in each case.

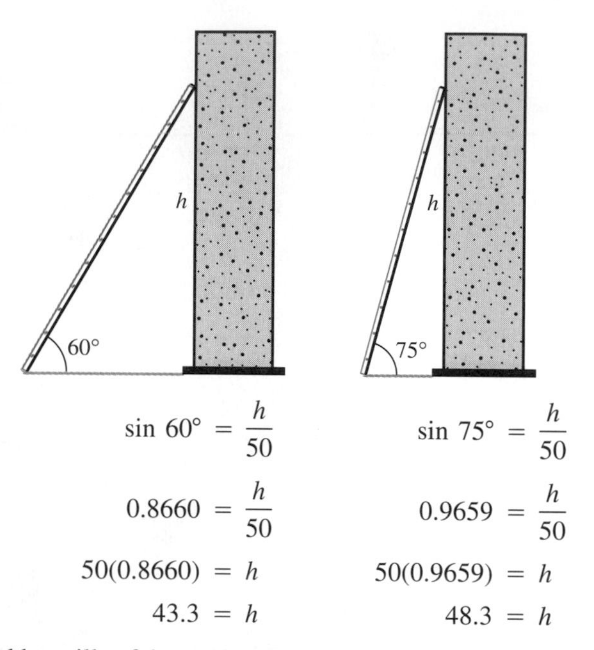


$$\sin 60^\circ = \frac{h}{50} \qquad \sin 75^\circ = \frac{h}{50}$$

$$0.8660 = \frac{h}{50} \qquad 0.9659 = \frac{h}{50}$$

$$50(0.8660) = h \qquad 50(0.9659) = h$$

$$43.3 = h \qquad 48.3 = h$$

Thus, the ladder will safely reach heights between 43.3 and 48.3 ft on the wall. ▪

## Example 2

A small airplane takes off from an airport at an angle of 32.3° with level ground. Three-fourths of a mile (3960 ft) from the airport is a 1500-ft peak in the flight path of the plane. If the plane continues that angle of ascent, find (a) its altitude when it is above the peak and (b) how far it will be above the peak.

**Solution:**

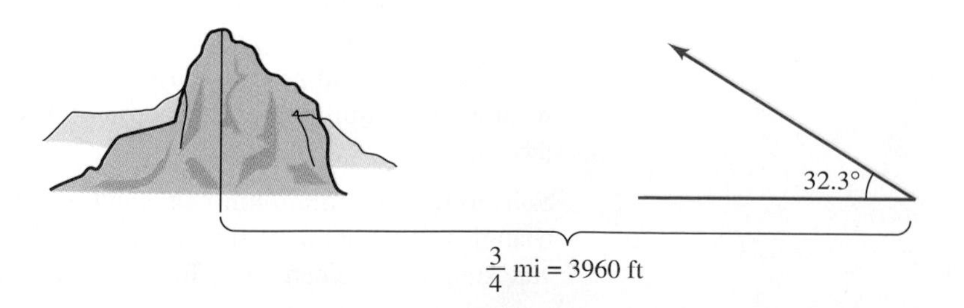

Let $a$ = the altitude of the plane

$$\tan 32.3° = \frac{a}{3960}$$

$$0.6322 = \frac{a}{3960}$$

$$3960(0.6322) = a$$

$$2503.5 = a$$

Thus, (a) the altitude of the plane is 2503.5 ft and (b) it is 2503.5 ft − 1500 ft = 1003.5 ft above the peak. ■

## Example 3

The swimming area at Shadow Cliffs Lake is roped off with floating markers as shown in the figure. If a section of the beach is 350 yd long and the ropes make a 90° angle at the platform, how long is the swim from one corner on the beach along the floating markers to the middle of the stationary platform and then to the other corner on the beach?

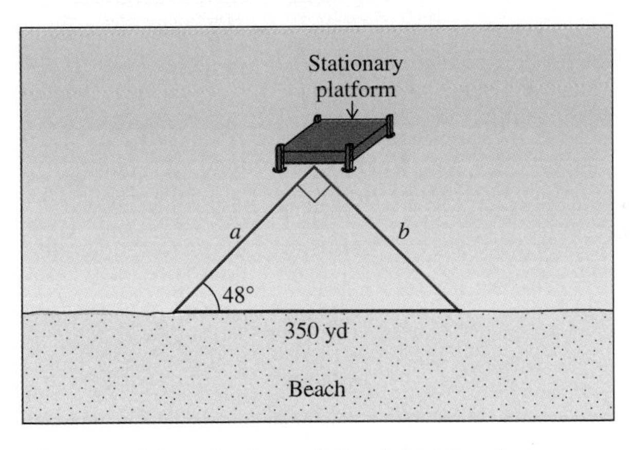


**Solution:** Let $a$ and $b$ = the legs of the right triangle.
We need to find $a + b$.

$$\cos 48° = \frac{a}{350} \qquad \sin 48° = \frac{b}{350}$$

$$0.6691 = \frac{a}{350} \qquad 0.7431 = \frac{b}{350}$$

$$350(0.6691) = a \qquad 350(0.7431) = b$$

$$234.2 = a \qquad 260.1 = b$$

Thus, the total distance is 234.2 + 260.1 = 494.3 yd. ■

## Angles of Elevation and Depression

Many problems with right triangles involve the angle made with an imaginary horizontal line. An angle between such a horizontal line and the line of sight to an object that is above the horizontal is called the **angle of elevation**, and the angle made between such a horizontal line and the line of sight to an object that is below the horizontal is called the **angle of depression**. Instruments such as transits and sextants can be used to measure such angles.

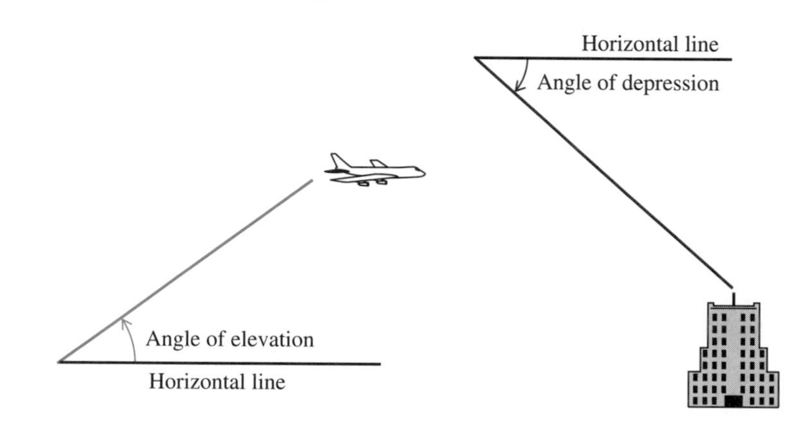


### Example 4

On level ground, at a point 75 ft from the base of a tree, the angle of elevation to the top of the tree is 70°. Find the height of the tree.

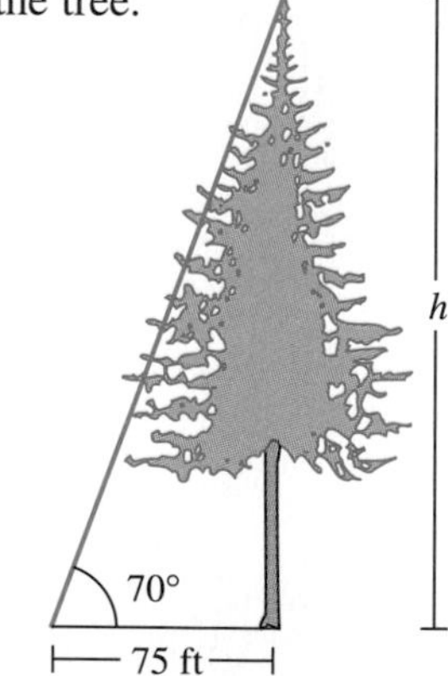


**Solution:** Let $h$ = the height of the tree in feet.

$$\tan 70° = \frac{h}{75}$$

$$2.7475 = \frac{h}{75}$$

$$75(2.7475) = h$$

$$206.1 = h$$

The height of the tree is 206.1 ft. ■

### Example 5

A point on the edge of the Grand Canyon in Arizona is 4600 ft above the Colorado River. The angle of depression to the middle of the canyon floor is 16°. Find, to the nearest tenth of a mile, the horizontal distance to a point directly above the middle of the canyon floor.

**Solution:** Let $x$ = the horizontal distance to the middle of the Grand Canyon floor from one side.

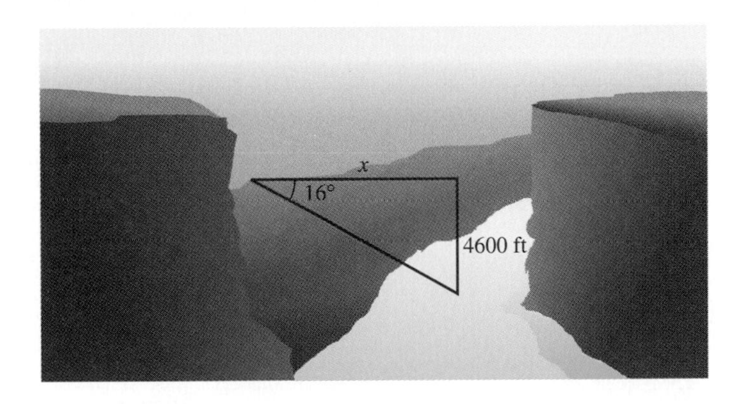


$$\tan 16^\circ = \frac{4600}{x}$$

$$0.2867 = \frac{4600}{x}$$

$$0.2867x = 4600$$

$$x = \frac{4600}{0.2867}$$

$$x = 16{,}044.6 \text{ ft} \approx 3.0 \text{ mi}$$

## Example 6

The sign on the side of a straight uphill stretch of highway reads, "9% GRADE NEXT 5 MI."

(a) Find the angle of elevation of the highway.
(b) Determine the change in altitude in feet after driving 5 mi.

**Solution:** A 9% grade indicates that the slope of the road is 9/100, which means the highway rises 9 ft vertically for every 100-ft change in the horizontal direction.

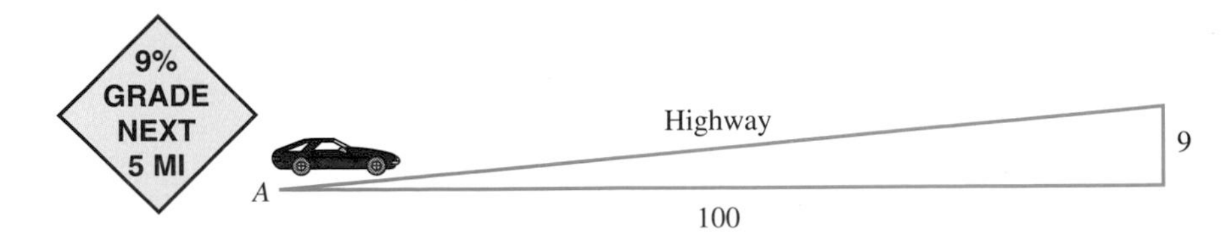

(a) To find the angle of elevation, find $\angle A$.

$$\tan A = \frac{9}{100} = 0.09$$

$$A = 5.1°$$

(b) Using the result from part (a), we get

$$5 \text{ mi} = 5(5280) = 26{,}400 \text{ ft}$$

$$\sin 5.1° = \frac{a}{26{,}400}$$

$$0.0889 = \frac{a}{26{,}400}$$

$$0.0889(26{,}400) = a$$

$$a = 2346.8$$

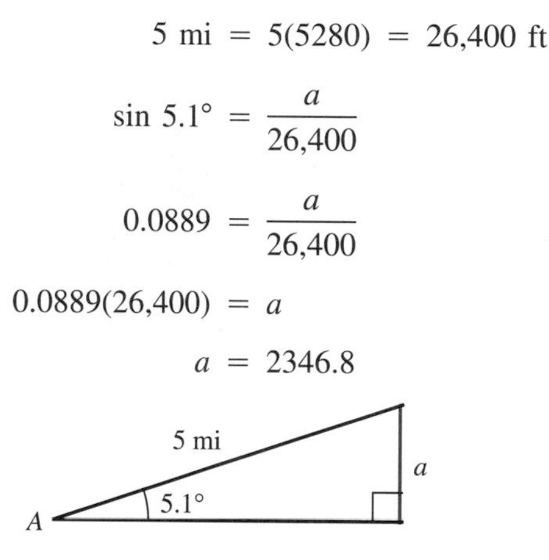


Thus, the angle of elevation of the highway is 5.1°, and after 5 mi of driving there is a change in altitude of 2346.8 ft. ■

## Example 7

To find the height of Mission Peak near the Ohlone College campus, a student went to the football field with a transit and found that the angle of elevation to the top of the peak was 9.8°. At 100 yd (300 ft) from that point and in line with the first measurement, the angle of elevation was 9.6°. How far above the football field is the top of Mission Peak?

**Solution:** The following figure shows Mission Peak along with the two measurements of the angles of elevation. We need to find the value of $a$ to find the height of Mission Peak above the football field. Both right triangles, $\triangle BCA$ and $\triangle BCD$, will be used in solving for $a$.

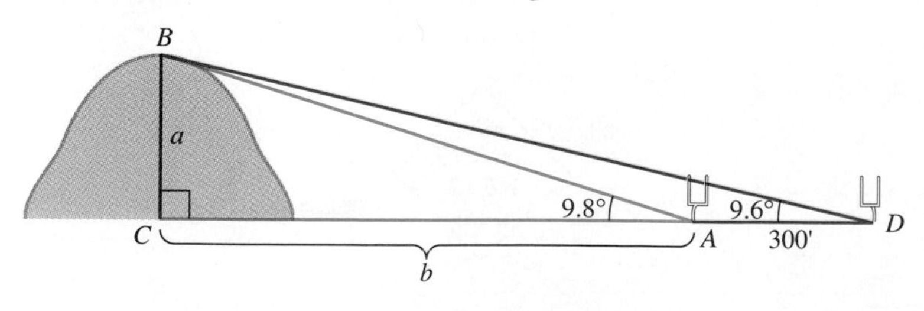

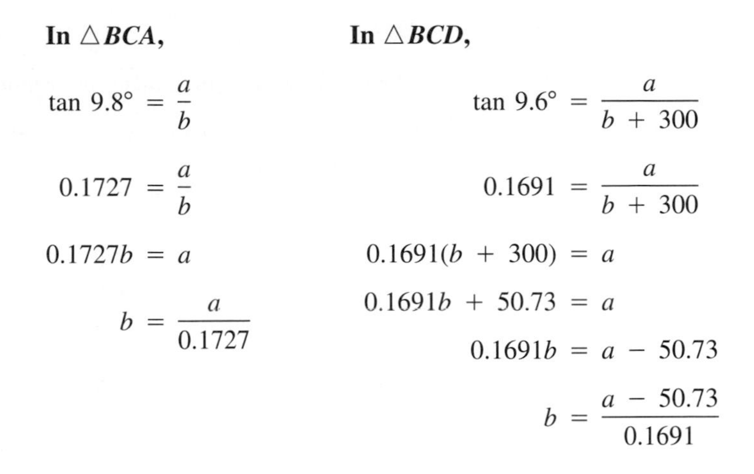

**In $\triangle BCA$,**

$$\tan 9.8° = \frac{a}{b}$$

$$0.1727 = \frac{a}{b}$$

$$0.1727b = a$$

$$b = \frac{a}{0.1727}$$

**In $\triangle BCD$,**

$$\tan 9.6° = \frac{a}{b + 300}$$

$$0.1691 = \frac{a}{b + 300}$$

$$0.1691(b + 300) = a$$

$$0.1691b + 50.73 = a$$

$$0.1691b = a - 50.73$$

$$b = \frac{a - 50.73}{0.1691}$$

Since we have found two different expressions for $b$, we can set them equal to each other and solve for $a$:

$$\frac{a}{0.1727} = \frac{a - 50.73}{0.1691}$$

$$0.1691a = 0.1727(a - 50.73)$$

$$0.1691a = 0.1727a - 8.7611$$

$$-0.0036a = -8.7611$$

$$a = 2433.6$$

Therefore, Mission Peak is approximately 2433.6 ft above the football field. ■

## Section 5.2 PROBLEMS

## *Explain ➡ Apply ➡ Explore*

### *Explain*

1. What four steps would be helpful in solving practical problems involving right triangles?
2. What is an angle of elevation?
3. What is an angle of depression?
4. Why can you not use the sine, cosine, or tangent functions on a triangle that has angles of 66°, 34°, and 80°?
5. Explain why the title of this chapter, "Trigonometry: A Door to the Unmeasurable," is appropriate in this section?

## *Apply*

In Problems 6–10, solve using the round-off rules established in Section 5.1.

**6.** An Eagle Scout taking trigonometry finds the distance across a river by doing the following:
(a) He stands on the bank of the river and chooses a boulder directly across the river as a marker.
(b) He paces off 25 yd along the bank.
(c) From that point he approximates that the angle back to the rock is 60°.
(d) Being well prepared, he takes out his calculator, uses trig, and determines that the river is 43.3 yd wide.

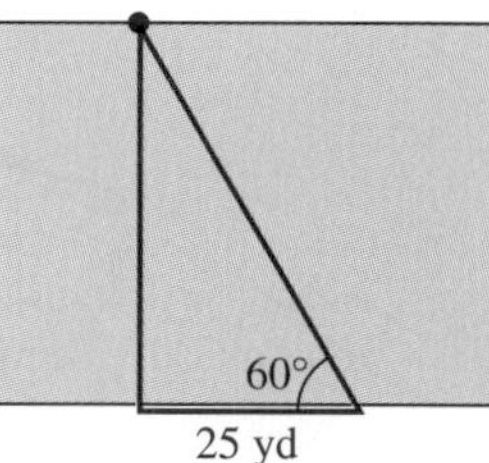


Show that the Eagle Scout was correct in his approximation.

**7.** A sign on a straight stretch of freeway reads, "10% GRADE altitude 3000 ft." If the next altitude sign you see states that the altitude is 4350 ft, how many miles have you driven since seeing the 10% grade sign?

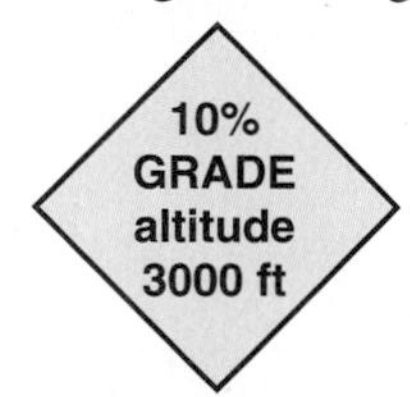


**8.** Due to road construction a 36-mi section of a highway is closed. The detour makes a 55° angle with the highway as shown. How many additional miles does one travel using the detour?

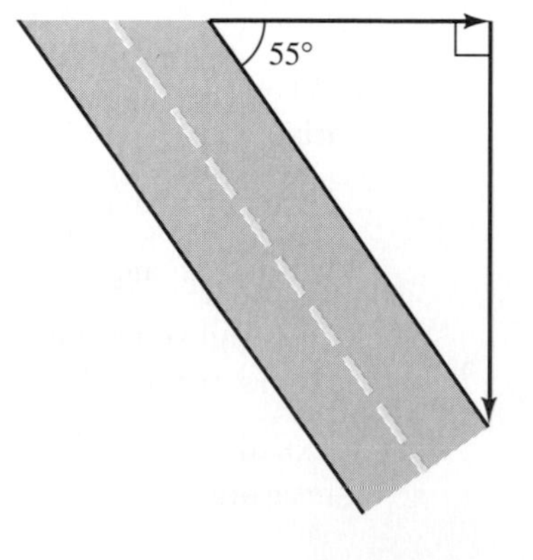

**9.** To find the distance across a lake, a tree on the opposite side of the lake was taken as a marker and 60 m was walked off, as shown in the figure. If the angle to the tree was 75°, how far is it across the lake?

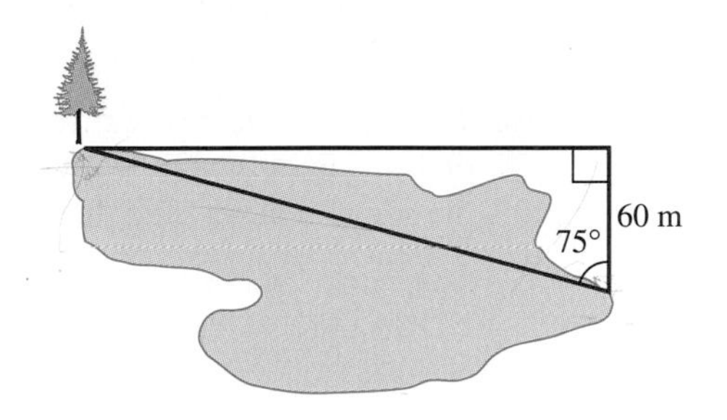


**10.** After lifting off from its launch site, a hot air balloon floats along over an open field with the top of its basket 250 ft above ground level. If the angle of depression to its take-off point is 2°, how far is the balloon from its take-off point?

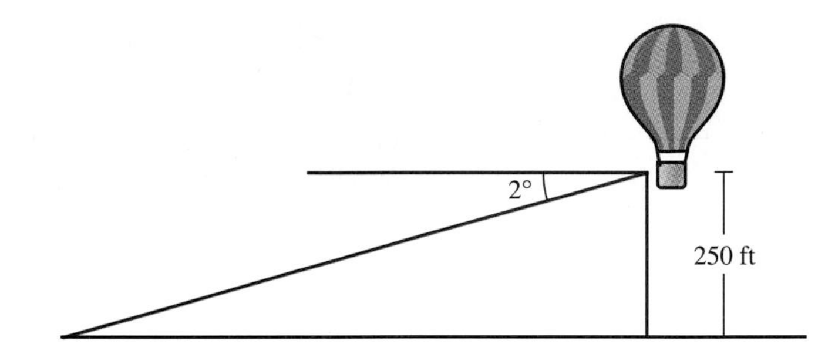


## *Explore*

In Problems 11–23, make a sketch of the situation and solve using the round-off rules established in Section 5.1.

**11.** A guy wire, which supports a vertical circus tent pole, makes a 76.5° angle with level ground. If the guy wire is secured to the ground 12 ft away from the bottom of the pole, how tall is the pole and how long is the guy wire?

**12.** The pitch of a roof is $5/12$. That means the roof rises 5 ft for each 12-ft change in the horizontal direction. What is the angle of elevation of the roof? If the actual length of the roof is 32 ft, how much does the roof rise in that distance?

**13.** How far up a vertical wall will a 32-ft ladder reach if it makes a 67° angle with the level ground?

**14.** Radar indicates that the distance to an approaching airplane is 8.6 mi and the angle of elevation to the plane is 26°. At what altitude is the plane flying?

**15.** From a point at eye level (5 ft off the ground) the angle of elevation to the top of a radio transmitting tower is 70°. If the person measuring this angle is 40 ft from the tower, what is the height of the tower?

**16.** The angle of depression from the roof of a 310-ft office building to the bottom of a statue in the Civic Center is 8.2°. If the ground between the building and the statue is level, find the distance from the building to the statue.

**17.** The angle of depression from the top of the largest Khufu Pyramid (height 482 ft) in Egypt to the bottom of a nearby marker measures 12.7°. If the ground is level, how far is the top of the Pyramid from the bottom of the marker?

**18.** A sea-to-air guided missile shot from a submarine leaves the water at an angle of elevation of 18.6° traveling at 480 ft/s. If the missile maintains a constant angle of ascent and the same speed, how far above sea level will it be after 30 s?

**19.** In Problem 18, how long will it take the missile to reach an altitude of 10,000 ft?

**20.** A ship at sea measures the angle of elevation to the top of a cliff on shore to be 14.8°. After traveling a half mile closer to the cliff, the angle of elevation is now 26.5°. How many feet is the top of the cliff above sea level?

**21.** A tennis player hits an overhead smash 35 ft from the net. If the ball is hit 10.5 ft off the ground and just clears the net at a height of 3 ft, how far from the net and at what angle will the ball strike the ground?

**22.** In *Tactics for Trout* the author Dave Hughes states that a trout has a 97° angle of vision in any direction. Furthermore, he states that because of the reflection of light a trout cannot see objects above the water surface within a 10° angle. The implication is that a trout has both a region within its range of vision and a region not in its range of vision.

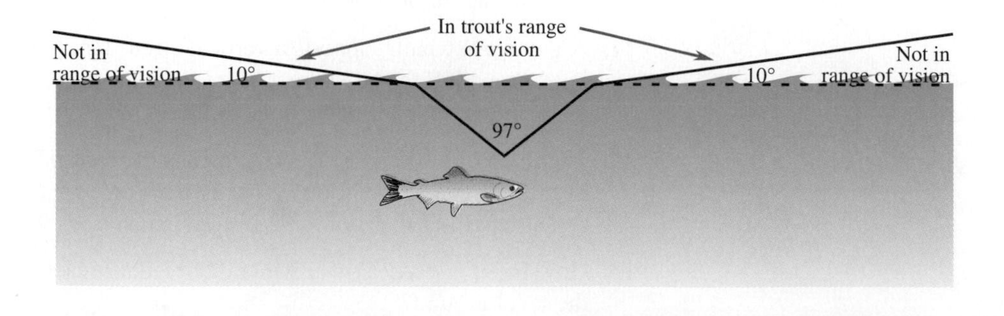

A knowledge of the range of vision of a trout can help fly fishermen determine how close he can get to a trout without being seen. If a trout is at a depth of 2 ft, how close can a 6-ft person get to the trout without being seen by the trout? (*Hint:* The angle at the trout's eye is 48.5° and you need to find the length of $d$.)

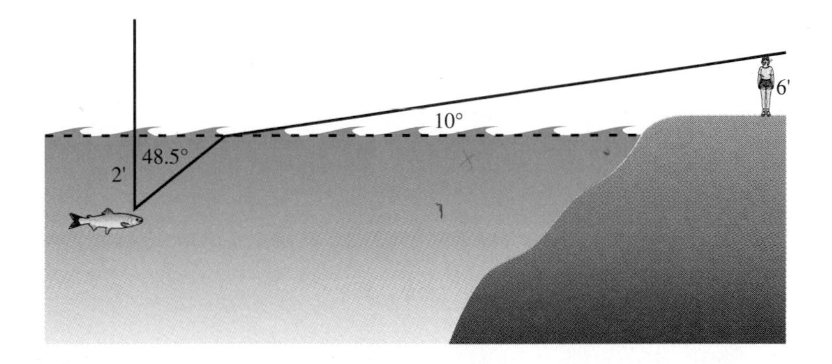


**23.** In Problem 22, if a trout is at a depth of 1 ft, how close can a 5-ft person get to the trout without being seen by the trout?

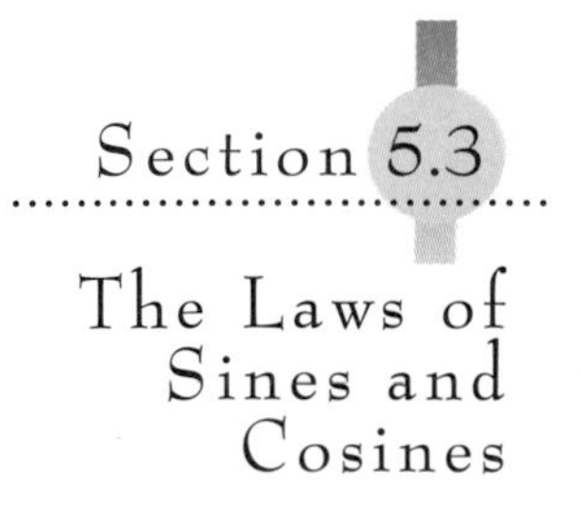

## Section 5.3 The Laws of Sines and Cosines

The three trigonometric functions studied in the previous sections were defined and used in reference to right triangles. The next two laws of trigonometry, the Law of Sines and the Law of Cosines, will allow us to solve triangles that are not right triangles. A triangle with each angle measuring less than 90° is called an **acute triangle**, and a triangle with one angle greater than 90° is called an **obtuse triangle**. The Laws of Sines and Cosines can be used to solve acute and obtuse triangles. However, in this survey of trigonometry we only apply them to acute triangles that have a unique solution.

### The Law of Sines

The **Law of Sines** gives a relationship between the sides and angles of a triangle. It states that the ratio of the length of the side of any triangle to the sine of the angle opposite that side is the same for all three sides of the triangle. That is, for any $\triangle ABC$:

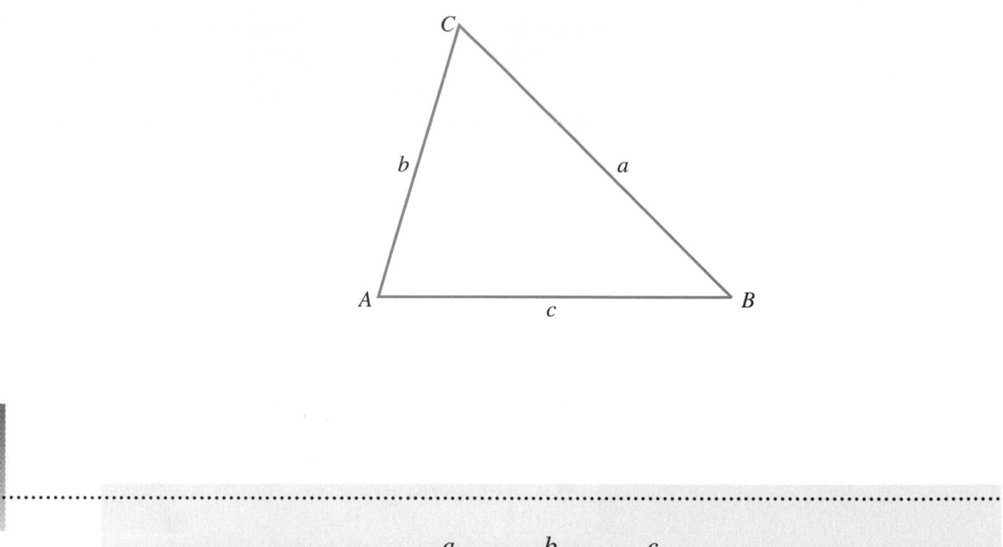


Law of Sines

$$\frac{a}{\sin A} = \frac{b}{\sin B} = \frac{c}{\sin C}$$

## Derivation of the Law of Sines

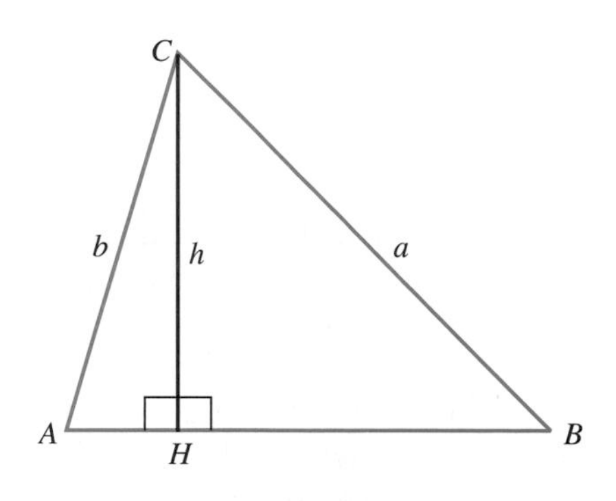


Let $h$ be the length of the segment from $C$ forming right angles with $\overline{AB}$ at $H$.

**In $\triangle ACH$,**

$$\sin A = \frac{h}{b}$$

$$b \sin A = h$$

**In $\triangle CHB$,**

$$\sin B = \frac{h}{a}$$

$$a \sin B = h$$

Since we have found two expressions for $h$, they must be equal. Setting them equal to each other, we get

$$a \sin B = b \sin A$$

$$\frac{a \sin B}{(\sin A)(\sin B)} = \frac{b \sin A}{(\sin A)(\sin B)} \quad \text{[dividing both sides by } (\sin A)(\sin B)\text{]}$$

$$\frac{a}{\sin A} = \frac{b}{\sin B}$$

If we use the same process with a segment from $B$ making right angles with $\overline{AC}$, we can show that the third ratio, $c/(\sin C)$, is equal to the two ratios given above.

The Law of Sines can be used to solve any triangle in which two angles and one side are known. In each example, you will notice that the ratio of the sine of a known angle and its opposite side will be set equal to a ratio that has either an unknown angle or an unknown side.

## Example 1

In $\triangle ABC$, $\angle A = 40°$, $\angle B = 60°$, and $b = 9$. Find the measure of $\angle C$ and the lengths of $a$ and $c$.

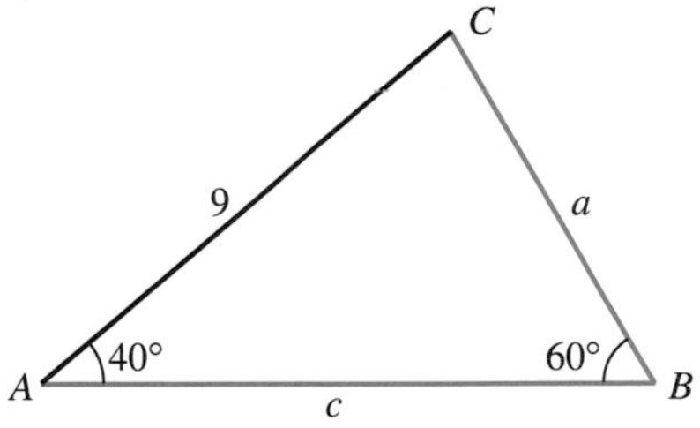

**Solution:** $\angle C = 80°$, since the sum of the angles of a triangle is 180° and $180° - 40° - 60° = 80°$.

To find sides $a$ and $c$, we use the Law of Sines.

$$\frac{a}{\sin A} = \frac{b}{\sin B} \qquad \frac{c}{\sin C} = \frac{b}{\sin B}$$

$$\frac{a}{\sin 40°} = \frac{9}{\sin 60°} \qquad \frac{c}{\sin 80°} = \frac{9}{\sin 60°}$$

$$\frac{a}{0.6428} = \frac{9}{0.8660} \qquad \frac{c}{0.9848} = \frac{9}{0.8660}$$

$$0.8660a = 9(0.6428) \qquad 0.8660c = 9(0.9848)$$

$$0.8660a = 5.7852 \qquad 0.8660c = 8.8632$$

$$a = 6.7 \qquad c = 10.2$$

**Note:** The Law of Sines should be used to solve any triangle in which two angles and one side are known. This situation is referred to as angle-angle-side (AAS) or angle-side-angle (ASA). The Law of Sines can also be used when two sides and an angle opposite one of the sides is known (SSA). Since this case may lead to more than one solution, it will not be discussed in this survey of trigonometry.

### Example 2

In $\triangle XYZ$, $\angle Y = 86.2°$, $\angle Z = 21.5°$, and $y = 110$. Find the measure of $\angle X$ and the lengths of sides $x$ and $z$.

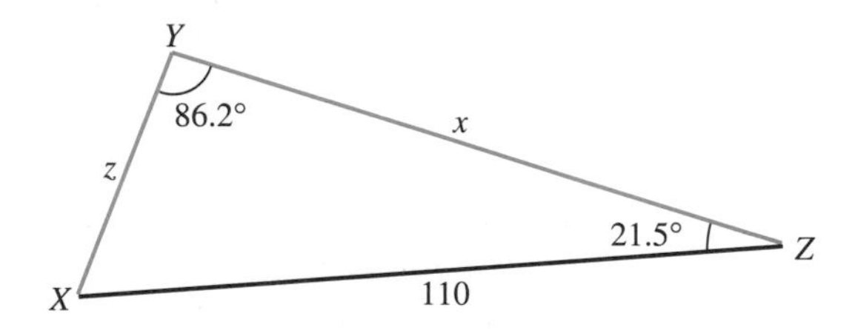


**Solution:** $\angle X = 180° - 86.2° - 21.5° = 72.3°$.
To find sides $x$ and $z$, we use the Law of Sines.

$$\frac{x}{\sin X} = \frac{y}{\sin Y} \qquad \frac{z}{\sin Z} = \frac{y}{\sin Y}$$

$$\frac{x}{\sin 72.3°} = \frac{110}{\sin 86.2°} \qquad \frac{z}{\sin 21.5°} = \frac{110}{\sin 86.2°}$$

$$\frac{x}{0.9527} = \frac{110}{0.9978} \qquad \frac{z}{0.3665} = \frac{110}{0.9978}$$

$$0.9978x = 110(0.9527) \qquad 0.9978z = 110(0.3665)$$

$$0.9978x = 104.797 \qquad 0.9978z = 40.315$$

$$x = 105.0 \qquad z = 40.4$$

## The Law of Cosines

As seen in the previous examples, the Law of Sines gives a very efficient means for solving acute triangles in which two angles and one side are known. However, it is not possible to use the Law of Sines to solve a triangle in which the three sides (SSS) or one angle and the two sides that form the angle (SAS) are known. The law of trigonometry that enables us to solve triangles with those conditions is called the

**Law of Cosines**. This law gives us a different relationship involving the sides and an angle of a triangle than does the Law of Sines. It states that for any $\triangle ABC$,

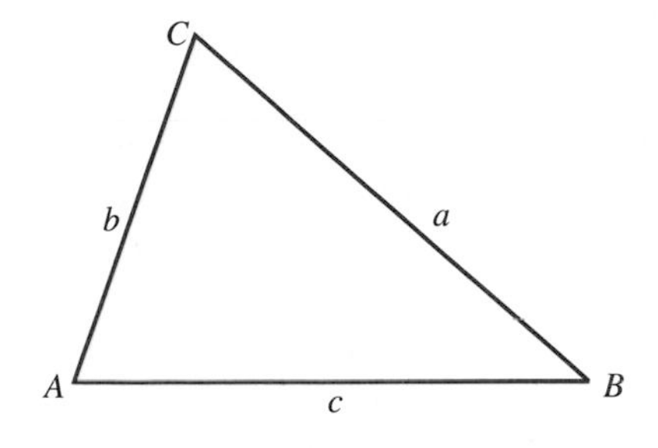


Law of Cosines

$$a^2 = b^2 + c^2 - 2bc \cos A$$

$$b^2 = a^2 + c^2 - 2ac \cos B$$

$$c^2 = a^2 + b^2 - 2ab \cos C$$

## Derivation of the Law of Cosines

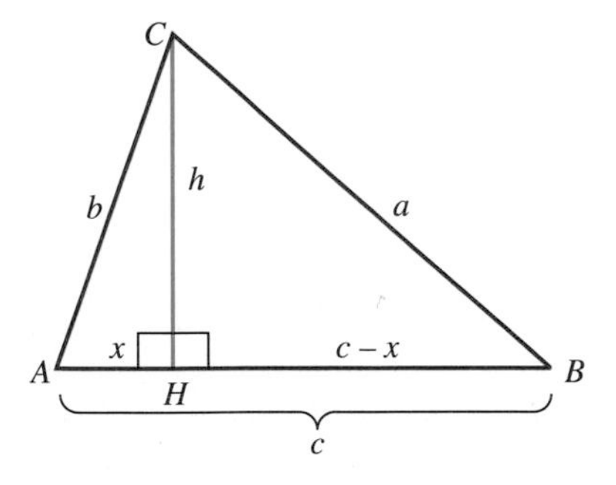


Let $h$ be the length of a segment from $C$ forming right angles with $\overline{AB}$ at $H$. Using the Pythagorean Theorem in $\triangle CHB$ and $\triangle CHA$, we get in right $\triangle CHB$,

$$a^2 = (c - x)^2 + h^2$$

$$a^2 = c^2 - 2cx + x^2 + h^2$$

But in right $\triangle CHA$, we have $b^2 = x^2 + h^2$, so

$$a^2 = c^2 - 2cx + b^2$$

$$a^2 = b^2 + c^2 - 2cx$$

Also in $\triangle CHA$, $\cos A = \frac{x}{b}$, so $b \cos A = x$.

$$\therefore a^2 = b^2 + c^2 - 2cb \cos A$$
$$a^2 = b^2 + c^2 - 2bc \cos A$$

If we use the same process, drawing segments from $A$ or $B$ that make right angles with the sides opposite those angles, we can derive the other forms of the Law of Cosines.

Notice that the variable of the triangle used on the left of the equal sign corresponds to the angle opposite that side on the right of the equal sign. That is, in $a^2 = b^2 + c^2 - 2bc \cos A$, $a^2$ on the left corresponds to $\cos A$ on the right side of the equation. Recognizing that this is true for all three forms of the Law of Cosines makes the task of remembering the Law of Cosines a little easier.

### Example 3

In $\triangle ABC$, $\angle A = 64°$, $b = 18$, and $c = 39$. Use the Law of Cosines to find (a) side $a$ and (b) $\angle B$.

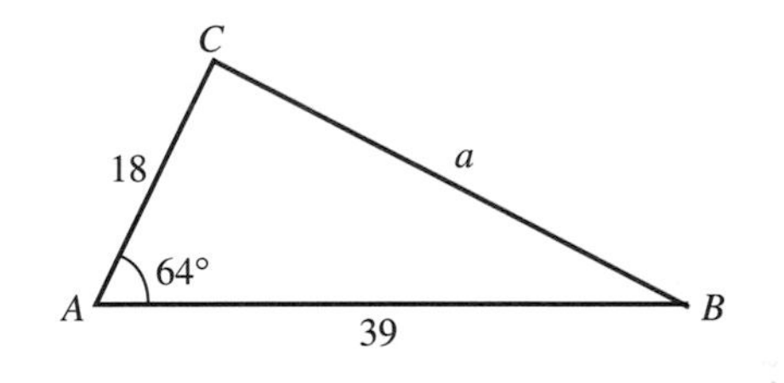


**Solution:**

(a) Since two sides and the included angle are known (SAS), this triangle can be solved by the Law of Cosines.

$$a^2 = b^2 + c^2 - 2bc \cos A$$
$$a^2 = 18^2 + 39^2 - 2(18)(39) \cos 64°$$
$$a^2 = 324 + 1521 - 1404(0.4384)$$
$$a^2 = 1229.4864$$
$$a = \sqrt{1229.4864} = 35.1$$

(b) Since we now know $a$, $b$, and $c$, we can find $\angle B$ from the Law of Cosines, where $B$ is the only unknown quantity.

$$b^2 = a^2 + c^2 - 2ac \cos B$$
$$18^2 = 35.1^2 + 39^2 - 2(35.1)(39) \cos B$$
$$324 = 1232.01 + 1521 - 2737.8 \cos B$$
$$-2429.01 = -2737.8 \cos B$$
$$0.8872 = \cos B$$
$$27.5° = B$$

**Note:** The Law of Cosines can be used to solve any triangle in which a side, an angle, and a side (**SAS**) or three sides (**SSS**) are known.

## Example 4

In $\triangle USA$, $u = 38$, $s - 42$, and $a = 29$. Find the measures of $\angle U$, $\angle S$, and $\angle A$.

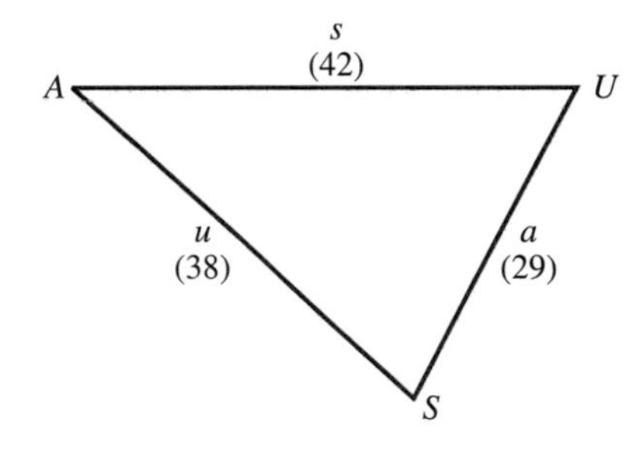


**Solution:** Since the three sides of the triangle are known, we first find $\angle U$ by writing the Law of Cosines, using $u$, $s$, $a$, and $U$.

$$u^2 = s^2 + a^2 - 2sa \cos U$$
$$38^2 = 42^2 + 29^2 - 2(42)(29) \cos U$$
$$1444 = 1764 + 841 - 2436 \cos U$$
$$-1161 = -2436 \cos U$$
$$0.4766 = \cos U$$
$$61.5° = U$$

We can now find $\angle S$ by using the Law of Cosines again. However, it would be easier to use the Law of Sines with $u$ and $\sin U$ forming the known ratio and $S$ being the only unknown quantity.

$$\frac{u}{\sin U} = \frac{s}{\sin S}$$

$$\frac{38}{\sin 61.5°} = \frac{42}{\sin S}$$

$$\frac{38}{0.8788} = \frac{42}{\sin S}$$

$$38 \sin S = 42(0.8788)$$

$$\sin S = 0.9713$$

$$S = 76.2°$$

Since we now know two of the angles of $\triangle USA$, we can find $\angle A$ by using the fact that the sum of the angles of a triangle equals 180°.

$$\angle A = 180° - 61.5° - 76.2° = 42.3°$$

## Section 5.3 PROBLEMS

## *Explain ➡ Apply ➡ Explore*

### *Explain*

1. What is the Law of Sines?
2. What is the Law of Cosines?
3. Explain when it is necessary to use the Law of Sines or the Law of Cosines.
4. In this section, what should be known about an acute triangle to use the Law of Sines?
5. In this section, what should be known about an acute triangle to use the Law of Cosines?

### *Apply*

In Problems 6–14, use the Law of Sines to find the designated side in $\triangle SUN$.

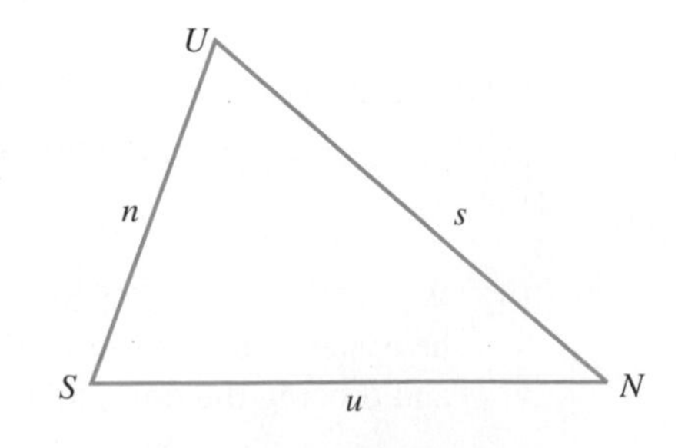

6. $\angle S = 36°$, $\angle U = 70°$, $s = 7$; find $u$.
7. $\angle S = 56.5°$, $\angle U = 62.9°$, $s = 25$; find $n$.
8. $\angle S = 9°$, $\angle U = 87°$, $n = 34.6$; find $s$.
9. $\angle S = 47°$, $\angle U = 60°$, $n = 450$; find $u$
10. $\angle S = 28.7°$, $\angle N = 71.5°$, $u = 56$; find $n$.
11. $\angle S = 66.6°$, $\angle N = 55.5°$, $s = 44.4$; find $u$.
12. $\angle S = 57°$, $\angle N = 57°$, $n = 7.5$; find $s$.
13. $\angle N = 12.5°$, $\angle U = 80.25°$, $u = 1240$; find $n$.
14. $\angle N = 75°$, $\angle U = 30°$, $s = 1.23$; find $u$.

In Problems 15–20, use the Law of Cosines to find the designated angle or side in $\triangle DOG$.

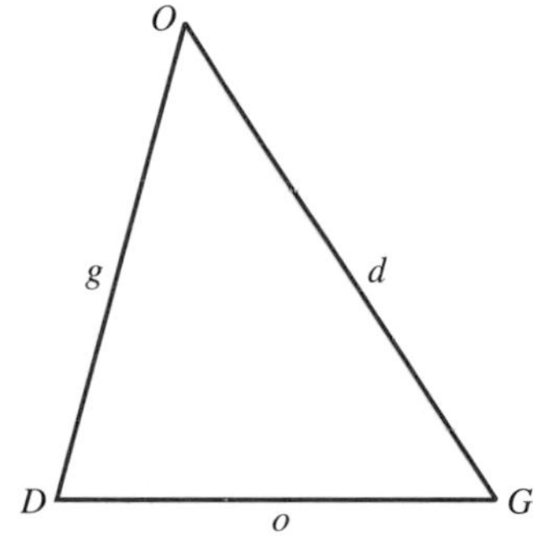


15. $\angle D = 73°$, $o = 12$, $g = 34$; find $d$.
16. $\angle O = 40.6°$, $g = 23$, $d = 45.5$; find $o$.
17. $\angle G = 37.25°$, $d = 5.5$, $o = 6.75$; find $g$.
18. $d = 15$, $o = 16$, $g = 17$; find $\angle D$.
19. $d = 9.4$, $o = 8.3$, $g = 5.8$; find $\angle O$.
20. $d = 189$, $o = 213$, $g = 220$; find $\angle G$.

## *Explore*

21. Solve $\triangle CAT$ if $\angle A = 56.7°$, $\angle T = 44.7°$, and $c = 32.1$.

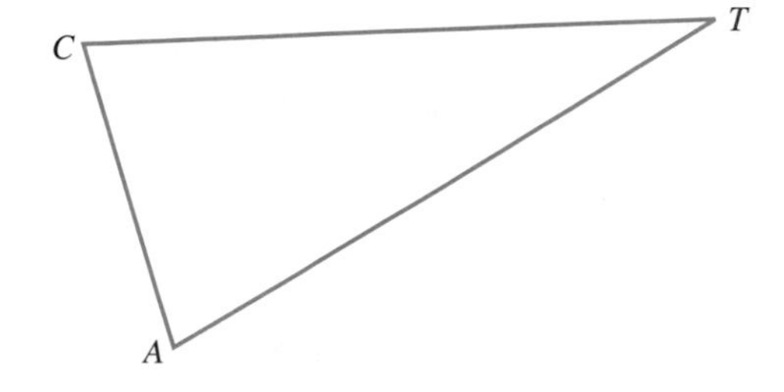

**22.** Solve $\triangle KEY$ if $\angle Y = 55°$, $\angle E = 55°$, and $k = 100$.

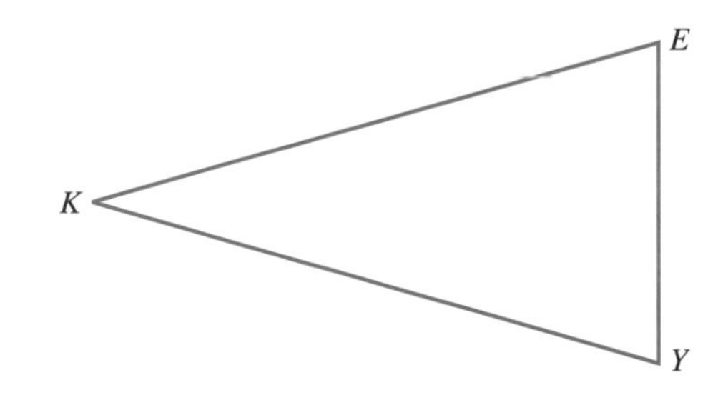


**23.** Solve $\triangle CUP$ if $c = 19.8$, $p = 20.6$, and $u = 14.4$.

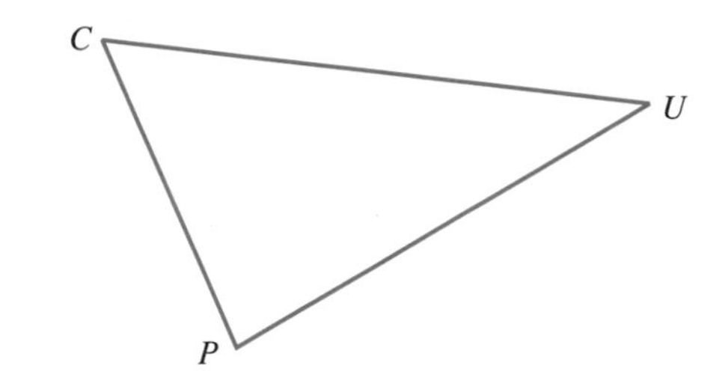


**24.** Solve $\triangle CAR$ if $c = 106$, $a = 155$, and $r = 127$.

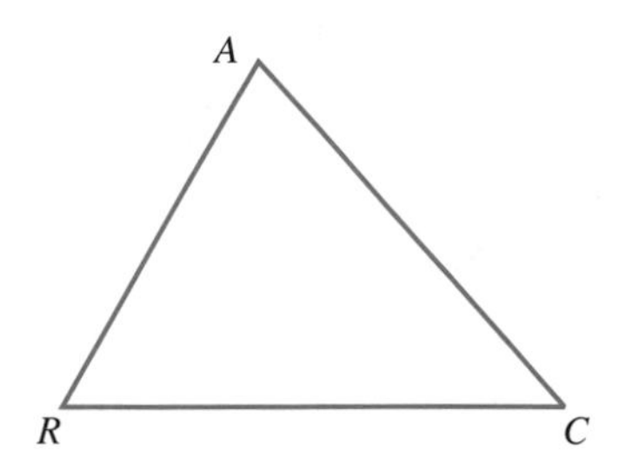


**25.** In $\triangle GET$, $\angle E = 67.5°$, $g = 23.5$, and $t = 45.3$. Find the measures of the parts of the triangle that are not given.

**26.** In $\triangle MET$, $t = 345.5$, $\angle E = 32.2°$, and $\angle M = 58.5°$. Find the measures of the parts of the triangle that are not given.

**27.** In $\triangle PET$, $p = 20$, $e = 23$, and $t = 28$. Find the measures of the parts of the triangle that are not given.

**28.** In $\triangle JET$, $j = 78.8$, $\angle E = 52.2°$, and $\angle J = 45°$. Find the measures of the parts of the triangle that are not given.

**29.** Verify that the Law of Sines applies to right triangles by determining the unknown sides and angles of $\triangle RAD$ where $\angle A = 90°$, $\angle R = 65°$, and $AR = 20$ using:
(a) the Law of Sines.
(b) right triangle trig.

**30.** Verify that the Law of Cosines applies to right triangles by determining the angles of a 3-4-5 triangle using:
(a) the Law of Cosines.
(b) right triangle trig.

***31.** Determine why it is impossible to have a triangle $\triangle ABC$ with sides $a = 10$, $b = 12$, and $c = 23$.

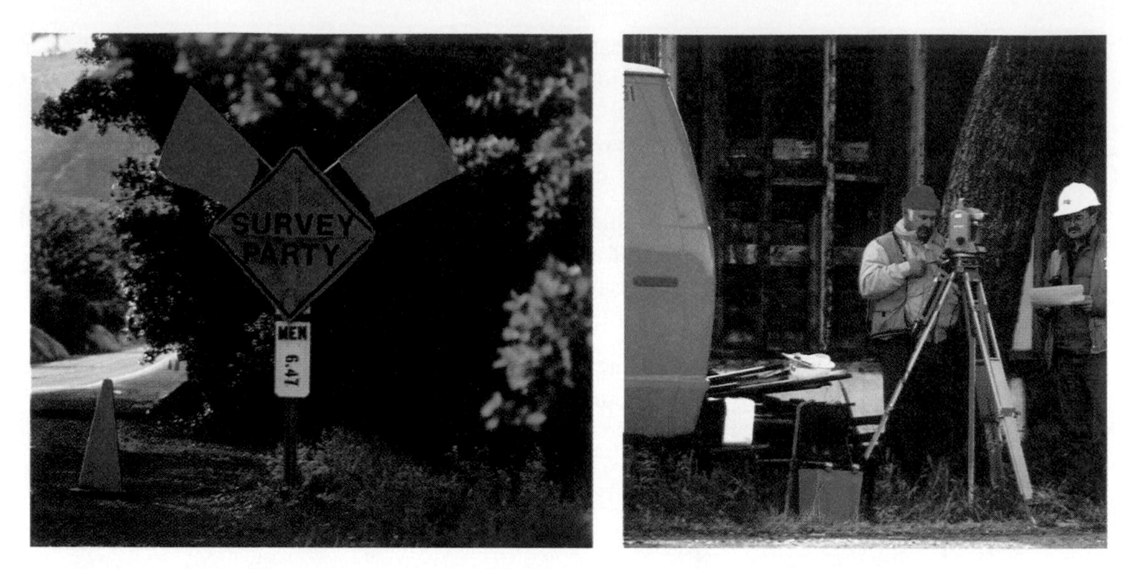


Surveyors make extensive use of trigonometry. (Courtesy of Kurt Viegelmann)

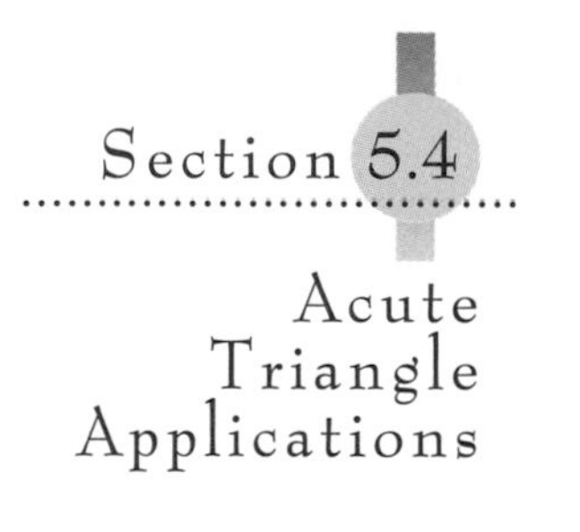

## Section 5.4 Acute Triangle Applications

In Section 5.2, we saw how the trigonometry of right triangles can be applied to many different kinds of problems. In Example 7 of that section, we showed how to solve a problem using two right triangles and a great deal of algebra. In this section, we will see how the Laws of Sines and Cosines can simplify this process.

### Example 1

Two fire-lookout stations are 15 mi apart, with station B directly east of station A. Both stations spot a fire on a mountain to the north. The line of sight to the fire from station A makes a 37.3° angle with a line running between the two stations (the east-west line), and the line of sight to the fire from station B makes a 54.2° angle with the east-west line. How far is the fire from station A?

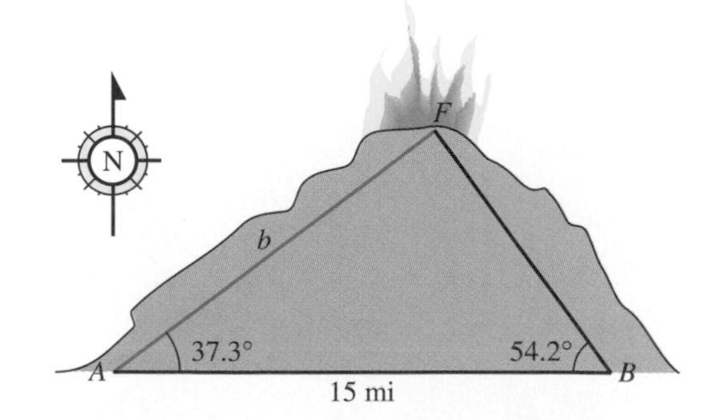

**Solution:** The figure shows the information given in the problem. Since $A = 37.3°$ and $B = 54.2°$, $F = 180° - 37.3° - 54.2° = 88.5°$. Since we now know $\angle F$ and the side opposite that angle, we can apply the Law of Sines to find side $b$.

Let $b$ = distance to the fire from $A$.

$$\frac{f}{\sin F} = \frac{b}{\sin B}$$

$$\frac{15}{\sin 88.5°} = \frac{b}{\sin 54.2°}$$

$$\frac{15}{0.9997} = \frac{b}{0.8111}$$

$$0.9997b = 15(0.8111)$$

$$b = 12.2$$

The distance to the fire from station A is approximately 12.2 mi.

## Example 2

The bottom of a hot air balloon is tethered at the top of a small hill with 200 ft of rope. Due to the wind blowing from the west, the angle of elevation to the balloon is 67.2°. At a point 56 ft down the hill, the balloon is directly overhead. If the hill makes an angle of 9° with the horizontal and the rope to the balloon is taut, how far above the ground is the bottom of the hot air balloon?

**Solution:**

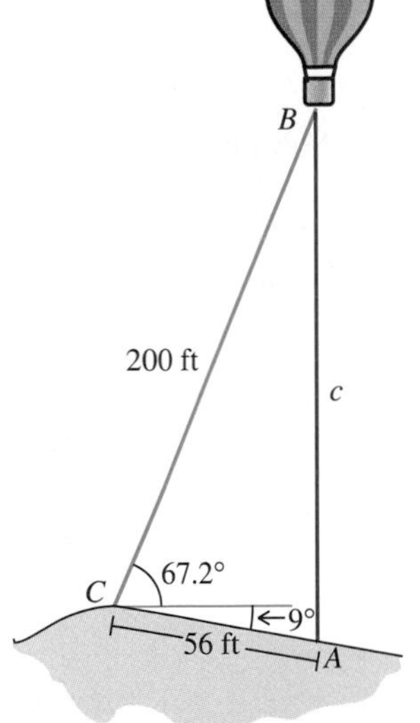

The hot air balloon along with the given information is shown in the figure. The angle at $C$ is 76.2° (67.2° + 9°). Now in $\triangle BCA$, a side, an angle, and a side are known (SAS). This allows us to use the Law of Cosines to find the height of the balloon above the ground. Let $c$ represent the height of the bottom of the balloon above the ground.

$$c^2 = a^2 + b^2 - 2ab \cos 76.2°$$

$$c^2 = 200^2 + 56^2 - 2(200)(56)(0.2385)$$

$$c^2 = 40{,}000 + 3136 - 5342.4$$

$$c^2 = 37{,}793.6$$

$$c = 194.4 \text{ ft}$$

## Example 3

A mining company digs a 750-yd horizontal mine shaft into a hill with an incline of 17°. How far up the hill should an air shaft, making an 87° angle with the hill, be drilled so that it will meet the end of the mine shaft? How long is the air shaft?

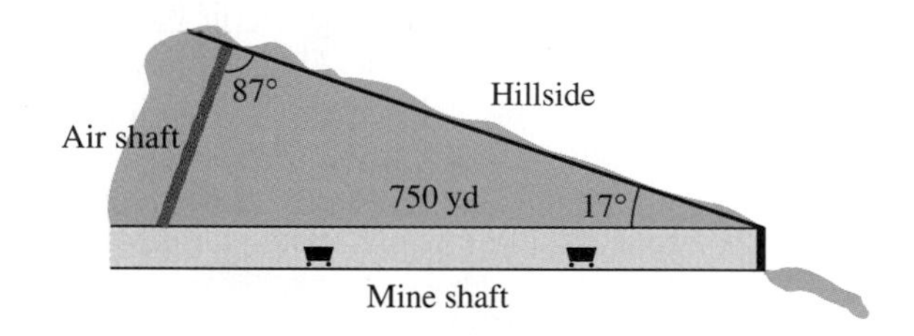


**Solution:** The angle where the air shaft meets the mine shaft is 76° (180° − 87° − 17°). Since two angles and a side are known (AAS), the Law of Sines can be used to find the distance up the hill where the air shaft is to be drilled and the length of the air shaft.

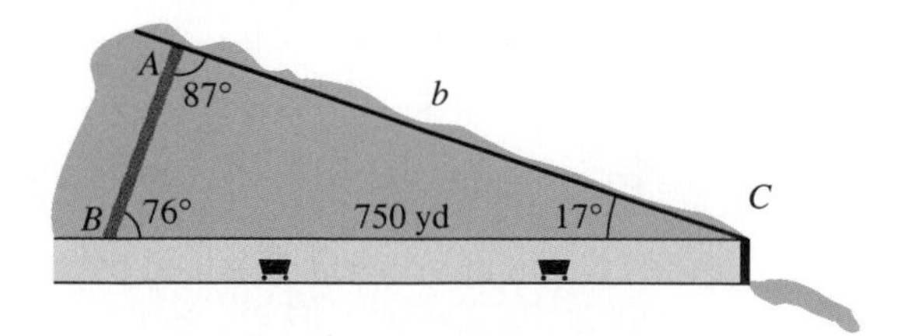

$$\frac{b}{\sin B} = \frac{a}{\sin A} \qquad \frac{c}{\sin C} = \frac{a}{\sin A}$$

$$\frac{b}{\sin 76°} = \frac{750}{\sin 87°} \qquad \frac{c}{\sin 17°} = \frac{750}{\sin 87°}$$

$$\frac{b}{0.9703} = \frac{750}{0.9986} \qquad \frac{c}{0.2924} = \frac{750}{0.9986}$$

$$0.9986b = 727.725 \qquad 0.9986c = 219.3$$

$$b = 728.7 \qquad c = 219.6$$

Therefore, the shaft should be drilled 728.7 yd up the hill and should be 219.6 yd long. ■

## *Navigation Problems*

In navigation problems, one of the common ways to give the course of a plane or ship is in terms of a **bearing**. A bearing is an acute angle measured from a north-south line toward either the east or the west. In this system either N (north) or S (south) is written, followed by an acute angle, then E (east) or W (west). The following examples show how this system is used to give a direction.

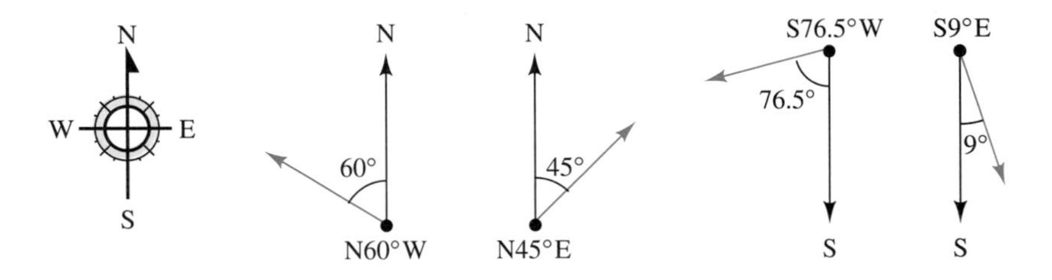


### Example 4

An airplane leaves an airport and flies directly north at 450 mph while a second airplane leaves the airport at the same time flying at 360 mph on a bearing of N70°E. How far apart are the airplanes after 2.5 hr?

**Solution:** Traveling at 450 mph after 2.5 hr, the plane flying directly north would have traveled 1125 mi ($2.5 \times 450$) and the plane flying N70°E at 360 mph would travel 900 mi ($2.5 \times 360$). Placing that information on the diagram shows that we have the SAS situation, so we will use the Law of Cosines.

Let $a$ = the distance between the planes after 2.5 hr.

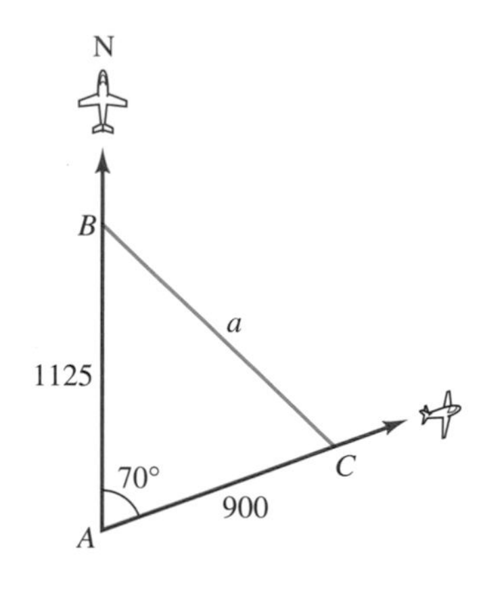


$$a^2 = b^2 + c^2 - 2bc \cos A$$
$$a^2 = 900^2 + 1125^2 - 2(900)(1125)\cos A$$
$$a^2 = 810{,}000 + 1{,}265{,}625 - 2{,}025{,}000 \cos 70°$$
$$a^2 = 2{,}075{,}625 - 692{,}590.8$$
$$a^2 = 1{,}383{,}034.2$$
$$a = 1176.0$$

Therefore, the planes are 1176 mi apart after 2.5 hr.

## Example 5

A sport-fishing boat leaves Bob's Pier heading directly east. After traveling for 50 mi, the captain hears a fishing report on the radio, which entices him to turn the boat and proceed on a bearing of S42°W for 27 mi. How far is the boat from Bob's Pier, and what bearing should the boat have originally taken to arrive at the fishing spot?

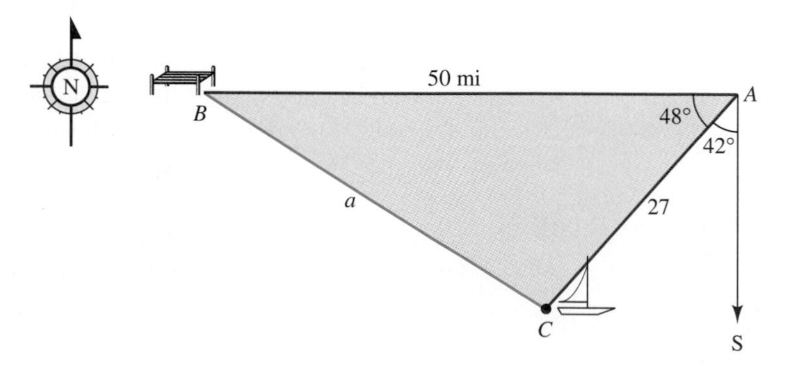

**Solution:** The figure shows the course taken by the fishing boat (starting at point $B$). Since a line going east forms a 90° angle with one going south, $\angle BAC = 90° - 42° = 48°$, and since we have an SAS situation, we can use the Law of Cosines to find the distance $a$ from Bob's Pier.

$$a^2 = b^2 + c^2 - 2bc \cos 48°$$

$$a^2 = 27^2 + 50^2 - 2(27)(50)(0.6691)$$

$$a^2 = 729 + 2500 - 1806.57$$

$$a^2 = 1422.43$$

$$a = 37.7 \text{ miles}$$

Since we know the three sides of the triangle (SSS), we will use the Law of Cosines to find the bearing from Bob's Pier to point $C$.

$$b^2 = a^2 + c^2 - 2ac \cos B$$

$$27^2 = 37.7^2 + 50^2 - 2(37.7)(50) \cos B$$

$$729 = 1421.29 + 2500 - 3770 \cos B$$

$$0.8468 = \cos B$$

$$32.1° = B$$

Since bearing is measured from the north-south line, we need to subtract 32.1° from 90° to find the angle made by the line from Bob's Pier to the fishing spot. Thus, the bearing angle is $90° - 32.1° = 57.9°$, and the actual bearing is S57.9°W. ■

In solving acute triangles, you will find it helpful to:

1. Sketch the situation and the triangle involved.
2. Write the known angles and sides on the triangle.
3. Use letters to represent the unknown angles and sides.
4. Find the unknown angle by subtracting the sum of the two angles from 180°, if two angles are known.
5. For triangles in which two angles and a side are known, use the Law of Sines.
6. For triangles in which a side, an angle, and a side (SAS) or three sides (SSS) are known, use the Law of Cosines.

The trigonometric methods shown in this chapter give you powerful tools for solving problems that contain triangles. The problems that follow will show you more applications that involve both right and acute triangles.

Section 5.4
PROBLEMS

## *Explain* ➡ *Apply* ➡ *Explore*

### *Explain*

1. What six steps are helpful in solving application problems involving acute triangles?

2. Explain how a bearing is measured in navigation problems.

3. Explain what a bearing of N22°E indicates. Make a sketch of that bearing.

4. Explain what a bearing of S75°W indicates. Make a sketch of that bearing.

5. Explain why the title of this chapter, "Trigonometry: A Door to the Unmeasurable," is appropriate in this section.

### *Apply*

6. A tall fir tree growing on the side of a hill makes a 70° angle with the hill. From a point 60 ft up the hill, the angle of elevation to the top of the tree is 65° and the angle of depression to the bottom of the tree is 20°. How tall is the fir tree?

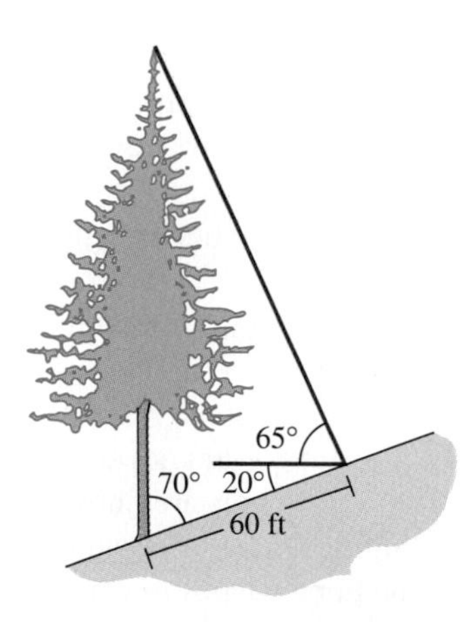


7. A large helium balloon advertising a sale at a local auto dealer is tethered by a rope at a west end of a level car lot. A wind blowing from the west causes the balloon to have an angle of elevation of 73°. If from the other end of the car

lot, 500 ft away, the angle of elevation to the balloon is 52°, how far above the car lot is the balloon?

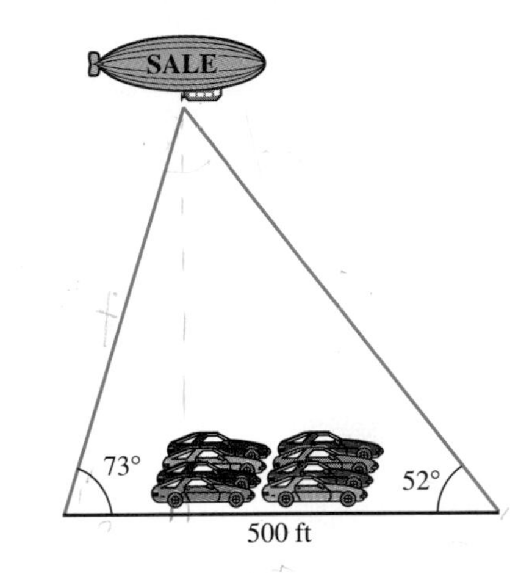


**8.** To find the distance between a fire-lookout station and a cabin on the other side of a canyon, a ranger uses a transit to measure angles at two different points as shown in the figure. Using the information shown, determine the distance from the fire-lookout station to the cabin.

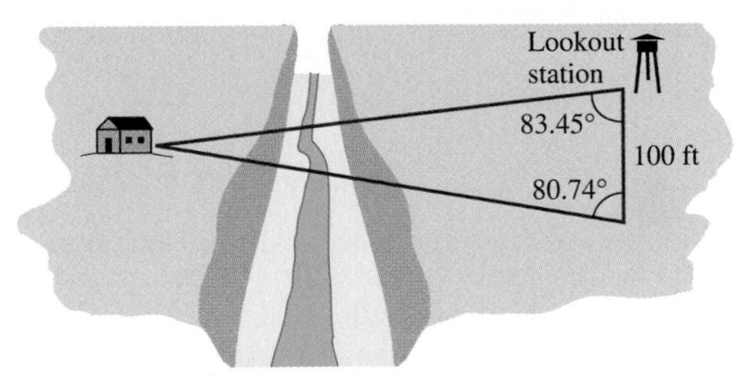


**9.** The walls of a trench form a 65° angle. To get out of the trench, highway workers place the bottom of a 25-ft plank on one side of the trench and lean the top of the plank on the other side. If the bottom of the plank makes an 80° angle with the wall of the trench, how far up the other side does the top of the plank reach?

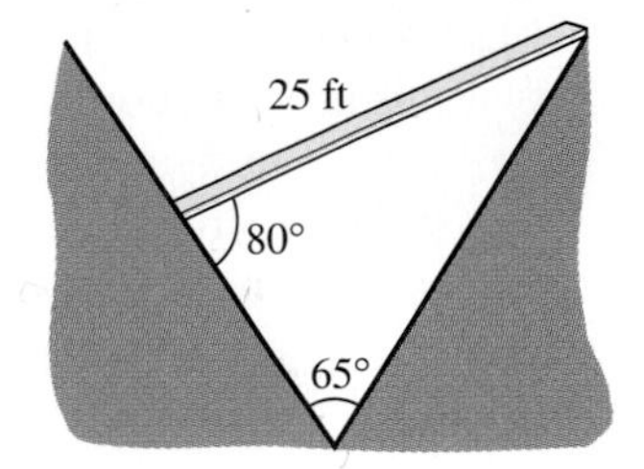

**10.** To find the distance across Mallard Cove at Bass Lake, a Girl Scout troop made the measurements in the diagram. What is the distance across Mallard Cove?

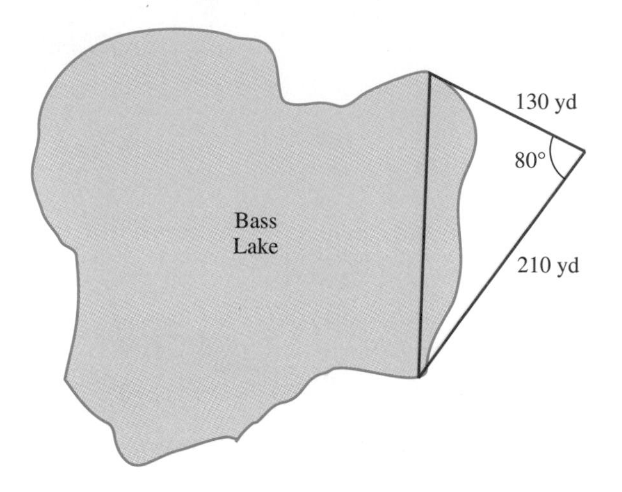


**11.** An A-frame mountain cabin is 30 ft wide. If the roof of the cabin makes a 63° angle with the base of the cabin, what is the length of the roof from ground level to the peak of the roof?

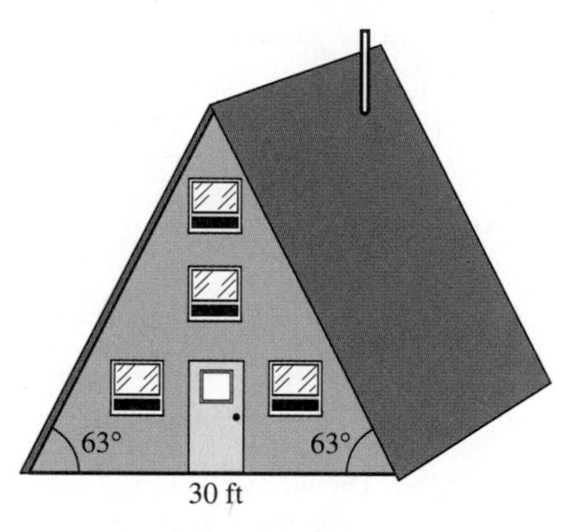


*Explore*

**12.** A pilot planned to fly a small plane from Kodiak, Alaska, to Tikchik Lakes, a distance of 300 mi. After flying for 2 hr at 150 mph, the pilot realized that she had been flying on a course that was off by 2°. How far from Tikchik Lakes was the airplane at this point? At what angle from the present course must the pilot turn to arrive at Tikchik Lakes?

**13.** Two cross-country skiers are at the bottom of a ravine. The sides of the ravine form a 72° angle. The first skier goes straight up one side of the ravine while the second one goes straight up the other side. After 20 min the first skier is 1600 m up one side of the ravine while the second skier is only 1000 m up the other side. How far apart are the two skiers?

**14.** A ship leaves Miami and cruises 74 mi on a bearing of S60°E. Another ship leaves at the same time and sails for 56 mi on a bearing of S20°E. How far apart are the two ships?

**15.** An airplane leaves Denver at 1:00 P.M. on a bearing of N20°E flying at 280 mph. At 1:30 P.M. a second airplane leaves on a bearing of N15°W flying at 375 mph. If the two airplanes continue flying on those courses, how far apart will they be at 3:30 P.M.?

**16.** How long is an escalator if it makes a 32° angle with the floor and carries people a vertical distance of 20 ft between floors?

**17.** From a point on the floor the angle of elevation to the top of a door is 48°, while the angle of elevation to the ceiling above the door is 59°. If the door is 8 ft tall, what is the vertical distance from the floor to the ceiling?

**18.** The pitch of a roof can be determined by taking the tangent function of the angle of elevation of the roof. Find the pitch of both sides of the roof in the figure, and determine the distance between the ground and the peak of the roof.

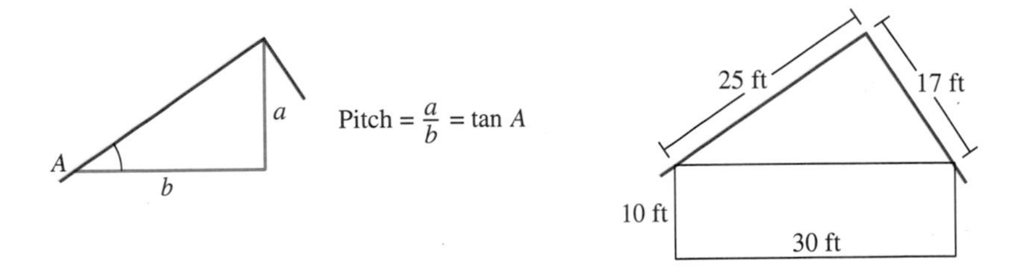


**19.** A triangular-shaped piece of property has sides that are 136 ft, 125 ft, and 178 ft. What are each of the angles formed by the property lines of this triangular lot?

**20.** A parallelogram-shaped lot has sides that are 120 m and 232 m. If the angle between the two sides is 58°, how long is the diagonal of the parallelogram that is opposite that angle?

**21.** When a kite is flying high in the sky, you will notice that the kite string does not make a straight line to the kite. Realizing this, find a method for determining the height above the ground for a kite that is flying in the air if (a) the ground is level and (b) the ground is not level.

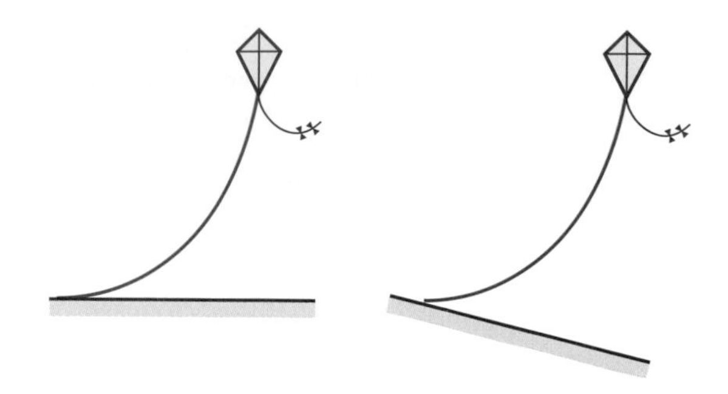

**22.** A rhombus is a parallelogram with equal sides, as shown. If $\angle A$ is an acute angle, find a formula for the length of the diagonal of the rhombus that is opposite $\angle A$. That is, if $d$ = the length of the diagonal, represent $d$ in terms of $x$ and $A$.

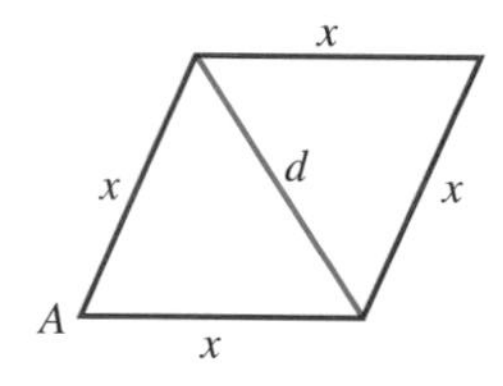


**23.** A hot air balloon is hovering over a lake. From one side of the lake the angle of elevation to the bottom of the balloon is 54°, and from the other side of the lake, 1660 ft away, the angle of elevation is 69°. How far is the bottom of the hot air balloon above the lake?

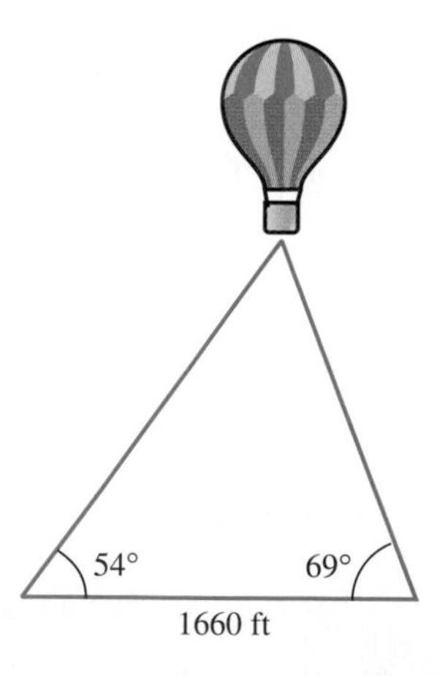


***24.** Explain how a person in a ship that is traveling parallel to the coast could use a lighthouse, the speed of the ship, a device for measuring angles, and a knowledge of trigonometry to determine the distance from the ship to the lighthouse.

# Chapter 5 • Summary

## Key Terms, Concepts, and Formulas

The important terms in this chapter are:

**Acute triangle:** A triangle with each angle measuring less than 90°. p. 365

**Angle of depression:** An angle made between a horizontal line and a line to an object that is below the horizontal line. p. 358

**Angle of elevation:** An angle made between a horizontal line and a line to an object that is above the horizontal line. p. 358

**Bearing:** A measure of direction relative to a north-south line. p. 379

**Cosine:** The ratio of the side adjacent to an angle to the hypotenuse of a right triangle. p. 344

**Hypotenuse:** The side opposite the 90° angle in a right triangle. p. 334

**Law of Cosines:** The relationship involving the sides and angles of a triangle, which states for any $\triangle ABC$,

$$a^2 = b^2 + c^2 - 2bc \cos A$$

$$b^2 = a^2 + c^2 - 2ac \cos B$$

$$c^2 = a^2 + b^2 - 2ab \cos C$$

p. 369

**Law of Sines:** The relationship involving the sides and angles of a triangle which states, for any $\triangle ABC$,

$$\frac{a}{\sin A} = \frac{b}{\sin B} = \frac{c}{\sin C}$$

p. 365

**Legs:** The sides of a right triangle that form the 90° angle. p. 334

**Obtuse triangle:** A triangle with one angle measuring greater than 90°. p. 365

**Pythagorean Theorem:** The relationship involving the sides of a right triangle which states that, if $a$ and $b$ are the legs and $c$ is the hypotenuse,

$$a^2 + b^2 = c^2$$

p. 335

**Pythagorean triple:** Three natural numbers that satisfy the Pythagorean Theorem. p. 339

**Right triangle:** A triangle with a 90° angle. p. 334

**Sine:** The ratio of the side opposite an angle to the hypotenuse of a right triangle. p. 344

**Solving triangles:** The process of finding unknown angles and sides of a triangle. p. 347

**Tangent:** The ratio of the side opposite an angle to the side adjacent to an angle of a right triangle. p. 344

After completing this chapter, you should be able to:

1. **1.** Define the sine, cosine, and tangent trigonometric functions with respect to right triangles. p. 344
2. **2.** Solve right triangles using the sine, cosine, and tangent functions. p. 347
3. **3.** Use trigonometric functions to solve various application problems involving right triangles. p. 355
4. **4.** Use the Law of Sines and the Law of Cosines to solve acute triangles. p. 365
5. **5.** Use the Law of Sines and Law of Cosines to solve various application problems involving acute triangles. p. 376

## • Summary Problems

Solve each of the triangles shown.

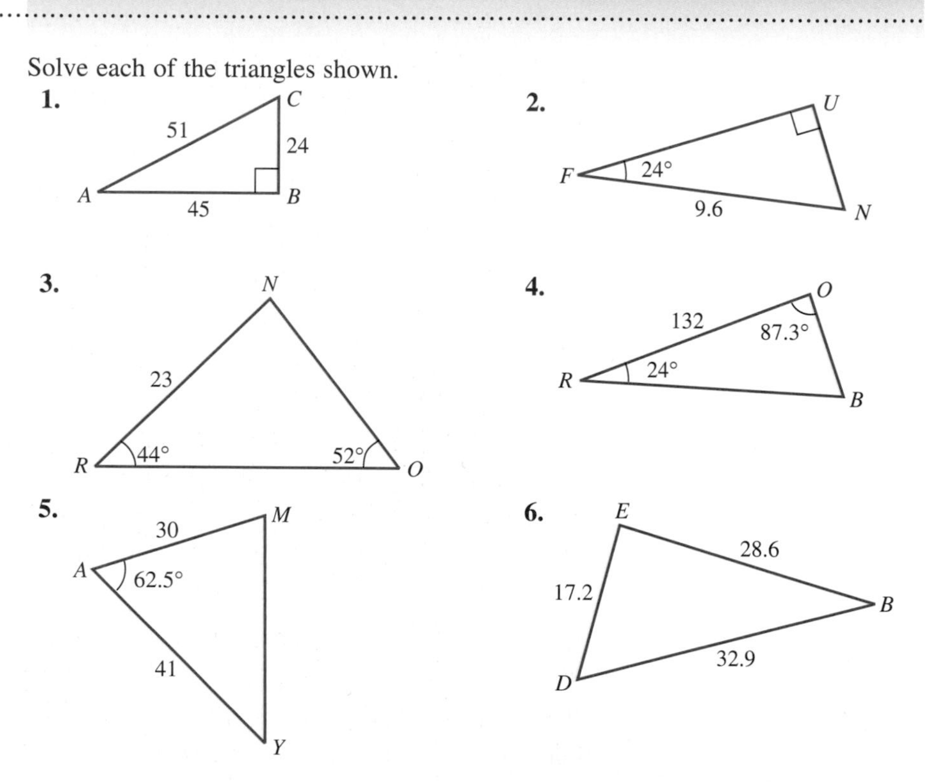


**7.** What angle does a 30-ft plank make with level ground if it is placed so that it reaches 20 ft up a vertical wall?

**8.** From a point 50 m from the bottom of a radio tower along level ground, the angle of elevation to the top of the tower is 67°. Find the height of the tower.

**9.** To determine the altitude of an approaching airplane, Debra and Charles measured the angle of elevation to the plane at the same time from locations that were 2000 ft apart, as shown in the figure. What is the altitude of the airplane?

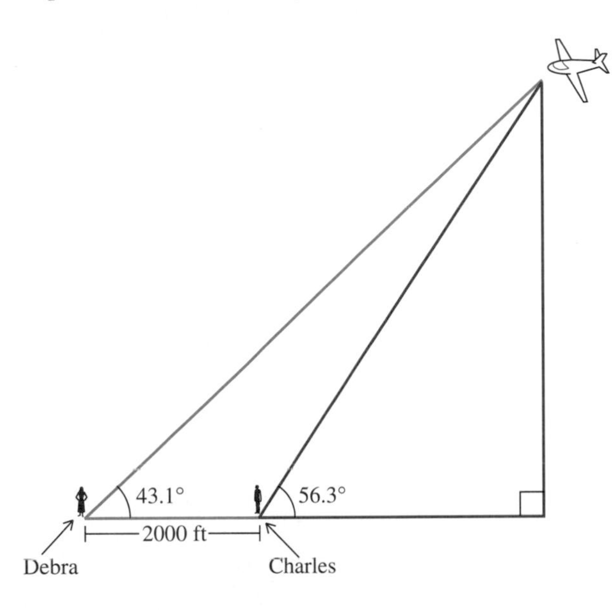


**10.** The Leaning Tower of Pisa in Italy is 177 ft tall and leans at an angle of about 84.5°.

(a) If a ball is dropped from the edge of the tower and strikes the ground at a 90° angle, how far from the base of the tower does the ball hit the ground and how far does the ball fall?

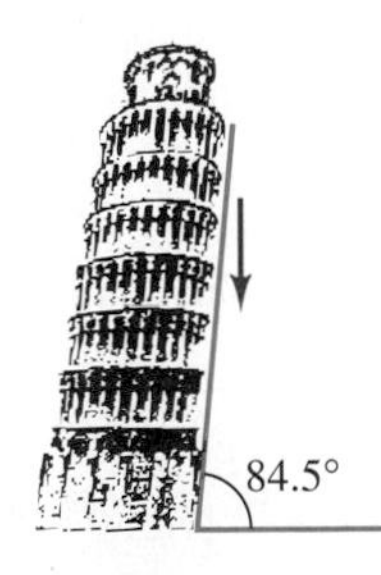


(b) If, because of the wind, a ball dropped from the tower takes a straight path to the ground and strikes the ground at an 80° angle of elevation to the top of the tower, how far from the base of the tower does it hit the ground and how far does the ball fall?

**11.** With the use of modern electronic equipment, the distance for various field events in a track meet can be measured without the use of measuring tapes. For example, in the discus throw, after a competitor has made a fair throw from the discus ring to point $D$, an electronic transmitter placed at $D$ sends a signal to a device in the official's booth above the track. The device then determines the angles at $D$ and at $B$. Since the distance from the booth to the center of the discus ring is a known distance, the length of the throw can be found using trigonometry.

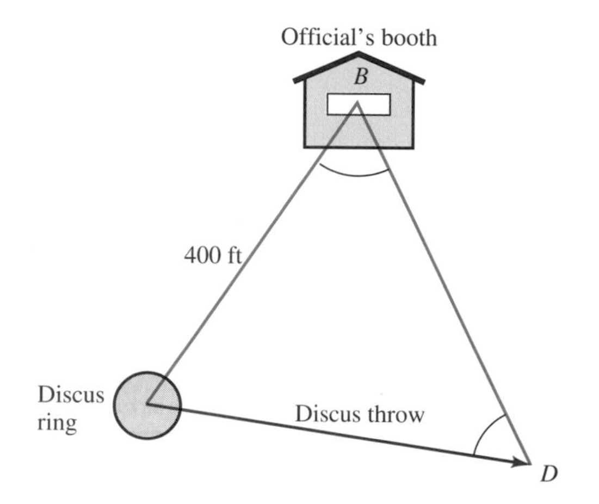


At an invitational track meet, the distance from the official's booth to the center of the discus ring is 400 ft, and the radius of the discus ring is 4 ft. If the angle at $D$ is 50.1° and the angle at $B$ is 24.6°, determine the length of a discus throw. Since a discus throw is measured from the outer edge of the discus ring, to get a final answer, be sure to subtract 4 ft from the distance you get from $D$ to the center of the ring.

**12.** Two lookout stations, which are 25 mi apart along the coast on a north-south line, spot an approaching yacht. One lookout station measures the direction to the yacht at N33°E, and the other station measures the direction to the yacht at S62°E, How far is the yacht from each lookout station? How far is the yacht from the coast?

**13.** Two airplanes leave San Francisco at 8:00 A.M. on different runways. One flies on a bearing of N64.5°W at 315 mph and the other plane on a bearing of S27°W at 295 mph. How far apart will the airplanes be at 10:00 A.M.?

**14.** In order to chart the movement of a polar bear, scientists attached a radio transmitter to its neck. Two tracking stations are monitoring the radio signals from the bear. Station B is 10 miles directly east of station A. On Monday, station A measured the direction to the bear at N43°E, and station B at N30°W. Three days later the directions to the bear from the two tracking stations were N24°E and N30°W, respectively. How much further from station B was the polar bear after those three days?

**15.** Three islands are located in the South Pacific. Island A is located 252 mi directly west of island B. Island C is located to the north of island A and island B. If the distance from island A to island C is 195 mi and the distance from island B to island C is 287 mi, on what bearing should one navigate to go from island A to island C and from island B to island C?

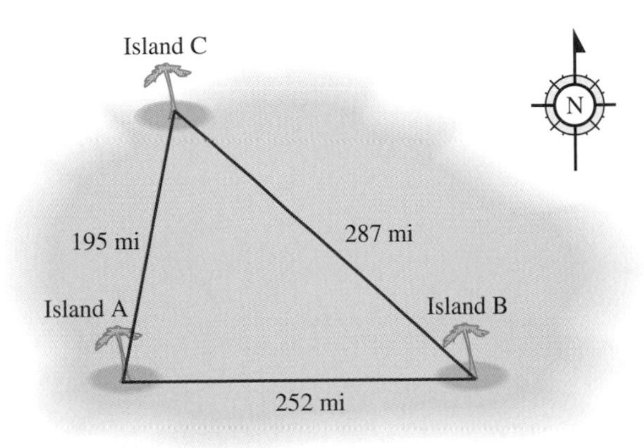


**16.** To determine the distance to an oil platform in the Pacific Ocean from both ends of a beach, a surveyor measures the angle to the platform from each end of the beach. The angle made with the shoreline from one end of the beach is 83°, from the other end 78.6°. If the beach is 950 yd long, what are the distances to the oil platform from both ends of the beach?

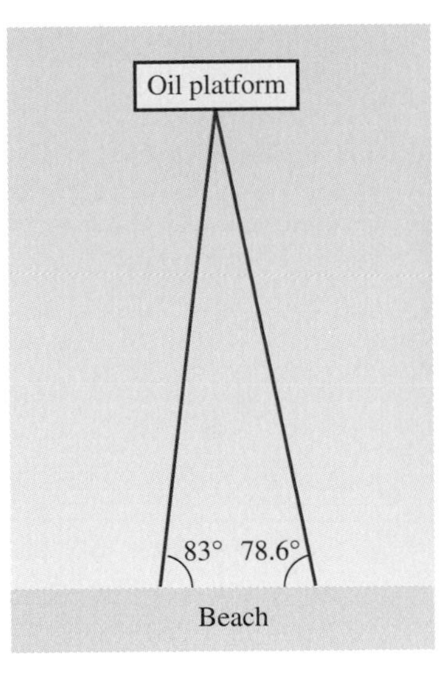

# 6 Finance Matters

Financial transactions, as seen in Masaccio's *The Rendering of the Tribute Money*, have always been a formal and highly regulated exchange. (Art Resources)

# Overview Interview

Handling financial dealings is one of the responsibilities of an adult in today's society. In this chapter, you will learn how to determine the amount of interest earned in various types of accounts, using both simple and compound interest. You will learn how to calculate the value of an annuity and the amount of a loan payment, and to see the benefits of paying off a loan early. If you ever plan to buy a car or save for retirement, this chapter will pique your interest.

## George Landavazo

GEORGE LANDAVAZO, FOUNDER OF LANDAVAZO BROTHERS, CONCRETE PUMPING, PLACING, AND FINISHING, HAYWARD, CALIFORNIA

TELL US ABOUT YOURSELF. I was born and raised in Gallup, New Mexico. After completing two years of military service, I relocated to Fremont, California, where I began working as a laborer. In 1988 I founded Landavazo Brothers Concrete Pumping, Placing, and Finishing. By 1992 the company had achieved $5.2 million in gross revenue, with a 420% increase in sales since 1988, making it the fifty-sixth fastest growing Hispanic company in the United States. My company has pumped and placed concrete in many large buildings in the San Francisco Bay area. One of the company's major projects was the pumping and placing of 110,000 yards of concrete in the 340,000-square-foot exhibit floor and convention facility in the Moscone Convention Center in San Francisco. Presently I am a consultant to Landavazo Brothers, Inc., overseeing the business transactions of the company and leaving its daily operations to my sons.

*George Landavazo in front of one of the cranes used to pump concrete in high rise buildings.*

HOW IS MATHEMATICS USED IN YOUR BUSINESS AND WHY IS IT IMPORTANT? The question is, how do I *not* use math in my business? If you consider the estimating, bidding, planning, scheduling, banking, bookkeeping, and accounting that it takes to run the business, you will see that a sound use of mathematics has been at the core of my success. If the mathematics involved in estimating and bidding, especially on large jobs, were not correct, I would not be in business today. If mistakes were made in finance matters, especially in loans and taxes, the company would not have succeeded. Knowing the expenses and profit-and-loss conditions of daily transactions is of the utmost importance. Believe me, mathematics is essential to my business. The correct use of mathematics has given me a sense of comfort and a feeling of success rather than the stress that frequently plagues the small businessman.

*The Interior of the Pantheon, Rome* by Giovanni Paolo Panini illustrates the classic architectural style that we associate with financial institutions. (National Gallery of Art)

# A Short History of Interest and Banking

## INTEREST

Interest has been charged for the use of money since Babylonian times (2000–500 B.C.). Ancient cuneiform tablets show that the Sumerians used both simple and compound interest. A tablet from 1700 B.C. contains a problem concerning how long it would take an amount of money to double if interest was compounded annually at a certain rate. Interest rates have varied greatly throughout history. Babylonian interest rates were 20%, and rates in Cicero's Rome (c. 60–43 B.C.) reached 48%. By the time of Justinian (483–565), rates were limited by law to 6%, but the Indian mathematician Bhāskara (c. 1150) mentions interest rates as high as 60%. In 1304, interest rates in Nuremberg were an astounding 220%!

Historically, lending money for a fee has been opposed. The Greek philosophers Plato and Aristotle wrote against the taking of interest. In fact, Aristotle believed that money is, by nature, "barren" and that the birth of money from money was "unnatural," and therefore he condemned the taking of interest.

Men such as Martin Luther and Thomas Aquinas argued that the Scriptures specifically forbade the taking of interest. The Roman Catholic Church was officially against **usury**, the charging of interest, until the 1830s, and penalties for disobeying Church law included being denied a Christian burial. Public feeling toward usury was so strong that in the Middle Ages it was believed that a prolonged rainstorm was caused by the burial of an Italian money lender in consecrated ground. In an attempt to stop the downpour, the body was disinterred from the grave and thrown into the Po River. In spite of the strong feelings against charging interest, lending still occurred. Powerful banks were founded in Venice in the late 1100s and in Genoa and Barcelona in the early 1400s. The success of these areas in commerce and the arts can be partially attributed to the availability of money from interest-charging banks.

Banks were also started by Jewish families. Since Jews were not associated with the Catholic Church, there was no conflict in charging interest. However, the stereotype that Jews charged unreasonably high interest rates during this time is not necessarily correct. Christian bankers operated under a much higher risk in the lending of money because of possible censure by the Church. Due to the risk involved, they often charged higher rates.

One method of avoiding the conflict with the Church over charging of interest was by calling it something else. A lender would agree to lend a sum of money at no charge if the money was returned within a specified time. If not, the borrower paid the lender an additional fee. The fee was computed by the lender (hence being as high as he chose) and was the difference between the lender's current financial standing and what his standing would have been had the money been repaid on time. The word ''interest'' comes from the Latin *Id quod inter est* or ''that which is between.''

Since the availability of money is necessary for the economy, negative attitudes toward interest have diminished. However, if the interest rate is extremely high, it is still called usury and is still viewed with disfavor.

## BANKING

The original bankers were the money changers. As long ago as the Roman Empire, there was a need to exchange the coins of one country for another. The money changers would sit in the plaza or in front of their shops, counting coins on a bench. It is from the Italian word for bench, *banco*, that we get the current word ''bank.''

As the Dark Ages ended and the amount of commercial activity in Europe increased, it became increasingly difficult to carry on all transactions in cash. The currency of the time was almost always coin. Bulky when carried in large amounts and very heavy, gold and silver coins were accepted by all merchants but were very impractical for the demands of the growing economy. Coupled with the necessity for increased amounts of cash, this situation created the need for an institution that would handle long-distance transactions and supply ready cash.

The powerful banks in Venice, Genoa, and Barcelona had branches in the major trading spots throughout Europe. These banks were able to complete transactions without the transfer of coin. All that was needed was a letter of credit that could be honored by any of the bank's branches. These powerful banks also were willing to defy the Church and lend money in return for the payment of interest.

Another activity of these banks was to act as a clearing house where money from different countries could be exchanged. As is true today, the rates of exchange fluctuated. The banks would exchange coins for their customers immediately, but would hold the foreign currency until a more favorable rate of exchange materialized. By this process of speculation in the currency markets, the banks were able to accumulate profits. The Church considered this activity a legitimate way of accumulating profits because a risk was involved.

Much of the modern system of banking is based on the arrangement started in England during the late 1600s. King Charles II of England was a profligate spender. In 1694, when Parliament refused to give him the money he needed to support his army, the king raided the Royal Mint. The merchants, whose money he stole, were outraged. To prevent a recurrence, the merchants decided to deposit their money in the vaults of the local goldsmiths. In return, the goldsmiths would issue a note saying that the merchant had a certain amount of gold or silver. Instead of handling all transactions in coin, merchants could now purchase goods by giving the seller a note. In time, a note attesting that John Doe had 20 ounces of gold coin kept with a certain goldsmith might pass through the hands of several people. None of these people ever saw the gold because the note was accepted by most everyone as money, and the possession of the note was far more convenient than keeping the actual coin.

When the goldsmiths realized that these notes of deposits were being used like real money, major changes ensued. First, the notes they issued to their depositors were no longer issued to the individual. Instead, the notes indicated that the person who possessed the note was the owner of the gold. This eliminated the difficulty of redeeming a note in someone else's name. The second major change was that the goldsmiths realized that only rarely did they need to redeem large numbers of these notes for the precious metals they represented. Usually, the daily deposits of coin were sufficient to supply the gold requested by customers that same day. Betting that they would never need to redeem all the outstanding notes, the goldsmiths started to print more notes. These additional notes were not backed by actual assets, but, if they were never redeemed, it didn't matter. The goldsmiths then used these notes to provide the public with loans. The person taking out the loan received the notes and promised to pay the goldsmith interest on the loan. The interest could be paid either with notes or with real money—gold or silver. If the loan was repaid in real money, the goldsmith could print more notes and issue more loans.

This was a sure method for the goldsmith to become wealthy as long as the public maintained confidence in him. As long as the noteholders believed the paper was worth real gold, they appreciated the convenience of not lugging around heavy sacks of coins. If public confidence in the goldsmith did falter, trouble ensued. Lack of confidence could cause all the noteholders to demand the gold that their certificates represented. Because there were more notes than actual coin, not all the customers would receive their money. A similar situation can be seen in modern times when a majority of the customers of a bank decide to withdraw their deposits. Called a ''run on the bank,'' this was a problem in the United States during the 1930s and the 1980s.

The goldsmiths' system developed into the Bank of England and the modern system of banking used today. As safeguards for both the depositors and the stockholders in the bank, regulations regarding the required amounts of reserves and the number of loans are strictly enforced.

## BANKING IN THE UNITED STATES

The First Bank of the United States was established in 1791 in Philadelphia, Pennsylvania, and had eight branches throughout the new country. This bank issued currency and provided loans to and accepted deposits from the federal government, business, and the public. Although the bank was considered a success, in 1811 Congress did not renew its charter due to pressure from the smaller, state-chartered banks.

To fill the void left by the demise of the Bank of the United States, 120 state-chartered banks were created within a year. Like the goldsmiths, many of the bankers saw a path to great wealth. Unfortunately, in 1812, the United States was at war with Great Britain and borrowed extensively from the state banks. By 1814, many of these banks stopped silver and gold payments, and public confidence dropped. Many believed this would not have occurred if a national bank had been in existence. Thus, in 1816, the Second Bank of the United States was created.

During its short life, the bank had a turbulent history. Run by Nicholas Biddle, a Philadelphia banker, the bank was opposed by the seventh president of the United States, Andrew Jackson (1829–1837). Biddle claimed that Jackson wanted to use a spoils system to appoint the directors of the bank. In return, Jackson claimed that Biddle was using the bank for private purposes and was opposed to government interests. Jackson won the battle. Instead of waiting for the charter to expire in 1836, Jackson ordered the Secretary of the Treasury to withdraw all federal funds from the bank in 1833. The funds were deposited in the state banks that Jackson favored. After 1836, the Second Bank of the United States tried to remain in business as a state-chartered bank but failed.

The next 30-year period is considered a disaster in U.S. banking history. Large numbers of state-chartered banks were quickly established only to fail due to mismanagement, greed, or insufficient reserves. These smaller banks issued their own currency and did not coordinate policy with either the government or other banks. Public distrust in this system led to the establishment of national banks in 1863 and the National Bank Act of 1864.

Having twice failed in attempts to create a national branch banking system, the government did not intend the new national banks to operate as a branch system. Instead, they operated in a manner similar to the state-chartered banks but had a uniform currency and were chartered by the federal government. The advantages of this system were that

1. The currency was widely recognized and, in the event of bank failure, backed by the federal government. Although this did not protect depositors, people holding cash were protected.
2. The number of bank failures was reduced due to the more conservative practices of the national banks.

The disadvantages of this system were that

1. Due to lack of centralization, the clearing of checks and other interbank transfers was costly and time-consuming.
2. Cash reserves were inadequate in times of a crisis, such as a run on the bank.
3. The total cash available did not fluctuate to meet the changes in demand that occurred at times such as the Christmas shopping season.
4. Because federal funds were deposited in these banks, bank operations were disrupted when the federal government made large deposits or withdrawals.

To help correct the drawbacks of the national banks, the Federal Reserve Act of 1913 set up the current system of banking in the United States. In addition to alleviating the problems of the national banks, the act created the Federal Reserve Bank. Further regulations regarding the Federal Reserve were passed in 1962. The actions of the Federal Reserve Bank appear frequently in the daily news and affect both business and the individual citizen.

## Check Your Reading

1. What was the general feeling about interest throughout history?
2. What is a "run on the bank"?
3. The late 1100s saw the arrival of tea in Japan from China while the Mayans of Central America were in the second stage of the development of their civilization. What event was occurring in banking history at this time?
4. While banks were being formed in Barcelona and Genoa, the Great Temple of the Dragon was being erected in Peking, China. During what time period did this happen?
5. Give a brief history of a national bank in the United States.
6. What role did goldsmiths play in the history of banking?
7. The period 46–44 B.C. saw Brutus and Cassius assassinate Julius Caesar and the Roman legions make Africa a Roman province. What were interest rates in Rome during this time?
8. In the early 1300s, Edward I of England was standardizing the units of measurement *yards* and *acres*. However, interest rates were far from being standardized. In what city did interest rates reach the amazing level of 220%?

9. Mozart's famous composition, *The Magic Flute*, had its first performance in 1791. What U.S. financial institution was being established during this same year?
10. When R. T. Laënnec invented the stethoscope and Nikolai Karamzin was writing *History of the Russian Empire*, the Second Bank of the United States was established. In what year did all three of these events occur?
11. In 1913, the U.S. Congress instituted the first federal income tax. What federal legislation was passed concerning banking during the same year, and what institution did it create?
12. When the first U.S. salmon cannery was being established in 1864, what piece of federal legislation was being passed?
13. Match each of the following names with the associated event or action.

| | |
|---|---|
| (a) Aquinas | Argued that the Scriptures forbade interest |
| (b) Aristotle | Condemned taking interest |
| (c) Bhāskara | Favored the Roman Catholic Church's policy against the taking of interest |
| (d) Biddle | Mentioned interest rates of 60% in India |
| (e) Charles II | Interest rates in his Rome reached 48% |
| (f) Cicero | Opposed deposits in the Bank of the United States |
| (g) Jackson | Raided the Royal Mint |
| (h) Luther | Ran the Second Bank of the United States |

## • Research Questions

To answer the following questions, you will need to refer to material not contained in the text. Possible sources of information are listed in the bibliography at the end of this book.

1. In modern television shows, the word "shylock" is used in reference to gangsters who charge exorbitant amounts of interest and enforce the penalties with violence. Determine the origin of the word "shylock." What was the interest charged on a loan by the original Shylock?
2. Find some common interest rates for Europe in A.D. 1400. Be sure to cite your references.
3. What connection did banking and interest rates play in the historic European rivalries between Christians and Jews?
4. What is meant by "unit banking"? What countries use this method? What is meant by "branch banking"? What countries use this method?
5. What is a federal reserve note? Have these notes always had the same physical size? How have these notes changed?
6. What is a silver certificate? During what years was it used? Why was it discontinued?

7. What are the main points of the Federal Reserve Act of 1913?
8. What is the Federal Reserve Bank? Who are its customers? What does the bank do? Where are the Federal Reserve Banks located? Who is the current chairman of the Federal Reserve?
9. What do the initials F.D.I.C. stand for? When was the organization started? What is the purpose of the organization?
10. Interview a stock broker to determine how commissions on different financial transactions are computed.
11. Discuss with a banker the conditions on a Visa™ or Mastercard™ credit card. What are the terms of repayment, the annual interest rate, and the minimum monthly payment? How are these values computed? What is a ''grace period''?

## • Projects

1. What interest rates were common in the 1960s? in the early 1970s? in the late 1970s? Compare these rates with those charged in Rome in 40 B.C. or in Germany in the 1300s. Discuss whether you consider the interest rates in the late 1970s to be high or low. What is the difference between usury and interest in modern times?
2. Find common interest rates for the United States and five other countries in each decade of the 20th century. Plot the information on a graph. Does there appear to be a trend?
3. There are many different terms used in the world of finance when you are borrowing money. Examples of such terms are points, APR, fixed rates, variable rates, balloon payments, closing costs, escrow account, and second mortgage. Give an organized report on the important terms associated with borrowing money. Give examples of each term.
4. There are many different terms used in the world of finance when you are saving or investing money. Examples of such terms are tax shelter, CD, stocks, bonds, APR, penalty for early withdrawal, commissions, interest, and minimum balance. Give an organized report on the important terms associated with the saving or investing of money. Give examples of each term.

In this section, we look at some of the algebra that will be used in solving problems in the mathematics of finance. We will discuss the use of a calculator, logarithms, and equations containing exponents.

## Using Your Calculator

This section will discuss keys needed in financial math.

---

**Note:** To understand this material, it is very important for you to have a calculator available when reading this section. As we do each example, you should try the example on your own calculator. We will demonstrate one method in the text. However, because of the differences among calculators, this method may not work with your calculator.

For further help, read your calculator instruction manual or ask your instructor.

---

When we want to indicate that a particular key is being used, the symbol is enclosed in a box. For example, for $3 \times 4 = 12$, we use the following format:

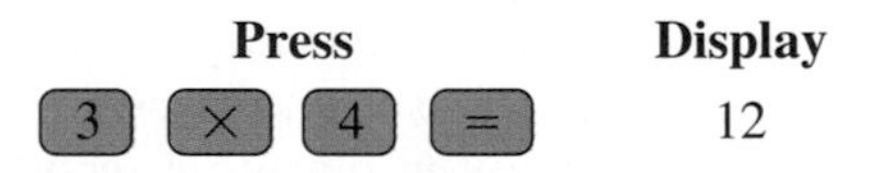


Since we pushed the keys for 3, 4, ×, and =, these characters are in boxes. The answer, 12, is not in a box since we did not push those keys.

The first key we discuss is the exponent key. Depending on your calculator, it may look like

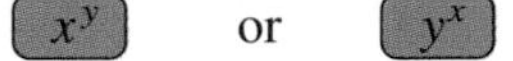


### Example 1

Use a calculator to find the value of $1.065^{-58}$.

**Solution:** The sequence of buttons pressed is

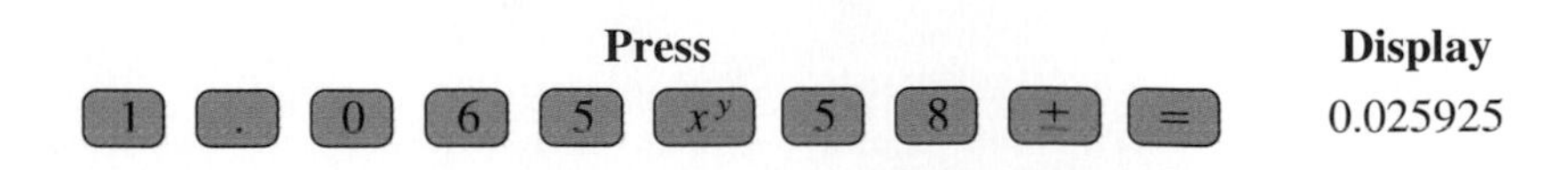

Because the exponent is negative, we used the ± symbol *after* entering 58. Entering this symbol before 58 may result in an incorrect answer. If you do not have a ± key, look for one labeled CHS (change sign). ■

The next key we will be using is the reciprocal key.

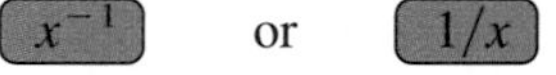


This key provides a shortcut for the division problem $1 \div x$.

From algebra, you may remember that fractional exponents represent roots. For example,

$$\sqrt{x} = \sqrt[2]{x} = x^{1/2} \qquad \text{and} \qquad \sqrt[5]{x} = x^{1/5}$$

Although most calculators have a key to calculate square roots, calculators do not have a fifth-root key. As a result, to compute most roots, it is necessary to use fractional exponents. To do this on a calculator, we use the reciprocal key.

In the next example, we combine the use of the reciprocal and exponent keys to perform calculations involving fractional exponents.

### Example 2

Calculate $85^{1/25}$.

**Solution:** Using the reciprocal key and exponent keys on the calculator gives

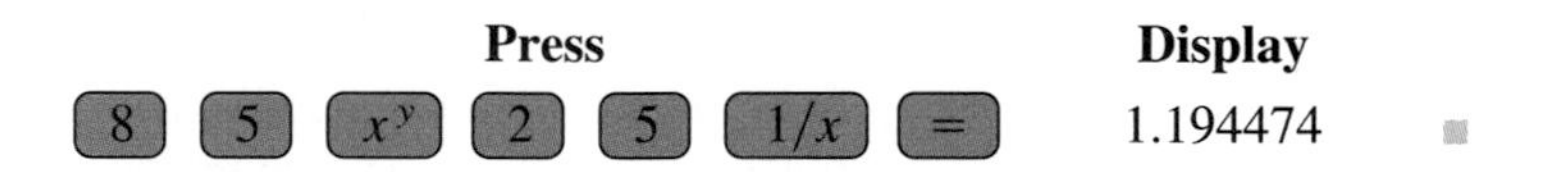

■

Throughout the chapter, we use six places after the decimal point unless we are talking about money. For greater accuracy, we could use more places, but this will usually not be necessary.

## Equations with Exponents

There are two types of equations involving exponents. The first type has a constant exponent. The second type has a variable exponent. The remainder of this section is a review on how to solve both types of equations.

To solve equations that have constant exponents, we recall one of the rules of exponents.

$$(x^m)^n = x^{m \times n}$$

For example, $(x^4)^3 = x^{4\times 3} = x^{12}$. We will use this rule to solve equations that have a constant exponent. Beginning with a problem with a known answer will allow us to verify that our method is correct.

### Example 3

Solve for $x$: $x^3 = 8$.

**Solution:** To find $x$, we eliminate the exponent from the variable. We do this by raising both sides to the one-third power.

$$x^3 = 8$$

$$(x^3)^{1/3} = 8^{1/3} \qquad \text{Now apply the rule of exponents.}$$

$$x^{3\times 1/3} = 8^{1/3}$$

$$x^1 = 8^{1/3}$$

$$x = 8^{1/3}$$

Using the following keys on a calculator gives the answer

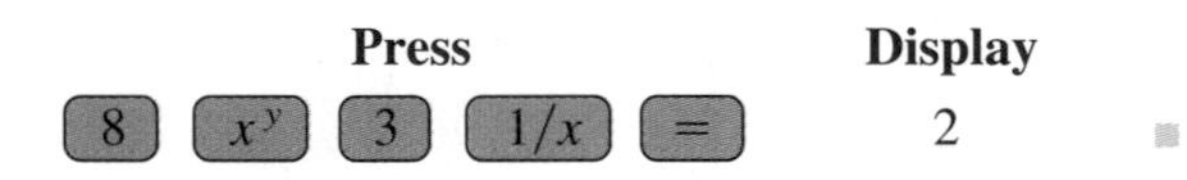


### Example 4

Solve find $x$: $x^{14} = 75$.

**Solution:** To find $x$, we eliminate the exponent from the variable. We do this by raising both sides to the $\frac{1}{14}$ power.

$$x^{14} = 75$$

$$(x^{14})^{1/14} = 75^{1/14}$$

$$x^{14\times 1/14} = 75^{1/14}$$

$$x = 75^{1/14}$$

Using a calculator, we get

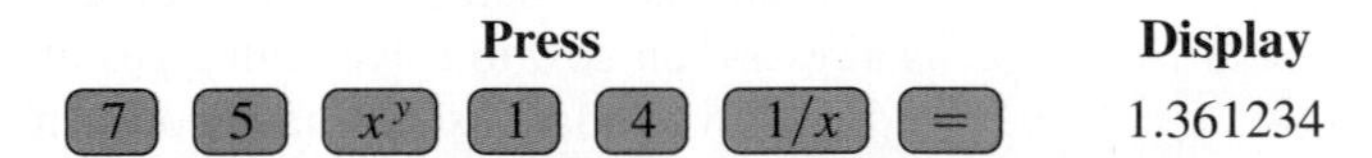


We can check this result by raising 1.361234 to the 14th power.

$$1.361234^{14} = 74.999615 \approx 75$$

The value was not exactly 75 because 1.361234 is an approximate value. For increased accuracy, use more decimal places.

The other type of equation is one in which the exponent is a variable. For example,

$$5^x = 25 \qquad \text{or} \qquad 2^x = 20$$

We know that the first equation is solved when $x = 2$ because $5^2 = 25$. The second equation is more difficult. Since $2^4 = 16$ and $2^5 = 32$, the answer must be between 4 and 5, but there is no simple method of finding it. To obtain the solution, we use logarithms.

**Note:** When solving equations in which the variable is in the exponent, use logarithms.

As you may remember, logarithms can be written with different bases. Throughout the chapter, we use natural logarithms. On a calculator, the key is

ln *x*

To calculate ln 5, press the following keys. Notice that we do *not* need to press the equals button.

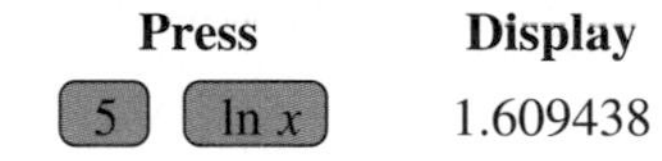

| Press | Display |
| --- | --- |
| 5 ln *x* | 1.609438 |

For the problems we will be solving, we will use the following rule of logarithms:

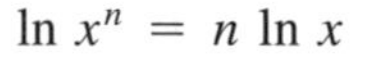

$$\ln x^n = n \ln x$$

For example, $\ln 5^3 = 3 \ln 5$. To find this value on a calculator, we press the following keys:

| Press | Display |
| --- | --- |
| 5 ln *x* × 3 = | 4.828314 |

The first example is a problem with an answer that is known.

### Example 5

Solve for $x$: $5^x = 25$.

**Solution:** Since the variable is in the exponent, we use logarithms to solve the equation.

$$5^x = 25 \quad \text{First, take the natural log of both sides.}$$

$$\ln (5^x) = \ln 25 \quad \text{Now, use the rule of logarithms.}$$

$$x \ln 5 = \ln 25 \quad \text{Finally, divide both sides by ln 5.}$$

$$x = \frac{\ln 25}{\ln 5}$$

Using the 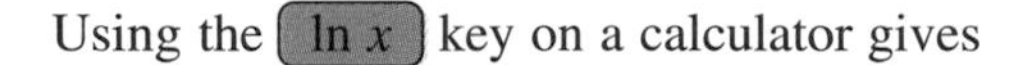 key on a calculator gives

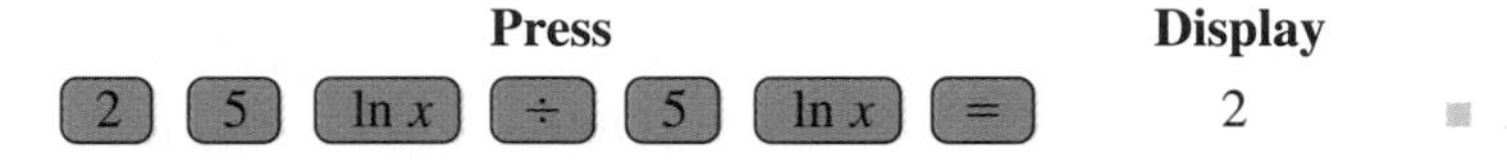


Note that this gives the expected answer. ■

Confident that the method of Example 5 will work, we can now attempt a problem whose answer is not as easily guessed.

### Example 6

Solve for $x$: $2^x = 20$.

**Solution:** Since the variable is in the exponent, we will again use logarithms.

$$2^x = 20$$

$$\ln (2^x) = \ln 20$$

$$x \ln 2 = \ln 20$$

$$x = \frac{\ln 20}{\ln 2} \approx 4.321928$$

■

### Example 7

Solve for $x$: $\dfrac{(1.07)^x - 1}{0.07} = 3.29$.

**Solution:** The equation contains a variable in the exponent, so we will use logarithms to solve the problem. First, however, we must isolate the term containing the exponent. To do this, multiply both sides of the equation by 0.07. This gives

$$(1.07)^x - 1 = 3.29(0.07) = 0.2303$$

Next, add 1 to both sides of the equation.

$$(1.07)^x = 1.2303$$

At this point, we can solve for $x$ as we did in Example 6.

$$x = \frac{\ln 1.2303}{\ln 1.07} \approx 3.063290$$

In summary, this section demonstrated the use of three keys on the calculator and recalled two formulas from algebra. It is important to master this material before attempting the rest of the chapter.

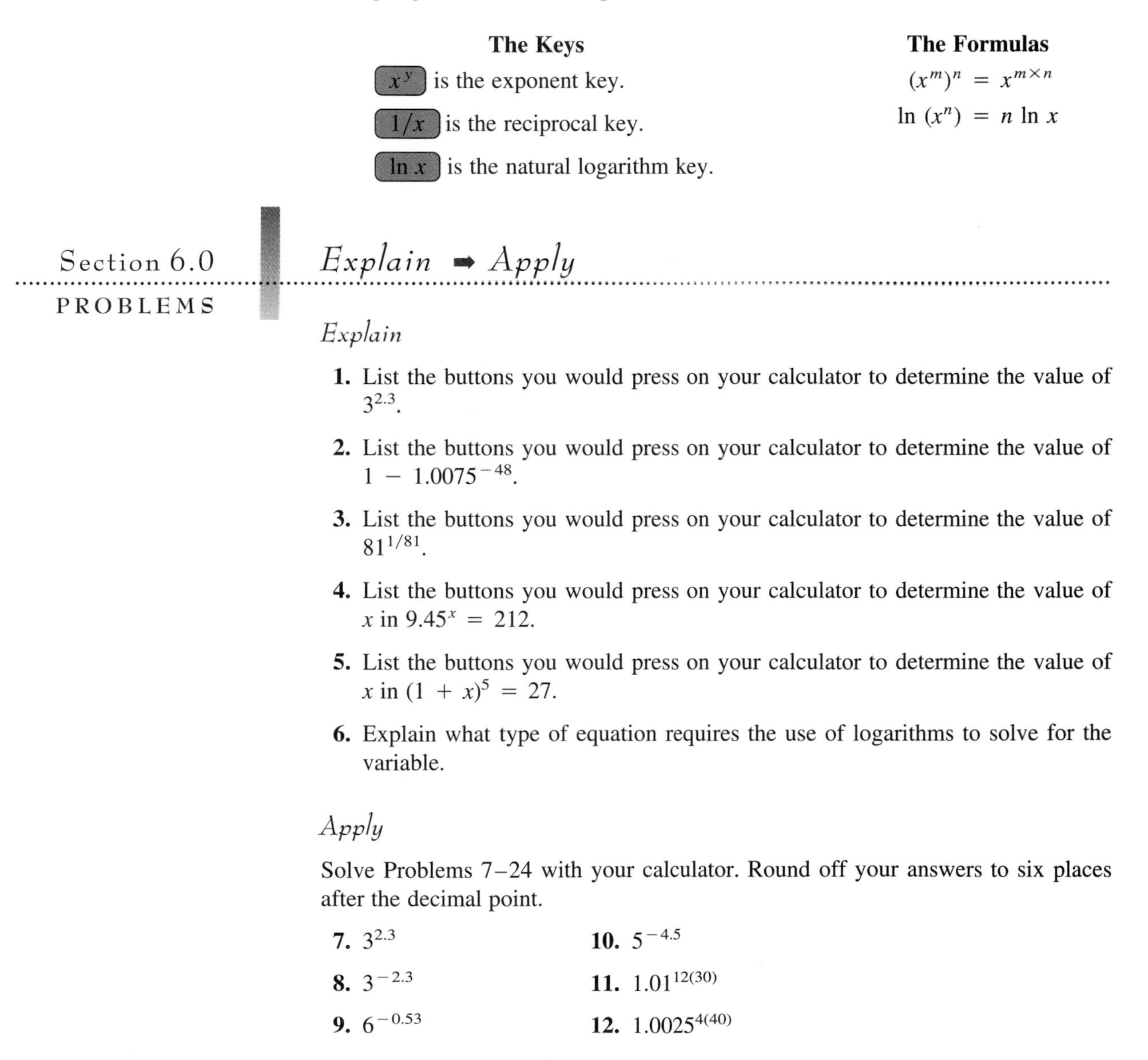

**The Keys**

[$x^y$] is the exponent key.

[$1/x$] is the reciprocal key.

[$\ln x$] is the natural logarithm key.

**The Formulas**

$(x^m)^n = x^{m \times n}$

$\ln(x^n) = n \ln x$

## Section 6.0 PROBLEMS

## *Explain ➡ Apply*

### *Explain*

1. List the buttons you would press on your calculator to determine the value of $3^{2.3}$.
2. List the buttons you would press on your calculator to determine the value of $1 - 1.0075^{-48}$.
3. List the buttons you would press on your calculator to determine the value of $81^{1/81}$.
4. List the buttons you would press on your calculator to determine the value of $x$ in $9.45^x = 212$.
5. List the buttons you would press on your calculator to determine the value of $x$ in $(1 + x)^5 = 27$.
6. Explain what type of equation requires the use of logarithms to solve for the variable.

### *Apply*

Solve Problems 7–24 with your calculator. Round off your answers to six places after the decimal point.

7. $3^{2.3}$
8. $3^{-2.3}$
9. $6^{-0.53}$
10. $5^{-4.5}$
11. $1.01^{12(30)}$
12. $1.0025^{4(40)}$

**13.** $1.005^{48} - 1$

**14.** $1.0075^{-360} - 1$

**15.** $1 - 1.0075^{-48}$

**16.** $1 - 1.009^{-120}$

**17.** $\dfrac{1.005^{48} - 1}{0.005}$

**18.** $\dfrac{1.007^{120} - 1}{0.007}$

**19.** $\dfrac{1 - 1.006^{-36}}{0.006}$

**20.** $\dfrac{1 - 1.01^{-120}}{0.01}$

**21.** $81^{1/81}$

**22.** $17^{-1/17}$

**23.** $17^{1/17}$

**24.** $81^{-1/81}$

Solve for $x$ in Problems 25–49. Round off your answers to six places after the decimal point.

**25.** $x^{10} = 108$

**26.** $x^{7} = 18$

**27.** $x^{-7} = 18$

**28.** $x^{-10} = 108$

**29.** $10^{x} = 108$

**30.** $7^{x} = 108$

**31.** $2.3^{x} = 95$

**32.** $7.56^{x} = 120$

**33.** $9.45^{x} = 212$

**34.** $43^{x} = 108$

**35.** $5 + 12^{x} = 108$

**36.** $4 + 5^{-x} = 27$

**37.** $5 + 8^{-x} = 53$

**38.** $17 + 83^{x} = 81$

**39.** $5 + 6^{x} = 21$

**40.** $17 + 30^{-x} = 125$

**41.** $4 + 8^{-x} = 36$

**42.** $(1 + x)^{5} = 27$

**43.** $(1 + x)^{7} = 39$

**44.** $(1 + x)^{12} = 7$

**45.** $(1 + x)^{20} = 31$

**46.** $\dfrac{1.06^{x} - 1}{0.06} = 2.56$

**47.** $\dfrac{1.03^{x} - 1}{0.03} = 4.29$

**48.** $\dfrac{1 - 1.005^{-x}}{0.005} = 32.87$

**49.** $\dfrac{1 - 1.0045^{-x}}{0.0045} = 92.57$

## Section 6.1 Simple Interest

**Interest** is the fee charged for the use of money. If we **deposit** money in a bank, the bank may use the money to provide loans for other customers. In return for the use of the money, the bank pays a certain percentage of the amount invested. In a similar manner, if we borrow money from a bank, we are required to pay interest to the bank in return for the privilege of using the money.

One type of interest is **simple interest**. Simple interest is earned when money is deposited in a bank and all the interest is paid only at the end of a specified time and is earned only on the principal. The formula used to calculate simple interest is

### Simple Interest

$$I = Prt \quad \text{where} \begin{cases} I = \text{interest} \\ P = \text{principal or amount deposited} \\ r = \text{interest rate} \\ t = \text{time} \end{cases}$$

### Example 1

If we deposit \$1500 in a bank for three years at an annual rate of 9%, find the amount of simple interest we will earn.

**Solution:** Since $I = Prt$,

$$I = (1500)(0.09)(3) = \$405$$

### Example 2

If we deposit \$1500 in the bank for three years and the bank is paying simple interest of 1.5% each month, find the amount of interest earned.

**Solution:** Because the time and the interest rate are not given in the same units of time, we cannot merely substitute the numbers into the formula as we did in Example 1. We first change the time, three years, into months by multiplying by 12. This gives $t = 3 \times 12 = 36$ months. We now have the interest rate per month and the time in months so we can substitute $P = \$1500$, $r = 0.015$, and $t = 36$ months into $I = Prt$.

$$I = (1500)(0.015)(36) = \$810.00$$

### Example 3

If Mr. Jackson earned \$500 on a \$12,000 investment that earned simple interest for 18 months, what was the annual interest rate?

**Solution:** Since we want to find the annual interest rate, convert time into years. This gives

$$t = 18 \div 12 = 1.5 \text{ years}$$

Now substituting $P = \$12{,}000$, $I = \$500$, and $t = 1.5$ into the formula $I = Prt$ gives

$$500 = (12{,}000)r(1.5)$$

$$500 = 18{,}000r$$

$$r = \frac{500}{18{,}000} \approx 0.0278 = 2.78\%$$

If interest is left in the bank account along with the principal, the amount in the account is given by

$$A = P + I$$

$$A = P + Prt \qquad \text{Substituting } I = Prt$$

$$A = P(1 + rt) \qquad \text{Factoring } P \text{ out of the right side}$$

**Amount in a Simple Interest Account**

$$A = P(1 + rt) \qquad \text{where} \begin{cases} A = \text{amount in the account (including interest)} \\ P = \text{principal} \\ r = \text{interest rate} \\ t = \text{time} \end{cases}$$

**Example 4**

If Ms. Wilson deposited \$1500 in a bank for three years and the bank pays her 9% simple interest per year, find the amount in the account after the interest has been added to the account.

**Solution:** Substituting $P = \$1500$, $r = 0.09$, and $t = 3$ into $A = P(1 + rt)$ gives

$$A = 1500[1 + 0.09(3)] = 1500(1.27) = \$1905$$

**Example 5**

Suppose a bank pays Bill 6% simple interest each year on the amount in the account at the beginning of each year. He deposits \$1000 on January 1. If Bill lets the interest accumulate in the account, how much is in the account after one year? If the simple interest for the second year is based on the amount in the account after the first year's interest is added to the account, how much is in the account after two years? Repeat this process to determine the amount in the account after three years.

**Solution:** Using the formula $A = P(1 + rt)$, with $P = \$1000$, $r = 0.06$, and $t = 1$ year, we have

$$A = 1000(1 + 0.06) = \$1060.00$$

Thus, during the second year, \$1060 is the principal in the account, so the amount at the end of the second year is

$$A = 1060(1 + 0.06) = \$1123.60$$

Similarly, during the third year, \$1123.60 is the principal in the account, so the amount at the end of the third year is

$$A = 1123.60(1 + 0.06) = \$1191.02$$ ■

As you can see, calculating the amount in the account after 30 years by this method would be a very tedious process. The amount in the account during each intermediate year must be calculated before you arrive at the final amount.

## Section 6.1 PROBLEMS

## *Explain → Apply → Explore*

### *Explain*

1. Explain what is meant by *principal.*
2. Explain how simple interest is calculated.
3. If you know the simple interest rate, the amount of interest earned, and the principal, explain how can you determine the length of time of the investment.
4. If you know the total amount in an account earning simple interest, the principal, and the time that the principal has been in the account, explain how you can determine the interest rate.
5. If the interest rate is given as a monthly rate and time is given in years, what step(s) must you complete to determine the simple interest?
6. If the interest rate is given as an annual rate and time is given in months, what step(s) must you complete to determine the simple interest?

### *Apply*

In Problems 6–28, use the simple interest formula $I = Prt$ and $A = P(1 + rt)$ and the given information to find the indicated value.

7. $P = \$2000$ 5% annually $t = 4$ years Find $I$.

| | | | |
|---|---|---|---|
| **8.** $P = \$3000$ | 4% annually | $t = 6$ years | Find $I$. |
| **9.** $P = \$25{,}000$ | 5% annually | $t = 3$ months | Find $I$. |
| **10.** $P = \$35{,}000$ | 4% annually | $t = 4$ months | Find $I$. |
| **11.** $P = \$2500$ | 0.75% monthly | $t = 3$ months | Find $I$. |
| **12.** $P = \$3500$ | 0.5% monthly | $t = 6$ months | Find $I$. |
| **13.** $P = \$1000$ | 0.0329% daily | $t = 1$ year | Find $I$. |
| **14.** $P = \$1000$ | 0.0247% daily | $t = 1$ year | Find $I$. |
| **15.** $P = \$3000$ | 4% annually | $t = 6$ years | Find $A$. |
| **16.** $P = \$2000$ | 5% annually | $t = 4$ years | Find $A$. |
| **17.** $P = \$35{,}000$ | 4% annually | $t = 4$ months | Find $A$. |
| **18.** $P = \$25{,}000$ | 5% annually | $t = 3$ months | Find $A$. |
| **19.** $P = \$3500$ | 0.5% monthly | $t = 6$ months | Find $A$. |
| **20.** $P = \$2500$ | 0.75% monthly | $t = 3$ months | Find $A$. |
| **21.** $P = \$1000$ | 0.0247% daily | $t = 1$ year | Find $A$. |
| **22.** $P = \$1000$ | 0.0329% daily | $t = 1$ year | Find $A$. |
| **23.** $A = \$3100$ | 4% annually | $t = 6$ years | Find $P$. |
| **24.** $A = \$6000$ | 5% annually | $t = 4$ years | Find $P$. |
| **25.** $P = \$3000$ | 4% annually | $A = \$3300$ | Find $t$. |
| **26.** $P = \$2000$ | 5% annually | $A = \$2350$ | Find $t$. |
| **27.** $P = \$2000$ | $A = \$3000$ | $t = 2$ years | Find $r$ (annually). |
| **28.** $P = \$2500$ | $A = \$2600$ | $t = 3$ months | Find $r$ (monthly). |

### *Explore*

**29.** Nina deposits \$3400 into a savings account earning simple interest at 6.3% annually. She intends to leave the money in the bank for eight months. How much money, including both principal and interest, can be withdrawn at the end of this time?

**30.** Diane deposits \$4700 into a savings account earning simple interest at 4.6% annually. She intends to leave the money in the bank for six months. How much money, including both principal and interest, can be withdrawn at the end of this time?

**31.** You win \$4700 in a charity drawing and decide to deposit it into a savings account earning simple interest at 5.51% annually. You intend to leave the money in the bank until the account is worth \$5000. How long must you wait?

**32.** Ed purchased 500 shares of stock for \$23.63 per share. After holding the stock for 18 months, he sells the stock for \$26.37 per share. Assuming that there are no commissions on either the purchase or sale of the stock, what was the annual simple interest rate earned on the investment?

**33.** Angelina purchases 2000 shares of stock for \$87.88 per share. After holding the stock for 18 months, she sells the stock for \$93.12 per share. Assuming that there are no commissions on either the purchase or sale of the stock, what is the annual simple interest rate earned on the investment?

**34.** You earn \$6350 for completing a special project at work and decide to deposit it into a savings account. What annual simple interest rate must be earned if you want to withdraw \$7000 in one year?

**35.** The Radoviches need \$20,000 as the down payment for a house. If they currently have \$18,500 in a bank account, what annual simple interest rate must the bank pay the Radoviches so that the account will have the total down payment after one year?

**36.** In 1993, Salvador purchased a 10-year bond with a face value of \$25,000. The purchase price was \$14,285.71. If the bond is redeemed for its face value in 2003, what simple interest rate will be earned on the bond?

**37.** Phuong Lan purchased a tax-free bond with a face value of \$50,000. The purchase price was \$31,250. If the bond is redeemed for its face value in ten years, what simple interest rate will be earned on the bond?

**38.** Stanley Chang is investing \$13,157.89 in bonds that can be redeemed for \$35,000. If the annual simple interest rate is 8.3%, how long must the bonds be held?

**39.** The Nishikawa family has purchased several bonds for \$126,582.29. The bonds will be redeemed for \$250,000. If the annual simple interest rate earned by the bonds is 6.5%, for how many years must the family hold the bonds?

**40.** Suppose you have a savings account that earns 5% interest annually, and you let the interest accumulate in your account until the end of each year. If the account initially has a balance of \$10,000, use the method of Example 5 to find how much is in the account at the end of each of the next three years.

**41.** Suppose you have a savings account that earns 10% interest annually, and you let the interest accumulate in your account until the end of each year. If the account initially has a balance of \$1000, use the method of Example 5 to find how much is in the account at the end of each of the next three years.

### Example 3

Carol is depositing \$1500 into a savings account with an annual rate of 9%, compounded monthly. How much money will be in the account after 25 years?

**Solution:** Using the formula $A = P(1 + r)^n$, with $P = \$1500$, $r = 0.09 \div 12 = 0.0075$, and $n = 25 \times 12 = 300$, gives

$$A = 1500(1 + 0.0075)^{300}$$
$$= 1500(9.408415) = \$14{,}112.62$$

Compare this result with Example 2. Notice that by compounding more frequently, the amount of interest earned has increased. What do you think will happen if the interest is compounded daily? ■

### Example 4

If \$1500 is deposited into an account earning 9%, compounded daily, how much money will be in the account after 25 years?

**Solution:** Using $r = 0.09 \div 365 = 0.0002466$ and $n = 25 \times 365 = 9125$, we have

$$A = 1500(1 + 0.0002466)^{9125} = 1500(9.4872386) = \$14{,}230.86$$

(*Note:* If values are not rounded off, $A = \$14{,}227.66$.) ■

Starting with the compound interest formula, we can solve for the amount in the account ($A$), the **present value** ($P$), the rate per period ($r$), or the number of times the account is compounded ($n$). The next problem involves solving for $P$, the present value of the account.

### Example 5

How much money must be deposited into an account that earns 6%, compounded monthly, so that \$20,000 can be withdrawn in seven years?

**Solution:** Using $r = 0.06 \div 12 = 0.005$ and $n = 7 \times 12 = 84$, we have

$$20{,}000 = P(1 + 0.005)^{84}$$
$$20{,}000 = P(1.520370)$$
$$P = 20{,}000 \div 1.520370 = \$13{,}154.69$$ ■

Most people don't have \$13,154.69 available to deposit into an account. Suppose we have only half this amount, \$6577.35. If this money is deposited into the account used in Example 5, will it take twice the time, 14 years, to reach the desired \$20,000?

### Example 6

If \$6577.35 is deposited into an account that earns 6%, compounded monthly, how long must we wait for the account to be worth \$20,000?

**Solution:** Using $r = 0.06 \div 12 = 0.005$, $P = \$6577.35$, and $A = \$20{,}000$, we have

$$20{,}000 = 6577.35(1 + 0.005)^n$$

$$\frac{20{,}000}{6577.35} = 1.005^n$$

$$3.040738 = 1.005^n$$

$$\ln (3.040738) = \ln (1.005^n)$$

$$\ln (3.040738) = n \ln (1.005)$$

$$n = \frac{\ln 3.040738}{\ln 1.005} = 223$$

Now we need to either think or worry. If $n = 223$ is measured in years, we will never live to see this money. The good news is that since the interest in the account was compounded monthly, $n$ is measured in months. Therefore, we will have the \$20,000 in 223 months, or $223 \div 12 = 18.58$ years, or 18 years, 7 months. The important thing to notice here is that you made one-half the original investment, but it required more than double the time to achieve the same account balance. ■

If we really wanted to have the \$20,000 in seven years, but we have only \$6577.35, we can meet our financial needs by securing a higher interest rate.

### Example 7

If \$6577.35 is deposited into an account, what annual rate, compounded monthly, is necessary to accumulate \$20,000 in the account after seven years?

**Solution:** Using $t = 7 \times 12 = 84$, $P = \$6577.35$, and $A = \$20{,}000$ gives

$$20{,}000 = 6577.35(1 + r)^{84}$$

$$\frac{20{,}000}{6577.35} = (1 + r)^{84}$$

$$3.040738 = (1 + r)^{84}$$

$$3.040738^{1/84} = [(1 + r)^{84}]^{1/84}$$

$$1.013327 = 1 + r$$

$$r = 0.0133$$

As in the previous example, this is a problem in which interest is compounded monthly. Thus, the annual interest rate is $0.0133 \times 12 = 0.1596$, or 15.96%. ■

## Interest Before Calculators

If you noticed that we relied heavily on a calculator while doing the problems, you may question how students of the 1960s, as well as the bankers of the 1600s, solved these problems. They did it through the extensive use of tables. Students and other users looked up the interest rate and number of periods in the table to find the appropriate value. Where did the tables come from? They were the result of long years of computations done by hand. To appreciate the magnitude of this effort, try computing $1.0075^5$ without the use of a calculator.

When people in the Middle Ages did computations, they did not merely use multiplication. The methods were very clever and involved algebra as well as arithmetic.

### Example 8

Compute the value of $1.04^5$ using the methods available in the Middle Ages.

**Solution:** First compute the value of the algebra expression

$$(1 + r)^2 = (1 + r)(1 + r) = (1 + 2r + r^2)$$

Multiplying this result by itself gives

$$(1 + r)^4 = (1 + 2r + r^2)(1 + 2r + r^2) = 1 + 4r + 6r^2 + 4r^3 + r^4$$

Multiply this result by $1 + r$. This gives

$$\begin{aligned}(1 + r)^5 &= (1 + 4r + 6r^2 + 4r^3 + r^4)(1 + r) \\ &= 1 + 5r + 10r^2 + 10r^3 + 5r^4 + r^5\end{aligned}$$

If we let $r = 0.04$ and calculate each term separately, we notice that several of the terms are very small.

$$\begin{aligned}(1 + 0.04)^5 &= 1 + 5(0.04) + 10(0.04)^2 + 10(0.04)^3 \\ &\quad + 5(0.04)^4 + (0.04)^5 \\ &= 1 + 0.20 + 0.016 + 0.00064 \\ &\quad + 0.0000128 + 0.0000001024 \\ &= 1.2166529024\end{aligned}$$

Notice that a good approximation to this value can be found by adding only $1 + 5r^2 + 10r^2$. In general, to estimate $(1 + r)^n$, you can use the formula

$$(1 + r)^n \approx 1 + nr + \frac{n(n - 1)r^2}{2}$$

Using this formula gives

$$(1 + 0.04)^5 \approx 1 + 5(0.04) + \frac{5(4)(0.04)^2}{2}$$

$$(1.04)^5 \approx 1 + 0.2 + 0.016$$

$$(1.04)^5 \approx 1.216$$

Notice that this is very close to the exact answer of 1.2166529024. ■

## Section 6.2 PROBLEMS

## *Explain ➡ Apply ➡ Explore*

### *Explain*

**1.** Explain what is meant by present value.

**2.** With respect to compound interest, explain what is meant by a period.

**3.** Explain what is meant by periodic interest rate.

**4.** An account with an initial amount of $1000, an annual interest rate of 10%, and a time period of two years will accumulate $200 in interest if the simple interest method is used. The same account will accumulate $210 if the interest is compounded annually. Explain why the compound interest method accumulates more interest.

**5.** Explain how to determine $r$ in the compound interest formula when the annual rate is 8.4% and the interest is compounded monthly.

**6.** Explain how to determine $r$ in the compound interest formula when the annual rate is 8.4% and the interest is compounded daily.

**7.** Explain how to determine $n$ in the compound interest formula when the interest is compounded monthly for five years.

**8.** Explain how to determine $n$ in the compound interest formula when the interest is compounded daily for five years.

### *Apply*

In Problems 9–32, use the compound interest formula $A = P(1 + r)^n$ and the given information to determine the indicated values. In all cases, the interest rates are given as annual rates.

**9.** $P = \$2000$ 6% compounded monthly $t = 4$ years Find $A$.

**10.** $P = \$3000$ 9% compounded monthly $t = 6$ years Find $A$.

**11.** $P = \$6000$ 5% compounded daily $t = 3$ years Find $A$.

12. $P = \$5000$ 10% compounded daily $t = 4$ years Find $A$.

13. $P = \$6000$ 5% compounded daily $t = 3$ months Find $A$.

14. $P = \$5000$ 10% compounded daily $t = 4$ months Find $A$.

15. $A = \$6000$ 6% compounded monthly $t = 2$ years Find $P$.

16. $A = \$5000$ 12% compounded monthly $t = 3$ years Find $P$.

17. $A = \$7500$ 6% compounded quarterly $t = 5$ years Find $P$.

18. $A = \$5000$ 7.2% compounded quarterly $t = 10$ years Find $P$.

19. $A = \$5000$ 5.5% compounded annually $t = 25$ years Find $P$.

20. $A = \$5000$ 3.2% compounded annually $t = 30$ years Find $P$.

21. $A = \$5000$ 9% compounded monthly $P = \$3500$ Find $t$.

22. $A = \$6000$ 6% compounded monthly $P = \$2000$ Find $t$.

23. $A = \$15{,}000$ 7.8% compounded quarterly $P = \$5000$ Find $t$.

24. $A = \$5500$ 4% compounded quarterly $P = \$1000$ Find $t$.

25. $A = \$10{,}000$ 6% compounded annually $P = \$5000$ Find $t$.

26. $A = \$12{,}000$ 5% compounded annually $P = \$2500$ Find $t$.

27. $A = \$5000$ $t = 2$ years $P = \$3500$ Find rate compounded annually.

28. $A = \$6000$ $t = 9$ years $P = \$2000$ Find rate compounded annually.

29. $A = \$10{,}000$ $t = 9$ years $P = \$5000$ Find rate compounded quarterly.

30. $A = \$12{,}000$ $t = 8$ years $P = \$6000$ Find rate compounded quarterly.

31. $A = \$75{,}000$ $t = 15$ years $P = \$25{,}000$ Find rate compounded monthly.

32. $A = \$60{,}000$ $t = 12$ years $P = \$20{,}000$ Find rate compounded monthly.

## *Explore*

33. Which is the better investment for a gift of \$1000, an 8% account compounded annually or a 7.8% account compounded daily? Each investment is for one year.

34. Which is the better investment for a gift of \$3000, an 8.5% account compounded annually or an 8.3% account compounded daily? Each investment is for one year.

35. Antonio deposits \$15,000 into a certificate of deposit that guarantees a 6.6% annual interest rate, compounded quarterly. How much will be in the account at the end of five years?

**36.** The Lee family has decided to invest $12,000 in an account that pays 7.3% interest, compounded daily. What is the value of the account after 12 years?

**37.** The Park family has $10,000 invested in a money market fund that pays 5.475% interest, compounded daily. The account has a three-year term. For tax purposes, the Parks must know the amount of interest earned during each year. Determine the amount of interest earned in each of the three years.

**38.** After moving to the United States from Eastern Europe, the Marvineviches invested their savings of $4000 into an account earning 7.2% interest, compounded monthly. How much interest was earned by the account during each of the first three months?

**39.** Suppose you invest $2500 in an account that earns 9% compounded monthly. After eight years you withdraw the entire amount and deposit it into an account that earns 10% compounded quarterly for six years. How much money have you accumulated at the end of those 14 years?

**40.** An inheritance of $7000 from your great aunt in Des Moines is deposited into an account that earns 7.5% compounded monthly. After eight years you withdraw the entire amount and deposit it into an account that earns 10% compounded quarterly. This second account requires you to keep the money in the account for seven years. How much money have you accumulated at the end of those seven years?

**41.** Uncle Bill and Aunt Marilyn plan on buying a vacation home in the future. They have $20,000 to invest and want to make a down payment of $40,000 on the home. If the best investment currently available is a 12% account compounded daily, how long do they have to wait until they have enough money to make the down payment?

**42.** You, being the wise parent, decide to invest $10,000 for your newborn child's college education. Setting your sights high, you aim for an education at Vine Covered University (VCU). Estimating the future costs at VCU, you arrive at a figure of $150,000. If you can get a 15% annual rate, compounded monthly, how old will your child be when you have the funds to send her to VCU?

**43.** After extensive negotiations, the Indians of Manhattan Island agree with the bankers of Amsterdam to invest $24 at 5% interest, compounded daily. After keeping this investment for 400 years, the Indians decide to cash in the account. What is the account balance?

**44.** Determine the value of a $100 deposit at the end of one year if the account earns 12% interest, compounded (a) monthly, (b) weekly, (c) daily, (d) every second. Compare these values to the value of $A = 100e^{0.12t}$, where $t = 1$ year.

In Problems 45–51, use the Middle Ages formula

$$(1 + r)^n \approx 1 + nr + \frac{n\,(n - 1)r^2}{2}$$

**45.** Compute the value of $1.0075^{12}$.

**46.** Compute the value of $1.005^{10}$.

**47.** Compute the value of $1.01^{24}$.

**48.** Compute the value of a $200 account earning 8.1%, compounded monthly, for two years.

**49.** Compute the value of a $300 account earning 9.6%, compounded monthly, for three years.

**50.** Compute the value of a $200 account earning 7.3%, compounded daily, for two years.

**51.** Compute the value of a $300 account earning 8.76%, compounded daily, for three years.

## Section 6.3 Annuities

As you were working the problems in Section 6.2, the situations may have seemed a little beyond your current financial status. For many, the idea of depositing $10,000 is not realistic. Most people are more likely to save a little money every month or every week rather than make one large deposit. An account in which money is deposited at the end of each period is called an **ordinary annuity**.*

Suppose you deposit $100 into an account at the end of every month for four months. If interest is to be compounded monthly and the annual interest rate is 12%, how much is in the account at the end of four months? To solve this problem, consider the compound interest earned by each deposit, using the formula $A = P(1 + r)^n$.

The first deposit is in the bank for three months, so its value is $\$100(1 + 0.01)^3$.

The value of the second deposit is $\$100(1 + 0.01)^2$.

The value of the third deposit is $\$100(1 + 0.01)^1$.

The value of the fourth deposit is $100 (since it has not earned any interest).

Adding these four terms gives

$$100 + 100(1.01)^1 + 100(1.01)^2 + 100(1.01)^3 = \$406.04$$

* An annunity that has payments made at the beginning of the period is called an annuity due. Only ordinary annuities will be discussed here.

This method certainly solved the problem, but suppose these monthly deposits continued for 30 years. Since this would mean 360 deposits, this process would soon become very tiresome. Instead, if you look at the calculations, you may recognize that this is the sum of the terms of a geometric series. Using the formula for the sum of a geometric series, we can find the formula for the sum of all the payments in an annuity. The derivation of the formula is not the intent of this book, but we will be using the formula for all the remaining problems. It will allow us to easily determine the sum of an annuity without repetitive calculations.

## Annuity Formula

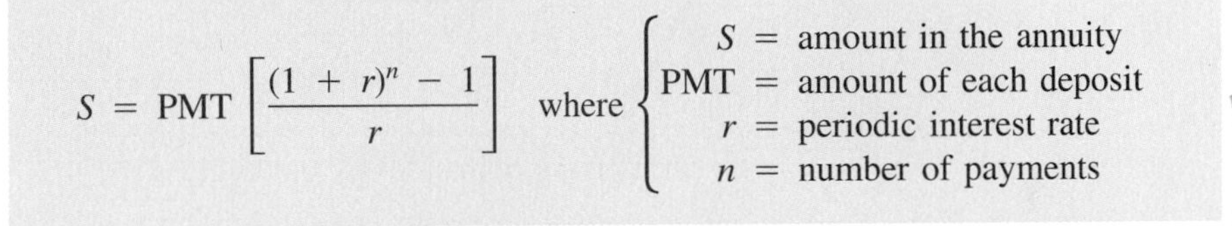

$$S = \text{PMT}\left[\frac{(1+r)^n - 1}{r}\right] \quad \text{where} \begin{cases} S = \text{amount in the annuity} \\ \text{PMT} = \text{amount of each deposit} \\ r = \text{periodic interest rate} \\ n = \text{number of payments} \end{cases}$$

We use PMT (which stands for payment) as a reminder that an annuity is different from an account with one deposit. You make payments into the account every period.

### Example 1

Suppose Mildred deposits \$100 every month for 20 years into an account earning 6%, compounded monthly.

(a) What is the value of the account after 20 years?
(b) How much of the total value of the account was paid through the deposits?
(c) How much interest was earned?

**Solution:**

(a) The amount of the annuity is to be found. Using PMT = 100, $r = 0.06 \div 12 = 0.005$, and $n = 20 \times 12 = 240$ gives

$$S = 100\left(\frac{(1 + 0.005)^{240} - 1}{0.005}\right)$$

$$= 100\left(\frac{3.310204 - 1}{0.005}\right) = \$46{,}204.09$$

(b) The total deposits were 240 payments of \$100 each, which gives \$24,000.00.
(c) The amount of interest earned is the difference between the amount in the annuity and the amount deposited. Therefore the interest earned is

$$\$46{,}204.09 - \$24{,}000 = \$22{,}204.09$$ ■

### Example 2

Suppose we deposit $100 every month into an account earning 6%, compounded monthly. How long must we continue making the deposits so that the account is worth $100,000?

**Solution:** The number of payments, $n$, is to be determined. Using $\text{PMT} = 100$, $r = 0.06 \div 12 = 0.005$, and $S = 100{,}000$, we have

$$100{,}000 = 100\left(\frac{(1 + 0.005)^n - 1}{0.005}\right)$$

Dividing both sides by 100 gives

$$1000 = \left(\frac{(1 + 0.005)^n - 1}{0.005}\right)$$

Multiplying both sides by 0.005 gives

$$5 = 1.005^n - 1$$

$$6 = 1.005^n$$

$$\ln 6 = \ln (1.005^n)$$

$$\ln 6 = n \ln 1.005$$

$$n = \frac{\ln 6}{\ln 1.005} = 359.2 \approx 360 \text{ months or 30 years}$$

There are two interesting things to notice here. The first is that even though the calculation for $n$ is 359.2, the answer is 360 months, not 359 months. The reason is that if we only made 359 deposits, the account would be slightly less than $100,000. By making 360 payments, the account is slightly over $100,000. The second important thing to notice is that while it took 20 years to accumulate the first $46,000, the account reached $100,000 in only ten additional years. ■

## Section 6.3 PROBLEMS — *Explain ➡ Apply ➡ Explore*

### *Explain*

**1.** What is an ordinary annuity?

**2.** What is the difference between an ordinary annuity and the type of account discussed in Section 6.2?

**3.** When solving for which variable in the formula for ordinary annuities will you use logarithms? Explain.

4. How can you determine the total amount of interest earned by an ordinary annuity?

5. How can you determine the total amount you deposit into an ordinary annuity?

6. Suppose you know that you require a certain amount of money for a purchase you will be making in the future, such as making the down payment on a house. How would the annuity formula be useful to you?

## *Apply*

$$S = \text{PMT}\left[\frac{(1 + r)^n - 1}{r}\right] \quad \text{where} \begin{cases} S = \text{amount in the annuity} \\ \text{PMT} = \text{amount of each deposit} \\ r = \text{periodic interest rate} \\ n = \text{number of payments} \end{cases}$$

In Problems 7–18 use the annuity formula and the given information to determine the indicated values. In all cases, interest rates are given as annual rates.

| | | | | |
|---|---|---|---|---|
| 7. | PMT = \$200 | 6% compounded monthly | $t$ = 5 years | Find $S$. |
| 8. | PMT = \$100 | 4% compounded quarterly | $t$ = 2 years | Find $S$. |
| 9. | PMT = \$150 | 6% compounded quarterly | $t$ = 25 years | Find $S$. |
| 10. | PMT = \$50 | 4% compounded monthly | $t$ = 30 years | Find $S$. |
| 11. | $S$ = \$20,000 | 4% compounded quarterly | $t$ = 10 years | Find PMT. |
| 12. | $S$ = \$35,000 | 12% compounded monthly | $t$ = 15 years | Find PMT. |
| 13. | $S$ = \$200,000 | 6% compounded monthly | $t$ = 20 years | Find PMT. |
| 14. | $S$ = \$435,000 | 9% compounded quarterly | $t$ = 30 years | Find PMT. |
| 15. | $S$ = \$20,000 | 9% compounded monthly | PMT = \$100 | Find $t$, in years. |
| 16. | $S$ = \$80,000 | 6% compounded quarterly | PMT = \$250 | Find $t$, in years. |
| 17. | $S$ = \$20,000 | 7.2% compounded monthly | PMT = \$50 | Find $t$, in years. |
| 18. | $S$ = \$80,000 | 8% compounded quarterly | PMT = \$250 | Find $t$, in years. |

## *Explore*

19. Sanjay has decided to invest \$500 each quarter into a retirement account that has annual earnings of 9.3%, compounded quarterly. If Sanjay continues his investments for a period of 25 years, how much money will he have in the retirement account?

20. Antonio will be retiring in 15 years. With his children grown and through with college, he now has the money to save a large amount of money towards his

retirement. He decides to deposit \$800 per month into a mutual fund that is earning \$6.72%, compounded monthly. How much will Antonio have in the account when he retires?

**21.** George listened when his banker told him to start saving money in an IRA account. Beginning when he was 22, George deposited \$100 every month into an account earning 9% compounded monthly.

(a) How much will be in the account when George retires at age 70?
(b) How much of this money did George deposit?
(c) How much of this money is interest?

George's brother Skippy decided to spend the first ten years buying himself toys. He reasoned that he could accumulate more money than George if he deposited \$200 every month starting at age 32. Skippy also plans to retire at age 70 and to use the same 9% account as George.

(d) How much money will be in Skippy's account?
(e) How much of this money did Skippy deposit?
(f) How much of this money is interest?

**22.** After hearing all the advertisements about becoming wealthy when you retire, you decide to contribute to a mutual fund that averages 13% per year.

(a) If you contribute \$2500 each year for the next 25 years, how much will be in the account?
(b) How much of this money did you deposit?
(c) How much of this money is interest?

**23.** When Jill was first hired by SemiTechCorp as a design engineer, she had sufficient income to deposit \$400 each month into an IRA paying 9% interest, compounded monthly. The monthly deposits lasted for ten years.

(a) How much was in the account at the end of ten years?
(b) Due to family responsibilities and investments in real estate, she was not able to continue with these IRA contributions. Instead, she deposited the entire IRA account into a 25-year certificate of deposit earning 11% compounded quarterly. What was the value of the account when it matured?

**24.** When Lisa was first hired by the Environmental Protection Agency as a research scientist, she had sufficient income to deposit \$600 each quarter into an IRA paying 10% interest, compounded quarterly. The quarterly deposits lasted for 12 years.

(a) How much was in the account at the end of 12 years?
(b) Because of the cost of maintaining her parents in a nursing home, Lisa was not able to continue these deposits. Instead, she deposited the entire IRA account into a 30-year certificate of deposit earning 12% compounded monthly. What was the value of the account when it matured?

**25.** In the hopes of driving a luxurious Belchfire 8088, Bert has been setting aside \$150 every month into an account earning 9%, compounded monthly. How long will it take to accumulate the necessary \$42,000 purchase price of the car?

**26.** Grandpa has decided to set up a college fund for his newborn grandson. How much should he deposit every month into an account paying 7.5% interest compounded monthly so that the account will be worth \$30,000 by the time his grandson is 18?

**27.** You are the owner of Edward's Printing Company and know that you will need to buy \$35,000 in new equipment seven years from now. To finance this purchase, you decide to deposit a fixed amount every month into an annuity. If the annuity pays 8.4% compounded monthly, how much is each deposit?

**28.** Anticipating the need for \$150,000 as a college fund for the children, the Cleavers deposit \$300 every month into an account earning 7.8% compounded monthly. How long will it take to accumulate the tentative college fund?

**29.** Akiko is making monthly deposits into an annuity that will be worth \$200,000 in 30 years. The annuity earns 7.8% annual interest, compounded monthly.

(a) What are the monthly payments into the annuity?
(b) What is the value of the annuity after 15 years?

**30.** Juanita is making quarterly deposits into an annuity that will be worth \$175,000 in 35 years. The annuity earns 8.4% annual interest, compounded quarterly.

(a) What are the quarterly payments into the annuity?
(b) What is the value of the annuity after 25 years?

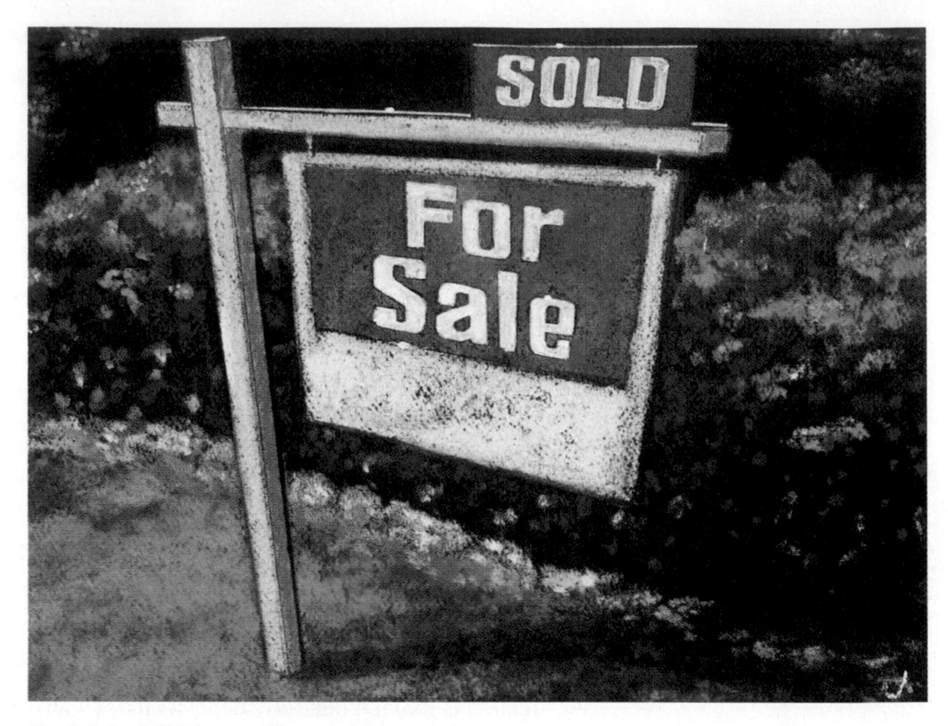


*For Sale* by Jeff Peterson. (Courtesy of the artist)

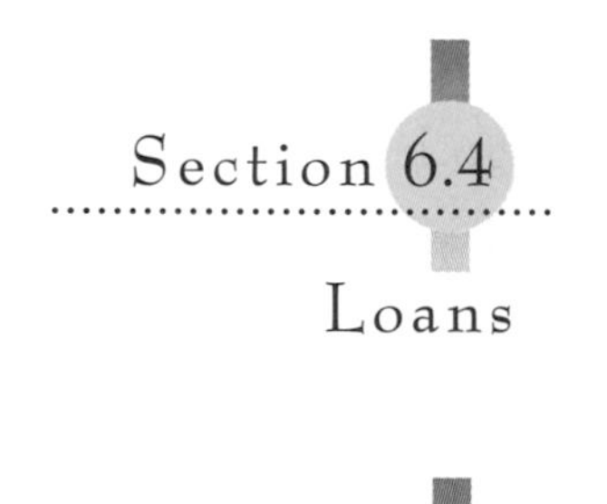

## Section 6.4 Loans

In Section 6.3, we discussed the idea of making repeated deposits into an account. The bank added interest, and the amount in the account accumulated rapidly. Banks acquire the money they pay out as interest from loans and other investments. In this section, we will discuss what happens when a loan is created and eventually paid off. Although we will not show the derivation, the formula can be derived by using the concepts of geometric series. The formula that gives loan payments is

### Loan Formula

$$L = \text{PMT}\left[\frac{1-(1+r)^{-n}}{r}\right] \quad \text{where} \quad \begin{cases} L = \text{the amount of the loan} \\ \text{PMT} = \text{the amount of each payment} \\ r = \text{periodic interest rate} \\ n = \text{the number of payments} \end{cases}$$

### Example 1

Suppose Bill receives a $9000 loan from his bank. He must make monthly payments for four years. The annual interest rate is 6%, compounded monthly. What are his monthly payments?

**Solution:** We are trying to find the payments PMT. Using $L = 9000$, $r = 0.06 \div 12 = 0.005$, and $n = 4 \times 12 = 48$, we have

$$9000 = \text{PMT}\left[\frac{1 - (1 + 0.005)^{-48}}{0.005}\right]$$

$$9000 = \text{PMT}\left[\frac{1 - 0.787098}{0.005}\right]$$

$$9000 = \text{PMT}\ [42.580318]$$

$$\text{PMT} = \$211.37 \text{ is the monthly payment.}$$

### Example 2

When buying a $180,000 home in Milpitas, Theresa made a down payment of $40,000 and took out a loan for the remaining $140,000. The loan has a 30-year term with monthly payments and an annual rate of 10.8%.

(a) What is the monthly payment?
(b) What is the total of the payments over the 30 years?
(c) How much interest will be paid on the loan?

**Solution:**

(a) Using $L = 140{,}000$, $r = 0.108 \div 12 = 0.009$, and $n = 30 \times 12 = 360$, we have

$$140{,}000 = \text{PMT}\left[\frac{1 - (1 + 0.009)^{-360}}{0.009}\right]$$

$$140{,}000 = \text{PMT}\left[\frac{1 - 0.039736}{0.009}\right]$$

$$140{,}000 = \text{PMT}\ [106.696041]$$

$$\text{PMT} = \$1312.14 \text{ is the monthly payment.}$$

(b) 360 payments of $1312.14 gives $360 \times \$1312.14 = \$472{,}370.40$.

(c) Since the interest is the difference between the total payments and the value of the loan, the total interest is

$$\$472{,}370.40 - \$140{,}000 = \$332{,}370.40$$

As we saw in Example 2, interest payments can exceed $300,000. Often when paying off a loan, people decide to **accelerate** the payments to lower the amount of interest paid. When you accelerate the payments, the actual loan payments are greater than what is required by the terms of the loan.

In the next example, we look at the effects of accelerating the payments on the mortgage of Example 2.

## Example 3

Theresa took out a loan for $140,000. The loan had a 30-year term with monthly payments and an annual rate of 10.8%. The loan payments were $1312.14. Instead, Theresa decided to pay $1500 each month.

(a) How long did it take to pay off the loan?
(b) What was the total amount paid on the loan?
(c) What was the total interest paid?

**Solution:**

(a) The question asks us to determine $n$. Using PMT $= 1500$, $L = 140{,}000$, and $r = 0.009$, we have

$$140{,}000 = 1500\left[\frac{1 - 1.009^{-n}}{0.009}\right]$$

Dividing by 1500 gives

$$93.3333 = \left[\frac{1 - 1.009^{-n}}{0.009}\right]$$

Multiplying by 0.009 gives

$$\begin{aligned}
0.84 &= 1 - 1.009^{-n}\\
1.009^{-n} &= 1 - 0.84\\
1.009^{-n} &= 0.16\\
\ln(1.009^{-n}) &= \ln 0.16\\
-n \ln 1.009 &= \ln 0.16\\
n &= \frac{\ln 0.16}{-\ln 1.009} = 204.54
\end{aligned}$$

so Theresa has to make 205 monthly payments (17 years and 1 month). Notice that this cuts almost 13 years off the duration of the loan!

(b) The total amount paid on the loan is given by 205 payments of $1500 each:

$$205 \times 1500 = \$307{,}500$$

(c) The total interest paid is the difference between the value of the loan and the total payments. So the total interest paid is

$$\$307{,}500 - \$140{,}000 = \$167{,}500$$

This looks very nice. By paying roughly $200 a month more than is required, total interest paid is reduced from $332,370.40 to $167,500. This is a savings of $164,870.40! ■

## Refinancing

When interest rates drop, a person who has a mortgage at a high interest rate will often want to refinance a loan. **Refinancing** a loan is the process of paying off an existing loan and replacing it with a new loan. In addition to refinancing to reduce the interest rate, some people also refinance their mortgage so they can draw on the value of their home to help meet other expenses, such as paying for an extensive remodeling job or college expenses for their children.

To refinance, the balance of the existing mortgage must be known. Although the balance is often available from the mortgage company, being able to calculate it yourself will help you determine if refinancing is a feasible idea in your specific situation. The next example shows how you can determine the balance of your existing loan.

### Example 4

Anita and Carlos originally purchased their home with a $20,000 down payment and a $180,000 loan. The loan has monthly payments for thirty years and has an annual interest rate of 9.6%, compounded monthly. After making payments for seven years, they are considering refinancing their mortgage at a lower rate of 6.6%, compounded monthly.

(a) What are the payments on the original loan?
(b) What is the balance of the original loan after seven years?
(c) If they refinance the remaining balance at 6.6%, compounded monthly, for 30 years, what will the new loan payments be?
(d) How much will the monthly payment decrease if Anita and Carlos refinance?

**Solution:**

(a) To determine the original loan payments, we use the loan formula with $L = 180{,}000$, $r = \frac{0.096}{12} = 0.008$, and $n = 30 \times 12 = 360$.

$$180{,}000 = \text{PMT}\left[\frac{1 - (1.008)^{-360}}{0.008}\right]$$

$$180{,}000 = \text{PMT}[117.902287]$$

$$\text{PMT} = \$1526.69$$

(b) To determine the loan balance after making payments for seven years, we use the loan formula again. This time, the amount of the payments is known, \$1526.69, and the length of time remaining on the loan is 23 years. Thus, using PMT = \$1526.69, $r = 0.008$, and $n = 23 \times 12 = 276$, we have

$$L = 1526.69\left[\frac{1 - (1.008)^{-276}}{0.008}\right]$$

$$L = 1526.69[111.138726]$$

$$L = \$169{,}674.38$$

This means that after making payments for seven years, the balance of the loan has decreased from \$180,000 to \$169,674.38.

(c) To determine the new loan payments, we use $L = 169{,}674.38$, $r = \frac{0.066}{12} = 0.0055$, and $n = 30 \times 12 = 360$.

$$169{,}674.38 = \text{PMT}\left[\frac{1 - (1.0055)^{-360}}{0.0055}\right]$$

$$169{,}674.38 = \text{PMT}[156.578125]$$

$$\text{PMT} = \$1083.64$$

(d) By refinancing, the monthly payments have been reduced from \$1526.69 to \$1083.64. This gives a monthly savings of

$$\text{Monthly savings} = \$1526.69 - \$1083.64 = \$443.05$$

■

## Amortization Schedules

An **amortization schedule** lists the amount of principal and interest included in each loan payment and the balance of the loan that remains after each payment. Such schedules are often included with the annual mortgage statement sent by mortgage companies to their customers.

### Example 5

Write an amortization schedule for a \$1200 loan that is being repaid in one year with quarterly payments. The annual interest rate is 13.2%, compounded quarterly.

**Solution:** To write an amortization schedule, the first step is to determine the mortgage payments. Using $L = 1200$, $n = 4$, and $r = 0.132 \div 4 = 0.033$, we have

$$1200 = \text{PMT}\left[\frac{1 - (1.033)^{-4}}{0.033}\right]$$

$$1200 = \text{PMT}[3.690585]$$

$$\text{PMT} = \$325.15$$

The next step is to create a table that includes the payment number, the amount of each payment, the interest included in each payment, the amount of principal included in each payment, and the current loan balance. At the start, our table is as follows:

| Payment Number | Payment | Interest Paid | Principal Repaid | Balance |
|---|---|---|---|---|
| | | | | $1200.00 |
| 1 | $325.15 | | | |
| 2 | | | | |
| 3 | | | | |
| 4 | | | | |

To determine the amount of interest included in each payment, we multiply the current balance by the interest rate for one period, 0.033. Thus, interest paid is given by

$$\text{Interest paid} = \$1200 \times 0.033 = \$39.60$$

To determine the amount of principal repaid, subtract the interest paid from the payment.

$$\text{Principal repaid} = \$325.15 - \$39.60 = \$285.55$$

To determine the new balance, subtract the repaid principal from the previous balance.

$$\text{Balance} = \$1200.00 - \$285.55 = \$914.45$$

Entering these values into the amortization schedule gives

| Payment Number | Payment | Interest Paid | Principal Repaid | Balance |
|---|---|---|---|---|
| | | | | $1200.00 |
| 1 | $325.15 | $39.60 | $285.55 | $ 914.45 |
| 2 | | | | |
| 3 | | | | |
| 4 | | | | |

Repeating this process for the next line, we have

Payment = $325.15
Interest = $914.45 × 0.033 = $30.18
Repaid principal = $325.15 − $30.18 = $294.97
Balance = $914.45 − $294.97 = $619.48

Updating the amortization schedule, we have

| Payment Number | Payment | Interest Paid | Principal Repaid | Balance |
|---|---|---|---|---|
| | | | | $1200.00 |
| 1 | $325.15 | $39.60 | $285.55 | $ 914.45 |
| 2 | $325.15 | $30.18 | $294.97 | $ 619.48 |
| 3 | | | | |
| 4 | | | | |

If we continue the process, we can complete the amortization schedule.

| Payment Number | Payment | Interest Paid | Principal Repaid | Balance |
|---|---|---|---|---|
| | | | | $1200.00 |
| 1 | $325.15 | $39.60 | $285.55 | $ 914.45 |
| 2 | $325.15 | $30.18 | $294.97 | $ 619.48 |
| 3 | $325.15 | $20.44 | $304.71 | $ 314.77 |
| 4 | $325.15 | $10.39 | $314.76 | $ 0.01 |

Notice that the last line of the amortization schedule does not give a zero ending balance to the loan. To fix this, we add one cent to the last payment so that the final balance is zero. The effects on the amortization schedule are shown in the following table.

| Payment Number | Payment | Interest Paid | Principal Repaid | Balance |
|---|---|---|---|---|
| | | | | $1200.00 |
| 1 | $325.15 | $39.60 | $285.55 | $ 914.45 |
| 2 | $325.15 | $30.18 | $294.97 | $ 619.48 |
| 3 | $325.15 | $20.44 | $304.71 | $ 314.77 |
| 4 | $325.16 | $10.39 | $314.77 | $ 0.00 |

From this final amortization schedule, we observe that the amount of principal included in each payment increases with each payment while the amount of interest decreases. ■

In summary, loan payments can be calculated quickly by using the formula presented in this section. Familiarity with the formula and the necessary algebra enables anyone to feel comfortable with the ideas behind borrowing money for a car or any other purchase. In addition, the actual costs of borrowing money can be computed. This will help you make intelligent choices when it comes time for major purchases.

Section 6.4
PROBLEMS

## *Explain* ➡ *Apply* ➡ *Explore*

### *Explain*

1. For a fixed rate loan, the payments remain the same each month. Does the amount of interest paid each month remain the same? Explain.
2. For a fixed rate loan, the payments remain the same each month. Does the amount of principal repaid each month remain the same? Explain.
3. What does it mean to accelerate the payments on a loan?
4. How can you determine the total amount of interest paid on a loan?
5. When solving for which variable in the loan formula will you need to use logarithms?
6. What is an amortization schedule?
7. When writing an amortization schedule, how do you determine the amount of interest paid each period?
8. When writing an amortization schedule, how do you determine the amount of principal repaid each period?

### *Apply*

$$L = \text{PMT}\left[\frac{1-(1+r)^{-n}}{r}\right] \quad \text{where} \begin{cases} L = \text{amount of the loan} \\ \text{PMT} = \text{amount of each deposit} \\ r = \text{periodic interest rate} \\ n = \text{number of payments} \end{cases}$$

In Problems 9–20, use the loan payment formula and the given information to determine the indicated values. In all cases, interest rates are given as annual rates.

9. PMT = \$200 6% compounded monthly $t = 5$ years Find $L$.
10. PMT = \$100 4% compounded quarterly $t = 2$ years Find $L$.

**11.** PMT = \$250 8% compounded quarterly $t = 5$ years Find $L$.

**12.** PMT = \$100 5.4% compounded monthly $t = 3$ years Find $L$

**13.** $L$ = \$20,000 4% compounded quarterly $t = 10$ years Find PMT.

**14.** $L$ = \$35,000 12% compounded monthly $t = 15$ years Find PMT.

**15.** $L$ = \$120,000 8% compounded quarterly $t = 20$ years Find PMT.

**16.** $L$ = \$235,000 9.3% compounded monthly $t = 30$ years Find PMT.

**17.** $L$ = \$20,000 9% compounded monthly PMT = \$200 Find $t$, in years.

**18.** $L$ = \$80,000 6% compounded quarterly PMT = \$2000 Find $t$, in years.

**19.** $L$ = \$180,000 8.4% compounded quarterly PMT = \$5000 Find $t$, in years.

**20.** $L$ = \$150,000 6.3% compounded monthly PMT = \$1200 Find $t$, in years.

## *Explore*

**21.** Bill and Judy are buying a seaside cottage in Bolinas. The mortgage will be \$109,000, to be repaid monthly within 15 years. Find the monthly payments if the interest rate is 10.5%, compounded monthly.

**22.** The Hopkins family is purchasing a Wavc Cruiser yacht costing \$75,000. If the down payment is \$60,000 and the loan is \$15,000, find the monthly payments on their five-year, 15% loan (compounded monthly).

**23.** You have decided to purchase a new Toyota Corolla, using your savings and an \$8000 loan. If the loan is at 13.2%, compounded monthly, and has monthly payments for four years, find the
(a) monthly payment
(b) total paid over four years
(c) total interest paid.

**24.** Olivia's Visa™ card has a balance of \$4250.00. She plans on paying it off in three years, using equal monthly payments. The interest rate is 20.4%, compounded monthly. Assuming no additional charges are made to the account, find the
(a) monthly payment
(b) total paid over three years
(c) total interest paid.

**25.** Lisa plans to accelerate the payments on her \$5000 car loan. The original loan had an interest rate of 13.5% compounded monthly for four years, and Lisa plans on paying \$250 per month.
(a) How much were the original loan payments?
(b) How long will it take Lisa to pay off the loan with her \$250 payments?
(c) How much will the accelerated payments save Lisa over the life of the loan?

**26.** Ron and Michelle plan to accelerate the payments on their $200,000 home loan. The original loan had an interest rate of 10.5% compounded monthly for 30 years.
(a) How much were the original loan payments?
(b) How long will it take to pay off the loan if they pay $100.00 extra each month?
(c) How much will the accelerated payments save over the life of the loan?

**27.** Phil and Angelika are refinancing their mortgage. The existing loan is a 30-year mortgage for $175,000 at 9.6%, compounded monthly. They have made payments on the loan for nine years. Find the remaining balance on the loan.

**28.** A corporation has a mortgage on its main facility. The original value of the mortgage was $1,200,000. The loan is a 30-year fixed rate loan at 8.4%, compounded monthly. After six years, the company has decided to refinance. Find the amount of the new loan.

**29.** Write an amortization schedule for a $2000 loan that is to be paid in one year if the loan has a 6.6% annual rate with quarterly compounding.

**30.** Write an amortization schedule for a $15,000 loan that is to be paid in one year if the loan has a 5.4% annual rate with quarterly compounding.

**31.** Write an amortization schedule for a $2000 loan that is to be paid in $\frac{1}{4}$ year if the loan has a 7.8% annual rate with monthly compounding.

**32.** Write an amortization schedule for a $2000 loan that is to be paid in $\frac{1}{4}$ year if the loan has a 7.2% annual rate with monthly compounding.

**33.** Mike and Melissa have a choice of two loans. The first loan has a 7.23% annual rate, compounded monthly, for 15 years. The other loan is a 30-year loan at 7.5%, compounded monthly. Either loan will be for $150,000.
(a) Calculate the payment for each loan.
(b) Calculate the total paid on each loan.
(c) Which loan is a better choice? Explain.

# Chapter 6 • Summary

## Key Terms, Concepts, and Formulas

The important terms in this chapter are:

**Accelerated payments:** Loan payments that are greater than the payments required by the terms of the loan. p. 430

**Amortization schedule:** A table that lists the amount of principal and interest included in each loan payment and the balance of the loan that remains after each payment. p. 432

**Annuity:** A type of account into which money is deposited on a periodic basis. p. 422

**Compound interest:** The interest paid on both the principal and the accumulated interest. p. 414

**Deposit:** The amount of money put into an account for the purpose of earning interest. p. 409

**Interest:** The fee charged for the use of money. p. 409

**Loan:** An amount of money that is borrowed. p. 428

**Period:** The interval of time between successive additions of interest to an account. p. 415

**Present value:** See Principal. p. 416

**Principal:** The beginning value in an account; the amount deposited. p. 414

**Refinancing:** The process of paying off an existing loan and replacing it with a new loan. p. 431

**Simple interest:** Interest that is added or charged to an account only once. p. 409

**Usury:** A practice of charging extremely high interest rates. p. 396

After completing this chapter, you should be able to:

**1.** Use the formulas

Simple interest $I = Prt \quad A = P(1 + rt)$ p. 409

Compounded interest $A = P(1 + r)^n$ p. 415

Annuities $S = \text{PMT}\left[\frac{(1 + r)^n - 1}{r}\right]$ p. 423

Loans $L = \text{PMT}\left[\frac{1 - (1 + r)^{-n}}{r}\right]$ p. 428

to calculate interest, interest rates, annuities, and loan payments.

**2.** Decide what type of situation (simple interest, compound interest, annuity, or loan) is being described in a problem and apply the appropriate formula(s).

**3.** Write an amortization schedule. p. 432

• Summary

## Problems

**1.** Alice invests $10,000 in an account that earns 8% simple interest. What is the value of the account after three years?

**2.** A certain investment has earned simple interest for five years. The initial investment was \$10,804.85. The account is now worth \$14,700. What is the annual interest rate?

**3.** Suppose \$1000 is deposited in an account earning 10% interest. Find the value of the account at the end of one year if the interest is compounded (a) annually, (b) quarterly, (c) monthly, (d) daily.

**4.** Grandma wants to create an account for her newborn grandchild's college fund. She wants the account to be worth \$25,000 in 18 years. If the account earns 8.1%, compounded monthly for the next 18 years, how much must be deposited?

**5.** A \$20,000 certificate of deposit will be worth \$26,500 in two years. If interest is compounded quarterly, what is the annual interest rate?

**6.** Joe wants to create an annuity that will ensure a comfortable retirement. He estimates that he will need \$500,000 to meet his retirement plans. How much must Joe deposit each month for 32 years into an account earning 7.65% to meet his goal?

**7.** After depositing \$400 per quarter for ten years into an account earning 8% compounded quarterly, Amanda is forced to stop making deposits.
(a) How much is in the account at the end of the ten years?
(b) How much interest has the account earned?
(c) If the amount from (a) is left in the account and the account continues to earn the same interest rate for another 25 years with no more deposits, how much will the account be worth?

**8.** A \$65,000 annuity was created by depositing \$100 every month into an account earning 6% interest compounded monthly. How many deposits were needed to create this annuity?

**9.** The purchase of a \$15,000 automobile is to be accomplished with a 20% down payment. The remainder will be financed at 11.4%, compounded monthly, for three years.
(a) What is the monthly payment?
(b) What is the total amount paid for the car?

**10.** A \$130,000 mortgage at 10.2%, compounded monthly, is to be paid with equal monthly payments of \$1600. How long will it take to pay off the loan?

**11.** Determine the balance of a 30-year \$200,000 mortgage at 8.4% interest, compounded monthly, if payments have been made on the loan for 12 years.

**12.** Write an amortization schedule for a \$3000 loan that is to be paid in one and one-half years. The interest rate is 6.8%, compounded quarterly.

# 7

# Probability and the Games People Play

*At the Races* by Edouard Manet. Horse racing is just one of the many sporting events in which the study of probability and odds play an important role. (National Gallery of Art)

# Overview Interview

In this chapter, you will examine some basic components of probability. You will learn techniques that determine the chance that an event will occur. This investigation will concentrate on the probability involved in the games that people play. These games include lotteries, Bingo, Poker, Roulette, Keno, Slot Machines, and Blackjack. You will determine the number of outcomes in such games and examine the corresponding odds. You will also discover that by looking at the probabilities, the amount wagered, and the amount won or lost on the wager, you can determine if a game is fair. If you like to play games of chance, you will find that this chapter will make you a more knowledgeable player.

## Frank Riolo

### FRANK RIOLO, VICE PRESIDENT OF CASINO OPERATIONS, TURNING STONE CASINO, VERONA, NY

TELL US ABOUT YOURSELF. I was raised in the Utica, New York area. In 1970, I moved to Las Vegas where I took an entry level position in a casino. I am now the Vice-President of Casino Operations at the Turning Stone Casino, owned by the Oneida Nation of Native Americans. I am responsible for all live gaming operations including the supervison of nearly 1,000 gaming employees at this first legal casino in the State of New York. The development of the business plan, preparation and monitoring of the casino operating budget, and identifying casino needs all fall under my realm of responsibility.

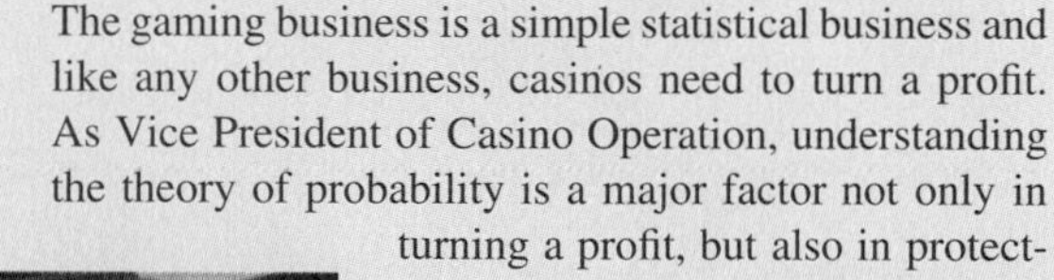

HOW IS PROBABILITY USED BY SOMEONE IN YOUR POSITION? The gaming business is a simple statistical business and like any other business, casinos need to turn a profit. As Vice President of Casino Operation, understanding the theory of probability is a major factor not only in turning a profit, but also in protecting the integrity of the games. Being able to calculate the probability of any particular outcome offers us the opportunity to maximize our profitability. There was a considerable amount of time spent on attaining the proper mix of games that would stimulate player interest and enable the casino to be profitable. Each game was studied as to its level of interest and payoff odds. In addition, games are analyzed daily as to their profitability. It is through this analysis that a proper balance is maintained at the casino.

*The Card Players* by Lucas van Leyden show a game in which probability theory has relevant application. (National Gallery of Art)

## A Short History of Probability

Probability is the science of determining the likelihood, or chance, that an event will occur. Combinatorics is the mathematical tool used to find the number of ways in which an event can occur. The histories of these two topics are interwoven, developing with people's interest in games and, later, science.

Originally, interest in probability arose from the study of games similar to dice and other modern pastimes. Evidence of this has been found in archaeological digs of Assyrian and Sumerian sites. Scorecards from games, tomb engravings, and dice-like implements called *tali* clearly demonstrate that the ancient Egyptians had an interest in games and gambling. A *talus*, or *astralagus*, is the heel bone of a running animal. Polished and engraved, tali were used by the Egyptians the way dice are used today. When thrown, a talus could land on any of four different sides. Since it was not uniformly shaped, each side had a different probability of landing face up.

Although many civilizations, such as the Greeks and the Chinese, developed advanced mathematics, modern probability theory did not begin to develop until the late 1500s. Since (as we shall see) probability depends heavily on arithmetic, it is believed that the development of probability was hampered by cumbersome systems of numeration. Little research has been conducted on the history of probability in India; however, an Indian text from A.D. 400 seems to indicate that the Indians possessed a greater knowledge of probability than did Westerners. It is believed that their arithmetic system, one which far surpassed other systems in ease of use, allowed for these advances in probability.

Early work on combinatorics has been contributed by writers from several civilizations. In China in 1100 B.C., permutations were mentioned in *I-Ching* [The Book of Changes] concerning the possible number of trigrams. The Latin writer Boethius (c. 510) gave a rule for selecting items two at a time from a large set. The Hindu mathematician Bhāskara gave rules for calculating permutations and combinations and discussed them as they related to such varied topics as medicine, music, and architecture. Although the Hebrew writer Rabbi ben Ezra did not provide a formula, he used combinations to discuss possible arrangements of Saturn and the other planets. Hérigone (1634) was the first to give the general formula for combinations (see Section 7.0) and Leonhard Euler used the notation

$$\left[\frac{p}{q}\right]$$

for combinations, a form close to the modern

$$\binom{p}{q}$$

Although probability was mentioned as early as 1477 in a commentary on Dante's *Divine Comedy*, it is said to have its origins in an unfinished dice game. Two gamblers were unable to complete a game of chance. They agreed to divide the stakes according to their respective chances of winning the game but could not decide what these chances were. The mathematician Blaise Pascal received a letter from his friend Chevalier de Méré requesting a solution to the problem (c. 1654). Pascal sent the problem to another French mathematician, Pierre de Fermat. Working together, they solved the problem and in the process began the development of modern probability theory.

As is always true in science, probability did not spring forth as a completely developed theory. There have been many who have made substantial contributions to the topics we will study. The Spanish alchemist Raymond Lulle (1234–1315) is credited as being the father of combinatorics. He wanted to find the symbols for all the chemical elements and then to write down all possible arrangements of these symbols. By doing so, he believed he would be able to construct every possible substance.

It was during the 16th and 17th centuries that European mathematicians developed combination theory and applied it to games of chance. In 1663, *Liber de ludo aleane* [The book on games of chance] was published. Written by the Italian Girolamo Cardano (1501–1576) nearly a century earlier, it was published after the works of Pascal and Fermat.

The Dutch astronomer Christiaan Huygens (1629–1695) wrote an introduction to dice games in *De ratiociniis in ludo aleane* [On reasoning in games of dice] in 1657. This treatise included the concept of expected value (see Section 7.4). Huygens was also one of the first to study probability from what is now considered the classical viewpoint. Whereas the investigations of Fermat and Pascal started with games, Huygens considered probability as the ratio of the number of successful outcomes of an event to the total number of possible outcomes.

The first substantial book on probability was published in 1713. Entitled *Ars conjectandi* [Art of conjecture], it was a posthumous work of the Swiss mathematician Jakob Bernoulli (1654–1705). Bernoulli developed the theory of probability, discussed the Law of Large Numbers (see Section 7.1), and provided a general theory of permutations and combinations. The work of Abraham De Moivre (1667–1754) played an important role in the development of actuarial mathematics and the theory of probability. The classical use of probability theory and combinatorics in the study of games of chance continued to develop throughout the 18th and 19th centuries. In the 1700s, probability was also used in courts of law to determine the validity of evidence and in the insurance industry to help calculate the proper rates to charge for an annuity.

In 1812, Pierre Simon de Laplace published *Théorie analytique des probabilités*. This gave the first complete theory of probability. Two years later, Laplace published *Essai philosophique sur les probabilités*, which was a compilation of the conceptual principles involved in probability. The arrival of these two works allowed probability to be used in the physical sciences. It quickly became one of the major mathematical tools of the 19th and 20th centuries.

In the 180 years since Laplace, probability theory has played a major role in science. From 1866 to 1887, Ludwig Boltzmann used probability to develop the kinetic theory of gases. In 1905, Albert Einstein used probability in developing the theory of Brownian motion. In 1907, A. A. Markov began the development of Markov chains. Probability theory was soon applied to epidemiology, sociology, quantum mechanics and population studies in emigration and immigration. Since the 1940s, combinatorics and probability have been used a new area—game theory. Game theory not only applies to the games of Fermat and Pascal, but it is also used in the study of economics, military strategy, politics, and psychology.

Today, probability is applied to many different fields. It is no longer merely a study of how the heel bone of a sheep will land. It is used to determine odds and payoffs by state lottery commissions, by gambling casinos, by racetrack handicappers, by managers of sports teams in making decisions during a game, by insurance companies in calculating insurance premiums, by advertising agencies in planning ad campaigns, meteorologists in predicting the weather, and by many others.

## Check Your Reading

1. What role did gamblers play in the development of probability theory?
2. What role did the study of alchemy play in the study of combinatorics?

3. Pigtails were introduced in the Prussian army during the same year that Jakob Bernoulli developed the theory of probability. What year was this?
4. In 1100 B.C., the Sun Pyramid was built in Mexico and silk fabrics were available in China. What event was occurring in the history of probability?
5. In 1634, Jean Nicolet explored Wisconsin while Anne Hutchinson arrived in Massachusetts. What was Hérigone doing this same year?
6. One eventful year saw all four of the following events: The drinking of chocolate was introduced to England; Velázquez painted *Las Hilanderas* (The Spinners); Parisian manufacturers produced fountain pens; Christiaan Huygens wrote an introduction to dice games. In what year did all these events occur?
7. In 1654, the Portuguese drove the Dutch out of Brazil, and Rembrandt painted *Portrait of Jan Six*. What work was being done on gambling during this year?
8. When Louisiana became a state and Beethoven wrote his seventh and eighth symphonies during 1812, what was Pierre Simon de Laplace doing?
9. In what year was probability mentioned in Dante's *Divine Comedy*?
10. In what year did neon lights appear, work progress on the Panama Canal, and Albert Einstein use probability theory to study Brownian motion?
11. How do gambling casinos make use of probability theory?
12. How do insurance companies make use of probability theory?
13. Match each of the following names with the correct event, idea, or occurrence.

| | |
|---|---|
| (a) Bernoulli | Development of modern probability theory |
| (b) Bhāskara | Father of combinatorics |
| (c) Boethius | First complete discourse on theory of probability |
| (d) Hérigone | First substantial book on probability |
| (e) Huygens | General formula for combinations |
| (f) Laplace | Reasoning in games of dice |
| (g) Lulle | Rules for calculating permutations and combinations |
| (h) Pascal/Fermat | Rule for selecting items two at a time |

## • Research Questions

To answer the following questions, you will need to refer to materials not contained in this book. Possible sources of information are listed in the Bibliography at the end of this book.

1. Investigate at least two areas that use probability. Write a paragraph or two about each area and how it uses probability. Possible areas are meteorology, sports, biology, genetics, insurance, psychology, advertising, and education.
2. Throughout history, people studying various topics have found the need to understand probability. Although only remotely connected to probability, these topics are of interest in themselves. Write a paragraph or two on at least two of the following areas.

(a) What is *I-Ching*? Who wrote it? What does it discuss?
(b) What are trigrams? What is their use and significance?
(c) What is an alchemist?
(d) What is the kinetic theory of gases?
(e) What is Brownian motion?

3. Discuss the Bernoulli family of mathematicians. What areas besides probability did they investigate?
4. What does a standard deck of playing cards consist of? What are the origins of playing cards?
5. Who was Hoyle and what is the *Book of Hoyle*?
6. Examine one of the state lotteries. How is the lottery played? What are the odds or probabilities of winning in the lottery?
7. What do the odds posted in horse or dog races indicate? What is pari-mutuel betting and how does it work?
8. Examine the statistics of popular sports like baseball, football, or basketball. Which of the statistics actually give the probability that a certain event will occur? Explain what such statistics mean in terms of determining the probability of an event.
9. Gambling casinos in cities such as Atlantic City, New Jersey; Las Vegas, Nevada; or Reno, Nevada, make extensive use of probability theory. Investigate the history of casinos and how profit is made in the games of chance played at the casinos.
10. Do some research on two of the mathematicians mentioned in the text. What interests did they have besides mathematics?

## Projects

1. The games of chance in gambling casinos make use of probability. Write a short history of at least three of the games listed below. Include a description of how the games are played and some of the probabilities or odds involved in the games.
   (a) Keno
   (b) Roulette
   (c) Craps
   (d) Slot Machines
   (e) Poker
   (f) Blackjack
2. Standard dice are cubes with six congruent square faces. Each face is marked with one to six dots. Cubes are not the only geometric solids (regular polyhedra) that could be used as "dice." What other polyhedra could be used? How many dots should be placed on the faces of these dice? If two of these dice are tossed, what is the minimum and maximum sum of the two dice? Make a sketch of the faces of these dice.

3. Some television game shows have wheels that are spun to determine prizes that contestants win. The *Wheel of Fortune* program is one of these shows. Study the wheel used on two TV game shows and give the probability (see Section 7.1) that the wheel will land on each of the prizes on the wheel. Which prize amount is the most likely to occur? Which is the least likely to occur? What implications do the results have for the contestants on the show?
4. According to the Law of Large Numbers (Section 7.1), experimental probability approaches the theoretical probability as the number of trials get larger. Examine events such as those described below to actually calculate the experimental probability of an event in a larger and larger number of trials.
   (a) Since the last digit of a phone number can be any one of the ten digits, the theoretical probability that the last digit of your phone number is a 5 is $1/10$. Use a phone book to tally the number of phone numbers that end in 5. Record the number of phone numbers ending in 5 for 100, 200, . . . , 1000 phone numbers. How do the experimental probabilities compare to the theoretical probability of $1/10$?
   (b) If you toss a pair of dice, you will see in Section 7.1 that the theoretical probability that the sum of the two dice is 11 is $2/36$ or about 0.0556. Toss a pair of dice and tally the number of sums that are 11 after 20, 40, 60, . . . , 200 tosses. How do the experimental probabilities compare to the theoretical probability of 0.0556?

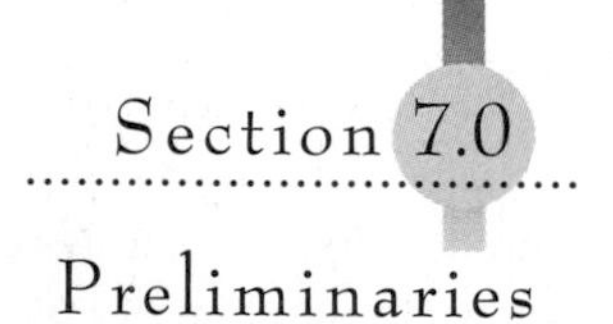

## Section 7.0 Preliminaries

Before discussing probability, we will examine a few algebraic concepts, namely factorials, permutations, and combinations.

### Factorials

The expression $n!$ is a special shorthand in mathematics. It is read ***n* factorial**. When a number is written with an exclamation point, such as 5!, it means to do the following calculation:

$$5! = 5 \times 4 \times 3 \times 2 \times 1 = 120$$

In general,

$$n! = n \times (n - 1) \times (n - 2) \times \cdots \times 3 \times 2 \times 1$$

where $n$ is any natural number.

**Note:** In the formula for $n!$, the "· · ·" indicates that there are factors not being written. For example, instead of writing

$$8! = 8 \times 7 \times 6 \times 5 \times 4 \times 3 \times 2 \times 1$$

we could write

$$8! = 8 \times 7 \times \cdots \times 2 \times 1.$$

Even though the numbers 3 through 6 are not written, they are still used in the calculation. Called an **ellipsis**, "· · ·" is used to abbreviate a long, repetitive mathematical expression.

The reason for this notation is that factorials can be very large numbers even though $n$ is relatively small. For example, 12 is not a very large number but 12! is

$$12! = 12 \times 11 \times 10 \times \cdots \times 3 \times 2 \times 1 = 479{,}001{,}600$$

It is much easier to write equations using 12! rather than 479,001,600.

We will now illustrate how to perform calculations using factorials.

### Example 1

Multiply $4! \times 3!$.

**Solution:**

$$\begin{aligned} 4! \times 3! &= (4 \times 3 \times 2 \times 1) \times (3 \times 2 \times 1) \\ &= 24 \times 6 = 144 \end{aligned}$$

■

**Note:** Although it may be tempting,

$$4! \times 3! \neq 12! = 12 \times 11 \times 10 \times \cdots \times 2 \times 1 = 479{,}001{,}600$$

This method gives an incorrect answer.

As seen in Example 1, there is no shortcut for multiplying factorials. However, Example 2 will show how **division of factorials** can be done without calculating the value of each factorial separately.

### Example 2

Find: 4! ÷ 3!.

**Solution:**

$$\frac{4!}{3!} = \frac{4 \times 3 \times 2 \times 1}{3 \times 2 \times 1} = \frac{4}{1} = 4$$

When dividing factorials, some factors in the numerator will cancel with factors in the denominator. ■

### Example 3

Find: 148! ÷ 146!.

**Solution:** Calculating 148! and 146! is a difficult task, even with a calculator. However, the division can be accomplished by using the methods of Example 2.

$$\frac{148!}{146!} = \frac{148 \times 147 \times 146 \times 145 \times \cdots \times 2 \times 1}{146 \times 145 \times \cdots \times 2 \times 1}$$

$$= 148 \times 147 = 21{,}756$$

Thus, by canceling, we are able to perform the division without actually calculating either of the factorials. ■

We conclude the review of factorials by discussing the following sequence:

$$\begin{aligned} 4! &= 24 \\ & \quad \div 4 \\ 3! &= 6 \\ & \quad \div 3 \\ 2! &= 2 \\ & \quad \div 2 \\ 1! &= 1 \\ & \quad \div 1 \\ 0! &= ? \end{aligned}$$

In this sequence, we see that as we decrease from 4! to 3!, we are actually dividing by 4. This process continues until we reach 0!. Two inferences can be drawn from this table. First, 0! is 1! ÷ 1 = 1. Thus, we define

$$0! = 1$$

The second inference is that it is not possible to discuss the factorial of a negative number. Since the pattern developed at the right indicates that $(-1)!$ must come from $0! \div 0$ and division by zero is not possible, $(-1)!$ is not allowed.

## Permutations

A **permutation** is one of the arrangements of a group of items. For example, three objects ▮ ◆ ✻ can be arranged in the following orders:

**Note:** Different arrangements of the same symbols count as different permutations.

These are the six possible permutations of the three objects. All six arrangements are different, and it is not possible to find any other arrangement.

Instead of listing all the arrangements, often we will be concerned only with the number of permutations. This will be true particularly when we deal with large numbers of objects. Thus, we want to develop a formula that will determine the number of permutations. To do this, reexamine the previous arrangement of three symbols. There are three choices for the first symbol. This leaves only two choices for the second symbol, which, in turn, leaves only one choice for the third symbol. If we compute $3 \times 2 \times 1 = 6$, we arrive at the number of permutations listed.

### Example 4

Calculate the number of ways to arrange two objects selected from the following group of four.

**Solution:** We expect that since there are four choices for the first object and three choices for the second object, there will be $4 \times 3 = 12$ possibilities. Listing the possibilities verifies this result.

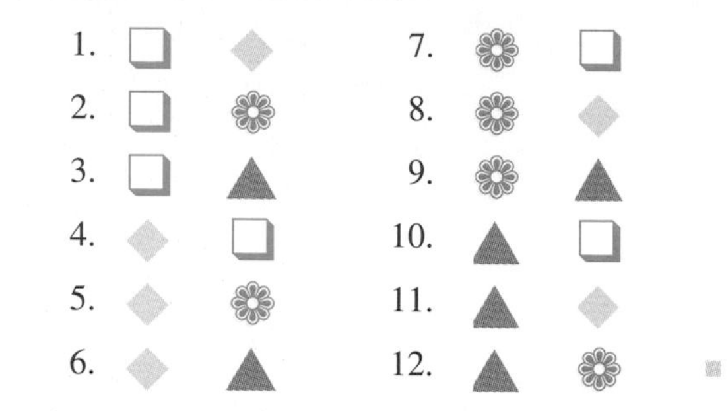

We can now use these examples to help determine a formula. If we have a set of ten objects and we want to find the number of ways to arrange a set of three, there are ten ways to pick the first item, nine ways to pick the second, and eight ways to pick the third. Multiplying gives $10 \times 9 \times 8 = 720$ ways. If we rewrite $10 \times 9 \times 8$ in the following way, we can make a connection between permutations and factorials:

$$10 \times 9 \times 8 = \frac{10 \times 9 \times 8 \times 7 \times 6 \times 5 \times 4 \times 3 \times 2 \times 1}{7 \times 6 \times 5 \times 4 \times 3 \times 2 \times 1}$$

$$= \frac{10!}{7!} = \frac{10!}{(10-3)!}$$

Thus, if we have a group of ten objects from which we want to arrange a set of three, there are $10!/(10-3)!$ ways in which this can be done.

In general, the number of permutations of $n$ objects taken $r$ at a time is

Permutation Formula

$$P_{n,r} = \frac{n!}{(n-r)!}$$

**Note:** The symbol $P_{n,r}$ is not the only symbol used for permutations. Other books may use ${}_nP_r$, or $P_r^n$, or $P(n,r)$.

$P_{n,r}$ is read as "the number of permutations of $n$ objects taken $r$ at a time."

**Example 5**

Find $P_{4,4}$.

**Solution:** Using the permutation formula with $n = 4$ and $r = 4$ gives

$$P_{4,4} = \frac{4!}{(4-4)!} = \frac{4!}{0!} = \frac{24}{1} = 24$$

We can see that the number of permutations of four objects taken four at a time is given by 4! Similarly, the number of permutations of $r$ objects taken $r$ at a time is given by $r$! We can also see that it is important that $0! = 1$. If $0! = 0$, we would not be able to use the formula to calculate $P_{4,4}$. ■

**Example 6**

Find $P_{48,3}$.

**Solution:** Using the formula above with $n = 48$ and $r = 3$ gives

$$P_{48,3} = \frac{48!}{(48-3)!} = \frac{48!}{45!}$$

$$\frac{48!}{45!} = \frac{48 \times 47 \times 46 \times 45 \times 44 \times \cdots \times 2 \times 1}{45 \times 44 \times \cdots \times 2 \times 1}$$

$$= 48 \times 47 \times 46 = 103{,}776$$

Thus, by canceling, we are able to perform the division without actually calculating either of the factorials. ■

Since permutations are arrangements of objects, **the order of the objects matters**. For example, (1) ▮ ◆ ✻ and (2) ▮ ✻ ◆ are considered two different permutations (arrangements) of the same three objects because the order of the objects is different.

## Combinations

Combinations are similar to permutations except that the **order of the objects does not matter**. In other words, although (1) ▮ ◆ ✻ and (2) ▮ ✻ ◆ are considered different permutations, they are the same combination of items. With combinations, we pick items from a set, but not arrange them.

## Example 7

Calculate the number of combinations of two objects selected from the following group of four.

**Solution:** We will do the problem by listing the possibilities.

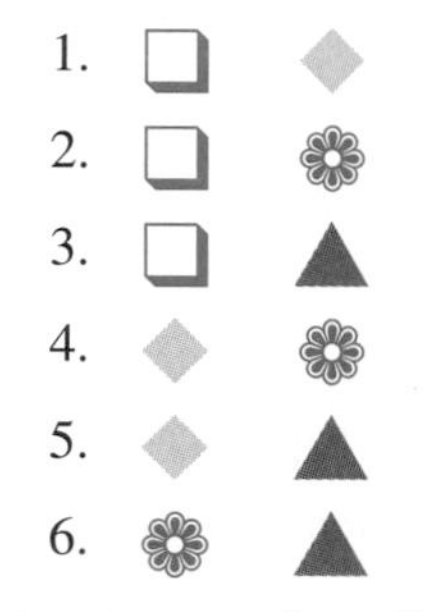

As long as order does not matter, we have listed all the possibilities. Thus, there are six combinations when selecting two items from a group of four. If we compare this to Example 4, we find that there are fewer combinations than there are permutations. ■

## Example 8

Calculate the number of combinations of four objects selected from the following group of four.

**Solution:** Since order does not matter, there is only one choice.

Thus, there is only one combination of four items selected from a group of four. ■

As we saw in Example 5, the number of arrangements of a group of $r$ objects is $r!$. Therefore, any set of $r$ objects can be arranged in $r!$ different ways. Since combinations do not depend on the order of the objects, dividing the number of permutations by $r!$ eliminates the repeated combinations. Thus, the number of combinations of $n$ objects taken $r$ at a time equals $P_{n,r} \div r!$. From this, we find that the formula for combinations is

**Combination Formula**

$$C_{n,r} = \frac{n!}{r!(n-r)!}$$

**Note:** The symbol $C_{n,r}$ is not the only symbol used for combinations. Other books may use $\binom{n}{r}$, ${}_nC_r$, $C_r^n$, or $C\,(n,r)$.

$C_{n,r}$ is read as ''the number of combinations of $n$ objects taken $r$ at a time.''

## Example 9

Given the objects ❑ ▼ ▮ ◆

(a) Find the number of combinations when two items are selected.
(b) List the combinations.
(c) Find the number of permutations when two items are selected.
(d) List the permutations.

**Solution:**

(a) For $n = 4$ and $r = 2$, the number of combinations is

$$C_{4,2} = \frac{4!}{2!(4-2)!} = \frac{4!}{2! \times 2!} = \frac{24}{2 \times 2} = 6$$

(b) The six combinations are

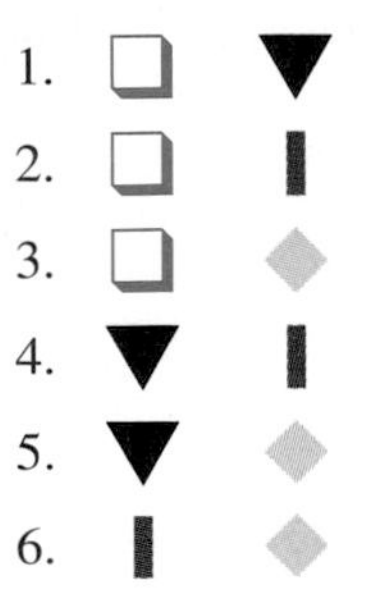

(c) For $n = 4$ and $r = 2$, the number of permutations is

$$P_{4,2} = \frac{4!}{(4-2)!} = \frac{4!}{2!} = \frac{24}{2} = 12$$

(d) The 12 permutations are

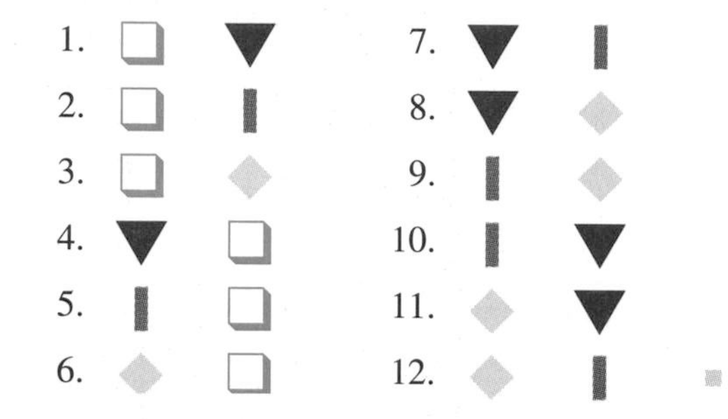

## Example 10

Compute the values of $P_{10,0}$ and $C_{10,0}$.

**Solution:** Applying the appropriate formulas gives

$$P_{10,0} = \frac{10!}{(10-0)!} = \frac{10!}{10!} = \frac{1}{1} = 1 \qquad C_{10,0} = \frac{10!}{0!\,(10-0)!} = \frac{10!}{0!\,10!} = \frac{1}{0!} = \frac{1}{1} = 1$$

Notice that whether we are determining permutations or combinations, when choosing groups containing zero items the result is 1. Intuitively, this makes sense. There is only one way to pick a group of zero items, and that is not to pick any of the items. ■

## Example 11

Find $C_{48,3}$.

**Solution:** Using the formula with $n = 48$ and $r = 3$ gives

$$C_{48,3} = \frac{48!}{3!\,(48-3)!} = \frac{48!}{3!\,45!}$$

$$\frac{48!}{3!\,45!} = \frac{48 \times 47 \times 46 \times 45 \times 44 \times \cdots \times 2 \times 1}{3 \times 2 \times 1 \times 45 \times 44 \times \cdots \times 2 \times 1}$$

$$= \frac{48 \times 47 \times 46}{3 \times 2 \times 1} = \frac{103{,}776}{6} = 17{,}296$$ ■

In summary, this section has discussed the following topics:

**Factorials** $n! = n \times (n - 1) \times (n - 2) \times \cdots \times 3 \times 2 \times 1$

**Permutations** $P_{n,r} = \dfrac{n!}{(n - r)!}$

**Combinations** $C_{n,r} = \dfrac{n!}{r!\,(n - r)!}$

**Permutations are used when order matters.**

**Combinations are used when order does *not* matter.**

## Pascal's Triangle

Pascal's triangle is an arrangement of numbers that has an interesting property. All the values in the triangle are combinations. The first nine rows are

| | |
|---|---|
| 1 | 0th row |
| 1 1 | 1st row |
| 1 2 1 | 2nd row |
| 1 3 3 1 | 3rd row |
| ***1 4 6 4 1*** | 4th row |
| 1 5 10 10 5 1 | 5th row |
| 1 6 15 20 15 6 1 | 6th row |
| 1 7 21 35 35 21 7 1 | 7th row |
| 1 8 28 56 70 56 28 8 1 | 8th row |

Suppose we look at the fourth row of the triangle. The values in this row are the same as the possible combinations of 4. In other words,

$$C_{4,0} = 1 \qquad C_{4,1} = 4 \qquad C_{4,2} = 6 \qquad C_{4,3} = 4 \qquad C_{4,4} = 1$$

Although it seems easy enough to write down a triangle full of combinations, the amazing aspect of Pascal's triangle is the pattern that can be used to construct the triangle. To find the terms in the triangle, add two consecutive terms in a row to get the term that is between these terms in the following row. The first and last terms in a row will always be 1. For example, let's look at the fourth and fifth rows:

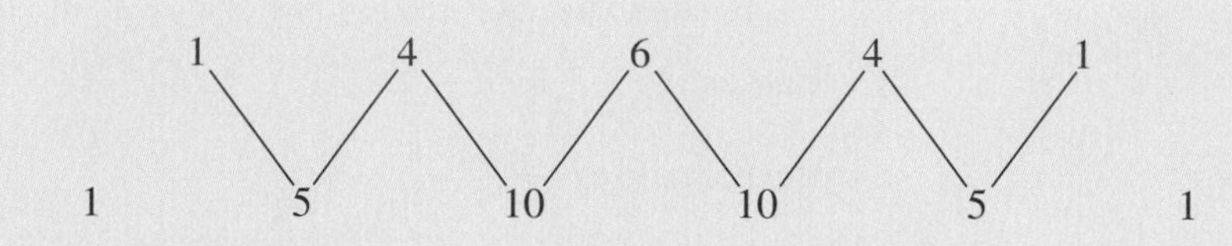


Writing the terms of Pascal's triangle in this way allows us to compute many combinations quickly. For example, to find $C_{8,3}$, we need to write the eighth row of the triangle. The combinations, beginning with $C_{8,0}$, will start at the left. Thus, $C_{8,3}$ will be the fourth term from the left, so $C_{8,3}$ equals 56.

### Example 12

Find $C_{9,5}$ from Pascal's triangle.

**Solution:** The ninth row of Pascal's triangle is

1 9 36 84 126 126 84 36 9 1

Thus, $C_{9,5}$, the sixth term in the row, is 126. ■

Section 7.0
PROBLEMS

## *Explain* ➡ *Apply*

### *Explain*

1. What is a factorial?
2. What are permutations?
3. What are combinations?
4. How are permutations and combinations different? Give an example.
5. Using the letters $a$, $b$, $c$, $d$, $e$, show the difference between $P_{5,2}$ and $C_{5,2}$ by actually listing the elements of each.
6. How are the formulas to determine the number of permutations and combinations different?
7. Without actually doing the calculations, why would you expect $P_{75,24}$ to be larger than $C_{75,24}$?

### *Apply*

In Problems 8–31, calculate the value of each expression.

**8.** 7!

**9.** 9!

**10.** 13!

**11.** 12!

**12.** $C_{7,3}$

**13.** $C_{8,3}$

**14.** $C_{71,3}$

**15.** $C_{68,3}$

**16.** $C_{7,7}$

**17.** $C_{8,8}$

**18.** $C_{71,0}$

**19.** $C_{68,0}$

**20.** $P_{7,3}$

**21.** $P_{8,3}$

**22.** $P_{71,3}$

**23.** $P_{68,3}$

**24.** $P_{7,7}$

**25.** $P_{8,8}$

**26.** $P_{7,0}$

**27.** $P_{6,0}$

**28.** $C_{7,3} \div C_{9,4}$

**29.** $C_{8,3} \div C_{10,5}$

**30.** $C_{7,3} \times C_{9,2}$

**31.** $C_{8,3} \times C_{10,2}$

**32.** Given the set of objects ☆ ×

(a) How many combinations of one object are there? List all of the combinations.

(b) How many combinations of two objects are there? List all of the combinations.

**33.** Given the set of objects ❤ → ✚

(a) How many combinations of one object are there? List all of the combinations.

(b) How many combinations of two objects are there? List all of the combinations.

(c) How many combinations of three objects are there? List all of the combinations.

**34.** Given the set of objects ☆ ×

(a) How many permutations of one object are there? List all of the permutations.

(b) How many permutations of two objects are there? List all of the permutations.

**35.** Given the set of objects ❤ → ✚

(a) How many permutations of one object are there? List all of the permutations.

(b) How many permutations of two objects are there? List all of the permutations.

(c) How many permutations of three objects are there? List all of the permutations.

**36.** Given the set of objects × ❑ ▼ ❙ ◆, how many permutations of two objects are there? List all of the permutations.

**37.** Given the set of objects × ❑ ▼ ❙ ◆, how many combinations of two objects are there? List all of the combinations.

**38.** How many three-letter code words can be made from the 26 letters of the alphabet if no letter can be used more than once in the code word?

**39.** How many different planning committees of three people can be selected from a club that has 56 members?

***40.** How many zeros are on the end of the expansion of 100!? For example, 5! = 120 has one zero on the end of the answer and 13! = 6,227,020,800 has two zeros on the end of the answer.

***41.** Arrange Pascal's triangle as shown below.

| | | | | | |
|---|---|---|---|---|---|
| 1 | | | | | |
| 1 | 1 | | | | |
| 1 | 2 | 1 | | | |
| 1 | 3 | 3 | 1 | | |
| 1 | 4 | 6 | 4 | 1 | |
| 1 | 5 | 10 | 10 | 5 | 1 |
| 1 | 6 | 15 | 20 | 15 | . . . |

Show how the Fibonacci numbers (1, 1, 2, 3, 5, 8, 13, 21, . . .) can be found by a systematic process adding ''diagonals'' of the triangle.

Odds play an intrinsic role in wagering on the outcome of an athletic event such as that depicted in *Club Night* by George Bellows. Unlike odds in card playing, in which the numbers of cards in the deck and in each hand determine the probability of subsequent draws, odds for athletic events are set by odds makers and often are based on complex systems for judging the relative quality of the competitors. (National Gallery of Art)

# Intuitive Concepts—Probability and Odds

## Probability

In everyday conversations, we frequently ask questions such as ''What's my chance of getting an A in this class?'' or ''What's the probability of my winning the drawing for the trip to Hawaii?'' or ''What is the chance that it will rain today?'' We are looking for a measure of the chance that the event will occur. We might say that there is a 90% chance that he will get an A in the class or the probability of my winning the Hawaii drawing is 1 in 20,000 or there is a 50% chance of rain today. The numbers measure the likelihood that the event will occur. For example, the probability of flipping a coin and having it land with the head side up is 1/2. Since the coin could land either heads or tails, heads is one of the two possible outcomes. In general, if outcomes are equally likely, the **probability** of an event is found by dividing the number of ways that an event can occur (number of *desired* outcomes) by the *total* number of possible outcomes. If we will let $P(E)$ represent the probability of an event $E$, the basic formula for determining the probability is

## Probability of an Event

$$P(E) = \frac{\text{number of ways an event can occur } (\textit{desired})}{\text{total number of possible outcomes } (\textit{total})} \quad \text{where } 0 \leq P(E) \leq 1$$

### Example 1

Find the probability of drawing an ace from a standard deck of 52 cards.

**Solution:** In this case, the event is drawing an ace. Because there are four aces in the standard deck, there are four ways to draw an ace out of a total of 52 possibilities.

$$P(\text{ace}) = \frac{4}{52} = \frac{1}{13}$$ ■

The probability of an event can be represented as a fraction, decimal, or percent. Each form gives a numerical way to analyze the chance that the event occurs. Thus, the probability of drawing an ace in Example 1 can be written

$$P(\text{ace}) = \frac{1}{13} \approx 0.077 = 7.7\%$$

### Example 2

A question on a multiple choice test has five answers. What is the probability that you guess the correct answer to the question?

**Solution:** There is only one correct (desired) answer out of the five possible (total) answers. Therefore,

$$P(\text{correct answer}) = \frac{1}{5} = 0.2 = 20\%$$ ■

### Example 3

Suppose you draw an ace from a standard deck of cards and do not return the ace to the deck before you draw from the deck a second time. What is the probability that you draw an ace on this second draw from the deck?

**Solution:** Since one ace was removed, there are now three aces in the deck. Similarly since one card was removed, there are 51 cards in the deck. Thus,

$$P(\text{ace on 2nd card}) = \frac{3}{51} = \frac{1}{17}$$ ■

## *What Does It Mean for $0 \leq P(E) \leq 1$?*

The inequality $0 \leq P(E) \leq 1$, tells us that the probability of an event $E$ has a value from zero to one. Furthermore, the closer the probability is to zero, the less of a chance there is that the event will occur. The closer the probability is to one, the more of a chance there is that the event will occur. If it is *impossible* for an event to occur, the probability of the event is 0. If an event is *certain* to happen, the probability of an event is 1. The next example demonstrates these concepts.

### Example 4

Two standard dice are tossed. (a) What is the probability that the sum of the two dice is 14? (b) What is the probability that the sum of the two dice is less than 13?

**Solution:**

(a) Since the maximum number of dots on a die is six, it is impossible to get a sum of 14. Therefore,

$$P(\text{sum of } 14) = 0$$

(b) Since the sum of two standard dice is always less than 13, the event is a certain event. The probability of an event that is certain is 1. Therefore,

$$P(\text{sum less than } 13) = 1$$

## *What Does Probability Really Tell Us?*

In the previous examples, we showed how to determine the probability of an event, but what does a given probability actually tell us? When the word probability is used, it can have either of two meanings—experimental probability or theoretical probability.

If we flip ten coins, we might find that we get seven heads. To say that the probability of getting a head is $7/10$ is an example of **experimental probability**. This means that we arrived at the value through an experiment rather than through mathematical calculations. Experimental probability is used to analyze events in which there is no mathematical way to determine the probability. For example, there is no mathematical way to determine the probability that the next person that walks into a restaurant will be wearing a red dress. However, if for an entire month we counted the number of people that entered the restaurant and the number that wore red dresses, we could give an approximation to the probability of that event.

**Theoretical probability** is what is predicted by mathematics. For example, we say that the probability of a coin landing heads is $1/2$, since the coin has two sides and only one of the sides is a head. When an event has a theoretical probability, let's say $1/2$, it means that in large number of trials, $1/2$ of the trials should result in the given event. The larger the number of trials, the closer the ratio of desired to total outcomes will be to $1/2$. For example, if 1000 coins were tossed with 523 landing heads, the experimental probability would be 523/1000. As we increase the number of coin tosses, the experimental probability will become close to the theoretical probability. The fact that the experimental probability is close to the theoretical probability when an experiment includes a large number of trials is known as the **Law of Large Numbers**.

## *Determining Probabilities of Single Events*

To determine the probability of an event we have two choices, an experimental approach or a theoretical approach. In an experimental approach we examine a number of trials, tally the number of occurrences of an event, and find the ratio of number of occurrences to the total number of trials examined. In the theoretical approach instead of examining a number of trials, we mathematically determine the number of ways the desired event can occur and the total number of possible outcomes. For example, if we wanted to determine the probability of obtaining a sum of seven when rolling a pair of dice, we could use either theoretical probability or experimental probability.

**Theoretical Probability** To find the theoretical probability, we could logically determine the possible outcomes for the sum of the two dice and determine the number of ways the sum could be seven. One way to do this is to systematically list all possible outcomes of two dice as shown on the next page.

There are 36 different ways to roll the dice and six of these ways have a sum of 7. Therefore, $P(\text{sum of } 7) = \frac{6}{36} = \frac{1}{6}$. This means that if we roll the dice many times, about $\frac{1}{6}$ of them would have a sum of 7. It does not mean that if we roll the dice six times, precisely one of the rolls will have a sum of seven.

**Experimental Probability** Toss a pair of dice 100 times and count the number of times the sum is 7. If we toss the dice 100 times and a sum of 7 occurred 20 times, we would say $P(\text{sum of } 7) = \frac{20}{100} = \frac{1}{5}$. We might then conclude that, on the average, one out of every five rolls would have a sum of 7.

The more we tally the results of tossing two dice, the more we would see that the experimental probability approaches the theoretical probability. Since experimental probability is a tedious matter of counting and recording results, we will focus most of our attention on determining the theoretical probability of various events.

### Example 5

There are 15 horses in a race. (a) Find the probability that you randomly pick the winner of the race. (b) Find the probability that you do not pick the winner of the race.

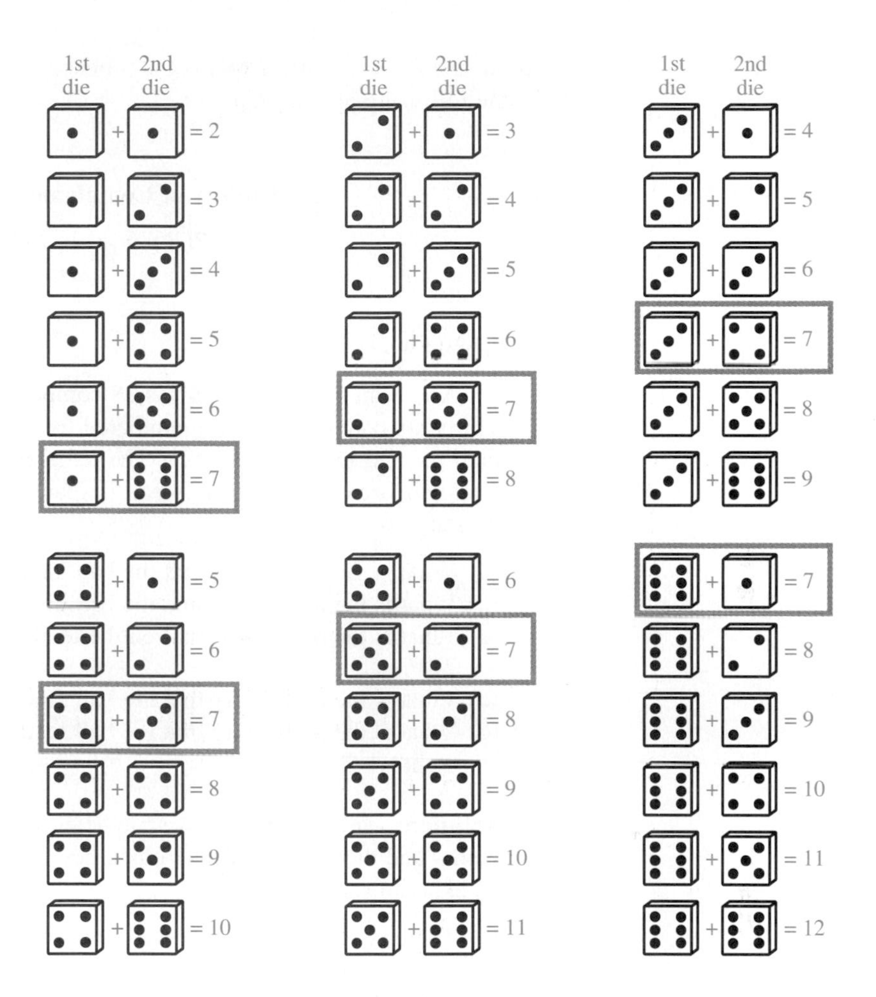


**Solution:**

(a) Let $P(W)$ be the probability that you will pick the winner. There are a total of 15 horses in the race and only one winning horse. Thus,

$$P(W) = \frac{1}{15}$$

(b) Let $\overline{W}$ be the event that you do not pick the winner. There are a total of 15 horses in the race and there are 14 losing horses. Thus,

$$P(\overline{W}) = \frac{14}{15}$$ ■

This example leads us to a very important rule of probability. If an event $E$ occurs, then its **complementary event** $\overline{E}$ is the nonoccurrence of $E$. If $P(E)$ is the probability that an event occurs, then $P(\overline{E})$ is the probability that the event does not occur, we get the following relationship.

**Probability of Complementary Events**

$$P(\overline{E}) = 1 - P(E)$$

## Example 6

An American Roulette wheel has 38 compartments around its circumference. Thirty-six compartments are numbered from 1 to 36, with half of them colored red and half colored black. The remaining two compartments, numbered 0 and 00, are colored green. A ball is spun and lands in one of the compartments.

(a) What is the probability that the ball lands on the number 27?
(b) What is the probability that the ball does not land on the number 27?
(c) What is the probability that on the tenth spin the ball lands on the number 27?
(d) What is the probability that the ball lands on an odd number?
(e) What is the probability that the ball lands on a green colored compartment?

**Solution:** In this problem, we can determine the number of desired outcomes and the number of total outcomes. Therefore, we will not need to use an experimental approach.

(a) There is one desired outcome out of the 38 total outcomes.

$$P = \frac{1}{38}$$

(b) Since this is the complementary event of Question (a),

$$P = 1 - \frac{1}{38} = \frac{37}{38}$$

(c) The probability of the ball landing on the number 27 for the tenth or any other spin is the same.

$$P = \frac{1}{38}$$

(d) There are 18 desired outcomes out of the 38 total outcomes.

$$P = \frac{18}{38} = \frac{9}{19}$$

(e) There are two desired outcomes out of the 38 total outcomes.

$$P = \frac{2}{38} = \frac{1}{19}$$

## Example 7

In the National Football League's Super Bowl XXIV, Joe Montana of the San Francisco 49ers had a completion rate of about 76%. If Montana threw 29 passes during the game, how many of them were complete?

**Solution:** In sporting events, the percentages given are arrived at by actually tallying the results. They are examples of experimental probability. Montana's 76% completion rate means that the probability that any given pass will be complete is $\frac{76}{100} = 0.76$. Since the probability that a pass is complete is 0.76 and 29 passes were thrown, the number of complete passes is $0.76 \times 29 = 22.04 \approx 22$.

*The Jockey* (*Le Jockey*) by Henri de Toulouse-Lautrec shows the age-old fascination with horse racing. (National Gallery of Art)

## Odds

In many situations, rather than probability, the **odds** for an event are given. For example, suppose that the odds of Patdancer winning the Derby are 1 to 35. This means that if the race is run 36 times, Patdancer is expected to win once and lose 35 times. Patdancer's probability of winning the race is 1/36. Alternatively, we can say the odds are 35 to 1 against Patdancer or there is a probability of 35/36 that Patdancer will not win the race. The odds for any event can be thought of as the ratio of the number of *desired* outcomes to the number of outcomes that are *not desired*, such as, wins to losses or successes to failures. Like probability, odds are also used to measure the likelihood of an event occurring. The following formulas can be used to convert from odds to probability and from probability to odds.

If the odds for event $E$ are $a : b$, then the probability is

$$P(E) = \frac{a}{a + b}$$

If the probability of event $E$ is $P(E)$, then the odds, $O(E)$ are

$$O(E) = \frac{\text{number of ways an event can occur } (\textit{desired})}{\text{number of ways an event fails to occur } (\textit{not desired})} = \frac{P(E)}{P(\overline{E})}$$

### Example 8

Suppose the San Francisco Giants have a 60% chance of winning the 1995 National League pennant. What are the odds of the Giants winning the pennant? What are the odds of the Giants not winning the pennant?

**Solution:** Since the probability is 60%, $P(E) = 0.60$ and $P(\overline{E}) = 1 - 0.60 = 0.40$.

The odds of the Giants winning the pennant are

$$O(E) = \frac{0.60}{0.40} = \frac{6}{4} = \frac{3}{2} = 3 : 2$$

The odds of the Giants not winning the pennant are 2 : 3. ■

### Example 9

The odds for winning the $49 prize in the California Lottery Gold Rush Scratcher game are listed at 1 : 83. What is the probability of winning a $49 prize?

**Solution:** Since the odds of winning are 1 : 83, use $a = 1$ and $b = 83$ in the formula that converts odds to probability.

$$P = \frac{a}{a+b} = \frac{1}{1+83} = \frac{1}{84}$$

## *House Odds*

The odds for various events as given by casinos, race tracks, or lotteries, are called **house odds**. They give the odds *against* an event occurring. For example, before the 1993 NCAA Basketball Tournament a Las Vegas Casino stated that the odds for Kansas State were 100 : 1. This really meant that odds against Kansas State winning the tournament were 100 to 1, or the probability of Kansas State losing was 100/101. House odds are given in such a manner to facilitate the betting of money that might accompany the event. Odds of 100 to 1 would mean that for every $1 you bet on Kansas State you would win $100 if Kansas State won the tournament. The odds for North Carolina in the same tournament were 5 : 2. This would mean that if North Carolina won the tournament you would win $5 for every $2 you bet. Odds are established to balance the amount won with the likelihood that the event occurs. The less probable an event is, the more money you win if the event does occur.

### Example 10

The odds for betting on a single number on an American roulette wheel are given by casinos as 35 : 1. (a) What does that mean in terms of the amount won when $1 is bet on a single number? (b) Are the odds fair? Explain.

**Solution:**

(a) House odds of 35 : 1 indicate that for every dollar bet on a single number the bettor wins $35.

(b) If $E$ is the event that the single number comes up,

$$P(E) = \frac{1}{38}, \quad P(\overline{E}) = 1 - \frac{1}{38} = \frac{37}{38}, \quad \text{and} \quad O(E) = \frac{\frac{1}{38}}{\frac{37}{38}} = \frac{1}{37}$$

Thus, the odds against the single number coming up are 37 : 1. The house odds are not fair. The casino is paying out $2 less than is mathematically correct. The discrepancy in the payoffs in this example gives the casino a margin of profit.

### Example 11

The odds makers in Atlantic City have stated that Jhun's odds are 7 to 1 in the title fight with Chavez. (a) What is the probability that Jhun wins the match? (b) If you bet \$50 on Jhun and Jhun beats Chavez, how much money will you win?

**Solution:**

(a) The house odds of 7 to 1 indicate that Jhun's odds for losing are 7 to 1. This means that the odds of Jhun winning are 1 to 7. Using $a = 1$ and $b = 7$ in the formula that converts odds to probability, we find that the probability of Jhun winning the fight is

$$P = \frac{1}{1 + 7} = \frac{1}{8}$$

(b) The house odds for Jhun of 7 : 1 mean that if Jhun wins, each \$1 bet on Jhun will generate \$7 in winnings. Thus, if you bet \$50, you will win $7 \times 50 = \$350$. ■

## Section 7.1 PROBLEMS

## *Explain* ➡ *Apply* ➡ *Explore*

### *Explain*

1. What is probability?
2. What are odds?
3. What are house odds and what do they tell you about money being bet on an event?
4. Explain what it means for $P(E) = 0$ and describe an event that has a probability of zero.
5. Explain what it means for $P(E) = 1$ and describe an event that has a probability of one.
6. Explain the difference between theoretical and experimental probability.
7. Explain if experimental or theoretical methods are more appropriate in determining the probability of each of the following events:
   (a) A serve by a tennis player will hit the net.
   (b) Three dice are thrown and the sum is 18.
   (c) A senior citizen will be riding a certain city bus on Friday.
   (d) The six numbers you picked for a lottery are all winning numbers.
   (e) Getting a jackpot when playing a slot machine in a casino.

**8.** Explain how you could determine the probability of randomly selecting a word from Shakespeare's play *Romeo and Juliet* and having it be the word "the."

**9.** Explain how you could determine the probability that a given household in your community is watching the 10:00 P.M. TV News on a Wednesday night.

**10.** Explain the Law of Large Numbers.

**11.** What are complementary events and how are the probabilities of complementary events related? Give an example of complementary events and their probabilities.

**12.** What is the meaning of a weather forecast that says there is a 40% chance of rain?

**13.** What does it mean when a political poll says there is a 63% chance that the candidate will win reelection?

## Apply

**14.** A card is cut (drawn) from a standard deck of 52 cards.

(a) What is the probability the 7 of spades is cut?
(b) What is the probability that a 7 is cut?
(c) What is the probability that a face card (king, queen, or jack) is cut?
(d) What is the probability that a heart is cut?
(e) What is the probability that a red card is cut?
(f) What is the probability that either an ace or an 8 is cut?

**15.** A card is cut (drawn) from a standard deck of 52 cards.

(a) What is the probability the 10 of diamonds is cut?
(b) What is the probability that a 10 is cut?
(c) What is the probability that a non–face card (ace, 2, 3, 4, . . . , 10) is cut?
(d) What is the probability that a spade is cut?
(e) What is the probability that a black card is cut?
(f) What is the probability that either a queen or a king is cut?

**16.** Two six-sided dice, with sides numbered 1 through 6, are rolled.

(a) What is the probability that the sum of the two dice is 8?
(b) What is the probability that the sum of the two dice is 1?
(c) What is the probability that exactly one of the two dice shows a 3?
(d) What is the probability that the sum of the two dice is 13?
(e) What is the probability that the sum of the two dice is less than 13?
(f) If the dice are rolled 9000 times, about how many times would you expect the dice to have a sum of 8?

**17.** In baseball, if a player has a batting average of 0.279 it means that for every 1000 official at bats, the player gets a hit 279 times. Ty Cobb of the Detroit

Tigers had a lifetime batting average of 0.367. How many hits would Cobb expect in a season if he had 530 official at bats?

**18.** In the course of a professional baseball season, the probability that a good lead-off hitter such as Lenny Dykstra will get on base is about 0.435. If Lenny Dykstra comes to bat 620 times during the season, how many times would you expect him to reach base?

**19.** If the probability of Anita making a 3-point shot in basketball is 0.15,

(a) What is the probability that Anita will not make a 3-point shot?
(b) What are the odds of Anita making a 3-point shot?
(c) What are the odds of Anita not making a 3-point shot?

**20.** If there is a 75% chance that Hector will get up when the alarm rings,

(a) What is the probability that Hector will not get up when the alarm rings?
(b) What are the odds that Hector will not get up when the alarm rings?
(c) What are the odds that Hector will get up when the alarm rings?

**21.** If a casino's house odds on Martina Navratilova in a tennis match are posted at 3 : 2,

(a) What is the probability that Martina wins the match?
(b) What is the probability that Martina loses the match?
(c) If you wager $10 on Martina and she wins the match, how much will you win?

**22.** If the probability of the Cleveland Browns winning the 1995 Super Bowl is $\frac{1}{25}$,

(a) What are the odds for the Browns winning the Super Bowl?
(b) What is the probability that the Browns lose the Super Bowl?
(c) What are the odds for the Browns losing the Super Bowl?
(d) What odds would a casino post for betting purposes?
(e) If you wager $10 on the Browns and they win the Super Bowl, how much will you win?

**23.** If the house odds for winning the prize in an Iowa lottery are posted as 28,560 : 1.

(a) What is the probability and what are the odds for winning the lottery?
(b) How much should you win if you buy a $2 ticket and win the lottery.

*Explore*

**24.** At some tables in Las Vegas casinos, the game of Blackjack is dealt from a five-deck-shoe, a device that contains five decks of cards. Suppose you are playing at such a table.

(a) What is the probability the 7 of spades is the first card dealt from the shoe?
(b) What is the probability that a 7 is the first card dealt from the shoe?

(c) What is the probability that a face card (king, queen, or jack) is the first card dealt from the shoe?
(d) What is the probability that a heart is the first card dealt from the shoe?
(e) What is the probability that either an ace or an 8 is the first card dealt?
(f) Suppose that the first card dealt from the shoe is an ace. What is the probability that the second card dealt from the shoe is an ace?
(g) Suppose the first and second cards dealt from the shoe are aces. What is the probability that the third card dealt from the shoe is an ace?

**25.** The game *Dungeons and Dragons* uses two regular dodecahedral "dice." Such dice are solids with 12 congruent pentagon-shaped faces. A pair of these dodecahedral dice with the numbers 1 to 12 on its faces are tossed. A possible result when such dice are tossed is shown.

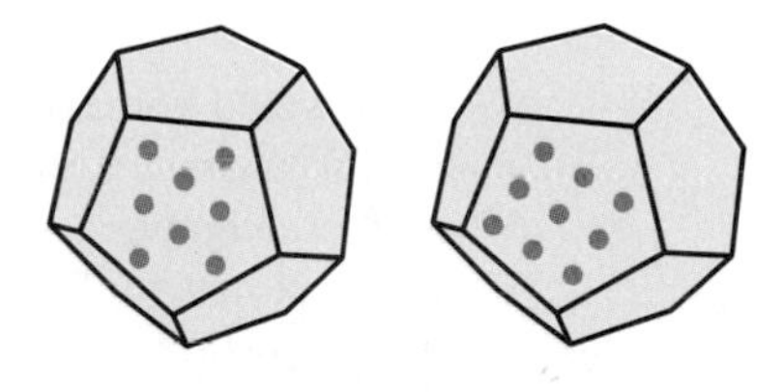

(a) List all the possible outcomes for tossing the two dice.
(b) What is the probability that the sum of the two dice is 24?
(c) What is the probability that the sum of the two dice is greater than 24?
(d) What is the probability that the sum of the two dice is 17?
(e) What is the probability that the sum of the two dice is greater than 17?
(f) What is the probability that the sum of the two dice is greater than 1?

**26.** A jar of pennies has 50 pennies dated 1993 and 150 pennies with other dates. You reach into the jar and pick a penny.

(a) What is the probability that you picked a 1993 penny?
(b) Suppose you picked a 1993 penny and did not put it back into the jar. What is the probability of now picking a 1993 penny from the jar?
(c) Suppose you picked ten 1993 pennies in a row and did not put them back into the jar. What is the probability of now picking a 1993 penny from the jar?

**27.** The results of the nine horse races held at Golden Gate Fields near Berkeley, California, on Friday, May 28, 1993, were reported in San Francisco Bay Area newspapers the day after the races were held. The winning horses for each of the nine races and the exact odds used to pay off those that bet on the winning horses are shown in the chart that follows.

| Race/Number | | Distance | Horse | Odds |
|---|---|---|---|---|
| 1 | g770 | $1\frac{1}{16}$ miles | Fair and Fabulous | 2.90 : 1 |
| 2 | g771 | 5 furlongs | Ramanujan | 6.40 : 1 |
| 3 | g772 | 1 mile | Chip Ahead | 3.20 : 1 |
| 4 | g773 | 6 furlongs | Somekindadeal | 2.20 : 1 |
| 5 | g774 | 6 furlongs | Bold Riparian | 2.20 : 1 |
| 6 | g775 | $1\frac{1}{16}$ miles | Miz Interco | 11.70 : 1 |
| 7 | g776 | 6 furlongs | Cassie June | 2.60 : 1 |
| 8 | g777 | $1\frac{1}{8}$ miles | Adorable Vice | 3.40 : 1 |
| 9 | g778 | 6 furlongs | Chosen Journey | 0.90 : 1 |

(a) If you bet $10 on Chip Ahead to win in the Third Race, how much did you win?

(b) How would you explain that Miz Interco in the Sixth Race had such a high payoff whereas Chosen Journey in the Ninth Race had such a low payoff?

(c) Suppose you were very lucky and placed a $2 bet on each of the horses that won that day at Golden Gate Fields, how much would you have won from those bets?

**28.** A Pinochle (pē′nuk′l) deck of cards consists of 48 cards—eight aces, eight kings, eight queens, eight jacks, eight 10s and eight 9s. Suppose you are cutting cards from a pinochle deck.

(a) What is the probability that you cut a king on the first card?

(b) Suppose you cut a king on the first card and do not return the king to the deck before you cut a second card. What is the probability of cutting a king?

(c) Suppose you cut another king on the second card and again you do not return the king to the deck before you cut the deck a third time. What is the probability that you now cut a king?

**29.** A BINGO card contains 25 small squares as shown. Five numbers from 1 to 15 are placed under the letter B, five numbers from 16 to 30 are placed under the letter I, four numbers from 31 to 45 along with the ''free space'' are placed under the letter N, five numbers from 46 to 60 are placed under the letter G, and five numbers from 61 to 75 are placed under the letter O.

When a regular BINGO game is played the numbers from 1 to 75 are randomly called until a player has a card with five numbers in a horizontal, vertical, or diagonal line. Suppose you are playing BINGO with the card on the next page.

(a) What is the probability that the first number called is on your card?

(b) What is the probability that G 59 is the first number called?

(c) What is the probability that the first number called is in your N row?

(d) Suppose the first number called is not on your card. What is the probability that the second number called is on your card?

| B | I | N | G | O |
|---|---|---|---|---|
| 3 | 17 | 40 | 52 | 73 |
| 8 | 21 | 42 | 56 | 63 |
| 1 | 20 | FREE | 47 | 69 |
| 13 | 19 | 45 | 59 | 71 |
| 10 | 26 | 34 | 46 | 68 |

(e) Suppose the first and second numbers called are not on your card. What is the probability that the third number called is on your card?

(f) After 20 numbers (seven from the B row) are called, you notice that you have B 3, B 13, B 8, and B 10 covered. You are one number away from getting a BINGO (five numbers in a vertical row). What is the probability that the next number called will give you a BINGO?

***30.** This problem repeats the experiment performed by French naturalist, Georges Buffon, in 1777.

Take a piece of paper with equally spaced lines in one direction and measure the distance between the lines. Call this distance $d$. Obtain a pin or a short nail and measure its length. Call this length $L$. (The length of the pin should be less than the distance between two lines. Toss the pin on the paper 100 times and count how many times the pin crosses a line. Calculate the probability $p$ that the pin crosses a line. Finally, perform the calculation $\frac{2L}{pd}$ and express it as a decimal. What value does this seem close to?

In a card game such as this one, *The Card Players* by Paul Cezanne, we can determine the likelihood of getting a particular card on the next draw by using theoretical probability. (The Metropolitan Museum of Art)

Every day people use the process of counting in many activities. Most of the time, the counting is fairly simple: "How many people are coming to dinner?" or "How many shopping days are left until Christmas?" When a more difficult problem arises, simple methods either fail or are overpowered by the problem. For example, how many different ways can the spark plug wires be connected to the distributor cap of your car? Although only one arrangement is correct, there are 720 ways for a six-cylinder car. This section will introduce you to some basic principles that can make counting an easier task. The ability to count the number of ways an event can occur will allow you to investigate a wider variety of real life applications of probability.

## Counting by Listing Possible Outcomes

The most straightforward method for determining the number of ways an event can occur is to simply list all the possible outcomes and count them. For example, if four coins are tossed, in how many different ways could they land with two heads and two tails facing up? If we list all the possible ways this could happen, we would see that there are six possibilities.

| 1st Coin | 2nd Coin | 3rd Coin | 4th Coin |
|---|---|---|---|
| H | H | T | T |
| H | T | T | H |
| H | T | H | T |
| T | H | H | T |
| T | H | T | H |
| T | T | H | H |

The listing of all desired outcomes is a legitimate way to count. However, it becomes very tedious if there are many possibilities. If you were to list all the possible five-card hands you could deal from a standard deck of 52 cards, you would have to list 2,598,960 different hands. To avoid this kind of task, we will investigate other methods of counting the number of ways an event can occur. However, the listing method is always a possibility in a given problem.

### *With or Without Replacement?*

Before examining other methods of counting, there is a concern that could affect the answer to a counting problem. Is the counting done with items being replaced or without being replaced? Counting is done **with replacement** if the same object can be used more than once. For example, suppose a person is buying a candy bar from a well-stocked candy machine. If the machine is working correctly and is well stocked, choosing a particular brand of candy bar does not prohibit the next person using the machine from choosing the same brand of candy. The candy bar chosen by the first person was replaced by another bar of the same brand. If the machine had only one bar of each brand, the candy would be chosen **without replacement**.

A standard deck of cards can be used to show the difference between an event done with replacement and without replacement. Suppose a player is to draw two cards from a standard deck of cards and the queen of hearts is drawn on the first card. If the second card is drawn without replacement (that is, the first card is not returned to the deck), it is not possible to draw another queen of hearts. On the other hand, suppose the second card is drawn after the first card has been replaced into the deck. It is now possible to draw the queen of hearts on the second card. When counting the number of ways an event can occur, it makes a difference if the event happens with or without replacement.

## Counting Using the Basic Counting Law

Besides listing outcomes, the number of ways an event can occur can be determined by a basic counting law. According to the **Basic Counting Law**, if there are ***n*** choices for the first item and $m$ choices for the second item, there are $n \times m$ ways in which to pick a set consisting of two items.

### Example 1

At the Produce Market Restaurant, the light lunch special consists of a choice of one of the four salads and one of the six types of fruit. How many different lunches are available?

**Solution:** Since there are four choices of salad and six choices of fruit, the Basic Counting Law gives a result of $4 \times 6 = 24$ different light lunch specials. ■

### Example 2

A group of three people decide to have the light lunch special at the Produce Market Restaurant. How many different ways can the group of three order lunches?

**Solution:** Assuming that there are enough supplies at the restaurant to accommodate many orders of the same item, this is a situation involving replacement. This means that all the people can order the same item. Using the results of Example 1, each of the people has 24 choices. Thus, by the Basic Counting Law, there are $24 \times 24 \times 24 = 13{,}824$ different orders. ■

### Example 3

For those whose appetites are not satisfied by the light lunch special, the restaurant has a dessert cart. There are ten different desserts on the cart, but only one of each type. In how many ways can our group of three each order a dessert?

**Solution:** In this case, it is not possible for the same dessert to be chosen by more than one person. The first person will have ten choices for dessert, but the second person will have only nine choices and the third person only eight choices. Thus, by the Basic Counting Law, there are $10 \times 9 \times 8 = 720$ different possibilities. ■

In Examples 1–3, the Basic Counting Law was used. In Examples 1 and 2, the selections were made with replacement, whereas in Example 3, the selections were made without replacement.

## Counting Using Permutations and Combinations

In some counting problems, you may be selecting a group of items from a larger set of items, like dealing five cards from a deck of 52 cards or picking two horses from a field of 12 horses or selecting three desserts from a tray containing ten desserts. Counting in situations such as these can be accomplished using the permutations or combinations discussed in Section 7.0. The number of permutations determined by $P_{n,r}$ indicates how many arrangements of $r$ items can be made from a set of $n$ items. The number of combinations determined by $C_{n,r}$ indicates how many groupings of $r$ items can be made from a set of $n$ items without regard for the order of the items. If we are selecting items without replacement, permutations are used if the order of the items matters and combinations are used if the order does not matter. With this in mind, let us take another look at the problem posed in Example 3. It can be reworded into the statement formulated in the next example.

### Example 4

How many ways can three people select a dessert from a dessert tray containing ten different desserts?

**Solution:** Since they are selecting desserts without replacement and the order of the selections gives each person a different dessert, we can use permutations to determine the number of ways this can be done. Using $n = 10$ and $r = 3$, the number of permutations of ten items, choosing three at a time, is

$$P_{10,3} = \frac{10!}{(10-3)!} = \frac{10!}{7!} = \frac{10 \times 9 \times 8 \times 7 \times \cdots \times 1}{7 \times 6 \times \cdots \times 1}$$

$$= 10 \times 9 \times 8 = 720$$

### Example 5

The lock on a safe is a combination lock with 50 numbers on the dial. Four numbers are needed to unlock the safe.

(a) How many four-number sequences are possible if a number can be used only once?
(b) How many four-number sequences are possible if a number can be used more than once?

**Solution:**

(a) If the numbers can only be used once, the selection of numbers is done without replacement. Since the order of the numbers is essential for opening the lock, permutations should be used. Thus, the number of four number sequences is

$$P_{50,4} = \frac{50!}{(50 - 4)!} = \frac{50!}{46!} = \frac{50 \times 49 \times 48 \times 47 \times 46 \times \cdots \times 1}{46 \times 45 \times \cdots \times 1}$$

$$= 50 \times 49 \times 48 \times 47 = 5{,}527{,}200$$

(b) If the numbers can be used more than once, the numbers are chosen with replacement. There are 50 choices for each of the four numbers in the combination. Therefore, by the Basic Counting Law, the number of possible four number sequences is

$$50 \times 50 \times 50 \times 50 = 6{,}250{,}000$$

## Example 6

A researcher has to select three out of a group of five rabbits to test a new dietary supplement. How many different sets of three rabbits can be selected?

**Solution:** Since she needs to select three rabbits from a group of five, the selection is done without replacement. Since the order in which they are selected does not matter, she can count the number of ways this can be done using combinations with $n = 5$ and $r = 3$.

$$C_{5,3} = \frac{5!}{3!(5 - 3)!} = \frac{5!}{3!2!} = \frac{5 \times 4 \times 3 \times 2 \times 1}{3 \times 2 \times 1 \times 2 \times 1} = \frac{5 \times 4}{2 \times 1} = 10$$

## Example 7

In the game of lotto, each player picks some numbers in the hope of matching the winning numbers. For example, in a state lotto game, seven different numbers between 1 and 49 are chosen. How many different sets of seven numbers are possible?

**Solution:** Because the problem states that seven different numbers are drawn, we cannot use the same number twice. Therefore, the sampling is done without replacement. Since we are concerned only about a set of numbers and not the order in which they are drawn, we should use combinations. Thus, the number of combinations of 49 numbers, taken seven at a time is

$$C_{49,7} = \frac{49!}{7!(49 - 7)!} = \frac{49!}{7!42!} = 85{,}900{,}584$$

## Example 8

A jar contains 75 balls, numbered 1 through 75. If balls are selected without replacement and the order of selection does not matter, are there more ways to select groups of two balls from the jar or ways to select groups of 73 balls?

**Solution:** Since we are selecting a group of balls from the 75 balls without replacement and the order in which the balls are selected does not matter, we compute the number of combinations in both cases.

$$C_{75,2} = \frac{75!}{2!(75-2)!} = \frac{75!}{2!73!} = 2775$$

$$C_{75,73} = \frac{75!}{73!(75-73)!} = \frac{75!}{73!2!} = 2775$$

The results are the same! After arriving at this answer, it is not hard to see why it is true. Imagine being told to select 73 items from a jar containing 75 items. Not wanting to go through the tedious chore of selecting 73 items, it would be simpler to select and discard two items while keeping the remaining 73 items. This means that there are as many groups of two as groups of 73 that can be chosen from the set of 75 items. ■

The preceding examples have introduced you to the following four methods used in counting the number of ways an event can occur.

1. Counting by listing all possibilities.
2. Counting with replacement using the Basic Counting Law.
3. Counting without replacement where the order of the objects matters. This can be done using permutations.
4. Counting without replacement where the order of the objects does not matter. This can be done using combinations.

If you are selecting a group of items and listing all possible outcomes is not feasible, the following chart can help you decide how to approach a given counting problem.

**Summary of Counting Principles**

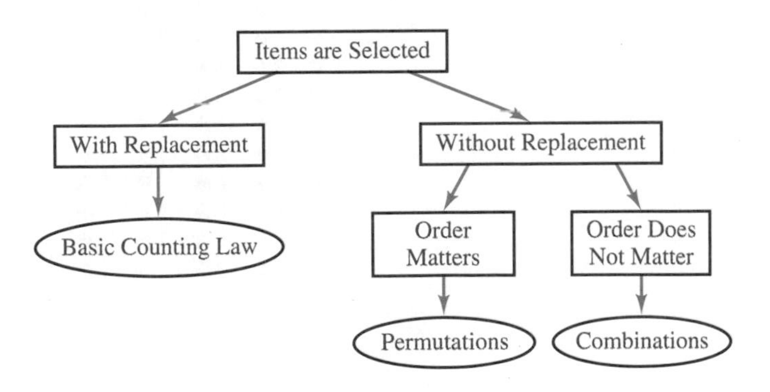


With these basic counting principles, we can examine applications of a more complex nature. We must state here that although these basic techniques will allow you to examine a large variety of applications, there are many counting problems that are beyond the scope of this text.

### Example 9

A standard deck of cards has 52 cards consisting of four suits of 13 cards each. In five-card poker, a hand with three cards of one rank and two cards of another rank is called a full house. How many ways are there to deal a five-card hand consisting of three 8's and two 7's?

**Solution:** Since the order of selection does not matter and the cards are dealt without replacement, we will use combinations. Since we want to get three of the four 8's in the deck, the number of ways to get three 8's is given by:

$$C_{4,3} = \frac{4!}{3!(4-3)!} = 4$$

Similarly, the number of ways to get two of the four 7's in the deck is given by

$$C_{4,2} = \frac{4!}{2!(4-2)!} = 6$$

Now, by the Basic Counting Law, the number of ways to get three 8's and two 7's is given by

$$C_{4,3} \times C_{4,2} = 4 \times 6 = 24$$

An actual listing of the 24 possible ways to get a full house of three 8's and two 7's is shown below. ■

| | | | | | | | | | |
|---|---|---|---|---|---|---|---|---|---|
| 8♥ | 8♦ | 8♣ | 7♥ | 7♦ | 8♥ | 8♣ | 8♠ | 7♥ | 7♦ |
| 8♥ | 8♦ | 8♣ | 7♥ | 7♣ | 8♥ | 8♣ | 8♠ | 7♥ | 7♣ |
| 8♥ | 8♦ | 8♣ | 7♥ | 7♠ | 8♥ | 8♣ | 8♠ | 7♥ | 7♠ |
| 8♥ | 8♦ | 8♣ | 7♦ | 7♣ | 8♥ | 8♣ | 8♠ | 7♦ | 7♣ |
| 8♥ | 8♦ | 8♣ | 7♦ | 7♠ | 8♥ | 8♣ | 8♠ | 7♦ | 7♠ |
| 8♥ | 8♦ | 8♣ | 7♣ | 7♠ | 8♥ | 8♣ | 8♠ | 7♣ | 7♠ |
| 8♥ | 8♦ | 8♠ | 7♥ | 7♦ | 8♦ | 8♣ | 8♠ | 7♥ | 7♦ |
| 8♥ | 8♦ | 8♠ | 7♥ | 7♣ | 8♦ | 8♣ | 8♠ | 7♥ | 7♣ |
| 8♥ | 8♦ | 8♠ | 7♥ | 7♠ | 8♦ | 8♣ | 8♠ | 7♥ | 7♠ |
| 8♥ | 8♦ | 8♠ | 7♦ | 7♣ | 8♦ | 8♣ | 8♠ | 7♦ | 7♣ |
| 8♥ | 8♦ | 8♠ | 7♦ | 7♠ | 8♦ | 8♣ | 8♠ | 7♦ | 7♠ |
| 8♥ | 8♦ | 8♠ | 7♣ | 7♠ | 8♦ | 8♣ | 8♠ | 7♣ | 7♠ |

### Example 10

How many ways are there to deal a full house of any type?

**Solution:** This problem is similar to Example 9 except that there are fewer conditions. Instead of having three 8's, we now want three cards of any rank.

Since there are 13 different ranks, there are 13 ways to choose the set of three of a kind. Similarly, there will be only 12 ranks from which to get the pair. Therefore, the number of ways to deal a full house is

$$13 \times C_{4,3} \times 12 \times C_{4,2} = 13 \times 4 \times 12 \times 6 = 3744$$ ■

## Example 11

(a) How many different five-card hands contain four aces?
(b) How many different five-card hands contain four cards of the same rank?

**Solution:**

(a) To get four aces it seems that it is only necessary to compute $C_{4,4}$, but it is not that simple. Because we have a five-card hand, we must also include the number of ways to select the fifth card. Since there are 48 cards in the deck that are not aces, the number of ways to select the fifth card is given by $C_{48,1}$. Thus, the Basic Counting Law gives

$$C_{4,4} \times C_{48,1} = 1 \times 48 = 48$$

(b) This is the same problem as part (a) except that we now have 13 choices for the rank (2's through aces). Therefore, the number of possible hands with four cards of the same rank is

$$13 \times C_{4,4} \times C_{48,1} = 13 \times 1 \times 48 = 624$$ ■

## Example 12

In the game Blackjack (also called Twenty-One or Vingt-et-Un), a player can win the hand by receiving a blackjack: an ace and a ten-point card in the first two cards. A ten-point card is a king, queen, jack, or 10. Find the number of ways a player can get a blackjack when two cards are dealt from a standard deck of 52 cards.

**Solution:** Since we want to get one of the four aces and one of the 16 ten-point cards, the number of ways to get a blackjack is

$$C_{4,1} \times C_{16,1} = 4 \times 16 = 64$$ ■

## Example 13

In casinos Blackjack is played with several decks of cards. Find the number of ways a player can draw a blackjack from three standard decks of 52 cards.

**Solution:** Since there are three decks of cards, each containing four aces, there is a total of 12 aces. This means that the number of ways to get one ace out of the 12 possible aces can be determined by $C_{12,1}$. Similarly, since there is a total of 48 ten-point cards in the three decks, the number of ways to get a ten-point card is given by $C_{48,1}$. Thus, the number of ways to get a blackjack is

$$C_{12,1} \times C_{48,1} = 12 \times 48 = 576$$

It seems odd that a casino would want to increase the number of winning hands. In the next section, we will investigate why a casino would do this.

### Example 14

KENO

| 1 | 2 | 3 | 4 | 5 | 6 | 7 | 8 | 9 | 10 |
|---|---|---|---|---|---|---|---|---|---|
| 11 | 12 | 13 | 14 | 15 | 16 | 17 | 18 | 19 | 20 |
| 21 | 22 | 23 | 24 | 25 | 26 | 27 | 28 | 29 | 30 |
| 31 | 32 | 33 | 34 | 35 | 36 | 37 | 38 | 39 | 40 |

| 41 | 42 | 43 | 44 | 45 | 46 | 47 | 48 | 49 | 50 |
|---|---|---|---|---|---|---|---|---|---|
| 51 | 52 | 53 | 54 | 55 | 56 | 57 | 58 | 59 | 60 |
| 61 | 62 | 63 | 64 | 65 | 66 | 67 | 68 | 69 | 70 |
| 71 | 72 | 73 | 74 | 75 | 76 | 77 | 78 | 79 | 80 |

In the game of Keno, 80 numbers are displayed on a board. Twenty of these numbers are randomly drawn and are considered the winning numbers. Suppose a person has selected nine numbers on a playing card. In how many ways can the person get exactly six winning numbers?

**Solution:** This problem is similar to the card problems. We are playing a game with 80 possibilities, 20 of which are considered the winning values. Since the order that the numbers are selected does not matter and we want to select six winning numbers from these 20, there are $C_{20,6}$ ways to pick the winning numbers. Since we are picking a total of nine numbers, we must also account for the three losing numbers. Since there are $80 - 20 = 60$ losing values, the three losing numbers can be chosen in $C_{60,3}$ ways.

Therefore, when playing a Keno card with nine numbers, the number of ways to pick exactly six of the 20 winning numbers is

$$C_{20,6} \times C_{60,3} = 38{,}760 \times 34{,}220 = 1{,}326{,}367{,}200$$

Section 7.2
PROBLEMS

## Explain → Apply → Explore

### Explain

1. What does counting by listing involve?
2. Why would you use other methods besides the listing method in determining the number of ways an event can occur?
3. What is meant by counting with replacement?
4. What is meant by counting without replacement?
5. What does the Basic Counting Law tell us about counting?
6. Under what conditions would combinations be used to count the number of ways an event can happen?
7. Under what conditions would permutations be used to count the number of ways an event can happen?
8. Under what conditions would the listing method be used to count the number of ways an event can happen?

### Apply

In Problems 9–21, determine if the number of ways for the event can be found using the Basic Counting Law, permutations, or combinations. Explain the reason for your choice. (There may be more than one correct answer.)

9. The number of ways eight runners in a 100-meter dash can finish 1st, 2nd, 3rd.
10. The number of ways three cards can be dealt from a standard deck of cards.
11. The number of ways to answer six true/false questions on a test.
12. The number of ways to pick five baseballs from a box that contains 50 baseballs.
13. The number of ways a group of five people can be selected from a class of 30 people.
14. The number of serial numbers that consist of a letter of the alphabet followed by five digits.
15. The number of ways you can dress if you have five pairs of shorts, six T-shirts, and two pairs of shoes to choose from.
16. The number of ways you can press three different keys on a basic calculator that has keys for the 10 digits and the $+$, $-$, $\times$, $\div$, $=$, and . keys.

**17.** The number of different four-letter code words that can be made from consonants if no letter can be used more than once.

**18.** The number of different code words that can be made from consonants if a letter can be used more than once.

**19.** The number of ways to plug in a black cord and a grey cord into an electrical strip that has six outlets in a row.

**20.** The number of ways to answer eight multiple choice questions on a test if each question has five choices.

**21.** The number of ways you can choose 4 of your 15 shirts to take on vacation.

### *Explore*

**22.** A slot machine consists of three wheels with 12 different objects on a wheel. How many different outcomes are possible?

**23.** A computer chip consists of four different switches. Each switch can be in either the off or the on position. What is the total number of arrangements of the chip's switches? Make a list of all the possible outcomes.

**24.** A VISA® credit card has a first digit of 4 followed by 15 other digits. How many different VISA accounts does this allow?

**25.** Social Security numbers have nine digits. How many different people can have distinct social security numbers?

**26.** How many 7-digit phone numbers are possible if the phone number does not begin with a 1 or 0?

**27.** A young couple has decided they will have three children. If all three births are single births, how many different ways could the couple have exactly two boys? Make a list of all cases.

**28.** An eight-cylinder car has eight wires running from the spark plugs to the distributor cap. A prankster has removed all the wires from the distributor. In how many possible ways can the wires be reattached?

**29.** Seven candidates are running for three positions on the local board of supervisors. The candidate with the most votes will be the board president while the second and third place finishers will have correspondingly lesser positions. How many different outcomes can the election have?

**30.** A pitcher knows how to throw five different pitches. He threw four pitches before the last batter was called out.

(a) Assuming he threw each pitch at most once, determine the number of possible arrangements of his pitches.
(b) Assuming he can use any pitch an unlimited number of times, determine the number of possible arrangements of his pitches.

**31.** Determine the number of ways a ten-question true-false test can be answered.

**32.** Determine the number of ways a ten-question multiple choice test can be answered if there are five possible answers to each question.

**33.** Determine the total number of five-card hands that can be dealt from a deck of 52 cards.

**34.** Determine the total number of 13-card bridge hands that can be dealt from a deck of 52 cards.

**35.** In poker, a straight is five cards in consecutive numerical order. The suit of the card does not matter.

(a) Find the number of ways to get a straight beginning with a 5 and ending with a 9.

(b) Find the number of ways to get any straight. (*Note:* An ace can be part of either A, 2, 3, 4, 5 or part of 10, J, Q, K, A but cannot be used in K, A, 2, 3, 4.)

**36.** In poker, a straight flush is five cards of the same suit in consecutive numerical order. If five cards are dealt,

(a) Find the number of ways to get a straight flush beginning with the 5 of hearts and ending with the 9 of hearts.

(b) Find the number of ways to get a straight flush beginning with a 5 and ending with a 9.

(c) Find the number of ways to get any straight flush. (*Note:* An ace can be part of either A, 2, 3, 4, 5 or part of 10, J, Q, K, A but cannot be used in K, A, 2, 3, 4.)

**37.** In bridge, a yarborough is a 13-card hand containing only cards numbered 2 through 9. Find the number of ways in which to get a yarborough.

**38.** A basketball league intends to add two more teams to the league. If 18 cities have applied for franchises, in how many ways can the league add two more cities?

**39.** In a dog show, a German Shepherd is supposed to pick the correct two objects from a set of 20 objects. In how many ways can the dog pick the two objects?

**40.** There are eight people on a committee. How many different subcommittees of two people can be formed?

**41.** An auditorium has scheduled three basketball games, two concerts, and four poetry meetings. You have a ticket allowing you to attend any three of the events. How many ways can you go to two of the poetry readings and one of the other events?

**42.** In a seven-card poker hand, find the number of hands containing four aces.

**43.** Find the total number of license plates that can be printed by the State of California using the format of a digit, followed by three letters, followed by three digits. Assume all arrangements of letters can be used.

**44.** Twenty people are at a party. If everyone at the party shakes the hand of everyone else at the party, determine the total number of handshakes.

**45.** Ten married couples are at a party. If each person at the party shakes the hand of everyone else except his/her spouse, determine the total number of handshakes at the party.

**46.** In the game of Keno, 80 numbers are displayed on a board. Twenty of these numbers are chosen to be the winning numbers. Suppose a person has selected seven numbers on a playing card. In how many ways can the person select exactly five of the winning numbers?

**47.** In the game of Keno, 80 numbers are displayed on a board. Twenty of these numbers are chosen to be the winning numbers. Suppose a person has selected six numbers on a playing card. In how many ways can the person select six of the winning numbers?

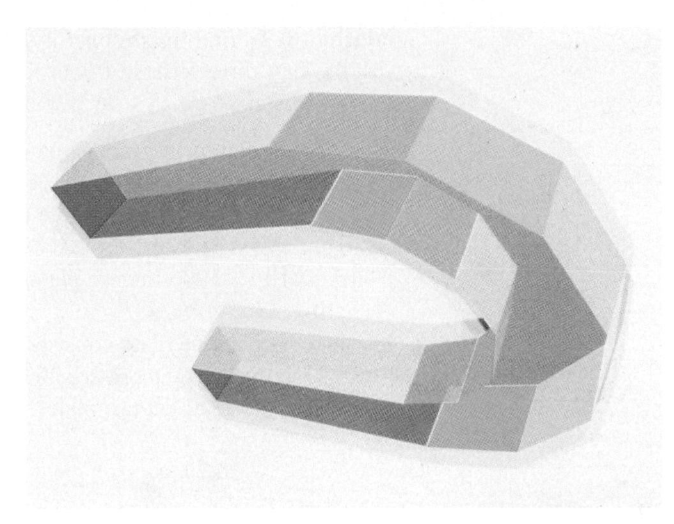

*Horseshoe* by Ron Davis shows the symbol of good luck valued by many a gambler. (Courtesy of the artist and The John Berrgruen Gallery)

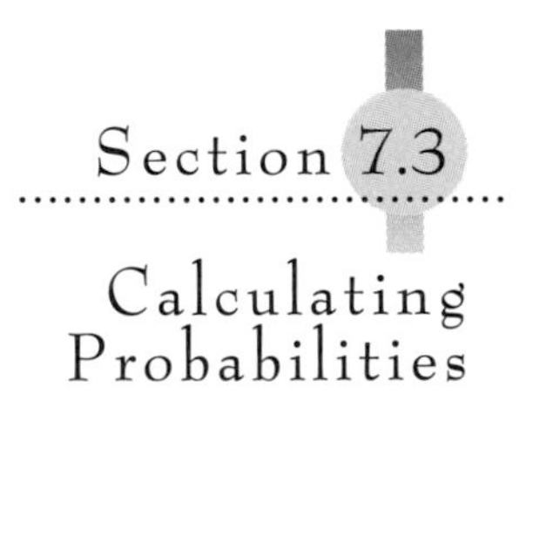

## Section 7.3 Calculating Probabilities

In Section 7.1 we introduced the basic terminology and formulas used in the study of probability. We are now ready to combine those concepts with the counting principles studied in Section 7.2 to further examine probabilities involved in real world situations such as games of chance. We will address the problems using the techniques of theoretical probability. Some problems will involve using permutations and combinations whereas others will require actually listing the outcomes of the event or using the basic counting law.

### Example 1

If an automobile's license plate consists of any three letters of the alphabet and three digits as shown below, what is the probability that the license plate assigned to your car has the first three letters of your name on it? What are the odds of this happening?

**Solution:** To find the probability, determine (a) the number of license plates with the first three letters of your name on it and (b) the total number of license plates consisting of three letters and three digits.

(a) Since the letters of your name must be in the first three positions of the license plate, each letter can occur in one way. Since any of the ten digits can be used in the three digits of the license plate, each digit could occur in ten ways. Thus, by the basic counting law there are $1 \times 1 \times 1 \times 10 \times 10 \times 10 = 1000$ license plates with the first three letters of your name on it.

(b) Since the license plate consists of three letters and three digits, by the Basic Counting Law there are $26 \times 26 \times 26 \times 10 \times 10 \times 10 =$ 17,576,000 total license plates. Therefore,

$$P(\text{3 letters of your name}) = \frac{1000}{17{,}576{,}000} = \frac{1}{17{,}576}$$

If $E$ is the event of having the first three letters of your name on your license plate, then the odds can be determined from the probability of the event as shown below.

$$O(E) = \frac{P(E)}{P(\overline{E})} = \frac{\dfrac{1}{17{,}576}}{1 - \dfrac{1}{17{,}576}} = \frac{\dfrac{1}{17{,}576}}{\dfrac{17{,}575}{17{,}576}}$$

$$= \frac{1}{17{,}575} = 1 : 17{,}575$$

## Example 2

If three standard dice are tossed, what is the probability that the sum of the numbers on the three dice is 15?

**Solution:** To determine the probability of the sum being 15, we must determine (a) the number of ways that we can get a sum of 15 and (b) the number of total outcomes for tossing three dice.

(a) By listing the ways the sum of three dice can be 15, we can count the number of ways that event can occur.

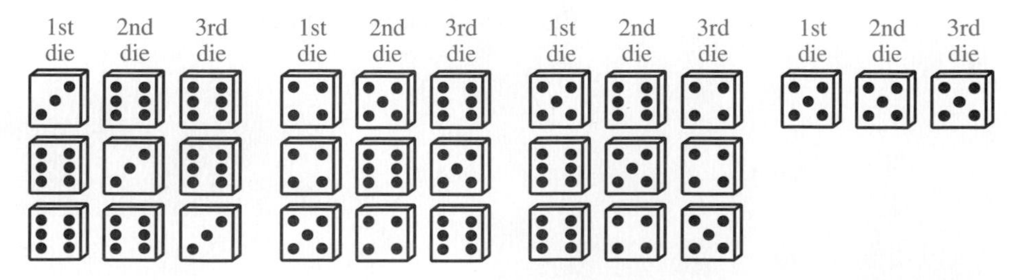

There are 10 different ways for three dice to have a sum of 15.

(b) Since the first die can land in six ways, and the second die can land in six ways, and the third die can land in six ways, by the basic counting law the total number of outcomes for the three dice is $6 \times 6 \times 6 = 216$. Therefore,

$$P(\text{sum of } 15) = \frac{10}{216} = \frac{5}{108} \approx \frac{1}{21.6} \approx 0.0463$$

(*Note:* Answers to a probability problem are often given as a fraction with one as its numerator. This simply makes the answer easier to read and understand.)

## Example 3

You discover that the six spark plug wires were removed from the distributor cap of your car. What is the probability of randomly reconnecting the wires in the correct order?

**Solution:** We must conncct all the wires in the correct order and each wire can be used only once. Since the order in which we connect the wires matters, we use permutations to find the number of total ways this can be done. Using $n = 6$ and $r = 6$, the number of different arrangements of the wires is given by $P_{6,6} = 6! = 720$. Since only one of these arrangements is correct, the probability of randomly selecting the correct order is $1/720$.

## Example 4

Find the probability of having four queens when dealing five cards from a standard deck of 52 cards.

**Solution:** There are four queens and 48 other cards in the deck. Since we want to get all four of the queens and one of the other 48 cards, the number of ways to draw the hand is given by

$$C_{4,4} \times C_{48,1} = 1 \times 48 = 48$$

The number of ways to draw five cards from a 52-card deck is given by

$$\begin{aligned}
C_{52,5} &= \frac{52!}{5!(52-5)!} = \frac{52!}{5!47!} \\
&= \frac{52 \times 51 \times 50 \times 49 \times 48 \times 47 \times \cdots \times 1}{5 \times 4 \times 3 \times 2 \times 1 \times 47 \times 46 \times 45 \times \cdots \times 1} \\
&= \frac{52 \times 51 \times 50 \times 49 \times 48}{5 \times 4 \times 3 \times 2 \times 1} \\
&= 52 \times 51 \times 10 \times 49 \times 2 = 2{,}598{,}960
\end{aligned}$$

Thus, the probability of drawing four queens in five cards from a standard 52-card deck is

$$P(4 \text{ queens}) = \frac{48}{2{,}598{,}960} = \frac{1}{54{,}145} \approx 0.00001847$$

This means, on the average, four queens will appear in a five-card poker hand once every 54,145 hands. ■

In the examples that follow, we will not show all the computations. Instead, we will give the answers in the following form:

$$\frac{C_{4,4} \times C_{48,1}}{C_{52,5}} = \frac{1 \times 48}{2{,}598{,}960} = \frac{1}{54{,}145}$$

In this way, the solutions and explanations can be presented without the page being filled with calculations. It is hoped that you have a calculator with combination and permutation keys to help simplify your computations.

## Example 5

(a) Find the probability of getting three queens and two aces (a full house of queens over aces) from a standard deck of 52 cards when seven cards are dealt.

(b) Find the probability of getting three cards of one rank and two cards of another rank (any full house) from a standard deck of 52 cards when seven cards are dealt.

**Solution:**

(a) Since the order in which the cards are received does not affect the results, we can use combinations to determine the number of ways the events can occur. The number of ways to get three queens out of the four queens in the deck is $C_{4,3}$. Similarly, there are $C_{4,2}$ ways to get two of the four aces. Since we are receiving a total of seven cards, we need to draw two more cards from the 44 cards remaining in the deck that are not queens or aces. There are $C_{44,2}$ ways to do this. We can now compute the desired probability.

$$
\begin{aligned}
P(\text{full house of queens over aces}) &= \frac{C_{4,3} \times C_{4,2} \times C_{44,2}}{C_{52,7}} \\
&= \frac{4 \times 6 \times 946}{133{,}784{,}560} \\
&= \frac{22{,}704}{133{,}784{,}560} \approx \frac{1}{5893} \approx 0.00017
\end{aligned}
$$

(b) Since there are 13 choices for the set of three cards and 12 choices for the set of two cards, the probability of getting a full house when being dealt seven cards is

$$
\begin{aligned}
P(\text{any full house}) &= \frac{(13 \times C_{4,3}) \times (12 \times C_{4,2}) \times C_{44,2}}{C_{52,7}} \\
&= \frac{(13 \times 4) \times (12 \times 6) \times 946}{133{,}784{,}560} \\
&= \frac{3{,}541{,}824}{133{,}784{,}560} \approx \frac{1}{37.8} \approx 0.0265
\end{aligned}
$$

■

In the game Twenty-One or Blackjack, casinos like to use several decks of cards at once. In the next example, we will investigate whether using more than one deck changes the probability of getting a Blackjack (21-point total consisting of two cards—an ace along with a ten-point card, king, queen, jack, or 10).

## Example 6

Find the probability of getting a Blackjack when drawing two cards from (a) one standard deck and (b) two standard decks.

**Solution:**

(a) Since the order in which the cards are received does not affect the results, we use combinations to determine the number of ways the events can occur. We want to get one of the four aces and one of the 16 ten-point cards that are in the deck. Thus, the number of desired hands is given by $C_{4,1} \times C_{16,1}$. The total number of possible two-card hands is given by $C_{52,2}$. Thus, the probability of getting a blackjack is

$$P(\text{blackjack}) = \frac{C_{4,1} \times C_{16,1}}{C_{52,2}} = \frac{4 \times 16}{1326} = \frac{64}{1326} \approx \frac{1}{20.7} \approx 0.04827$$

(b) When playing with two decks of cards, there are 104 cards with eight aces and 32 ten-point cards. Thus, the probability of drawing a Blackjack is

$$P(\text{blackjack}) = \frac{C_{8,1} \times C_{32,1}}{C_{104,2}} = \frac{8 \times 32}{5356}$$

$$= \frac{256}{5356} \approx \frac{1}{20.9} \approx 0.04780$$

When more than one deck is used, the probability of getting a blackjack decreases. It also makes it more difficult for the players to keep track of which cards have been played.

## Example 7

KENO

| 1 | 2 | 3 | 4 | 5 | 6 | 7 | 8 | 9 | 10 |
|---|---|---|---|---|---|---|---|---|---|
| 11 | 12 | 13 | 14 | 15 | 16 | 17 | 18 | 19 | 20 |
| 21 | 22 | 23 | 24 | 25 | 26 | 27 | 28 | 29 | 30 |
| 31 | 32 | 33 | 34 | 35 | 36 | 37 | 38 | 39 | 40 |

| 41 | 42 | 43 | 44 | 45 | 46 | 47 | 48 | 49 | 50 |
|---|---|---|---|---|---|---|---|---|---|
| 51 | 52 | 53 | 54 | 55 | 56 | 57 | 58 | 59 | 60 |
| 61 | 62 | 63 | 64 | 65 | 66 | 67 | 68 | 69 | 70 |
| 71 | 72 | 73 | 74 | 75 | 76 | 77 | 78 | 79 | 80 |

In the game of Keno, 80 numbers are displayed on a board. A player marks from 1 to 20 numbers on a Keno card. Twenty of the 80 numbers are randomly drawn as the winning numbers. Suppose a person selects seven numbers on a Keno card. What is the probability that the person gets exactly five winning numbers out of the seven marked on the Keno card? What are the odds for getting exactly five winning numbers out of seven numbers marked?

**Solution:** This is a game with 80 numbers, 20 of which are winning ones and 60 are losing ones. Since the order in which the numbers are selected does not affect the results, use combinations to determine the number of ways the events can occur. The number of ways to select five out of the 20 winning numbers is $C_{20,5}$. The number of ways to select two out of the 60 losing numbers is $C_{60,2}$. Finally, there are $C_{80,7}$ total ways in which any seven numbers can be selected out of the 80 numbers. This gives the probability of picking exactly five winning numbers from a Keno card of seven marked numbers as

$$P(5 \text{ of } 7 \text{ marked}) = \frac{C_{20,5} \times C_{60,2}}{C_{80,7}} = \frac{15{,}504 \times 1770}{3{,}176{,}716{,}400} \approx \frac{1}{115.8} \approx 0.00864$$

Let $E$ be the event of getting exactly five winning numbers out of the seven numbers selected. The odds would be determined as follows:

$$O(E) = \frac{P(E)}{P(\overline{E})} \approx \frac{\frac{1}{115.8}}{\frac{114.8}{115.8}} = \frac{1}{114.8} = 1 : 114.8$$

Lotteries are held in many states of the United States. Millions of people wager a dollar each week hoping that they will become millionaires. What are their chances of "striking it rich"? The following example will show you that the probability of becoming a millionaire in a state run lottery is very small. In fact, there is a greater chance of a person being struck by lightning than winning the State of California SuperLotto.

### Example 8

In the California SuperLotto, a player selects six numbers (from 1 to 51) and wagers $1. Every Wednesday and Saturday evening at 7:56 P.M., six winning numbers are selected for the SuperLotto. If the numbers a player chooses match the six winning numbers, the player wins the SuperLotto grand prize for that game. What is the probability that the player selects the six winning numbers?

**Solution:** The game has 51 numbers, six of which are the winning numbers. Since the order in which the numbers are selected does not matter, use combinations to determine the number of ways the events can occur. There are $C_{6,6}$

ways to pick the winning numbers. There are $C_{51,6}$ total ways in which any six winning numbers can be drawn. This gives the probability of picking six numbers correctly in the SuperLotto as

$$P(6 \text{ out of } 6) = \frac{C_{6,6}}{C_{51,6}} = \frac{1}{18{,}009{,}460} \approx 0.0000000555$$

## Example 9

A certain "jackpot only $1 slot machine" has three wheels on it as shown below. Each wheel rotates and lands on the pay line with either a BAR symbol or a blank.

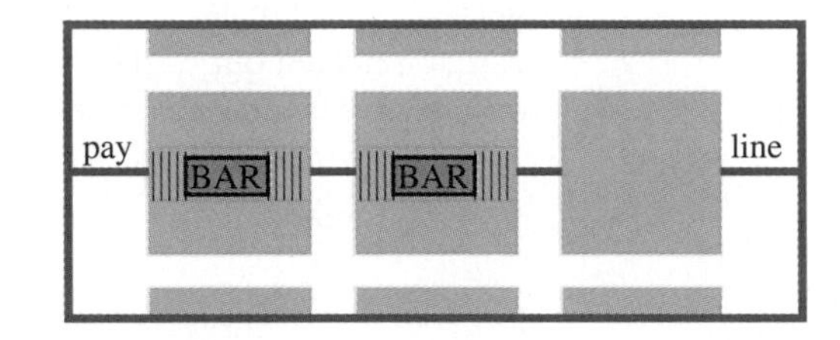


The first wheel contains six BAR symbols and 14 blanks, the second wheel contains five BAR symbols and 15 blanks, and the third wheel contains three BAR symbols and 17 blanks. After depositing $1 and spinning the wheels, you win the jackpot when each wheel lands on the pay line with a BAR symbol showing. What are the probability and the odds of winning the jackpot?

**Solution:** The desired outcome is to have a BAR symbol on each wheel. There are six BAR symbols on the first wheel, five on the second wheel, and three on the third wheel. By the basic counting law, a BAR symbol on all three wheels can occur in $6 \times 5 \times 3 = 90$ ways. Since there are 20 symbols on each wheel, by the basic counting law the total number of possible ways the wheels can land on the pay line is $20 \times 20 \times 20 = 8000$. Thus, the probability of the jackpot is

$$P(\text{jackpot}) = \frac{6 \times 5 \times 3}{20 \times 20 \times 20} = \frac{90}{8000} = \frac{9}{800}$$

If $E$ is the event of winning the jackpot, the odds can be determined by using the formula: $O(E) = \dfrac{P(E)}{P(\overline{E})}$. Since there are 90 winning combinations, there are $8000 - 90 = 7910$ losing combinations. Thus,

$$P(\overline{E}) = \frac{7910}{8000} \quad \text{and} \quad O(E) = \frac{\frac{90}{8000}}{\frac{7910}{8000}} = \frac{9}{791} \approx 1:88$$

Since the slot machine only pays off when there is a BAR symbol on each wheel, these odds indicate that if the slot machine was to be perfectly fair, the house odds would be about 88 : 1 and it would pay off about $88 for a dollar bet. ■

In this section, we have determined the probability and odds for various events using four methods: listing possibilities, the Basic Counting Law, combinations, and permutations. Armed with these principles you should be able to determine the probability and odds in games of chance.

Section 7.3
PROBLEMS

## Explain ➡ Apply ➡ Explore

### Explain

1. What are four methods used in determining the number of ways an event can occur? Under what conditions is each one used?

2. In the card game of Blackjack what constitutes a blackjack?

3. In the game of Keno, why are you not allowed to select over 20 numbers?

4. What does it mean to have a full house in playing cards?

5. In Example 9, the house odds for the "jackpot only $1 slot machine" were 88 : 1. Why would a casino pay you less than $88 for hitting a jackpot on the slot machine?

6. At Harrah's Casino in Lake Tahoe, Nevada, if you get five out of seven winning numbers on a $2 keno ticket, you will win $34. Is this what you would mathematically expect based on the results of Example 7? How can the difference be explained?

### Apply

7. Three standard dice are tossed. Find the probability and the odds for obtaining
   (a) a sum of 18 on the three dice
   (b) a sum of 4 on the three dice
   (c) a sum of 16 on the three dice

8. Five cards are dealt from a standard deck of 52 cards.
   (a) What is the probability of being dealt three kings?
   (b) What is the probability of being dealt three of any rank?

9. Five cards are dealt from a standard deck of 52 cards.
   (a) What is the probability of being dealt an ace and four kings?
   (b) What is the probability of being dealt an ace and four more cards of another rank?

**10.** Five cards are dealt from two standard decks mixed together.

(a) What is the probability of being dealt an ace and four kings?
(b) What is the probability of being dealt an ace and four more cards of another rank?

**11.** Five cards are dealt from two standard decks mixed together.

(a) What is the probability of being dealt three kings?
(b) What is the probability of being dealt three cards of any rank?

**12.** In poker, a full house is a five-card hand with three cards of one rank and two cards of another rank. Suppose you are dealt five cards from a standard deck of 52 cards.

(a) What is the probability of drawing a full house with three kings and two queens?
(b) What is the probability of drawing a full house with three kings and two cards of another rank?
(c) What is the probability of drawing any full house?

**13.** In horse racing, an ''exacta'' occurs when you correctly pick the horses which finish first and second. It is important to pick the correct order. If eight horses are in the race, what is the probability of correctly guessing the exacta?

**14.** In horse racing, a ''trifecta'' occurs when you correctly pick the order of the first three horses. If six horses are in the race, what is the probability of correctly guessing the trifecta?

**15.** In the game of Keno, 80 numbers are displayed on a board. Twenty of these numbers are chosen to be the winning numbers. Suppose a person has selected five numbers on a playing card.

(a) What is the probability of selecting no correct numbers?
(b) What is the probability of selecting exactly five correct numbers?
(c) What is the probability of selecting exactly four correct numbers?

**16.** In the game of Keno, 80 numbers are displayed on a board. Twenty of these numbers are chosen to be the winning numbers. Suppose a person has selected ten numbers on a playing card.

(a) What is the probability of selecting no correct numbers?
(b) What is the probability of selecting exactly six correct numbers?

**17.** A certain lottery has 49 numbers, six of which are the winning numbers for a particular game. To play the game each participant chooses six numbers. What is the probability of choosing exactly

(a) six correct numbers?
(b) five correct numbers?
(c) four correct numbers?

**18.** A certain lottery has 49 numbers, six of which are the winning numbers for a particular game. To play the game each participant chooses six numbers. What is the probability of choosing exactly

(a) three correct numbers?
(b) two correct numbers?
(c) one correct number?
(d) zero correct numbers?

**19.** A "Wild 7's" slot machine has three wheels on it. On each wheel there are four 7 symbols and six blanks. Find the probability and the odds of hitting the jackpot by getting a 7 symbol on the pay line on each wheel of the slot machine.

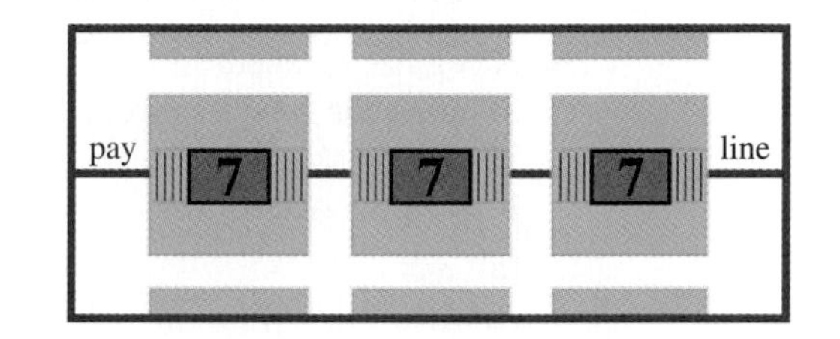


*Explore*

**20.** Suppose the NBA (National Basketball Association) has decided to add two more teams to the league. If 18 cities have applied for franchises, what is the probability that in a random drawing the league selects the two westernmost cities?

**21.** In a dog show, a Beagle is supposed to pick the correct two objects from a set of 20 objects. What is the probability that the dog picks the two objects?

**22.** Eight people on a committee are randomly selecting a subcommittee of two people. What is the probability that the two oldest committee members are selected?

**23.** An auditorium has scheduled three basketball games, two concerts, and four poetry meetings. You have a ticket allowing you to attend any three of the events. What is the probability that, by selecting three events at random, you attend two of the poetry readings and one of the other events?

**24.** Fourteen paintings are to be picked at random and placed along a wall. What is the probability that the paintings will be placed on the wall in alphabetical order, according to their titles? Assume that no two paintings have the same title.

**25.** License plates are printed by the State of California using the format of a digit, followed by three letters, followed by three digits. Assume all arrangements of letters can be used. What is the probability that the first digit of a plate is 5?

**26.** Ten people are on a bus. There are three stops until the end of the line. Assume that it is equally likely for a person to get off at any of the three stops. What is the probability that all the people get off the bus at the last of the three stops?

**27.** There are 30 people in your speech class. For the last ten days of the semester, three students are randomly selected to give their final speech for the semester.

(a) What is the probability that you are selected to give your speech on the first of the ten days?
(b) If you were not selected to give your speech during the first five days, what is the probability that you are selected to give it on the sixth day?
(c) If you were not selected to give your speech during the first nine days, what is the probability that you are selected to give it on the tenth day?

**28.** A regular octahedron is a solid with eight faces that are equilateral triangles. Suppose the numbers from 1 to 8 were placed on the faces of the octahedron. If three octahedrons are tossed, what is the probability that the sum of the numbers on the octahedrons is

(a) 3
(b) 15
(c) 24

**29.** In horse racing a ''Pick Six'' bet requires that you pick the first place horse in six designated races. Suppose that on a certain day at Churchill Downs in Louisville, Kentucky, there are nine horses in the first, ten horses in the second, 12 horses in the third, 15 horses in the fourth, eight horses in the fifth, and seven horses in the sixth Pick Six race. What are the probability and the odds of randomly guessing the winner in all of the six races?

**30.** Find the probability of getting a BINGO in the B column of a card when only five numbers have been called. (See Section 7.1, Problem 29, for an explanation of the game of BINGO.)

**31.** Find the probability of getting a BINGO in the N column of a card when only four numbers have been called. (See Section 7.1, Problem 29, for an explanation of the game of BINGO.)

**32.** There are 40 people in your math class. Four people are chosen at random to write their solution to Problem 28 on the chalkboard.

(a) What is the probability that you are one of those selected?
(b) What are the odds that you are not selected?

**33.** A ''25¢ Super 7'' slot machine has four wheels on it. The first wheel has two red 7's, four white 7's, six blue 7's, and eight blanks. The second wheel has four red 7's, six white 7's, two blue 7's, and eight blanks. The third wheel has six red 7's, two white 7's, four blue 7's, and eight blanks. The fourth wheel has five red 7's, five white 7's, five blue 7's, and five blanks. Find the probability and the odds for the following jackpots.

(a) getting a red 7 on the pay line of each wheel
(b) getting a white 7 on the pay line of each wheel
(c) getting a blue 7 on the pay line of each wheel
(d) The sum of probabilities of parts (a), (b), and (c) would give the probability of any jackpot. What are the probability and odds for getting any jackpot?
(e) Use the results of part (d) to answer the following question. If the "25¢ Super 7" slot machine was a mathematically fair one, how much should you receive on a 25¢ wager?

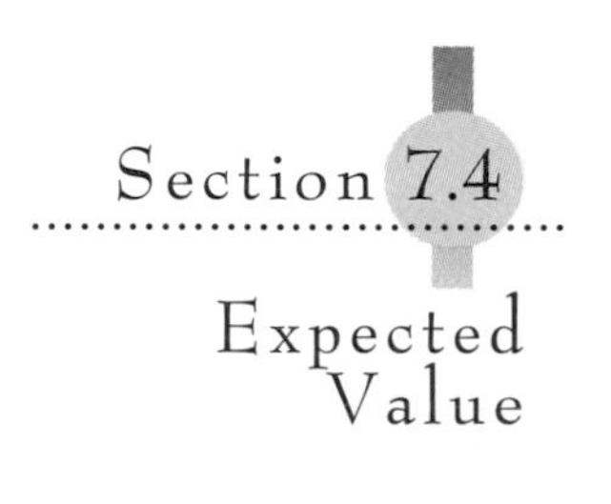


**Expected value** is the term used to describe the expected winnings from a contest or a game. Expected value is used throughout the world to determine prizes in contests and premiums on insurance policies. It is also used in the mathematical field called decision theory. In this section, we will show how an expected value can be calculated.

Suppose a game is played with one six-sided die. If the die is rolled and lands on 1, 2, or 3, the player wins nothing. If the die lands on 4 or 5, the player wins $3. If the die lands on 6, the player wins $12. The following table summarizes this information including the probability of each situation.

| Event | *P*(event) | Winnings |
|---|---|---|
| 1, 2, or 3 | 3/6 | $0 |
| 4 or 5 | 2/6 | $3 |
| 6 | 1/6 | $12 |

If you play this game, how much can you expect to win? What is the expected value of this game?

**Expected Value**

To compute the expected value of a game, find the sum of the products of the probability of each event and the amount won or lost if that event occurs.

For this game, the expected value (E.V.) is given by,

$$\text{E.V.} = \frac{3}{6} \times 0 + \frac{2}{6} \times 3 + \frac{1}{6} \times 12 = \$3$$

An expected value of \$3 means we would expect to win an average of \$3 for each game played. This means that if we played 1000 games, we would expect to win \$3000. However, as is true for probability, this is true only for a large numbers of games. If we played three games, although the expected winnings are \$9, we could win between \$0 and \$36.

Suppose the operator of the game charges \$1 to play the game. The amount won by the customer would then be reduced by \$1. By using the following table, we can compute the expected winnings for a player, including the cost of the game.

| **Event** | ***P*(event)** | **Winnings** |
|---|---|---|
| 1, 2, or 3 | 3/6 | −\$1 |
| 4 or 5 | 2/6 | \$2 |
| 6 | 1/6 | \$11 |

For this game, the expected value is

$$\text{E.V.} = \frac{3}{6}(-1) + \frac{2}{6}(2) + \frac{1}{6}(11) = \$2$$

As might be expected, a charge of \$1 to play the game reduces the expected value by \$1, from \$3 down to \$2.

## Example 1

A state lottery has 49 numbers, six of which are the winning numbers for a particular game. The cost of playing the lottery is \$1. To play the game, the player must pick six numbers. If a player picks three winning numbers, the payment is \$20. Similarly, picking four numbers pays \$100, picking five numbers pays \$10,000, and picking six winning numbers pays \$1,000,000. Find the expected value of this game, including the cost of playing.

**Solution:** First, we calculate the probability of correctly guessing the winning numbers. Using the methods of the previous sections, this is summarized in the following table along with the winnings at each level. As an example, the probability of selecting three numbers correctly is given by

$$P(3) = \frac{C_{6,3} \times C_{43,3}}{C_{49,6}} = \frac{20 \times 12{,}341}{13{,}983{,}816} = 0.0176504$$

Similar computations give the following table. Notice that the amounts paid are reduced by \$1 to account for the cost of playing the game.

| Event (number correct) | P(event) | Winnings |
|---|---|---|
| 0 | $\frac{C_{6,0} \times C_{43,6}}{C_{49,6}} = 0.4359650$ | −$1 |
| 1 | $\frac{C_{6,1} \times C_{43,5}}{C_{49,6}} = 0.4130195$ | −$1 |
| 2 | $\frac{C_{6,2} \times C_{43,4}}{C_{49,6}} = 0.1323780$ | −$1 |
| 3 | $\frac{C_{6,3} \times C_{43,3}}{C_{49,6}} = 0.0176504$ | $19 |
| 4 | $\frac{C_{6,4} \times C_{43,2}}{C_{49,6}} = 0.0009686$ | $99 |
| 5 | $\frac{C_{6,5} \times C_{43,1}}{C_{49,6}} = 0.0000184$ | $9999 |
| 6 | $\frac{C_{6,6} \times C_{43,0}}{C_{49,6}} = 0.0000001$ | $999,999 |

From this table, we can calculate the expected value.

$$\begin{aligned} \text{E.V.} &= (-1) \times 0.4359650 + (-1) \times 0.4130195 + (-1) \\ &\quad \times 0.1323780 + 19 \times 0.0176504 \\ &\quad + 99 \times 0.0009686 + 9999 \times 0.0000184 \\ &\quad + 999{,}999 \times 0.0000001 \\ &\approx -\$0.27 \end{aligned}$$

By doing this calculation, we see that we expect to lose an average of 27¢ every time we play the lottery. This means that the operators of the lottery expect to earn 27¢ every time someone buys a ticket. ■

## Example 2

A game is called **fair** if the expected value of the game is zero. Suppose a certain game is fair and costs $3 to play. The probability of winning is 0.6 and the probability of losing is 0.4. How much should you win for the game to be fair?

**Solution:** Since the game is fair, we can set up an equation for the expected value of the game with E.V. = 0. We will use $W$ to represent the prize for winning. Since it costs \$3 to play the game, the amount won if we win the game is $W - 3$ and the amount won if we lose the game is $-3$.

$$\text{E.V.} = \begin{pmatrix}\text{amount won if}\\ \text{you win game}\end{pmatrix} \times P(\text{winning}) + \begin{pmatrix}\text{amount won if}\\ \text{you lose game}\end{pmatrix} \times P(\text{losing})$$

$$0 = (W - 3) \times 0.6 + (-3) \times 0.4$$

$$0 = 0.6W - 1.8 - 1.2$$

$$0 = 0.6W - 3$$

$$-0.6W = -3$$

$$W = 5$$

Therefore, since the game is fair, a player should receive \$5 if he wins. Note that this means a real winnings of only \$2 since it costs \$3 to play the game. ■

## Example 3

Suppose a certain game is fair and costs \$3 if we lose and has a net payoff of \$10 if we win. The only possible outcomes of the game are winning and losing. What is the probability of winning?

**Solution:** Let $p$ be the probability of winning. Because there are only two choices, winning and losing, the probability of losing must be $1 - p$. Since the game is fair, we can set up an equation for the expected value of the game with E.V. = 0.

$$\text{E.V.} = (\text{amount won}) \times P(\text{winning}) + (\text{amount lost}) \times P(\text{losing})$$

$$0 = 10p + (-3)(1 - p)$$

$$0 = 10p - 3 + 3p$$

$$0 = 13p - 3$$

$$-13p = -3$$

$$p = \frac{3}{13}$$

Therefore, the probability of winning the game is 3/13. ■

Insurance companies determine the premiums for a policy by examining the risk involved. The risk is calculated by looking at statistics involving the situations covered by the policy. A person in a high-risk category pays more for insurance than someone in a low-risk category.

### Example 4

The Lagomorph Insurance Company has a customer, Mr. Roger Abbit, with a \$250,000 fire insurance policy on his art collection. The company estimates that there is a 1% chance that the art will be destroyed by fire. If the insurance company tries to maintain an expected value of \$200 on each policy, what should Mr. Abbit's premiums be?

**Solution:** If $a$ = amount of the premiums, then $a - 250{,}000$ is the amount the insurance company will lose if Roger's art collection is destroyed by fire. Since the company estimates there is a 1% chance of loss by fire, there is a 99% chance that there will be no loss.

$$\text{E.V.} = \begin{pmatrix}\text{amount company}\\ \text{stands to lose}\end{pmatrix} \times P\begin{pmatrix}\text{losing}\\ \text{the art}\end{pmatrix} + \begin{pmatrix}\text{amount paid}\\ \text{in premiums}\end{pmatrix} \times P\begin{pmatrix}\text{not losing}\\ \text{the art}\end{pmatrix}$$

$$200 = (a - 250{,}000) \times 0.01 + a \times 0.99$$

$$200 = 0.01a - 2500 + 0.99a$$

$$200 = a - 2500$$

$$2700 = a$$

This means that the Lagomorph Insurance Company should charge \$2700 for this policy. ■

## Decision Theory

Expected values can also be used to help make decisions that have financial implications. The process of weighing the risks versus the benefits of two or more alternatives is called **decision theory**. In decision theory, the expected value for each possible outcome is computed and a decision is made by analyzing the results.

### Example 5

An engineering firm, Sasselli Satellites, is expanding its facilities and needs some electrical work done within a week. The firm has received three bids on the work. The first contractor, Acme Parts, says that they will charge \$10,000. The second company, Yablok Electric, will charge \$11,000 if they finish within one week or \$9000 if they cannot finish the job within one week. The third company, Zak Communications, has the low bid of \$8500, but they want an extra \$4500 if they can complete the job in less than one week. Sasselli researches the history of all three contractors and finds that they all do very good work. Sasselli also finds out that Yablok Electric completes its work as scheduled 85% of the time and Zak Communications finishes ahead of schedule 20% of the time. Which contractor should Sasselli Satellites choose to do their electrical work if the primary concern is to keep costs low?

**Solution:** Since all three contractors have a reputation for high-quality workmanship, and since cost is the primary consideration, Sasselli Satellites should choose the contractor with the lowest expected cost.

For Acme Parts, the cost will always be $10,000.

For Yablok Electric, the cost will be $11,000 with a probability of 0.85 and $9000 with a probability of 0.15 ($1 - 0.85 = 0.15$). Therefore, the expected costs for Yablok Electric are

$$11{,}000 \times 0.85 + 9000 \times 0.15 = \$10{,}700$$

For Zak Communications, the cost will be $8500 with a probability of 0.80 and $13,000 ($8500 + $4500 = $13,000) with a probability of 0.20. Therefore the expected costs for Zak Communications are

$$8500 \times 0.80 + 13{,}000 \times 0.20 = \$9400$$

From this information, Sasselli Satellites should use Zak Communications. ■

## Example 6

A bank account is guaranteed to earn a fixed rate of 9% on a $10,000 deposit over the next year. A speculative investment offers the possibility of 15% earnings on the $10,000 if the investment succeeds and a loss of 5% of the $10,000 if the investment fails. Determine the probability of success necessary for the speculative investment to be the better choice.

**Solution:** To choose the better way to invest the money, we find the expected gain from each investment over the next year. Since the bank account offers a 9% gain with no possibility of loss, the bank account has an expected value of

$$10{,}000 \times 0.09 = \$900$$

To find the expected gain from the speculative investment, we must know the probability of a success. Since this is unknown, we will assign it a variable, $p$. The probability of failure is then $1 - p$. The expected gain from the speculative account is then given by

$$(10{,}000 \times 0.15) \times p + [(-10{,}000) \times 0.05] \times (1 - p)$$

Since we want to determine when the speculative account is the better investment, its expected value must be greater than the expected value of the bank account. If we solve the following inequality, we will find the value of $p$.

$$(10{,}000 \times 0.15) \times p + [(-10{,}000) \times 0.05] \times (1 - p) > 900$$
$$1500p - 500(1 - p) > 900$$
$$1500p - 500 + 500p > 900$$
$$2000p > 1400$$
$$p > 0.70$$

Thus, for the speculative venture to be the more lucrative investment, the probability of success must be greater than 70%. ■

## Section 7.4 PROBLEMS

## *Explain → Apply → Explore*

### *Explain*

1. What is expected value and how is it calculated?
2. In terms of expected value, when is a game considered to be fair?
3. What does it mean for the expected value of a \$1 slot machine to be −\$0.15?
4. What does it mean for the expected value of a \$1 lotto game to be \$0.15?
5. What is decision theory?
6. Why is your expected value for games of chance in a gambling casino negative? Explain.
7. Over a long period of time, what could you predict for the owners of a carnival game if the expected value for a player is a positive amount? Explain.

### *Apply*

8. In a certain game, the probability of winning is 0.3 and the probability of losing is 0.7. If a player wins, the player will collect \$50. If the player loses, the player will lose \$5. What is the expected value of this game? If the game is played 100 times what are the expected winnings (or losses) of the player?
9. In a game of dice, the probability of rolling a 12 is 1/36. The probability of rolling a 9, 10, or 11 is 9/36. The probability of rolling any other number is 26/36. If the player rolls a 12, the player wins \$5. If the player rolls a 9, 10, or 11, the player wins \$1. Otherwise, the player loses \$1. What is the expected value of this game? If the game is played 100 times what are the expected winnings (or losses) of the player?
10. In a game of dice, the probability of rolling a 12 is 1/36. The probability of rolling a 9, 10, or 11 is 9/36. The probability of rolling any other number is

26/36. If the player rolls a 12, the player wins $8. If the player rolls a 9, 10, or 11, the player wins $2. How much should the player lose when the player rolls any other number if the game is fair?

**11.** In a certain game, the probability of winning is 0.2 and the probability of losing is 0.8. If the player loses, the player will lose $5. How much does the player collect when the player wins if the game is fair?

**12.** In the game of Keno, 80 numbers are displayed on a board. Twenty of these numbers are chosen to be the winning numbers. Suppose a person has selected five numbers on a playing card. The probabilities and winnings, are as follows:

| Event | Probability | Amount Won |
|---|---|---|
| 0 winning numbers | 0.227184 | −$1 |
| 1 winning number | 0.405686 | −$1 |
| 2 winning numbers | 0.270457 | −$1 |
| 3 winning numbers | 0.083935 | $1 |
| 4 winning numbers | 0.012092 | $10 |
| 5 winning numbers | 0.000645 | $250 |

(a) What is the expected value of the game?
(b) If the game costs $1, how much money should the player expect to win or lose after playing 1000 games?

**13.** Many charities and other organizations use lotteries or other similar marketing devices to acquire funds. (Publisher's Clearinghouse or *Reader's Digest* may come to mind.) By law in many states, it is required to post the odds of winning the various prizes on the back of the tickets or in some other conspicuous spot.

(a) *Fancy That Poultry* magazine is running a contest with the following odds and prizes.

| Odds | Prize |
|---|---|
| 1 to 49 | $10 (in back issues) |
| 1 to 9999 | $50 (in poultry feed) |
| 1 to 99,999 | $2000 (in rare ducks) |

If tickets are free, find the expected value of a "winning ticket."
(b) Is the drawing worth the price of a first class stamp?
(c) At what postage rate would the drawing be considered fair?

**14.** Dennis is in charge of designing a game for the school fund raiser. Participants will be paying \$2 for each game. There will be three prizes. The lowest has a value of \$0.50, the second has a value of \$1 and the third has an undetermined value. The probability of winning the lowest prize is 0.35, the probability of winning the second prize is 0.15, and the probability of winning the third prize is 0.01. The probability of not winning any prize is 0.49. If the school wants an expected value of \$1 per ticket, what should Dennis choose as the value of the third prize?

**15.** The batting average of a baseball player gives the probability that the player will get a hit in the next at bat. The table gives the batting averages for the San Francisco Giants during the 1993 season and the number of official at bats each player has in a particular game. Determine the expected number of hits for the team during this game.

| Player | Batting Avg. | At Bats |
|---|---|---|
| Lewis cf | 0.257 | 5 |
| Thompson 2b | 0.321 | 5 |
| Clark 1b | 0.284 | 3 |
| Williams 3b | 0.303 | 4 |
| Bond 1f | 0.336 | 4 |
| McGee rf | 0.298 | 4 |
| Clayton ss | 0.283 | 3 |
| Manwaring c | 0.275 | 4 |
| Swift p | 0.270 | 4 |

**16.** The table gives a partial listing of mortality rates for guinea pigs. It gives the probability of a guinea pig living to a certain age. Assuming that all guinea pigs die by age 5, determine the life expectancy of guinea pigs.

| Age | Probability |
|---|---|
| 1 | 0.14 |
| 2 | 0.07 |
| 3 | 0.26 |
| 4 | 0.29 |
| 5 | 0.24 |

**17.** At many Bingo parlors, the operators sell ''pull tabs,'' which are very similar to slot machines. Each pull-tab card has rows of symbols that are covered by paper tabs. If the paper tabs are removed and three of the same symbols are in a straight line, you win a designated amount. A summary of a typical pull-tab game where each card costs $0.50 is shown below.

| **Symbols** | **Probability** | **Amount Won** |
|---|---|---|
| Three diamonds | 2/2783 | $149.50 |
| Three rubies | 2/2783 | 74.50 |
| Three pearls | 4/2783 | 9.50 |
| Three coins | 20/2783 | 2.50 |
| Three stars | 250/2783 | 0.50 |
| Three moons | 400/2783 | 0 |
| Any other combination | 2105/2783 | −0.50 |

Determine the expected value of this pull-tab game.

*Explore*

**18.** In the game of Roulette, players bet that a ball will land on a certain number. A player can choose any number from 1 through 36, 0, or 00. It costs $1 to play the game. If the player correctly selects the number, the $1 is returned and the player receives an additional $35. What is the expected value of this game? Suppose a casino has 100 players, each of whom plays ten times each hour for 24 hours. Each player bets $1. What is the casino's expected profit?

**19.** In the game of Roulette, a player can bet that a ball will land on any one of the numbers in a square of four numbers by placing a bet at the juncture of four numbers as shown below.

If a player bets $1 on a square of four numbers and one of the numbers comes up, the $1 is returned and the player receives an additional $8. What is the expected value of such a bet?

**20.** In the California State Lottery game of Keno, 20 winning numbers are randomly selected from a total of 80 numbers. Suppose you play the "3-Spot Game" in which you choose three numbers on your Keno card and bet $2. If zero or one of your numbers is correct, you lose your bet. If two of your numbers are correct, you win $2. If all three of your numbers are correct, how much should you win to make this a fair game? Why does the California State Lottery only give you winnings of $38 when you match three numbers on the 3-Spot Game?

**21.** In a ''5-spot Game'' of Keno at a casino, you select five numbers on your card. The number of winning numbers out of the 20 winners you get and the amount you bet determine the amount you win. The payoffs are shown in the chart that follows.

| Number of Winners Picked | $1 Ticket ($) | $2 Ticket ($) | $3 Ticket ($) |
|---|---|---|---|
| 5 | 820 | 1640 | 2460 |
| 4 | 10 | 20 | 30 |
| 3 | 1 | 2 | 3 |

If you pick fewer than three winning numbers, you lose the amount you bet. The amounts given in the chart are the actual amounts paid to you if you win. Deduct the amount bet to get the actual amount won in each case. Find the expected value for each of the three tickets shown in the chart. From your results, which is the best ticket to play?

**22.** A television game show contestant has current prizes worth $12,500. If the contestant participates in the next round of competition, he will have $50,000 if he wins and $0 if he loses. The contestant uses expected values to determine that he should participate in the next round. What did the contestant determine his probability of winning the next round to be?

**23.** An engineer has provided a customer with the choice of two different procedures for extending the lifetime of a certain structure. The first procedure has a success rate of 93% and will extend the life of the building by eight years. The second procedure is still experimental and has not been perfected. It will extend the life of the structure by 15 years if the procedure works. However, there is only a 47% success rate. Assume that the customer can afford to use only one procedure and that if a procedure fails, the building will last two more years. Which procedure should be used to maximize the expected life of the building?

**24.** In planting a playing field, a park manager must decide between planting seed or sod. If seed is used, there is a 33% chance that the grass lawn will grow with one seeding and a 67% chance that it will need two seedings. If the lawn is seeded once, it will cost $60. If the lawn needs two seedings, the cost will be $400. Planting sod will cost $300 and has a 100% success rate. Which method is more cost effective?

**25.** The circles have radii of 2, 8, 16, and 24 in. respectively.

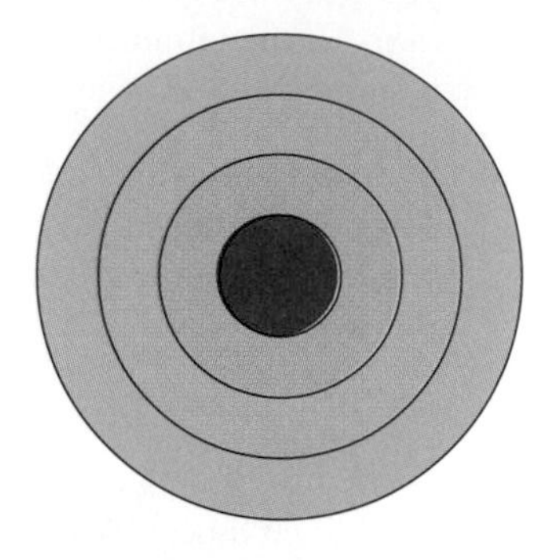

In the game of skeeball, a player rolls a ball up a ramp and wins \$10 if the ball lands in the center circle, wins \$2 if the ball lands in the second band, wins nothing if it lands in the third band, and loses \$1 if it lands in the outer band. If the ball misses the target, the ball is rolled again. The probability that the ball lands in a certain region is the same as the area of that region divided by $576\pi$. The areas of the regions are given in the table.

| Region | Area |
|---|---|
| Center | $4\pi$ sq in. |
| 2nd band | $60\pi$ sq in. |
| 3rd band | $192\pi$ sq in. |
| Outside band | $320\pi$ sq in. |

(a) Find the expected value of the game.
(b) Assuming everything else is unchanged, what should the prize be for the inner circle so that the game is fair?

**26.** The squares have sides of 4, 8, 16, and 32 in. respectively.

Suppose in the game of skeeball, a ball is rolled up a ramp and wins \$10 if it lands in the center square, wins \$5 if it lands in the second band, wins \$1.50 if it lands in the third band, and loses \$1 if it lands in the outer band. If the ball misses the target, the ball is rolled again. The probability that the ball lands in a certain region is equal to the area of the region divided by 1024. The areas of the regions are given in the table.

| Region | Area |
|---|---|
| Center | 16 sq in. |
| 2nd band | 48 sq in. |
| 3rd band | 192 sq in. |
| Outside band | 768 sq in. |

(a) Find the expected value of the game.
(b) Assuming that everything else is unchanged, what should the prize be for the inner square so that the game is fair?

# Chapter 7 • Summary

## Key Terms, Concepts, and Formulas

The important terms in this chapter are:

**Basic Counting Law:** The principle which states that if there are $n$ choices for one item and $m$ choices for a second item, there are $n \times m$ ways to pick both items. p. 477

**Combination:** A group of objects in which the order of the objects does not matter. p. 453

**Complementary events:** If an event $E$ occurs, then its complementary event $\overline{E}$ is the nonoccurrence of $E$. p. 465

**Decision theory:** The use of expected values to determine choices. p. 505

**Expected value:** The average outcome of an event; this could be the expected amount of money won when playing a game or the anticipated cost of a project based on the associated risks involved in the project. p. 501

**Experimental probability:** The probability of an event that has been approximated by analyzing a number of trials rather than by using mathematical calculations. p. 463

**Factorial:** The mathematical expression given by $n! = n \cdot (n - 1) \cdot (n - 2) \cdot \cdots \cdot 2 \cdot 1$. p. 448

**Fair game:** A game in which the expected value is zero. p. 501

**Law of Large Numbers:** The principle which states that as the number of trials increases the experimental probability approaches the theoretical probability. p. 464

**Odds:** Another way of expressing the chance that an event will occur. If the probability of an event is $P(E)$ and the probability of its complement is $P(\overline{E})$, the odds that the event will occur are given by $O(E) = \dfrac{P(E)}{P(\overline{E})}$. p. 468

**Permutation:** An arrangement of objects in which the order of the objects matters. p. 451

**Probability:** A measure of the chance that an event will occur. p. 461

**Theoretical probability:** The probability of an event that has been predicted through mathematical calculations rather than by an experiment. p. 463

**With replacement:** A phrase which indicates that an object can be used more than once. p. 477

**Without replacement:** A phrase which indicates that an object cannot be used more than once. p. 477

After completing this chapter, you should be able to:

1. Explain the difference between combinations and permutations. p. 453
2. Determine the probability and odds of single events using experimental and theoretical methods. p. 461
3. Explain the difference between using "with replacement" and using "without replacement" in determining the probability of an event. p. 477
4. Determine the number of ways an event can occur by applying the Basic Counting Law, listing possibilities, and using formulas for

   Combinations $C_{n,r} = \dfrac{n!}{r!(n-r)!}$ p. 455

   Permutations $P_{n,r} = \dfrac{n!}{(n-r)!}$ p. 452
5. Apply the concepts of probability, odds, and expected value. p. 501

• Summary

## Problems

1. Find the value of the following:

   (a) $C_{7,4}$
   (b) $P_{7,4}$
   (c) $C_{73,4}$
   (d) $P_{73,4}$

2. What is wrong with the following argument? Since the probability of a flipped coin landing heads up is 0.5, flipping a fair coin ten times will result in five heads and five tails.

3. If the house odds for a team winning the championship are 99 : 1, what is the probability that the team will win the championship? What do the house odds tell you about wagering on that team?

4. A Pioneer® compact disk player has a cartridge that holds 5 CDs and can randomly select and play any song from the cartridge. Suppose in the cartridge the first CD has 12 songs, the second CD has 13 songs, the third CD has 15 songs, the fourth CD has 11 songs, and the fifth CD has 14 songs. How many ways can three songs be selected and played with the random feature,
   (a) if a song can be repeated?
   (b) if no song is repeated?

5. Dodecahedral dice have 12 sides instead of the standard six sides. The sides are numbered 1 through 12. What is the probability of rolling two dodecahedral dice and having the sum of the face-up sides equal 8?

6. If there are six candidates for the president of an organization, three candidates for secretary, and four candidates for treasurer, how many different executive teams of a president, secretary, and treasurer are possible?

7. (a) Find the number of ways in which three aces and two kings can occur in a six-card hand from a standard 52-card deck.
   (b) Find the number of ways in which three aces and a pair of any rank occur in a six-card hand from a standard 52-card deck.
   (c) Find the number of ways in which three cards of one rank and two cards of another rank occur in a six-card hand from a standard 52-card deck.

8. What are the probability and the odds of correctly guessing the top three finishers in order in a 100m swimming race with 10 competitors?

9. There are 14 players on a college basketball team. Before each game the coach randomly draws the names of two players to meet with game officials and players from the opposing team. What is the probability that the center, Pat Smith, is one of the two players selected for the
   (a) first game of the season?
   (b) tenth game of the season?

10. A certain game is played with two standard six-sided dice. The player will win an amount equal to the sum of the spots on both dice if the sum is greater than 7. The player will lose an amount equal to the sum if the sum is less than or equal to 7. What is the expected value of this game?

**11.** In the picture there are three equilateral triangles. The largest has sides of 2 ft, the next has sides of 1 ft, and the smallest has sides of 0.5 ft. A player rolls a ball up a ramp towards the target and earns 5 points if the ball lands in the center region, 2 points if the ball lands in the middle region, and 1 point if the ball lands in the outer region. If the ball completely misses the target, the player rolls the ball again. If each probability is equal to the area of the region divided by $\sqrt{3}$, and we are given the following information about the areas of the three regions:

The area of the innermost triangle is $\sqrt{3}/16$.

The area of the middle band is $3\sqrt{3}/16$

The area of the outer band is $3\sqrt{3}/4$.

(a) Find the expected value of the game.
(b) Find the total number of points a player will expect to accumulate in ten tosses that land in the target.

**12.** (a) In the game of keno as described in this chapter, what are the probability and the odds of selecting four numbers on your playing card and all four of the numbers are winning numbers?
(b) What are the probability and odds of not selecting all four winning numbers?
(c) If you wager $1 and actually win $250 if all four of the numbers you choose are winning numbers and lose your $1 wager if you do not select all four winning numbers, what is the expected value of the game.
(d) By examining the results of part (c), what can you say about this game? If you play this game for a long period of time what can you expect?

***13.** The following is a famous problem known as the birthday problem.
(a) In how many ways can 20 days of the year be selected if the same day may be used more than once? Leave your answer in terms of a formula.
(b) In how many ways can 20 days be selected if the same day may be used only once? Leave your answer in terms of a formula.
(c) Calculate the value of the result of part (b) divided by the result of part (a).
(d) Explain why the answer to part (c) is the probability that no two people from a set of 20 are born on the same day of the year.
(e) Explain why the probability that at least two people from a set of 20 are born on the same day is given by 1 minus the result from part (c).

# 8
# Statistics and Voting

Politicians, like those shown in *The Senate* by William Gropper, use apportionment and voting patterns to set up their campaign strategy—where should they concentrate their campaigning for the biggest ''vote-getting'' impact. (The Museum of Modern Art, New York)

# Overview Interview

In this chapter, we will introduce you to some of the basic concepts of statistics. We will show you how data can be organized, displayed, and analyzed using various techniques. Our focus in this chapter will be on the statistics that can be applied in areas associated with some aspects of voting. Many of our examples and problems will investigate areas such as political affiliation, voter characteristics, legislative apportionment, voter registration, and opinion polls.

## Blanche M. Lambert

BLANCHE M. LAMBERT, CONGRESSWOMAN, HOUSE OF REPRESENTATIVES, CONGRESS OF THE UNITED STATES, 1ST DISTRICT, ARKANSAS

TELL US ABOUT YOURSELF. I was born and raised in Helena, Arkansas, the seventh generation of a cotton and rice farming family. After graduating with a Bachelor of Science degree from Randolph-Macon Women's College in Lynchburg, Virginia, I moved to Washington, D.C., where I spent ten years in congressional research. In 1992, the people of the First Congressional District of Arkansas elected me to represent their rural interests of farming, job creation, and improving access to health care.

HOW IS STATISTICS USED BY SOMEONE IN YOUR POSITION? Elected officials use statistics in a variety of ways. Statistics assist them in determining the concerns of the electorate. For example, if statistics show that violent crime is increasing 50% faster in rural areas than in urban areas, members of Congress might be persuaded to allocate more money for community policing in rural areas. If statistics reveal that 82% of the people in prisons are high school dropouts, lawmakers might determine that focusing on education is one of the best bets for deterring criminal activity. Elected officials and legislators must understand statistical data in order to determine what areas can be improved through legislation.

The feeling of being merely a statistic in an urban jungle can be sensed in the painting *Urban Freeways* by Wayne Thiebaud. (Courtesy of The Campbell-Thiebaud Gallery, San Francisco)

## A Short History of Statistics

The word *statistik* was first used in 1749 by Gottfried Achenwall. It comes from the Latin word *statisticus*, meaning "of the state." This is appropriate because, until the 1850s, statistics almost always referred to information about a country or other political body. This information could refer to political, social, or economic conditions and was usually given in the form of tables of numerical data.

When the Great Plague hit England in 1665, mortality rates soared. John Graunt of England (1620–1674), known as the father of vital statistics, published *Natural and Political Observations Made upon the Bills of Mortality*. By studying the English death records, Graunt found patterns in the number of deaths by suicide, disease, and accidents in English cities. He also found that the number of male births exceeded the number of female births. Graunt's work, along with that of William Petty, established life expectancy tables used by the life insurance companies to determine the premiums that the companies should charge their policyholders. By the mid-1700s, all Western nations were compiling information from periodic censuses, recording such information as age of death, cause of death, and the male-female ratio of births. The availability and use of this information allowed life insurance companies to determine the rates by scientific methods.

In 1763, the Englishman Thomas Bayes' (1702–1761) posthumous publication *Essay Towards Solving a Problem in the Doctrine of Chances* became one of the first works to argue from a small sample of information to what could be expected from the population as a whole. This was the forerunner of the activities of George Gallup and the Gallup poll. This process involves asking a small percentage of a population their opinions or status on a question and then using this small sample to predict the position of the population as a whole.

In 1806, Adrien-Marie Legendre of France (1752–1833) introduced the method known as "least squares" in the supplement to his *New Methods for Determination of a Comet's Orbit*. This method determines the equation of a line that passes through a cluster of points with the least error. While the derivation of the method requires calculus, it is easily applied in many areas of study, from economic forecasting to sociology as well as in nearly all the sciences. The method of least squares was also discovered by the great German mathematician, Carl Friedrich Gauss (1777–1855) in 1794, but was not published by Gauss until he released his work in *Theoria motus* in 1809. In this book, Gauss also discussed the bell-shaped curve called the Gaussian, or normal, distribution. This is the distribution used with many standardized testing systems.

Adolphe Quetelet (1796–1874) of Belgium constructed the first statistical breakdown of a national census in 1829. He examined the census information to determine possible connections between age of death and various other variables such as season of the year, occupation, age, and economic status.

In 1889, the Englishman Francis Galton (1822–1911) published *National Inheritance*. This was the compilation of 15 years of work on statistical relationships in the area of genetics. One interesting fact determined by Galton was that of regression. Regression is a process that occurs when a biological characteristic reverts to a simpler or more general form. An example of this can be seen with the heights of successive generations in a family. The children of two exceptionally tall people will probably be taller than the average person, but the children's heights usually will not exceed the heights of the parents. In the same way, the children of two exceptionally short people will be taller than the parents, and their heights will become closer to the height of an average person rather than the heights of the parents.

As was true with probability, statistics developed rapidly in the twentieth century. In the 1920s Ronald Fisher (1890–1962) studied the problems involved in sampling. Sampling is the procedure followed when selecting a few individuals to determine information about the population as a whole. Fisher's studies showed that the results of a poll are more reliable if the sample is chosen at random.

In 1929 and 1930, A. K. Erlang and J. F. Steffensen studied the phenomenon of the extinction of family names. Certain family names that once were common were slowly but inevitably dying out. After analyzing the data, Erlang and Steffensen concluded that it was a form of Darwin's theory of survival of the fittest. Families that were wealthy and in good health were able to raise large numbers of children and to provide these children with the means to succeed in later life. Families in desperate conditions were often in poor health, with many children dying before they reached maturity. The less advantaged families had fewer family members to carry on the family name.

New discoveries continue up to the present day. In 1950, Gertrude Mary Cox published the first work of note on statistical design and analysis of complicated experiments in *Experimental Designs*. Many new statistical distributions have been put into use during the 20th century. Others, like Fisher, found that the Gaussian distribution (the standard bell-shaped curve) did not correctly model many real-life situations. Today, magazines and newspapers have statistical presentations about a myriad of facts. Stock averages, the latest cancer research, and the status of the national debt are some ever-present examples.

The availability of computers allows anyone to accumulate and arrange numerical data in many ways. It is up to each person to be able to interpret the data and its presentation to determine the validity of the position the data is being used to support.

## Check Your Reading

1. What did statistics refer to until the 1850s?
2. What is sampling?
3. When John Graunt was publishing his mortality statistics, Caleb Cheeshateaumuck became the first American Indian to receive a Bachelor of Arts degree at Harvard College. What year was this?
4. The Indian war near what would later be called Detroit, Michigan, occurred in 1763. What work of Thomas Bayes was published during this year? What was important about this book?
5. What historical figure was the forerunner to George Gallup?
6. Cite some areas where statistics is used in present-day society.
7. During the 1920s, the Soviet states formed the Union of Soviet Socialist Republics and the Cuban chess player José Raoul Capablanca won the world chess championship. What was the statistician Ronald Fisher doing during this period?
8. While Vincent Van Gogh was painting *Landscape with a Cypress Tree* and North Dakota, South Dakota, Montana, and Washington were becoming states, Francis Galton published a statistical work related to genetics. In what year did this occur?
9. The British chemist James Smithson bequeathed funds to found the Smithsonian Museum in Washington, D.C., and Adolphe Quetelet constructed a statistical breakdown of the Belgian national census. In what year did this occur?
10. In 1929–1930, Pablo Picasso was painting *Woman in an Armchair*, and Alketta Jacobs became the first woman physician in Holland. What were A. Erlang and J. Steffensen studying during this period?

11. Carl Gauss published a work on bell-shaped curves and Charles Darwin and Abraham Lincoln were born during what year?
12. The year 1806 marked the official end of the Roman Empire. What was Frenchman Adrien-Marie Legendre doing during this year?
13. Match each of the following names with the correct event, idea, or occurrence.

| | |
|---|---|
| (a) Achenwall | Father of vital statistics |
| (b) Bayes | First to argue from a small sample of information |
| (c) Cox | First to use the word *statistik* |
| (d) Fisher | First statistical breakdown of a national census |
| (e) Galton | Method of least squares |
| (f) Gauss | Normal distribution (bell-shaped curves) |
| (g) Graunt | Sampling procedures |
| (h) Legendre | Statistical relationships in area of genetics |
| (i) Quetelet | Pioneer in statistical design |

## • Research Questions

To answer the following questions, you will need to refer to material not contained in the text. Possible sources of information are listed in the Bibliography at the end of this book.

1. Many of the mathematicians and statisticians mentioned in this short history of statistics are also known for other mathematical endeavors or had other interests besides mathematics. Do some research on two of the mathematicians mentioned in this section. Tell something about their lives, their achievements, and their interests.
2. The phenomenon of regression is mentioned in the *Short History of Statistics*. Find other examples of regression. The examples can come from studies of humans, animals, or other living organisms.
3. What is the A.C. Nielsen Co.? What does the company do? How are its rating statistics found and used?
4. Look in the reference section of the library and three different books that have statistical information. Give a short summary of the contents of each book.
5. What is the Dow Jones average? What does it represent? What is its history?
6. What was the Great Plague? Find some information about the cause of the plague, the number of deaths caused by the plague, and what countries were affected by the plague. What is its connection to statistics?
7. What is demography? What are its uses? Cite some examples.
8. The statistical graphs studied in Section 8.1 are used in newspapers and magazines to display data. Find some examples of statistical graphs and explain some implications or conclusions one would reach by examining the graphs.

9. What is the U.S. Bureau of the Census? What information does it collect? When, why, and how does it collect information?
10. Statistics are used in elections and voting. Cite some of the statistics that are kept for presidential elections. What do these tell you about the voters in the election?

## • Projects

1. What is the Gallup poll? How are this and other polls used in elections? What do experts say about these polls and the way they influence voters? How do politicians use these polls during an election?
2. What is the meaning of the expression "to lie with statistics"? Give some examples of this and explain how each of the examples is a "lie."
3. Statistics are used extensively in sports such as baseball, football, basketball, tennis, golf, bowling, horse racing, and car racing. Examine the statistics from one of these areas and describe the kinds of statistics that are kept and how these statistics are used.
4. Many "fast food" chains have pamphlets that state the amount of fat, sodium, calories, and vitamins contained in their menu items. Examine literature from several "fast food" chains and determine which food item would be the best to eat,
   (a) if you wanted to lessen your fat intake.
   (b) if you wanted to lessen your sodium intake.
   (c) if you wanted to lessen your calorie intake.
   (d) if you wanted to increase your intake of certain vitamins and minerals.

   Construct graphs of the types shown in Section 8.1.

## Section 8.0 Preliminaries

The character $\Sigma$ is the Greek letter uppercase sigma and is called a **summation** symbol. Like the factorial symbol used in the previous chapter, it has a special use in mathematics and is intended to make expressions more compact. $\Sigma$ is used to indicate addition of a sequence of terms. For example,

$$\sum_{i=1}^{6} i = 1 + 2 + 3 + 4 + 5 + 6$$

The equation is read "the sum of $i$ where $i$ goes from 1 to 6." This means that we want to add all the integer values from 1 through 6. The terms above and below the summation sign are called the **indices** and indicate the values that begin and end the sum. The expression that follows $\Sigma$ indicates the operation that is to be performed. For example:

$$\sum_{i=1}^{6} i^2 = 1^2 + 2^2 + 3^2 + 4^2 + 5^2 + 6^2$$

Since the expression to the right of $\Sigma$ indicates that the values of $i$ must be squared, each of the values 1 through 6 must be squared before they are added.

## Example 1

Compare the values of $\sum_{i=1}^{4} i^2$ and $\left(\sum_{i=1}^{4} i\right)^2$.

**Solution:** The value of $\sum_{i=1}^{4} i^2$ is

$$\sum_{i=1}^{4} i^2 = 1^2 + 2^2 + 3^2 + 4^2$$
$$= 1 + 4 + 9 + 16$$
$$= 30$$

The value of $\left(\sum_{i=1}^{4} i\right)^2$ is

$$\left(\sum_{i=1}^{4} i\right)^2 = (1 + 2 + 3 + 4)^2$$
$$= 10^2$$
$$= 100$$

Notice that the values are not the same! It is important to pay attention to whether the square is included in the summation or whether the result of the summation is squared. ■

If the expression following $\Sigma$ is enclosed in parentheses, the values of the entire expression must be summed. We see this in the following examples.

### Example 2

Find the value of $\sum_{i=3}^{7} (2i + 1)$.

**Solution:** Noticing that the values of $i$ start at $i = 3$, we have

$$\sum_{i=3}^{7} (2i + 1) = (2 \cdot 3 + 1) + (2 \cdot 4 + 1) + (2 \cdot 5 + 1) + (2 \cdot 6 + 1) + (2 \cdot 7 + 1)$$

$$= 7 + 9 + 11 + 13 + 15 = 55$$

### Example 3

Find the value of $\sum_{i=1}^{4} (3i + 2)^2$.

**Solution:** Letting $i$ take on the values from 1 through 4 gives

$$\sum_{i=1}^{4} (3i + 2)^2 = 5^2 + 8^2 + 11^2 + 14^2 = 406$$

One use of summations is to indicate that we want to add a long list of numbers. To do this, we introduce the notation $x_i$. Read *x sub i*, $x_i$ means the $i$th measurement of the variable $x$. For example, if the $x_i$ are the heights of trees, the statement "$x_5 = 62$ ft" means that the height of the fifth tree is 62 ft. The subscript 5 refers only to which tree is being measured, not to the height of a tree.

### Example 4

A group of bird-watchers counted the number of Cooper's hawks that passed the Fort Baker Lookout on five consecutive days. On the first day, the count was 47, the second day the count was 32, the third day 26, the fourth day 28, and the fifth day 25. Use summation notation and $x_i$ to represent the number of hawks that passed the Fort Baker Lookout.

**Solution:** Let $x_i$ be the number of Cooper's hawks that passed on the $i$th day. This gives

$$x_1 = 47, \quad x_2 = 32, \quad x_3 = 26, \quad x_4 = 28, \quad \text{and} \quad x_5 = 25$$

The total number of birds that passed the observations point is

$$\sum_{i=1}^{5} x_i = 47 + 32 + 26 + 28 + 25 = 158$$

## Operations with Summations

To conclude this section, we will present three rules of summations. First, suppose we have

$$\sum_{i=1}^{5} 3 = 3 + 3 + 3 + 3 + 3 = 5 \times 3$$

Notice that this result is merely the value of the upper index multiplied by the constant inside the summation sign. In general,

$$\sum_{i=1}^{n} c = n \times c \qquad \text{where } c \text{ is any constant}$$

Now, look at the problem

$$\sum_{i=1}^{5} 3i = 3(1) + 3(2) + 3(3) + 3(4) + 3(5)$$

$$= 3 \times (1 + 2 + 3 + 4 + 5) = 3 \sum_{i=1}^{5} i$$

This implies that if a variable inside the summation sign is multiplied by a constant, the constant can come out of the summation sign. In general,

$$\sum_{i=1}^{n} cx_i = c \sum_{i=1}^{n} x_i \qquad \text{where } c \text{ is any constant}$$

For the third rule, let's examine the problem

$$\sum_{i=1}^{3} (x_i + y_i) = (x_1 + y_1) + (x_2 + y_2) + (x_3 + y_3)$$

$$= (x_1 + x_2 + x_3) + (y_1 + y_2 + y_3)$$

$$= \sum_{i=1}^{3} x_i + \sum_{i=1}^{3} y_i$$

This implies that if several terms are added together inside of a summation sign, the summation sign can be used on each term separately. In general,

$$\sum_{i=1}^{n} (x_i + y_i) = \sum_{i=1}^{n} x_i + \sum_{i=1}^{n} y_i$$

### Example 5

Use the formulas given to simplify the expression $\sum_{i=1}^{n} (2x_i + 5)$.

**Solution:**

$$\sum_{i=1}^{n} (2x_i + 5) = \sum_{i=1}^{n} 2x_i + \sum_{i=1}^{n} 5 \quad \text{(by using the third rule)}$$

$$= 2\sum_{i=1}^{n} x_i + \sum_{i=1}^{n} 5 \quad \text{(by using the second rule)}$$

$$= 2\sum_{i=1}^{n} x_i + 5(n) \quad \text{(by using the first rule)}$$

## Section 8.0 PROBLEMS

## *Explain* ➡ *Apply*

### *Explain*

1. What is the summation symbol and how is it used?
2. What are the indices in summation notation and what are they used for?
3. What is the difference between $\sum_{i=1}^{5} i$ and $\sum_{i=1}^{5} x_i$? Write out the terms of both summations.
4. What are three rules for operating with summations?
5. What is the difference between $\sum_{i=1}^{5} 3i^2$ and $\left(\sum_{i=1}^{5} 3i\right)^2$?
6. What rule is used incorrectly in this simplification of the summation expression, $\sum_{i=1}^{n} 3x_i^2 = 3 + \sum_{i=1}^{n} x_i^2$? What is the correct simplification?
7. What rule is used incorrectly in this simplification of the summation expression, $\sum_{i=1}^{n} 3 = 3i$? What is the correct simplification?

*Apply*

In Problems 8–17, find the value of each summation.

**8.** $\sum_{i=1}^{5} (2i - 3)$

**9.** $\sum_{i=3}^{8} (i^2 - 3)$

**10.** $\sum_{i=1}^{5} (2i^2 + 3)$

**11.** $\sum_{i=3}^{8} (i^2 - 2i + 1)$

**12.** $\sum_{i=1}^{5} i^3$

**13.** $\left(\sum_{i=1}^{5} i\right)^3$

**14.** $3\sum_{i=1}^{10} i + \sum_{i=2}^{6} i^2$

**15.** $\sum_{i=0}^{4} (3i^2 - 5) - \sum_{i=1}^{4} (5 - i)$

**16.** $\dfrac{\sum_{i=1}^{6} 3i}{\sum_{i=1}^{6} i}$

**17.** $\dfrac{\sum_{i=1}^{6} 3i + 4}{\sum_{i=1}^{6} i}$

In Problems 18–21, write out the terms of each summation and simplify.

**18.** $\sum_{i=1}^{5} 3x_i + 4$

**19.** $\sum_{i=1}^{5} 3x_i - 4$

**20.** $\sum_{i=0}^{2} (x_i - 1)^2$

**21.** $\sum_{i=0}^{2} (3 + x_i)^2$

In Problems 22–26, simplify using the rules of summations.

**22.** $\sum_{i=1}^{n} 5x_i$

**23.** $\sum_{i=1}^{n} (5x_i + 3y_i)$

**24.** $\sum_{i=1}^{n} (5x_i + 3)$

**25.** $\dfrac{\sum_{i=1}^{n} (3x_i)}{12}$

**26.** $\dfrac{\sum_{i=1}^{n} (2x_i - 6)}{2}$

In Problems 27–32, write each expression using $\Sigma$ notation.

**27.** $1 + 2 + 3 + 4 + 5$

**28.** $3 + 4 + 5 + 6 + 7 + 8 + 9 + 10$

**29.** $2 + 4 + 6 + 8 + 10$

**30.** $3 + 6 + 9 + 12 + 15 + 18$

**31.** $1 + 4 + 9 + 16 + 25 + 36$

**32.** $1 + 3 + 5 + 7 + 9 + 11 + 13 + 15$

**33.** The term ''average'' is generally taken to mean the sum of a sequence of terms divided by the number of terms.

(a) Find the average of 7, 9, 8, and 12.

(b) If there are $n$ terms, $x_1$ to $x_n$, use $\Sigma$ notation to write a formula for the average.

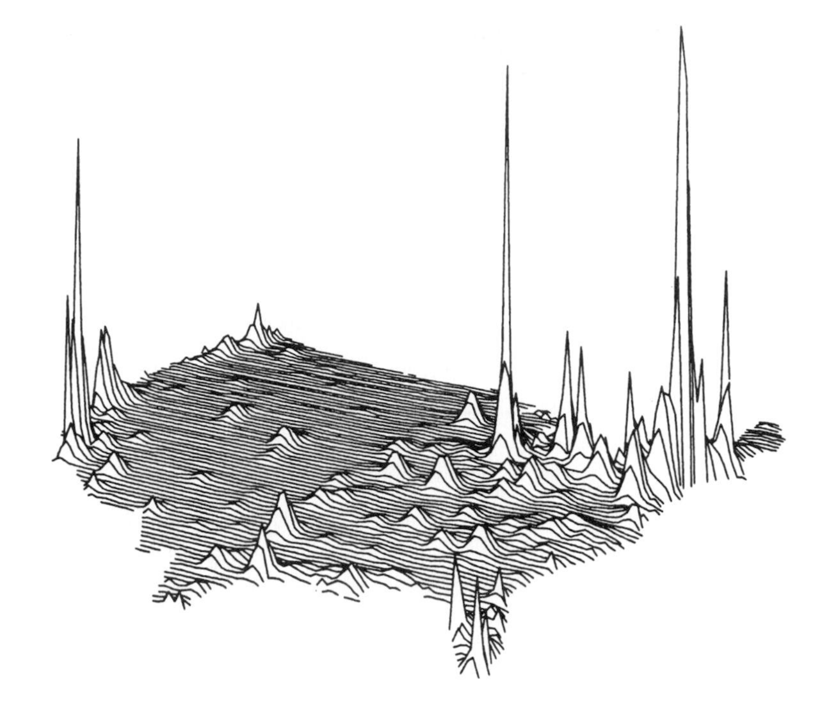

The population-density map (here of the United States) is a unique method of arranging information. (Produced by the Aspex Program at the Harvard Laboratory for Computer Graphics and Spatial Analysis)

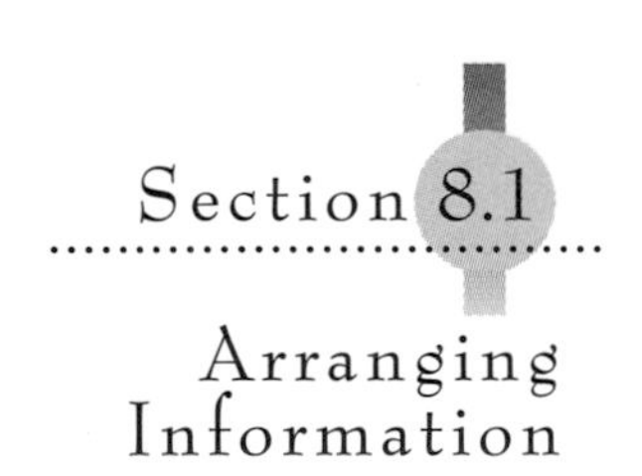

## Section 8.1 Arranging Information

Statistics are numerical data assembled in such a way as to present significant information about a subject. You come into contact with statistics on a daily basis. Credit bureaus use your payment history to determine your credit rating. Schools use standardized tests to determine student placement. Executives accumulate and organize data for a presentation. Government officials present findings regarding pollution, demographics, and economics. Magazines and newspapers contain articles with charts and data in almost every issue. In this section, we discuss methods of graphically presenting data.

### The Data

When a study or a poll is made, much of the information is in numerical form. Merely listed, the results of a survey can be overwhelming. Suppose 100 people respond to a survey with ten questions, each with numerical answers. There will be 10 numbers from each survey, giving a total of 1000 numbers. Most people will not be able to make any pertinent observations when the data is in this form. The following will show how to overcome such a difficulty.

Displayed here are the test scores of 70 students who have taken a math placement test. The test has a possible low score of 0 and a possible high score of 25.

| | | | | | | |
|---|---|---|---|---|---|---|
| 12 | 23 | 25 | 5 | 9 | 5 | 24 |
| 14 | 14 | 15 | 21 | 2 | 18 | 13 |
| 22 | 16 | 17 | 15 | 19 | 23 | 14 |
| 11 | 8 | 7 | 16 | 11 | 10 | 19 |
| 24 | 16 | 11 | 22 | 20 | 14 | 17 |
| 11 | 13 | 18 | 9 | 6 | 15 | 4 |
| 6 | 23 | 20 | 13 | 9 | 7 | 15 |
| 16 | 14 | 21 | 10 | 20 | 3 | 16 |
| 11 | 22 | 7 | 10 | 11 | 18 | 14 |
| 15 | 12 | 19 | 25 | 23 | 2 | 21 |

As we look at the data, there does not appear to be any pattern nor can we make any conclusions about the test scores. In order to examine the data more carefully, one technique is to put the data into categories.

When data are divided into categories, the different categories are called **classes**. The number and size of the classes we choose are determined by the goals of the investigation. Often, studies have predetermined classes. For example, many college courses use the standard groupings 90–100 for an A, 80–89 for a B, 70–79 for a C, 60–69 for a D, and 0–59 for an F. Economic studies concerned with household income have classes such as ''below the poverty level (less than $14,000 income per year),'' and ''lower middle income ($9000 to $16,000).''

A second criterion for choosing the classes is common sense. It would not be very wise to pick only one class because all the data will fall into that class. On the other hand, creating 30 classes for our data would mean that some of the classes would be empty and most of the classes would contain only a few scores. Finally, if we want the number of items in a class to be meaningful, the size of each class should be the same.

For our test data, we do not have any predetermined classes. Since the data vary from 0 through 25, it is convenient to pick five classes. This number was chosen because it allows us to create nearly equal size classes. The classes are 0 through 5, 6 through 10, 11 through 15, 16 through 20, and 21 through 25.

Now that we have chosen the classes, the next step is to determine the number of data values in each class. To do this, we have created the following chart, called a **frequency distribution**. The column labeled Tally is used to count the number of scores that fall into a class. The column labeled Frequency contains the number of tallies for that class. The entries in the column labeled Relative Frequency are found by dividing the frequency for each class by the total number of items, in this case 70. The final column is called Percentage Frequency. Its entries are found by multiplying the relative frequency by 100. They are the percentages of the data that fall into each category.

*Frequency Distribution*

| Class | Tally | Frequency | Relative Frequency | Per centage Frequency |
|---|---|---|---|---|
| 0–5 | 𝍸 \| | 6 | 0.09 | 9% |
| 6–10 | 𝍸 𝍸 \|\| | 12 | 0.17 | 17% |
| 11–15 | 𝍸 𝍸 𝍸 𝍸 \|\| | 22 | 0.31 | 31% |
| 16–20 | 𝍸 𝍸 𝍸 \| | 16 | 0.23 | 23% |
| 21–25 | 𝍸 𝍸 \|\|\|\| | 14 | 0.20 | 20% |

With the information presented in this way, we can see that of the 70 test scores, 22 were in the class from 11 through 15. This group comprised 31% of those who took the test.

Having the data organized in this fashion is a great improvement, but it is still a bit intimidating. Many people do not like to look at tables of numbers. Information can often be presented with greater impact if it is in graphical form. For our discussion, we will examine three of the many types of graphical representations. The first of these is called a **bar graph**. A bar graph uses one axis to represent the different classes while the other axis is used to indicate the frequency for each class.

Bar graphs may be oriented horizontally or vertically and may use either the frequencies or the relative frequencies as the lengths of the bars. The choice of display is left to the person creating the graphs. With a bar graph, it is easy to see which class contains the largest number of test scores. The numbers at the ends of the bar permit the reader to determine the actual number of scores in each class.

Three different bar graphs of our data are shown.

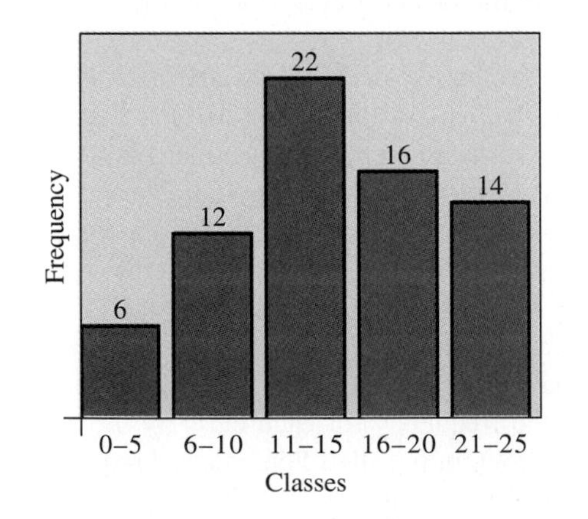


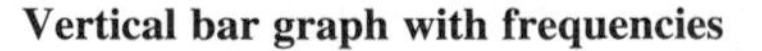
**Vertical bar graph with frequencies**

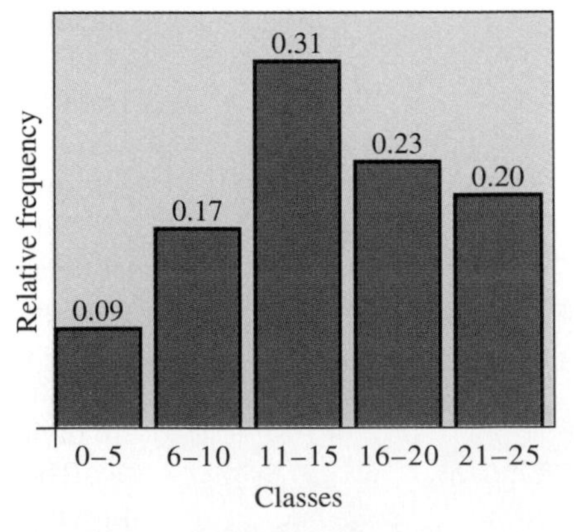


**Vertical bar graph with relative frequencies**

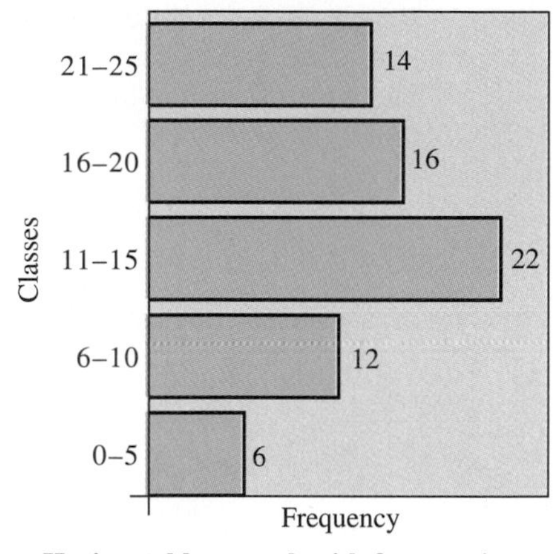

**Horizontal bar graph with frequencies**

A second way that information can be displayed is with a **line graph**, also called a frequency polygon or broken line graph. In a line graph, the bars of a bar graph are replaced with dots that are connected by line segments. A line graph that represents the frequency of scores on the math placement test is shown on the figure.

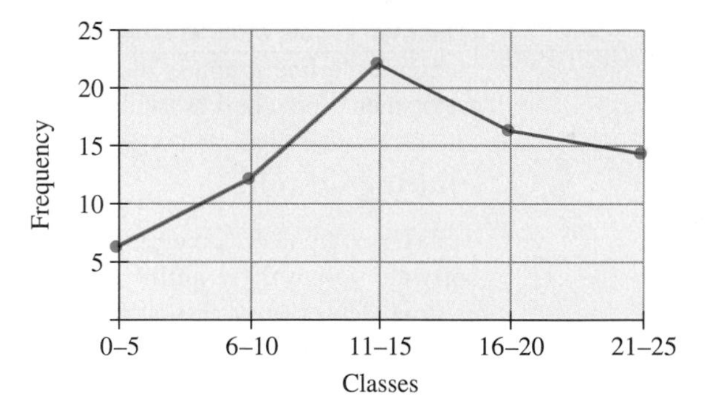


A line graph allows you to quickly see the increases and decreases in the number of scores in each class. For example, you can see that there is a large jump between the number that scored from 6–10 and the number that scored from 11–15. On the other hand, you can see that there is only a slight decrease from the number that scored from 16–20 and the number that scored from 21–25.

A third way that information can be presented is through the use of a **pie chart**, also called a circle graph. A pie chart is a circular diagram divided into sectors (wedge-shaped pieces) where the areas of the sectors are used to represent percentage frequencies. Each sector represents a part or percentage of a whole. The percentage frequency determines the angle of the sector, with the entire circle representing 100%. The percentage frequencies of the math placement test can be displayed in the pie chart that follows. From the pie chart you can visually determine the relative size of each class. You can quickly see which class had the largest number of scores and which class had the smallest number of scores.

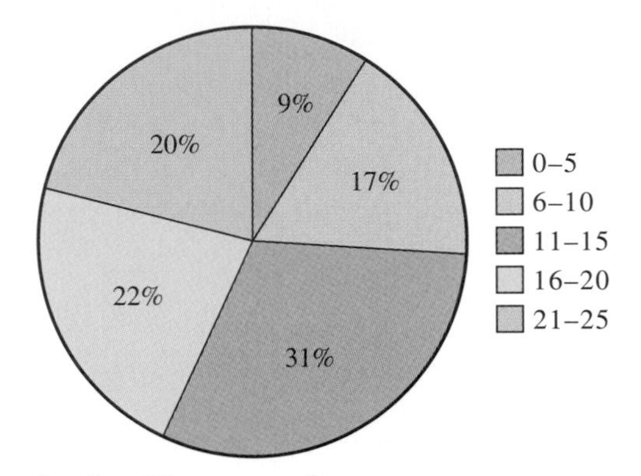


### *Which Type of Graph Should Be Used?*

Since there are three basic types of statistical graphs, which one should be used for a given set of data? Depending on what you are trying to emphasize, the data can be displayed with any of these three statistical graphs. If you want to clearly show which category contains the largest or the least number of items, a bar graph is effective. If you want to emphasize the patterns of change (rise and fall) of various categories, a line graph is useful. If you want to show how a whole is divided into parts, then a pie chart is the best one to use.

## Making Graphs

Today's computers accomplish the drawing of these graphs by using chart-drawing software. Often these abilities are incorporated into large software packages, such as spreadsheet programs and word processors. However, it may be necessary to create a graph without the aid of a computer. The remaining part of this section describes how to draw these basic statistical graphs.

Suppose you wanted to compare the voters from the states of Florida and New Hampshire. According to the Committee for the Study of the American Electorate in 1990, the registered voters in those states had the following political affiliations.

| **Florida** | | **New Hampshire** | |
|---|---|---|---|
| Democrats | 52% | Democrats | 29% |
| Republicans | 41% | Republicans | 39% |
| Other parties | 7% | Other parties | 32% |

The data can be represented by either a bar graph, pie chart, or line graph.

## *Bar Graphs*

To make a bar graph, determine the following:

**1.** The categories to be placed along the vertical and horizontal axes.
**2.** The scale used to encompass the numerical data.
**3.** The style of bars used and the length of each bar.

One possible way to draw the bar graph comparing the voter registration data for Florida and New Hampshire is to have the vertical axis show the percentage of the registered voters and the horizontal axis list the political parties. Since the data consist of percentages that range from 7% to 52%, by letting the vertical scale be 0.5 in. for every 10%, the graph will fit in a reasonable area. To find the length of each bar we divide each percentage by 10 and multiply the result by 0.5 in. Thus, the length of the bar for each percentage is as follows:

$$52\% \rightarrow \frac{52}{10} \times 0.5 = 2.6 \text{ in.} \qquad 29\% \rightarrow \frac{29}{10} \times 0.5 = 1.45 \text{ in.}$$

$$41\% \rightarrow \frac{41}{10} \times 0.5 = 2.05 \text{ in.} \qquad 39\% \rightarrow \frac{39}{10} \times 0.5 = 1.95 \text{ in.}$$

$$7\% \rightarrow \frac{7}{10} \times 0.5 = 0.35 \text{ in.} \qquad 32\% \rightarrow \frac{32}{10} \times 0.5 = 1.6 \text{ in.}$$

Finally, since we have two states to compare, we can use a solid bar for Florida and a striped bar for New Hampshire. Using vertical bars of the given lengths, we get the following graph.

**Registered Voters 1990**

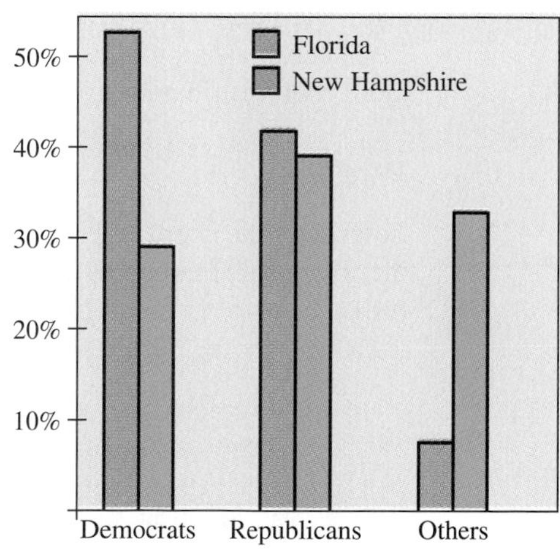


The bar graph allows us to visualize the data. By simply looking at the bar graph, we can make conclusions about the registered voters in Florida and New Hampshire. For example, the percentage of Republicans in each state is similar.

## *Line Graphs*

Using the same values as calculated from the previous bar graph, we can create a line graph that compares the voter registration for Florida and New Hampshire. We can simply place dots on the vertical scale for Florida at 2.6 in. (Democrats), 2.05 in. (Republicans), 0.35 in. (Others) and for New Hampshire at 1.45 in. (Democrats), 1.95 in. (Republicans), 1.6 in. (Others) and then connect those dots with line segments.

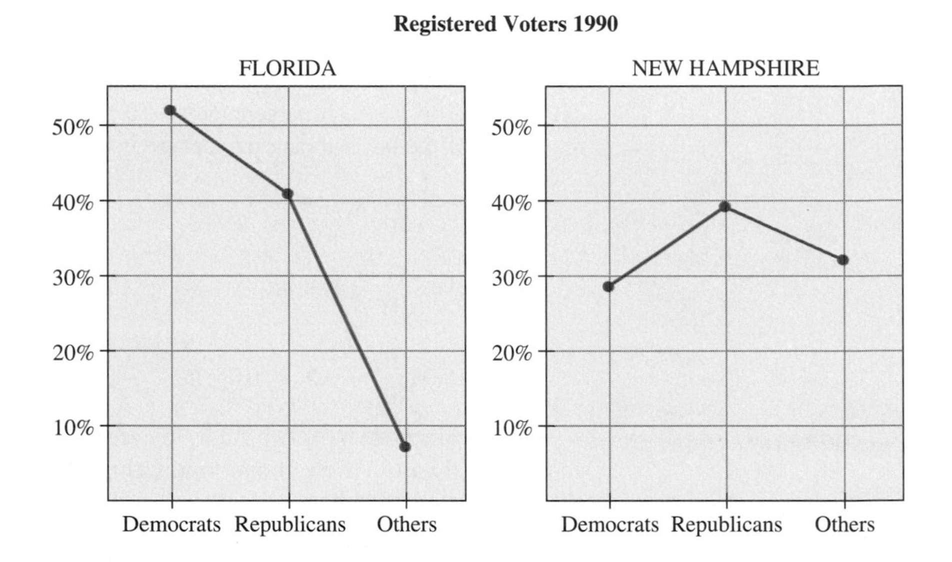


From the line graphs we can quickly see that in Florida there is a large drop from Democratic voters to other political affiliations while in New Hampshire there is a more even distribution of voters.

## *Pie Charts*

To make a pie chart, determine the size of each sector that will be used to represent the percentage frequency. Since there are 360° around the center of a circle, the angle for each sector is found by multiplying the percentage by 360°. Thus, for the registered voters for Florida and New Hampshire the angle of each sector is as follows:

| Florida | New Hampshire |
|---|---|
| $52\% \rightarrow 0.52 \times 360° = 187.2°$ | $29\% \rightarrow 0.29 \times 360° = 104.4°$ |
| $41\% \rightarrow 0.41 \times 360° = 147.6°$ | $39\% \rightarrow 0.39 \times 360° = 140.4°$ |
| $7\% \rightarrow 0.07 \times 360° = 25.2°$ | $32\% \rightarrow 0.32 \times 360° = 115.2°$ |

Using a protractor to measure the desired angles, divide each circle into three sectors.

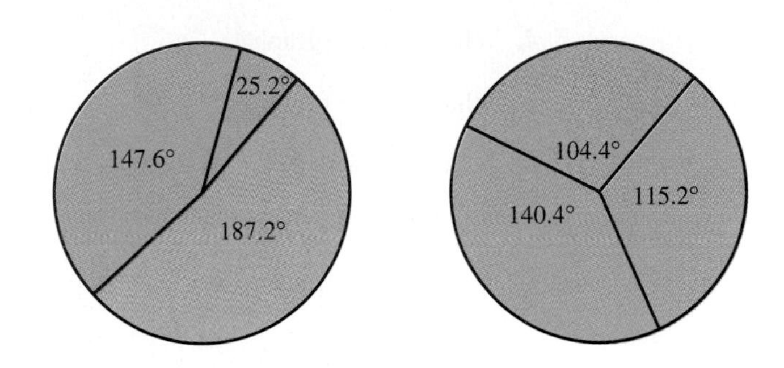


Using a different pattern for each political party and including labels, we get the following pie charts for the 1990 registered voters in Florida and New Hampshire.

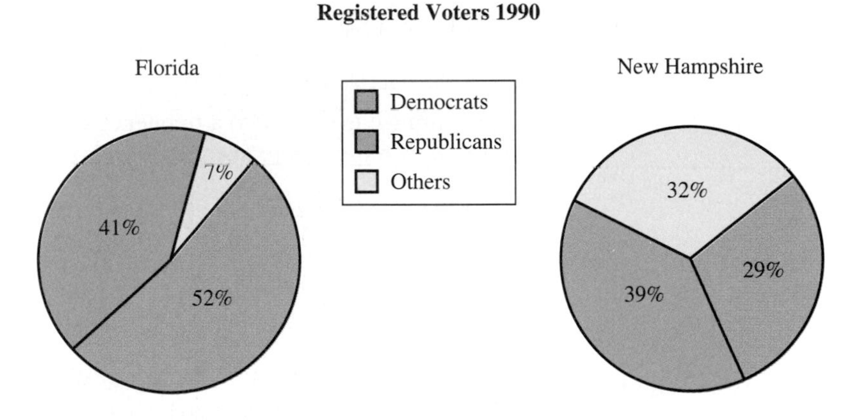


As we did with the bar and line graphs, we can draw conclusions about the registered voters in Florida and New Hampshire. If you examine newspapers, magazines, business reports, advertising material, and demographic studies, you will see that a variety of data is displayed using various forms of these three basic statistical graphs.

## Section 8.1 PROBLEMS

## *Explain ➡ Apply ➡ Explore*

### *Explain*

1. What are statistics?
2. What is a bar graph?

3. What is the difference between the frequency, relative frequency, and percentage frequency distribution?

4. What is a line graph?

5. What is a pie chart?

6. What are the three major steps in making a bar graph?

7. How is the angle of each sector of a pie chart determined?

8. When would it be effective to display data using a bar graph? line graph? pie chart?

## *Apply*

9. For the data 2, 4, 6, 1, 7, 9, 5, 3, 7, 6:
   (a) Arrange the data into a frequency distribution with three classes.
   (b) Draw a bar graph, line graph, and pie chart.

10. For the data 2, 4, 3, 7, 2, 8, 6, 3, 7, 5:
   (a) Arrange the data into a frequency distribution with four classes.
   (b) Draw a bar graph, line graph, and pie chart.

11. For the data 10, 12, 14, 13, 17, 12, 18, 16, 13, 17, 15, 17:
   (a) Arrange the data into a frequency distribution with five classes.
   (b) Draw a bar graph, line graph, and pie chart.

12. For the data 20, 29, 21, 21, 28, 28, 23, 23, 23, 26, 26, 26, 25, 25, 25, 25, 25:
   (a) Arrange the data into a frequency distribution with five classes.
   (b) Draw a bar graph, line graph, and pie chart.

13. For the data 42, 64, 75, 82, 96, 93, 77, 82, 67, 78, 88, 90, 80, 72, 71, 80, 81, 98, 61, 75:
   (a) Arrange the data into a frequency distribution with six classes.
   (b) Draw a bar graph, line graph, and pie chart.

14. The bar graph shown gives the number of seeds of a particular type of cactus that germinated under greenhouse conditions within a specified number of weeks after planting. At the end of six weeks, the experiment was discontinued.
   (a) Determine the total number of germinated seeds.
   (b) What percentage of the germinating seeds sprouted during each week?
   (c) Make use of the results of part (b) to create a line graph and pie chart displaying the percentage of seeds sprouting in each of the six weeks.

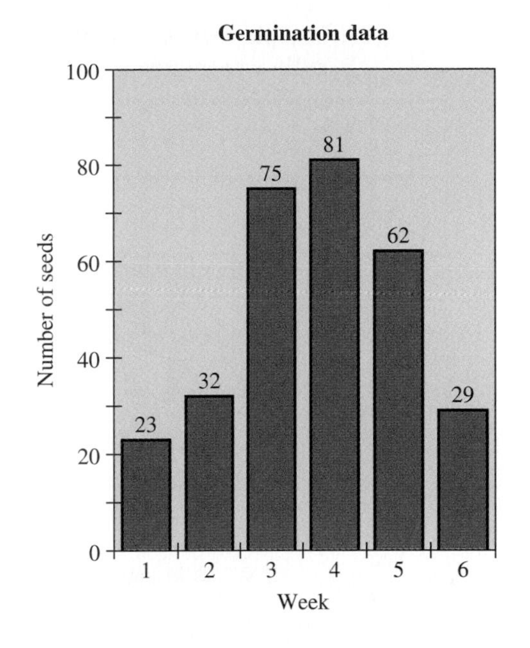


**15.** The bar graph shows the 1990 voting-age population in the five Pacific states of Alaska (AK), Washington (WA), Oregon (OR), California (CA), and Hawaii (HI). (Source: U.S. Bureau of the Census)

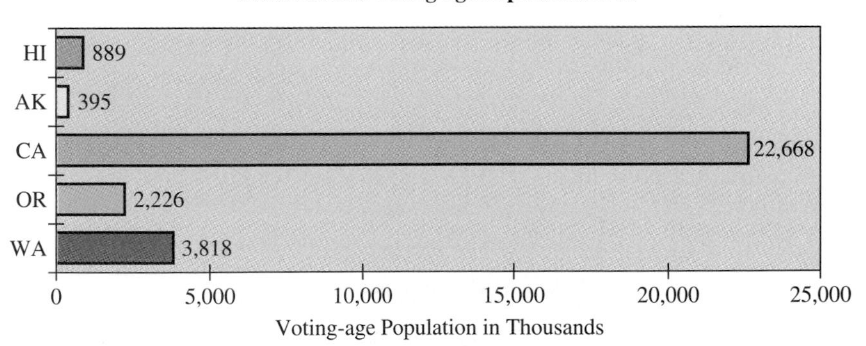


(a) How many more people of voting age are there in California than Alaska?
(b) What percent of the voting-age population in the Pacific states resided in each of the five states in 1990?
(c) Use the results of part (b) to make a pie chart showing the percentage of the voting-age population in the Pacific states in 1990.

**16.** The pie chart shown gives the percentage of 1990 U.S. voters in each of six different age-groups. (Source: U.S. Bureau of the Census)
In 1990, there were 182,100,000 U.S. citizens eligible to vote.

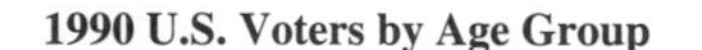


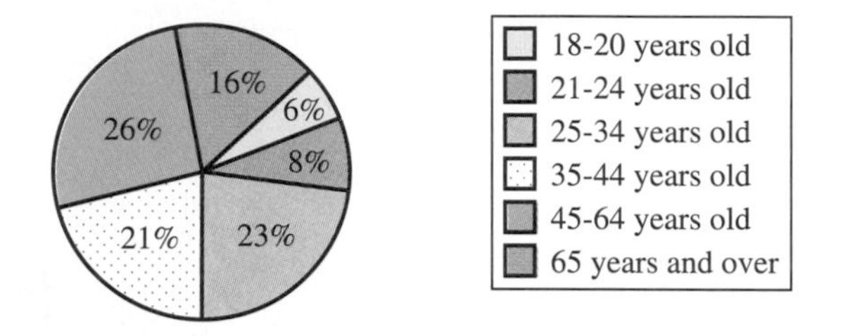


(a) How many people were in each age-group?
(b) Use the results of part (a) to make a bar and line graph showing the number of voters in each age-group.

**17.** The following pie chart shows beverage consumption of the typical U.S. resident in 1990. (Source: The U.S. Department of Agriculture Economic Research Service)

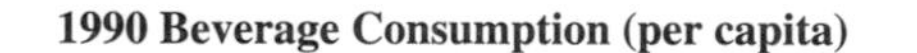


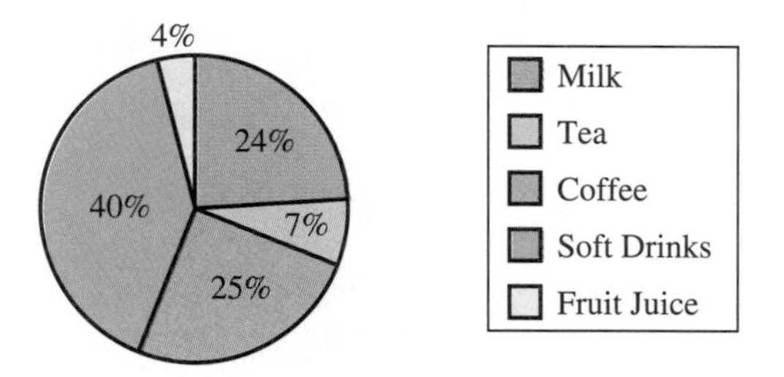


In 1990, the typical person consumed 106 gallons of these beverages.
(a) How many gallons of each beverage did the typical person consume?
(b) Use the results of part (a) to construct a bar graph.

## *Explore*

**18.** The following table gives the average monthly salaries for people 18 years or older at various levels of education. (Source: U.S. Bureau of the Census, *Current Population Reports*)

| Education Level | 1990 Average Monthly Income Males | Females |
|---|---|---|
| Elementary | $1,599 | $1,110 |
| High School (1–3 years) | $1,880 | $1,282 |
| High School (4 years) | $2,337 | $1.580 |
| College (1–3 years) | $2,849 | $1,888 |
| College (4 years) | $3,713 | $2,409 |
| College (5 or more years) | $4,653 | $2,986 |

(a) Create a bar graph to display the data for both sexes.
(b) Create a line graph to display the data for both sexes.
(c) Using the graphs, make three pertinent observations about the data.

**19.** The number of congressional bills vetoed by the presidents of the United States from 1961 to 1992 is as follows. (Source: *Congressional Quarterly*)

| President | Period | Total Vetoes |
|---|---|---|
| Kennedy (D) | 1961–63 (≈3 yr) | 21 |
| Johnson (D) | 1963–69 (≈5 yr) | 30 |
| Nixon (R) | 1969–74 (5 yr) | 42 |
| Ford (R) | 1974–77 (3 yr) | 72 |
| Carter (D) | 1977–81 (4 yr) | 31 |
| Reagan (R) | 1981–89 (8 yr) | 78 |
| Bush (R) | 1989–93 (4 yr) | 57 |

(a) Make a bar graph for this data where the number of vetoes are on the vertical axis and the president is on the horizontal axis.
(b) From the graph in part (a), which president vetoed the most and vetoed the least?
(c) Create a similar bar graph using the number of vetoes each president enacted per year.
(d) From the graph in part (c), which president vetoed the most and which president vetoed the least?
(e) Use the above results to explain how graphs can be misleading.
(f) Create a pie chart to show a comparison of Republican and Democratic presidents in using the veto. Explain why this pie chart may be misleading.

**20.** The number of U.S. governors by political party affiliation is given in the chart that follows. (Source: National Governors Association, Washington, DC)

| Year | Democratic | Republican | Independent |
|---|---|---|---|
| 1986 | 34 | 16 | 0 |
| 1987 | 26 | 24 | 0 |
| 1988 | 27 | 23 | 0 |
| 1989 | 28 | 22 | 0 |
| 1990 | 29 | 21 | 0 |
| 1991 | 28 | 20 | 2 |
| 1992 | 27 | 21 | 2 |

(a) Make a bar graph to display the data by using different coded bars for each political party at each year.
(b) Make one line graph to display the data by using a different style of dots for each political party at each year
(c) What do the graphs show about the political party affiliation of governors from 1986 to 1992?
(d) If the data from 1986 is omitted, what do the graphs show about the political affiliation of governors?

**21.** The table below shows the number of women who held statewide elective executive offices or who were members of state legislatures in 1992. (Source: Center for American Women and Politics)

| | | | | | | | | | | | |
|---|---|---|---|---|---|---|---|---|---|---|---|
| AL | 9 | AK | 14 | AZ | 33 | AR | 12 | CA | 24 | CO | 34 |
| CT | 45 | DE | 9 | FL | 31 | GA | 34 | HI | 21 | ID | 37 |
| IL | 34 | IN | 28 | IA | 25 | KS | 47 | KY | 8 | LA | 12 |
| ME | 60 | MD | 44 | MA | 37 | MI | 23 | MN | 44 | MS | 12 |
| MO | 32 | MT | 33 | NE | 11 | NV | 15 | NH | 131 | NJ | 15 |
| NM | 17 | NY | 27 | NC | 25 | ND | 25 | OH | 21 | OK | 16 |
| OR | 25 | PA | 26 | RI | 26 | SC | 23 | SD | 28 | TN | 15 |
| TX | 26 | UT | 12 | VT | 56 | VA | 18 | WA | 49 | WV | 28 |
| WI | 33 | WY | 25 | | | | | | | | |

(a) Arrange the data into a frequency distribution with seven classes, the first one being 1–20.

(b) Draw a line graph for the classes.
(c) Draw a bar graph for the classes.
(d) Make a pie chart for the classes.

**22.** The table gives the median household income in 1990 for each state and the District of Columbia. Income is given in constant 1990 dollars. (Source: U.S. Bureau of the Census, *Current Population Reports*)

| | | | | | | | | | | | |
|---|---|---|---|---|---|---|---|---|---|---|---|
| AL | 23,397 | AK | 39,298 | AZ | 29,224 | AR | 22,786 | CA | 33,290 | CO | 30,733 |
| CT | 38,870 | DE | 30,804 | DC | 27,392 | FL | 26,685 | GA | 27,561 | HI | 38,921 |
| ID | 25,305 | IL | 32,542 | IN | 26,928 | IA | 27,288 | KS | 29,917 | KY | 24,780 |
| LA | 22,405 | ME | 27,464 | MD | 38,857 | MA | 36,247 | MI | 28,937 | MN | 31,465 |
| MS | 20,178 | MO | 27,332 | MT | 23,375 | NE | 27,482 | NV | 32,023 | NH | 40,805 |
| NJ | 38,734 | NM | 25,039 | NY | 31,591 | NC | 26,329 | ND | 25,264 | OH | 30,013 |
| OK | 24,384 | OR | 29,281 | PA | 29,005 | RI | 31,968 | SC | 28,735 | SD | 24,571 |
| TN | 22,592 | TX | 28,228 | UT | 30,142 | VT | 31,098 | VA | 35,073 | WA | 32,112 |
| WV | 22,137 | WI | 30,711 | WY | 29,460 | | | | | | |

(a) Arrange the data into a frequency distribution with five classes, the first one being 20,001–25,000.
(b) Draw a bar graph for the classes.
(c) Draw a pie chart for the classes.

**23.** The fees collected for Vehicle Licenses and Registration in the State of California for the budget year 1993–94 were distributed as follows: (Source: California Department of Motor Vehicles Document, DMV 77 B REV 8/93)

| Distribution of License and Registration Fees (1993–94) | |
|---|---|
| City and County, Local Governments | $3044 million |
| California Highway Patrol | $ 661 million |
| Dept. of Motor Vehicles | $ 513 million |
| State Highways | $ 381 million |
| State's General Fund | $ 450 million |
| Various State Agencies* | $ 121 million |
| Total | $5170 million |

*Includes Dept. of Justice, Air Resources Board, and Environmental Agencies.

(a) Which type of graph would be effective in displaying the data? Explain.
(b) Draw the graph.

**24.** Another graphical representation for data displays cumulative frequencies. The cumulative frequency for each class is determined by adding the frequency of the class to the combined frequency of the previous classes. In such a graph the cumulative frequency for the last class must be the same as the total number of values in the survey. These graphs are called ogives (pronounced *oh'jives*). Ogives can be used to compare data that use time as the variable for the classes. For example, two marathons, run in two different cities, have the finishing times presented in the two ogives shown.

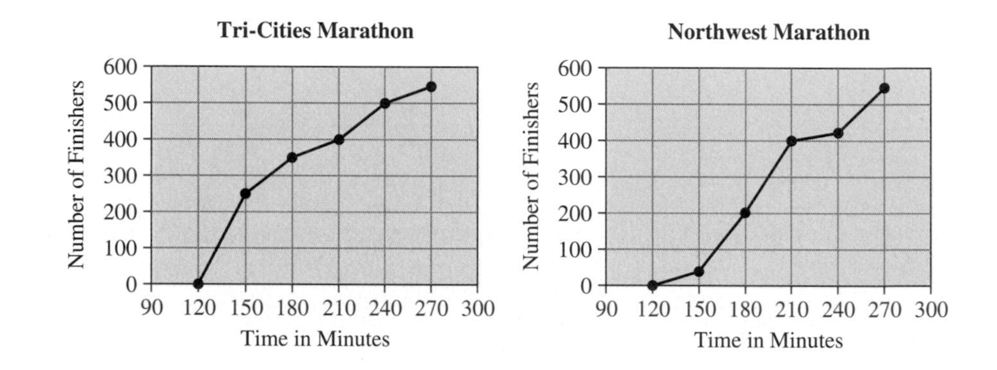


(a) How many runners finished in 4 hours or less in each marathon?
(b) How many runners finished in 3 hours or less in each marathon?
(c) What percentage of the runners finished in 3 hours or less in each marathon?
(d) If we assume that the abilities of the runners in each marathon were similar, which marathon was an easier race? Explain your reasoning.

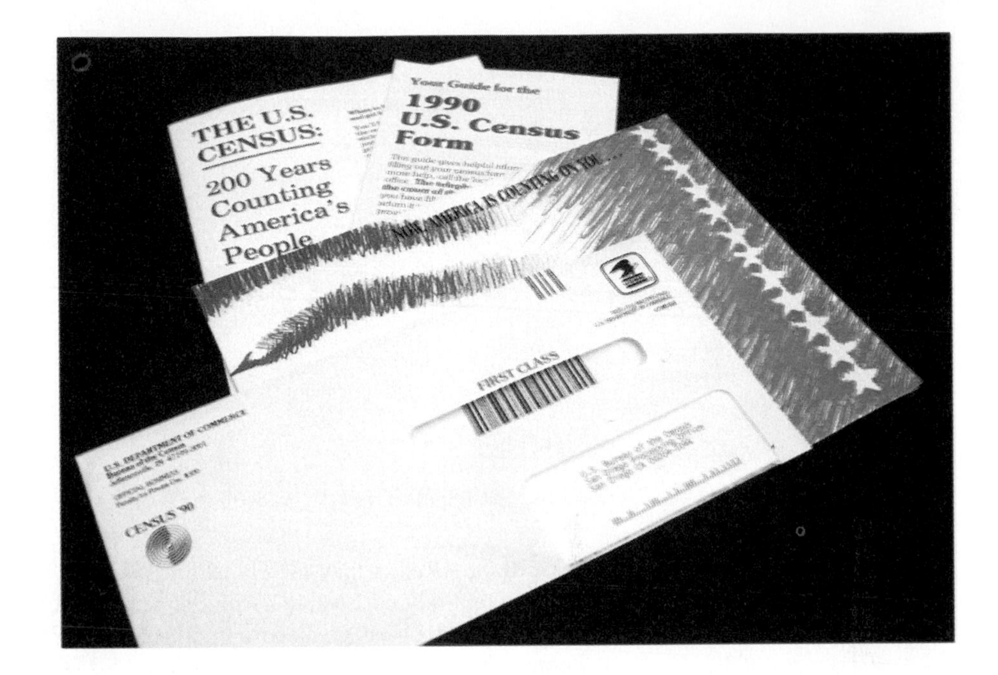


Census forms are an efficient means of collecting the information that is used in the statistical analysis of many facets of American life. (Courtesy of Kurt Viegelmann)

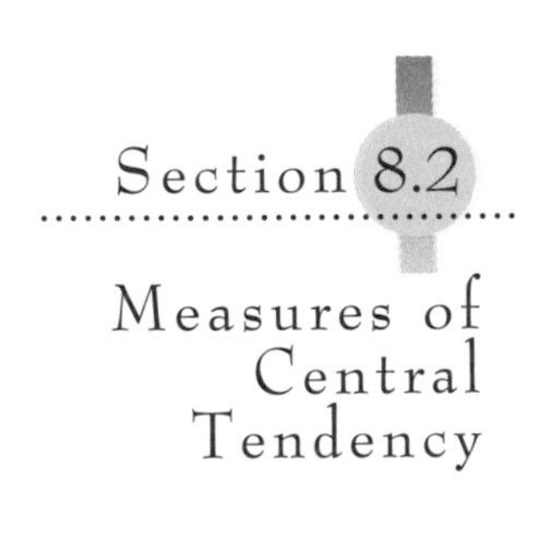

## Section 8.2 Measures of Central Tendency

Finding an average is a computation that requires only arithmetic skills. It has, however, very important uses in real-world problems. An average allows us to find one value that represents the middle of a set of data. As we shall see, however, the middle of a set of data can be described in different ways. Collectively, these ways of describing the middle value are called **measures of central tendency**. In this section, we introduce three measures of central tendency and describe situations in which one of the three measures is the most appropriate to use.

### The Mean

The **mean** or **arithmetic** (pronounced *ar ith met′ ic*) **mean** of a set of values is what most people are referring to when they say "average." It is found by adding up all the data and then dividing by the number of data values. Statisticians use the symbol $\mu$ (read *mu*) for the mean and define it with the formula

$$\text{mean} = \mu = \frac{\sum_{i=1}^{n} x_i}{n} \qquad \text{where} \begin{cases} n \text{ is the number of data values} \\ x_i \text{ represent the data values} \end{cases}$$

### Example 1

Find the mean of the numbers 3, 6, 8, 5, 4.

**Solution:**

$$\mu = \frac{\sum_{i=1}^{5} x_i}{5} = \frac{3 + 6 + 8 + 5 + 4}{5} = \frac{26}{5} = 5.2$$

### Example 2

Find the mean of 1, 1, 1, 1, 3, 3, 5, 5, 6, 6, 6, 6, 6.

**Solution:** Rather than merely adding all these numbers, it is more convenient to multiply each value by its frequency and then add. This total is then divided by the total frequency. In other words, rather than doing the computation

$$\mu = \frac{1 + 1 + 1 + 1 + 3 + 3 + 5 + 5 + 6 + 6 + 6 + 6 + 6}{13}$$

$$= \frac{50}{13} = 3.85$$

it is easier to write

$$\mu = \frac{4(1) + 2(3) + 2(5) + 5(6)}{13} = \frac{50}{13} = 3.85$$

In general, when a set of data has values that appear several times or when the data are grouped into classes, we can use the formula

$$\text{mean for grouped data} = \mu = \frac{\sum_{i=1}^{n} f_i x_i}{\sum_{i=1}^{n} f_i} \quad \text{where} \begin{cases} n \text{ is the number of classes} \\ f_i \text{ is the frequency of each class} \\ x_i \text{ represent the data values} \end{cases}$$

## Example 3

Find the mean of the following frequency distribution.

| Class | Frequency | Class Midpoint $x_i$ |
|---|---|---|
| 1–5 | 6 | 3 |
| 6–10 | 12 | 8 |
| 11–15 | 22 | 13 |
| 16–20 | 16 | 18 |
| 21–25 | 14 | 23 |

**Solution:** In the table, there is a column that we have not previously used. The **class midpoint** is the mean of the boundary values of each class. For example, the boundary values for the first class are 1 and 5. The mean of these two numbers is 3. This gives the midpoint for the class. We use the midpoints because we do not know the actual values of the data in each class. For example, we know that there are six values in the 1–5 class, but we do not know what those values are. As an estimate of these values, we will use the class midpoints. Now, using the formula for the mean of grouped data, we get

$$\begin{aligned}\mu &= \frac{\sum_{i=1}^{5} f_i x_i}{\sum_{i=1}^{5} f_i} \\ &= \frac{6(3) + 12(8) + 22(13) + 16(18) + 14(23)}{6 + 12 + 22 + 16 + 14} \\ &= \frac{1010}{70} \\ &= 14.43\end{aligned}$$

■

Another use of the formula for the mean of grouped data is to compute a **weighted average**. A weighted average is the mean of a group of numbers in which certain values have more importance, or weight, than do other values. When computing a weighted average, we **use the weights of each value as frequencies in the mean of grouped data formula**. An example of this type of situation can be seen in the next problem, concerning a student's grade.

### Example 4

A student is trying to calculate his grade in an English class. His total scores are

| | |
|---|---|
| Midterms | 82 |
| Homework | 87 |
| Final | 92 |

The course syllabus says that the midterms count for 60% of the grade, homework for 10% of the grade, and the final for 30% of the grade. Compute the student's total score in the class.

**Solution:** Using the percentages as the weights of the scores in each category, we have

$$\mu = \frac{0.60(82) + 0.10(87) + 0.30(92)}{0.60 + 0.10 + 0.30} = \frac{85.5}{1.00} = 85.5$$

## The Median

### Example 5

Five houses are listed for sale in a real estate broker's advertisement. The prices are \$269,000, \$256,000, \$249,000 \$235,000, and \$749,000. Find the average price of the houses listed by the agent.

**Solution:**

$$\mu = \frac{\sum_{i=1}^{5} x_i}{5} = \frac{269{,}000 + 256{,}000 + 249{,}000 + 235{,}000 + 749{,}000}{5}$$

$$= \frac{1{,}758{,}000}{5}$$

$$\mu = \$351{,}600$$

Notice that the average is higher than all but one of the house prices. The reason is that the price of the most expensive house is distorting the results. The mean price of these five houses is \$351,600, but the mean does not accurately reflect the typical or ''average'' price.

To solve this difficulty, we will use a measure of central tendency called the **median**. The median is found by listing the data in increasing order and choosing the middle value. If there is an even number of items, the median is found by taking the mean of the two middle values. The median is frequently used when there are a few extreme values in the data that will greatly alter the value of the mean.

### Example 6

Find the median of the prices listed in Example 5.

**Solution:** Listing the data in increasing order gives

$235,000, $249,000, $256,000, $269,000, $749,000

The median price is $256,000 because this is the middle value. This value is much closer to what we would call the ''average'' price of these five houses. Also notice that increasing the price of the most expensive house will have no effect on the value of the median. ■

### Example 7

Find the median of the values 0, 3, 4, 9, 16, 90.

**Solution:** The values are listed in order, so we need only find the middle value. Since there is an even number of values, the median is found by calculating the mean of the center pair of values.

$$\text{median} = \frac{4 + 9}{2} = 6.5$$ ■

## The Mode

The **mode** of a set of data is the value that occurs most frequently. It is the only measure of central tendency that can be used with nonnumerical as well as numerical data. For example, if a design consultant wanted to determine the most popular color for exterior house paint, he or she would conduct a survey to find out how many people liked each color. It would not be possible to use the mean or median.

### Example 8

Determine the mode of the following sets of values.

(a) 1, 2, 2, 3, 4, 6, 6, 6, 8, 9
(b) 1, 2, 2, 3, 4, 6, 6, 7, 8, 9
(c) 1, 2, 2, 3, 4, 4, 6, 6, 8, 9

**Solution:**

(a) In the first group of data, the number 6 appears three times, and all the other numbers appear at most twice. Therefore, the mode is 6.

(b) In the second group, the numbers 2 and 6 both appear twice. No other number appears more than once. In a situation like this, there are two modes, 2 and 6. When a set of data has two modes, it is called **bimodal**.

(c) In the third set of data, the numbers 2, 4, and 6 all appear twice. When more than two values have the highest frequency, we say that the set of data does not have a mode. ■

## Example 9

In 1990, there were 66,090,000 families in the United States. Of these families, 33,801,000 had no children, 13,530,000 families had one child, 12,263,000 had two children, 4,650,000 families had three children, and 1,846,000 families had four or more children.

(a) Determine the modal number of children per family.
(b) Determine the mean number of children per family.

**Solution:**

(a) Since the class with the highest frequency is the class with no children, the modal number of children is 0.

(b) Finding an accurate mean of the data will not be possible. Since the final category consists of families with four or more children, we do not know the actual number of children in these families. However, we can use the method of finding the mean for grouped data to give an estimate.

$$\begin{aligned}\mu &= [33{,}801{,}000(0) + 13{,}530{,}000(1) + 12{,}263{,}000(2) \\ &\quad + 4{,}650{,}000(3) + 1{,}846{,}000(4)] \div 66{,}090{,}000 \\ &= \frac{59{,}390{,}000}{66{,}090{,}000} \approx 0.90\end{aligned}$$

Since the number of children in the last class may be greater than four per family, we can say that the mean number of children per family is greater than or equal to 0.90. ■

## Example 10

From data published in the *Congressional Directory*, the makeup of the 1991 U.S. Congress by age is as shown in the bar graphs.

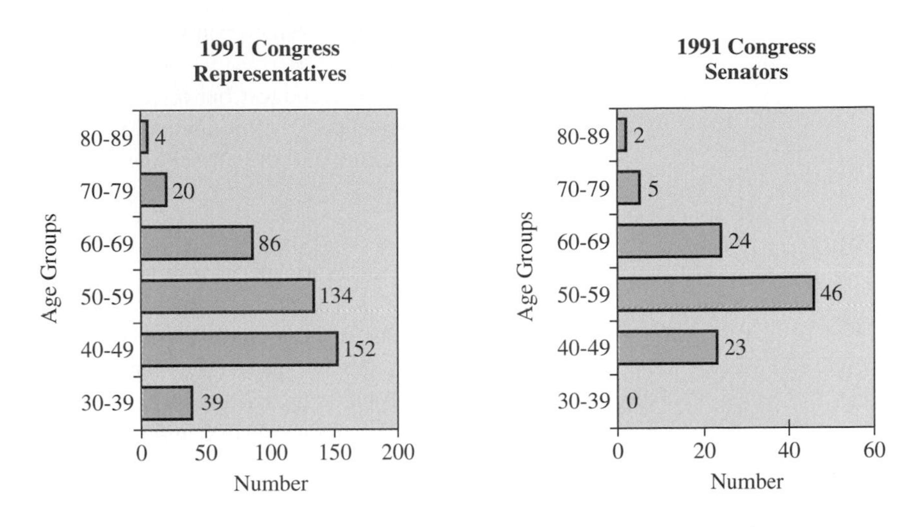


(a) Find the mean, median, and mode of the ages of the representatives and senators of the 1991 Congress.

(b) Which of the three measures gives a better approximation of the ''average'' age of a representative or a senator? Explain.

**Solution:**

(a) *Mean:* There are 435 representatives. Using the class midpoints and corresponding frequencies, the mean age of the representatives is

$$\mu = \frac{39(34.5) + 152(44.5) + 134(54.5) + 86(64.5) + 20(74.5) + 4(84.5)}{435}$$

$$= \frac{22{,}787.5}{435} \approx 52.4$$

There are 100 senators. Using the class midpoints and corresponding frequencies, the mean age of the senators is

$$\mu = \frac{0(34.5) + 23(44.5) + 46(54.5) + 24(64.5) + 5(74.5) + 2(84.5)}{100}$$

$$= \frac{5620}{100} = 56.2$$

*Median:* Since we do not have a list of all the ages of either the representatives or the senators, we cannot find the median of the data. However, we can conclude that the middle value of the representatives (218th one) lies in the 50–59 age-group. For the senators the middle value also lies in the 50–59 age-group. Thus, the median age of both houses of Congress is in the 50s.

*Mode:* The longest bar in each graph shows the class with the greatest frequency. Thus, the modal age of the representatives is the 40–49 year-old age-group and the modal age for senators is the 50–59 year-old age-group.

(b) The mean seems to give the best approximation for the "average" age of representatives and senators since there are no very extreme ages that are affecting its value. The mean shows that there is a difference between the two groups while including the effect of all the age classes. On the other hand, the median just tells us the "average" age is from 50–59. The mode only tells us that the largest number of representatives are in the 40–49 year-old age-group and the largest number of senators are in the 50–59 year-old age-group. Neither the median nor the mode gives us a good indication of the "average" age of the Congress. ■

## Section 8.2 PROBLEMS

## *Explain ➡ Apply ➡ Explore*

### *Explain*

**1.** What are measures of central tendency?

**2.** What is the mean for a set of data and how is the mean determined?

**3.** What is the weighted average? When is it used?

**4.** What is the median for a set of data?

**5.** How is the median determined for an odd number of data? an even number of data?

**6.** What conditions in the data would make the median an effective measure of central tendency?

**7.** What is the mode for a set of data?

**8.** How is the mode determined?

**9.** For what kind of data would the mode be an effective measure of central tendency?

**10.** What does it mean for a set of data to be bimodal?

### *Apply*

**11.** For the set of values 2, 4, 7, 2, 1, 8, 9, 10, 9, 6:

(a) Find the mean.

(b) Find the median.
(c) Find the mode.

**12.** For the set of values, 3, 8, 4, 2, 4, 6, 7, 1, 5, 0:

(a) Find the mean.
(b) Find the median.
(c) Find the mode.

**13.** For the data 20, 29, 21, 21, 28, 28, 23, 23, 23, 26, 26, 26, 25, 25, 25, 25, 25:

(a) Find the mean.
(b) Find the median.
(c) Find the mode.

**14.** For the data 42, 64, 75, 82, 96, 93, 77, 82, 67, 78, 88, 90, 80, 72, 71, 80, 81, 98, 61, 75:

(a) Find the mean.
(b) Find the median.
(c) Find the mode.

**15.** Due to health reasons, many people watch their sodium and cholesterol intake. The Taco Bell® restaurant chain publishes a nutrition pamphlet that among other things contains the amount of sodium and cholesterol in its menu items. (Source: Taco Bell Corp. PFS#3943)

| Menu Item | Sodium (mg) | Cholesterol (mg) |
|---|---|---|
| Taco | 276 | 32 |
| Soft Taco | 554 | 32 |
| Tostada | 596 | 16 |
| Chicken Soft Taco | 615 | 52 |
| Taco Supreme | 276 | 32 |
| Bean Burrito | 1148 | 9 |
| Beef Burrito | 1311 | 57 |
| Chicken Burrito | 880 | 52 |
| Burrito Supreme | 1181 | 33 |
| Combo Burrito | 1136 | 33 |

(a) Find the mean and median amounts of sodium and cholesterol in the ten menu items.
(b) Find the separate mean and median amounts of sodium and cholesterol in tacos and burritos.
(c) If you were watching your sodium intake, should you buy tacos or burritos? Explain.

(d) If you were watching your cholesterol intake, should you buy tacos or burritos? Explain.

**16.** The top five scorers for the Chicago Bulls in the 1993 NBA Championship series are given in the table. Included are the number of field goal attempts and the field goal percentage for each of the five players.

| Player | Attempts | Percentage |
|---|---|---|
| Jordan | 199 | 51% |
| Pippen | 123 | 44% |
| Armstrong | 63 | 51% |
| Grant | 53 | 53% |
| Paxson | 21 | 62% |

In parts (a) and (b), find the average shooting percentage by computing

(a) The weighted mean shooting percentages, using the number of attempts as the weights.

(b) The mean of the shooting percentages.

(c) Which of the averages computed gives a better representation of the team shooting percentage? Explain.

**17.** A student had seven 100-point tests during the semester. The scores on his tests were 90, 92, 59, 65, 94, 73, and 94. The instructor has given him the option of determining his semester grade by using the mean, median, or mode of his scores. Which would be the fairest selection? Explain.

**18.** A student is computing her cumulative grade point average (GPA). During six semesters, she has received 5 C's, 11 B's, and 8 A's. If a grade of C is worth 2 points, a B worth 3 points, and an A worth 4 points, and all classes have the same number of units, compute the student's GPA. (*Hint:* Use the number of grades as the weights.)

## *Explore*

**19.** According to *Sport* magazine, June 1991, the top ten prize winners in the three major women's professional sports in 1991 were as follows:

| Golf | | Tennis | | Bowling | |
|---|---|---|---|---|---|
| B. Daniel | $863,578 | S. Graf | $1,921,853 | T. Johnson | $94,420 |
| P. Sheehan | 732,618 | M. Seles | 1,637,222 | L. Barrette | 91,390 |
| B. King | 543,844 | M. Navratilova | 1,330,794 | L. Wagner | 58,055 |
| C. Gerring | 487,326 | G. Sabatini | 975,490 | D. Miller-Mackie | 57,805 |
| P. Bradley | 480,018 | J. Novotna | 645,500 | N. Gianulias | 57,087 |
| R. Jones | 353,832 | Z. Garrison | 602,203 | W. MacPherson | 54,245 |
| A. Okamoto | 302,865 | H. Sukova | 562,715 | K. Terrell | 50,027 |
| N. Lopez | 301,262 | M. Fernandez | 518,366 | R. Romeo | 45,645 |
| D. Ammaccapane | 300,231 | A. Vicario | 517,662 | A. Sill | 44,867 |
| C. Rarick | 259,163 | N. Zvereva | 462,770 | L. Nichols | 38,372 |

(a) Find the mean and median salaries of the top ten in each sport.
(b) Find the mean and median salaries for the entire group of 30 athletes.
(c) Is the median salary for the entire group of 30 athletes a realistic measure of the ''average'' salary of top female professional athletes? Explain.
(d) Is the mean salary for the entire group of 30 athletes a realistic measure of the ''average'' salary of top female professional athletes? Explain.

**20.** The following data is from the 1990 congressional elections in the five districts in the state of Kansas. (Source: Elections Research Center, *America Votes*)

| District | Total Votes Cast | Democratic (%) | Republican (%) |
|---|---|---|---|
| 1 | 164,000 | 37 | 63 |
| 2 | 158,000 | 63 | 37 |
| 3 | 148,000 | 40 | 60 |
| 4 | 158,000 | 71 | 29 |
| 5 | 153,000 | 41 | 59 |

In parts (a) and (b), determine what percentage voted Democratic and Republican by

(a) Computing the weighted mean percentage using the number of votes cast as the weights.
(b) Computing the mean of the percentages for each political party.
(c) Why are the results of parts (a) and (b) close to each other?

**21.** Based on a survey by the U.S. Bureau of the Census published in *Current Population Reports*, the number of people who reported that they voted in the 1990 congressional election by age-group is as follows.

| Age-Group | Number Voted (million) |
|---|---|
| 18–20 | 2.0 |
| 21–24 | 3.1 |
| 25–34 | 14.4 |
| 35–44 | 17.0 |
| 45–64 | 26.2 |
| 65–99 | 18.0 |

(a) Find the mean age of those who reported they voted.
(b) In what age groups do the median and mode lie?
(c) Use the result of parts (a) and (b) to describe the ''average'' voter in the congressional election of 1990.

**22.** The table gives the median household income in 1990 for each state and the District of Columbia. Income is given in constant 1990 dollars. (Source: U.S. Bureau of the Census, *Current Population Reports*)

| | | | | | |
|---|---|---|---|---|---|
| AL 23,397 | AK 39,298 | AZ 29,224 | AR 22,786 | CA 33,290 | CO 30,733 |
| CT 38,870 | DE 30,804 | DC 27,392 | FL 26,685 | GA 27,561 | HI 38,921 |
| ID 25,305 | IL 32,542 | IN 26,928 | IA 27,288 | KS 29,917 | KY 24,780 |
| LA 22,405 | ME 27,464 | MD 38,857 | MA 36,247 | MI 28,937 | MN 31,465 |
| MS 20,178 | MO 27,332 | MT 23,375 | NE 27,482 | NV 32,023 | NH 40,805 |
| NJ 38,734 | NM 25,039 | NY 31,591 | NC 26,329 | ND 25,264 | OH 30,013 |
| OK 24,384 | OR 29,281 | PA 29,005 | RI 31,968 | SC 28,735 | SD 24,571 |
| TN 22,592 | TX 28,228 | UT 30,142 | VT 31,098 | VA 35,073 | WA 32,112 |
| WV 22,137 | WI 30,711 | WY 29,460 | | | |

(a) Find the median of the data.
(b) Arrange the data into a frequency distribution with five classes, the first one being $20,001–$25,000.
(c) What income bracket is the mode?
(d) Find the mean of the grouped data.

(e) Which state has a household income that is closest to the mean?
(f) Even though a state like Ohio does not have the highest household income, what positive statement about its household income can it make?

**23.** The table below shows the number of women who held statewide elective executive offices or who were members of state legislatures in 1992. (Source: Center for American Women and Politics)

| | | | | | | | | | | | |
|---|---|---|---|---|---|---|---|---|---|---|---|
| AL | 9 | AK | 14 | AZ | 33 | AR | 12 | CA | 24 | CO | 34 |
| CT | 45 | DE | 9 | FL | 31 | GA | 34 | HI | 21 | ID | 37 |
| IL | 34 | IN | 28 | IA | 25 | KS | 47 | KY | 8 | LA | 12 |
| ME | 60 | MD | 44 | MA | 37 | MI | 23 | MN | 44 | MS | 12 |
| MO | 32 | MT | 33 | NE | 11 | NV | 15 | NH | 131 | NJ | 15 |
| NM | 17 | NY | 27 | NC | 25 | ND | 25 | OH | 21 | OK | 16 |
| OR | 25 | PA | 26 | RI | 26 | SC | 23 | SD | 28 | TN | 15 |
| TX | 26 | UT | 12 | VT | 56 | VA | 18 | WA | 49 | WV | 28 |
| WI | 33 | WY | 25 | | | | | | | | |

(a) Find the median of the data.
(b) Arrange the data into a frequency distribution with seven classes, the first one being 1–20.
(c) Find the mode of the grouped data.
(d) Find the mean of the grouped data.
(e) Is the set of data given sufficient to justify a claim that women are underrepresented in high political offices? If not, what additional information would you need? Explain.

## Section 8.3 Measures of Dispersion

Suppose the mean, mode, and median for two sets of data are identical. Does this suggest that the data are the same? Consider the two sets of data: 1, 1, 100, 100, 100, 199, 199 and 99, 99, 100, 100, 100, 101, 101. For each set, the mean, mode, and median all equal 100, yet, the data are not the same. Not only are the data not the same, the first set has values that range between 1 and 199, whereas the second set is closely clustered around 100. From this, we can see that we need more tools to help describe a distribution of numbers. Collectively, the tools used to do this are called **measures of dispersion**. A measure of dispersion will provide a tool to determine to what extent the data in a set differ from a central value. In this section, we will discuss two of these measures, range and standard deviation.

## The Range

The **range** of a set of values is the difference between the highest and lowest values in the set.

### Example 1

Find the range of each of the following sets of numbers.

(a) 1, 1, 100, 100, 100, 199, 199
(b) 99, 99, 100, 100, 100, 101, 101

**Solution:**

(a) The range of the first set of data is 199 − 1 = 198.
(b) The range of the second set of data is 101 − 99 = 2. ■

We can see that the range will provide some help in analyzing the difficulty mentioned in the introductory remarks. However, it will not completely solve the problem. Consider the following two sets of data: 1, 1, 1, 100, 199, 199, 199 and 1, 100, 100, 100, 100, 100, 199. In both sets, the range is 198, but the first set of data has most of its values at the extreme ends, whereas most of the data in the second set have the value of 100. Only two of the numbers are at the extremes. What is needed is a method to determine the average of the distance between each data value and the mean. If this average is high, then the data are spread out. If the average is low, then the data are clustered together.

## Standard Deviation

The measure we will use to determine how closely the data are clustered around the mean is called the **standard deviation**. The standard deviation is the square root of the average of the squares of the differences between the data values and the mean. The standard deviation is represented by the Greek letter $\sigma$ (sigma) and can be determined using the formula,

Standard Deviation Formula

$$\sigma = \sqrt{\frac{\sum_{i=1}^{n} (x_i - \mu)^2}{n}} \quad \text{where} \begin{cases} x_i \text{ are the data values} \\ \mu \text{ is the mean of the data} \\ n \text{ is the number of data} \end{cases}$$

### Example 2

Find the standard deviation of the numbers 4, 6, and 11.

**Solution:** To find the standard deviation, first find the mean.

$$\mu = \frac{4 + 6 + 11}{3} = 7$$

Next, subtract the mean from each of the data values and then square the results. Using the formula gives

$$\sigma = \sqrt{\frac{(4 - 7)^2 + (6 - 7)^2 + (11 - 7)^2}{3}}$$

$$= \sqrt{\frac{9 + 1 + 16}{3}}$$

$$= \sqrt{\frac{26}{3}} = 2.94$$

Although this process does not seem too difficult, let's try to find the standard deviation of the numbers 2, 5, 7, 8, 10, 6, and 7. We first find the mean:

$$\mu = \frac{2 + 5 + 7 + 8 + 10 + 6 + 7}{7} = \frac{45}{7} = 6.429$$

Now, finding the standard deviation will involve subtracting 6.429 from each of these values and then squaring the results. Even with the help of a calculator, this is tedious. Therefore, before we continue with this problem, we will simplify the equation for the standard deviation. This derivation will involve the use of algebra and the three rules of summations that were presented in Section 8.0. For your convenience, these rules are given along with the derivation of the simplified equation for standard deviation.

## Derivation of the Computing Formula for Standard Deviation

The three rules of summations given in Section 8.0 are

1. $\sum_{i=1}^{n} c = n \times c$ where $c$ is any constant
2. $\sum_{i=1}^{n} cx_i = c \sum_{i=1}^{n} x_i$ where $c$ is any constant
3. $\sum_{i=1}^{n} (x_i + y_i) = \sum_{i=1}^{n} x_i + \sum_{i=1}^{n} y_i$

Starting with our current formula for standard deviation, we have

$$\sigma = \sqrt{\frac{\sum_{i=1}^{n} (x_i - \mu)^2}{n}}$$

By expanding the term inside the square, we get

$$\sigma = \sqrt{\frac{\sum_{i=1}^{n} \left((x_i)^2 - 2x_i\mu + \mu^2\right)}{n}}$$

Now, using the third rule of summations to distribute the summation sign gives

$$\sigma = \sqrt{\frac{\sum_{i=1}^{n} (x_i)^2 - \sum_{i=1}^{n} 2\mu x_i + \sum_{i=1}^{n} \mu^2}{n}}$$

Since the $2\mu$ in the second term is a constant, by the second rule of summations it can be brought outside the summation sign.

$$\sigma = \sqrt{\frac{\sum_{i=1}^{n} (x_i)^2 - 2\mu \sum_{i=1}^{n} x_i + \sum_{i=1}^{n} \mu^2}{n}}$$

Since the third summation contains only constants, we can use the first rule of summations to get

$$\sigma = \sqrt{\frac{\sum_{i=1}^{n} (x_i)^2 - 2\mu \sum_{i=1}^{n} x_i + n\mu^2}{n}}$$

Since $\mu = \frac{\sum_{i=1}^{n} x_i}{n}$, $\sum_{i=1}^{n} x_i = n\mu$:

$$\sigma = \sqrt{\frac{\sum_{i=1}^{n} (x_i)^2 - 2\mu n\mu + n\mu^2}{n}}$$

Multiplying gives

$$\sigma = \sqrt{\frac{\sum_{i=1}^{n} (x_i)^2 - 2n\mu^2 + n\mu^2}{n}}$$

Finally, combining the like terms gives

$$\sigma = \sqrt{\frac{\sum_{i=1}^{n} (x_i)^2 - n\mu^2}{n}}$$

This gives the formula we will be using to compute the standard deviation.

## Computing Formula for Standard Deviation

$$\sigma = \sqrt{\frac{\sum_{i=1}^{n} (x_i)^2 - n\mu^2}{n}} \qquad \text{where} \begin{cases} x_i \text{ are the data values} \\ \mu \text{ is the mean of the data} \\ n \text{ is the number of data values} \end{cases}$$

In the following example, we use the computing formula to finish the problem begun earlier.

### Example 3

Find the standard deviation of 2, 5, 7, 8, 10, 6, and 7.

**Solution:** We had determined that the mean was 6.429. However, we did not want to do all the work of subtracting this value from each of the given numbers. The new formula for standard deviation does not require this. Instead, we simply square the original data and the mean:

Since

$$\begin{aligned}\sum_{i=1}^{7} (x_i)^2 &= 2^2 + 5^2 + 7^2 + 8^2 + 10^2 + 6^2 + 7^2 \\ &= 4 + 25 + 49 + 64 + 100 + 36 + 49 \\ &= 327\end{aligned}$$

We have

$$\sigma = \sqrt{\frac{\sum_{i=1}^{7} (x_i)^2 - n\mu^2}{n}} = \sqrt{\frac{327 - 7(6.429)^2}{7}}$$

$$= \sqrt{\frac{37.676}{7}} = 2.32 \quad ■$$

The computing formula works nicely for individual data, but we have discovered that much of the data found in the real world is organized in classes. Therefore, we need a formula for standard deviation of grouped data. The derivation of this formula is similar to the derivation of the computing formula for standard deviation and is left as a starred(*) problem.

## Computing Formula for Standard Deviation of Grouped Data

$$\sigma = \sqrt{\frac{\sum f_i(x_i)^2 - n\mu^2}{n}}$$

**Example 4**

| Class | Frequency |
|---|---|
| 1–5 | 6 |
| 6–10 | 12 |
| 11–15 | 22 |
| 16–20 | 16 |
| 21–25 | 14 |

These data are the test scores used in Example 3 of Section 8.2. Find the mean and standard deviation of these scores.

**Solution:** Since this is grouped data, we use the midpoint of each class as the $x_i$. To help with the solutions. The calculations are displayed in tabular form.

| Class | Frequency $f_i$ | $x_i$ | $f_i x_i$ | $(x_i)^2$ | $f_i(x_i)^2$ |
|---|---|---|---|---|---|
| 1–5 | 6 | 3 | 18 | 9 | 54 |
| 6–10 | 12 | 8 | 96 | 64 | 768 |
| 11–15 | 22 | 13 | 286 | 169 | 3718 |
| 16–20 | 16 | 18 | 288 | 324 | 5184 |
| 21–25 | 14 | 23 | 322 | 529 | 7406 |
| | $\Sigma f_i = 70$ | $\Sigma x_i = 65$ | $\Sigma f_i x_i = 1010$ | $\Sigma(x_i)^2 = 1095$ | $\Sigma f_i(x_i)^2 = 17{,}130$ |

$$\mu = \frac{\sum f_i x_i}{n} = \frac{1010}{70} = 14.43$$

$$\sigma = \sqrt{\frac{\sum f_i(x_i)^2 - n\mu^2}{n}} = \sqrt{\frac{17{,}130 - 70(14.43)^2}{70}} = \sqrt{36.49} = 6.04$$

## How is Standard Deviation Used?

We have now seen how to compute the standard deviation of either grouped or ungrouped data but, so far, we have no real feeling for what standard deviation means or how it can be used. In the remainder of this section, we discuss the meaning of standard deviation in terms of a frequency distribution. In Section 8.4 we discuss how standard deviation can be used.

### Example 5

Consider the following sets of scores from two intermediate algebra classes.

| Morning Class | Afternoon Class |
|---|---|
| 78 65 83 91 98 25 67 88 81 77 | 34 87 81 93 99 24 77 62 98 100 |
| 53 76 80 72 75 69 64 62 85 93 | 57 34 81 72 61 59 68 74 77 94 |
| 70 44 85 73 75 63 | 56 71 70 81 78 83 25 94 31 |

(a) Find the mean and standard deviation of each class.
(b) Construct the frequency distribution of each class.
(c) Draw a bar chart for each class.
(d) Use the bar chart to draw conclusions about the meaning of standard deviation.

**Solution:**

(a) To find the mean of each class, use the formula

$$\mu = \frac{\sum_{i=1}^{n} x_i}{n}$$

For the morning class, this gives

$$\mu = \frac{\sum_{i=1}^{26} x_i}{26} = \frac{1892}{26} = 72.77 \approx 73$$

For the afternoon class, the mean is

$$\mu = \frac{\sum_{i=1}^{29} x_i}{29} = \frac{2021}{29} = 69.69 \approx 70$$

To find the standard deviation, we use the computing formula for ungrouped data,

$$\sigma = \sqrt{\frac{\sum_{i=1}^{n} (x_i)^2 - n\mu^2}{n}}$$

For the morning class, this gives

$$\sigma = \sqrt{\frac{\sum_{i=1}^{26} (x_i)^2 - 26(72.77)^2}{26}} = \sqrt{\frac{143{,}784 - 26(5295.4729)}{26}} = 15.32$$

and the afternoon class has a standard deviation of

$$\sigma = \sqrt{\frac{\sum_{i=1}^{29} (x_i)^2 - 29(69.69)^2}{29}} = \sqrt{\frac{154{,}939 - 29(4856.6961)}{29}} = 22.05$$

(b) Completing the frequency distributions for each class gives the following:

| Morning Class | | Afternoon Class | |
|---|---|---|---|
| **Class** | **Frequency** | **Class** | **Frequency** |
| 0–9 | 0 | 0–9 | 0 |
| 10–19 | 0 | 10–19 | 0 |
| 20–29 | 1 | 20–29 | 2 |
| 30–39 | 0 | 30–39 | 3 |
| 40–49 | 1 | 40–49 | 0 |
| 50–59 | 1 | 50–59 | 3 |
| 60–69 | 6 | 60–69 | 3 |
| 70–79 | 8 | 70–79 | 7 |
| 80–89 | 6 | 80–89 | 5 |
| 90–100 | 3 | 90–100 | 6 |

(c) The bar charts for both classes are shown here.

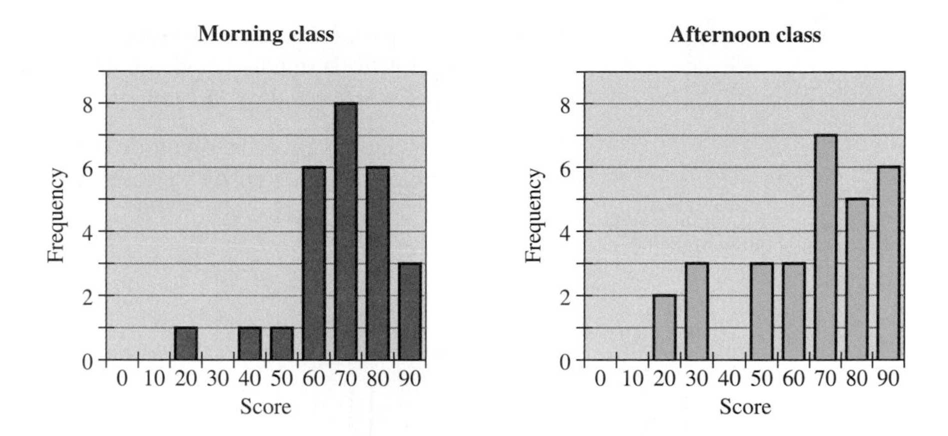


(d) We now come to the important part of this problem. We have taken two sets of data, found the mean and standard deviation of each set, and drawn the bar graphs for each set of data. Both classes have means close to 70. This is seen in the bar charts by noting that most of the shading is clustered around the bar representing the 70–79 category. The standard deviation of the afternoon class is 22.05 versus 15.32 for the morning class. In the bar graphs, this can be seen by noticing that the afternoon class has more scores 20 units or more away from 70. This observation that the higher standard deviation corresponds to the bar graph showing the greater spread from the mean confirms the discussion at the beginning of this section. At that time, we introduced standard deviation as an indicator of how widely the data values were dispersed from the mean. ■

Section 8.3

PROBLEMS

## *Explain ➡ Apply ➡ Explore*

### *Explain*

1. What is a measure of dispersion?
2. What is the range of a set of data?
3. What is the standard deviation of a set of data?
4. What is the computing formula for the standard deviation?
5. What is the computing formula for the standard deviation of grouped data? When should it be used?
6. You are comparing data from two similar experiments that have a mean of 34.5. The data in experiment A has a range of 150 and the data for experiment B has range of 20. What does that tell you about the data in the experiments?

**7.** You are comparing data from two similar experiments that have a mean of 34.5. Experiment A has a standard deviation of 10.1 and experiment B has a standard deviation of 5.5. What does this tell you about the data in the experiments?

**8.** Suppose two frequency distributions are represented by the following bar graphs. Which distribution has a greater standard deviation? Explain.

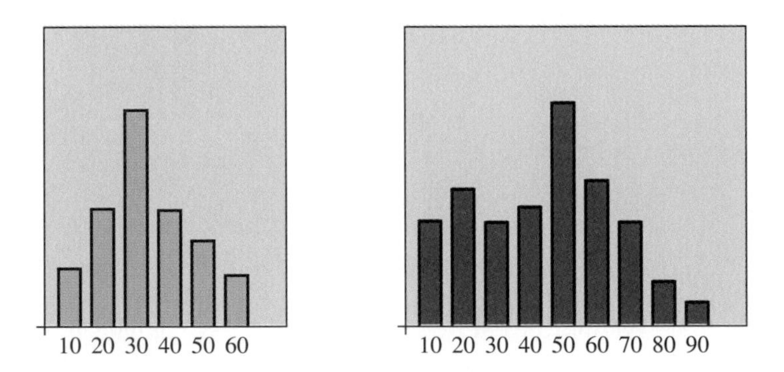


**9.** If a set of test scores has a standard deviation of zero, what can be said about the scores?

## *Apply*

**10.** For the set of numbers 3, 6, 10, 14, 17:

(a) Find the range.

(b) Find the standard deviation, using the formula $\sigma = \sqrt{\dfrac{\sum_{i=1}^{n} (x_i - \mu)^2}{n}}$.

(c) Find the standard deviation, using the computing formula.

(d) Suppose that the original data are changed. The 3 is replaced by 2, and the 17 is replaced by 18. Without doing any calculations, what effect does this have on the standard deviation?

**11.** For the set of numbers 2, 6, 11, 12, 14:

(a) Find the range.

(b) Find the standard deviation, using the formula $\sigma = \sqrt{\dfrac{\sum_{i=1}^{n} (x_i - \mu)^2}{n}}$

(c) Find the standard deviation, using the computing formula.

(d) Suppose that the original data are changed. The 2 is replaced by 5, and the 14 is replaced by 12. Without doing any calculations, what effect does this have on the standard deviation?

**12.** For the set of values 2, 4, 7, 2, 1, 8, 9, 10, 9, 6:

(a) Find the range.

(b) Find the standard deviation, using the formula $\sigma = \sqrt{\dfrac{\sum_{i=1}^{n} (x_i - \mu)^2}{n}}$.

(c) Find the standard deviation, using the computing formula.

**13.** For the set of values 3, 8, 4, 2, 4, 6, 7, 1, 5, 0:

(a) Find the range.

(b) Find the standard deviation, using the formula $\sigma = \sqrt{\dfrac{\sum_{i=1}^{n} (x_i - \mu)^2}{n}}$.

(c) Find the standard deviation, using the computing formula.

**14.** For the data 20, 29, 21, 21, 28, 28, 23, 23, 23, 26, 26, 26, 25, 25, 25, 25, 25:

(a) Find the range.

(b) Find the standard deviation, using the formula $\sigma = \sqrt{\dfrac{\sum_{i=1}^{n} (x_i - \mu)^2}{n}}$.

(c) Find the standard deviation, using the computing formula.

**15.** For the data 42, 64, 75, 82, 96, 93, 77, 82, 67, 78, 88, 90, 80, 72, 71, 80, 81, 98, 61, 75:

(a) Find the range.

(b) Find the standard deviation, using the formula $\sigma = \sqrt{\dfrac{\sum_{i=1}^{n} (x_i - \mu)^2}{n}}$.

(c) Find the standard deviation, using the computing formula.

## *Explore*

**16.** The table given in Problem 23 in Section 8.2 shows the number of women who held statewide elective executive offices or who were members of state legislatures in 1992. (Source: Center for American Women and Politics) Use the results of that problem and the computing formula for the standard deviation of grouped data to find $\sigma$.

**17.** Use the results of Problem 15 in Section 8.2 to find the standard deviation for both sodium and cholesterol in the ten menu items from the Taco Bell® chain of "fast food" restaurants.

**18.** An accounting firm plans to buy a large number of cartridges for laser printers. The cartridge is available from two different suppliers. The first supplier says that the expected lifetime is 3000 pages with a standard deviation of 100 pages. The second supplier says their cartridge has an expected lifetime of 3000 pages with a standard deviation of 400 pages. If you are the buyer for the accounting firm, which supplier would you choose? Explain your reasoning.

**19.** According to *Sport* magazine, June 1991, the top ten salaries in the four major professional team sports in 1991 were as follows:

| **Baseball** | | **Football** | |
|---|---|---|---|
| D. Strawberry | $3,800,000 | J. Montana | $4,000,000 |
| W. Clark | 3,750,000 | J. Kelley | 3,300,000 |
| K. Mitchell | 3,750,000 | J. George | 2,768,000 |
| J. Carter | 3,666,667 | W. Moon | 2,700,000 |
| M. Davis | 3,625,000 | J. Everett | 2,400,000 |
| E. Davis | 3,600,000 | H. Walker | 2,225,000 |
| W. Magee | 3,562,000 | E. Dickerson | 2,066,667 |
| M. Langdon | 3,500,000 | B. Kosar | 1,942,857 |
| J. Canseco | 3,500,000 | J. Elway | 1,878,541 |
| T. Raines | 3,500,000 | R. Cunningham | 1,703,571 |

| **Basketball** | | **Hockey** | |
|---|---|---|---|
| J. Williams | $5,000,000 | W. Gretzky | $3,000,000 |
| P. Ewing | 4,600,000 | M. Lemieux | 2,333,000 |
| H. Olajuwon | 4,062,451 | B. Hull | 1,366,000 |
| C. Barkley | 2,900,000 | S. Yzerman | 1,300,000 |
| I. Thomas | 2,720,000 | R. Bourque | 1,194,000 |
| M. Malone | 2,506,000 | D. Savard | 1,125,000 |
| C. Mullen | 2,500,000 | P. Roy | 1,080,000 |
| R. Parrish | 2,500,000 | S. Stevens | 1,058,000 |
| M. Jordan | 2,500,000 | C. Chelios | 990,000 |
| D. Ferry | 2,450,000 | R. Cunningham | 962,000 |

(a) Find the range of the 40 salaries.
(b) Find the median of the 40 salaries.
(c) Find the mode of the 40 salaries.
(d) Arrange the data into a frequency distribution with five classes, the first one being $1–$1,000,000.
(e) Make a bar graph of the grouped data.

(f) Find the mean of the grouped data.
(g) Find the standard deviation of the grouped data.
(h) Which measure would you use to describe the ''average'' salary of these athletes? Explain your choice.

***20.** Starting with the formula $\sigma = \sqrt{\dfrac{\sum_{i=1}^{n} (x_i - \mu)^2}{n}}$ where $n = \sum f_i$, derive the computing formula for grouped data that is given in the text.

## Section 8.4 The Normal Distribution

The following curve is the **normal distribution**, also called a **bell curve** or the **Gaussian distribution**. Using this curve, we will connect concepts of probability and statistics.

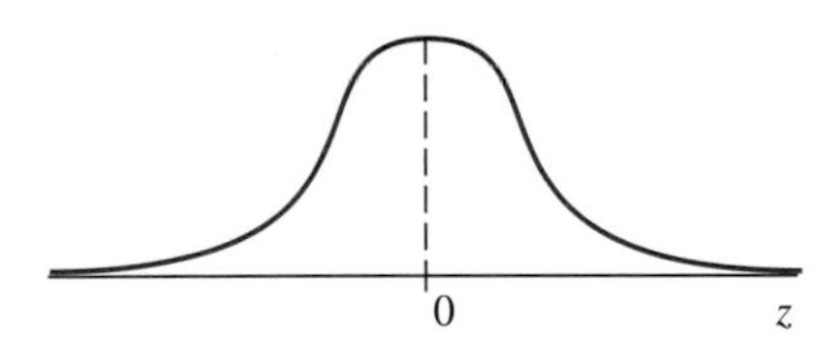


First, let's reexamine the results of the first problem discussed in Section 8.1. The frequency distribution and the bar graph are given again here.

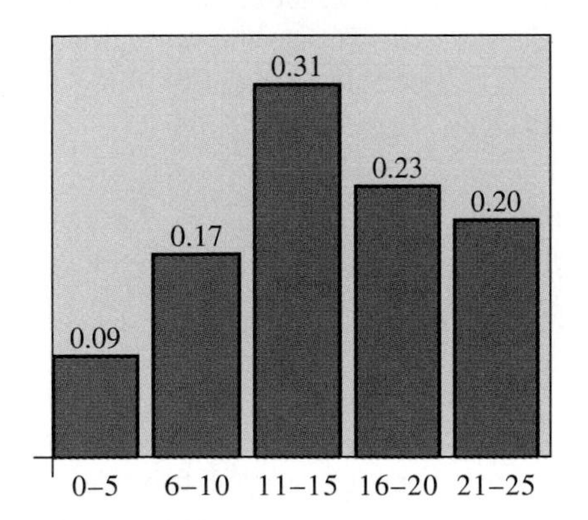

*Frequency Distribution*

| Class | Relative Frequency |
|---|---|
| 0–5 | 0.09 |
| 6–10 | 0.17 |
| 11–15 | 0.31 |
| 16–20 | 0.23 |
| 21–25 | 0.20 |

Notice that the height of each rectangle in the bar graph also represents the probability of a data value being in a certain class. For example, the probability that a randomly picked test score will be in the 6–10 class is 0.17. This means that, on the average, 17 out of 100 test scores will be in the class 6–10. This is true because both the probability and the relative frequency for a certain class are calculated by taking the number of items in a group and dividing by the total number of items. Thus, there is a connection between relative frequency, bar graphs, and probability.

A set of data is said to be normal if it follows a very definite set of mathematical criteria that is beyond our discussion. Informally, a set of data is said to be normal if 95% of all data values are within two standard deviations of the mean and 50% of the data is on each side of the mean. The normal distribution is used for many mathematical models, ranging from test scores, to the number of defective items coming off a production line, to the analysis of sociological surveys.

Using the normal distribution requires being able to find the area between the $x$ axis and the graph of the normal curve. The equation of the normal curve is $y = (1/\sqrt{2\pi})e^{-x^2/2}$. Even using the powerful tools of calculus, this is a difficult problem. To avoid this difficulty, it is standard practice to use a table of values, called the normal table. The normal table is given in the Appendix. In a normal table, the letter $z$ is used as the independent variable and as the label for the horizontal axis. To use the normal table, look up the desired value of $z$ and read the corresponding probability in the adjacent column.

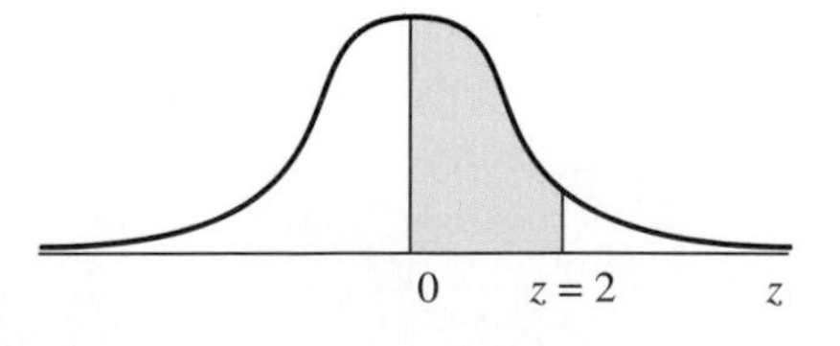


The values in the table give the area between the $z$ axis and the normal curve, bounded on the left by the line $z = 0$ and bounded on the right by the $z$ value being used in the problem. For example, for $z = 2$ the table gives a value of 0.4772. This means that the area of the shaded region in the figure is 0.4772. It also means that the probability that $z$ is between 0 and 2 is 0.4772. Symbolically, this is written as

$$P(0 < z < 2) = 0.4772$$

This means that 47.72% of the area under the curve is between $z = 0$ and $z = 2$.

As you can see in the diagram, $z = 0$ is the middle value for the normal distribution. For most real-life problems, the middle or mean is usually a value other than 0. Also, the normal distribution has a standard deviation of 1, which is probably not the case in most situations. Therefore, to make the normal distribution useful, we must transform information that is given in a problem into $z$ values. To do this we use the formula

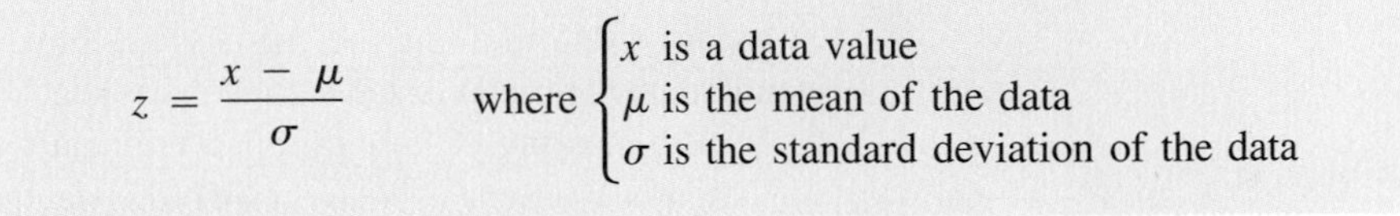

$$z = \frac{x - \mu}{\sigma} \quad \text{where} \begin{cases} x \text{ is a data value} \\ \mu \text{ is the mean of the data} \\ \sigma \text{ is the standard deviation of the data} \end{cases}$$

At this introductory level, if the bar graph has a large number of rectangles and has a ''bellshape,'' we will assume that the bar graph can be approximated by a normal distribution.

The $z$ value gives the number of standard deviations between the mean and a particular data value.

## Example 1

The heights of 1000 students are measured and found to have a mean of 70 in. and a standard deviation of 3 in. Assuming the heights of the students are normally distributed, what is the probability that a student, chosen at random, has a height between 70 and 73 in.?

**Solution:** The first step is to draw a normal distribution with a mean of 70 and shade the desired area. Since the mean is 70, that is the value at the center of the distribution. We want to find $P(70 < x < 73)$, where $x$ represents the height of the student. Therefore, the region between 70 and 73 is shaded on the diagram.

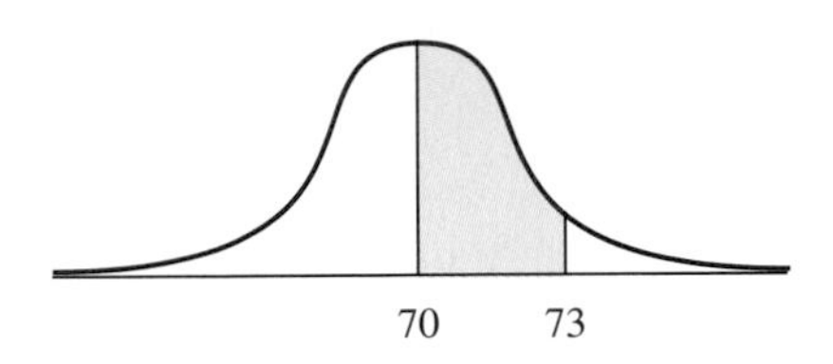

To find $P(70 < x < 73)$, convert 73 into a $z$ value. The formula gives

$$z = \frac{73 - 70}{3} = 1$$

In the normal table, for $z = 1$, the area is 0.3413. This can be interpreted in two ways. We can say that the probability that the student is between 70 in. and 73 in. tall is 0.3413, or we can say that approximately 34% of the students are between 70 and 73 in. tall. ■

## Example 2

Suppose the problem is similar to that in Example 1, except that the standard deviation is 4 in. rather than 3 in. Find $P(70 < x < 73)$.

**Solution:** Before doing the calculation, think about the physical meaning of standard deviation. We stated in the previous section that the standard deviation was a measure of how far apart the data were spread. Since the standard deviation in Example 2 is larger than the standard deviation in Example 1, the data are more spread out. Because of this, we should expect $P(70 < x < 73)$ to now be smaller.

We can verify this by doing the calculations. First, we need to find the $z$ value. This is

$$z = \frac{73 - 70}{4} = 0.75$$

Looking up $z = 0.75$ in the $z$ table, we get 0.2734. Therefore, $P(70 < x < 73) = 0.2734$, which, as expected, gives a smaller value than in Example 1. ■

## Example 3

A study of the educational aspirations of military women under age 21 was undertaken. Results of the study showed that these women hoped to achieve, on the average, 15.301 years of schooling, with a standard deviation of 1.893 years. Assuming a normal distribution,

(a) What percentage of the women hoped to have more than 15.301 years of school?
(b) What percentage hoped to have between 12 and 15.301 years of school?
(c) What percentage desired to have no more than 12 years of school?
(d) What percentage of the women wanted to have between 16 and 18 years of school?
(e) What percentage of the women wanted to have between 14 and 16 years of school?

**Solution:**

(a) Since 15.301 is the mean, we expect half of the respondents wanted to have more than 15.301 years of education while half did not want more. Therefore, 50% of the women desired to have more than 15.301 years of school.

For the remaining questions, we will sketch a graph of the normal distribution.

(b) To find the percentage of women who hoped to have between 12 and 15.301 years of school, look at the graph.

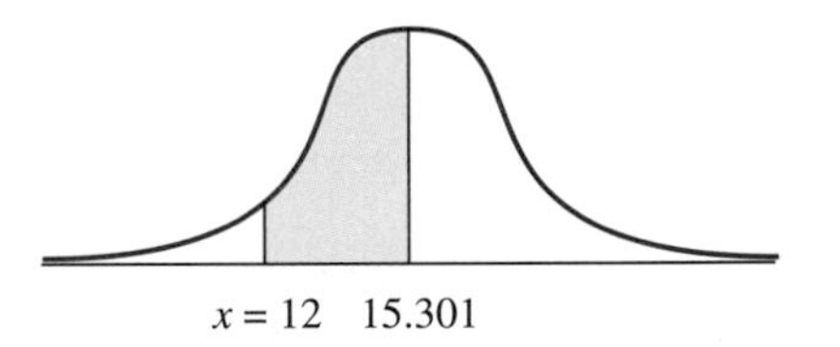


As we did in previous examples, we will find the $z$ value:

$$z = \frac{12 - 15.301}{1.893} = -1.74$$

This $z$ value is negative, but the normal table includes only positive values. However, since the graph of the normal distribution is symmetric with respect to the mean, we can find the area between 12 and 15.301 by using $z = +1.74$. This gives the value $A = 0.4591$. This means that 45.91% of the women surveyed hoped to have between 12 and 15.301 years of school.

(c) In this part of the problem, we are interested in the percentage of women who wanted 12 or fewer years of schooling, so we want to find the area to the left of 12.

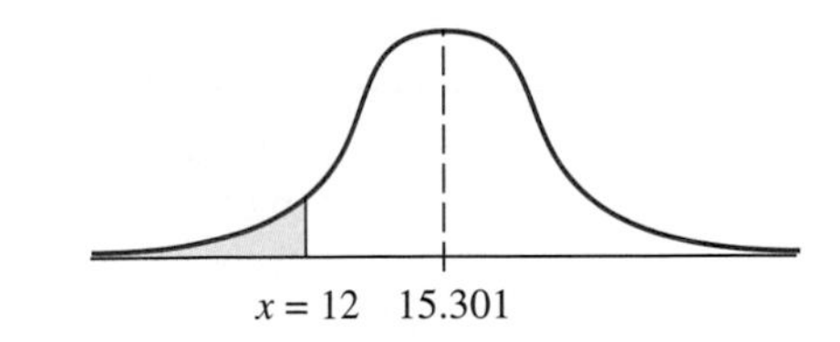


Since the normal table only gives areas that are adjacent to the center line of the distribution, we need to do this problem in another way. We use the fact that the area to the left of 15.301 is 0.5000 and the area between 12 and 15.301 is 0.4591. Subtracting these gives $A = 0.5000 - 0.4591 = 0.0409$. Therefore, only 4.09% of the women wanted the equivalent of a high school education or less.

(d) To determine the percentage of women who wanted to have between 16 and 18 years of schooling, we again examine a picture.

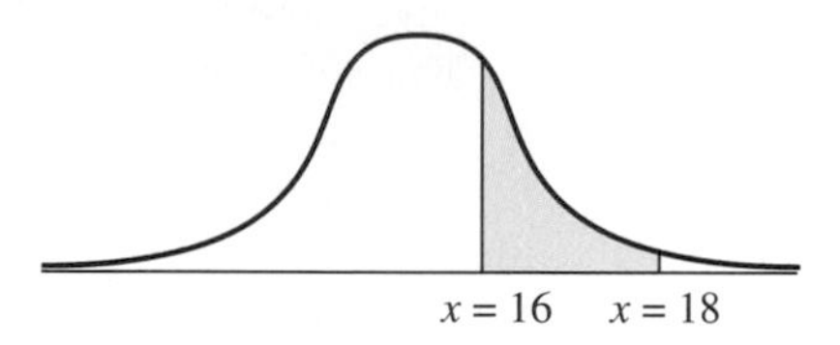


As was true in part (c), we cannot get the desired area directly, since the area is not adjacent to the middle of the distribution. As we did in part (c), we look at two separate areas and subtract the smaller from the larger to get the final answer.

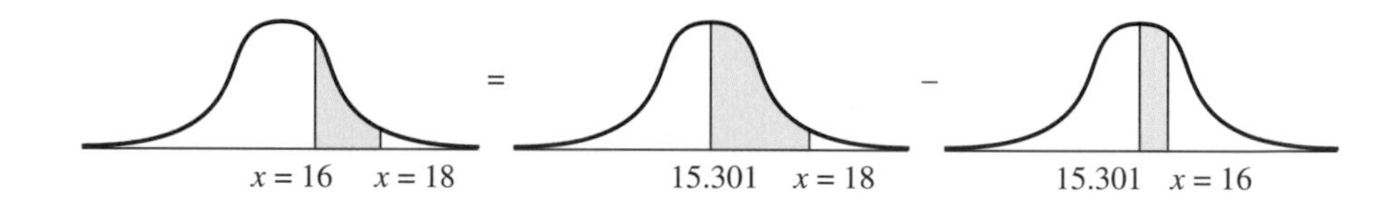


With the help of some pictures, we can see that if we want to find the area between 16 and 18, we can do it by finding the area between 15.301 and 18 and then subtracting the area between 15.301 and 16. We will need two $z$ values. For $x = 18$,

$$z = \frac{18 - 15.301}{1.893} = 1.43$$

which gives an area of 0.4236. For $x = 16$,

$$z = \frac{16 - 15.301}{1.893} = 0.37$$

giving an area of 0.1443. Therefore, the probability that a randomly selected woman surveyed wanted between 16 and 18 years of education is $0.4236 - 0.1443 = 0.2793$. This means that 27.93% of the women hoped to have between 16 and 18 years of schooling.

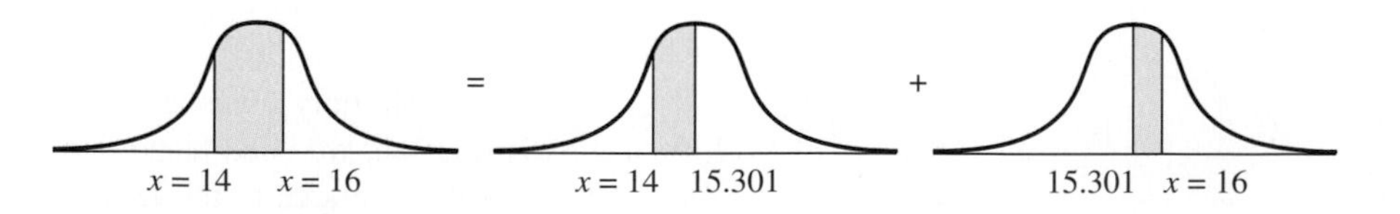


(e) For the final part of this problem, we want to determine the percentage of women who wanted to have between 14 and 16 years of schooling.

As we can see from the pictures, this problem is similar to part (d) except that since the mean is between the values of 14 and 16, we need to add the two areas to find the area of the combined regions. For $x = 14$,

$$z = \frac{14 - 15.301}{1.893} = -0.69$$

which gives an area of 0.2549. For $x = 16$,

$$z = \frac{16 - 15.301}{1.893} = 0.37$$

giving an area of 0.1443. Therefore, since $0.2549 + 0.1443 = 0.3992$, we can say that 39.92% of the women surveyed aspired to have between 14 and 16 years of education. ■

To summarize, we used the normal table to determine the probability of an event occurring. Even though the normal table only gives $P(0 < z < c)$, where $c$ is some number, we found that using the formula $z = \dfrac{x - \mu}{\sigma}$ allows us to determine the probability of an event if we know the mean and the standard deviation.

The final topic for this section is the use of the normal table to reverse the above process. In this type of problem, we know the percentage or probability but we want to determine the value of $z$ or $x$ that gives this probability. The next two examples will demonstrate what we mean.

## Example 4

Find the values of $c$ that make the following statements true.

(a) $P(0 < z < c) = 0.2580$ (b) $P(c < z < 0) = 0.2580$

(c) $P(0 < z < c) = 0.3000$ (d) $P(z > c) = 0.0563$

**Solution:** As we have done in previous examples, we will draw a picture for each problem. The picture will help us understand what we must do.

(a) The shaded part of the following diagram represents the area or the probability. The problem is to determine the value of $c$ that makes $P(0 < z < c) = 0.2580$. Look in the $z$ table for the $z$ value that gives a corresponding area of 0.2580. This gives $z = 0.70$. This means $P(0 < z < 0.70) = 0.2580$.

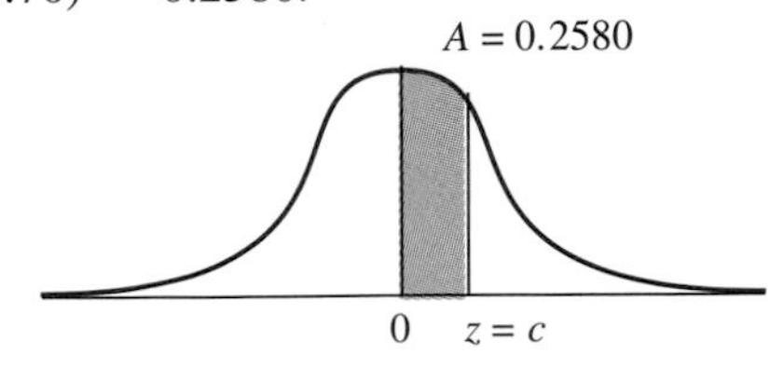

(b) To find the solution to $P(c < z < 0) = 0.2580$, we know that the value of $c$ must be negative because $c$ is less than 0. Since the value of the area is the same as it was in part (a) and we know that $c$ must be negative, we know that $P(-0.70 < z < 0) = 0.2580$.

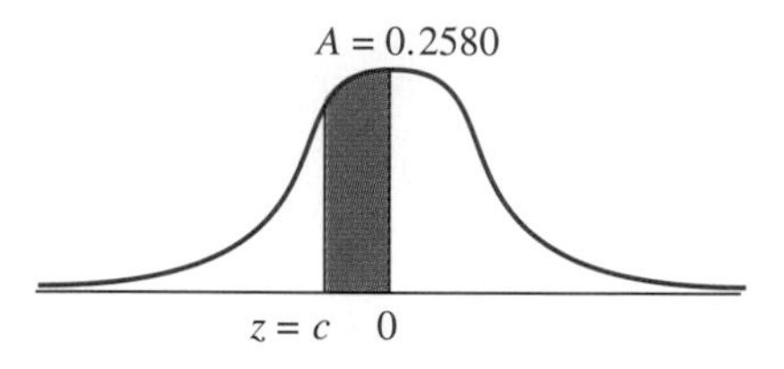


(c) The problem $P(0 < z < c) = 0.3000$ is very similar to part (a). The difference arises when we try to find the value 0.3000 in the area column of the $z$ table. For $z = 0.84$, $A = 0.2995$, and $z = 0.85$ gives $A = 0.3023$. Since neither of these values is 0.3000, we need to make some compromises. The choices are to find a table that has more digits of accuracy, use a mathematical technique called interpolation, or choose the value of $A$ that is closest to 0.3000. Because this is only a survey course, we choose the easiest method, picking the closest value, which is $A = 0.2995$. Therefore, we will say $P(0 < z < 0.84) \approx 0.3000$.

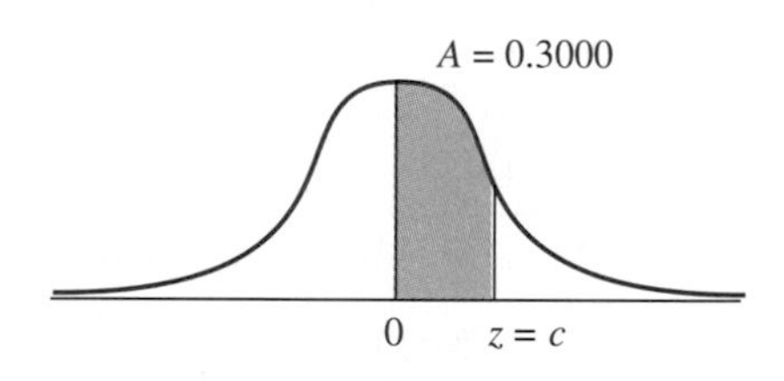


(d) The problem $P(z > c) = 0.0563$ is different from parts (a), (b), and (c), since the specified area is given by $z > c$, which means that the area is at the far right tail (light-colored region). Since the normal table only gives areas that are adjacent to the mean, using the dark-colored area gives $0.5000 - 0.0563 = 0.4437$. The normal table now gives $P(z > 1.59) = 0.0563$. ■

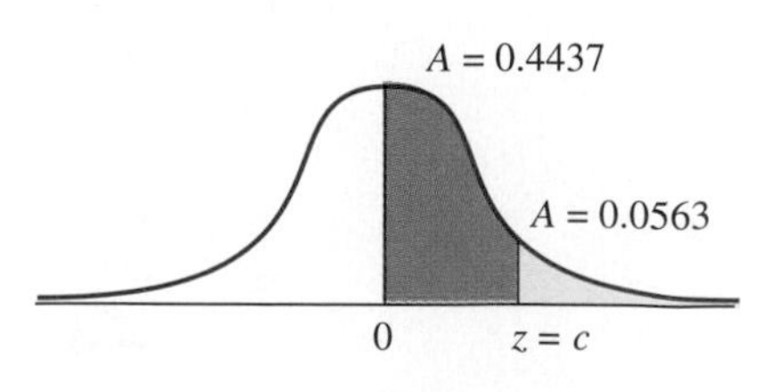

## Example 5

A certain college instructor has a fixed grading policy. He always gives the top 8% of the students A's; the next 15% receive B's, 54% receive C's, 15% receive D's, and 8% receive F's. Assuming that the scores are normally distributed with a mean of 72 and a standard deviation of 12, determine the scores that receive each grade.

**Solution:** Because there are five different grades, we must divide the normal distribution into five different regions, as shown in the diagram. Since the A's

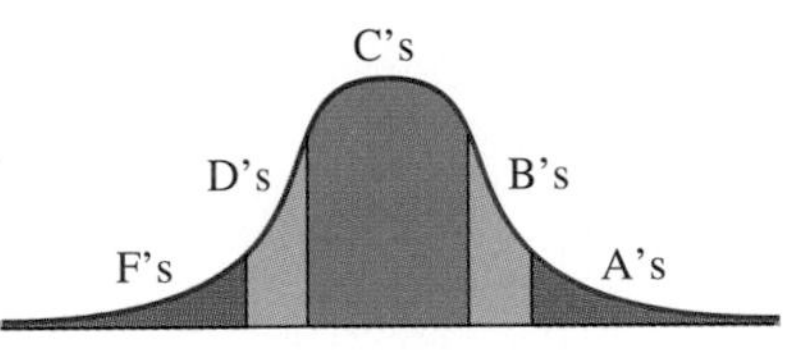


occupy the top 8%, we have the diagram shown. As we did in part (d) of the previous example, look at the area between the mean and some value of $x$, labeled $c$. Using $A = 0.5000 - 0.0800 = 0.4200$, we find $z = 1.41$.

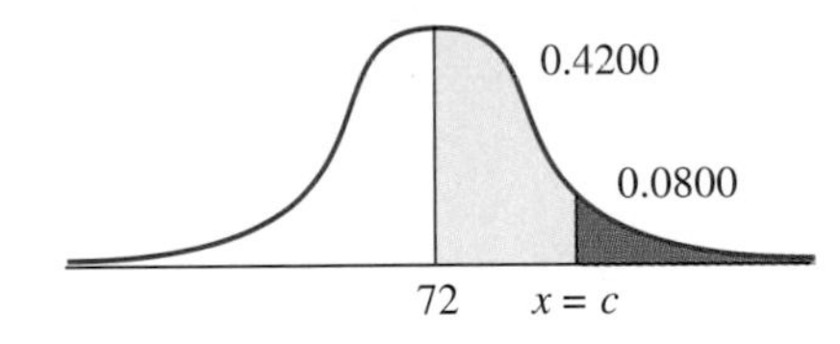


Since $\mu = 72$ and $\sigma = 12$, use the formula $z = (x - \mu)/\sigma$. This gives

$$1.41 = \frac{x - 72}{12}$$

$$16.9 = x - 72$$

$$88.9 = x$$

This means that all students scoring 88.9 or higher will receive a grade of A.

Since 15% of the students will receive B's, the scores that will receive a grade of B are shown in the following diagram in the light-colored region marked 0.1500. To find the area of the dark region, we use the fact that 50% of the scores must be greater than the mean. Since 8% receive A's and 15% receive B's, we find the dark area to be $0.5000 - 0.0800 - 0.1500 = 0.2700$.

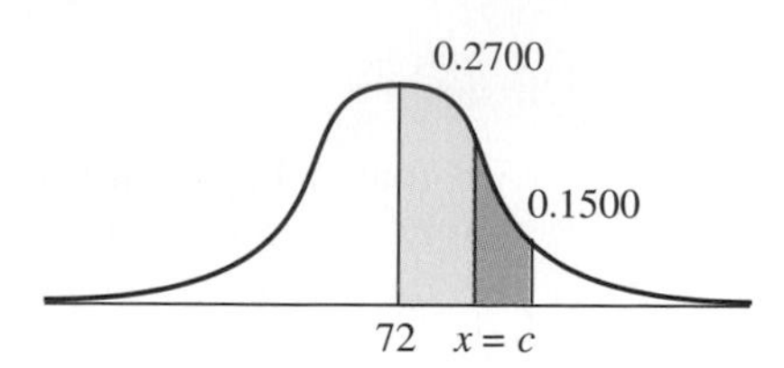

Using $A = 0.2700$, we find $z = 0.74$. Using the formula

$$z = \frac{x - \mu}{\sigma}$$

$$0.74 = \frac{x - 72}{12}$$

$$8.9 = x - 72$$

$$80.9 = x$$

Thus, students with scores between 80.9 and 88.9 will receive B's.

To find the scores for the other three grades, we use the fact that the normal distribution is symmetric about the mean.

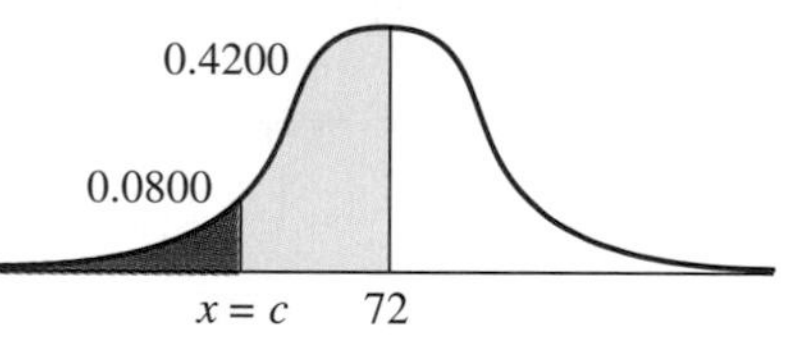


To find the scores receiving a grade of F, look at the lowest 8% of the scores. Since the diagram is the mirror image of the picture used to find the A grades, we know that $z = -1.41$.

$$-1.41 = \frac{x - 72}{12}$$

$$-16.9 = x - 72$$

$$55.1 = x$$

This means that all students scoring 55.1 or lower will receive a grade of F.

The scores earning D's are found in a similar way. From our work on the B's, we have $z = -0.74$. This gives

$$-0.74 = \frac{x - 72}{12}$$

$$-8.9 = x - 72$$

$$63.1 = x$$

This means that all students scoring between 55.1 and 63.1 will receive a grade of D.

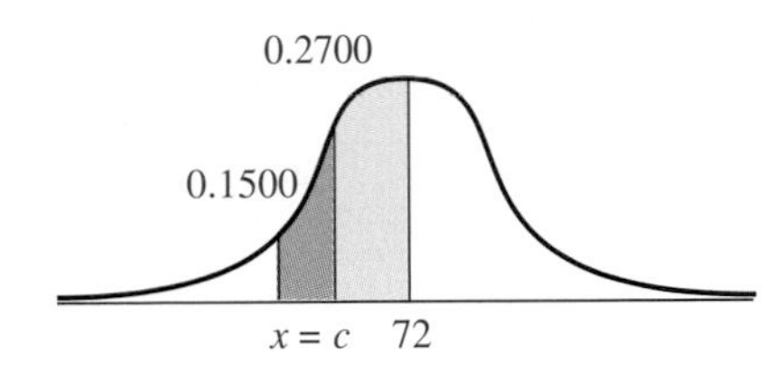


At this point we have only to determine the scores that will earn a grade of C. Since the highest score that will earn a D is 63.1 and the lowest score that will earn a B is 80.9, all scores between 63.1 and 80.9 will receive a grade of C. ■

## Section 8.4 PROBLEMS

## *Explain* ➡ *Apply* ➡ *Explore*

### *Explain*

1. What is a normal distribution?

2. When is a set of data said to be normal?

3. What does a $z$ value for a score give you?

4. In a normal distribution, if $z = 1.53$, what does this mean in terms of an area?

5. In a normal distribution, if $z = 1.53$, what does this mean in terms of a probability?

6. In a normal distribution, if $z = -0.76$, what does this mean in terms of probability?

7. In a normal distribution, if $z = -0.76$, what does this mean in terms of an area?

8. In a normal distribution with $\mu = 25$ and $\sigma = 5$, if a score $x$ has a $z$ value of 1.2, what does that tell you about $x$?

9. In a normal distribution with $\mu = 25$ and $\sigma = 5$, what does it mean if for score $x$, $P(x > 30) = 0.159$?

10. In a normal distribution with $\mu = 25$ and $\sigma = 5$, what does it mean for score $x$ to have an area of $A = 0.4332$?

*Apply*

In Problems 11–18, find the indicated area (probability) under the normal curve.

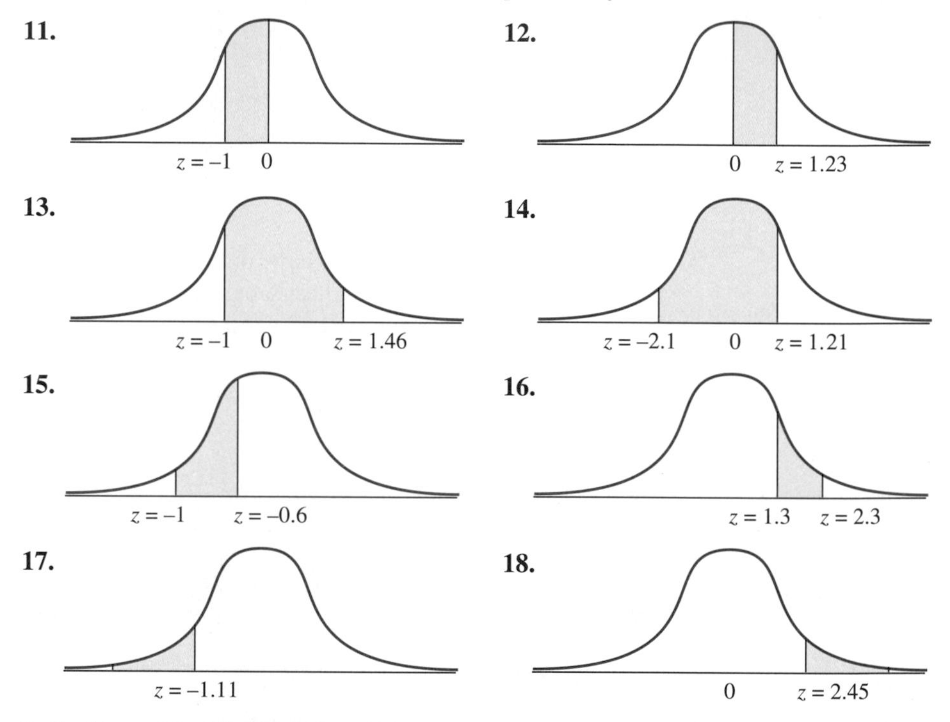


In Problems 19–24, determine the value of $c$ that will give the indicated area (probability) under the normal curve.

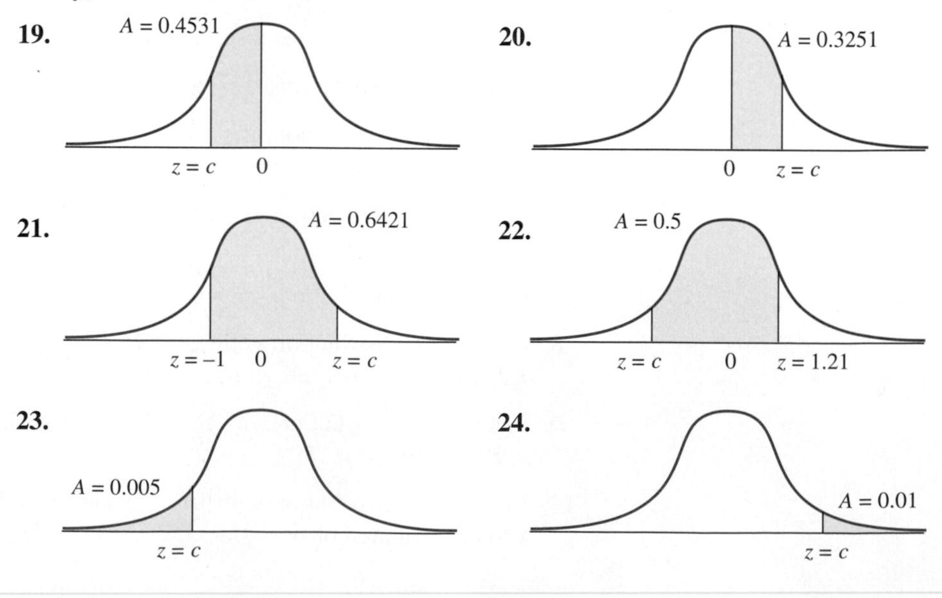

*Explore*

**25.** A survey of blue-collar workers found that the workers had a mean of 17.34 years of experience, with a standard deviation of 11.14 years. Assume a normal distribution.

(a) What percentage of the blue-collar workers in the survey had between 11 and 15 years of experience?
(b) Find the probability that a blue-collar worker in the survey, selected at random, had more than 15 years of experience.

**26.** A study was made of black families with a female as the head of the household. Among working mothers of age 30 or more, the mean number of hours worked per year was 1582.2, with a standard deviation of 728.5 hr. Assume a normal distribution.

(a) What percentage of the women in the study worked between 1000 and 1582.2 hr per year?
(b) What percentage of the women in the study worked less than 1000 hr per year?
(c) Assuming a 40-hr work week, how many full work weeks are represented by 1582.2 hr?

**27.** A survey of women aged 14–21 determined that the mean travel time to work was 17 min with a standard deviation of 13 min. Assume a normal distribution.

(a) What is the probability that a woman from this survey, chosen at random, traveled more than 20 min to reach work?
(b) What percentage of the women traveled between 10 and 20 min to get to work?
(c) What percentage of the women traveled between 5 and 10 min to get to work?

**28.** A survey of men aged 14–21 determined that the mean travel time to work was 19.7 min with a standard deviation of 20 min. Assume a normal distribution.

(a) What is the probability that a man from this survey, chosen at random, traveled more than 30 min to reach work?
(b) What percentage of the men traveled between 10 and 30 min to get to work?
(c) What percentage of the men traveled between 5 and 10 min to get to work?

**29.** A study of men aged 35–57 at increased risk of coronary disease found that the average number of cigarettes smoked per day was 21.7 with a standard deviation of 20.5. Assuming a normal distribution, determine the number of cigarettes smoked by the 10% of the group who were the heaviest smokers.

**30.** The All American University accepts, unconditionally, students who score in the top 10% nationally on the SAT exam. For those students in the next 15% on the SAT exam, the university accepts the students on the basis of their grades in the senior year of high school. Assume the SAT average is 950 with a standard deviation of 280.

(a) Determine the lowest SAT score a student may have and be accepted to the All American University unconditionally.

(b) Determine the SAT scores that will allow a student to enter All American University dependent on his or her senior-year grades.

**31.** Pick one of the following topics (or decide on one of your own) for a statistical investigation. Collect 40 or more data values for your topic. Arrange the data in a frequency distribution and draw a bar graph for the data. Find the mean and standard deviation for the data. Finally, determine the cutoff for the top 10% of the data.

(a) Season-ending batting averages of baseball players.

(b) Bowling scores of a bowling league.

(c) Lengths of the most popular songs on the radio according to the *Billboard* charts.

(d) Winning times in the 100-m dash at the collegiate level for track meets across the country.

(e) Weights of chickens sold in the grocery store.

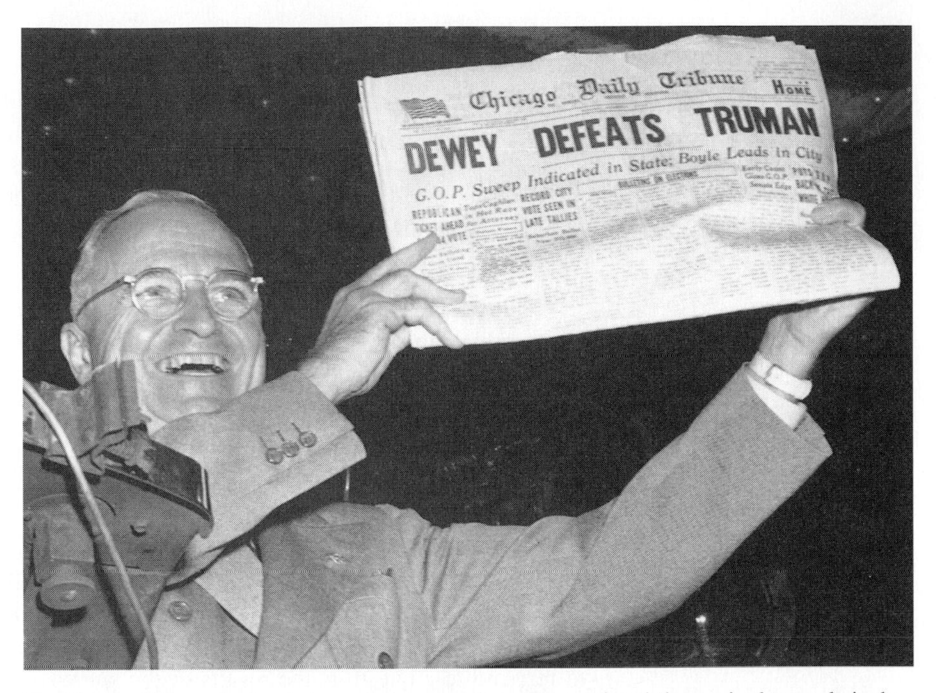


*The Chicago Tribune*, basing their conclusion on voting returns and statistics received too early in the night, picked the wrong winner of the 1948 Presidential election. (UPI/Bettman Newsphotos)

# Section 8.5 Polls and the Margin of Error

One of the most visible uses of statistics and the normal curve can be found in the public opinion surveys. In these polls, a relatively small segment of the population is asked about a certain issue and the responses are used to predict the views of the entire population. In this section, we will examine how the results of these polls can be interpreted and the accuracy of the predictions determined.

## Inferential Statistics and Confidence Intervals

Inferential statistics are the statements about a population that are derived from information about a small segment of a population. In other words, inferential statistics are exactly the types of information that can be derived from a poll or survey. A particular type of inferential statistic is the confidence interval. Before defining this term, consider the following example.

### Example 1

Two thousand registered voters have been asked to respond to a questionnaire about whether the state's members of the House of Representatives should be subject to term limitations of eight years. The results of the survey show that 1120 of the 2000 people feel that the Representatives should be limited to a maximum of eight years in office. What does this survey say about the entire voting population of the state?

**Solution:** Since 1120 of the 2000 voters endorse term limitations, it is appropriate to say that approximately $1120 \div 2000 = 0.56$ or 56% of voters endorse term limitations. However, we cannot say that 56% of all the voters endorse the proposal. If we took another survey of a different group of 2000 people, we might find that 1160 or 1070 people endorse term limitations. All we can say is that approximately 56% of the people endorse term limitations of eight years. ■

Now that we recognize that any survey will only produce an estimate of an actual value, the next step is to determine what factors contribute to the difference between the estimate and an actual value. One obvious factor that will affect the accuracy of the estimate is the size of the sample. If the survey asks a large number of people to respond to a question, the results of the survey should be more accurate than if the survey asks only a small number of people. With an increase in the size of the sample comes an increase in the accuracy of the estimate and an increase in the confidence we have in the validity of our estimate.

An additional consideration comes from the theory of statistics. Suppose that we conduct the poll described in Example 1 one hundred times, using a different sample of the population in each poll. Each poll will have a resulting estimate of the percentage of the people who are in favor of term limitations. By drawing a graph of these proportions, it can be shown that the percentages are distributed in the shape of the normal curve. As was true with the situations described in Section 8.4, the exact shape and position of the normal distribution will depend on the mean and the standard deviation of the sample percentages.

The above discussion leads to two important concepts in statistics, confidence intervals and confidence levels. A **confidence interval** is an interval centered around the estimate generated by the survey. For instance, in Example 1, the results of the survey stated that 56% of the voters endorsed term limitations. If we allow an error of four percentage points, we anticipate that the true percentage in favor of term limitations could be as low as 52% ($56 - 4$) and as high as 60% ($56 + 4$). A confidence interval for this situation would then be 52% to 60%.

A **confidence level** is a statement of the probability that the actual percentage being studied is contained within the confidence interval. With all other factors remaining equal, the higher the confidence level, the larger the confidence interval. With these considerations in mind, it can be shown that the formula for the margin of error in a poll or survey is given as follows.

Margin of Error

$$M = \frac{z}{2\sqrt{n}} \qquad \text{where} \begin{cases} M = \text{margin of error} \\ z = \text{value determined by the confidence level and the normal distribution} \\ n = \text{sample size} \end{cases}$$

## Confidence Levels and the Normal Distribution

Before continuing with our discussion of estimates, it is important to understand what a confidence level means in terms of the normal distribution. Suppose that we have a 95% confidence level. This means that the probability that the confidence interval contains the actual percentage is 0.95. Using the methods of Section 8.4 we can now calculate the $z$ value used to determine the margin of error.

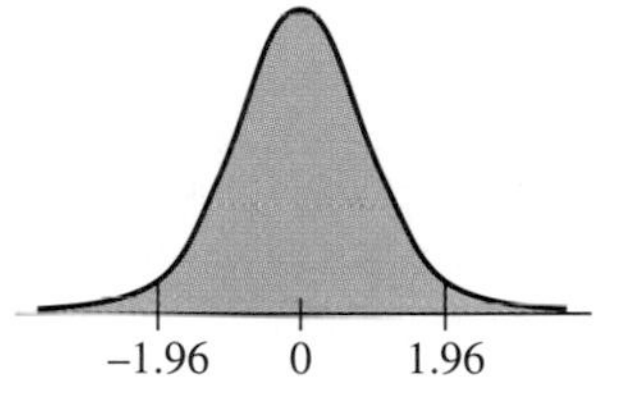


Since the confidence interval is centered around the estimate, the standard normal distribution is centered around zero. Therefore, dividing 0.95 by 2 gives 0.475. Using area = 0.475, we find $z = 1.96$. Notice that this calculation is done without consideration of the survey or the number of people responding to the survey. Therefore, any 95% confidence interval will use $z = 1.96$.

### Example 2

Determine the $z$ value for a 90% confidence interval.

**Solution:** Dividing 0.90 by 2 gives area = 0.45. Looking up this area in the normal table gives $z = 1.645$. Thus, a 90% confidence interval will use $z = 1.645$. ■

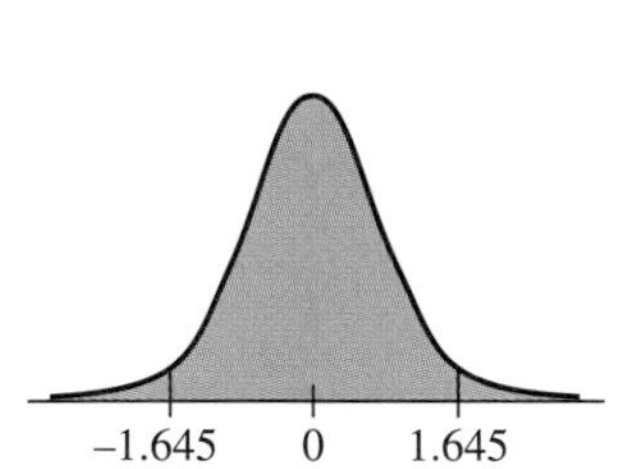


Since a particular confidence interval will result in a particular $z$ value, we give the common confidence levels in the following table.

| Confidence Level | $z$ value |
|---|---|
| 90% | 1.645 |
| 95% | 1.96 |
| 98% | 2.327 |
| 99% | 2.575 |

## Margin of Error and Surveys

Now that we have completed the background material, we can apply the margin of error formula to a variety of situations.

### Example 3

In August 1992, *Time*/CNN commissioned a poll that included the following question.*

Do you have a favorable impression of George Bush?

The response was that 52% of 836 voters surveyed had a favorable opinion of Bush. If the survey is conducted with a 98% confidence level, what is the margin of error of the survey?

**Solution:** Since we have a 98% confidence level, $z = 2.327$. Using this $z$ value and $n = 836$, we have

$$M = \frac{2.327}{2\sqrt{836}} = 0.04$$

Thus the margin of error is 4%. This means that the percentage of people that had a favorable opinion of George Bush is somewhere between 48% and 56%. ■

### Example 4

Suppose that the pollster wants to decrease the margin error from 4% to 1% while maintaining the same level of confidence. How large a sample should the pollster use?

**Solution:** Using $M = 0.01$ and $z = 2.327$, we have

$$\begin{aligned} 0.01 &= \frac{2.327}{2\sqrt{n}} \\ 0.02\sqrt{n} &= 2.327 \\ \sqrt{n} &= 116.35 \\ n &= 116.35^2 \\ n &\approx 13{,}538 \text{ people} \end{aligned}$$

■

Notice that a reduction in the margin of error requires a very large increase in the sample size. This is an important consideration for pollsters. While it is desirable to

* All polls mentioned in these examples were published by *Time* magazine in the September 14, 1992 issue.

decrease the margin of error, the resulting increase in the sample size could cause a large increase in the cost of the poll.

### Example 5

As part of another poll commissioned by *Time*/CNN, 755 voters in St. Louis County, Missouri, were asked if they planned to vote for the Republican or Democratic congressional candidate in their district. Of those responding, 39% stated that they planned to vote for the Democratic candidate. The margin of error in the survey was 3.5%

(a) What is the confidence level of this survey?
(b) What is the confidence interval for the percentage of voters planning to vote for the Democratic congressional candidate?

**Solution:**

(a) Using $n = 755$ and $M = 0.035$, we have

$$0.035 = \frac{z}{2\sqrt{755}}$$

$$0.035 = \frac{z}{54.9545}$$

$$z = 0.035 \times 54.9545$$

$$z \approx 1.92$$

Since this value is not in the table of common confidence levels, we must use the normal table. Using a $z$ value of 1.92, the normal table returns an area of 0.4726. Because this is the area on only the right side of the normal distribution, we must double the result. This gives a value of $0.4726 \times 2 = 0.9452$. Thus, the confidence level is 94.52% or approximately 95%.

(b) To determine the confidence interval, we construct an interval with 39% at its center and include all values within the margin of error. Since the margin of error is 3.5%, the confidence interval is

$$(39 - 3.5)\% \quad \text{to} \quad (39 + 3.5)\% \qquad \text{or} \qquad 35.5\% \quad \text{to} \quad 42.5\%$$

Thus we have a 95% assurance that the percentage of voters planning to vote for the Democratic candidate is somewhere between 35.5% and 42.5%. ■

In this section, we have seen one application of the normal distribution to polling and margins of error, a topic that can be found in almost every issue of the national news magazines. In the next section, we again see how statistics can be applied to a practical situation, the allocation of voting power.

Section 8.5
PROBLEMS

## *Explain* ➡ *Apply* ➡ *Explore*

### *Explain*

1. What is a confidence interval?
2. What is a confidence level?
3. What is a margin of error?
4. What is a sample size?
5. What happens to the margin of error if the sample size increases? Explain.
6. What happens to the margin of error if the confidence level increases? Explain.
7. A survey estimates that 25% of the viewing audience watched the Cotton Bowl on January 1, 1994. If there is a 5% margin of error, what does this say about the percentage of the viewing audience that watched the Cotton Bowl?
8. A poll estimates that 63% of the public wants to see an increase in handgun control laws. The margin of error is 4%. What does this say about the percentage of the public that wants to see an increase in the number of handgun control laws?

### *Apply*

9. Determine the $z$ value for an 80% confidence level.
10. Determine the $z$ value for a 70% confidence level.
11. Determine the $z$ value for an 85% confidence level.
12. Determine the $z$ value for a 75% confidence level.
13. A poll has a result of 63% with a margin of error of 4%. What is the confidence interval for this poll?
14. A poll has a result of 52% with a margin of error of 3%. What is the confidence interval for this poll?
15. A poll has a result of 47.6% with a margin of error of 2.4%. What is the confidence interval for this poll?
16. A poll has a result of 39.2% with a margin of error of 3.1%. What is the confidence interval for this poll?
17. A poll of 750 people is taken and the confidence level is 98%. What is the margin of error?

**18.** A poll of 1200 people is taken and the confidence level is 95%. What is the margin of error?

**19.** A poll of 1000 people is taken and the confidence level is 86%. What is the margin of error?

**20.** A poll of 800 people is taken and the confidence level is 99.9%. What is the margin of error?

**21.** A poll of 2000 people has a margin of error of 4%. What is the confidence level for this poll?

**22.** A poll of 600 people has a margin of error of 2.5%. What is the confidence level for this poll?

**23.** A poll of 800 people has a margin of error of 3.5%. What is the confidence level for this poll?

**24.** A poll of 1000 people has a margin of error of 3%. What is the confidence level for this poll?

**25.** A customer requests that a polling company produce a survey with a 98% confidence level and a margin of error of 2%. How many people must respond to the survey?

**26.** A customer requests that a polling company produce a survey with a 99% confidence level and a margin of error of 1%. How many people must respond to the survey?

**27.** A customer requests that a polling company produce a survey with an 80% confidence level and a margin of error of 3%. How many people must respond to the survey?

**28.** A customer requests that a polling company produce a survey with an 85% confidence level and a margin of error of 2%. How many people must respond to the survey?

## *Explore*

**29.** According to a poll commissioned by Time, Inc., August 25–30, 1992, 55% of those polled in DeKalb County, Georgia, stated that they planned to vote for Bill Clinton in the 1992 presidential election. If 578 likely voters were polled and the margin of error was 4%,

(a) What is the confidence level of the poll?
(b) How large a sample would be needed to have a confidence level of 99% while maintaining a 4% margin of error?
(c) How large a sample would be needed to decrease the margin of error to 2% while maintaining a 99% confidence level?

**30.** According to a poll commissioned by Time, Inc., August 25–30, 1992, 49% of those polled in Middlesex County, New Jersey, stated that they planned to vote for Bill Clinton in the 1992 presidential election. If 595 likely voters were polled and the margin of error was 4%,

(a) What is the confidence level of the poll?
(b) How large a sample would be needed to have a confidence level of 99% while maintaining a 4% margin of error?
(c) How large a sample would be needed to decrease the margin of error to 2% while maintaining a 99% confidence level?

**31.** A poll of 994 people was conducted by the Field Institute on May 14–22, 1993. Eighty-one percent of the respondents said that the California economy was heading on the wrong track. The margin of error is 3.2%.

(a) What is the confidence level?
(b) What is the confidence interval for this survey?

**32.** A poll of 994 people was conducted by the Field Institute on May 14–22, 1993. Sixty-three percent of the respondents said that the U.S. economy was heading on the wrong track. The margin of error is 3%.

(a) What is the confidence level?
(b) What is the confidence interval for this survey?

**33.** According to a *USA Today*/CNN Gallup poll conducted October 13–18, 1993, 53% of a survey of 314 people favored the toughening of parole possibilities for violent criminals.

(a) If the margin of error is 6%, what is the confidence level?
(b) How large a sample would be needed to have a confidence level of 99%?
(c) How large a sample would be needed to decrease the margin of error to 4% while maintaining the original confidence level?

**34.** According to a *USA Today*/CNN Gallup poll conducted October 13–18, 1993, 46% of a survey of 314 people favored making it more difficult for arrested violent crime suspects to get out on bail.

(a) If the margin of error is 6%, what is the confidence level?
(b) How large a sample would be needed to have a confidence level of 90%?
(c) How large a sample would be needed to decrease the margin of error to 3% while maintaining the original confidence level?

**35.** According to a *USA Today*/CNN Gallup poll conducted October 13–18, 1993, 38% of a survey of 870 people favored the enactment of tougher gun control laws. If the margin of error is 4%, what is the confidence level?

**36.** According to a *USA Today*/CNN Gallup poll conducted October 13–18, 1993, 48% of a survey of 870 people favored making prison sentences more severe for all crimes. If the margin of error is 4%, what is the confidence level?

**37.** According to a *USA Today*/CNN poll, 79% of a survey of 500 people said they knew how to change the oil in a car. If the confidence level of the survey is 90%

(a) What is the margin of error?

(b) Is it possible to keep the same margin of error and to increase the confidence level to 100%? If so, describe how. If not, explain why not.

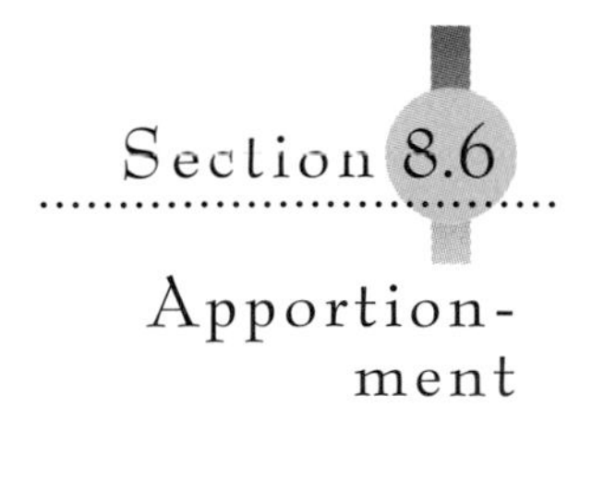

## Section 8.6 Apportionment

Our discussions in this chapter have shown us some of the statistical measurements that can be used on a set of data and how these measurements can be used to make decisions and predictions based on that data. In this section, we continue this investigation by analyzing how applying a different set of assumptions to the same set of data can lead to different conclusions. While this section uses none of the statistical techniques of the previous sections, it does demonstrate how various methods of manipulating data can produce different results.

In particular, we will investigate the way a representative body of leaders can be selected in a democratic system by different methods of apportionment. **Apportionment** is the process by which a given amount of a resource can be divided into parts. In this section, the apportionment will involve determining the number of representatives from each of the five counties of Rhode Island to that state's Senate.

The following table gives the populations of the different counties in Rhode Island.

| County | Population |
|---|---|
| Bristol | 48,859 |
| Kent | 161,135 |
| Newport | 87,194 |
| Providence | 596,270 |
| Washington | 110,006 |
| Total | 1,003,464 |

The first assumption is that each seat in the legislature should provide representation for approximately the same number of people. Thus, since there are 50 seats, one way to determine the apportionment is to determine the percentage of the state's population that is in each region and multiply that percentage by 50. For example, for Bristol County we have

$$\frac{48,859}{1,003,464} \times 50 \approx 2.43$$

The results of the calculations for each region are presented in the Quotas column of Table 8.6.1. Since it is not possible to have a decimal number of seats apportioned to a county, we determine the number of seats by rounding off each quota to the nearest whole number.

**Table 8.6.1** *Number of Seats Using Rounding Method*

| County | Population | Quotas | No. Seats |
|---|---|---|---|
| Bristol | 48,859 | 2.43 | 2 |
| Kent | 161,135 | 8.03 | 8 |
| Newport | 87,194 | 4.35 | 4 |
| Providence | 596,270 | 29.71 | 30 |
| Washington | 110,006 | 5.48 | 5 |
| Totals | 1,003,464 | 50.00 | 49 |

Notice that although the total number of seats should be 50, only 49 of the seats have been allocated. While rounding off the quotas seems to be a reasonable method, it fails to allocate all of the available seats. When the government of the United States was being established in the late 1780s, similar difficulties were encountered when setting up the apportionment for the seats in the House of Representatives. In search of a solution, several methods of apportionment were devised. We will look at three of these methods, those devised by Alexander Hamilton, Daniel Webster, and Thomas Jefferson.

The **Hamilton method** assigns an initial number of seats by taking the integer part of each quota. This may give a total number of seats that is fewer than the actual number of seats. In our example, the initial total number of seats is 48 (see Table 8.6.2). Since this allows for an additional two seats, the Hamilton method assigns those seats to the counties with the greatest decimal parts in their quotas. Since Providence (decimal part 0.71) and Washington (0.48) have the greatest decimal parts in their quotas, they receive the additional seats.

**Table 8.6.2** *Number of Seats Using Hamilton's Method*

| County | Population | Quotas | Initial No. of Seats | No. Seats |
|---|---|---|---|---|
| Bristol | 48,859 | 2.43 | 2 | 2 |
| Kent | 161,135 | 8.03 | 8 | 8 |
| Newport | 87,194 | 4.35 | 4 | 4 |
| Providence | 596,270 | 29.71 | 29 | 30 |
| Washington | 110,006 | 5.48 | 5 | 6 |
| Totals | 1,003,464 | 50.00 | 48 | 50 |

## Divisor Methods

The apportionment methods of Jefferson and Webster are called divisor methods. **Divisor methods** apportion a resource by dividing the number in each group by a divisor and then rounding off the result to a nearby integer.

**Jefferson's method** works as follows. Since we have a total of 1,003,464 people in the state and 50 seats, each seat should represent $\frac{1{,}003{,}464}{50} \approx 20{,}069$ voters. This number 20,069 is our divisor. The number of representatives for each county is calculated by dividing the county population by the divisor and rounding down to the nearest whole number. This is shown in Table 8.6.3.

**Table 8.6.3** *Number of Seats Using Jefferson's Method*

| County | Population | Quotas | No. of Seats |
|---|---|---|---|
| Bristol | 48,859 | 2.43 | 2 |
| Kent | 161,135 | 8.03 | 8 |
| Newport | 87,194 | 4.35 | 4 |
| Providence | 596,270 | 29.71 | 29 |
| Washington | 110,006 | 5.48 | 5 |
| Totals | 1,003,464 | 50.00 | 48 |

Since the total number of seats used, 48, is fewer than the 50 seats that are available, the divisor is decreased until the number of seats apportioned has increased to 50. The first divisor that achieves this goal is 19,234 instead of the initial 20,069. (See Table 8.6.4.)

**Table 8.6.4** *Number of Seats Using Jefferson's Method*

| County | Population | Pop./Divisor | No. of Seats |
|---|---|---|---|
| Bristol | 48,859 | 2.54 | 2 |
| Kent | 161,135 | 8.38 | 8 |
| Newport | 87,194 | 4.53 | 4 |
| Providence | 596,270 | 31.00 | 31 |
| Washington | 110,006 | 5.72 | 5 |
| Totals | 1,003,464 | | 50 |

Webster's method is another divisor method. With **Webster's method**, you proceed as with Jefferson's method but round off to the nearest whole number rather than always rounding down. An example is given in Table 8.6.5.

**Table 8.6.5** *Number of Seats Using Webster's Method*

| | | Divisor = 20,069 | | Divisor = 20,001 | |
|---|---|---|---|---|---|
| **County** | **Population** | **Quotas** | **No. of Seats** | **Pop./Divisor** | **No. of Seats** |
| Bristol | 48,859 | 2.43 | 2 | 2.44 | 2 |
| Kent | 161,135 | 8.03 | 8 | 8.06 | 8 |
| Newport | 87,194 | 4.35 | 4 | 4.36 | 4 |
| Providence | 596,270 | 29.71 | 30 | 29.81 | 30 |
| Washington | 110,006 | 5.48 | 5 | 5.50 | 6 |
| Totals | 1,003,464 | 50.00 | 49 | | 50 |

## Why So Many Methods?

We have introduced three methods that provide a clear way of determining the apportionment of a legislature if the number of available seats is known. There are other methods such as the quota method, the Hill-Huntington method, and Lowndes' method. While the Hill-Huntington method is currently being used to apportion the U.S. House of Representatives, a discussion of it is beyond the scope of this course.

You may wonder why there are so many methods and which method is the best to use. The reason that there are many methods is that all of the methods have biases, some obvious and some hidden. As these problems were discovered, the various methods fell out of favor. For example, in the apportionment of the U.S. House of Representatives, Jefferson's method was used until 1842. Hamilton's method was used until 1901 when it was replaced by Webster's method. This was replaced, in turn, by the current method, Hill-Huntington, in 1941. In the remainder of this section, we will examine some of the problems associated with some of these methods.

### *The Bias in Jefferson's Method*

Because the apportionment of the legislature requires some sort of rounding off of the quotas, one of the guidelines in setting up an apportionment procedure is given by the quota rule. The **quota rule** states that the number of seats apportioned to a group should be a whole number within one of the quota. For example, if a county's quota is 1.95, it should receive an apportionment of either 1 or 2.

Jefferson's method tends to favor states or counties with large populations. If we reexamine the apportionment of the Rhode Island Legislature using Jefferson's method with respect to the quota rule, we see that Providence county has a quota of 29.71 (see Table 8.6.3), but it receives an apportionment of 31 seats, thus violating the quota rule (Table 8.6.4).

## *Two Paradoxes with Hamilton's Method*

A problem that arose with Hamilton's method is now known as the Alabama paradox. In 1880, Hamilton's method was used to apportion the U.S. House of Representatives. In the **Alabama paradox**, the House was apportioned according to Hamilton's method. Using the same method with the same population figures, the size of the legislature was increased and the House reapportioned. The paradox arose because Alabama's apportionment decreased even though the total number of seats had increased and there was no change in the population. We can examine this paradox in the following example.

### Example 1

Consider the populations of the following three counties in a state with a 60-seat legislature. Using Hamilton's method, we get

| County | Population | Quotas | Initial No. of Seats | No. Seats |
|---|---|---|---|---|
| Anderson | 3,010 | 30.21 | 30 | 30 |
| Black Hills | 2,710 | 27.20 | 27 | 27 |
| Coldwater | 258 | 2.59 | 2 | 3 |
| Totals | 5,978 | 60.00 | 59 | 60 |

What happens to the apportionment if the number of seats is increased to 61?

**Solution:** If the number of seats increases, the quota for each county must be recalculated. For Anderson County, we find

$$\frac{3010}{5978} \times 61 \approx 30.71$$

Repeating this procedure for the other two counties and computing the corresponding number of seats in the legislature, we have

| County | Population | Quotas | Initial No. of Seats | No. Seats |
|---|---|---|---|---|
| Anderson | 3,010 | 30.71 | 30 | 31 |
| Black Hills | 2,710 | 27.65 | 27 | 28 |
| Coldwater | 258 | 2.63 | 2 | 2 |
| Totals | 5,978 | 61.00 | 59 | 61 |

Notice that though there was no change in the population of any of the three counties, Coldwater County lost one of its three seats. ■

A second paradox that affects Hamilton's method is called the population paradox. The **population paradox** occurs when reapportionment takes place and a region with a greater rate of increase in population loses seats to another region with a slower growth rate.

### Example 2

The following table gives the population figures for five counties, with populations given in thousands, and the associated apportionment as calculated by Hamilton's method.

*Seats Computed Using Hamilton's Method*

| County | Population | Quotas | Initial No. of Seats | No. Seats |
|---|---|---|---|---|
| Anderson | 1,732 | 9.60 | 9 | 10 |
| Black Hills | 1,503 | 8.33 | 8 | 8 |
| Coldwater | 784 | 4.34 | 4 | 4 |
| Shasta | 2,049 | 11.35 | 11 | 11 |
| Tehachapi | 2,957 | 16.38 | 16 | 17 |
| Totals | 9,025 | 50.00 | 48 | 50 |

Suppose the population of Anderson County increases by 5% while the population of Tehachapi County increases by 0.3%. The populations of the other three counties remain the same. Use Hamilton's method to calculate the new apportionment.

**Solution:** Since the population of Anderson County changed by 5%, the population increased by $1{,}732{,}000 \times 0.05 = 87{,}000$ or 87 thousand people. Therefore the new population is $1732 + 87 = 1819$ thousand people. Similarly multiplying the population of Tehachapi County by 0.003 and adding this amount to the original population results in a new population of 2966 thousand people.

Using Hamilton's method on these figures, we arrive at the following table.

*Seats Computed Using Hamilton's Method*

| County | Population | Quotas | Initial No. of Seats | No. Seats |
|---|---|---|---|---|
| Anderson | 1,819 | 9.97 | 9 | 10 |
| Black Hills | 1,503 | 8.24 | 8 | 8 |
| Coldwater | 784 | 4.30 | 4 | 5 |
| Shasta | 2,049 | 11.23 | 11 | 11 |
| Tehachapi | 2,966 | 16.26 | 16 | 16 |
| Totals | 9,121 | 50.00 | 48 | 50 |

Notice what has happened. Coldwater County's apportionment has increased from four seats to five seats despite having no growth in its population. In addition, Tehachapi County has suffered a decrease in the number of its representatives despite having a small increase in its population. ■

After seeing how Hamilton's method is subject to both the Alabama paradox and the population paradox, and that Jefferson's method is biased for regions with larger populations, you may wonder if there are similar difficulties with Webster's method or any other method. The answer is a definitive "Yes." In 1980, Michel L. Balinski and H. Peyton Young stated and proved what is known as the **Balinski and Young impossibility theorem**. The theorem states that it is mathematically impossible for any apportionment method to be flawless. If a method satisfies the quota rule, it will occasionally produce paradoxes. On the other hand, if the method does not produce paradoxes, in some situations it will violate the quota rule.

As a result of the impossibility theorem, you can see that an analysis of any apportionment method will result in the discovery of a flaw. Currently, the apportionment of the U.S. House of Representatives is accomplished using the Hill-Huntington method. However, because of the impossibility theorem, the debate continues which method is best. Currently, mathematicians are debating the merits of Webster's method versus those of the Hill-Huntington method. Webster's method is considered by some to be the best method. It is theoretically possible that it will violate the quota rule. However, if it had been applied in every apportionment of the House of Representatives since 1790, no such violation would have occurred. Other mathematicians believe that the Hill-Huntington method is best because it minimizes the relative differences in size between districts. Will the method change? If it does, the U.S. Constitution says that it would require an act of Congress.

From the discussions in this section, you have observed how different analyses of the same set of data can lead to different conclusions and decisions. This points out the importance of understanding how statistics are computed and interpreted. Without this underlying understanding, you will not be able to fully use the power of statistics.

## Section 8.6 PROBLEMS

## *Explain → Apply → Explore*

### *Explain*

**1.** What is meant by ''apportionment''?

**2.** Describe Hamilton's method.

**3.** Describe Jefferson's method.

**4.** Describe Webster's method.

**5.** What is the Alabama paradox?

**6.** What is the population paradox?

**7.** What is the quota rule?

**8.** What is the impossibility theorem?

### *Apply*

**9.** Rhode Island's Assembly has 100 members. Use the population statistics given in the text and Hamilton's method to determine the apportionment of the assembly.

**10.** Rhode Island's Assembly has 100 members. Use the population statistics given in the text and Webster's method to determine the apportionment of the assembly.

**11.** Rhode Island's Assembly has 100 members. Use the population statistics given in the text and Jefferson's method to determine the apportionment of the assembly.

Delaware has three counties with the populations given in the table. The State Assembly has 41 seats.

| County | Population |
|---|---|
| Kent | 110,993 |
| New Castle | 441,946 |
| Sussex | 113,229 |
| Total | 666,168 |

**12.** Use Hamilton's method to determine the apportionment of the Assembly.

**13.** Use Webster's method to determine the apportionment of the Assembly.

**14.** Use Jefferson's method to determine the apportionment of the Assembly.

## *Explore*

In Problems 15–17, use the following information.

A school has the resources to offer 20 sections of math courses each semester. It is expected that there will be a total of 600 students wanting to take math courses:

| **Course** | **Number of Students** |
|---|---|
| College Algebra | 120 |
| Calculus | 80 |
| Math for Business | 235 |
| Liberal Arts Math | 165 |

**15.** Use Hamilton's method to determine the distribution of the number of sections of each course.

**16.** Use Webster's method to determine the distribution of the number of sections of each course.

**17.** Use Jefferson's method to determine the distribution of the number of sections of each course.

In Problems 18–20, use the following information.

A city has 45 garbage trucks that are in use five days a week. This provides for 225 different routes. The city is divided into six areas because of geographical and transportation limitations. The populations of these six areas are given in the chart.

| **District** | **Population** |
|---|---|
| Bayview | 25,456 |
| Downtown | 32,723 |
| East End | 27,568 |
| Greenhaven | 16,475 |
| Ingleside | 8,696 |
| Riverview | 11,458 |
| Total | 122,376 |

**18.** Use Hamilton's method to determine the number of garbage trucks in each district.

**19.** Use Webster's method to determine the number of garbage trucks in each district.

**20.** Use Jefferson's method to determine the number of garbage trucks in each district.

In Problems 21–23, use the following information.

The state of Iowa has 6 congressional districts. The number of people that voted in the 1990 congressional election in each district is given in the chart.

| District | Voters |
|---|---|
| 1 | 90,000 |
| 2 | 166,000 |
| 3 | 102,000 |
| 4 | 131,000 |
| 5 | 147,000 |
| 6 | 156,000 |

Suppose that 800 voting sites are to be used in the 1996 elections.

**21.** Use Hamilton's method to allocate the number of voting sites in each district.

**22.** Use Webster's method to allocate the number of voting sites in each district.

**23.** Use Jefferson's method to allocate the number of voting sites in each district.

In Problems 24–26, use the following information.

The state of Arizona has 5 congressional districts. The number of people that voted in the 1990 congressional election in each district is given in the chart.

| District | Voters |
|---|---|
| 1 | 167,000 |
| 2 | 116,000 |
| 3 | 237,000 |
| 4 | 231,000 |
| 5 | 215,000 |

Suppose that a candidate plans to spend 150 days campaigning for the 1996 elections.

**24.** Use Hamilton's method to allocate the number of days that should be spent campaigning in each district.

**25.** Use Webster's method to allocate the number of days that should be spent campaigning in each district.

**26.** Use Jefferson's method to allocate the number of days that should be spent campaigning in each district.

## Chapter 8 • Summary

### Key Terms, Concepts, and Formulas

The important terms in this chapter are:

**Apportionment:** The process by which a given amount of a resource can be divided into parts. p. 591

**Bar graph:** A graphical representation of data using rectangles. p. 532

**Central tendency:** A general term describing the middle value of a set of data; usually refers to mean, mode, or median. p. 545

**Class:** A category into which data are distributed; used in frequency distributions. p. 531

**Class midpoints:** The mean of the boundary values for a class. p. 547

**Confidence interval:** An interval centered around the estimate of the desired statistic. p. 584

**Confidence level:** A statement of the probability that the actual percentage being studied is contained within the confidence interval. p. 584

**Dispersion:** A general term describing how far data are spread from a central value. p. 557

**Divisor method:** Apportionment of a resource by dividing the number in each group by a divisor and rounding off the result to an integer. p. 593

**Frequency:** The number of times a data value or a class value occurs. p. 531

**Line graph:** A graphical representation of data formed by dots connected by line segments. p. 533

**Margin of error:** The distance from the center of a confidence interval to each end point. p. 585

**Mean:** The arithmetic mean; the sum of the data values divided by the number of data values. p. 545

**Median:** The middle value of a set; found by listing the data in increasing numerical order and selecting the middle value; if there is an even number of data values, it is the arithmetic mean of the middle pair. p. 548

**Mode:** The data value that occurs most frequently. p. 549

**Normal distribution:** A statistical distribution that models many real-world situations; the bell-shaped curve. p. 569

**Pie chart:** A graphical representation of percentage frequencies using sectors of a circle. p. 533

**Quota rule:** Principle of apportionment that states that the number of seats apportioned to a group should be a whole number within one of the quota. p. 594

**Range:** The difference between the highest and lowest data values. p. 558

**Standard deviation:** A number that measures the dispersion of data. p. 558

**Weighted average:** The mean of a group of numbers in which certain values have more importance, or weight, than do other values. p. 547

After completing this chapter, you should be able to:

1. Organize data into tables, charts, and graphs. p. 530
2. Use and explain the concepts of central tendency and dispersion. p. 545
3. Decide which measure of central tendency is most appropriate for a situation and how to find that value. p. 580
4. Find the standard deviation of either grouped or nongrouped data. p. 558
5. Use the mean and standard deviation of two sets of data to compare the distributions of the two sets. p. 563
6. Analyze the normal distribution using the normal table. p. 569
7. Interpret the meaning of margin of error and confidence interval in a survey. p. 585
8. Apportion a resource using the Hamilton, Jefferson, and Webster methods. p. 592

## Summary Problems

1. Given the numbers, 2, 4, 6, 8, 3, 6, 7, 9, 1, 3, find:
   (a) The mean.
   (b) The median.

(c) The mode.
(d) The range.
(e) The standard deviation.

**2.** In 1990, there were 182.1 million people eligible to vote in the United States. The data organized according to the years of school they completed are as follows. (Source: U.S. Bureau of the Census, *Current Population Reports*)

| School Years Completed | Voters (million) |
|---|---|
| 8 or less | 17.7 |
| 1–3 high school | 21.0 |
| 4 high school | 71.5 |
| 1–3 college | 36.3 |
| 4 or more college | 35.6 |

(a) Draw a pie chart for this data.
(b) Draw a bar graph for this data.

**3.** The recording industry uses various recording media such as phonograph records, cassettes, and compact discs. The following data give the net number of units shipped for various media from 1975–1991. (Source: Recording Industry Association of America)

*Recording Media Shipments in Millions*

| | 1975 | 1980 | 1985 | 1990 | 1991 |
|---|---|---|---|---|---|
| Phono. Records | 421.0 | 487.1 | 287.7 | 39.3 | 28.6 |
| Cassettes | 16.2 | 110.2 | 339.1 | 442.2 | 360.1 |
| Compact Discs | 0 | 0 | 22.6 | 285.6 | 333.3 |

(a) Make a line graph that displays all three types of recording media on the same graph.
(b) What are some conclusions you can make from the graph?

**4.** The following table gives the number of people that voted in the 34 Congressional Districts in the state of New York in the 1990 U.S. House of Representatives election. (Source: Election Research Center, *America Votes*)

| Dst | No. Voters | Dst | No. Voters |
|---|---|---|---|
| 1 | 134,000 | 18 | 41,000 |
| 2 | 102,000 | 19 | 75,000 |
| 3 | 137,000 | 20 | 131,000 |
| 4 | 130,000 | 21 | 140,000 |
| 5 | 132,000 | 22 | 139,000 |
| 6 | 61,000 | 23 | 183,000 |
| 7 | 51,000 | 24 | 178,000 |
| 8 | 78,000 | 25 | 109,000 |
| 9 | 55,000 | 26 | 97,000 |
| 10 | 76,000 | 27 | 151,000 |
| 11 | 39,000 | 28 | 151,000 |
| 12 | 43,000 | 29 | 141,000 |
| 13 | 59,000 | 30 | 165,000 |
| 14 | 98,000 | 31 | 160,000 |
| 15 | 90,000 | 32 | 124,000 |
| 16 | 57,000 | 33 | 110,000 |
| 17 | 98,000 | 34 | 129,000 |

(a) Find the range, the median, and the mode for the data.
(b) Arrange the data into a frequency distribution with six classes, with the first class being 30,001–60,000. Draw a bar graph for the grouped data.
(c) Find the mean of the grouped data.
(d) Find the standard deviation of the grouped data.

**5.** Suppose you need to purchase 500 resistors rated at 10 ohms. It is critical that the resistance not vary substantially from the rated value. Two companies have told you they can supply the part for the same low, low price. Both companies claim their resistors are rated at 10 ohms. However, company A says the standard deviation of their resistors is 0.6 ohm, and company B says the standard deviation of their resistors is 1.8 ohms. Which company should you choose as your supplier? Explain your answer.

**6.** A survey of white-collar workers found that the workers had a mean of 18.14 years of experience, with a standard deviation of 10.08 years. Assume the experience level is normally distributed.

(a) What percentage of the white-collar workers had between 10 and 20 years of experience?
(b) What percentage of the workers had between 5 and 10 years of experience?

(c) Find the probability that a white-collar worker selected at random had more than 15 years of experience.
(d) Find the probability that a white-collar worker selected at random had more than 25 years of experience.

**7.** An algebra class averaged 73 on the first midterm with a standard deviation of 14.2. If the top 10% received A's, the next 20% received B's, the next 40% received C's, the next 20% D's, and the final 10% F's, use the normal distribution to determine which scores receive which grades.

**8.** As part of a poll conducted for Time, Inc., on August 25–30, 1992, 658 residents of Contra Costa County, California, were asked the following question: "Which candidate, Bush or Clinton, will win the most votes in your county?" Fifty-one percent of the respondents replied that Bill Clinton would win more votes. Assume that the margin of error is 4%.

(a) What is the confidence interval for this information? What does the confidence interval say about the anticipated votes for Bush and Clinton?
(b) What is the confidence level?
(c) How large a sample would be needed to have a confidence level of 99% while maintaining a 4% margin of error?
(d) How large a sample would be needed to decrease the margin of error to 2% while maintaining a 99% confidence level?

**9.** Delaware has three counties with the populations given in the table. The State Senate has 21 seats.

| County | Population |
|---|---|
| Kent | 110,993 |
| New Castle | 441,946 |
| Sussex | 113,229 |
| Total | 666,168 |

(a) Use Hamilton's method to determine the apportionment of the Senate.
(b) Use Webster's method to determine the apportionment of the Senate.
(c) Use Jefferson's method to determine the apportionment of the Senate.

# 9

# From Fingers to the Computer

Rod Thomas illustrates what hieroglyphics would have looked like if computers were in ancient Egypt. (Courtesy of the artist)

# Overview Interview

In this chapter, we will trace the development of calculating devices. The investigation begins with a look at the use of fingers to record numbers and perform basic operations and continues to other devices such as abacuses, "Napier's Bones," slide rules, and mechanical calculators. We look at some of the mathematics behind these devices and learn how they actually calculate. We then enter the electronic age and examine computers and hand-held calculators, taking a look at their history and uses. Our goal in this chapter is to give you a basic understanding of the origins and applications of present-day calculators and computers.

## Michael Fallon

MICHAEL FALLON, ENVIRONMENTALIST, ENVIRONMENTAL AND ENERGY STUDY INSTITUTE (EESI), WASHINGTON, D.C.

TELL US ABOUT YOURSELF. I attended West Virginia University, where my various interests led me to take course work in journalism, computers, and environmental sciences while majoring in business. As a volunteer in the Mountain Stream Monitors, a community organization dedicated to preserving West Virginia's river systems, I became interested in environmental issues. I presently work for the Environmental and Energy Study Institute in Washington, D.C., which promotes policies and programs that lead to solutions of environmental problems. At EESI, I manage a newsletter that reports on Congressional environmental activities, maintain the hardware and software of our computer network system, help publish environmental research results and policy recommendations, and am active in the marketing department.

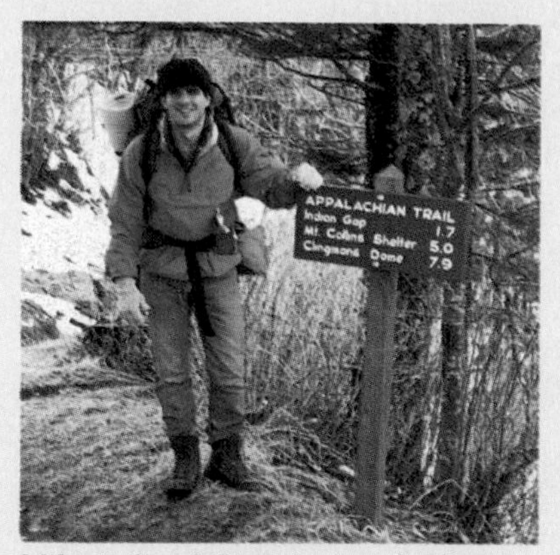


*Mike Fallon hiking the Appalachian Trail.*

HOW ARE COMPUTERS USED BY AN ENVIRONMENTALIST? The computer's power, adaptability, and widespread use has made it the ideal tool for the environmentalist. Its ability to instantly transmit and access correspondence, news reports, official documents, and libraries enables me to take part in environmental debates and decisions from virtually anywhere. The computer's power can process information from experiments, surveys, and research in minutes instead of days. Results can then be presented in a clear and compelling way using today's powerful publishing software.

Information once stored in paper files is now kept in computer databases. This information can be used to make models that predict the growth of cities, transportation patterns, or sensitive ecological areas. These models and their predictions are crucial for avoiding environmental problems and disasters. Computer databases and the speed at which the information can be accessed enable me to target key decision makers, such as the U.S. Congress, with the information needed to make informed decisions that could preserve our planet.

Calculating devices from the abacus to the electronic calculator have been used by students to do computation. The personal computer has given students even more power, as is seen in the computer-generated graphics creation, *An Interpretation of Picasso's Woman* by Daniel Lim, Ohlone College student. (Courtesy of the artist)

## A Short History of Computing Devices

After humans developed the ability to count and formulate number systems, they began to combine numbers with the operations of addition, subtraction, multiplication, and division. Through the centuries, different methods were developed to perform these calculations. The first aids to computation were the fingers. Many different schemes for finger reckoning were developed. Along with finger techniques various ''pen and paper'' computational schemes were also designed. The methods we use today were techniques to do the basic operations developed during the Renaissance in Europe. These techniques rely on the work of humans and are subject to human speed and human error. To increase the speed and accuracy of computation, mechanical devices were invented.

The abacus was the first of these mechanical computing devices. With an abacus, calculations are performed by moving stones placed in grooves or beads strung on wires. For a well-trained user of an abacus, its efficiency and speed of calculation is astounding. On November 11, 1946, a competition was held between Kiyoshi Matsuzake of the Japanese Ministry of Postal Administration, using a Japanese soroban abacus, and Private Thomas Wood of the U.S. Occupation Forces, using a 1940s rotary calculator. Even though Private Wood was an experienced calculator user, Kiyoshi Matsuzake won four out of five calculation contests.

Besides the abacus, other devices were invented to help in calculation. ''Napier's Bones'' developed in the early 1600s simplified multiplication. The slide rule, which uses logarithmic scales, was invented by William Oughtred in about 1622 and was used until the late 1960s. Mechanical calculators that performed calculations with rotating wheels were invented beginning in the 1620s. In the 20th century, the progress of calculating devices took giant steps. The advancements in electronics such as electromagnetic relays, vacuum tubes, transistors, integrated circuits, and microprocessors led to the development of the present-day calculators and computers.

A summary of the important dates, inventors, and events connected with the development of computing devices follows.

- c. 400 B.C. A primitive abacus was used in Greece for computation.
- c. 100 B.C. Evidence that basic counting-board abacus was used in Europe.
- c. 1200 Suan-pan abacus was used in China.
- c. 1600 Soroban abacus was used in Japan.
- 1617 John Napier published *Rabdologiae*, which described ''Napier's Bones'' and how they were used to simplify multiplication.
- 1621 William Oughtred invented the slide rule as an aid to computation.
- 1623 William Schickard designed the first mechanical calculating machine. Unfortunately it was destroyed in a fire before completion and Schickard died of the plague before he could rebuild it.
- 1624 Henry Briggs published 14-place common logarithm tables used for computation.
- 1642 Blaise Pascal invented and demonstrated a calculating machine that performed addition and subtraction.
- 1674 Gottfried Leibniz constructed a stepped-drum gear calculating machine.
- 1820 The first commercial mechanical calculating machine, the *Arithmometer*, was made available by Frenchman Thomas de Colmar.
- 1823 Charles Babbage designed a steam-powered digital calculator, the *Difference Engine*.
- 1843 Ada Augusta Byron, Countess of Lovelace, introduced computer programming techniques of looping and recursion when she wrote a computer program for Charles Babbage's *Analytic Engine*.
- 1890 Herman Hollerith's electromechanical tabulating machines read punched cards to analyze 1890 U.S. census data.
- 1892 William Burroughs introduced an adding-subtracting machine for business purposes with a built-in printer.
- 1893 Swiss engineer Otto Steiger designed and marketed the first practical multiplication-table calculator, the *Millionaire*.

- 1924 Computing-Tabulating-Recording Corporation became International Business Machines (IBM) under the leadership of Thomas J. Watson.
- 1938 Konrad Zuse of Germany produced *Z1*, the first of his automatic binary calculating machines.
- 1942 Jon V. Atanasoff and Clifford Berry introduced the *ABC*, one of the first electronic calculating machines using electron vacuum tubes.
- 1943 The code-breaking computer *COLOSSUS*, based on designs by Alan Turing of England, became operational.
- 1943 Contracts made with the U.S. Army and University of Pennsylvania's Moore School of Engineering allowed John Mauchly and J. Presper Eckert to begin construction of *ENIAC*, the first general-purpose electronic digital computer.
- 1945 A computer team working on a *Mark II* computer at the Bureau of Ordnance Computation Project at Harvard used the term ''bug'' to refer to a computer glitch when a moth caused computer malfunctions after it actually got inside a *Mark II* computer.
- 1948 The transistor was invented by Bell Labs, signaling the end of the vacuum tube.
- 1952 Grace Brewster Hopper pioneered the writing of computer commands in English.
- 1958 Jack Kilby of Texas Instruments developed the integrated circuit, which within 10 years brought an end to electrically driven mechanical calculators and began the microcomputer age.
- 1963 The Bell Punch Co. of England produced the first fully transistorized four-function calculator.
- 1968 U.S. engineer Gilbert Hyatt invented the first microprocessor chip. The U.S. Patent Office did not issue a patent until July 17, 1990.
- 1971 Ted Hoff of Intel designed the microprocessor chip that is used in personal computers.
- 1971 U.S. manufacturers introduced four-function hand-held electronic calculators to the retail market.
- 1971 American Edward Roberts began making personal computers, using microchip technology.
- 1974 Hewlett-Packard Company introduced the first programmable hand-held calculator.
- 1976 Steven Jobs and Stephen Wozniak developed the *Apple I* computer in Jobs' garage.
- 1976 The first supercomputer was designed by Seymour Cray of the United States. The *Cray-1* contained 200,000 integrated circuits and did 150,000,000 calculations per second.

- 1979 *VisiCalc* (*VISI*ble *CALC*ulator) was the first business accounting software package designed for a personal computer.
- 1981 IBM entered the microcomputer market with its 16-bit *PC*(*P*ersonal *C*omputer).
- 1983 IBM became the dominant microcomputer producer. IBM compatibles, clones, and related models became available.
- 1984 Apple Corporation released its user friendly machine, the *Macintosh.*
- 1987 IBM introduced the *PC/2* with the software package *Windows* to compete with the Apple *Macintosh.*
- 1989 The *Cray X-MP* supercomputer was capable of 300 million calculations per second; the Japanese electronics firm NEC announced its *SX-X* supercomputer with an expected speed of 20 billion calculations per second.
- 1993 A supercomputer after 2500 hours of computing time found F22, the largest known prime number. It contained 1,262,612 digits and took 169 pages with 7500 digits on a page to display.

## THE HUMAN CALCULATORS

Advances in the development of calculating devices have been phenomenal. However, the first and most common calculating device is still the human mind. Over the centuries, human minds have performed calculations on billions of digits. In fact, there have been men and women that have been called ''living calculators'' or calculating prodigies. They have been able to do computations without using ''pen-and-paper'' algorithms. Their ability to memorize numbers and quickly perform mental calculations is astounding. Some of the prominent people with these powers, along with some of their feats in computation, are listed here.

- Thomas Fuller (b. 1710), a black slave in Virginia, could mentally multiply four- and five-digit numbers and calculate the number of seconds in any period of time.
- Johann Dase (b. 1824) of Germany was a professional calculator by the age of 15. He could mentally multiply two 20-digit numbers in 6 minutes and extract the square root of a number with 100 digits in 52 minutes.
- Truman Safford (b. 1836) of Vermont mentally raised 365 to the 11th power in less than a minute.
- Gottfried Rückle (b. 1880) of Germany could recite all the factors of each integer less than 1000 by the age of 12.
- Alexander Aitken (b. 1895) of New Zealand memorized the first 1000 digits of $\pi$, and Hans Eberstark (b. 1929) of Austria recited the first 11,944 digits of $\pi$. Eberstark's feat was bettered by Hideaki Tomoyori of Japan and, later, by Creighton Carvello, who memorized the first 20,013 digits of $\pi$.

- Salo Finkelstein (b. 1896) of Russia could memorize 30 digits of a number after looking at it for only 3 seconds.
- Wim Klein (b. 1912) of Holland mentally multiplied two five-digit numbers in 44 seconds.
- Shyam Marathe (b. 1931) of India could mentally raise single-digit numbers up to the 20th power.
- Shahuntala Dévi (b. 1940) of India extracted the 23rd root of a 501-digit number in 50 seconds.

There are many other men and women throughout the world that have displayed this uncanny ability to work with numbers. However, the majority of us will continue to rely on the marvel of technology, the hand-held electronic calculator.

## Check Your Reading

1. Give evidence that the abacus is really a fast and accurate calculating device.
2. What sequence of developments in electronics made the present-day calculators and computers a reality?
3. In 1614, Sir Walter Raleigh of England wrote *The History of the World.* What was John Napier working on at about this time?
4. In 1621, when the potato was introduced in Germany, what calculating device was introduced?
5. When the carrier pigeon was used to deliver messages in Greece in 400 B.C., what calculating device was in use?
6. What is important about Steven Jobs' garage?
7. What are Ada Byron's (Countess of Lovelace) and Grace Hopper's contributions to computers?
8. What were the feats of India's mental calculators Shyam Marathe and Shahuntala Dévi?
9. What is the origin of the computer programming term "bug"?
10. When the Ford Motor Company was producing its ten-millionth automobile in 1924, what was Thomas J. Watson forming?
11. By 1959, Alaska became the 49th state in the United States, Lorraine Hansberry wrote *A Raisin in the Sun*, and the cha-cha was a popular dance. What happened in electronics just before these events?
12. In 1823, as Mexico was becoming a republic, what was Charles Babbage doing?
13. What are some of the feats connected with memorizing the value for $\pi$?
14. What did IBM do in 1981?
15. What was released by the Apple Corporation in 1984?
16. What did a supercomputer accomplish in 1993?

17. Match each of the following names with the correct mathematical discovery or event.

| | |
|---|---|
| (a) Babbage | Adding machine for business purposes |
| (b) Burroughs | Apple I computer |
| (c) Carvello | Early American mental calculator |
| (d) Cray | Electromechanical tabulating machine |
| (e) Fuller | Programming techniques |
| (f) Hollerith | Integrated circuit |
| (g) Kilby | Mechanical calculators |
| (h) Jobs, Wozniak | Memorized 20,013 digits of $\pi$ |
| (i) Pascal, Schickard, Leibniz | Steam-powered digital calculator |
| (j) Ada Byron, Grace Hopper | Supercomputers |

## • Research Questions

To answer the following questions, you will need to refer to material not contained in the text. Possible sources of information are listed in the bibliography at the end of this book.

1. Do some research on one of the mathematicians or computer pioneers mentioned in this section. Explain that person's contribution to calculating devices or computers. What other interests did the individual have?
2. What is a modem? What is "E-mail"? Explain how each is used.
3. What do services like *Prodigy, CompuServe*, and *America On Line* provide? What do you need to use these services?
4. Discuss the speed of present-day computers in doing calculations.
5. What are some of the methods used by the mental calculators in performing their computational feats?
6. What is an algorithm? Why is it important in the study of computer programming?
7. What is a programmable calculator? How are programmable calculators different from the scientific and graphing calculators described in Section 9.3?
8. Explain how cassette tapes, compact disks, video tapes, and video disks store information using principles that are used by computers.
9. What are computer viruses? How are they treated?
10. What are CAD and CAI? Who uses them?
11. What are RAM and ROM? How are they different? What is a CD-ROM?
12. The three types of printers used today are dot matrix printers, inkjet printers, and laser printers. How does each type of printer work? What causes the differences in letter quality in each type of printer?
13. What are computer "networks"? Give a brief description of the following networks: *BITNET, MEDLINE, LEXIS, FEDWIRE, INTERNET*, and *Holidex*.

14. What are computer ''hackers''? What do they do? What has happened to some infamous ''hackers''?
15. Computers are subject to errors, some because of machine errors and others because of human errors. What are some of the basic machine and human errors?

## Projects

1. There are various methods for representing numbers and doing calculations on fingers. Do some research on some of these methods. Show how numbers are recorded and give examples of how calculations are performed.
2. There have been many different calculating devices besides those mentioned in this chapter. Do some research on some of these devices, giving information on the inventors and how the devices work.
3. The computer is used in almost every area of our present society. Find some new and interesting uses of computers. Give details on how and why they are being used in the given area.
4. Computers have used various schemes for storing data. What are some of the methods of storing data both external to the computer and inside the computer?
5. There have been many computer languages developed since programmable computers were invented. Investigate some of the languages such as *FORTRAN*, *BASIC*, *Pascal*, *COBOL*, *C*, *LISP*, *LOGO*, and *Ada*. What are the uses and limitations of each language?

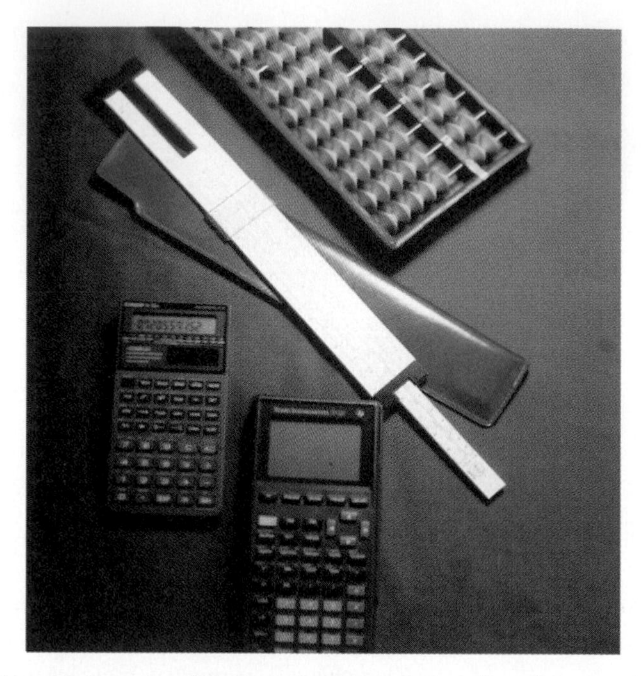

The abacus, slide rule, electronic calculator, and graphing calculator show the progression of calculating devices used by mathematics students throughout history.

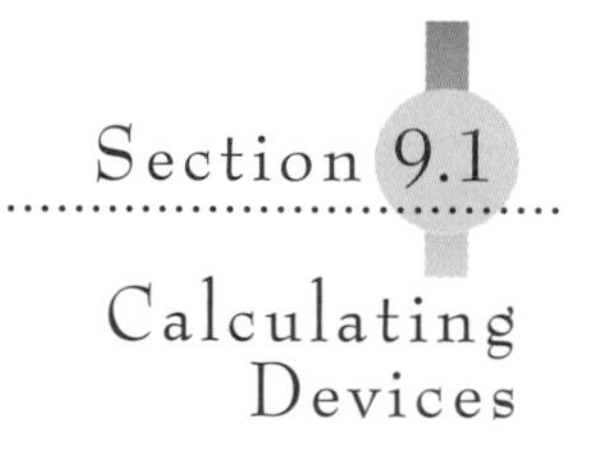

## Section 9.1 Calculating Devices

Numbers and computation have been a concern to mankind through the centuries. Many devices have been developed to record and calculate numbers. In this section, we will take a look at some of the calculating devices invented before present-day electronic calculators.

### Fingers

The first calculating devices were, quite simply, fingers. Fingers have been and continue to be used as an aid in counting, addition, and subtraction. Many different **finger reckoning** schemes, which use the fingers to represent numbers and perform computations, have been developed. The works of the Greek historian, Herodotus, indicate that a system of finger notation for numbers was in use as early as the 5th century B.C. It is documented that the Romans had a system of finger reckoning as well, but the earliest complete description of finger reckoning came from an English Benedictine monk, Venerable Bede (673–735). In *De Temporium Ratione*, Bede described an intricate set of finger signs and hand positions to represent the numbers from 1 to 1,000,000. Using these signs, the fingers could be used as a memory device for performing calculations. For example, if you wanted to add three numbers, you could record the first number on your fingers, add the second number mentally, and record that sum on your fingers. Finally, you could add the third number mentally to the previous sum and record the final answer on your fingers.

The Chinese also developed a system of representing the numbers from 1 to 9999 by using the fingers of the right hand to point at positions on the finger joints of the left hand. In 15th-century Europe, multiplication of numbers from $5 \times 5$ to $30 \times 30$ could be accomplished with a finger reckoning scheme of extending fingers from clenched left and right hands. In 1757, a Mexican priest, Buenaventura Francisco de Ossorio, developed a finger system to determine the dates of all movable Christian feasts. His methods enabled him to do calculations that would have required him to have access to 35 different almanac tables. These and other schemes of finger reckoning, however, were limited. They simply did not allow men and women to perform calculations quickly and accurately.

## The Abacus

An **abacus** is a device used to perform arithmetic by moving markers placed on grooves or strung on wires. Abacuses were used by many different cultures throughout history. There is some evidence that the Egyptians (c. 1500 B.C.) made use of the principles of an abacus. Greek (c. 400 B.C.) and Roman (c. 100 B.C.) abacuses consisted of pebbles placed on lines inscribed on stone tables. Chinese (c. 1200), Korean (c. 1400), and Japanese (c. 1600) abacuses had beads strung on wires. The natives of North and South America also used abacus-like manipulations. In 1590, Joseph de Acosta, a Jesuit priest, recorded that the Incas performed difficult computations by moving kernels of grain placed on the ground. The **counting board abacus** was used in Europe from ancient Roman times into the 17th century. This type of abacus utilized markers moved on a table inscribed with lines for the digits of Roman numerals. Markers were added to the table and moved so that amounts could be recorded and calculations performed.

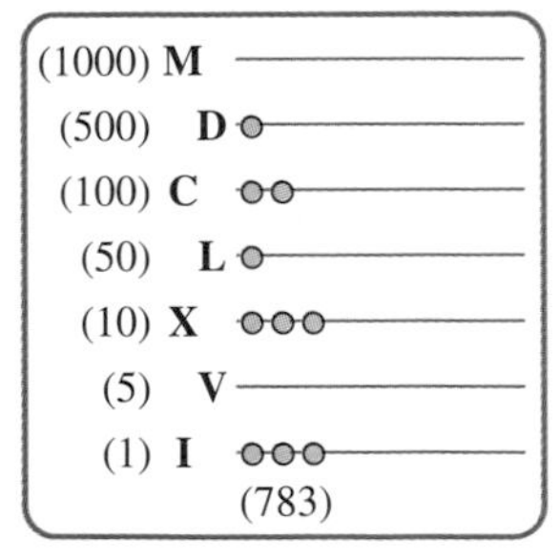


Suppose a 13th-century Italian merchant wanted to add DCCLXXXIII + XLIII (783 + 43). (*Note:* Roman numerals are explained in Section 1.1.) The first number is put on the counting board by placing coin-like markers as shown. To add the number 43, more markers are placed on the board and the markers are regrouped.

1. Three markers are added to the I row, making six markers in the I row. Since five 1's equals V (5), one marker is placed in the V row and one is left in the I row.
2. Four markers are added to the X row, making seven markers in the X row. Since five 10's equal L (50), one marker is placed in the L row and two are left in the X row.
3. Now the L row has two markers. Since two 50's equal C (100), one marker is placed in the C row, and then no markers are left in the L row.

**4.** Thus, the sum (826) can be written in Roman numerals, DCCCXXVI, by looking at the markers in each row starting at the top of the board.

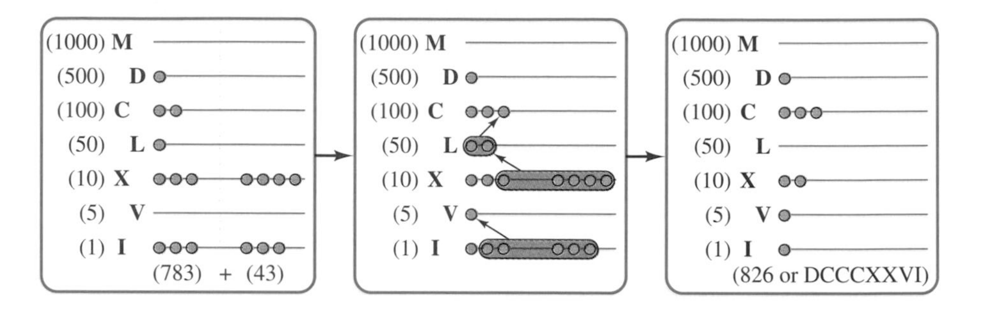


The previous example demonstrates a fundamental principle for computing on an abacus.

**Abacus Regrouping Principle**

If the value of a group of markers in one row equals a marker of the next row, the group is removed and one marker is added to the next row.

## Example 1

Use a counting-board abacus to find the sum of MCDXXXV (1435) and CCLXXIX (279). Give the sum in both Roman numerals and Hindu-Arabic numerals.

**Solution:** Place the first number MCDXXXV (1435) on the counting board, and add CCLXXIX (279) to the board starting with the markers in the I row. Using the abacus regrouping principle, we get

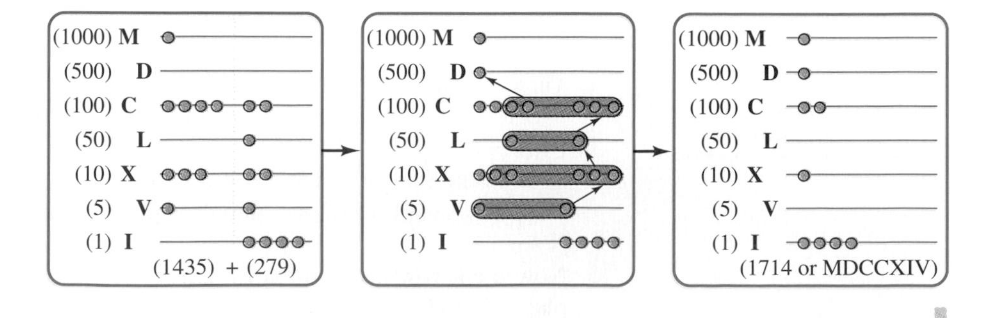

Other types of abacuses work on the same principles as the counting-board abacus just described. The "bead-and-wire" Chinese **suan-pan** and the Japanese **soroban** are still used today. The abacus, in the hands of a skilled person, is a very efficient calculating device for all the basic operations of arithmetic.

## Napier's Bones

The manual dexterity and the mental concentration required by the abacus left the average person without an easy means to make quick calculations. Even when the acceptance of the Hindu-Arabic numerals simplified written computation in the 15th century, mathematicians looked for even easier mechanical ways to perform computation. Scottish mathematician John Napier (1550–1617) spent a large part of his life devising schemes to do just that. One of his best-known schemes involved placing numbered rods next to each other to perform the operation of multiplication. Since these rods were usually made of animal horn, bone, or ivory, they were called **Napier's Bones**. Napier's Bones contained the multiplication facts from 0 × 0 to 9 × 9 as shown in Figure 9.1.

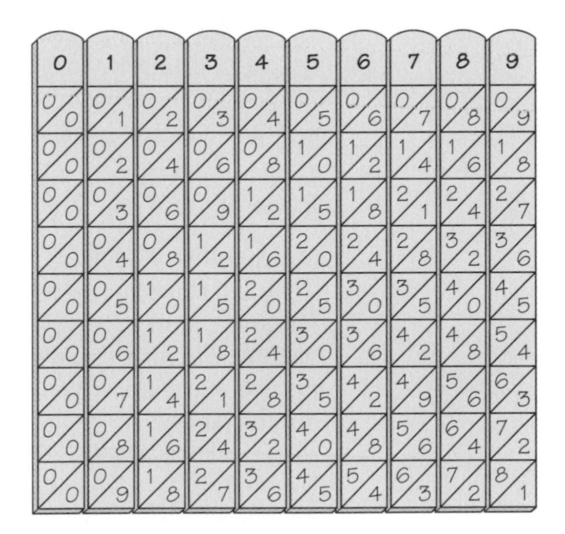


**Figure 9.1**

The scheme he devised imitates the long multiplication process we still use today. To multiply numbers using Napier's Bones, the rods are placed next to each other and digits on the diagonals between adjacent squares are added to find partial products. The sum of the partial products gives the desired answer. The following example shows how Napier's Bones are used to calculate 42 × 276 by determining 2 × 276 and 40 × 276 and then adding these products.

To find 42 × 276,

1. Place the rods for 2, 7, and 6 next to each other.
2. To find the digits of 2 × 276, move from right to left and find the total of each diagonal in the "2" row as shown in Figure 9.2.
3. To find the digits of 40 × 276, move from right to left and find the total of each diagonal in the "4" row. If the sum of a diagonal is 10 or more, the tens digit is carried to the diagonal on the left. Since the multiplier is 40, attach a zero to the result.
4. Add those partial products to obtain the answer of 42 × 276.

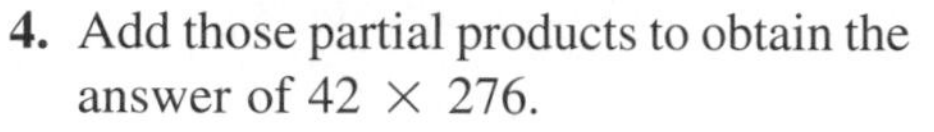

552 + 11,040 = 11,592

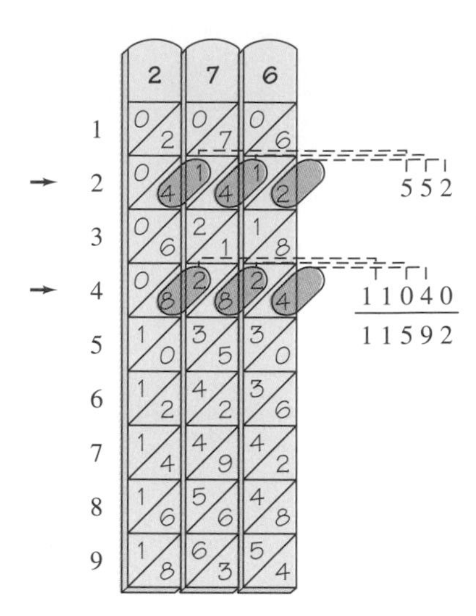


**Figure 9.2**

## Example 2

Use Napier's Bones to find 879 × 276.

**Solution:** As shown in Figure 9.3, the product is obtained by adding the partial products of 800 × 276, 70 × 276, and 9 × 276.

1. To find 800 × 276, move from right to left and find the total of the diagonals in the "8" row. Notice the carrying that occurred in the second and third diagonal. Since the multiplier is 800, attach two zeros and get 220,800.
2. To find 70 × 276, move from right to left and find the total of the diagonals in the "7" row. Since the multiplier is 70, attach one zero and get 19,320.
3. To find 9 × 276, move from right to left and find the total of the diagonals in the "9" row. The result is 2,484.
4. Add those partial products and get the answer of 242,604. ■

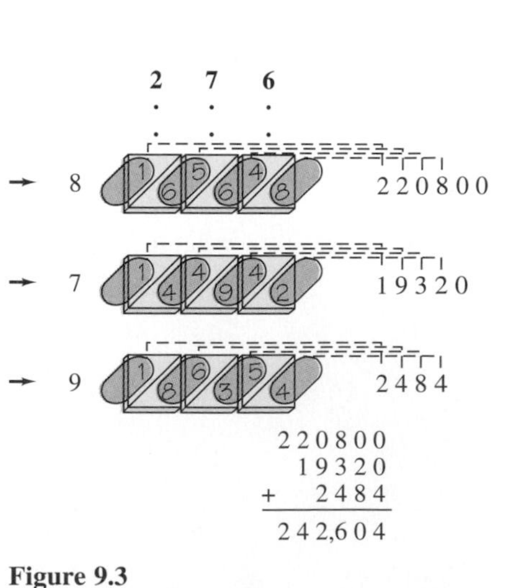


**Figure 9.3**

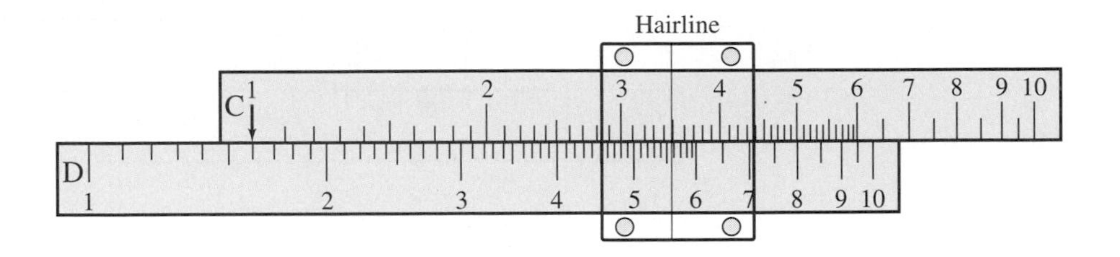


**Figure 9.4**

## The Slide Rule

Like the abacus, Napier's Bones had drawbacks. The user still had to do mental calculations and add columns of numbers to obtain final results. It also did not aid in performing division or advanced calculations. In the early 1600s, Napier's work on the concept of a logarithm and the work of two professors at Gresham College in London, Henry Briggs in logarithms and Edmund Gunter in calculating instruments, led to the development of a new calculating device, the slide rule. The **slide rule** is a calculating device invented by William Oughtred that consists of pieces of wood with logarithmic scales marked along the edges. Products and quotients can be determined by sliding these pieces of wood.

The basic slide rule consists of two sliding logarithmic scales called the C-scale and the D-scale and a sliding hairline marker as shown in Figure 9.4. The slide rule in Figure 9.4 is positioned to find $1.6 \times 3.5$.

1. The pointer on the left of the C-scale is positioned above the first factor 1.6 on the D-scale.
2. The hairline is positioned on the 3.5 on the C-scale.
3. The product is read on the hairline from the D-scale. Thus, $1.6 \times 3.5 = 5.6$.

The same positioning of the slide rule can also be used to find $5.6 \div 3.5$. Since division is the inverse of multiplication, we simply reverse the steps.

1. Place the divisor 3.5 on the C-scale above 5.6 on the D-scale.
2. The quotient is read from the D-scale below the pointer on at the end of the C-scale.
3. Thus, $5.6 \div 3.5 = 1.6$.

The diagrams in Figure 9.5 show you how basic multiplication and division are done for numbers P, Q, and R positioned on the slide rule.

Multiplying on a Slide Rule

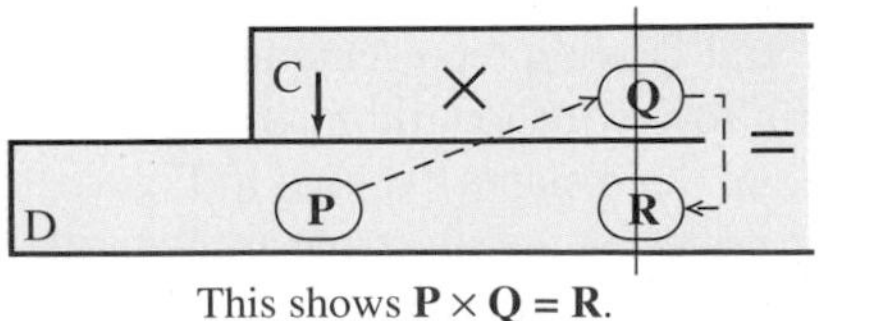


This shows **P × Q = R**.

Dividing on a Slide Rule

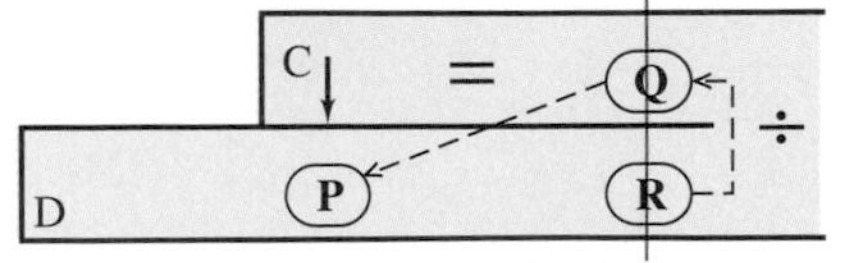


This shows **R ÷ Q = P**.

**Figure 9.5**

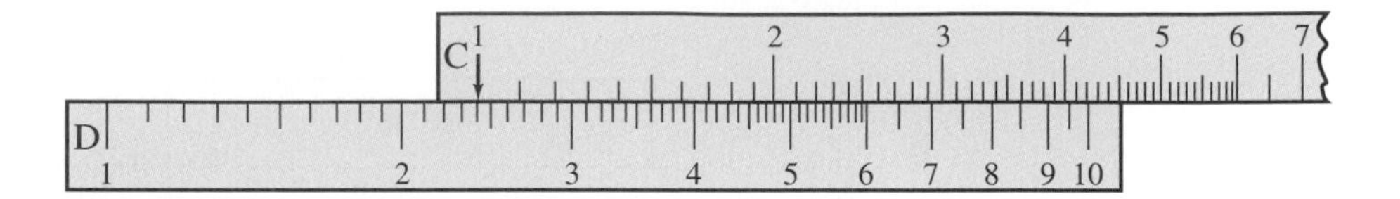


Figure 9.6

### Example 3

In the slide rule set up in Figure 9.6, find the answers to the following problems: (a) $2.4 \times 3.6$ and (b) $7.4 \div 3.1$.

**Solution:**

(a) Since the end of the C-scale is positioned over 2.4, simply slide the hairline over 3.6. The result is about 8.6. We get

$$2.4 \times 3.6 \approx 8.6$$

(b) The scales are positioned so that 3.1 is positioned over 7.4. Thus, the end of the C-scale points to the quotient of about 2.4. Thus,

$$7.4 \div 3.1 \approx 2.4$$

■

You will notice that the scales on the slide rule only have the numbers from 1 to 10. To multiply or divide numbers that are larger than ten or less than one, the numbers are written in scientific notation, that is, as the product of a power of 10 and a number between 1 and 10. For example,

$$3400 \times 0.00028 = (3.4 \times 10^3) \times (2.8 \times 10^{-4}) = 3.4 \times 2.8 \times 10^{-1}$$

The slide rule is used to find $3.4 \times 2.8$. Then the decimal is positioned in the answer using the power of 10. The accuracy of the slide rule is based on the number of dividing lines that are placed between whole numbers. The accuracy of the slide rule used here is only two digits. Thus, $3.41 \times 2.804$ yields the same result as $3.4 \times 2.8$. Although we will not go into the details for using scientific notation or discuss other necessary considerations in using a slide rule, we now give an explanation of why pieces of wood with scales marked along their edges actually give the answers to multiplication.

## *Why Does the Slide Rule Work?*

What did the slide rule really do when we found that $2.4 \times 3.6 \approx 8.6$ in Example 3? The slide rule is designed so that the distance from one to each number marked on the C-scale and D-scale represents the common logarithm of the number. (See Section 2.5 for a review of logarithms.) When the C-scale was positioned over the D-scale, we set up the distances shown in Figure 9.7.

Because the scales on the slide rule are logarithmic, a line at 2.4 on the slide rule represents the value log 2.4. We arrive at the value log 8.6 by adding the lengths

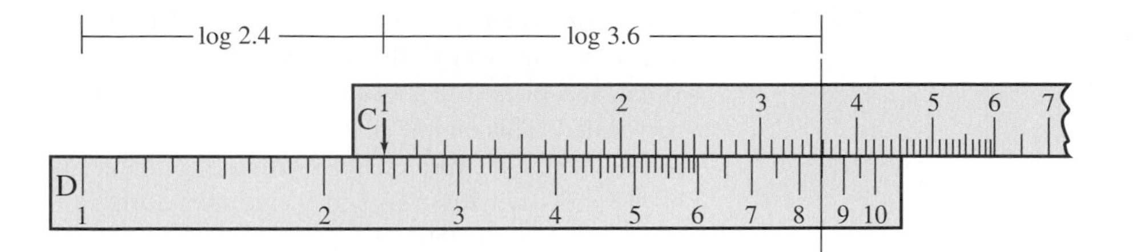


Figure 9.7

log 2.4 and log 3.6. However, the product rules of logarithms states that log 2.4 + log 3.6 = log(2.4 × 3.6). Thus, 8.6 is the product of 2.4 and 3.6.

## Mechanical Calculators

Further developments in the slide rule allowed the user to do computation involving exponential, logarithmic, and trigonometric functions. The slide rule became a widespread, hand-held calculating device used by scientists, mathematicians, and students into the 1960s. The accuracy of a slide rule, however, depends on the number of dividing lines that are placed between whole numbers. The standard slide rule has only three digits of accuracy. This limited accuracy was the slide rule's major drawback. Furthermore, it still did not calculate automatically. The users did most of the work. They had to use scientific notation and the laws of exponents and carefully move and position different scales. Mathematicians and scientists wanted more. They wanted an accurate device that would compute automatically without much human intervention. Early pioneers in the development of automatic computation such as Wilhelm Schickard (1592–1635), Blaise Pascal (1623–1662), Gottfried Leibnitz (1646–1716), and Samuel Moreland (1625–1695) used mechanical levers, gears, wheels, and cranks to replace the human element. These early calculating devices were essentially ''automatic abacuses'' where regrouping and carrying were done mechanically. The system of three simple gears shown in Figure 9.8 gives an indication of how these operations can be accomplished.

In Figure 9.8, the wheel on the left represents the tens digit and the wheel on the right represents the units digit. When the units-wheel turns ten notches in the

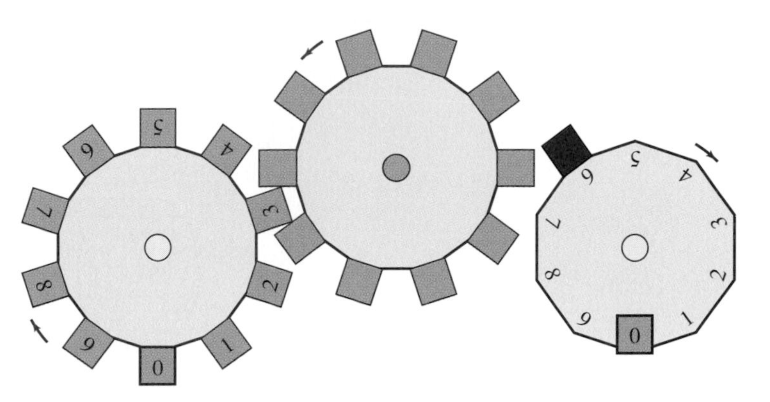

Figure 9.8

clockwise direction, the gear in the center turns one notch in the counterclockwise direction. This causes the tens-wheel to turn one notch in the clockwise direction. When the ones-wheel has recorded ten ones, the carry function is automatically performed. This simple device can actually be used to add numbers. For example, to find $6 + 9$ we turn the ones-wheel six notches so that the number 6 is displayed in the view window. To add 9, we then turn the ones-wheel nine more notches. When the pickup tooth on the ones-wheel activates the center gear, ten 1's are carried into the tens-wheel and a final result of 1 5 appears in the view windows.

These basic principles are still in use today. Look at the odometer in your car. The number representing the miles driven appears in the view window. Every time you travel $\frac{1}{10}$ mile, the right-hand wheel moves ahead one digit. When you travel 1 mile, the next wheel moves ahead one digit, and so on. In your car's odometer, the tooth on the 9 on each wheel is designed to catch the next wheel as it rotates. This is really an automatic application of the abacus regrouping principle. When a given wheel has rotated through ten digits, one unit is added to the next wheel and the given wheel is set back to zero (see Figure 9.9).

Using this basic principle, very ingenious systems of gears, levels, wheels, and cranks were devised to perform not only addition but also subtraction, multiplication, and division. Amounts were entered into these machines by pulling levers, and computation was done by turning cranks. In 1820, Frenchman Thomas de Colmar introduced the first reputable commercial calculating machine, the *Arithmometer*. In the late 1870s, Frank S. Baldwin (U.S.) and T. Odhner (Sweden) produced variable toothed-gear machines that were more reliable and greatly decreased in size. Their four-function calculators could fit on the corner of a desk rather than take up the entire surface of the desk. In 1892, William Burroughs introduced an adding-

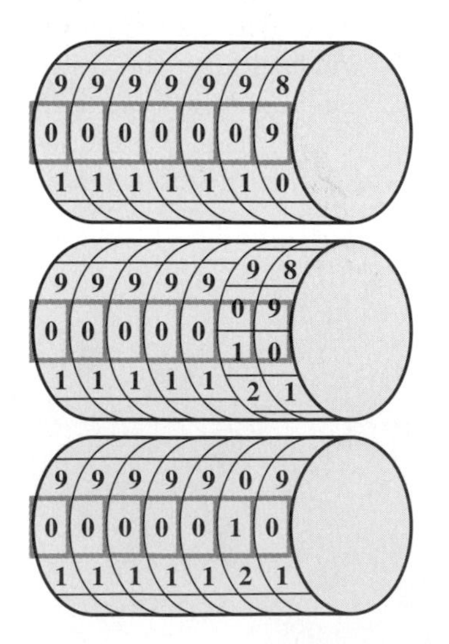


Figure 9.9

subtracting machine for business purposes with a built-in printer. In 1893, Swiss engineer Otto Steiger designed and marketed the first practical multiplication-table calculator, the *Millionaire.* Between 1900 and 1910, key-driven machines called *Comptometers* were common place. Instead of turning cranks, computation was done by pushing numbered keys. By 1922, for example, the Burroughs Corporation was manufacturing completely portable *Comptometers* in the United States. With the availability of electric power in the 1920s, the mechanical calculators began to be driven by electricity. Electric calculators could be found in companies, banks, accounting departments, and universities. These calculators had increased speed and greater accuracy, but they could only perform basic arithmetic. But still, the mathematicians and scientists wanted more.

## Section 9.1 PROBLEMS

## *Explain ➡ Apply ➡ Explore*

### *Explain*

1. What is finger reckoning? What are some of the drawbacks with using finger reckoning to calculate?
2. What is an abacus? What are some of the drawbacks with using an abacus to calculate?
3. What is the abacus regrouping principle?
4. What is the difference between a counting-board abacus and the suan-pan or soroban?
5. What are Napier's Bones? What are some of the drawbacks with using Napier's Bones to calculate?
6. How and why does a slide rule give the product of two numbers?
7. What is a slide rule? What are some of the drawbacks in using a slide rule to calculate?
8. How does the odometer in your car demonstrate the abacus regrouping principle?
9. What four qualities were mathematicians and scientists looking for in calculating devices?
10. Why can early mechanical calculators be considered to be ''automatic abacuses''?

*Apply*

In Problems 11–16, set up a counting-board abacus to find the following sums. Give the answers in both Roman numerals and Hindu-Arabic numerals.

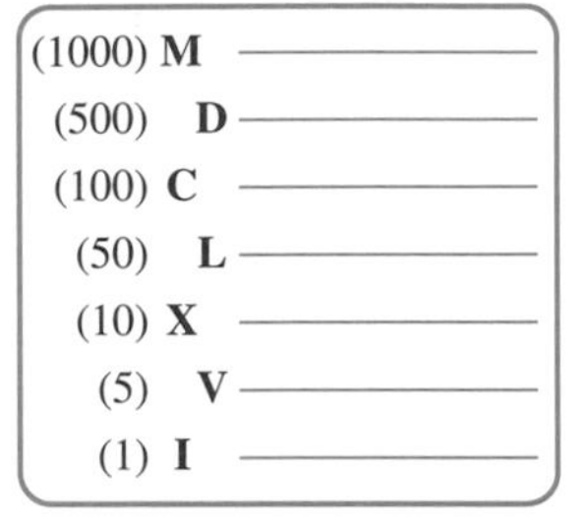


**11.** 67 + 23

**12.** 1467 + 384

**13.** 1234 + 2678

**14.** DCLXXXVII + LVIII

**15.** MCCXLIII + MMDCLXXIV

**16.** CMLXXIX + CDXXXVI

In Problems 17–22, use the set of Napier's Bones to perform each multiplication.

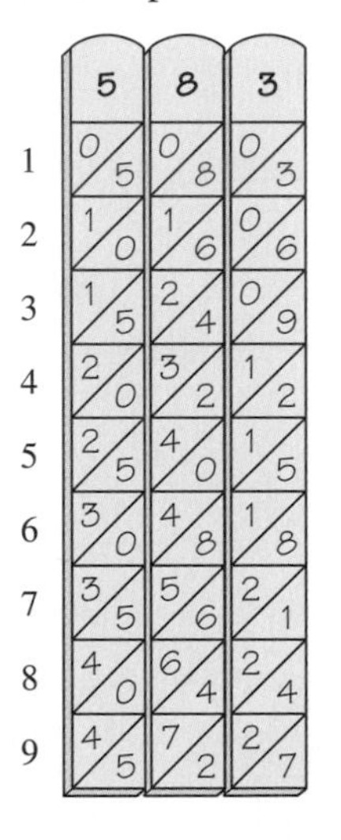


**17.** 57 × 583

**18.** 94 × 583

**19.** 162 × 583

**20.** 861 × 583

**21.** 5477 × 583

**22.** 9866 × 583

In Problems 23–30, use the slide rule shown to find each product or quotient.

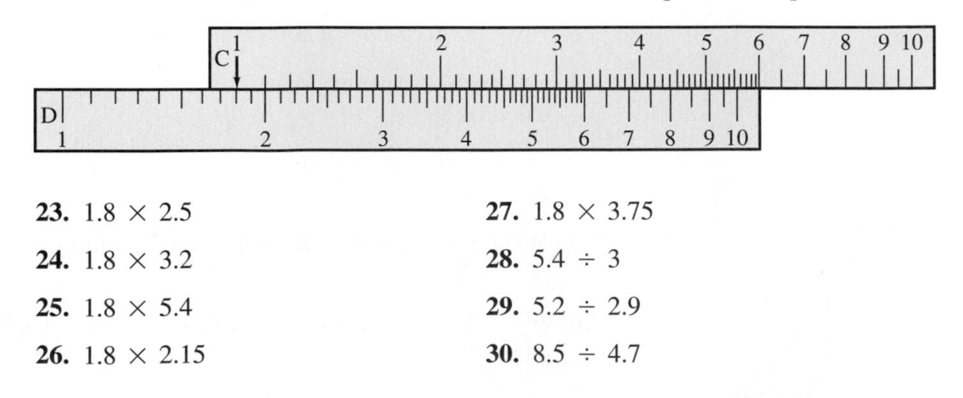


**23.** 1.8 × 2.5

**24.** 1.8 × 3.2

**25.** 1.8 × 5.4

**26.** 1.8 × 2.15

**27.** 1.8 × 3.75

**28.** 5.4 ÷ 3

**29.** 5.2 ÷ 2.9

**30.** 8.5 ÷ 4.7

In Problems 31–33, explain what happens to gears *A*, *B*, and *C* in the figure when each sum is calculated.

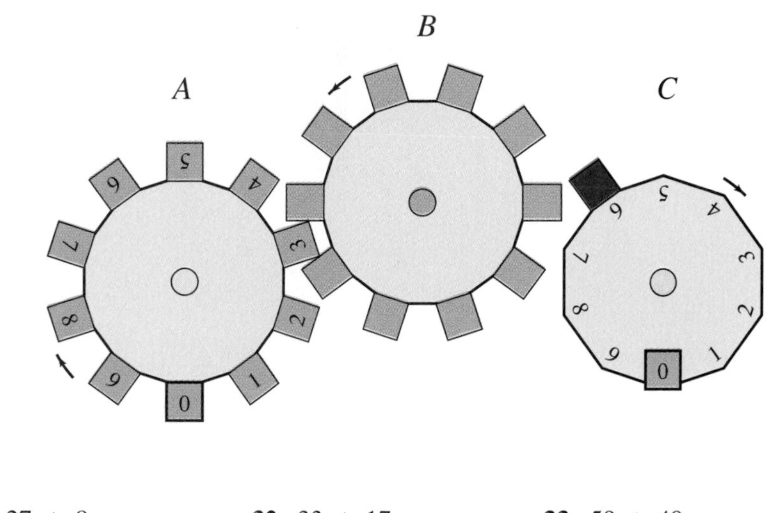


**31.** 27 + 8 **32.** 33 + 17 **33.** 59 + 49

## *Explore*

Alan Vorwald, in his book *Computers! From the Sand Table to Electronic Brain*, speculates that the Egyptians placed pebbles into grooves in the sand to represent numbers and moved these pebbles to perform computation. For example, the Egyptian hieroglyphic numeral 𓆼𓍢𓍢𓍢𓍢𓍢𓍢𓎆𓎆𓎆𓏺𓏺𓏺𓏺𓏺 (1635) would be represented by pebbles in the sand as shown in Figure 9.10.

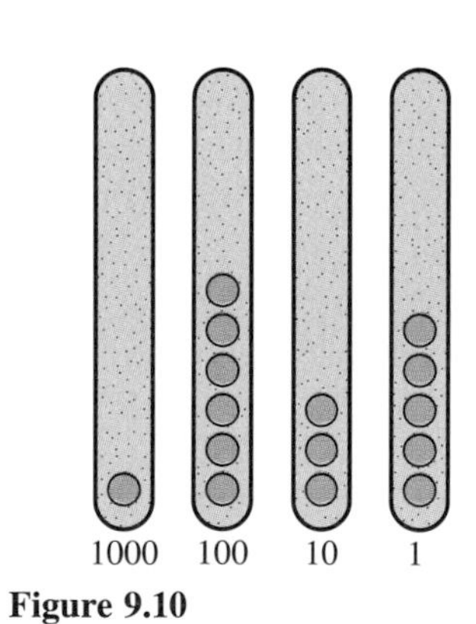


**Figure 9.10**

In Problems 34–36, represent each numeral as pebbles in the sand.

**34.** 56 **35.** 372 **36.** 4718

In Problems 37–39, explain how the sums can be found using such an abacus.

**37.** 56 + 9 **38.** 372 + 48 **39.** 4718 + 357

The soroban from Japan is an abacus that has upright wires with five counters on each wire. The top counter is separated from the bottom four counters by a crosspiece. The upper part above the crosspiece is called ''Heaven,'' and a counter in Heaven is a 5-unit bead. The lower part below the crosspiece is called ''Earth,'' and a counter in Earth is a 1-unit bead. The upright wires have place values that are the same as the place values in the Hindu-Arabic System. A soroban is cleared or in home position when no counter is touching the crosspiece. Numbers are entered on

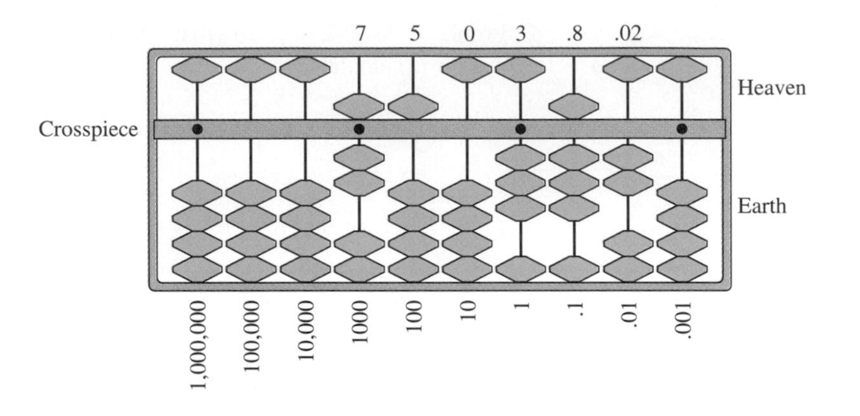


**Figure 9.11**

a soroban by moving counters toward the crosspiece. For example, the soroban in Figure 9.11 is displaying the number 7503.82.

In Problems 40–43, represent each number on a soroban.

**40.** 12,345.6

**41.** 298.765

**42.** 455,112

**43.** 5,234,985

Addition on the soroban can be performed by moving the beads in a manner that is very similar to the way markers were moved on the counting-board abacus described in this section. If the total value of the counters on an upright is 10 or more, the abacus regrouping principle is put into use. That is, the amount greater than 10 remains on the upright and a 1-unit bead is moved up toward the crosspiece on the upright to the left.

In Problems 44–47, explain how the sums can be found using a soroban.

**44.** 16 + 14

**45.** 476 + 98

**46.** 5.67 + 8.99

**47.** 38.4 + 64.7

**48.** In this section, the odometer on your car was used as an example of a present-day device that is really an "automatic abacus." Find other such devices and explain how the abacus regrouping principle is automatically performed.

**49.** When an automobile's odometer reads 30,000.0 miles, how many complete revolutions has each wheel of the odometer made?

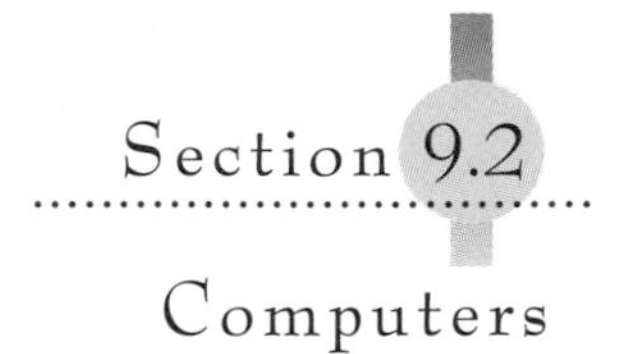


As some mathematicians, scientists, and inventors developed mechanical devices that performed basic arithmetic, others had grander ideas. They wanted machines that did more, were faster, and had greater accuracy. One such mathematician-inventor was Charles Babbage. In 1822, he received a grant from the English government to build his *Difference Engine*, a machine that would automatically calculate with a high degree of accuracy. Babbage hoped that his machine would find and correct the errors in mathematical tables of the time. As is true today, the government and military also had an interest. They believed that these accurate calculations could be used for navigation and ballistics. However, difficulties in constructing parts and getting continued financing caused the abandonment of the project. This setback did not discourage Babbage. In 1833, using the input mechanism of punched cards invented by Joseph Jacquard for a weaving loom, Babbage designed a general-purpose computing device. His proposed steam-powered *Analytic Engine* with its complex arrangement of gears and linkages included the main components of present-day computers. These components are

1. An **input device** to enter data and instructions.
2. A **control unit** to direct the flow of information and operations.
3. A **processor** to perform calculations.
4. A **memory unit** to store data.
5. An **output device** to display results.

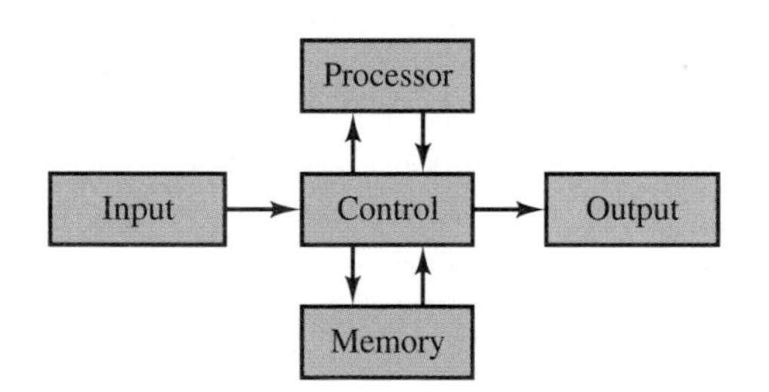


Although Charles Babbage never saw the construction of his *Analytical Engine*, his design served as a model for the computers to come.

While Babbage and others were looking for more accurate calculating devices, Herman Hollerith of the United States was designing a machine to tabulate the 1890 U.S. census. Using tabulation methods then available, the 1880 census took eight years to complete. Because of the increasing population, it was feared that the data from the 1890 census would not be completely analyzed before the census of 1900. Herman Hollerith's punched-card tabulating machine solved the problem. With this machine, the census data were processed more thoroughly and in far less time. Hollerith's work led to the formation of the Tabulating Machine Company. Hollerith's tabulating machines were used by the Bureau of the Census for the 1900 census and by large companies such as the Chicago department store Marshall Field & Co., the New York Central Railroad Co., and the Pennsylvania Steel Co. to handle paperwork. Various mergers with Hollerith's Tabulating Machine Co. eventually created a major player in today's computer industry—International Business Machines (IBM).

The development of computers that performed more than the basic operations of arithmetic was hastened by the military interests of World War II. In the 1940s, Konrad Zuse of Germany developed machines that used electromagnetic telephone relays to accurately calculate the wing adjustments on radio controlled flying bombs. In the early 1940s, England's Alan Turing helped design *COLOSSUS* to break secret German codes. This single-purpose computer used vacuum tubes instead of the slower electromagnetic relays and used punched-paper tape for input.

During the same years, computers were being developed in the United States. With backing from Thomas Watson of IBM, Howard H. Aiken created the *Harvard Mark I.* This computer used electromagnetic telephone relays to perform addition and subtraction in 0.3 seconds, multiplication in 6 seconds, and evaluate trigonometric, exponential, and logarithmic functions in 60 to 65 seconds. The *Harvard Mark I* was a very large machine. It was 51 feet long, 8 feet high, and 6 feet wide and contained over 500 miles of electrical wire. In 1945, John W. Mauchly and J. Presper Eckert unveiled *ENIAC*, the first general-purpose computer. This computer used punched cards for input and output, had a memory of twenty 10-digit numbers, and performed multiplication in 0.003 seconds. Its sequences of instructions were given by wiring circuits by hand. The *ENIAC* was about twice as long as the *Harvard Mark I*, contained 18,000 vacuum tubes, and weighed about 30 tons. The interest in these computers led to more advanced computers in the late 1940s. In England, Maurice Wilkes developed *EDSAC* (*E*lectronic *D*elay *S*torage *A*utomatic *C*omputer), and in the United States, John von Neumann developed *EDVAC* (*E*lectronic *D*elay *V*ariable *A*utomatic *C*omputer). These were the first electronic digital computers to be controlled by internally stored sets of step-by-step instructions called **programs**.

By the 1950s, laboratories, big business, and governments needed computers to do more than basic arithmetic. Electronic machines used stored programs to perform complex calculations, analyze all types of data, and perform accounting procedures. In 1952, when CBS used a *UNIVAC I* computer to predict Dwight Eisenhower's victory over Adlai Stevenson, the American public became aware of the power of the computer and the computer revolution had begun. The rapid development of computers that occurred in the next 40 years centered around the discoveries in electronics. These developments have been classified into four generations.

## First Generation (1951–1958)

Using the concepts and technology developed by the early computer pioneers from 1800 to 1950, companies such as Remington Rand, IBM, Honeywell, Burroughs, and General Electric produced machines such as the *UNIVAC I*, *IBM 650*, and the *IBM 700* series that

- used thousands of vacuum tubes
- accomplished input and output with punched cards and magnetic tape
- used teletype-like printers
- stored programs in memory using binary codes
- were programmed using binary machine language that was very time intensive
- were so large that a single computer system would a fill a small room
- used large amounts of electricity and produced large amounts of heat

## Second Generation (1959–1964)

In 1948, **transistors** were invented by Bell Lab scientists. These tiny solid-state electronic devices were used to control current flow but were $\frac{1}{100}$ the size of a vacuum tube and also produced less heat and had lower power requirements. By 1959, transistors were produced at relatively low costs and began replacing the vacuum tubes in computers. This technology led to computers such as the *IBM 1400* series, *GE 2*, *CD6600*, and *UNIVAC LARC*, which, when compared with first-generation computers,

- were smaller, faster, and more reliable
- used better peripheral devices
- were easier to program because they used programming languages such as *FORTRAN* (*FOR*mula *TRAN*slation) and *COBOL* (*CO*mmon *B*usiness-*O*riented *L*anguage)
- generated less heat and used less electricity

## Third Generation (1965–1970)

In 1958, Jack Kilby of Texas Instruments developed the integrated circuit (IC). **Integrated circuits** are tiny chips of silicon that contain many interconnected electronic circuits. By 1965, integrated circuits with 1000 circuits in a chip had replaced transistors. This technology led to computers such as the *IBM 360*, *DEC PDP 8*, and *HP 2000*, which, when compared with second-generation computers,

- were smaller, faster, more powerful, and less expensive
- incorporated better peripheral devices such as magnetic disks and printers
- allowed for more programming languages
- had more software packages

## Fourth Generation (1971–Present)

In 1970, using the technology that allowed more and more circuits to be placed into a silicon chip, Ted Hoff of Intel Corporation designed the microprocessor. A

**microprocessor** is an integrated circuit that is the entire processing unit of the computer. This technology led to computers such as the *IBM 370*, *DEC VAX 11/785*, *IBM PC/XT/AT/PC/2*, *Apple II*, and *Macintosh II/SE/Quadra/Power Book*, which

- were smaller, faster, more powerful, and less expensive
- incorporated better peripherals such as floppy disks, color screens, laser printers, pointing devices (''mice''), and CD-ROMs
- had more software available in different areas such as education, games, accounting, graphics, word processing, and data management
- allowed for the use of high-level languages
- were easier to use. Software became *user friendly*.

The number of uses of computers has increased tremendously. The June 1963 issue of *Computers and Automation* listed only 500 areas of computer applications. Today almost every area of our society is affected by the computer. The development of even better computers, software, and peripherals is just around the corner. As new technologies are developed, the computer scientists incorporate them into the next wave of computers.

Now that you are familiar with the origin of the personal computer, it is important to have some idea of how computers work—how they read and store data and how they perform calculations.

## How Do Computers Read and Store Data?

Present day computers read and store data using the fact that electric current can exist in one of two states, *on* or *off*. The *on* state can be achieved by a positive (+) magnetic charge and the *off* state by a negative (−) magnetic charge. These two states are associated with the digits of the binary system that were introduced in Section 1.3. The digit 1 is used to indicate *on* [(+) charge] and the digit 0 is used to indicate *off* [(−) charge]. Each 1 or 0 is called a **bit**, and a sequence of bits can be read electronically as (+) or (−) charges. The number of binary digits that can be processed at one time varies from computer to computer. For example, an 8-bit computer reads and stores data as a sequence of eight 1's or 0's. Thus, when such a computer reads information from a floppy disk, it reads eight bits at a time, and a letter pressed on its key board is transformed into a sequence of eight 1's and 0's. For example, the binary sequence 0101 1010 represents the letter Z. It would be represented by a sequence of magnetized dots that are − + − + + − + − on the floppy disk.

Each letter of the alphabet, digit, and special character, such as a period, comma, or hyphen, has a binary code. This code is called **ASCII**, **A**merican **S**tandard **C**ode for **I**nformation **I**nterchange. When a computer reads or sends information, it is processed in ASCII code. When it sends data to a device such as a printer, the printer decodes the binary information and converts it to the corresponding character.

Figure 9.12

| ASCII (American Standard Code for Information Interchange | | | | | | | | | | | |
|---|---|---|---|---|---|---|---|---|---|---|---|
| Chr | Hex | Chr | Hex | Chr | Hex | Chr | Hex | Chr | Hex | Chr | Hex |
| space | 20 | 0 | 30 | @ | 40 | P | 50 | ` | 60 | p | 70 |
| ! | 21 | 1 | 31 | A | 41 | Q | 51 | a | 61 | q | 71 |
| " | 22 | 2 | 32 | B | 42 | R | 52 | b | 62 | r | 72 |
| # | 23 | 3 | 33 | C | 43 | S | 53 | c | 63 | s | 73 |
| $ | 24 | 4 | 34 | D | 44 | T | 54 | d | 64 | t | 74 |
| % | 25 | 5 | 35 | E | 45 | U | 55 | e | 65 | u | 75 |
| & | 26 | 6 | 36 | F | 46 | V | 56 | f | 66 | v | 76 |
| ' | 27 | 7 | 37 | G | 47 | W | 57 | g | 67 | w | 77 |
| ( | 28 | 8 | 38 | H | 48 | X | 58 | h | 68 | x | 78 |
| ) | 29 | 9 | 39 | I | 49 | Y | 59 | i | 69 | y | 79 |
| * | 2A | : | 3A | J | 4A | Z | 5A | j | 6A | z | 7A |
| + | 2B | ; | 3B | K | 4B | [ | 5B | k | 6B | { | 7B |
| , | 2C | < | 3C | L | 4C | \ | 5C | l | 6C | \| | 7C |
| - | 2D | = | 3D | M | 4D | ] | 5D | m | 6D | } | 7D |
| . | 2E | > | 3E | N | 4E | ^ | 5E | n | 6E | ~ | 7E |
| / | 2F | ? | 3F | O | 4F | _ | 5F | o | 6F | delete | 7F |

Since sequences of binary numerals are difficult to read, write, and remember, ASCII codes are usually written as base 8 or base 16 numerals. The ASCII chart in Figure 9.12 uses HEX codes. The abbreviation HEX is for hexadecimal or base 16. The ability to convert between binary, octal, and hexadecimal forms is useful in working with and understanding computer codes. The following examples will show you how to convert between binary, octal, and hexadecimal numerals.

## *Converting Binary, Octal, and Hexadecimal Numerals*

In Section 1.3, the place-value system for any base was discussed. For example, the binary number, $10\,110_2$ means you have the following:

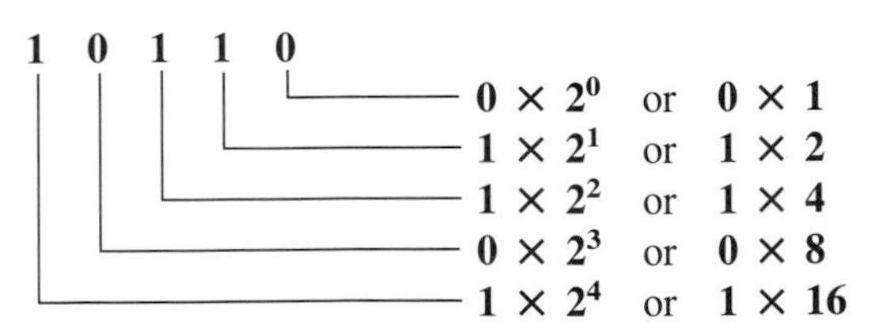


Furthermore, since $8 = 2^3$, an octal digit can be represented by a three-digit binary numeral, and since $16 = 2^4$, a hexadecimal digit can be represented by a four-digit binary numeral. If the digits of a binary numeral are written in groups of three or four digits, the binary numeral can be easily converted into an octal numeral or a hexadecimal numeral. If you add up the place values for each nonzero binary digit given in Figure 9.13, you get the corresponding octal or hexadecimal numeral.

| Binary (place values) 4 2 1 | Octal |
|---|---|
| 0 0 1 | 1 |
| 0 1 0 | 2 |
| 0 1 1 | 3 |
| 1 0 0 | 4 |
| 1 0 1 | 5 |
| 1 1 0 | 6 |
| 1 1 1 | 7 |

| Binary (place values) 8 4 2 1 | Hex |
|---|---|
| 0 0 0 1 | 1 |
| 0 0 1 0 | 2 |
| 0 0 1 1 | 3 |
| 0 1 0 0 | 4 |
| 0 1 0 1 | 5 |
| 0 1 1 0 | 6 |
| 0 1 1 1 | 7 |
| 1 0 0 0 | 8 |
| 1 0 0 1 | 9 |
| 1 0 1 0 | A(10) |
| 1 0 1 1 | B(11) |
| 1 1 0 0 | C(12) |
| 1 1 0 1 | D(13) |
| 1 1 1 0 | E(14) |
| 1 1 1 1 | F(15) |

**Figure 9.13**

## Example 1

Write 10 110 $111_2$ as (a) an octal numeral and (b) a hexadecimal numeral.

**Solution:**

(a) Octal: Separate the binary numeral into groups of three digits starting from the right of the numeral. Find the octal digit for each group of three digits by adding the nonzero place values shown in Figure 9.13.

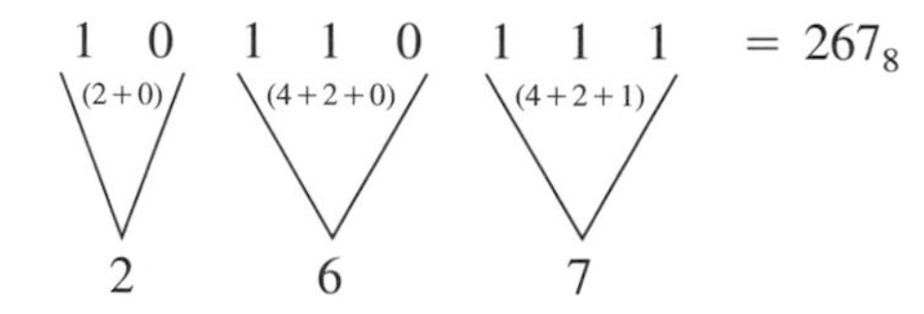


(b) Hexadecimal: Separate the binary numeral into groups of four digits and find the hexadecimal digit for each group of four digits adding the nonzero place values shown in Figure 9.13.

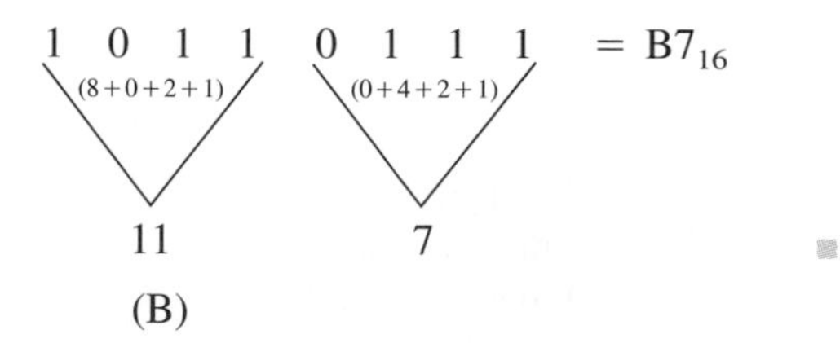


## Example 2

Using the ASCII chart, find the eight-digit binary code for A, Z, and >.

**Solution:** Since every hexadecimal digit can be represented by four binary digits, find the four-digit binary numeral for each digit in the hexadecimal representation of A, Z, and >.

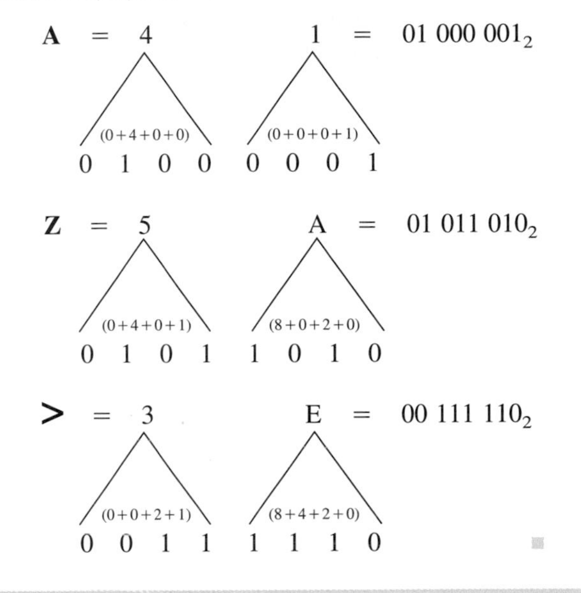


**Note:** By convention, binary numerals are written in groups of three digits.

## How Are Calculations Done by a Computer?

The last aspect of computers we will investigate is how computers do calculations, particularly the operations of addition and multiplication. In our base-10 system, to add or multiply efficiently, you need to know the 100 addition facts from $0+0$ to $9+9$ and the 100 multiplication facts from $0\times0$ to $9\times9$. These facts along with the process of "carrying" allow you to add and multiply. However, in the binary system there are only four addition facts and four multiplication facts.

### *Binary Arithmetic Facts*

| Addition | Multiplication |
|---|---|
| $0+0=0$ | $0\times0=0$ |
| $0+1=1$ | $0\times1=0$ |
| $1+0=1$ | $1\times0=0$ |
| $1+1=10$ | $1\times1=1$ |

Because we are accustomed to thinking in base 10, there is an entry in the binary arithmetic facts chart that seems strange, $1 + 1 = 10$. You must understand that the "10" is in base 2 and $10_2$ means $1 \times 2 + 0 \times 1 = 2$. Thus, the chart entry actually corresponds to our understanding of addition. With these binary arithmetic facts and the process of "carrying," we can explain how a computer performs addition and multiplication quickly and efficiently.

## *Binary Addition and Multiplication*

To add or multiply decimal numerals, the computer first converts them into binary numerals. The addition or multiplication is done electronically by the computer using the binary numerals. Before the answer is displayed, it is converted back into its decimal form. Let's show how this is done by finding $28 + 13$ and $28 \times 13$.

**1.** The computer converts the 28 and 13 into binary numerals. Paper-and-pencil methods for doing that are given in Section 1.3.

$$28 = 11\ 100_2 \qquad \text{and} \qquad 13 = 1\ 101_2$$

**2.** Addition and multiplication is done electronically using the basic binary arithmetic facts and a "carrying" procedure. We will use the process for addition and multiplication that we use with decimal numerals to demonstrate the process.

Addition:

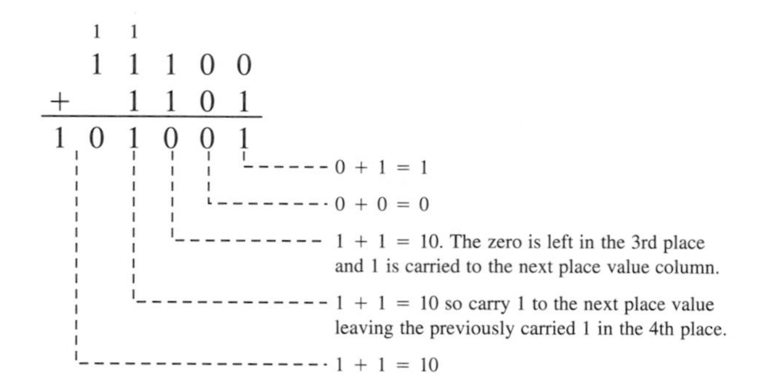

1 1
1 1 1 0 0
+ 1 1 0 1
1 0 1 0 0 1
0 + 1 = 1
0 + 0 = 0
1 + 1 = 10. The zero is left in the 3rd place and 1 is carried to the next place value column.
1 + 1 = 10 so carry 1 to the next place value leaving the previously carried 1 in the 4th place.
1 + 1 = 10


Check: $101\ 001_2 = 1\times32+0\times16+1\times8+0\times4+0\times2+1\times1 = 41$.
Using decimal numerals, $28 + 13 = 41$.

Multiplication:

```
              1 1 1 0 0
          ×     1 1 0 1
          -------------
              1 1 1 0 0----- Using the multiplier in the "1's" place, we get 1 × 11100 = 11100.
            0 0 0 0 0------- Using the multiplier in the "2's" place, we get 0 × 11100 = 00000.
          1 1 1 0 0--------- Using the multiplier in the "4's" place, we get 1 × 11100 = 11100.
        1 1 1 0 0----------- Using the multiplier in the "8's" place, we get 1 × 11100 = 11100.
  -----------------------
  1 0 1 1 0 1 1 0 0--------- Add the digits in each place value column. Remember to
                             carry "1" to the next column whenever you get 1 + 1.
```

Check: $101\ 101\ 100_2 =$
$1\times256+0\times128+1\times64+1\times32+0\times16+1\times8+1\times4+0\times2+0\times1 = 364.$
Using decimal numerals, $28 \times 13 = 364$.

3. To output the answers, the computer converts them back into decimal form. Paper-and-pencil methods for doing that are given in Section 1.3. Thus, we would see $28 + 13 = 41$ and $28 \times 13 = 364$.

Using only 1's and 0's simplifies computation in general. In the integrated circuits of a computer, the 1's and 0's are manipulated at great speed. If you are not concerned about the decimal answer to a problem and simply work in binary, addition and multiplication can be done quite easily.

## Example 3

Find the sum of the binary numbers, 101 111 101 and 10 011 010.

**Solution:** The binary numbers are positioned above each other, and the digits in each place-value column are added using the binary arithmetic facts and the normal ''carrying'' procedure as noted. Proceeding from the 1's place and working to the left gives the following result.

$$\begin{array}{r} {\scriptstyle 1\ \ 111\ 11\ \ } \\ 101\ 111\ 101 \\ +\quad 10\ 011\ 010 \\ \hline 1\ 000\ 010\ 111 \end{array} = 1\ 000\ 010\ 111_2$$

*Note:* Writing the numbers in decimal form, we have $381 + 154 = 535$.

## Example 4

Find the product of the binary numbers, 10 011 and 1 110.

**Solution:** The binary numbers are positioned above each other, and the first factor is multiplied by each digit of the second factor. The process starts with the 1's place and moves one place-value column to the left with each new multiplication. The digits in each place-value column are added using the binary arithmetic facts and the normal ''carrying'' procedure.

$$\begin{array}{r} 1\,0\,0\,1\,1 \\ \times\quad 1\,1\,1\,0 \\ \hline 0\,0\,0\,0\,0 \\ 1\,0\,0\,1\,1\phantom{0} \\ 1\,0\,0\,1\,1\phantom{00} \\ 1\,0\,0\,1\,1\phantom{000} \\ \hline 1\,0\,0\,0\,0\,1\,0\,1\,0 \end{array} = 100\ 001\ 010_2$$

*Note:* In decimal form, we have $19 \times 14 = 266$.

Section 9.2
PROBLEMS

## *Explain ➡ Apply ➡ Explore*

### *Explain*

**1.** What are basic components of a computer as designed by Charles Babbage?

**2.** What reasons did Charles Babbage have for building the *Difference Engine*?

**3.** What problem was Herman Hollerith trying to solve with his tabulating machine?

**4.** What World War II concerns were Konrad Zuse's computer and Alan Turing's *COLOSSUS* designed to solve?

**5.** How were the Harvard Mark I and the *ENIAC* alike? How were they different?

**6.** What "first" was accomplished by the *EDSAC* and *EDVAC* computers?

**7.** Describe the computers of the first generation (1951–1958).

**8.** Describe the computers of the second generation (1959–1964).

**9.** Describe the computers of the third generation (1965–1970).

**10.** Describe the computers of the fourth generation (1971–present).

**11.** What three advances in technology generated major changes in the computer?

**12.** What is a bit? What is an 8-bit machine?

**13.** What is ASCII? What is it used for?

**14.** How is data read and stored by a computer?

**15.** Why is $1 + 1 = 10$ in the binary system?

**16.** How does a computer use binary numbers in addition?

**17.** How does a computer use binary numbers in multiplication?

### *Apply*

In Problems 18–23, write each binary numeral in (a) octal and (b) hexadecimal.

**18.** $11\ 011_2$

**19.** $10\ 101_2$

**20.** $111\ 111_2$

**21.** $1\ 100\ 100_2$

**22.** $111\ 110\ 101\ 110_2$

**23.** $101\ 110\ 101\ 110_2$

In Problems 24–29, use the ASCII chart in Figure 9.12 to find the 8-bit binary code for each character.

**24.** The letter M.

**25.** The letter e.

**26.** The letter n.

**27.** The percent sign (%).

**28.** The equals sign (=).

**29.** The plus sign (+).

In Problems 30–37, give the answers to each problem as a binary numeral.

**30.** $1\ 010_2 + 1\ 110_2$

**31.** $110\ 011_2 + 10\ 110_2$

**32.** $10\ 011\ 111_2 + 111\ 011_2$

**33.** $111\ 000\ 101_2 + 101\ 011\ 001_2$

**34.** $110_2 \times 101_2$

**35.** $10\ 100_2 \times 1\ 101_2$

**36.** $101\ 111_2 \times 1\ 011_2$

**37.** $111\ 000\ 111_2 \times 111_2$

## *Explore*

To transfer data from one word processing program to another word processing program, the data are usually stored and read in binary ASCII format. If each line of binary code represents one character, change each line to hexadecimal and find what characters were being transferred in Problems 38 and 39.

**38.** 0100 0111
0110 1111
0110 1111
0110 0100
0010 0000
0100 0100
0110 1111
0110 0111

**39.** 0101 0100
0110 1111
0010 0000
0110 0010
0110 0101
0010 0000
0110 1111
0111 0010
0010 0000
0110 1110
0110 1111
0111 0100
0011 1111

In Problems 40–44, use the ASCII code to write each word in binary.

**40.** LOVE

**41.** HOPE

**42.** M*A*S*H

**43.** Mom

**44.** YES!

**45.** This section gives a method to convert binary numerals into octal (base 8) and hexadecimal (base 16). Based on those methods, how could you convert binary numerals into base 32 numerals or base 32 numerals into binary numerals? Give some examples of your method. (*Hint:* $2^5 = 32$.)

The power of the personal computer can be seen in these fractal images. They required over 800 billion calculations and 52 hours of computer time. If you could do one calculation per second, nonstop, it would take you over 25,000 years to complete the calculations. The fractal images were created using the software program, *Fractint 17.2*, on an IBM 486 DX-33 PC and an HP-500C printer. (Courtesy of Tom Falbo)

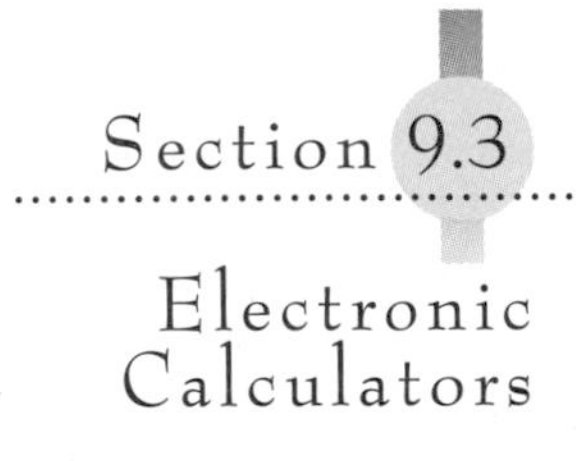

# Section 9.3 Electronic Calculators

With the increasing miniaturization of electronics during the 1960s, the room-size computers of the 1950s evolved into computers that were more compact and more powerful. This led to the obvious result—a portable, hand-held device capable of computing with speed and accuracy.

## Four-Function Calculators

Four-function calculators are those calculators whose primary purpose is restricted to addition, subtraction, multiplication, and division. These were the first hand-held calculators that were available to the public, costing about $100 in 1971. Today, this type of calculator costs less than $10 and is available as a built-in feature in check-books and watches. Despite the simplicity of these calculators, they gave the average person access to some of the power of the computer. Although other devices such as the mechanical calculator and the slide rule had been available, had not provided the speed and accuracy of the electronic calculator. Shortly after the introduction of the hand-held calculator, the era of the slide rule and the mechanical calculator came to an end.

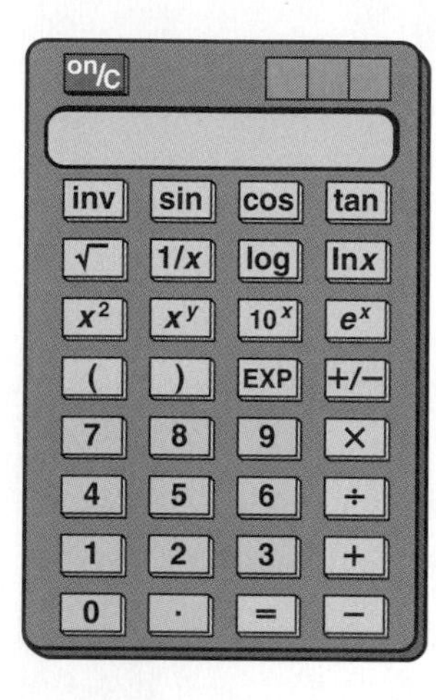


## Scientific Calculators

A scientific calculator is one that includes the keys for trigonometry, logarithms, and exponents. It may also contain keys for statistical functions such as the *mean* and *standard deviation.* Scientific calculators are often available for less than $20. They can be found at discount stores as well as at department stores and electronics shops.

You may be familiar with the basic keys on a calculator, but there may be a few you rarely use. This section will discuss some of those keys. When we want to indicate that a particular key is being used, the symbol for that key will be enclosed in a box. For example, for $3 \times 4 = 12$, we use the keys

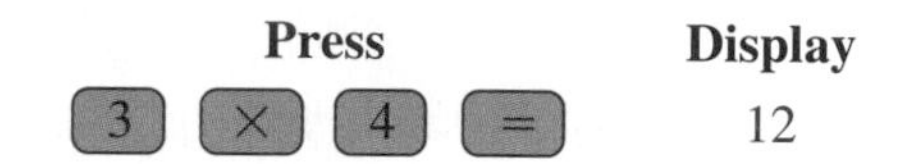


To understand the material in this section, it is very important that you have a scientific calculator available. As we do each example, you should try the example on your own calculator. We will demonstrate one method in the text. However, because of the differences between calculators, this method may not work with your calculator. For further help, read your calculator instruction manual or ask your instructor.

In this introduction to the scientific calculator, we will look at a few keys that you may find useful in complex calculations.

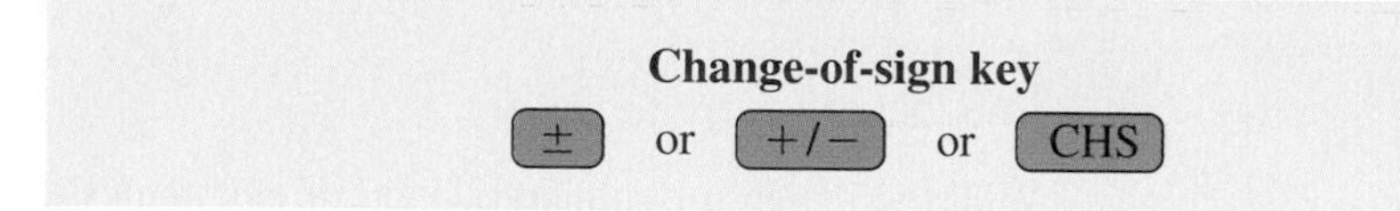


This key is used to create negative numbers in the calculator. On most calculators, to enter a negative value the negative sign must be entered after the number. For example, to enter the value $-6$, the keys must be pressed in the order

### Example 1

Use a calculator to find the value of $15 \div (-3)$.

**Solution:** We know that the answer should be $-5$. Using a calculator, we get

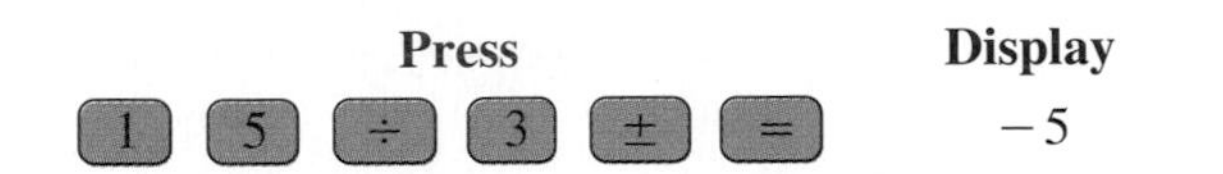

| Press | Display |
|---|---|
| 1 5 ÷ 3 ± = | −5 |

Notice that the parentheses around −3 are not entered into the calculator. ■

**Reciprocal key**

$x^{-1}$ or $\frac{1}{x}$

This key provides a shortcut for the division problem $1 \div x$.

### Example 2

Find the decimal value of $\frac{1}{5}$.

**Solution:** Using the reciprocal key on the calculator gives

| Press | Display |
|---|---|
| 5 $x^{-1}$ | 0.2 |

Notice that we did not need to press the equals key. ■

The reciprocal key is especially useful when you add fractions with numerators that are 1's.

### Example 3

Find the decimal value of $\frac{1}{5} + \frac{1}{7} - \frac{1}{6}$.

**Solution:** Using the reciprocal key on the calculator gives

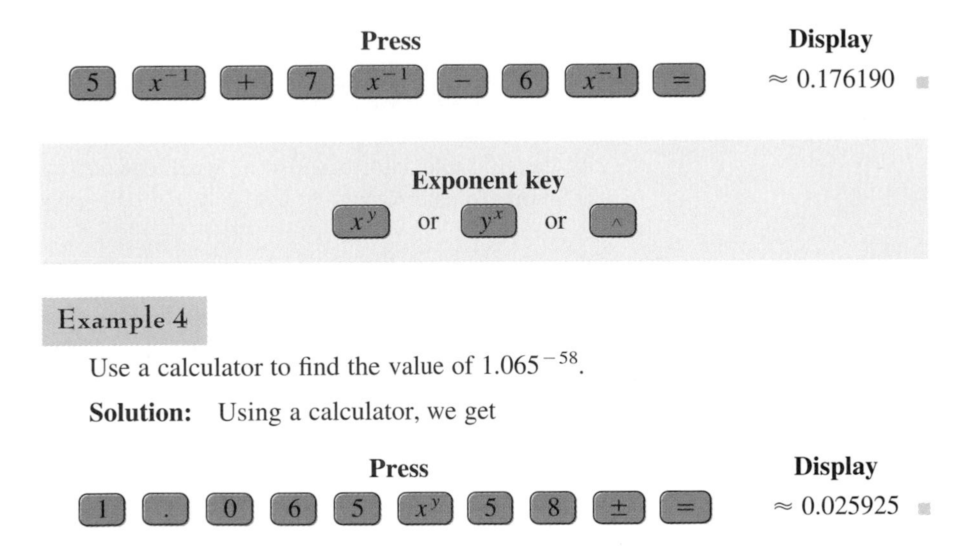

| Press | Display |
|---|---|
| 5 $x^{-1}$ + 7 $x^{-1}$ − 6 $x^{-1}$ = | ≈ 0.176190 ■ |

**Exponent key**

$x^y$ or $y^x$ or ^

### Example 4

Use a calculator to find the value of $1.065^{-58}$.

**Solution:** Using a calculator, we get

| Press | Display |
|---|---|
| 1 . 0 6 5 $x^y$ 5 8 ± = | ≈ 0.025925 ■ |

From algebra, you may remember that fractional exponents represent roots. For example,

$$\sqrt{x} = x^{1/2} \qquad \text{and} \qquad \sqrt[5]{x} = x^{1/5}$$

Although most calculators have a key to calculate square roots, calculators do not have a fifth-root key. As a result, to compute most roots, it is necessary to use fractional exponents. In the next example, we want to combine the use of the reciprocal and exponent keys to perform calculations involving fractional exponents.

### Example 5

Calculate $\sqrt[3]{8} = 8^{1/3}$.

**Solution:** We know that the answer is 2 since the cube root of 8 is 2. Using the reciprocal and exponent keys, we have

**Press** **Display**

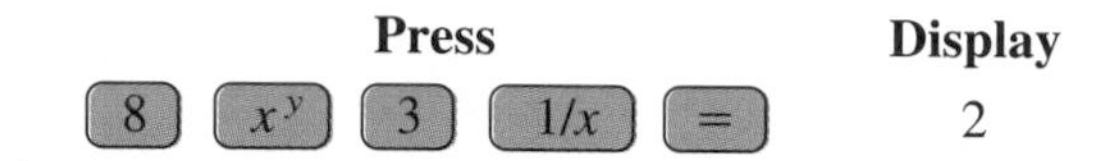


2

Push the keys again and watch what happens in the display. When the reciprocal key is pushed after hitting 3, the display shows 0.333333333. This is the reciprocal of 3, not the cube root of 8. It is necessary to hit the equals key to complete the calculation. ■

### Example 6

Calculate $85^{1/25}$.

**Solution:** Using the reciprocal and exponent keys, we have

**Press** **Display**

$\approx 1.194474$ ■

A scientific calculator can also be used to work with numbers written in scientific notation. To enter a number like $4 \times 10^6$ on a calculator, use a button labeled

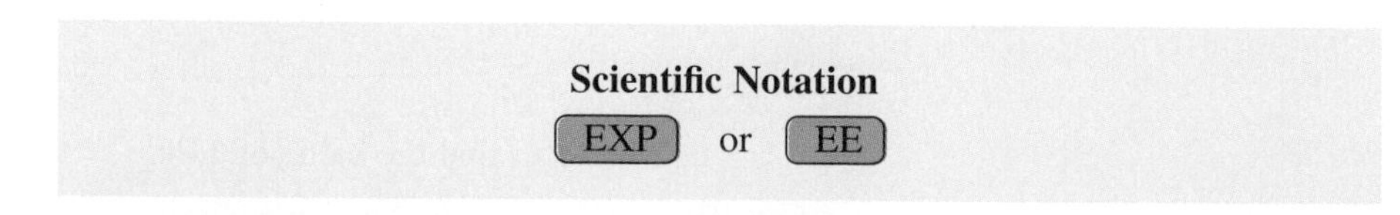


### Example 7

Enter the number $4.23 \times 10^{19}$ on a calculator.

**Solution:** Using the *EXP* button gives

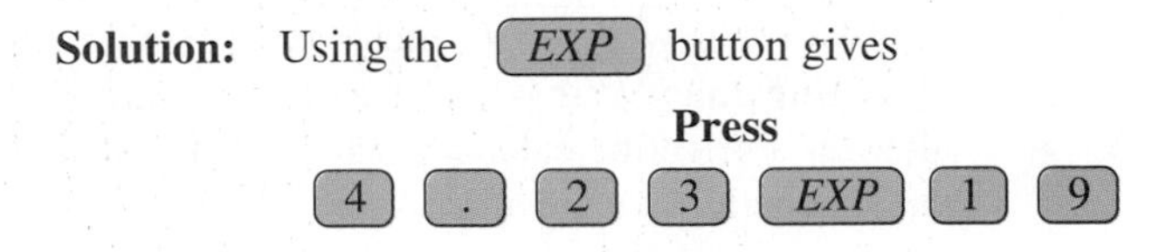

The calculator should now display the numbers 4.23 and 19, with the gap between the 4.23 and the 19 indicating that the calculator is using scientific notation. ■

## Example 8

Use a calculator to find the value of $(2 \times 10^{19}) \div (8 \times 10^{29})$.

**Solution:** Using the *EXP* button gives

**Press** **Display**

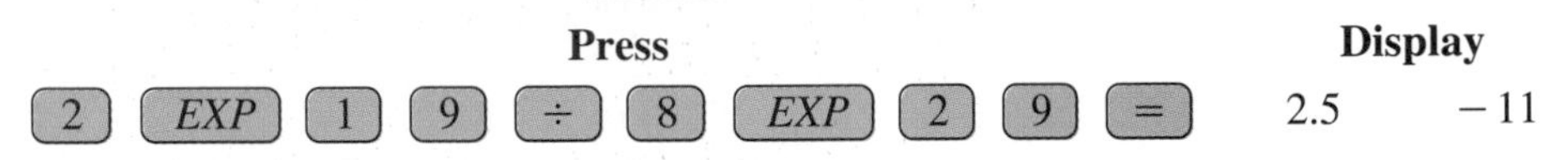

2.5 −11

Notice that the display of your calculator has a −11 off to the right side of 2.5. Since this is how a calculator displays scientific notation, the answer is $2.5 \times 10^{-11}$. ■

There are many other keys on a scientific calculator that are used for other mathematical functions. In Chapter 2, the exponential and logarithm functions, in Chapter 5, the sine, cosine and tangent functions, and in Chapter 7, factorials, combinations, and permutations are used. We have provided you with a good start in using a scientific calculator. The instruction manual for your calculator can provide assistance for these and other topics.

## Graphing Calculators (Optional)

A graphing calculator. (Reprinted by permission of Texas Instruments)

In 1990, a new type of calculator became available, the graphing calculator. A **graphing calculator** is the size of a standard scientific calculator, has all the functions of a scientific calculator, and has the capability to display the graph of one or more equations on the built in screen. The graphs given below are the actual displays from one model of these calculators, the TI 85.

## Example 9

Use a graphing calculator to display the graph of $y = x^2 - 6x + 4$.

**Solution:** Shown here is the graph as displayed by a TI 85 calculator. The range of values for $x$ and $y$ that are plotted are chosen by the person using the calculator. On the $x$ axis, the values vary from $-17$ to 17, while on the $y$ axis, the values vary from $-10$ to 10.

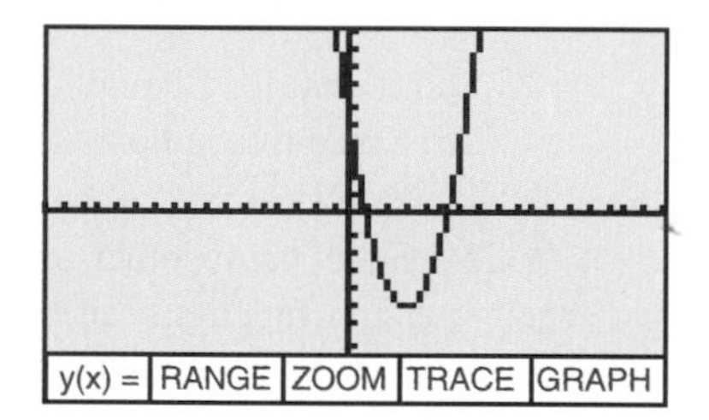


Parabolas such as the one given in Example 9 are familiar to you if you have studied Chapter 2 of this text. However, to help you appreciate the power of a graphing calculator, we give you a few more complicated examples.

## Example 10

Use a graphing calculator to display the graph of $y = 3x^4 + 4x^3 - 12x^2 + 9$ using $-10 \leq x \leq 10$ and $-10 \leq y \leq 10$.

**Solution:** Entering the equation into the calculator gives us the following graph. Notice that the bottom portion of the graph seems to be cut off.

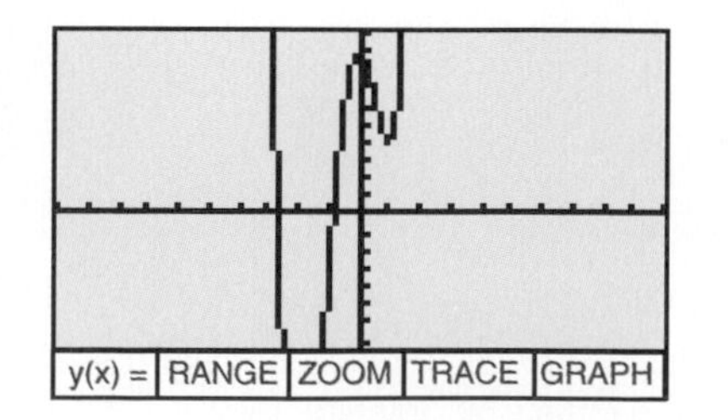

If we use the RANGE button on the calculator, we can change the $y$ values shown on the screen to include the interval $-35 \leq y \leq 15$. This gives the following picture.

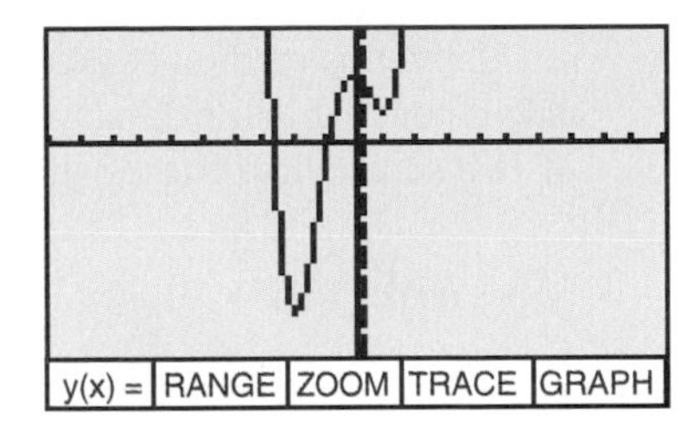


In our next example, we use a graphing calculator to sketch the graph of a trigonometric function you may have studied in Chapter 5.

## Example 11

Use a graphing calculator to sketch the graph of $y = \sin x$ using the values $-360° \leq x \leq 360°$ and $-2 \leq y \leq 2$.

**Solution:** After entering the equation and the desired values of $x$ and $y$ into the calculator, we have the following graph.

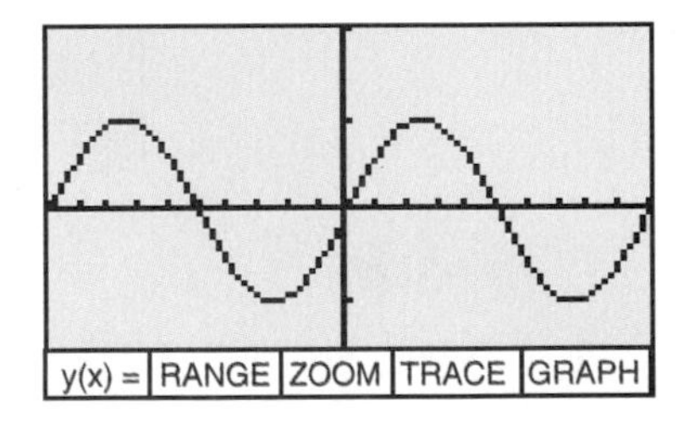


For our next two examples, we introduce equations that you may not have seen in previous courses, parametric equations. Equations such as $y = x^2 - 4x$ express a relationship between $x$ and $y$. **Parametric equations** are equations that express the values of $x$ and $y$ in terms of a third variable, usually $t$. When graphing parametric equations, we use the values of $t$ to determine the values of $x$ and $y$ and then graph the resulting $(x, y)$ pairs.

## Example 12

Given the set of parametric equations $\begin{cases} x = t^2 \\ y = t - 5 \end{cases}$, find the values of $x$ and $y$ when $t = 3$.

**Solution:** Substituting $t = 3$ into the parametric equations gives

$$x = 3^2 = 9$$

$$y = 3 - 5 = -2$$

Using this method of determining values of $x$ and $y$, we can now use a graphing calculator to graph parametric equations. We conclude this section with the graphs of two such pairs of parametric equations.

## Example 13

Graph the parametric equations

$$x = 8 \sin 2t \cos t$$

$$y = 8 \sin 2t \sin t$$

for values of $t$ contained in the interval $0° \leq t \leq 360°$.

**Solution:** Setting the calculator to graph parametric equations, we obtain the following graph.

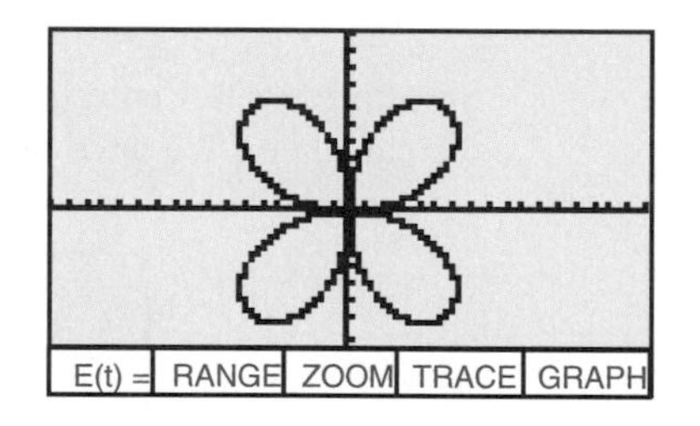


## Example 14

Graph the parametric equations

$$x = 3(\cos t)^3$$

$$y = 3(\sin t)^3$$

for values of $t$ contained in the interval $0° \leq t \leq 360°$.

**Solution:** Setting the calculator to graph parametric equations, we obtain the following graph.

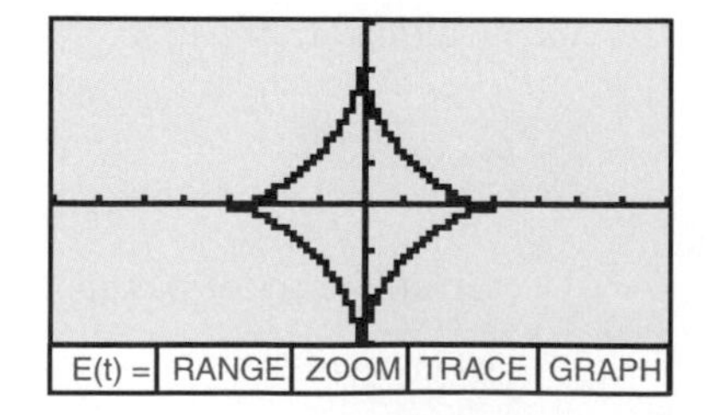

Graphing calculators are extremely useful in visualizing mathematical functions. Like the computer, hand-held scientific and graphing calculators will remain an intrinsic component of mathematics and science, at least until they are replaced by a more advanced technology.

Section 9.3
PROBLEMS

## Explain ➡ Apply ➡ Explore

### Explain

1. When and why did the four-function calculator replace the slide rule as a calculating device?
2. What is a scientific calculator?
3. What is a graphing calculator?
4. How are the four-function calculator, the scientific calculator, and the graphing calculator alike? How are they different?
5. What keys should be used to calculate the value of $7^{1/7}$?
6. Explain the difference between the [EXP] and the [$x^y$] button on your calculator.
7. Can the calculation $23^{12}$ be performed by using the keystrokes [2] [3] [EXP] [1] [2] [=]? Why or why not?
8. Perform the calculation 8,000,000 × 20,000,000 on a scientific calculator. Why does the calculator display the answer in scientific notation?

### Apply

In Problems 9–24, use a scientific calculator. If necessary, round off your answers to six decimal places.

9. $\frac{1}{5}$
10. $\frac{1}{4}$
11. $\frac{1}{27}$
12. $\frac{1}{17}$
13. $(-6)(-8)$
14. $(-12)(-4)$
15. $(-6.23)(-8.3)$
16. $(-1.56)(-2.7)$
17. $3^{2.3}$
18. $3^{-2.3}$
19. $6^{-0.53}$
20. $5^{-4.5}$
21. $81^{1/81}$
22. $17^{-1/17}$
23. $17^{1/17}$
24. $81^{-1/81}$

In Problems 25–28, solve using the scientific notation key on your calculator.

**25.** $(2 \times 10^{-9})(5 \times 10^{21})$

**26.** $(3.2 \times 10^{15}) \div (8 \times 10^{19})$

**27.** $(5 \times 10^{-9}) - (3 \times 10^{-10})$

**28.** $(2 \times 10^{-30}) + (8 \times 10^{-29})$

In Problems 29–32, solve using the reciprocal key on your calculator.

**29.** $\frac{1}{6} - \frac{1}{5}$

**30.** $\frac{1}{9} + \frac{1}{7}$

**31.** $\frac{1}{21} + \frac{1}{3} + \frac{1}{5}$

**32.** $\frac{1}{8} + \frac{1}{13} - \frac{1}{5}$

## *Explore*

In addition to performing calculations, a scientific calculator can be used to demonstrate non-trivial examples of algebraic properties. In Problems 33–38, use a scientific calculator to verify that both sides of the equation give the same result. Write the algebraic property that is being demonstrated by each example,

**33.** $4.7^3 \times 4.7^5 = 4.7^8$

**34.** $(1.75^2)^{5.6} = 1.75^{11.2}$

**35.** $\dfrac{\pi^8}{\pi^3} = \pi^5$

**36.** $\sqrt[4]{\sqrt[3]{5280}} = \sqrt[12]{5280}$

**37.** $\log(37.9 \times 55.556) = \log 37.9 + \log 55.556$

**38.** $2^{-\pi} = \dfrac{1}{2^{\pi}}$

In Problems 39–42, write the result shown in the display of your calculator after attempting each calculation. Explain why the calculator is displaying that result.

**39.** $173{,}821.6 \div 0$

**40.** $\ln(-5)$

**41.** $\log(14^2 - 13^2 - 3^3)$

**42.** $\sqrt{15.6^2 - 4(26.7)(53.3)}$

The following problems require the use of a graphing calculator.

In Problems 43–46, sketch the graph using a graphing calculator. Use the following values for $x$ and $y$: $-10 \le x \le 10, \quad -10 \le y \le 10$.

**43.** $y = x^2 - 6x + 9$

**44.** $y = x^2 + 4x - 2$

**45.** $y = -x^2 + 6x$

**46.** $y = x^2 - 2x - 3$

In Problems 47–56, sketch the graph using a graphing calculator. Use the indicated values of $x$ and $y$

**47.** $y = 2x^3 + 3x^2 - 5x + 6, \quad -5 \le x \le 5, \quad -10 \le y \le 20$

**48.** $y = -2x^3 - 4x^2 + 8x + 6, \quad -5 \le x \le 5, \quad -15 \le y \le 20$

**49.** $y = -x^4 - 4x^3 + 6, \quad -5 \le x \le 5, \quad -20 \le y \le 40$

**50.** $y = 3x^4 - 8x^2, \quad -5 \le x \le 5, \quad -20 \le y \le 40$

**51.** $y = 500\,(1.1)^x, \quad 0 \le x \le 30, \quad 0 \le y \le 7000$

**52.** $y = 14.7\,(10)^{-0.000018x}, \quad 0 \le x \le 20{,}000, \quad 0 \le y \le 20$

**53.** $y = 29 + 48.8 \log (x + 1), \quad 0 \le x \le 20, \quad 0 \le y \le 100$

**54.** $y = -172.7 + 25 \ln x, \quad 1000 \le x \le 2000, \quad -1 \le y \le 20$

**55.** $y = \cos 2x, \quad -360° \le x \le 360°, \quad -2 \le y \le -2$

**56.** $y = \tan x, \quad -472.5° \le x \le 472.5°, -20 \le y \le 20$

In Problems 57–62, sketch the following sets of parametric equations using a graphing calculator.

**57.** $x = 3t,\ y = t^2$ with $-3 \le t \le 3,\ -10 \le x \le 10,\ -10 \le y \le 10$

**58.** $x = 2t,\ y = \sqrt{t}$ with $0 \le t \le 3,\ -6 \le x \le 6,\ -2 \le y \le 5$

**59.** $x = \sin t,\ y = \cos t$ with $0° \le t \le 360°,\ -3.4 \le x \le 3.4,\ -2 \le y \le 2$

**60.** $x = 3 \cos t,\ y = \sin t$ with $0° \le t \le 360°,\ -3.4 \le x \le 3.4,\ -2 \le y \le 2$

**61.** $x = \dfrac{t}{t^2 + 1},\ y = \dfrac{2t}{t^2 + 1}$ with $-3 \le t \le 3,\ -2 \le x \le 2,\ -2 \le y \le 2$

**62.** $x = \sqrt{t + 4},\ y = \sqrt{t + 4}$ with $-4 \le t \le 4,\ -3 \le x \le 3,\ -3 \le y \le 3$

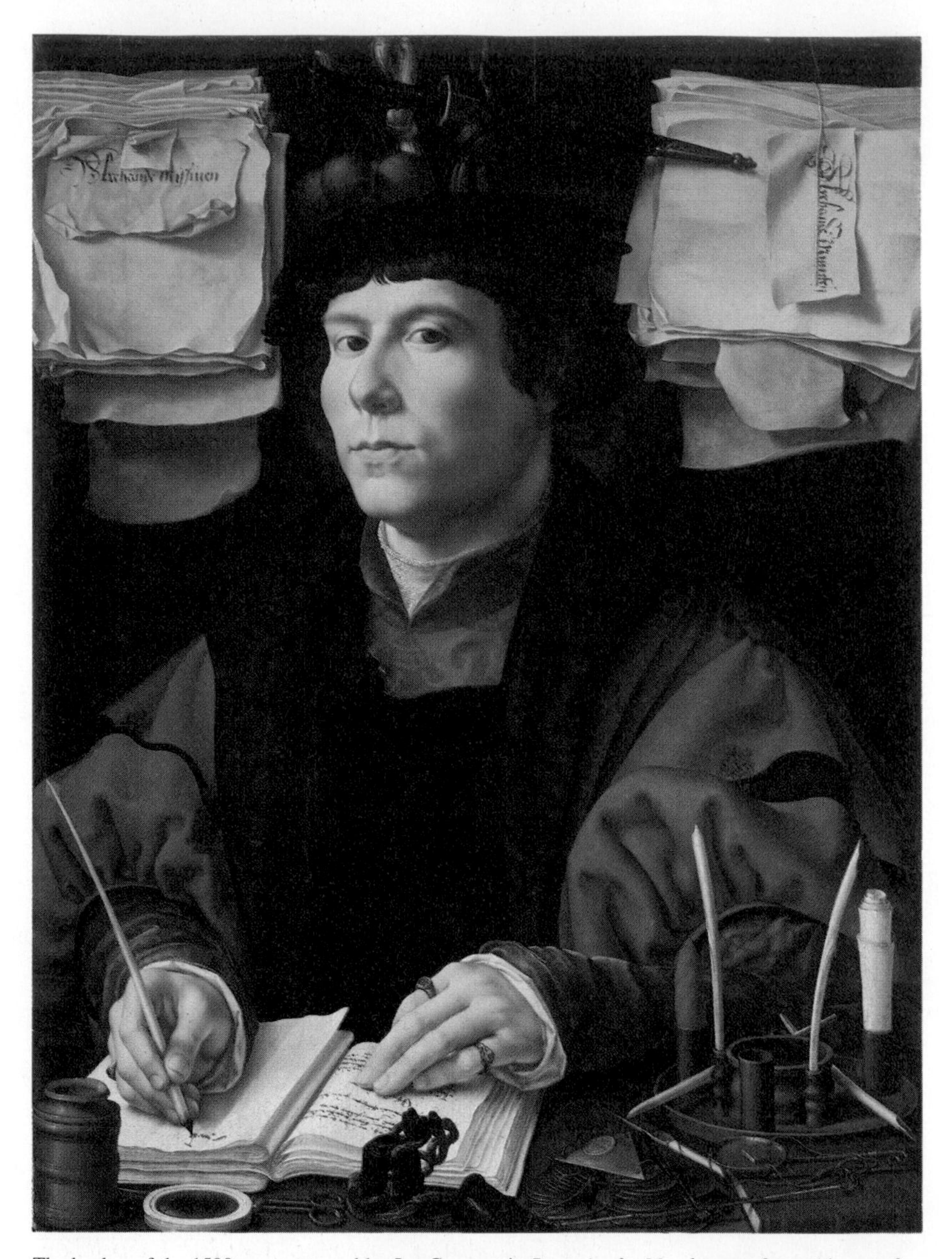

The banker of the 1500s, as portrayed by Jan Gossaert in *Portrait of a Merchant* performed hours of quill and paper computation and bookkeeping that would take the personal computer using basic accounting software only a few minutes. (National Gallery of Art)

## Section 9.4 Uses of Personal Computers

After much research, you have just purchased your very own personal computer. The **hardware**, all the components of your computer system, sits on your desk ready to be used. However, unless you have some **software**, programs to direct the operation of your computer, the system will be an *unused* marvel of technology. So in this section, we describe several types of software that you might want to buy for that computer system—word processors, spreadsheets, databases, and educational/entertainment software.

### Word Processors

**Word processors** are software programs that enable a computer to serve as an elaborate typewriter. If you use a computer at home or work, this is probably the most frequently used program. Because some people require a computer only for the purpose of word processing, it is possible to buy a computer whose sole function is word processing. Such a word processor does not require the software needed in a standard personal computer.

A word processor can be used for short notes, outlining, or even complete books. The manuscript of this book, including most of the illustrations, was originally typed on a word processor. This included different size and style typefaces, page numbers, and footnotes. Examples of commercial word processing programs include Microsoft Word® and WordPerfect®.*

One of the greatest advantages of word processors over typewriters is the capability to edit your writing on the screen without having to retype material. For example, if you are working on a document that has three main sections and decide that the placement of two of the sections should be reversed, using a typewriter you would be required to retype much and maybe all of those two sections. Using a word processor you would only use a few strokes on the keyboard or movements of the computer's mouse to accomplish the same revision.†

Other features that are available on many word processors are spelling checkers and grammar checkers. A spelling checker compares what has been typed with the words in a dictionary that has been stored in the program. If the checker finds a word such as "teh," it may suggest the words "the," "tech," "tee," "tea," "Ted," or "ten" as alternate correct spellings. The user then determines which word should replace "teh." However, there are some limitations with spelling checkers.

The following error will not be noticed by a spelling checker: "The package of tea arrived form Liverpool." Since the word "form" is correctly spelled, the spelling checker will not notice that the intended word is "from." Therefore, even if your

* Microsoft Word is a registered trademark of the Microsoft Corporation and WordPerfect is a registered trademark of WordPerfect, Inc.

† A cursor marks the position on the computer screen that is currently being used. A mouse is a device used by some computers to move the cursor to a different position on the screen.

word processor contains a spelling checker, it is important to carefully proofread any document you produce.

Grammar checkers are similar to spelling checkers in that they compare what has been typed to a set of grammatical rules stored in the computer. Given the sentence, ''She purchased three gifts.'' One grammar checker suggested that the word ''purchased'' be changed to a verb like ''buy'' and the word ''gift'' be changed to the word ''gifts.'' The first suggestion is to simplify the sentence and is not a necessary change. The second suggestion is necessary if the writer wants to conform to correct English grammar. As is true with spelling checkers, grammar checkers are not perfect. For example, the grammar checker mentioned above did not detect any errors in the sentence, ''I walks to the science lab.'' Again, it is important to realize that a grammar checker, like a spelling checker, is a useful tool, but it does not relieve the writer from carefully proofreading the document.

## Spreadsheets

A **spreadsheet** program has the capability of displaying numbers and text on a grid and performing calculations with the numbers. The resulting grid is called a spreadsheet. Spreadsheets are often used in businesses to keep track of the costs and quantities of items in inventory or production. But a spreadsheet can also be used to keep track of any set of numerical data that occurs in everyday life, such as the family checkbook, a teacher's grades, your bowling scores, or statistics for a baseball team. Examples of commercial spreadsheet programs include Microsoft Excel® and Lotus 1-2-3®.*

When a spreadsheet program is started, the screen is a grid of empty cells, as shown here.

| | A | B | C | D | E | F | G |
|---|---|---|---|---|---|---|---|
| 1 | | | | | | | |
| 2 | | | | | | | |
| 3 | | | | | | | |
| 4 | | | | | | | |
| 5 | | | | | | | |
| 6 | | | | | | | |
| 7 | | | | | | | |
| 8 | | | | | | | |
| 9 | | | | | | | |
| 10 | | | | | | | |
| 11 | | | | | | | |
| 12 | | | | | | | |
| 13 | | | | | | | |
| 14 | | | | | | | |

*Microsoft Excel is a registered trademark of the Microsoft Corporation and Lotus 1-2-3 is a registered trademark of Lotus, Inc.

A person using the spreadsheet constructs the spreadsheet by entering either text, a number, or a formula into each cell.

## Example 1

One typical use of a spreadsheet is to keep track of checkbook entries.

| | A | B | C | D | E | F | G |
|---|---|---|---|---|---|---|---|
| 1 | date | check | item | withdrawal | deposit | balance | |
| 2 | 4/23/94 | | initial deposit | | $920.56 | $920.56 | |
| 3 | 4/28/94 | 100 | Lousiana Mortgage Corp. | $758.65 | | $161.91 | |
| 4 | 5/7/94 | | deposit | | $920.56 | $1,082.47 | |
| 5 | 5/13/94 | 101 | Northern Telephone | $52.26 | | $1,030.21 | |
| 6 | 5/13/94 | 102 | Mastercharge | $251.68 | | $778.53 | |
| 7 | 5/13/94 | 103 | Municipal Water | $51.17 | | $727.36 | |
| 8 | 5/15/94 | | cash withdrawal | $200.00 | | $527.36 | |
| 9 | 5/21/94 | | deposit | | $920.56 | $1,447.92 | |
| 10 | 5/23/94 | 104 | Central Insurance Co. | $172.00 | | $1,275.92 | |
| 11 | 5/26/94 | 105 | Louisana Mortgage Corp. | $758.65 | | $517.27 | |
| 12 | 5/27/94 | | cash | $200.00 | | $317.27 | |
| 13 | | | | | | | |
| 14 | | | | | | | |

In this example, labels have been entered into the cells in the first row. For example, in cell 1C the word ''item'' has been entered. In each subsequent row, a transaction such as a check or a deposit is listed along with the date and the amount.

In the balance column, the spreadsheet calculates the new balance by subtracting withdrawals and adding deposits to the previous balance. This is accomplished through the use of formulas. For example, to compute the balance in the third row (cell F3), the formula could be

$$\text{F2} - \text{D3}$$

In a similar fashion, to determine the balance in cell F4, the formula could be F3 + E4 ■

Many spreadsheet programs also have the capability of drawing charts to illustrate information.

## Example 2

The following spreadsheet lists the number of violent crimes committed in the United States for a 13-year period. The software can be used to create a chart that gives a pictorial view of the information. (Source: *Time*, August 23, 1993)

| | A | B |
|---|---|---|
| 1 | | # of violent crimes |
| 2 | year | per 100,000 people |
| 3 | 1980 | 597 |
| 4 | 1981 | 594 |
| 5 | 1982 | 571 |
| 6 | 1983 | 538 |
| 7 | 1984 | 539 |
| 8 | 1985 | 557 |
| 9 | 1986 | 618 |
| 10 | 1987 | 610 |
| 11 | 1988 | 637 |
| 12 | 1989 | 663 |
| 13 | 1990 | 732 |
| 14 | 1991 | 758 |
| 15 | 1992 | 758 |

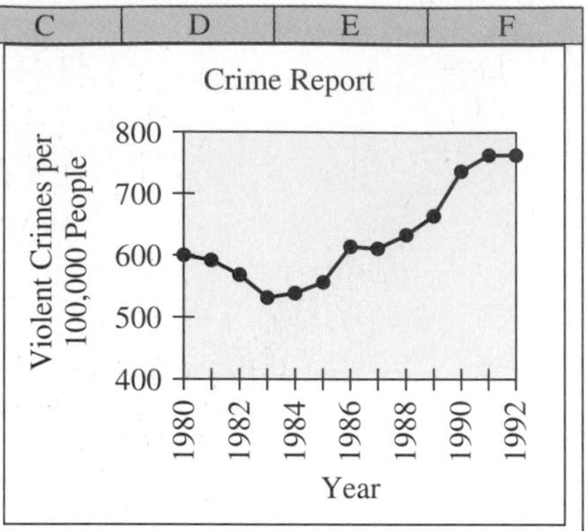

Crime Report
Violent Crimes per 100,000 People
800
700
600
500
400
1980 1982 1984 1986 1988 1990 1992
Year

The line graph was produced by the spreadsheet using the data in columns A and B. ■

## Databases

A **database** program enables the user to create a list of information and then manipulate the information. The collection of information is called a database. With database software, the information can be expanded, updated, and retrieved rapidly. For example, a business might create a database containing the name, address, zip code, and phone number of each of its customers. Since the post office charges lower bulk mailing rates when a business presorts its outgoing mail by zip code, a business will often use the database program to sort its mailing list by zip codes.

Databases created by government agencies and businesses are rapidly becoming available to the general public in libraries, on the INTERNET, and on other computer networks. One such database, that of information available in the periodical section of the library, is illustrated in the next example.

### Example 3

Suppose a local library has a computerized database of magazine and newspaper articles written in the past ten years. Although the database includes those articles published in 700 different periodicals, the library subscribes to only 200 of these. You want to read reviews on utility vehicles such as the Ford Explorer and Toyota 4 Runner that have been written in the past two years and are in periodicals available in the library. What information should you give the computer so that you receive the appropriate items from the database?

**Solution:** If you request all information on the database regarding utility vehicles, you will receive a long list that includes articles in magazines not carried by your library and whose information is more than two years old.

Most library databases have commands that allow you to specify the dates to be included in the search. So if you request information on utility vehicles that is at most two years old, you will meet the second of your criteria.

Since most periodical databases are compiled by outside companies and not by individual libraries, the list received from the computer includes articles that are not available to you. To further refine your search, you can specify the periodicals that are available in your library. By doing this you will receive information that is both current and useful. For example, if this is done at a local library, you can also specify a list of articles from *Consumer Reports* and *Car and Driver* magazines. ■

## Educational/Entertainment Software

Although some programs are readily identified as educational or entertainment, some programs now available combine the benefits of both categories. Educational programs at all levels from elementary school through graduate school are currently available. At the elementary level, there are programs in which a children's book such as *Cinderella* has both text and illustration displayed on the screen. The child is allowed to read the large print and then check if he is reading the material correctly by using the computer's mouse to highlight the words. The computer will then "speak" the selected text aloud.

At the junior high and high school levels, there are "games" in which the player is asked to participate in the decision-making process of a historical event. For example, in a program that describes the travels of a wagon train in the mid-1800s, the player plays the role of wagon master. The decisions made by the player determine the next series of events that take place.

In a game called SimCity,™* the player designs and builds a city. Decisions involve the placement of roads, railways, residential, commercial, and industrial areas, and major facilities such as airports and power plants. The player must then cope with difficulties such as pollution, crime, congested traffic, and natural disasters. This game has been used in graduate schools as a tool to simulate the problems associated with city planning.

---

* SimCity is a registered trademark of the Maxis company.

## Other Software

Other major categories of software include communications programs, graphics packages, and integrated software packages. **Communications software** enables computers in different locations to ''talk'' to each other through the use of telephone lines or other wiring systems. A communications program enables business travelers to transfer computer information back to the home office and researchers to access universities and libraries around the world.

**Graphics software** enables the user to create and modify illustrations on a computer. Specialized programs enable a person to edit photographs that have been entered into the computer or to create architectural drawings. The advantage of using graphics software is that changes are easily made without to redrawing the entire illustration.

**Integrated software** programs combine the features of more than one of the above categories of software. A typical integrated program will include word processing, spreadsheets, graphics, and communications programs. Although the integrated packages often do not contain all the features of the individual applications, they have the most frequently used features of the separate programs. In addition, the integrated packages are available for a fraction of the combined costs of the included applications.

## Section 9.4 PROBLEMS

## *Explain ➡ Apply ➡ Explore*

### *Explain*

1. What is hardware? What is software?
2. What are some advantages of a word processor over a typewriter?
3. What is a spreadsheet program?
4. What is a database?
5. What does communications software do?
6. What is graphics software used for?
7. What is an integrated software package?

### *Apply*

8. Give three examples of databases you have used or on which you are listed.
9. Give three examples of situations in which you could use a spreadsheet.
10. Explain why a spelling checker might miss the error in the following sentence: ''I purchased too hamburgers.''

**11.** Explain why a spelling checker might miss the error in the following sentence: "Juanita went to there house for dinner."

**12.** Give two of your own examples where a spelling checker may fail to detect an error.

**13.** Given is a portion of a spreadsheet used by the purchasing department of a pizza restaurant.

| | A | B | C | D |
|---|---|---|---|---|
| 1 | item purchased | amount purchased | cost for each item | total cost of item |
| 2 | tomato sauce | 24 cans | $5.99/can | |
| 3 | cheese | 25 pounds | $2.59/pound | |
| 4 | mixed spices | 6 jars | $8.49/jar | |
| 5 | flour | 6 fifty-pound bags | $17.50/bag | |
| 6 | pepperoni | 10 pounds | $2.00/pound | |
| 7 | sausage | 20 pounds | $2.75/pound | |
| 8 | | | **Grand Total** | |

(a) State in words the way the spreadsheet will calculate the value in cell D2.
(b) What formula should be entered in cell D2 to perform this calculation?
(c) State in words the way the spreadsheet will calculate the value in cell D8.
(d) What formula should be entered in cell D8 to perform this calculation?

**14.** To comply with state environmental requirements, your company is required to maintain records regarding the gas mileage of each vehicle in the company fleet. Given is a portion of a spreadsheet used by the vehicle pool of your company.

| | A | B | C | D |
|---|---|---|---|---|
| 1 | vehicle | miles driven | gallons of gas used | miles per gallon |
| 2 | delivery truck 1 | 167 | 22.1 | |
| 3 | executive sedan | 357 | 11.2 | |
| 4 | messenger car | 458 | 13.8 | |
| 5 | delivery truck 2 | 253 | 35.2 | |
| 6 | delivery truck 1 | 212 | 28.1 | |
| | Total Miles | | | |
| | | | | |

(a) State in words the way the spreadsheet will calculate the value in cell D2.
(b) What formula should be entered in cell D2 to perform this calculation?
(c) State in words the way the spreadsheet will calculate the value in cell B7.
(d) What formula should be entered in cell B7 to perform this calculation?

## *Explore*

**15.** An author uses a computer extensively for writing and occasionally for other uses such as communications and drawings. What software should the author purchase?

**16.** You are using a computer package that will enable you to keep records for your income taxes. Which of the categories of software might be helpful? Explain your reasoning.

**17.** A dentist has hired you to create and mail a newsletter to her clients. The newsletter will include both text and illustrations. What categories of software can help you create a newsletter and organize the mailing at the most economical rate?

**18.** You are using a computer package that will enable you to enter a list of ingredients and search for a recipe that uses these ingredients. Which of the categories of software would likely be helpful? Explain your reasoning.

**19.** You want to put all your favorite recipes on your computer so that you can easily access a particular type of recipe (e.g., chocolate cake that contains coconut). What type of software would you use for this application? Why?

**20.** What is desktop publishing software? What is its purpose? Give three examples of popular desktop publishing software.

**21.** What is a scanner? What is OCR (optical character recognition) software and how is it used?

**22.** SimCity was given as an example of an educational/entertainment software. Find the names of three other examples of educational/entertainment software and write a description of each.

**23.** What is meant by multimedia? What software is used to create a multimedia presentation? Why is a CD ROM disk drive important for multimedia?

In Problems 24–32, you will need the use of a computer with certain software and may require the assistance of someone with knowledge of the software.

**24.** Obtain access to a computer with a word processing program. Type in the first two paragraphs of this section. Interchange the order of the words, ''spreadsheets,'' and ''databases'' in the third sentence without retyping them. Make the second paragraph the first paragraph without retyping it. If the software has a spelling checker and a grammar checker, use them on the paragraphs. Get a printout of the modified paragraphs.

**25.** Obtain access to a computer with a word processing program. Type in any paragraph of your choosing. Interchange the last two sentences without retyping them. If the software has a spelling checker and a grammar checker, use them on the paragraph. Get a printout of the modified paragraph.

**26.** Obtain access to a computer with a spreadsheet program. Use the program, to keep track of the amounts you spent on lunch from Monday through Friday for two consecutive weeks. The amounts you spent are as follows: M–$3.29, T–$4.56, W–$2.89, Th–$4.88, F–$1.82, M–$4.51, T–$0.95, W–$1.78, Th–$2.84, and F–$1.99. Use the spreadsheet program to display the day, amount spent for that day, total spent each week, and total for the two weeks. Get a printout of your results.

**27.** Obtain access to a computer with a spreadsheet program. Use the program to keep track of your quiz scores in a math class. Each week for ten weeks you had a 25-point quiz. Your scores were as follows: 23, 20, 25, 19, 18, 22, 21, 23, 24, 20. Use the spreadsheet program to display the week, number score for the week, total points scored, and your average for the ten weeks. Get a printout of your results.

**28.** Obtain access to a computer with a graphics program. Use the program, draw a large rectangle that has a small gray circle and a pair of parallel line segments within the rectangle. Get a printout of your results.

**29.** Obtain access to a computer with a graphics program. Use the program, draw a rectangle, a circle, and a triangle. Shade the interiors with different patterns. Complete the drawing by adding two intersecting lines. Get a printout of your results.

**30.** Go to a library that has a database system for cataloging the books in the library. Use the system to find a list of books on ''endangered species in North America.'' Get a printout of the list.

**31.** Go to a library that has a database system for cataloging journal or magazine articles in the library. Use the system to find a list of articles on ''investigations on computer hackers.'' Get a printout of the list.

**32.** Obtain access to a service like *Prodigy*, *CompuServe*, or *America On Line*. Use the communications capabilities of the computer to get information about some area that interests you. Get a printout of the information.

## Chapter 9 • Summary

### Key Terms, Concepts, and Formulas

The important terms in this chapter are:

**Abacus:** Device used to perform arithmetic by moving markers placed in grooves or strung on wires. p. 617

**Abacus regrouping principle:** Principle that governs the movement of markers on an abacus: If the value of a group of markers in one row equals a marker in the next row, the group is removed and one marker is added to the next row. p. 618

**ASCII:** American Standard Code for Information Interchange. p. 632

**Bit:** A 1 or 0 that can be read electronically as *on* or *off* or as a positive (+) or negative (−) charge. p. 632

**Communications software:** Computer programs that enable computers in different locations to "talk" to each other through the use of telephone lines or other wiring systems. p. 658

**Control unit:** Computer component that directs the flow of information and operations of a computer. p. 629

**Counting-board abacus:** Calculating device that uses markers on a table inscribed with lines for the digits of Roman numerals. p. 617

**Database:** Computer software that enables the user to create and manipulate lists of information. p. 656

**Finger reckoning:** Methods for using the fingers to record numbers and perform calculations of a computer. p. 618

**Graphics software:** Computer programs that enable the user to create and modify illustrations. p. 658

**Hardware:** The components of a computer system—computer and peripherals. p. 653

**Input device:** Device used to enter data and instructions into a computer. p. 629

**Integrated circuit:** Silicon chips that contain many interconnected electronic circuits. p. 631

**Integrated software:** Computer programs that combine the features of different types of computer software packages. p. 658

**Memory unit:** Computer component used to store data and instructions. p. 629

**Microprocessor:** Small silicon chip with integrated circuits forming the entire processing unit of a computer. p. 632

**Napier's Bones:** Calculating device consisting of rods inscribed with numbers and used for multiplication. p. 619

**Output device:** Device used to display results of processing by a computer. p. 629

**Parametric equations:** Equations that express the values of $x$ and $y$ in terms of a third variable. p. 647

**Processor:** Computer component where calculations and logic are performed. p. 629

**Program:** Step-by-step set of instructions written in the language of a computer that controls the operation of the computer. p. 630

**Slide rule:** Calculating device consisting of two pieces of material with logarithmic scales marked along their edges. p. 621

**Software:** A set of instructions that directs the operation of a computer. p. 653

**Soroban:** Japanese version of a "bead-and-wire" abacus. p. 619

**Spreadsheet:** Computer software that enables the user to display numbers and text on a grid and perform calculations with the numbers. p. 654

**Suan-pan:** Chinese version of a ''bead-and-wire'' abacus. p. 619

**Transistor:** Tiny solid-state electronic device used to control current flow. p. 631

**Word processor:** Computer software that enables the user to easily modify the written word. p. 653

After completing this chapter you should be able to:

1. Demonstrate how to perform calculations on an abacus, Napier's Bones, and a slide rule. p. 617
2. Explain how a mechanical calculator performs the ''carrying'' function in addition. p. 623
3. Describe the basic components of a computer and the advances in electronics that stimulated the development of microcomputers. p. 629
4. Convert binary numerals to octal and hexadecimal numerals and vice versa. p. 633
5. Perform binary addition and multiplication using the basic binary arithmetic facts. p. 636
6. Use a scientific calculator to find roots and powers, perform operations with reciprocals, and compute with scientific notation. p. 642
7. Use a graphing calculator to graph functions and parametric equations. p. 646
8. Explain the uses of the commonly used computer software programs—word processors, spreadsheets, databases, educational/entertainment software, communication software, graphics software, and integrated software. p. 653

• Summary

## Problems

1. Find the following sums on a counting-board abacus. Give the answers in both Roman numerals and Hindu-Arabic numerals.
   (a) $247 + 194$
   (b) MMDLXXI + DCXLVIII

2. Place Napier's Bones for 4 and 9 next to each other. Use them to find the following products. Explain how results were obtained.
   (a) $38 \times 49$
   (b) $567 \times 49$

**3.** Use the slide rule shown in the figure to find the following products or quotients. Explain how the results were obtained.

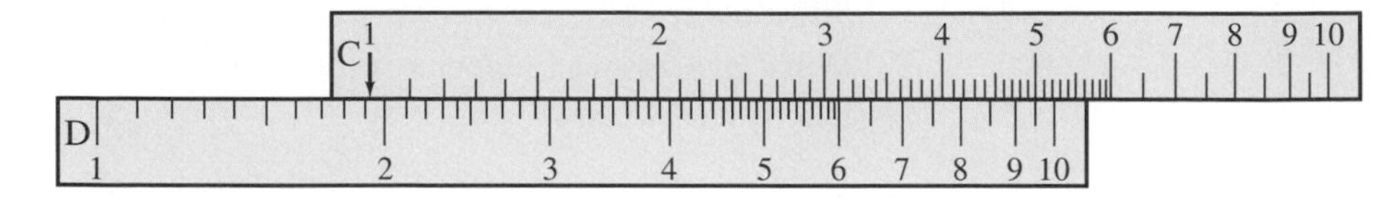


(a) $1.9 \times 3.6$
(b) $1.9 \times 4.7$
(c) $5.9 \div 3.1$

**4.** What three advancements in electronics stimulated the development of the microcomputer? Explain.

**5.** Why do computers store and use binary numbers? Explain.

**6.** Use the ASCII chart to find 8-bit binary codes for each character of the phrase, ''Just do it!''

**7.** Convert $101\ 110\ 011_2$ into both octal and hexadecimal.

**8.** Give the answers in binary.

(a) $101\ 110\ 011_2 + 1\ 100\ 111_2$
(b) $10\ 101_2 \times 1\ 011_2$

**9.** Find the following answers using a scientific calculator. If necessary, round off results to four decimal places.

(a) $\frac{1}{6} + \frac{1}{9} - \frac{1}{8}$
(b) $4.76^5$
(c) $\sqrt[5]{4.76}$
(d) $\frac{(3.46 \times 10^4)(5.39 \times 10^3)}{2.45 \times 10^{-3}}$

**10.** If you have access to a graphing calculator, graph the following:

(a) $y = x^3 - 6x^2 + 6x + 6$, with $-5 \le x \le 5$ and $-10 \le y \le 10$
(b) $x = t + 1,\ y = t^{1/3}$ with $-10 \le t \le 10,\ -9 \le x \le 11,\ -5 \le y \le 3$
(c) $y = \cos x$, with $0° \le x \le 720°$ and $-2 \le y \le 2$

**11.** Consider the sentence, ''I has for pears of sandals four ewe.''

(a) Explain why a computer spell-checking program might miss the errors in the sentence.
(b) The grammar checker in the word processor used to write this book suggested the ''has'' be changed to ''have'' and the ''four ewe'' be changed to ''four ewes.'' Is this correct? Does it correct all the errors in the sentence? Explain.

**12.** Explain which type of computer software would be the best to use for each of the applications in parts (a)–(i): word processor, spreadsheet, database, graphics software, educational software, or communications software.

(a) Writing poetry.
(b) Keeping the statistics for your basketball team.
(c) Cataloging the songs in your CD collection.
(d) Getting the latest stock prices.
(e) Making a floor plan for the setup of tables at a wedding.
(f) Keeping track of the badges earned by your daughter's Bluebird group.
(g) Writing a letter to the parents of your son's Cub Scout den.
(h) Addressing the envelopes for 500 flyers.
(i) Getting practice in the multiplication facts.

# 10

# Calculus and the Infinitesimal

A modern representation of a Mobius Strip can be seen in *Endless Ribbon* by Max Bill. (Musée National d'Art Moderne, Paris)

# Overview Interview

You may have known people who have taken a course in calculus and wondered, "What is calculus?" or "What do they study in calculus?" In this chapter, we will try to remove some of the mystery that surrounds calculus. You will investigate two of the fundamental processes in calculus, integration and differentiation. You will come to an understanding of what these terms mean and discover their importance in mathematics. You will see that integration allows you to find the area under a curve by examining very small quantities called infinitesimals. You will see that differentiation allows you to find the slope of a curve by letting distances approach zero. This chapter will show you that calculus is "no big thing" because it is really a study of the infinitesimal.

## Debbie Blumberg

DEBBIE BLUMBERG, RESEARCH BIOLOGIST, UNIVERSITY OF CALIFORNIA SANTA BARBARA, NEUROSCIENCE RESEARCH INSTITUTE

TELL US ABOUT YOURSELF. I was born and raised in New York and received my bachelor's degree from Stanford University and doctorate from the University of Wisconsin in Madison. I am presently a postdoctoral research fellow at the University of California at Santa Barbara. I study the molecular action of "growth factors," proteins produced by the body that are critical in establishing the proper pattern of nerve connections during embryonic development and that are needed throughout one's life for the maintenance of these connections. I hope that my research into these molecules will be useful in finding therapy to aid in the recovery of the nervous system following injury caused by disease or trauma.

*Research biologist Debbie Blumberg.*

WHY IS MATHEMATICS, FROM ARITHMETIC TO CALCULUS, IMPORTANT TO THE RESEARCH BIOLOGIST? Research biologists use mathematics on a daily basis. For example, basic mathematics is used in making up solutions and reagents. Arithmetic computations are made to determine how much of a given chemical to add to a solution, based on molecular weights and desired concentrations. Algebra and powers of 10 are used in making conversions from one unit to another and calculating dilutions. Statistics and probability theory are used in analyzing data from experiments and making predictions from the data. Calculus is the basis for the equations used to define certain biological events, such as enzyme kinetics and the binding of molecules. Even though I use calculators and computers to do calculations and analyze data, the answers are more meaningful if I understand the math behind the answers. Mathematics is essential to the quantitative analysis that I do as a research biologist. I could not imagine carrying out my research without using mathematics. I would fail miserably if I were "math-phobic."

Calculus can be used to analyze the curves, volumes, and areas of structures such as the *Interior of Saint Peter's, Rome* by Giovanni Paolo Panini. (National Gallery of Art)

# A Short History of Calculus

Mathematics is one of the subjects considered by many as an essential part of the training of the well-educated person. Like speaking a foreign language, reading a work by Shakespeare, or reciting poetry, understanding some mathematics was (and still is) the sign of a well-versed person. For many, calculus is the culmination of such mathematical training.

**Calculus** is the study of two mathematical processes called differentiation and integration. Both involve functions and the concept of very small increments of the variables in the functions. Because understanding the full scope of calculus requires a good grasp of algebra, geometry, and trigonometry, many students have difficulty with calculus. However, the basic concepts of calculus are not difficult. Some of the ideas in calculus have been studied for over 2000 years, long before modern algebra was invented. In this chapter, we examine the two basic concepts of calculus and some of their applications.

The first concept of calculus is the determination of areas bounded by curves or lines. The ancient Babylonians and Greeks developed formulas to determine areas of regular geometric figures such as triangles and rectangles, but the areas of circles and ellipses presented a greater challenge. Because the Greeks did not have a clear understanding of $\pi$, they could only determine the approximate area of a given circle.

Their knowledge of the area of a circle was that the ratio of the areas of two circles was equal to the ratios of the squares of their diameters.

By the fifth century B.C., to find the area of a circle, the concept of an **infinitesimal** was used. An infinitesimal is a very small quantity. For example, the number 0.0000 . . . 0001 is infinitesimally small. As another example, imagine a very small square, divide it into four square sections, and you now have a smaller square. By repeating this process enough times, you have an infinitesimally small square.

The use of infinitesimals led Eudoxus of Cnidus (c. 370 B.C.), to discover the *method of exhaustion*. He used this method to calculate the area of circles or other curved figures in the following manner. Start by inscribing a square inside a circle. Drawing four radii, one to each corner of the square, will create four triangles. Bisect each of these triangles with four more radii of the circle. Connect the end points of the radii to form an octagon consisting of eight triangles. Continue the process to create 16-, 32-, 64-, . . . sided polygons. As the number of sides increases, so does the number of triangles. As the number of triangles gets very large, the area of each triangle becomes very small. The area of each triangle is infinitesimally small and the sum of the areas of the triangle approaches the area of the circle. As can be seen from the inscribed square and octagon, the areas of the successive polygons will soon fill the circle, eventually exhausting the uncovered portion of the circle.

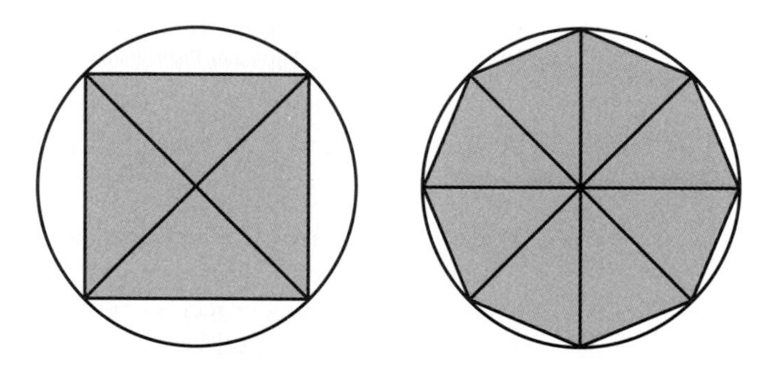

This method is the central idea behind the topic in calculus called **integration**. Integration is a method for finding areas of regions by filling the desired region with shapes of known area. By creating a large enough number of smaller pieces, the area of the unknown region can be found by adding the areas of the smaller pieces. This method can also be used for finding volumes of solids and lengths of curves.

Archimedes of Syracuse (c. 225 B.C.), one of the greatest mathematicians and scientists of ancient history, also used infinitesimals. After using an inductive process to discover formulas for areas of curved regions and volumes of solids, Archimedes used the method of exhaustion to derive formulas for the following geometric figures:

Surface area of spheres and spherical segments

Volumes of spherical segments and the segment of a hyperboloid of revolution

Areas of spirals and parabolic segments

In terms of the circles just discussed, Archimedes would argue as follows: By inscribing a series of polygons inside the circle, the area of the circle must be greater than the area of a polygon with a large number of sides. Similarly, by circumscribing polygons around the circle, the area of the circle must be smaller than the area of a polygon with a large number of sides. As the number of sides increases, the difference between the area of the circle and the area of the polygon decreases.

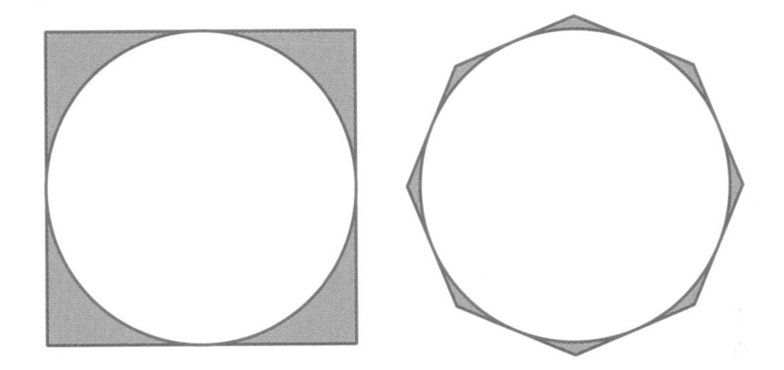

The method of exhaustion was an exceptional achievement for the Greek mathematicians. It allowed them to accurately determine the areas of curved regions. However, there are two difficulties with the method of exhaustion. The first is that the method is cumbersome because it requires a large number of calculations. The second difficulty is that a many-sided polygon is not a circle. No matter how many sides the inscribed polygon has, it will never become a circle. The method makes sense intuitively and gives accurate solutions.

As has been true in many other areas of mathematics, little progress in the development of calculus was made between the era of the Greeks and the Renaissance of the 16th century. For calculus, the 17th century was the beginning of an era of great discoveries.

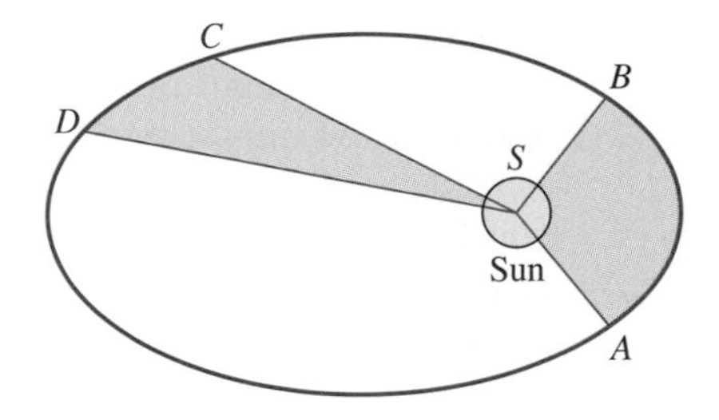


In 1609, Johann Kepler gave his three laws of planetary motion. One of these was that planets sweep out equal areas in equal times as they travel along their orbits. This means that as a planet travels along its elliptical path around the sun, if the areas of the regions $ASB$ and $CSD$ are equal, then the time required to travel from $A$ to $B$ is the same as the time required to travel from $C$ to $D$. This relationship required that Kepler determine the areas of elliptical sectors. He found the areas by a method called the *sum of the radii*.

The next major event in the history of integration is the work of Bonaventura Cavalieri, *Geometria Indivisibilibus*, first published in 1635. In this work, Cavalieri, influenced by Kepler, described a term called *indivisibles*. Indivisibles are objects that cannot be broken into smaller pieces. Like infinitesimals, indivisibles divide an object into many small pieces. Cavalieri thought of a line as an infinite collection of points; the points were the indivisibles for the line. The indivisibles for a surface were lines, and the indivisibles for a solid were planes. By combining the indivisibles, Cavalieri determined the length, area, or volume of the original figure. For example, think about a salami that has been thinly sliced. The total volume of the salami is the sum of the volumes of the slices. The slices of the salami are the indivisibles that, when added together, form the entire salami.

The second central idea in calculus, finding the slope of a tangent line to a curve, requires a process called **differentiation**. Finding the slope of a line tangent to a curve at a point $P$ can be used to determine the minimum or maximum point on a curve.

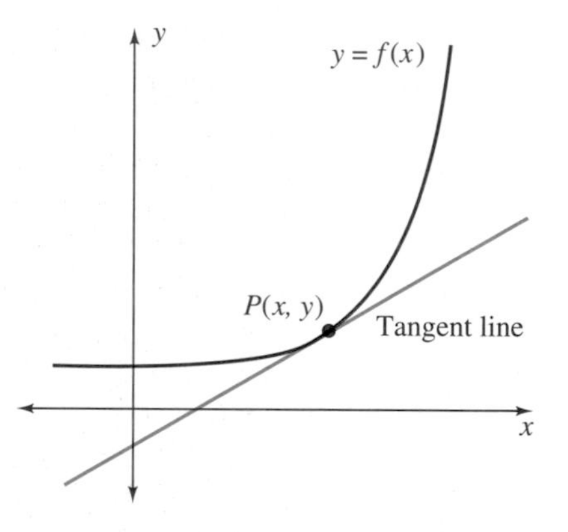


Like integration, differentiation has been studied for more than 2000 years. Though it was not investigated as intensively as integration, differentiation was studied by many mathematicians, such as Aristotle (384–322 B.C.), Pierre de Fermat (1601–1665), and Isaac Barrow (1630–1677). It was Fermat who recognized the connection between the maximum/minimum problem and differentiation.

In 1663, Isaac Barrow, Sir Isaac Newton's predecessor and mentor at Cambridge University in England, was the first to recognize and prove the startling fact that, like addition and subtraction, the operations of integration and differentiation are inverses of each other. With this work, the stage was set for the appearances of Sir Isaac Newton of England (1642–1727) and Gottfried Leibniz of Germany (1646–1716).

Leibniz and Newton are considered the co-inventors of calculus. Though they communicated with each other during the early part of their work, they soon disagreed over who first discovered certain aspects of calculus. This started a major conflict between the mathematicians of England and continental Europe. Today, it is agreed that Leibniz and Newton each formulated the rules of calculus independently, with each man inventing his own terminology and symbolism.

After reading of the gradual development of calculus over 2000 years, it may seem odd to say that Leibniz and Newton are the co-inventors. They are given credit for inventing calculus because they built on the ideas developed through the ages and constructed a set of rules and formulas that could be applied to many functions. Although Archimedes was able to determine the area bounded by a spiral, Newton and Leibniz determined a method that could be applied to many functions without using the tedious method of exhaustion. Barrow had proved that integration and differentiation were inverses of each other by using a cumbersome geometrical argument. Leibniz showed how this could be done with algebra. The work of Newton and Leibniz allowed calculus to be applied with great success to problems in science and engineering. During the 18th and 19th centuries, many of the great discoveries of science were precipitated by the work of these two men.

Due to the work of Newton and Leibniz, calculus became a very powerful mathematical tool. Algebra and geometry could deal with constant speeds, but calculus could help study the motion of objects with changing speeds. Algebra could determine the maximum value of a quadratic function by using the vertex of a parabola, but calculus could be used to determine the maximum value of many types of functions. Geometry could be used to find the areas and volumes of simple shapes, but calculus could be used to find areas between two functions, such as a logarithmic function and a quadratic function, or to determine the volume of curved solids in space.

Despite the significance and power of calculus, the new method was not complete. Leibniz and Newton concentrated on how the method worked and how it could be applied. Some of the details in the foundations had been neglected. This led to conclusions like $1 - 1 + 1 - 1 + \cdots = 1/2$. (Is it possible to add and subtract integers and get a result that is a fraction?) Many questioned whether infinitesimals actually existed and whether the new method performed division by zero, a most undesirable technique. As a statement against these methods, Bishop George Berkeley in 1734 wrote a treatise called *The Analyst or a Discourse Addressed to an Infidel Mathematician Wherein It Is Examined Whether the Object, Principles, and Inferences of the Modern Analysis Are More Distinctly Conceived, or More Evidently Deduced, than Religious Mysteries and Points of Faith. . . .* The ''infidel mathematician'' referred to Newton's friend Edmund Halley.

Although the methods in calculus produced great results throughout the 18th and 19th centuries, the discussion over the logical foundations of the methods was not settled until the end of the 19th century. It was only through the work of Augustin-Louis Cauchy of France (1789–1857) and Karl Weierstrass of Germany (1815–1897) that the great invention was put on a solid foundation.

The study of mathematics did not stop at placing calculus on a solid foundation. The 20th century brought its own mathematicians. Some of them continued the study of pure mathematics, some involved themselves with the application of mathematics to other sciences, others investigated using computers in mathematical research, and still others concentrated on the teaching of mathematics. These men and women of mathematics are a very diverse group with a wide range of intellectual interests and personal characteristics. We must keep in mind that they are people just like you and me. They have, however, found mathematics to be an interesting, challenging, enjoyable, and living subject. We are indebted to all the mathematicians from all the cultures of the world for their contributions to the development of mathematics.

## • Check Your Reading

1. Why are Newton and Leibniz considered the co-inventors of calculus?
2. What is the method of exhaustion? Explain and give examples.
3. While the Great Wall of China was being built in 275 B.C., who was proving geometry formulas?
4. In 1609, the artist Rubens was painting a self-portrait while the artist El Greco was painting *Brother Paravicino*. During this same year, who was investigating the laws of planetary motion?
5. What are integration and differentiation?
6. What are infinitesimals and indivisibles?
7. Bonaventura Cavalieri worked on indivisibles in the same year that Giulio Alenio published the first biography of Christ written in Chinese. What year was this?
8. In 370 B.C., catapults were used as weapons of war, and there were trumpet playing competitions in Greece. What was Eudoxus of Cnidus doing during this time?
9. In 1734, the first official horse race took place in America and George Sale translated the *Koran* into English. What was Bishop Berkeley publishing during this year?
10. While the Persian Wars (c. 460 B.C.) were being fought, what elements of calculus were the Greeks studying?
11. What were Cauchy and Weierstrass working on during the 1800s?
12. The end of a century saw the Great Plague hit London, the completion of the cathedral in Mexico, and the work of Newton and Leibniz relating to calculus. What century was this?
13. Match each of the following names with the correct idea, event, or description.

| | |
|---|---|
| (a) Archimedes | Co-inventor of calculus (*use twice*) |
| (b) Barrow | Indivisibles |
| (c) Cauchy | Logical foundations of calculus |
| (d) Cavalieri | Method of exhaustion |
| (e) Eudoxus | Derivation of many formulas using method of exhaustion |
| (f) Kepler | |

(g) Leibniz    Teacher of Newton
(h) Newton    Three laws of planetary motion

## Research Questions

To answer the following questions, you will need to refer to material not contained in the text. Possible sources of information are listed in the Bibliography at the end of this book.

1. Interview someone who has taken a calculus course or teaches a calculus course. Find out what kinds of problems can be solved with the knowledge of calculus. Find out what difficulties are inherent in learning and teaching calculus.
2. Investigate some of the inventions and discoveries of Archimedes.
3. Investigate the development of calculus in Japan during the 17th and 18th centuries.
4. Aside from calculus, what other inventions or discoveries are credited to Sir Isaac Newton?
5. Many of the mathematicians mentioned in this section had other interests besides mathematics. Describe some of these other interests. What anecdotes have been recorded about these mathematicians?
6. Galileo Galilei is called the father of modern science. Why is he given that title? What did he discover about the trajectory of horizontally fired cannonballs?
7. It was mentioned in the text that Archimedes was able to determine the volume and areas of certain geometrical objects. Investigate the following questions. What is an ellipsoid? What is a hyperboloid? What is a paraboloid? What is a spherical segment? How are volumes and surface areas of these objects determined?
8. One of Kepler's laws of planetary motion was described in this section. What are the other two laws of planetary motion according to Kepler? Make sketches illustrating these laws.
9. Calculus is used in many fields of engineering. Interview an engineer and find out why engineers study calculus and how calculus is used in engineering.
10. Some people associated with modern mathematics are George Pólya, Constance Reid, Garrett Birkhoff, Shiing-shen Chern, Persi Diaconis, Palageya Polubainava-Kochina, Paul Erdos, Raymond Smullyan, Henry Pollak, Olga Taussky-Todd, Martin Gardner, David Blackwell, Irmgard Flügge-Lotz, H.M.S. Coxeter, Paul Halmos, Morris Kline, Benoit Mandelbrot, Brother Alfred Brousseau, F.S.C., Ernest Wilkins, Donald Knuth, Mina Rees, Srinavasa Ramanujan, Grace Chisholm Young, Emmy Noether, and Anthony Barcellos. Do some research on one of these mathematicians or another 20th-century mathematician. Besides investigating his or her contributions to mathematics, discuss the person's interests, personal life, and so on.

## Projects

1. Investigate the history of the conflict over the priority of the discovery of calculus between Sir Isaac Newton and Gottfried Leibniz. Present arguments on who should be given credit for the "discovery" of calculus. What are some basic differences in the notation used by each in their development of calculus.
2. In this section, we have traced the development of differential calculus from the point of view of the tangent line to a curve. Another idea critical to the development of calculus is that the velocity of a particle can be found by examining the derivative of the function giving the position of the particle at any given time. Research this aspect of the history of calculus.
3. The derivative in calculus is a useful tool in determining the graph of a third degree function, $f(x) = ax^3 + bx^2 + cx + d$, where $a$ $b$, $c$, and $d$ are integers. Show how the derivative can help determine the graph of such a functions. Give examples.
4. What is a brachistochrone? What is its significance in mathematics? Build a model of a brachistochrone and compare it to a straight line model.

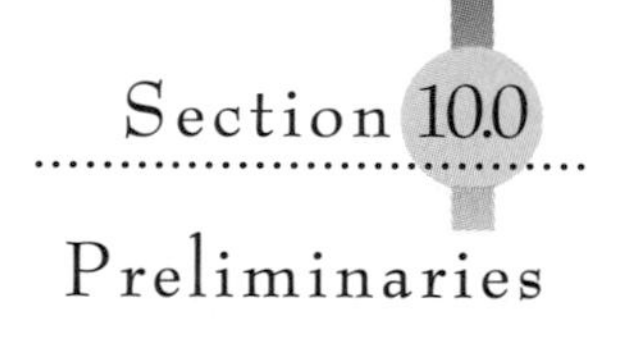


For our overview of calculus, we need an understanding of three concepts from algebra: area, slope, and function. We assume all students know that the area of a rectangle is length times width. We will briefly review the slope of a line. If you need a more detailed review of slopes, see Section 2.0. Most of this section contains a discussion of functions.

### Functions

A **function** is a set of ordered pairs $(x, y)$ such that for each value of $x$ there is exactly one value of $y$. In what are perhaps more familiar terms, for a function $y = f(x)$ (read "$f$ of $x$"), $x$ is the abscissa, $y$ is the ordinate, and $y$ is said to be a function of $x$.

**Example 1**

Determine if the following relationships are functions of $x$.

(a) $y = 3x + 4$
(b) $2x + 5y = 7$
(c) $x^2 + y^2 = 4$
(d) The situation where $x$ represents the date and $y$ represents the high temperature for the day in Minneapolis, Minnesota.

(e) The situation where $x$ represents the high temperature for the day in Minneapolis, Minnesota, and $y$ represents the date.

**Solution:**

(a) For $y = 3x + 4$, each value of $x$ will give us only one value of $y$. For example, if $x = 7$, then $y = 25$. Since each $x$ value has at most one $y$ value, $y = 3x + 4$ is a function.
(b) For $2x + 5y = 7$, for each value of $x$ we get only one $y$ value. For example, if $x = 3$, we find that $y = 1/5$. Therefore, $2x + 5y = 7$ is a function. *It is considered a function even though the equation is not written in the form* $y = \ldots$.
(c) $x^2 + y^2 = 4$ is not a function. This can be seen by picking a value for $x$, let's say $x = 0$. Substituting $x = 0$ into the equation gives $y^2 = 4$, which implies that $y$ can be either 2 or $-2$. Since one value of $x$ gives two values of $y$, $x^2 + y^2 = 4$ is not a function.
(d) The relationship in which $x$ represents the date and $y$ represents the high temperature for the day in Minneapolis, Minnesota, is a function because for any date, there can only be one high temperature.
(e) The relationship in which $x$ represents the high temperature for the day in Minneapolis, Minnesota, and $y$ represents the date is not a function. To see this, we choose a value of $x$, let's say 75°. Since there have been many days where 75° has been the high temperature in Minneapolis, the value $x = 75$ gives us many values of $y$. Therefore, this situation is not a function. ■

Since we are primarily concerned with mathematical functions, we now look at the notation that is used in mathematics, $y = f(x)$. In the previous example, we could write $y = 3x + 4$ as a function by writing

$$f(x) = 3x + 4$$

The only change is in the left side of the equation. Instead of $y$, functions use the notation $f(x)$. The notation $f(x)$ does not mean to multiply $f$ and $x$. It is indicating that if we assign a value to $x$, this value must be substituted for each occurrence of $x$ in the expression on the right side of the equal sign. We can see this in the following examples.

## Example 2

Find the value of $f(6)$, where $f(x) = 2x^2 - 5x + 8$.

**Solution:** To find $f(6)$, substitute 6 for each occurrence of $x$ in the equation. This gives

$$\begin{aligned} f(6) &= 2(6)^2 - 5(6) + 8 \\ &= 2(36) - 30 + 8 \\ &= 50 \end{aligned}$$ ■

### Example 3

Find the value of $f(6)$, where $f(x) = 18$.

**Solution:** To find $f(6)$, substitute 6 for each occurrence of $x$ in the equation. Since $x$ does not occur in the equation, this gives $f(6) = 18$. Note that since $x$ does not occur in the equation, it does not matter which value of $x$ we use. In terms of symbols, $f(6) = 18$, $f(3) = 18$, and $f(321) = 18$. Any value of $x$ will return the same value, 18. ■

In a function, it does not matter what we substitute for $x$. In mathematics, we are accustomed to substituting numbers for $x$. However, there is no reason why we cannot substitute any expression for $x$, including other variables or more complicated expressions.

### Example 4

For the function $f(x) = 2x^2 - 5x + 8$, find the following:

(a) $f(3)$
(b) $f(q)$
(c) $f(2 + h)$
(d) $f(x + h)$

**Solution:**

(a) As in Example 2, to find $f(3)$ substitute 3 for each occurrence of $x$. This gives $f(3) = 2(3)^2 - 5(3) + 8 = 11$.
(b) $f(q)$ means replace each occurrence of $x$ with $q$. Changing each occurrence of $x$ to $q$, we have $f(q) = 2(q)^2 - 5(q) + 8$.
(c) If we want to substitute more complicated expressions for $x$, we can. $f(2 + h)$ means replace each occurrence of $x$ with the expression $2 + h$. This gives

$$f(2 + h) = 2(2 + h)^2 - 5(2 + h) + 8$$

$$f(2 + h) = 2(4 + 4h + h^2) - 10 - 5h + 8$$

$$f(2 + h) = 8 + 8h + 2h^2 - 10 - 5h + 8$$

$$f(2 + h) = 6 + 3h + 2h^2$$

(d) Replacing $x$ with $x + h$ gives

$$f(x + h) = 2(x + h)^2 - 5(x + h) + 8$$

$$f(x + h) = 2(x^2 + 2xh + h^2) - 5x - 5h + 8$$

$$f(x + h) = 2x^2 + 4xh + 2h^2 - 5x - 5h + 8$$ ■

## The Slope of a Line

The last topic for this section is the slope of a line. As you may recall, the slope of a line measures the steepness of the line. The slope of a line through two points $(x_1, y_1)$ and $(x_2, y_2)$ is given by the expression

$$m = \frac{y_2 - y_1}{x_2 - x_1}$$

We can use this expression to find the slope of a line through two points.

**Example 5**

Find the slope of the line through the points (2, 6) and (−5, 8).

**Solution:** Using the slope formula, we find that the slope of the line through these two points is

$$m = \frac{8 - 6}{-5 - 2} = -\frac{2}{7}$$

This concludes the preliminaries for this chapter. In the rest of the chapter we use the slope and function concepts to investigate some of the intuitive ideas behind calculus.

Section 10.0 PROBLEMS

## *Explain ➡ Apply*

### *Explain*

1. What is a function?
2. What does the slope of a line measure?
3. What is the formula for the slope of a line through the points $(x_1, y_1)$ and $(x_2, y_2)$?
4. Describe in your own words how to find the slope of a line.
5. The notation $f(x)$ does not mean to multiply $f$ and $x$. What does it actually mean?
6. If $f(x) = 2x + 6$, what does $f(q)$ tell you to do?
7. If $g(x) = x^3 - 4$, what does $g(x + h)$ tell you to do?
8. Explain why $x^2 + y^2 = 25$ is not a function.
9. Explain why $x^2 + y = 25$ is a function.

**10.** If you wanted to create a function $y = f(x)$ that gives a correspondence between student Social Security Numbers and scores earned on a particular test, should you use $x$ or $y$ for the score? Explain.

**11.** If you wanted to create a function $y = f(x)$ that gives a correspondence between the number of points scored by players on a basketball team and their uniform numbers, should you used $x$ or $y$ for the uniform numbers? Explain.

## *Apply*

Find the value of the following functions at the specified point or expression.

$f(x) = 8$ $\quad g(x) = 3x - 6$ $\quad k(x) = x^2 + 7x$

**12.** (a) $f(4)$ (b) $g(4)$ (c) $k(4)$

**13.** (a) $f(-3)$ (b) $g(-3)$ (c) $k(-3)$

**14.** (a) $f(Q)$ (b) $g(Q)$ (c) $k(Q)$

**15.** (a) $f(M)$ (b) $g(M)$ (c) $k(M)$

**16.** (a) $f(3 + h)$ (b) $g(3 + h)$ (c) $k(3 + h)$

**17.** (a) $f(1 + h)$ (b) $g(1 + h)$ (c) $k(1 + h)$

**18.** (a) $f(x + h)$ (b) $g(x + h)$ (c) $k(x + h)$

**19.** (a) $f(x - h)$ (b) $g(x - h)$ (c) $k(x - h)$

Determine if the following are functions of $x$.

**20.** $y = 3x - 4$

**21.** $y = 3x + 9$

**22.** $3x + 3y = 7$

**23.** $5x - 4y = 9$

**24.** $y = 2x^2$

**25.** $y = 3x^2$

**26.** $x = 3y^2$

**27.** $x = 2y^2$

**28.** $x^2 + y^3 = 1$

**29.** $x^3 + y^2 = 1$

**30.** The line $y = 6$

**31.** The line $y = 7$

**32.** The line $x = 8$

**33.** The line $x = 7$

Find the slope of the line through the following pairs of points.

**34.** (a) $(-2, 3)$ and $(-4, 13)$
(b) $(2, 7)$ and $(-4, 5)$

**35.** (a) $(-3, 2)$ and $(-1, 6)$
(b) $(1, 7)$ and $(-5, 9)$

**36.** Determine if the following are functions of $x$.

(a) A situation where $x$ represents the time of sunset in Detroit and $y$ represents the date.

(b) A situation where $x$ represents the date and $y$ represents the time of sunset in Detroit.

**37.** Determine if the following are functions of $x$.

(a) A situation where $x$ represents a manufacturer's suggested retail price and $y$ represents the name of the item being sold.

(b) A situation where $x$ represents the name of the item being sold and $y$ represents the suggested retail price.

**38.** Determine if the following are functions of $x$.

(a) A correspondence where $x$ represents an employee's year of birth and $y$ represents the number of years a person has been employed by a company.

(b) A correspondence where $x$ represents the number of years a person has been employed by a company and $y$ represents the employee's year of birth.

**39.** For the function $y = f(x)$, let $(x, f(x))$ and $((x + h), f(x + h))$ be two points. Find an expression for the slope of the line through these two points.

**40.** For $f(x) = 6x + 1$, find and simplify each of the following.

(a) $f(2)$

(b) $f(2 + h)$

(c) $f(2 + h) - f(2)$

(d) $\dfrac{f(2 + h) - f(2)}{h}$

**41.** For $f(x) = 3x^2$, find and simplify each of the following.

(a) $f(2)$

(b) $f(2 + h)$

(c) $f(2 + h) - f(2)$

(d) $\dfrac{f(2 + h) - f(2)}{h}$

**42.** For $f(x) = 5x - 2$, find and simplify each of the following.

(a) $f(x + h)$

(b) $f(x + h) - f(x)$

(c) $\dfrac{f(x + h) - f(x)}{h}$

**43.** For $f(x) = x^2$, find and simplify each of the following.

(a) $f(x + h)$

(b) $f(x + h) - f(x)$

(c) $\dfrac{f(x + h) - f(x)}{h}$

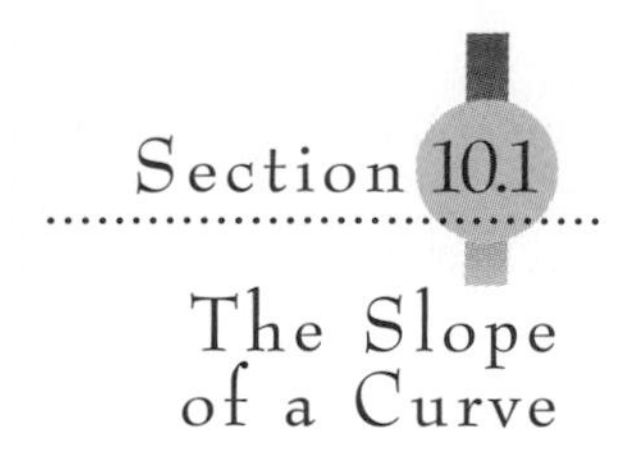


## The Slope of a Curve

In algebra, some of the many topics studied are slopes of lines and the maximum or minimum values of parabolas. The graphs of many different functions, such as semicircles and exponential functions, are also studied. In this section, we discuss how to find the slope of a curve. In Section 10.2, we will look at how we can use the slope of any curve to determine its maximum or minimum points.

Let's look at the graph of the equation $y = 3x^2 + 1$ and draw a line that intersects the graph in two points: at point $P$, where $x = 0$ and at point $Q$, where $x = h$. **The letter $h$ is used to represent any value except zero**. Now, find the slope of the line through points $P$ and $Q$.

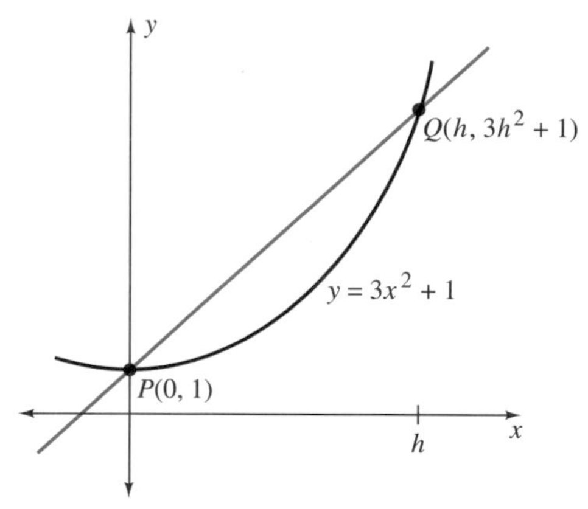


Since the slope of a line is given by $m = \dfrac{y_2 - y_1}{x_2 - x_1}$, to find the slope of the line segment $PQ$ first find the $y$ coordinates of the points $P$ and $Q$. This can be done by substituting the $x$ values into the equation $y = 3x^2 + 1$. For $x = 0$, $y = 1$, and $x = h$ gives $y = 3h^2 + 1$.

Using the points (0, 1) and $(h, 3h^2 + 1)$, we can find the slope of the line through the points $P$ and $Q$.

$$m = \frac{y_2 - y_1}{x_2 - x_1} = \frac{3h^2 + 1 - 1}{h - 0} = \frac{3h^2}{h} = 3h \qquad (\text{provided } h \neq 0)$$

We now know that the slope of the line through $P$ and $Q$ is given by $m = 3h$. This means that the slope of the line will depend on the value of $h$.

Examine the following series of diagrams that show point $Q$ moving along the curve toward $P$. What happens to the slope of the line $PQ$? If you determined that the slope was getting close to zero as point $Q$ moved closer to point $P$, you were correct. As $h$ takes on smaller values, the distance between the points decreases and the slope of the line segment $PQ$ approaches 0. This can also be seen by looking again at the slope formula and letting $h$ get close to 0. Since $m = 3h$, if $h$ approaches zero, then the slope of the line $PQ$ approaches zero. In addition, as $h$ gets close to 0, the line will intersect the graph at two points, $P$ and $Q$, that are very close together.

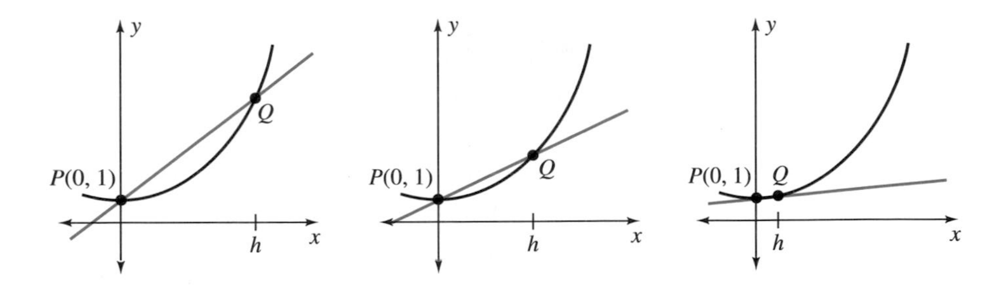


When, in a region very near to point $P$, the line intersects the curve only at the point $P$, the line is called the **tangent line to the curve** at the point $P$.*

The following pictures give three examples of tangent lines.

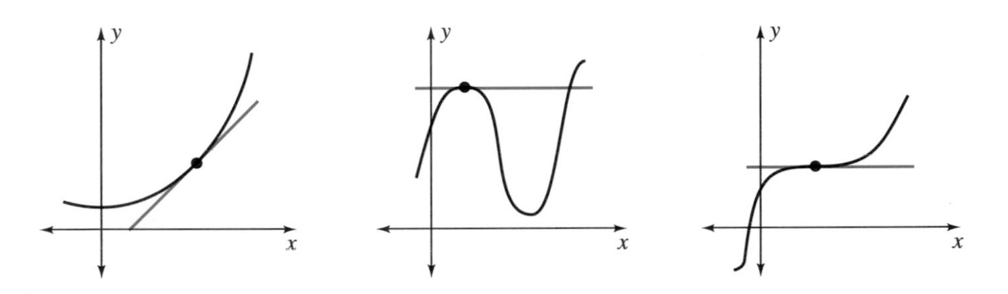


The following pictures give examples of lines that are not tangent to the curve.

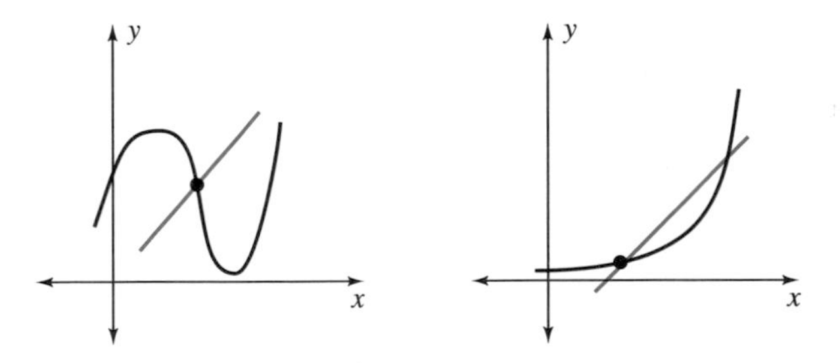


We are discussing the slope of the tangent line because the **slope of the curve at point *P*** is defined as the slope of the tangent line at point $P$. The slope of the curve is such an important topic that mathematicians have a specific name for it. The slope of a curve, or function, at a point $P$ is called the **derivative** of the function at the point $P$, and the process for finding a derivative is called **differentiation**.

---

*This statement is true only for functions whose graphs do not include any sharp points or jumps in the graph. Functions whose graphs contain sharp points or jumps will not be discussed here.

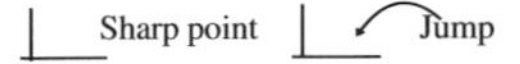

**Example 1**

Find the slope of the tangent line to the curve $y = x^2 + 4x$ at the point (2, 12).

**Solution:** We start by sketching a graph of the function. As we did before, to find the slope we use two points $P$ and $Q$. Point $P$ is the point (2, 12). We choose point $Q$ a small distance from point $P$, so we let the $x$ coordinate of $Q$ as $2 + h$. To find the $y$ coordinate of $Q$, substitute $2 + h$ for $x$ in the function $y = x^2 + 4x$. This gives

$$\begin{aligned} y &= (2 + h)^2 + 4(2 + h) \\ &= 4 + 4h + h^2 + 8 + 4h \\ &= 12 + 8h + h^2 \end{aligned}$$

Thus, point $Q$ is given by $(2 + h, 12 + 8h + h^2)$.

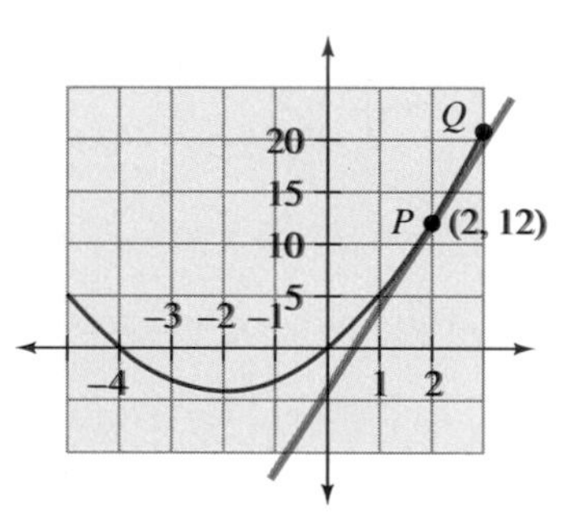


We can now use points $P$ (2, 12) and $Q(2 + h, 12 + 8h + h^2)$ to find the slope of the line through points $P$ and $Q$.

$$m = \frac{(12 + 8h + h^2) - 12}{(2 + h) - 2} = \frac{8h + h^2}{h} = 8 + h \qquad (\text{provided } h \neq 0)$$

To find the slope of the tangent line to $y = x^2 + 4x$ at $x = 2$, let the points $P$ and $Q$ get close together. In other words, let $h$ get close to 0. As $h$ takes on a value close to 0, the formula $m = 8 + h$ becomes $m = 8$. This implies that, at $x = 2$, the slope of the tangent line to the curve is 8. ■

In the two previous problems, we found the slope of the tangent line at a particular point. For instance, we found at the point (2, 12) that the slope of the tangent line to $y = x^2 + 4x$ is 8. We also stated that the slope of the tangent line at a point is the same as the slope of the curve at that point. However, as we can see from the diagram, the slope of the curve changes, depending on the point being examined.

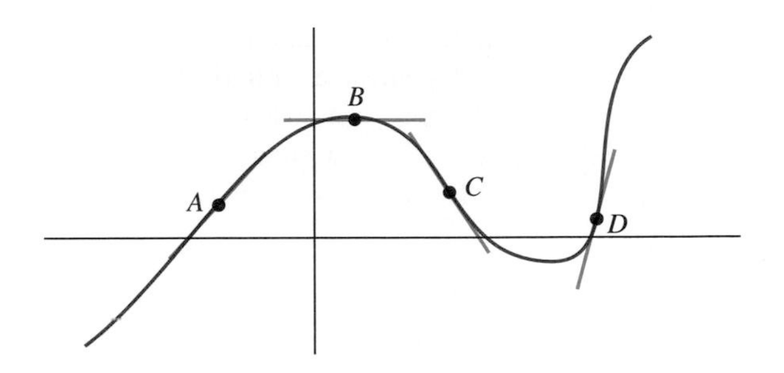


At points $A$ and $D$, the slope of the tangent line is positive, with the curve being steeper at point $D$ than at $A$. At point $C$, the slope is negative, while the slope of the tangent line is approximately zero at point $B$. Since a **slope** represents the steepness of a line and since the steepness of a curve depends on the point being considered, the slope of a curve is not a constant. Instead, the slope of a curve is changing and can be found by a formula.

In the previous example, we chose two points to find the slope. One $x$ value, $x = 2$, was given, and the other value was chosen to be $x = 2 + h$. Suppose we are given an arbitrary function, $y = f(x)$, and want to find the slope of the curve at some point $(x, f(x))$. The second point will be chosen as $(x + h, f(x + h))$. Now that the points are chosen, use the points to calculate the slope.

$$m = \frac{y_2 - y_1}{x_2 - x_1} = \frac{f(x + h) - f(x)}{(x + h) - x} = \frac{f(x + h) - f(x)}{h}$$

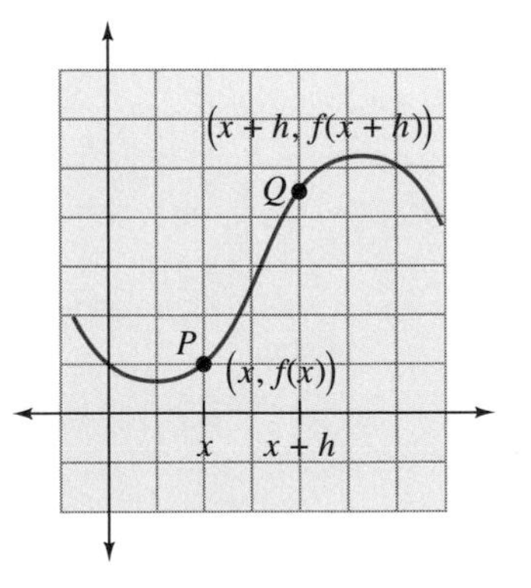


After finding the slope, slide point $Q$ toward point $P$ along the curve. This is done by letting $h$ approach 0. The notation used to indicate this is

$\lim\limits_{h \to 0}$ which is read "limit as $h$ gets close to 0."

Though this notation looks peculiar, it simply means that the value of $h$ becomes smaller and smaller until it is very close (but never equal) to zero. This has the effect of keeping the first point stationary while sliding the second point on the curve toward the first point.

This discussion leads to the definition of the derivative of a function $f(x)$.

## The Derivative

The derivative of a function $y = f(x)$ is

$$\frac{dy}{dx} = \lim_{h \to 0} \frac{f(x + h) - f(x)}{h}$$

### Example 2

Find the derivative of the function $y = x^2 + 4x$.

**Solution:** Notice that this is the same function that was discussed in Example 1. In Example 1, we determined the derivative (slope of the tangent line) when $x = 2$. Now, we find the value of the derivative at any point. Therefore, we will repeat the work done in Example 1 using the variable $x$ rather than the number 2.

The first point is $(x, f(x))$. For the second point, the $x$ coordinate is $x + h$ and the $y$ coordinate is $f(x + h)$, where

$$\begin{aligned} f(x + h) &= (x + h)^2 + 4(x + h) \\ &= x^2 + 2xh + h^2 + 4x + 4h \end{aligned}$$

This means that the derivative of $y = x^2 + 4x$ is

$$\begin{aligned} \frac{dy}{dx} &= \lim_{h \to 0} \frac{f(x + h) - f(x)}{h} \\ &= \lim_{h \to 0} \frac{x^2 + 2xh + h^2 + 4x + 4h - (x^2 + 4x)}{h} \\ &= \lim_{h \to 0} \frac{2xh + 4h + h^2}{h} \\ &= \lim_{h \to 0} (2x + 4 + h) \\ &= 2x + 4 \end{aligned}$$ ■

Notice that this gives a formula for the slope of the curve. The equation

$$\frac{dy}{dx} = 2x + 4$$

means that the slope of the curve at any value of $x$ can be determined by substituting the value of $x$ into the equation. For instance, in Example 1, we found that at the point (2, 12), the slope of the curve is 8. We can verify this by substituting $x = 2$ into the equation $dy/dx = 2x + 4$. As expected, this gives the slope as $2(2) + 4 = 8$. Similarly, at the point (1, 5), the slope of the curve is $2(1) + 4 = 6$.

## Example 3

Find the $x$ value of the point on the curve $y = x^2 + 4x$ where the slope is 0.

**Solution:** From Example 2, we know that the slope of the curve at any point is $dy/dx = 2x + 4$. Since we want to find the point where the slope of the curve is 0, set the equation for the slope equal to 0 and solve for $x$.

$$2x + 4 = 0$$

$$2x = -4$$

$$x = -2$$

This means that, at $x = -2$, the slope of the curve, or the slope of the tangent line to the curve, is zero. To visualize this, we can sketch the graph of $y = x^2 + 4x$ by plotting a few points. Notice that $x = -2$ is the vertex of the parabola, and the tangent line is horizontal.

| $x$ | $y$ |
|---|---|
| $-4$ | 0 |
| $-3$ | $-3$ |
| $-2$ | $-4$ |
| $-1$ | $-3$ |
| 0 | 0 |
| 1 | 5 |

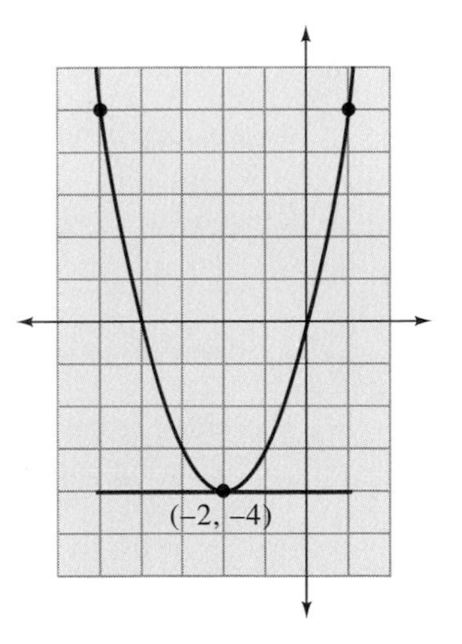


In this section, we discussed one of the two main topics in calculus, the derivative. The derivative gives the slope of a curve at any point. In particular, we found how we can determine the slope of a curve and how we can determine the point(s)

where the derivative is equal to zero. The next section shows how the slope of the curve can help us sketch the graph of a function and how the derivative can be applied to problems involving maximum and minimum points. The next section also gives a method for finding derivatives without resorting to the algebraic process that was used here.

## Section 10.1 PROBLEMS

## Explain ➡ Apply ➡ Explore

### Explain

1. What is a tangent line to a curve?
2. What is meant by the slope of the curve at a point?
3. What does the derivative of a function at a point give?
4. What is differentiation?
5. What does $\lim_{h \to 0}$ mean?
6. What is $\frac{dy}{dx}$ and what formula is used to calculate it?
7. Draw a curve where $\frac{dy}{dx}$ is positive at $x = 3$. Draw the tangent line to the point on the curve where $x = 3$.
8. Draw a curve where $\frac{dy}{dx}$ is negative at $x = 3$. Draw the tangent line to the point on the curve where $x = 3$.
9. Draw a curve where $\frac{dy}{dx} = 0$ at $x = 3$. Draw the tangent line to the point on the curve where $x = 3$.

### Apply

10. The graph represents the function $y = f(x)$. Determine which point(s) satisfy the following conditions.

(a) $\frac{dy}{dx} = 0$

(b) $\frac{dy}{dx}$ is positive

(c) $\frac{dy}{dx}$ is negative

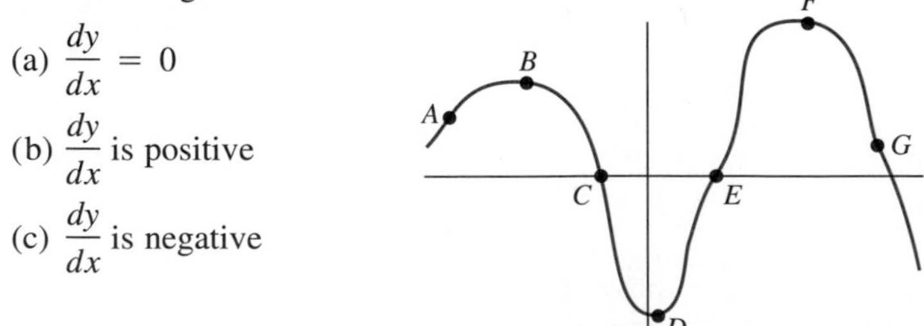

**11.** The graph represents the function $y = f(x)$. Determine which point(s) satisfy the following conditions.

(a) $\dfrac{dy}{dx} = 0$

(b) $\dfrac{dy}{dx}$ is positive

(c) $\dfrac{dy}{dx}$ is negative

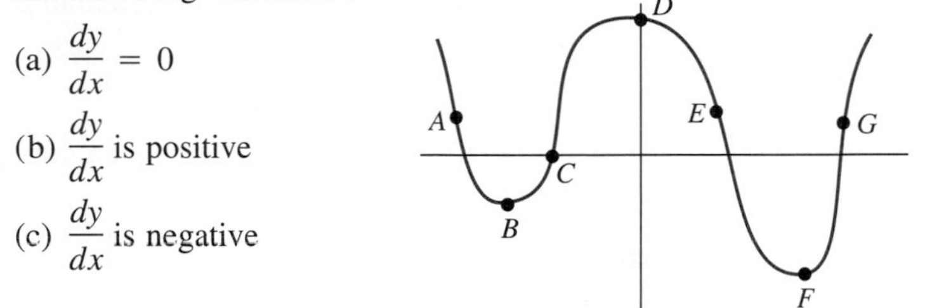


**12.** On the graph, the value of the derivative at $x = 1$ is 0.5, and the derivative at $x = 4$ is 2. What can you say about the value of the derivative at $x = 2$?

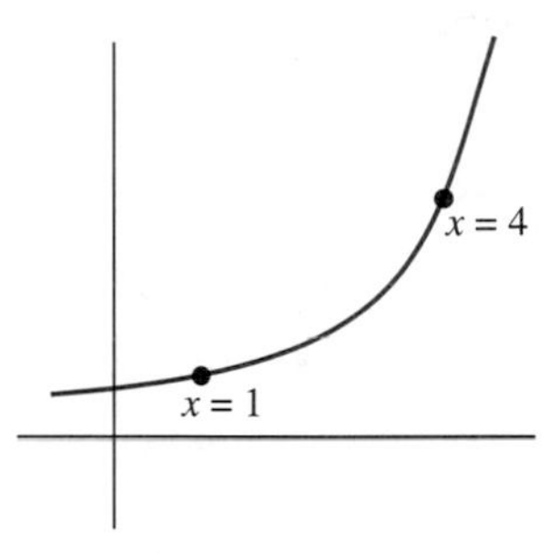


**13.** On the graph, the value of the derivative at $x = 1$ is $-3$, and the value of the derivative at $x = 4$ is $-1$. What can you say about the value of the derivative at $x = 3$?

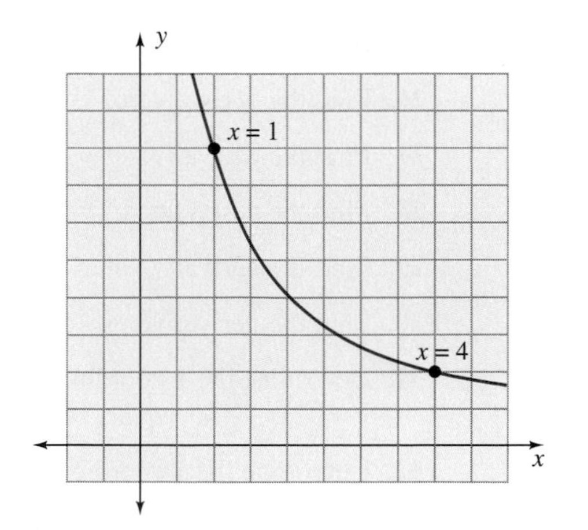

**14.** For the line $y = 6x + 2$:

(a) Find the slope of the line, by examining the equation.
(b) Find the derivative.

**15.** For the line $y = 5x - 1$:

(a) Find the slope of the line, by examining the equation.
(b) Find the derivative.

**16.** For the line $y = \frac{2}{3}x - 7$:

(a) Find the slope of the line, by examining the equation.
(b) Find the derivative.

**17.** For the line $y = -\frac{4}{3}x + 2$:

(a) Find the slope of the line, by examining the equation.
(b) Find the derivative.

**18.** Find the derivative of $y = -x^2$ at the point $(3, -9)$.

**19.** Find the derivative of $y = -x^2 + 4$ at the point $(0, 4)$.

**20.** Find the derivative of $y = x^2 - 4x - 3$ at the point $(2, -7)$

**21.** Find the derivative of $y = 2x^2 - 6x$ at the point $(1, -4)$.

**22.** Find the derivative of $y = 3x^2 - x$ at the point $(1, 2)$.

**23.** Find the derivative of $y = 3x^2 + 2$ at the point $(1, 5)$.

*Explore*

**24.** Find the derivative of $y = 5x^2$ at any point on the curve.

**25.** Find the derivative of $y = 2x^2$ at any point on the curve.

**26.** Find the derivative of $y = -x^2 + 15$ at any point on the curve.

**27.** Find the derivative of $y = x^2 - 4$ at any point on the curve.

**28.** Find the derivative of $y = -3x^2 + 4$ at any point on the curve.

**29.** Find the derivative of $y = -2x^2 + 3x$ at any point on the curve.

**30.** Find the derivative of $y = x^2 - 4x + 8$ at any point on the curve.

**31.** Determine the $x$ coordinate of the point where the derivative of $y = 2x^2 - 4x$ equals zero.

**32.** Determine the $x$ coordinate of the point where the derivative of $y = 5x^2 + 20x$ equals zero.

**33.** Determine the $x$ coordinate of the point where the derivative of $y = x^2 - 4x + 7$ equals zero.

**34.** Determine the $x$ coordinate of the point where the derivative of $y = x^2 + 6x - 5$ equals zero.

**35.** Determine the $x$ coordinate of the point where the derivative of $y = 5x^2 + 60x - 7$ equals zero.

**36.** Determine the $x$ coordinate of the point where the derivative of $y = 12 - 3x^2$ equals zero.

***37.** Find the derivative of each of the following:

(a) $y = x^2$
(b) $y = x^3$
(c) $y = x^4$
(d) Based on the answers to parts (a), (b), and (c), what do you think the derivative of $y = x^{23}$ is?

*Restaurant by the Sea* by John Register shows a young man in deep contemplation, the kind of contemplation that is commonly engaged in by calculus students. (Courtesy of Magnolia Editions)

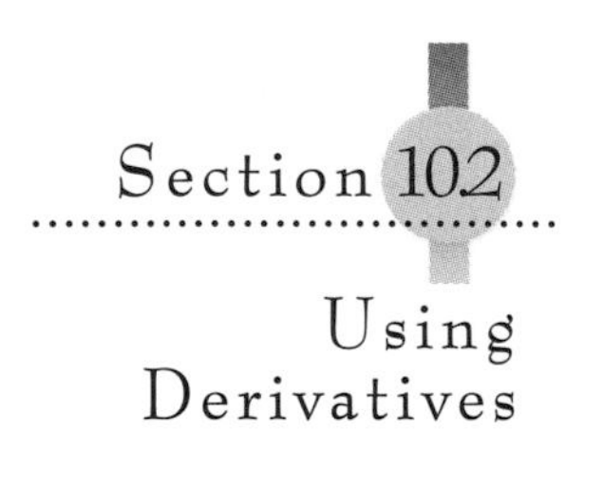

## Section 102 Using Derivatives

In the previous section, we found derivatives of linear and quadratic functions. However, the algebraic process was tedious. Therefore, before discussing how derivatives are used, we will want to reexamine how to find a derivative.

### Shortcuts for Finding Derivatives

If you worked on Problem 37 in the previous section, you may have guessed how certain differentiations can be performed more quickly. Problem 37 asked you to compute the derivatives of three different functions and to guess the derivative of a fourth function. The results were:

| **Function** | **Derivative** |
|---|---|
| $f(x) = x^2$ | $\frac{dy}{dx} = 2x$ |
| $f(x) = x^3$ | $\frac{dy}{dx} = 3x^2$ |
| $f(x) = x^4$ | $\frac{dy}{dx} = 4x^3$ |
| $f(x) = x^{23}$ | $\frac{dy}{dx} = 23x^{22}$ |

You were expected to grind out the first three results, whereas the fourth result was to be obtained using inductive reasoning (guessing). Notice that each of the first three functions has a derivative with a coefficient that is the same as the exponent in the original function. The derivative has an exponent that is 1 less than the original exponent. If you noticed this pattern, you probably were able to guess correctly the derivative of $f(x) = x^{23}$. If we continue this inductive reasoning, we arrive at the following formula:

**Power Rule for Derivatives**

If $f(x) = x^n$, then the derivative of $f(x)$ is $\frac{dy}{dx} = nx^{n-1}$.

Although we have not proven it, the power rule is true for any constant exponent. The exponent can be a fraction, a negative number, or even an irrational number.

### Example 1

Use the power rule to find the derivatives of the following functions:

(a) $f(x) = x$
(b) $f(x) = x^{-3}$
(c) $f(x) = x^{1/2}$

**Solution:**

(a) For $f(x) = x = x^1$,

$$\frac{dy}{dx} = 1x^0 = 1$$

(b) For $f(x) = x^{-3}$,

$$\frac{dy}{dx} = -3x^{-3-1} = -3x^{-4}$$

(c) For $f(x) = x^{1/2}$,

$$\frac{dy}{dx} = \frac{1}{2}x^{1/2-1} = \frac{1}{2}x^{-1/2}$$ ■

### Example 2

Find the derivative of the line $y = 5$.

**Solution:** The line $y = 5$ does not contain any $x$'s, but we can find the derivative by realizing that the line $y = 5$ is a horizontal line and that the slope of a horizontal line is 0. Therefore, since finding the slope is equivalent to finding the derivative, the derivative of $y = 5$ must equal 0. ■

**(1)** If $f(x) = k$, where $k$ is any constant, $\frac{dy}{dx} = 0$.

Since this is true for every horizontal line, we can generalize this with the following formula:

Each of the problems in Example 1 consisted of a single term with a coefficient of 1. To determine how to find the derivative of a function that contains more than one term or a function whose coefficient is not 1, we will look at the derivatives that were completed in the examples and odd-numbered problems from the previous section.

| Original Problem | Functions | Derivatives |
|---|---|---|
| Section 1, Example 2 | $y = x^2 + 4x$ | $\frac{dy}{dx} = 2x + 4$ |
| Section 1, Problem 24 | $y = 5x^2$ | $\frac{dy}{dx} = 10x$ |
| Section 1, Problem 31 | $y = 2x^2 - 4x$ | $\frac{dy}{dx} = 4x - 4$ |

In the first of these, the derivative of $x^2$ is, as expected, $2x$. The derivative of $4x$ is 4. Since Example 1 of this section showed the derivative of $x$ is 1, it appears that the derivative of $4x$ is merely 4 times the derivative of $x$. Since Problem 24 showed that the derivative of $5x^2$ is $10x$, and power rule states that the derivative of $x^2$ is $2x$, it appears that the derivative of $5x^2$ can be calculated as $5 \times 2x = 10x$. This pattern repeats itself in the third example and leads to the following two rules for derivatives.

**(2)** If $f(x) = kx^n$, then $\frac{dy}{dx} = knx^{n-1}$.

**(3)** The derivative of $f(x) \pm g(x)$ equals the sum (difference) of the derivatives of $f(x)$ and $g(x)$.

**Example 3**

Find the derivatives of the following functions.

(a) $f(x) = 8x^5$
(b) $f(x) = 4x^3 - 6x^2 + 5x + 1$

**Solution:**

(a) For $f(x) = 8x^5$, we use rule 2 to get $dy/dx = 8 \cdot 5x^4 = 40x^4$.
(b) For $f(x) = 4x^3 - 6x^2 + 5x + 1$, we use rule 2 for the derivatives of the first three terms and rule 1 for the derivative of the fourth term. From rule 2, the derivative of $4x^3$ is $4 \cdot 3x^2 = 12x^2$, the derivative of $6x^2$ is $6 \cdot 2x^1 = 12x$, and the derivative of $5x$ is 5. By rule 1, the derivative of 1 is 0. Therefore, by rule 3, if $f(x) = 4x^3 - 6x^2 + 5x + 1$,

$$\frac{dy}{dx} = 12x^2 - 12x + 5$$

In summary, to find the derivatives of polynomial functions, we no longer need to use the algebraic process of the previous section. Instead, we can find derivatives of polynomials by using the following three rules.

**(1)** If $f(x) = k$, where $k$ is any constant, $\frac{dy}{dx} = 0$.

**(2)** If $f(x) = kx^n$, then $\frac{dy}{dx} = knx^{n-1}$.

**(3)** The derivative of $f(x) \pm g(x)$ equals the sum (difference) of the derivatives of $f(x)$ and $g(x)$.

## Applications of the Derivative

The first application of the derivative that we will consider is how the derivative can aid in the graphing of functions. In algebra, the graphing of parabolas is facilitated by finding the location of the vertex of the parabola. The vertex gives the location of the maximum or minimum point of the parabola. We now discuss how to use derivatives to determine the maximum or minimum points of any curve.

As can be seen in the diagram on page 696, a curve may have several maximum and minimum points. If we use the common English usage of the word maximum, there can only be one maximum or highest point for a graph. In mathematics, this is called an absolute maximum.

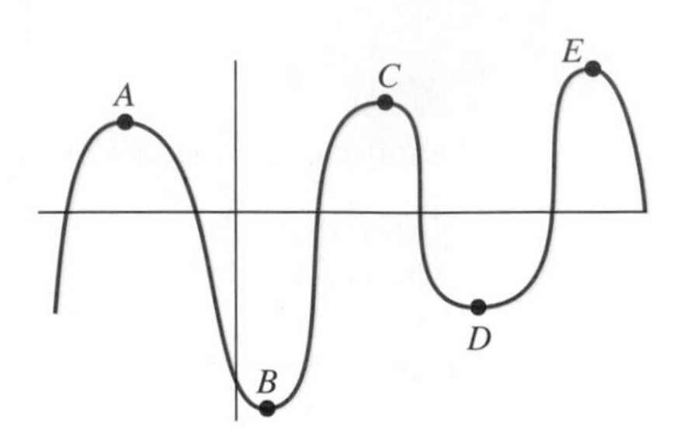


To indicate that we are concerned with all the points that are at the top of a curve, the term **relative maximum** is used. A relative maximum is a point that is higher than all the points very close by. Similarly, a **relative minimum** is a point that is lower than all the points very close by. The graph shows three relative maximums, located at points $A$, $C$, and $E$. There are also two relative minimums, located at $B$ and $D$.

If we redraw the diagram, including the tangent lines at the relative maximum and relative minimum points, we will be able to see the role that the derivative plays in determining relative maximum and relative minimum points. Notice that each of

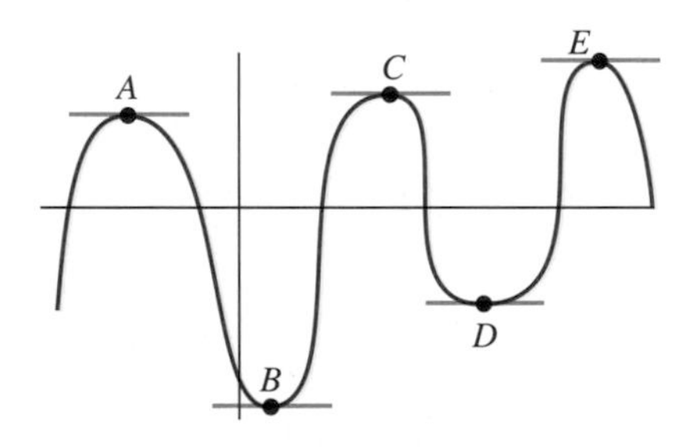


the five tangent lines in the drawing is horizontal. Since the slope of a horizontal line is zero, **we can determine the relative maximum and relative minimum points of a curve by setting the derivative equal to zero**.

## Example 4

Find the relative maximum and minimum points of the curve $y = 6x^3 - 8x + 1$.

**Solution:** To find the relative maximum and minimum points, we first sketch the graph for integer values of $x$. From the graph, we can see that there is a relative maximum point near $x = -1$ and a relative minimum point near $x = 1$. However, we cannot tell from the graph exactly where the relative maximum and relative minimum points are located.

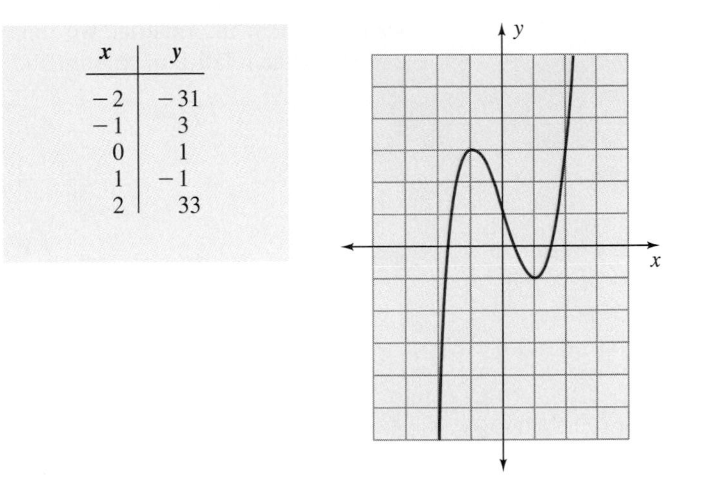

| $x$ | $y$ |
|---|---|
| −2 | −31 |
| −1 | 3 |
| 0 | 1 |
| 1 | −1 |
| 2 | 33 |



The relative maximum and relative minimum points occur where the slope of the tangent line is equal to zero. Therefore, we find the locations of the relative maximum and relative minimum points by setting the derivative equal to 0 and comparing the resulting values of $x$ with the information from the graph.

From our rules for derivatives, we know that if $y = 6x^3 - 8x + 1$, $dy/dx = 18x^2 - 8$. We determine the location of the relative maximum and minimum points by setting the derivative equal to 0.

$$18x^2 - 8 = 0$$

$$18x^2 = 8$$

$$x^2 = 4/9$$

$$x = \pm 2/3$$

Since we have two solutions for $x$, we look at the graph to determine which point is the maximum and which is the minimum. From the graph we find that the relative maximum point was located near $x = -1$. Therefore, we know that the curve reaches its relative maximum at $x = -2/3$.

To determine the $y$ value at the relative maximum point, we substitute the value $x = -2/3$ into the equation $y = 6x^3 - 8x + 1$. This gives

$$y = 6\left(\frac{-2}{3}\right)^3 - 8\left(\frac{-2}{3}\right) + 1 = \frac{41}{9} \approx 4.56$$

In the same way, we find that the relative minimum occurs at $x = 2/3$. This gives a $y$ value of

$$y = 6\left(\frac{2}{3}\right)^3 - 8\left(\frac{2}{3}\right) + 1 = \frac{-23}{9} \approx -2.56$$

Using this new information, we can re-draw the graph including the exact locations of the relative maximum and relative minimum points.

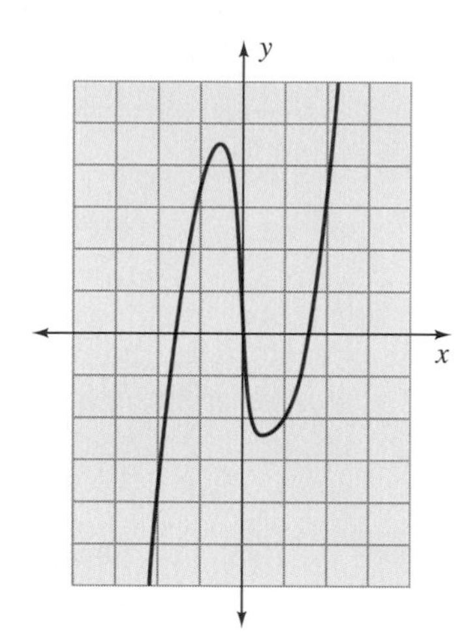


Calculus is a very powerful mathematical tool and has many applications. The standard calculus course requires two or three semesters to learn all its techniques and uses. Because of this, we cannot possibly study the many areas in which calculus plays a role. We do however, present some typical applications.

## Example 5

Find the dimensions of the rectangular region with maximum area that can be fenced with 120 ft of wire mesh if the fencing forms the perimeter of the region and one interior fence divides the area into two equal sections. (We know there will be a maximum since there is a given amount of available fencing.)

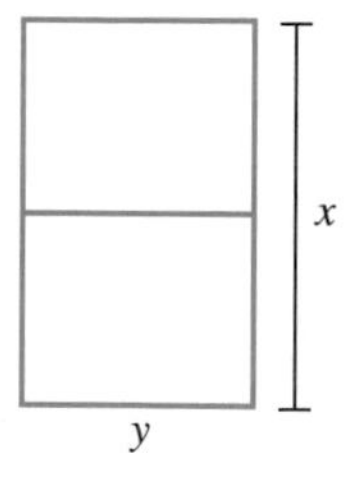


**Solution:** We start by drawing the rectangle and writing equations that model the situation. The total length of fencing is given by

$$2x + 3y = 120$$

We also know that the total area of the rectangle is

$$A = xy$$

Since we want to find the maximum area, we express the area as a function of one variable. Solving the first equation for $y$ we have

$$2x + 3y = 120$$
$$3y = 120 - 2x$$
$$y = \frac{120 - 2x}{3} = 40 - \frac{2}{3}x$$

Substituting $y$ into the equation for area gives

$$A = x\left(40 - \frac{2}{3}x\right) = 40x - \frac{2}{3}x^2$$

We now find the derivative and use it to determine the maximum value of the area.

$$\frac{dA}{dx} = 40 - \frac{2}{3}(2x) = 40 - \frac{4}{3}x$$

Setting the derivative equal to zero and solving for $x$ gives

$$40 - \frac{4}{3}x = 0$$
$$40 = \frac{4}{3}x$$
$$40\left(\frac{3}{4}\right) = x$$
$$30 = x$$

Since $y = 40 - (2/3)x$, substituting $x = 30$ gives $y = 20$. This means that the maximum area for the region fenced with the 120 ft of fencing is wire mesh is $30 \times 20 = 600$ sq ft. ■

## Example 6

A closed rectangular box is to be constructed from 2400 sq cm of cardboard. If the box has a square base of side $x$ and height $y$, find the values of $x$ and $y$ that form the box of maximum volume. (We know there will be a maximum since there is a given amount of available cardboard.)

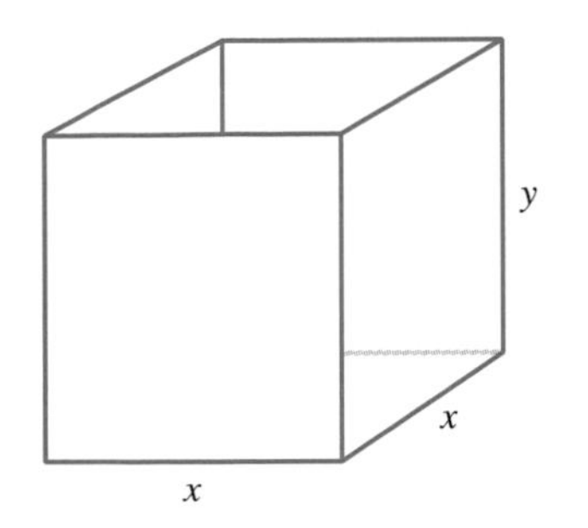

**Solution:** Since the box has a square base, the area of the top and the bottom of the box is given by $x^2$. The four sides of the box each have area $xy$, so the total surface area (S.A.) of the box is

$$\text{S.A.} = 2x^2 + 4xy$$

Since the total amount of cardboard available is 2400 sq cm, we have the equation

$$2400 = 2x^2 + 4xy$$

The problem asks us to find the maximum volume of the box. Since the volume of a box equals length × width × height, we have

$$V = x^2y$$

Since this expression has two variables, we use the surface area equation to solve for $y$ in terms of $x$, and substitute for $y$ in the volume equation.

$$2400 = 2x^2 + 4xy$$

$$2400 - 2x^2 = 4xy$$

$$\frac{2400 - 2x^2}{4x} = y$$

Substituting this value of $y$ into the expression for the volume gives

$$V = x^2y$$

$$V = x^2\left(\frac{2400 - 2x^2}{4x}\right)$$

$$V = 600x - \frac{x^3}{2}$$

To determine the maximum, find the derivative, set the derivative equal to 0, and solve for $x$.

$$\frac{dV}{dx} = 600 - \frac{3x^2}{2}$$

$$0 = 600 - \frac{3x^2}{2}$$

$$\frac{3x^2}{2} = 600$$

$$x^2 = 400$$

$$x = 20$$ (Since $x$ represents a dimension of the box, $x$ must be positive.)

Substituting $x = 20$ into $y = \frac{2400 - 2x^2}{4x}$ gives $y = 20$. Therefore, the box of maximum volume for a given surface area is a cube. In this case, the cube has dimensions $20 \times 20 \times 20$. ■

In summary, this section has introduced two topics. The first topic was a study of some of the shortcuts used to find a derivative. The second topic was how the derivative can be used to determine the relative maximum or relative minimum value of a function. The problem set that follows will give you some practice in working with both topics.

## Section 10.2 PROBLEMS

## *Explain* ➡ *Apply* ➡ *Explore*

### *Explain*

**1.** What is the power rule for finding dervatives?

**2.** What does it mean for $f(x)$ to be a constant function?

**3.** If $f(x)$ is a constant function, what is its derivative and what does that mean?

**4.** What is a relative minimum point for a curve?

**5.** What is a relative maximum point for a curve?

**6.** What steps should you take in order to determine the relative minimum points for a curve?

**7.** What steps should you take in order to determine the relative maximum points for a curve?

### *Apply*

Find the derivatives of the functions in Problems 8–22.

**8.** $f(x) = 2x + 5$

**9.** $f(x) = 2x - 7$

**10.** $f(x) = x^2 + 5x$

**11.** $y = -x^2 + 9$

**12.** $f(x) = x^3 - 6x$

**13.** $y = 3x^3 - 5x - 7$

**14.** $f(x) = -2x^3 + 8x^2$

**15.** $f(x) = x^4 + x^3 + x^2 + x + 1$

**16.** $f(x) = -4x^4 + 3x^2 - 16$

**17.** $y = 4x^5$

**18.** $y = 3x^8$

**19.** $y = 4x^5 - 5x^2$

**20.** $y = 2x^5 - 3x^2$

**21.** $f(x) = x^5 - 3x^2 + 7x + 5$

**22.** $f(x) = x^5 - 7x^2 + 3x + 4$

**23.** (a) Sketch the graph of the line $y = 3x + 7$. Does this graph have any relative maximum or relative minimum points?
(b) Find the derivative of $y = 3x + 7$. Can the derivative equal 0? If so, find the value(s) of $x$ where $dy/dx = 0$. If not, explain.

**24.** (a) Sketch the graph of $y = x^2 + 2x$. Does this graph have any relative maximum or relative minimum points?
(b) Find the derivative of $y = x^2 + 2x$. Find the point where $dy/dx = 0$.
(c) Does the value found in part (b) give a relative maximum or a relative minimum point?

**25.** (a) Sketch the graph of $y = -x^2 + 4x$. Does this graph have any relative maximum or relative minimum points?
(b) Find the derivative of $y = -x^2 + 4x$. Find the value(s) of $x$ where $dy/dx = 0$.
(c) Does the value found in part (b) give a relative maximum or a relative minimum point?

**26.** (a) Sketch the graph of $y = -x^2 + 6x$. Does this graph have any relative maximum or relative minimum points?
(b) Find the derivative of $y = -x^2 + 6x$. Find the value(s) of $x$ where $dy/dx = 0$.
(c) Does the value found in part (b) give a relative maximum or a relative minimum point?

**27.** (a) Plot points to sketch the graph of $y = 2x^3 - 6x$.
(b) Find the derivative of $y = 2x^3 - 6x$.
(c) Find the value(s) of $x$ where $dy/dx = 0$.
(d) Determine which value gives a relative maximum and which gives a relative minimum.

**28.** (a) Plot points to sketch the graph of $y = 2x^3 - 6x + 12$.
(b) Find the derivative of $y = 2x^3 - 6x + 12$.
(c) Find all the points where $dy/dx = 0$.
(d) Determine which value gives a relative maximum and which gives a relative minimum.

## *Explore*

**29.** The path traveled by a frog is given by $h = \frac{-1}{98}x^2 + \frac{6}{7}x$, where $x$ is the horizontal distance traveled and $h$ is the height in inches at any point along the path.
(a) Determine the $x$ value that gives the maximum height.
(b) Determine the maximum height.

**30.** A ball is thrown vertically in the air with an initial speed of 139.33 ft/s from an initial height of 9 ft. The equation describing the height of the ball at any time is $h = -16t^2 + 139.33t + 9$, where $t$ is the time in seconds and $h$ is the height in feet.

(a) Determine the time when the ball reaches its maximum height.
(b) Determine the maximum height.

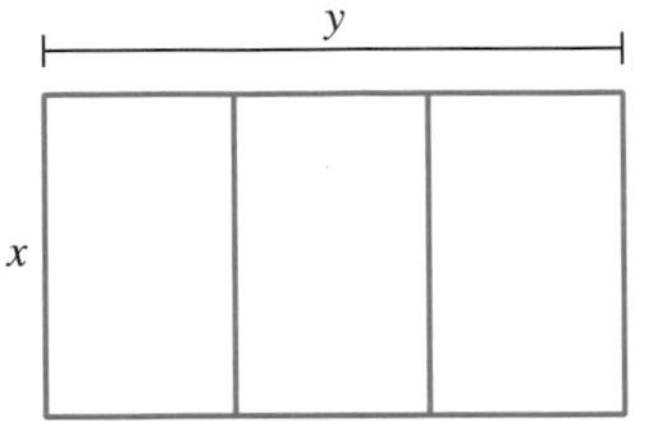


**31.** Use calculus to solve the following problem. A chicken farmer wants to build a fenced area for her free-roaming chickens. Because she wants to separate the different breeds, she plans to have three adjacent pens, as shown in the diagram. If 1200 ft of fencing is available, what is the largest total area that can be fenced off for the chickens?
(*Hint:* Area $= xy$ and total fencing $= 4x + 2y$.)

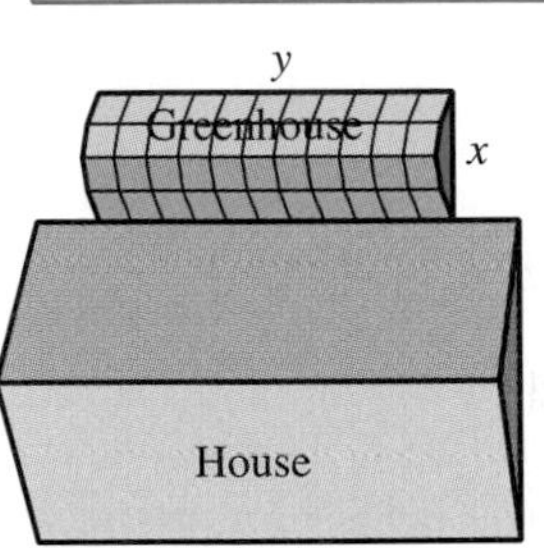


**32.** Use calculus to solve the following problem. An orchid fancier is building a fiberglass enclosure next to his house, as shown in the diagram. Since the house will be used for one side of the enclosure, only three sides will need to be enclosed by the fiberglass. The roof of the greenhouse is to be built of some other material. If 60 ft of the fiberglass walls are available, what is the maximum rectangular area that can be enclosed?
(*Hint:* Area $= xy$ and total length of walls $= 2x + y$.)

**33.** Use calculus to solve the following problem. A contractor wants to build a covered box that is twice as long as it is wide. The height is to be determined by the amount of wood available. If the contractor has 48 sq ft of plywood available, find the dimensions of the box with maximum volume.
(*Hint:* Volume $= 2x^2h$ and total plywood $= 2(2x^2) + 2(2xh) + 2(xh)$.)

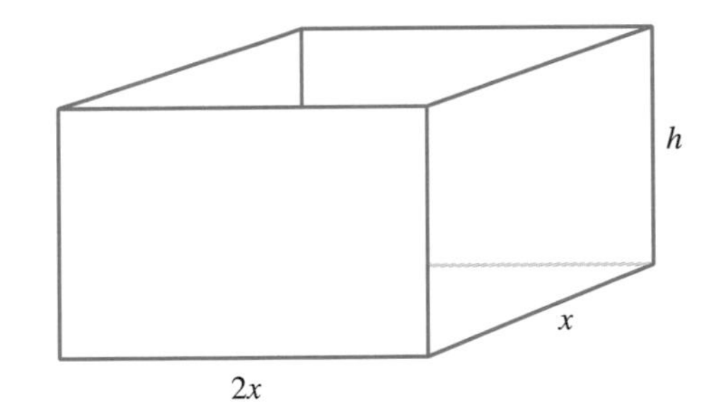

**34.** A rectangle is drawn with sides along the $x$ and $y$ axis and one vertex on the line $y = -2x + 16$ as shown in the figure below.
What is the area of the largest rectangle that can be drawn satisfying these conditions?
(*Hint:* The area of the rectangle is $A = xy$.)

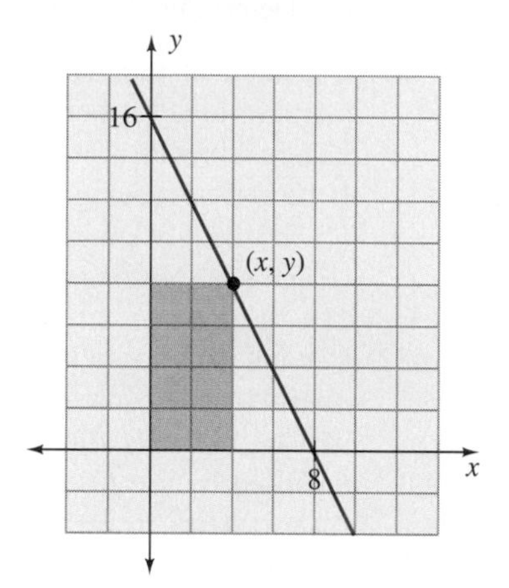


**35.** A manufacturer of aerobics shoes determines that the average cost for producing a pair of shoes is given by
$C = 80{,}000x^{-1} + 0.0002x + 8.2$ where $x$ is the number of shoes manufactured.
(a) Find the production level that minimizes the average cost.
(b) What is the minimum average cost?

**36.** A wood sculptor that creates custom headboards determines that the profit from selling the headboards can be approximated by
$P = 6x^2 - 0.02x^3$ where $x$ is the number of headboards sold.
(a) Find the number of headboards that must be sold in order to maximize the profit.
(b) What is the maximum profit?

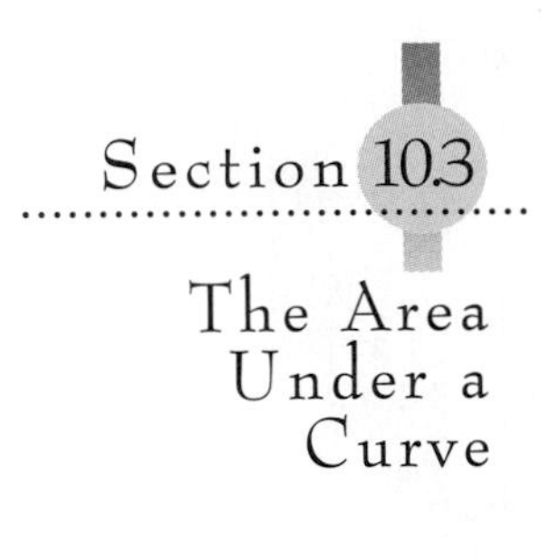


# Section 10.3 The Area Under a Curve

The next topic we consider is **integration**. Integration is a method for finding the area of a region by filling the region with shapes of known area. Integration has many uses in calculus; however, we will restrict our investigation to finding the area bounded by a curve in a plane. As we saw in the history section, many methods, such as the Eudoxian method of exhaustion and Cavalieri's indivisibles, have been used to find areas. We will use the method of rectangles, developed by Gilles Persone de Roberval (1602–1675) of France. It is essentially the method used to introduce integration to a modern calculus class.

The problem is to find the area of the region bounded by the lines $y = 2x$, $y = 0$, and $x = 4$. Using the formula for the area of a triangle, the area is $(1/2)(4)(8) = 16$. However, we will use a method involving rectangles so that we can arrive at a process for finding any area.

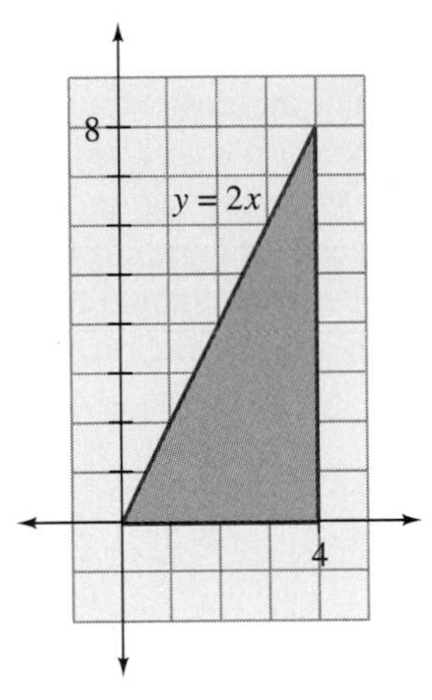


The first step in this method is to divide the region into a number of rectangles. The more rectangles we use, the greater the accuracy of our result. As in the method of exhaustion, we can put the rectangles inside or outside the triangle. In the first diagram there are four rectangles inside the triangle (one of them has a height of 0), whereas the second diagram has the rectangles extending outside of the triangle. From here, we can anticipate that the area of the lower rectangles will be less than the area of the region, and the area of the upper rectangles will exceed the area of the region.

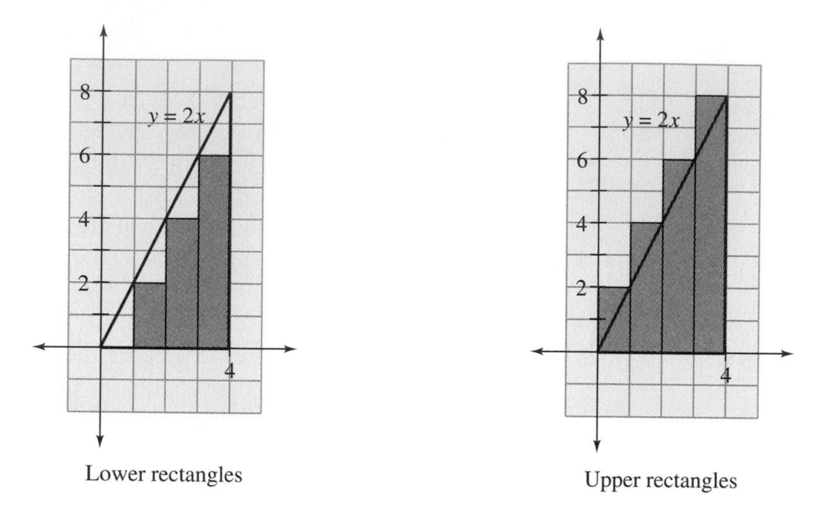


Lower rectangles

Upper rectangles

An improvement on this method is to use rectangles that approximate the region even more closely. Instead of using rectangles that are obviously smaller or larger than the desired area, we use rectangles that intersect the curve at the midpoint of their top edges, as shown in the diagram. The upper left corner of each rectangle is outside the region and therefore will provide an overestimate of the area. However, each rectangle has some white space above it, underestimating the area of the region. The result should be a value that gives a more accurate approximation of the actual area of the region.

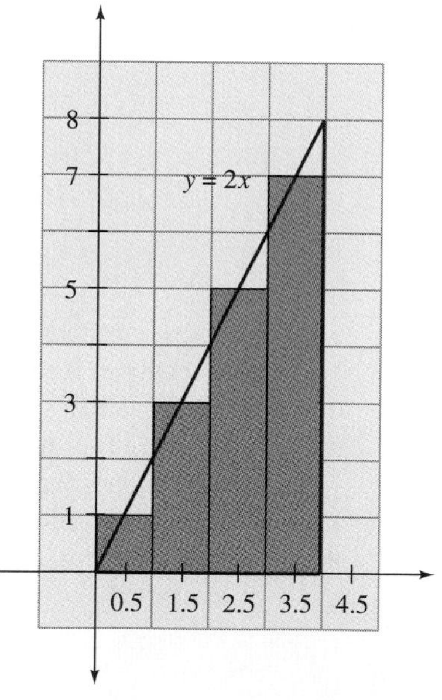


| $x$ | $y$ |
|---|---|
| 0.5 | 1 |
| 1.5 | 3 |
| 2.5 | 5 |
| 3.5 | 7 |

To estimate the area of the region, find the area of the four rectangles. Each rectangle has width 1. We need only find the heights of the rectangles. To do this, we use the fact that the line $y = 2x$ crosses the top of each rectangle at its midpoint. Therefore, the $x$ values are 0.5, 1.5, 2.5, and 3.5. Thus, the heights of the rectangles can be found by substituting these $x$ values into the equation $y = 2x$ as shown in the table. The area of the four rectangles is

$$A = \text{width} \times \text{height of the four rectangles}$$
$$= 1 \times 1 + 1 \times 3 + 1 \times 5 + 1 \times 7 = 16$$

Through the use of four rectangles, intersecting the line $y = 2x$ at the midpoints of their upper edges, we were able to find the correct area. Although we will usually only be able to find the approximate area, the rectangle method appears to work and can be applied to other regions.

In summary, the rectangle method can be used to estimate the area bounded by a curve by dividing the desired area into rectangles with the midpoint of the top edge of each rectangle intersecting the curve. The area of the region is estimated by calculating the sum of the areas of the rectangles.

## Example 1

Approximate the area of the region bounded by the $x$ axis, the line $x = 4$, and the curve $y = x^2/4$.

(a) Use the rectangle method with four rectangles.
(b) Use the rectangle method with eight rectangles.

**Solution:**

(a) Our first step will be to sketch the desired region. As before, the region has been divided into four rectangles. The midpoints of the rectangles are 0.5, 1.5, 2.5, and 3.5. Substituting each of these $x$ values into the equation $y = x^2/4$ gives the following $y$ values:

| $x$ | $y$ |
|---|---|
| 0.5 | 0.0625 |
| 1.5 | 0.5625 |
| 2.5 | 1.5625 |
| 3.5 | 3.0625 |

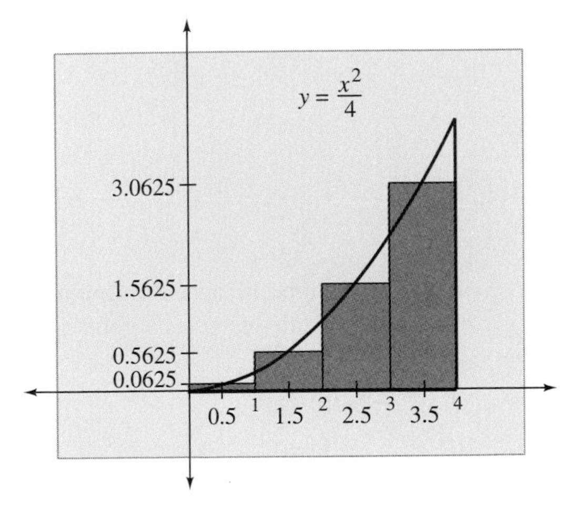

Since the width of each of the rectangles is 1, the sum of the areas of the rectangles is

$$A = \text{width} \times \text{height of the four rectangles}$$
$$= 1 \times 0.0625 + 1 \times 0.5625 + 1 \times 1.5625 + 1 \times 3.0625 = 5.25$$

(b) We now repeat this process with eight rectangles. Since we want to create eight rectangles in the interval 0 to 4, the width of each rectangle is 1/2, since $(4 - 0)/8 = 1/2$. We can now find the midpoint of each rectangle and its corresponding height, the $y$ value. Since the width of the first rectangle is 1/2 or 0.5, the midpoint of the first rectangle is 0.25. Adding 0.5 gives the midpoint of the second rectangle as 0.75. Repeating this process gives the midpoints of each of the eight rectangles. Using the width of 0.5 and the heights as given by the $y$ values, we have

| $x$ | $y$ |
|---|---|
| 0.25 | 0.0156 |
| 0.75 | 0.1406 |
| 1.25 | 0.3906 |
| 1.75 | 0.7656 |
| 2.25 | 1.2656 |
| 2.75 | 1.8906 |
| 3.25 | 2.6406 |
| 3.75 | 3.5156 |

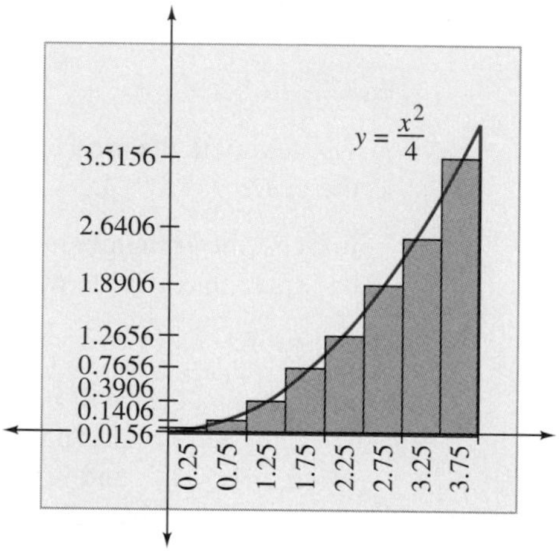

$$A = \text{width} \times \text{height of the four rectangles}$$
$$= 0.5 \times 0.0156 + 0.5 \times 0.1406 + 0.5 \times 0.3906 + 0.5 \times 0.7656 + 0.5 \times 1.2656 + 0.5 \times 1.8906 + 0.5 \times 2.6406 + 0.5 \times 3.5156$$
$$= 5.3124$$

Using more advanced techniques, it can be shown that the area is $5\frac{1}{3}$. Thus, using eight rectangles gives a good approximation of the area.

### Example 2

A lawn fertilizer company suggests that Weed O Burn be applied at a rate of 2 lb per 1000 sq ft. Because of the potency of the fertilizer, Chauncy, the gardener, wants to accurately determine the area of the irregularly shaped lawn, shown in the figure. His only tool is a 100-ft tape measure. How can Chauncy determine the area? What is the approximate area?

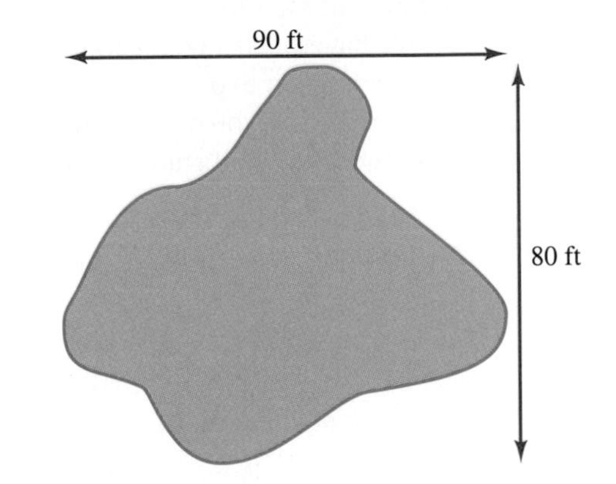


**Solution:** To determine the area, Chauncy decides to divide the lawn into 10-ft-wide strips and measure the length of each strip. This will give him a set of rectangles whose combined areas will approximate the area of the lawn. There will be eight rectangles ($80 \div 10$), each being 10 ft wide. Chauncy measured the length of each rectangle. His results are given in the figure. Using the formula for the area of a rectangle, we can now find the approximate area of the lawn.

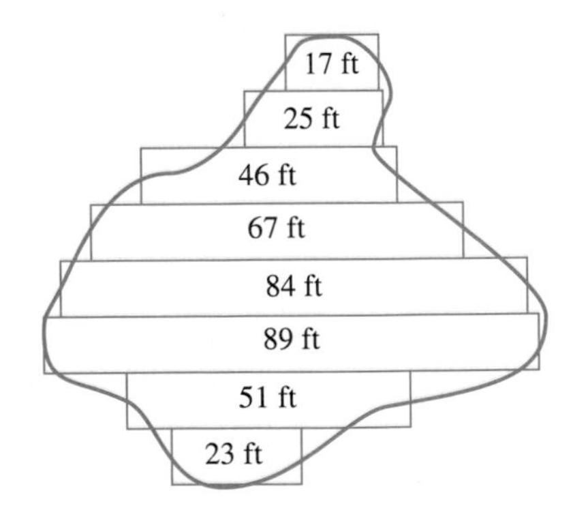


$$\begin{aligned} A &= \text{width} \times \text{length} \\ &= 10 \times 17 + 10 \times 25 + 10 \times 46 + 10 \times 67 + 10 \times 84 \\ &\quad + 10 \times 89 + 10 \times 51 + 10 \times 23 \\ &= 4020 \text{ sq ft of lawn} \end{aligned}$$

This section concludes our short discussion of calculus. In this chapter, we have shown that differentiation gives the slope of a curve and how, with the help of some shortcuts, the derivative can be applied to certain mathematical problems. We have also shown that integration gives a method for determining the areas of regions bounded by curves. As is true for differentiation, there are shortcuts for doing integration, but that investigation is beyond the scope of this chapter.

We have presented a very brief overview of calculus. We have not attempted to give a complete explanation of the power and applicability of calculus. However, it is our hope that you have seen some mathematics that is intriguing enough for you to pursue additional studies in this area.

## Section 10.3 PROBLEMS

## *Explain* ➡ *Apply* ➡ *Explore*

### *Explain*

**1.** What is integration?

**2.** When calculating an area, what are ''upper'' rectangles? Explain what happens to the approximation of the area under a curve as the number of upper rectangles increases.

**3.** When calculating an area, what are ''lower'' rectangles? Explain what happens to the approximation of the area under a curve as the number of lower rectangles increases.

**4.** When calculating an area, what are ''midpoint'' rectangles? Explain what happens to the approximation of the area under a curve as the number of midpoint rectangles increases.

### *Apply*

In Problems 5–12, sketch the given area and use the midpoint rectangle method to approximate the specified area.

**5.** The area bounded by the lines $y = -2x + 24$, the $x$ axis, and the $y$ axis, using three rectangles.

**6.** The area bounded by the lines $y = -3x + 24$, the $x$ axis, and the $y$ axis, using four rectangles.

**7.** The area bounded by the lines $y = 2x + 4$, the $x$ axis, the $y$ axis, and the line $x = 8$, using four rectangles.

**8.** The area bounded by the lines $y = 3x + 12$, the $x$ axis, the $y$ axis, and the line $x = 12$, using three rectangles.

**9.** The area bounded by the $x$ axis, the line $x = 5$, and the curve $y = x^2$, using five rectangles.

**10.** The area bounded by the $x$ axis, $y$ axis, the line $x = 5$, and the curve $y = x^2 - 2x + 3$, using five rectangles.

**11.** The area bounded by the $x$ axis and the curve $y = 36 - x^2$, using six rectangles.

**12.** The area bounded by the $x$ axis and the curve $y = 12x - x^2$, using six rectangles.

### *Explore*

**13.** In the sketch of a small pond, the measurements are 25 ft apart and start 12.5 ft from the ends of the pond. Use the measurements to estimate the area of the pond.

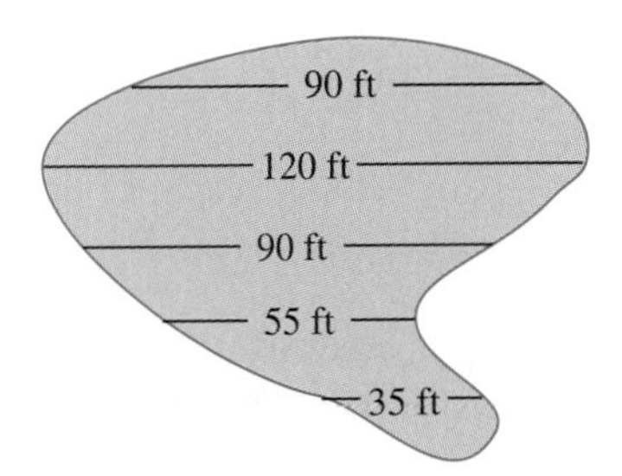


**14.** The following figure is drawn to scale. Trace the figure onto your paper. Use a metric ruler and the rectangle method with six horizontal rectangles to find the area of the figure in square centimeters.

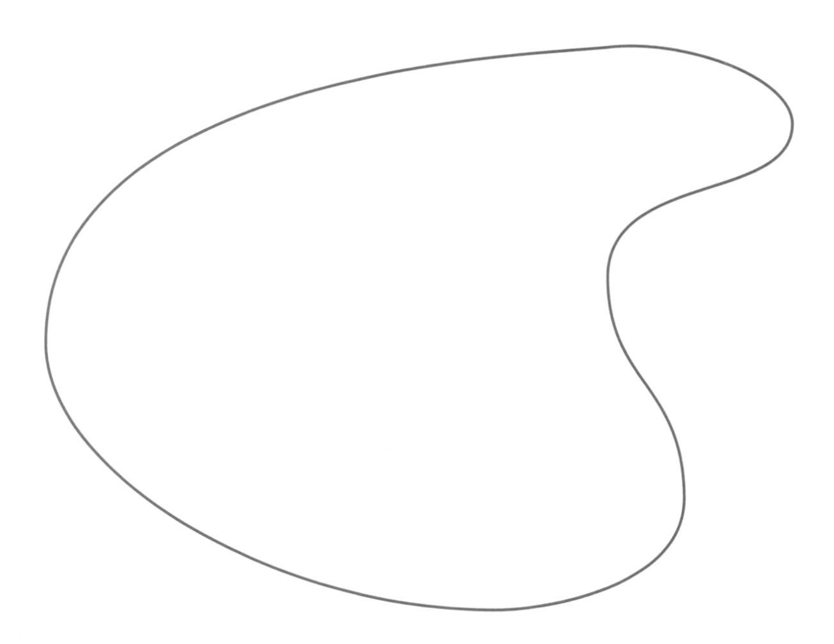

**15.** Consider the region bounded by the curve $y = x^3$, the $x$ axis, and the line $x = 4$.

(a) Draw a sketch of the region.
(b) Approximate the area of that region using four rectangles.

**16.** Consider the region bounded by the $x$ axis, the line $x = 4$, and the curve $y = \sqrt{x}$.

(a) Draw a sketch of the region.
(b) Approximate the area of that region using four rectangles.

**17.** Consider the region bounded by the $x$ axis, the line $x = 1$, the line $x = 4$, and the curve $y = \dfrac{1}{x}$.

(a) Draw a sketch of the region.
(b) Approximate the area of that region using three rectangles.

**18.** The equation $y = \sqrt{9 - x^2}$ gives the graph of the semicircle with a radius of 3 that is shown below.

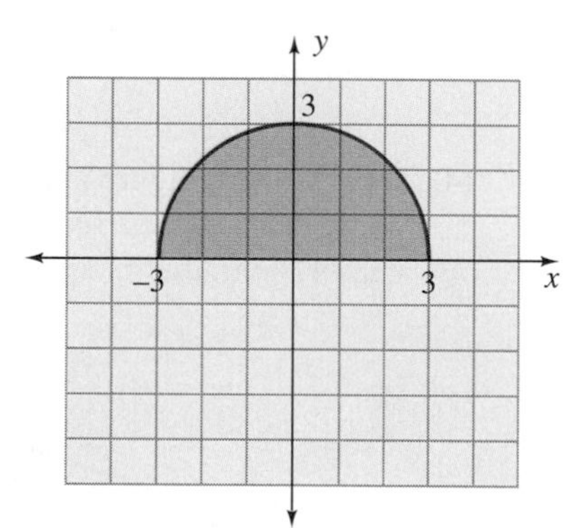


(a) Approximate the area of the semicircle using six rectangles.
(b) How does that approximation compare to the actual area of a semicircle with radius $r = 3$ given by the formula $A = \frac{1}{2}\pi r^2$.

**19.** The area of the ellipse, $\dfrac{x^2}{a^2} + \dfrac{y^2}{b^2} = 1$, is given by $A = \pi ab$. Consider an ellipse given the equation $\dfrac{x^2}{64} + \dfrac{y^2}{25} = 1$.

(a) Find the area of the ellipse by using the formula. Distances should be measured in centimeters. Give the answer accurate to two decimal places.
(b) Find the area of the ellipse using 8 ''midpoint'' rectangles with a width of 2 centimeters. The lengths of the rectangles will be determined by the equation. Round off your answer to two decimal places.
(c) How accurate is the area determined by the rectangle method?

**20.** The area of a circle with radius $r$ is given by $A = \pi r^2$. Consider a circle with a radius of 10 centimeters.

(a) Find the area of the circle by using the formula. Give the answer accurate to two decimal places.

(b) Draw the circle on an $xy$ coordinate system with the center of the circle at the origin. Find the area of the circle by using 10 "midpoint" rectangles with a width of 2 centimeters. The lengths of the rectangles will be determined by the equation $x^2 + y^2 = 100$. Round off your answer to two decimal places.

(c) How accurate is the area determined by the rectangle method?

## Chapter 10 • Summary

### Key Terms, Concepts, and Formulas

The important terms in this chapter are:

**Calculus:** An area of mathematics concerned with the study of derivatives and integrals. p. 669

**Derivative:** The slope of a curve at a point. p. 683

**Differentiation:** A process in calculus used to find the slope of a tangent line (derivative). p. 672

**Function:** A relation between two variables, often written as $y = f(x)$, such that each $x$ value is associated with at most one $y$ value. p. 676

**Infinitesimal:** A very small quantity; a very small change in $x$ or $y$. p. 670

**Integration:** A process in calculus used to find areas of regions in a plane. p. 670

**Relative maximum point:** A point on a curve with a $y$ value that is greater than the $y$ values of all points immediately to the left and to the right of the point. p. 696

**Relative minimum point:** A point on a curve with a $y$ value that is less than the $y$ values of all points immediately to the left and to the right of the point. p. 696

**Slope:** A measure of the steepness of a line or curve. p. 685

After completing this chapter, you should be able to:

**1.** Use the formula $\dfrac{dy}{dx} = \lim\limits_{h \to 0} \dfrac{f(x + h) - f(x)}{h}$

and shortcuts to find the derivative of a function. p. 688

2. Explain the concepts of tangent lines, slopes of curves, relative maximums and relative minimums, and areas bounded by curves. pp. 684, 707
3. Perform the necessary calculations to find slopes of curves, relative maximums and relative minimums, and areas bounded by curves. pp. 692, 705

• Summary

## Problems

1. Find the value of the following if $f(x) = 2x^2 - 3x + 2$.
   (a) $f(4)$
   (b) $f(m)$
   (c) $f(3 + h)$
   (d) $f(x + h)$
2. Find the value of the following if $f(x) = 3x^2 + 5x - 1$.
   (a) $f(-2)$
   (b) $f(w)$
   (c) $f(5 + h)$
   (d) $f(x + h)$
3. Using the formula for $dy/dx$, find the derivative of $y = 3x - 8$.
4. Using the derivative shortcuts, find the derivative of $y = x^3 + 6x - 1$.
5. Find the slope of the tangent line to the curve $f(x) = 3x^2 + 5x - 1$ at $x = 3$.
6. Find the slope of the tangent line to the curve $f(x) = 2x^2 - 3x + 2$ at $x = -2$.
7. Use a derivative to find the relative maximum point of $f(x) = -4x^2 - 16x + 2$.
8. Use a derivative to find the relative minimum point of $f(x) = 4x^2 - 24x + 9$.
9. A set of six equal pigpens is to be constructed out of 120 ft of fencing. The arrangement of the pens is given in the diagram. Use calculus to find the maximum total area of the pens. (*Hint:* $A = xy$ and $4x + 3y = 120$.)

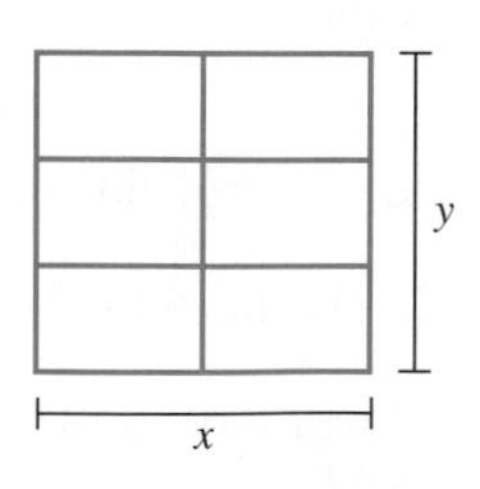

**10.** A field, adjacent to a barn, is to be divided into four pens, as shown in the following diagram. Because the field is adjacent to the barn, only three outside fences need to be built along with the interior fences. If 3000 ft of fencing is available, use calculus to find the outer dimensions of the field that will give the maximum total area. (*Hint:* $A = xy$ and $3x + 2y = 3000$.)

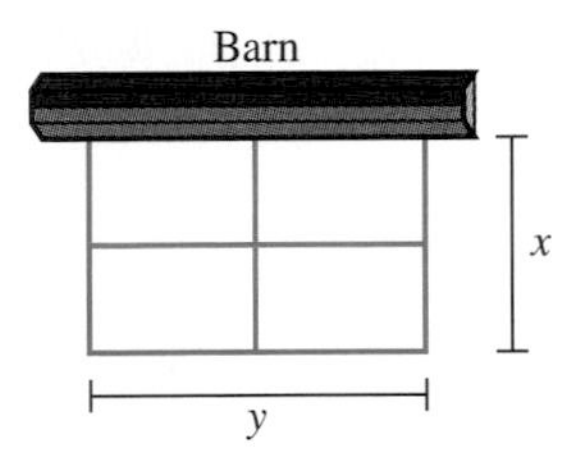


**11.** Use the rectangle rule with four rectangles to find the area bounded by the curve $y = -2x^2 + 16x$ and the $x$ axis.

**12.** Use the rectangle rule with three rectangles to find the area bounded by the curve $y = -3x^2 + 18x$ and the $x$ axis.

**13.** The lawn shown needs to receive 1 lb of fertilizer per 1000 sq ft. Use the given measurements to estimate the fertilizer requirements for the lawn. The measurements across the lawn are 16 ft apart and start 8 ft from the top and end 8 ft from the bottom of the lawn.

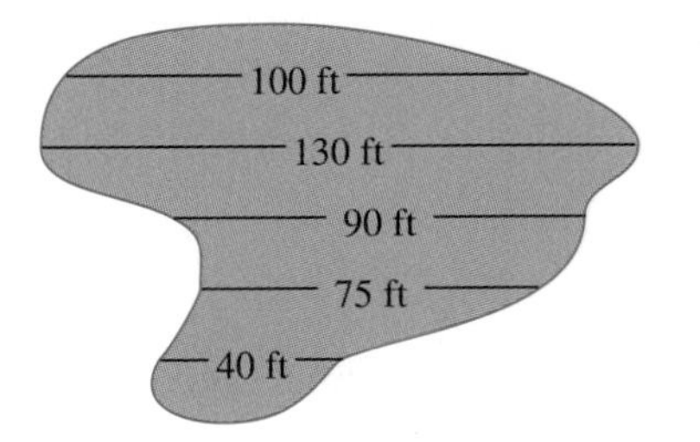

# Appendix 1

## Normal Table

This table gives the area under the standard normal curve, between $z = 0$ and the given value of $z$.

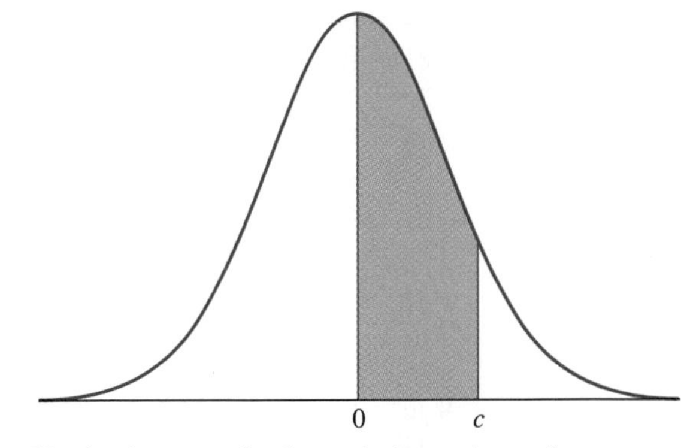


The values in the table give the area under the standard normal curve between $z = 0$ and $z = c$ where $z \geq 0$.

| $z$ | Area | $z$ | Area | $z$ | Area | $z$ | Area |
|---|---|---|---|---|---|---|---|
| 0.00 | 0.0000 | 0.07 | 0.0279 | 0.14 | 0.0557 | 0.21 | 0.0832 |
| 0.01 | 0.0040 | 0.08 | 0.0319 | 0.15 | 0.0596 | 0.22 | 0.0871 |
| 0.02 | 0.0078 | 0.09 | 0.0359 | 0.16 | 0.0636 | 0.23 | 0.0910 |
| 0.03 | 0.0120 | 0.10 | 0.0398 | 0.17 | 0.0675 | 0.24 | 0.0948 |
| 0.04 | 0.0160 | 0.11 | 0.0438 | 0.18 | 0.0714 | 0.25 | 0.0987 |
| 0.05 | 0.0199 | 0.12 | 0.0478 | 0.19 | 0.0753 | 0.26 | 0.1026 |
| 0.06 | 0.0239 | 0.13 | 0.0517 | 0.20 | 0.0793 | 0.27 | 0.1064 |

| $z$ | Area | $z$ | Area | $z$ | Area | $z$ | Area |
|---|---|---|---|---|---|---|---|
| 0.28 | 0.1102 | 0.68 | 0.2517 | 1.08 | 0.3599 | 1.48 | 0.4306 |
| 0.29 | 0.1141 | 0.69 | 0.2549 | 1.09 | 0.3621 | 1.49 | 0.4319 |
| 0.30 | 0.1179 | 0.70 | 0.2580 | 1.10 | 0.3643 | 1.50 | 0.4332 |
| 0.31 | 0.1217 | 0.71 | 0.2611 | 1.11 | 0.3665 | 1.51 | 0.4345 |
| 0.32 | 0.1255 | 0.72 | 0.2642 | 1.12 | 0.3686 | 1.52 | 0.4357 |
| 0.33 | 0.1293 | 0.73 | 0.2673 | 1.13 | 0.3708 | 1.53 | 0.4370 |
| 0.34 | 0.1331 | 0.74 | 0.2704 | 1.14 | 0.3729 | 1.54 | 0.4382 |
| 0.35 | 0.1368 | 0.75 | 0.2734 | 1.15 | 0.3749 | 1.55 | 0.4394 |
| 0.36 | 0.1406 | 0.76 | 0.2764 | 1.16 | 0.3770 | 1.56 | 0.4406 |
| 0.37 | 0.1443 | 0.77 | 0.2794 | 1.17 | 0.3790 | 1.57 | 0.4418 |
| 0.38 | 0.1480 | 0.78 | 0.2823 | 1.18 | 0.3810 | 1.58 | 0.4429 |
| 0.39 | 0.1517 | 0.79 | 0.2852 | 1.19 | 0.3830 | 1.59 | 0.4441 |
| 0.40 | 0.1554 | 0.80 | 0.2881 | 1.20 | 0.3849 | 1.60 | 0.4452 |
| 0.41 | 0.1591 | 0.81 | 0.2910 | 1.21 | 0.3869 | 1.61 | 0.4463 |
| 0.42 | 0.1628 | 0.82 | 0.2939 | 1.22 | 0.3888 | 1.62 | 0.4474 |
| 0.43 | 0.1664 | 0.83 | 0.2967 | 1.23 | 0.3907 | 1.63 | 0.4484 |
| 0.44 | 0.1700 | 0.84 | 0.2995 | 1.24 | 0.3925 | 1.64 | 0.4495 |
| 0.45 | 0.1736 | 0.85 | 0.3023 | 1.25 | 0.3944 | 1.65 | 0.4505 |
| 0.46 | 0.1772 | 0.86 | 0.3051 | 1.26 | 0.3962 | 1.66 | 0.4515 |
| 0.47 | 0.1808 | 0.87 | 0.3079 | 1.27 | 0.3980 | 1.67 | 0.4525 |
| 0.48 | 0.1844 | 0.88 | 0.3106 | 1.28 | 0.3997 | 1.68 | 0.4535 |
| 0.49 | 0.1879 | 0.89 | 0.3133 | 1.29 | 0.4015 | 1.69 | 0.4545 |
| 0.50 | 0.1915 | 0.90 | 0.3159 | 1.30 | 0.4032 | 1.70 | 0.4554 |
| 0.51 | 0.1950 | 0.91 | 0.3186 | 1.31 | 0.4049 | 1.71 | 0.4564 |
| 0.52 | 0.1985 | 0.92 | 0.3212 | 1.32 | 0.4066 | 1.72 | 0.4573 |
| 0.53 | 0.2019 | 0.93 | 0.3238 | 1.33 | 0.4082 | 1.73 | 0.4582 |
| 0.54 | 0.2054 | 0.94 | 0.3262 | 1.34 | 0.4099 | 1.74 | 0.4591 |
| 0.55 | 0.2088 | 0.95 | 0.3289 | 1.35 | 0.4115 | 1.75 | 0.4599 |
| 0.56 | 0.2123 | 0.96 | 0.3315 | 1.36 | 0.4131 | 1.76 | 0.4608 |
| 0.57 | 0.2157 | 0.97 | 0.3340 | 1.37 | 0.4147 | 1.77 | 0.4616 |
| 0.58 | 0.2190 | 0.98 | 0.3365 | 1.38 | 0.4162 | 1.78 | 0.4625 |
| 0.59 | 0.2224 | 0.99 | 0.3389 | 1.39 | 0.4177 | 1.79 | 0.4633 |
| 0.60 | 0.2257 | 1.00 | 0.3413 | 1.40 | 0.4192 | 1.80 | 0.4641 |
| 0.61 | 0.2291 | 1.01 | 0.3438 | 1.41 | 0.4207 | 1.81 | 0.4649 |
| 0.62 | 0.2323 | 1.02 | 0.3461 | 1.42 | 0.4222 | 1.82 | 0.4656 |
| 0.63 | 0.2357 | 1.03 | 0.3485 | 1.43 | 0.4236 | 1.83 | 0.4664 |
| 0.64 | 0.2389 | 1.04 | 0.3508 | 1.44 | 0.4251 | 1.84 | 0.4671 |
| 0.65 | 0.2422 | 1.05 | 0.3531 | 1.45 | 0.4265 | 1.85 | 0.4678 |
| 0.66 | 0.2454 | 1.06 | 0.3554 | 1.46 | 0.4279 | 1.86 | 0.4686 |
| 0.67 | 0.2486 | 1.07 | 0.3577 | 1.47 | 0.4292 | 1.87 | 0.4693 |

| $z$ | **Area** | $z$ | **Area** | $z$ | **Area** | $z$ | **Area** |
|---|---|---|---|---|---|---|---|
| 1.88 | 0.4699 | 2.28 | 0.4887 | 2.68 | 0.4963 | 3.08 | 0.49897 |
| 1.89 | 0.4706 | 2.29 | 0.4890 | 2.69 | 0.4964 | 3.09 | 0.49900 |
| 1.90 | 0.4713 | 2.30 | 0.4893 | 2.70 | 0.4965 | 3.10 | 0.49903 |
| 1.91 | 0.4719 | 2.31 | 0.4896 | 2.71 | 0.4966 | 3.11 | 0.49907 |
| 1.92 | 0.4726 | 2.32 | 0.4898 | 2.72 | 0.4967 | 3.12 | 0.49970 |
| 1.93 | 0.4732 | 2.33 | 0.4901 | 2.73 | 0.4968 | 3.13 | 0.49913 |
| 1.94 | 0.4738 | 2.34 | 0.4904 | 2.74 | 0.4969 | 3.14 | 0.49916 |
| 1.95 | 0.4744 | 2.35 | 0.4906 | 2.75 | 0.4970 | 3.15 | 0.49918 |
| 1.96 | 0.4750 | 2.36 | 0.4909 | 2.76 | 0.4971 | 3.16 | 0.49921 |
| 1.97 | 0.4756 | 2.37 | 0.4911 | 2.77 | 0.4972 | 3.17 | 0.49924 |
| 1.98 | 0.4761 | 2.38 | 0.4913 | 2.78 | 0.4973 | 3.18 | 0.49926 |
| 1.99 | 0.4767 | 2.39 | 0.4916 | 2.79 | 0.4974 | 3.19 | 0.49929 |
| 2.00 | 0.4772 | 2.40 | 0.4918 | 2.80 | 0.4974 | 3.20 | 0.49931 |
| 2.01 | 0.4778 | 2.41 | 0.4920 | 2.81 | 0.4975 | 3.21 | 0.49934 |
| 2.02 | 0.4783 | 2.42 | 0.4922 | 2.82 | 0.4976 | 3.22 | 0.49936 |
| 2.03 | 0.4788 | 2.43 | 0.4925 | 2.83 | 0.4977 | 3.23 | 0.49938 |
| 2.04 | 0.4793 | 2.44 | 0.4927 | 2.84 | 0.4977 | 3.24 | 0.49940 |
| 2.05 | 0.4798 | 2.45 | 0.4929 | 2.85 | 0.4978 | 3.25 | 0.49942 |
| 2.06 | 0.4803 | 2.46 | 0.4931 | 2.86 | 0.4979 | 3.26 | 0.49944 |
| 2.07 | 0.4808 | 2.47 | 0.4932 | 2.87 | 0.4979 | 3.27 | 0.49946 |
| 2.08 | 0.4812 | 2.48 | 0.4934 | 2.88 | 0.4980 | 3.28 | 0.49948 |
| 2.09 | 0.4817 | 2.49 | 0.4936 | 2.89 | 0.4981 | 3.29 | 0.49950 |
| 2.10 | 0.4821 | 2.50 | 0.4938 | 2.90 | 0.4981 | 3.30 | 0.49952 |
| 2.11 | 0.4826 | 2.51 | 0.4940 | 2.91 | 0.4982 | 3.31 | 0.49953 |
| 2.12 | 0.4830 | 2.52 | 0.4941 | 2.92 | 0.4982 | 3.32 | 0.49955 |
| 2.13 | 0.4834 | 2.53 | 0.4943 | 2.93 | 0.4983 | 3.33 | 0.49957 |
| 2.14 | 0.4838 | 2.54 | 0.4945 | 2.94 | 0.4984 | 3.34 | 0.49958 |
| 2.15 | 0.4842 | 2.55 | 0.4946 | 2.95 | 0.4984 | 3.35 | 0.49960 |
| 2.16 | 0.4846 | 2.56 | 0.4948 | 2.96 | 0.4985 | 3.36 | 0.49961 |
| 2.17 | 0.4850 | 2.57 | 0.4949 | 2.97 | 0.4985 | 3.37 | 0.49962 |
| 2.18 | 0.4854 | 2.58 | 0.4951 | 2.98 | 0.4986 | 3.38 | 0.49964 |
| 2.19 | 0.4857 | 2.59 | 0.4952 | 2.99 | 0.4986 | 3.39 | 0.49965 |
| 2.20 | 0.4861 | 2.60 | 0.4953 | 3.00 | 0.4987 | 3.40 | 0.49966 |
| 2.21 | 0.4864 | 2.61 | 0.4955 | 3.01 | 0.49869 | 3.41 | 0.49968 |
| 2.22 | 0.4868 | 2.62 | 0.4956 | 3.02 | 0.49874 | 3.42 | 0.49969 |
| 2.23 | 0.4871 | 2.63 | 0.4957 | 3.03 | 0.49878 | 3.43 | 0.49970 |
| 2.24 | 0.4875 | 2.64 | 0.4959 | 3.04 | 0.49881 | 3.44 | 0.49971 |
| 2.25 | 0.4878 | 2.65 | 0.4960 | 3.05 | 0.49886 | 3.45 | 0.49972 |
| 2.26 | 0.4881 | 2.66 | 0.4961 | 3.06 | 0.49889 | 3.46 | 0.49973 |
| 2.27 | 0.4884 | 2.67 | 0.4962 | 3.07 | 0.49893 | 3.47 | 0.49974 |

| $z$ | **Area** | $z$ | **Area** | $z$ | **Area** | $z$ | **Area** |
|---|---|---|---|---|---|---|---|
| 3.48 | 0.49975 | 3.62 | 0.49985 | 3.75 | 0.49991 | 3.88 | 0.49995 |
| 3.49 | 0.49976 | 3.63 | 0.49986 | 3.76 | 0.49992 | 3.89 | 0.49995 |
| 3.50 | 0.49977 | 3.64 | 0.49986 | 3.77 | 0.49992 | 3.90 | 0.49995 |
| 3.51 | 0.49978 | 3.65 | 0.49987 | 3.78 | 0.49999 | 3.91 | 0.49995 |
| 3.52 | 0.49978 | 3.66 | 0.49987 | 3.79 | 0.49993 | 3.92 | 0.49996 |
| 3.53 | 0.49979 | 3.67 | 0.49988 | 3.80 | 0.49993 | 3.93 | 0.49996 |
| 3.54 | 0.49980 | 3.68 | 0.49988 | 3.81 | 0.49993 | 3.94 | 0.49996 |
| 3.55 | 0.49981 | 3.69 | 0.49989 | 3.82 | 0.49993 | 3.95 | 0.49996 |
| 3.56 | 0.49982 | 3.70 | 0.49989 | 3.83 | 0.49994 | 3.96 | 0.49996 |
| 3.57 | 0.49982 | 3.71 | 0.49990 | 3.84 | 0.49994 | 3.97 | 0.49996 |
| 3.58 | 0.49983 | 3.72 | 0.49990 | 3.85 | 0.49994 | 3.98 | 0.49997 |
| 3.59 | 0.49986 | 3.73 | 0.49990 | 3.86 | 0.49994 | 3.99 | 0.49997 |
| 3.60 | 0.49984 | 3.74 | 0.49991 | 3.87 | 0.49995 | 4.00 | 0.49997 |
| 3.61 | 0.49985 | | | | | | |

# Bibliography

## General


Aleksandrov, A. D., *Mathematics—Its Content, Methods, and Meaning,* M.I.T. Press, Cambridge, MA, 1963.

Asimov, Isaac, *Asimov's Biographical Encyclopedia of Science and Technology,* Doubleday, New York, 1982.

Asimov, Isaac, *Asimov's New Guide to Science,* Basic Books, New York, 1984.

Ball, W. W. Rouse, *A Short Account of the History of Mathematics,* Dover Publications, New York, 1960.

Bell, E. T., *Men of Mathematics,* Simon and Schuster, New York, 1937.

Bell, E. T., *Mathematics Queen and Servant of Science,* Mathematics Association of America, Washington, DC, 1987.

Boyer, Carl B., *A History of Mathematics,* John Wiley and Sons, New York, 1968.

Cajori, Florian, *The Early Mathematical Sciences in North and South America,* Gorham Press, Boston, 1928.

Cajori, Florian, *A History of Mathematics,* Chelsea Publishing, New York, 1918.

Campbell, Douglas and Higgins, John, eds., *Mathematics: People, Problems, Results,* Wadsworth International, Belmont, CA, 1984.

Durant, Will, *Our Oriental Heritage,* Simon and Schuster, New York, 1963.

Eves, Howard, *Great Moments in the History of Mathematics,* Mathematics Association of America, Washington, DC, 1982.

Eves, Howard, *In Mathematical Circles,* vols. 1, 2, Prindle, Weber, & Schmidt, Boston, 1969.

Eves, Howard, *An Introduction to the History of Mathematics,* Saunders College Publishing, Philadelphia, 1990.

Gillispie, Charles C., *Dictionary of Scientific Biography,* vols. I–XVI, Charles Scribner's Sons, New York, 1970.

Grinstein, Louise and Campbell, Paul, eds., *Women of Mathematics,* Greenwood Press, New York, 1987.

Heath, Sir Thomas, *A History of Greek Mathematics,* Dover Publications, New York, 1981.

Hogben, Lancelot, *Mathematics in the Making,* Doubleday, New York, 1960.

Kline, Morris, *Mathematics in Western Culture,* Oxford University Press, New York, 1953.

Marcorini, Edgardo, ed., *The History of Sci-*

*ence and Technology, A Narrative Chronology,* Facts on File, New York, 1988.

Menninger, Karl, *Number Words and Number Symbols,* M.I.T. Press, Cambridge, MA, 1958.

Mount, Ellis and List, Barbara, *Milestones in Science and Technology: The Ready Reference Guide to Discoveries, Inventions and Facts,* Oryx Press, Phoenix, 1987.

Newell, Virginia, ed. *Black Mathematicians,* Dorrance and Co., Ardmore, PA, 1980.

Newman, James, *The World of Mathematics,* Simon and Schuster, New York, 1956.

Osen, Lynn, *Women in Mathematics,* M.I.T. Press, Cambridge, MA, 1974.

Parkinson, Claire L., *Breakthrough, A Chronology of Great Achievements in Science and Mathematics,* G. K. Hall, Boston, 1985.

Peterson, Ivars, *The Mathematical Tourist,* W. H. Freeman, New York, 1988.

Resnikoff, H. L. and Wells, R. O., *Mathematics in Civilization,* Dover Publications, New York, 1984.

Smith, David E., *History of Mathematics,* Dover Publications, New York, 1958.

Smith, David and Mikami, Yoshio, *A History of Japanese Mathematics,* The Open Court Press, Chicago, 1914.

Temple, Robert, *The Genius of China,* Simon and Schuster, New York, 1986.

## Chapter 1

Andrews, W. S., *Magic Squares and Cubes,* Dover Publications, New York, 1960.

Beckman, Petr, *A History of Pi,* Martin Press, New York, 1971.

Dantzig, Tobias, *Number: The Language of Science,* Macmillan Company, New York, 1930.

Friend, J. Newton, *Numbers: Fun & Facts,* Charles Scribner's Sons, New York, 1954.

Ifah, Georges, *From One to Zero,* Viking Penguin, New York, 1985.

Zaslavsky, Claudia, *Africa Counts,* Chicago Review, Chicago, 1979.

## Chapter 3

Barnsley, Michael, *Fractals Everywhere,* Academic Press, New York, 1988.

Gardner, Martin, *Penrose Tiles to Trapdoor Ciphers,* W. H. Freeman, New York, 1988.

Gleick, James, *Chaos,* Viking Press, New York, 1987.

Heath, Sir Thomas, *Euclid's Elements,* vol. 1, Dover Publications, New York, 1956.

Holt, Michael, *Mathematics in Art,* Van Nostrand Reinhold, New York, 1971.

Mandelbrot, Benoit, *The Fractal Geometry of Nature,* W. H. Freeman, New York, 1982.

Peitgen, H. O. and Richter, P. H., *The Beauty of Fractals,* Springer-Verlag, New York, 1986.

Peitgen, H. O. and Saupe, D., eds., *The Science of Fractal Images,* Springer-Verlag, New York, 1988.

Prenowitz, Walter and Jordan, Meyer, *Basic Concepts of Geometry,* Blaisdell Publishing, Waltham, MA, 1965.

Seymour, Dale and Britton, Jill, *Introduction to Tessellations,* Dale Seymour Publications, Palo Alto, CA, 1989.

## Chapter 4

Hurley, Patrick, J., *A Concise Introduction to Logic,* Wadsworth Publishing Co., Belmont, CA, 1988.

## Chapter 6

Durant, Will, *The Age of Faith,* Simon and Schuster, New York, 1950.

Hildreth, Richard, *The History of Banks,* Augustus M. Kelley Publishers, New York, 1968.

Klise, Eugene, *Money and Banking,* 5th ed., Cincinnati South-Western Publishing, Cincinnati, 1972.

Shaw, W. A., *The History of Currency,* Augustus M. Kelley Publishers, New York, 1967.

## Chapter 7

Hacking, Ian, *The Emergence of Probability,* Cambridge University Press, Cambridge, Great Britain, 1975.

Stewart, Ian, *Does God Play Dice?,* Basil Blackwell, New York, 1989.

## Chapter 8

Balinski, Michael and Young, H. Peyton, *Fair Representation,* Yale University Press, New Haven, CT, 1982.

Eagles, Charles W., *Democracy Delayed,* University of Georgia Press, Athens, 1990.

Hoffman, Paul, *Archimedes Revenge,* W. W. Norton and Company, New York, 1988.

Kruskal, William and Tanur, Judith, eds., *International Encyclopedia of Statistics,* The Free Press, New York, 1978.

Schmeckebier, Laurence F., *Congressional Apportionment,* Greenwood Press Publishers, Westport CN, 1941.

Stigler, Stephen M., *The History of Statistics,* Belknap Press, Cambridge, MA, 1986.

## Chapter 9

Augarten, Stan, *Bit by Bit—An Illustrated History of Computers,* Ticknor & Fields, New York, 1984.

Burke, James, *Connections,* Little, Brown and Company, Boston, 1978.

Burke, James, *The Day the Universe Changed,* Little, Brown and Company, Boston, 1985.

Hartung, Maurice L., *A Teaching Guide for Slide Rule Instruction,* Pickett & Eckel, Inc., Chicago, 1960.

Kershner, Helene G., *Introduction to Computer Literacy,* D. C. Heath and Company, Lexington, MA, 1990.

Moreau, R., *The Computer Comes of Age: The People, the Hardware, and the Software,* M.I.T. Press, Cambridge, MA, 1984.

Pullan, J. M., *The History of the Abacus,* Frederick A. Praeger Publishers, New York, 1969.

Ritchie, David, *The Computer Pioneers: The Making of the Modern Computer,* Simon and Schuster, New York, 1986.

Tani, Yukio, *The Magic Calculator—The Way of Abacus,* Japan Publications Trading Company, New York, 1967.

Vorwald, Alan and Clark, Frank, *COMPUTERS! From the Sand Table to the Electronic Brain,* McGraw-Hill Book Company, New York, 1964.

Williams, Michael R., *A History of Computing Technology,* Prentice-Hall, Englewood Cliffs, N.J., 1985.

Yoshino, Yozo, *The Japanese Abacus Explained,* Dover Publications, Inc., New York, 1963.

## Chapter 10

Boyer, Carl B., *A History of the Calculus,* Dover Publications, New York, 1959.

# Selected Answers

## Chapter 1

### *Section 1.1*

**11.** 121,000 **13.** 1,000,000 + 10,000 + 100 + 1 = 1,010,101
**15.** 1106 **17.** 734
**19.** 40,000 + 1000 + 1000 + 400 + 50 + 5 + 1 + 1 = 42,457
**21.** 7502 **23.** 37
**25.** 2 × 10,000 + 5 × 1000 + 60 + 2 = 20,000 + 5000 + 60 + 2 = 25,062
**27.** 11 **29.** (1)(360) + 0 + 3 = 363
**31.** 19 × 2,880,000 + 16 × 144,000 + 6 × 7200 + 4 × 360 = 57,068,640
**33. a.** × HHHH △△△△
**b.** 一千四百四十
**c.** ≺ ≺ ⋎ ⋎ ⋎ ⋎ ⊲
**35. a.** $\overline{\text{X}}$MMCCCXV
**b.** 一万二千三百一十五

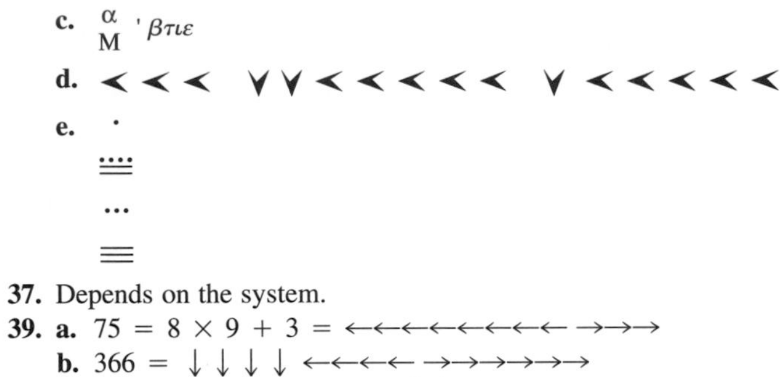

**37.** Depends on the system.

**39. a.** $75 = 8 \times 9 + 3 =$ ←←←←←←←←← →→→
**b.** $366 =$ ↓↓↓↓ ←←←←← →→→→→→
**c.** $5280 =$ ↑↑↑↑↑↑↑ ↓↓ ← →→→→→→→
**d.** $1000 =$ ↑ ↓↓↓ ←←←← →

**41.** Answers will vary.

**43. a.** In Egyptian, as is done in Hindu Arabic, powers of ten will carry.

| | | | |
|---|---|---|---|
| | 𓍢 | 𓎆 | |
| | 𓍢 | 𓎆𓎆𓎆𓎆𓎆𓎆𓎆 | \|\|\|\|\|\|\|\| |
| + | | 𓎆𓎆𓎆𓎆 | \|\|\|\|\|\|\| |
| | 𓍢𓍢 | 𓎆𓎆 | \|\|\|\|\| |

**b.**

| | | | | | |
|---|---|---|---|---|---|
| | C | L | X | V | |
| | C | L | XX | V | III |
| + | | | XL | V | II |
| | CC | | XX | V | |

**c.**

| | | | | | | |
|---|---|---|---|---|---|---|
| 一 | | 一 | | | | 三 |
| 百 | | 百 | | | | 百 |
| 一 | + | 七 | + | 四 | = | 三 |
| 十 | | 十 | | 十 | | 十 |
| | | 八 | | 七 | | 五 |

## *Section 1.2*

**7.** $1 \times 10^2 + 3 \times 10^1 + 9 + 10^0$
**9.** $4 \times 10^2 + 3 \times 10^1 + 7 \times 10^0 + 1 \times 10^{-1} + 5 \times 10^{-2}$
**11.** $3 \times 10^{-1} + 1 \times 10^{-2} + 4 \times 10^{-3}$
**13.** $5 \times 10^5 + 4 \times 10^4 + 3 \times 10^3 + 8 \times 10^2 + 6 \times 10^1 + 7 \times 10^0$

**15.** $5 \times 10^1 + 3 \times 10^0 + 1 \times 10^{-1} + 7 \times 10^{-2} + 1 \times 10^{-3}$
**17.** $6 \times 10^{-1} + 2 \times 10^{-2} + 1 \times 10^{-3} + 9 \times 10^{-4} + 3 \times 10^{-5}$
**19.** Tens, hundredths, tens, ten-thousands, tenths, hundredths, thousands, hundredths, ones, ones, hundred-thousandths, thousandths
**21.** 1/10 + 1/4 = 7/20 **23.** 1/1000 + 1/300 + 1/20 = 163/3000
**25.** 12/60 = 1/5 **27.** 10/60 + 10/3600 + 10/216,000 = 3661/21,600
**29.** 1/3 **31.** 3/4 **33.** 1/63
**35.** The numerator is $\mu\theta = 49$. The denominator is $\phi\alpha = 501$.
Thus, $\mu\theta\ \phi\alpha'\ \phi\alpha' = 49/501$.
**37.** (12)(60) + 23 + 12/60 = 743 1/5 **39.** 101 1/34
**41.** 5/12 = 4/12 + 1/12 = 1/3 + 1/12 = ◎ over |||   ◎ over ∩||
**43.** 8/15 = 5/15 + 3/15 = 1/3 + 1/5 = ◎ over |||   ◎ over |||||
**45.** ◎ over |||   ◎ over ||
**47.** ···· + S = S ····
**49.** XII × III S . . . = XLV
**51.** $\rho\kappa \times \gamma\ \gamma\delta'\delta' = \upsilon\nu$ **53.** Answers will vary.

## *Section 1.3*

**7.** 77 **9.** 5.625 **11.** 2952 **13.** 1132.25
**15.** 136 **17.** 15 **19.** $3101_5$ **21.** $1736_8$
**23.** $2160_7$ **25.** $10B0_{12}$ **27.** $81_{16}$ **29.** $313126_9$
**31.** $11\,011\,110_2$ **33.** $2431_5$ **35.** $556_8$ **37.** $7D0_{16}$
**39.** $224\,000\,000_5$ **41.** $3\,641\,100_8$ **43.** $F4\,240_{16}$ **45.** $254_9$
**47.** $3033_4$ **49.** $150_7$ **51.** 8
**53. a.** $45 = 1200_3$, $100 = 10201_3$, $200 = 21102_3$
$45 = 50_9$, $100 = 121_9$, $200 = 242_9$
**b.** As the base increases, the number of digits decreases.
**c.** Taken in pairs, the digits in the base 3 representation give the base 9 representation. For example, writing $21102_3$ as $2_3\ 11_3\ 02_3$, we have 2, 4, and 2. These are the digits in the base 9 representation of 200.
**55.** Answers will vary. **57.** Answers will vary.

## *Section 1.4*

**13. a.** $\sqrt{100}$, 19
**b.** 0, $\sqrt{100}$, 19
**c.** $-7$, $-\sqrt{64}$, 0, $\sqrt{100}$, 19
**d.** $-14{,}785$, $-7$, $-\sqrt{64}$, $\frac{-5}{16}$, 0, $5\frac{7}{8}$, 9.76555. . . , $\sqrt{100}$, 19
**e.** $\sqrt[5]{19}$, $\sqrt{8}$, $\pi$
**f.** $-14.785$, $-7$, $-\sqrt{64}$, $\frac{-5}{16}$, 0, $\sqrt[5]{19}$, $\sqrt{8}$, $\pi$, $5\frac{7}{8}$, 9.76555. . . , $\sqrt{100}$, 19

**g.** $-\sqrt{-25}, \sqrt{-8}$
**h.** $\pi$

**15.**

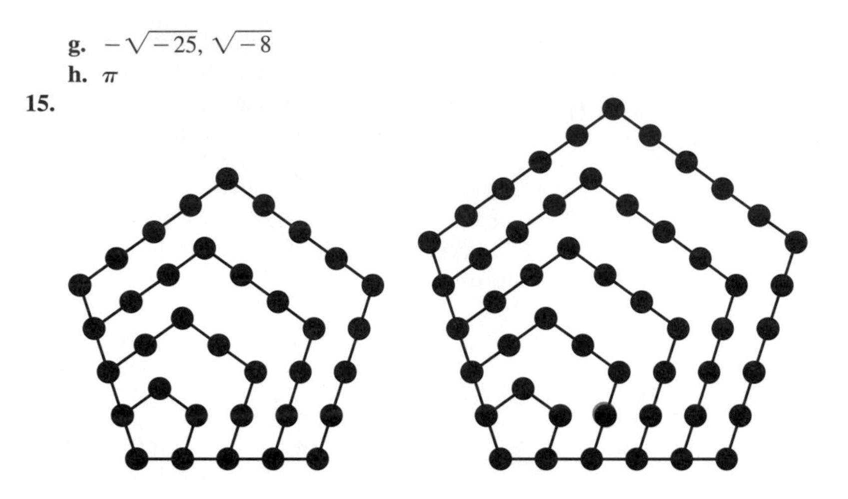

The fifth pentagonal number is 35. The sixth pentagonal number is 51.

**17. a.** Prime, deficient
**b.** Composite, deficient—proper factors are 1, 7, 11.
**c.** Composite, deficient—proper factors are 1, 5, 29.
**d.** Composite, abundant—proper factors are 1, 2, 4, 7, 14, 28, 71, 142, 284, 497, 994.
**e.** Composite, perfect—proper factors are 1, 2, 4, 8, 16, 32, 64, 127, 254, 508, 1016, 2032, 4064.
**f.** Composite, abundant—proper factors are 1, 2, 3, 4, 5, 6, 8, 9, 10, 12, 15, 18, 20, 24, 25, 30, 36, 40, 45, 50, 60, 72, 75, 90, 100, 120, 125, 150, 180, 200, 225, 250, 300, 360, 375, 450, 500, 600, 750, 900, 1000, 1125, 1500, 1800, 2250, 3000, 4500.

**19.** No, compare the diagonals; yes; no, compare the diagonals.

**23.**

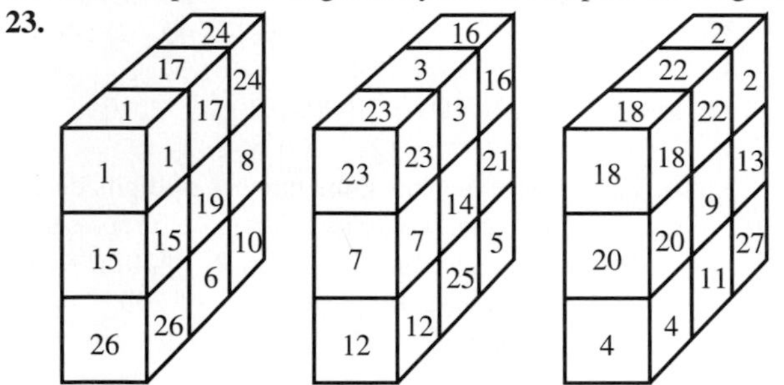


**25.** The number 1 has no proper factors.

## *Chapter Summary*

**1. a.** 𓍢𓎆𓎆𓏺𓏺𓏺
**b.** CXXIII

**c.** 一<br>百<br>二<br>十<br>三

**d.** ρκγ

**e.** 123 = 2(60) + 3 = 𒁹𒁹 𒁹𒁹𒁹

**f.** ·<br>—<br>···

**g.** 123

**h.** $\begin{array}{r} 1 \\ 64\overline{)123} \\ \underline{64} \\ 59 \end{array}$ $\begin{array}{r} 1 \\ 32\overline{)59} \\ \underline{32} \\ 27 \end{array}$ $\begin{array}{r} 1 \\ 16\overline{)27} \\ \underline{16} \\ 11 \end{array}$ $\begin{array}{r} 1 \\ 8\overline{)11} \\ \underline{8} \\ 3 \end{array}$ $\begin{array}{r} 0 \\ 4\overline{)3} \\ \underline{0} \\ 3 \end{array}$ $\begin{array}{r} 1 \\ 2\overline{)3} \\ \underline{2} \\ 1 \end{array}$ $\begin{array}{r} 1 \\ 1\overline{)1} \\ \underline{1} \\ 0 \end{array}$ $1\ 111\ 011_2$

**i.** $\begin{array}{r} 4 \\ 25\overline{)123} \\ \underline{100} \\ 23 \end{array}$ $\begin{array}{r} 4 \\ 5\overline{)23} \\ \underline{20} \\ 3 \end{array}$ $\begin{array}{r} 3 \\ 1\overline{)3} \\ \underline{3} \\ 0 \end{array}$ $443_5$

**j.** $\begin{array}{r} 1 \\ 64\overline{)123} \\ \underline{64} \\ 59 \end{array}$ $\begin{array}{r} 7 \\ 8\overline{)59} \\ \underline{56} \\ 3 \end{array}$ $\begin{array}{r} 3 \\ 1\overline{)3} \\ \underline{3} \\ 0 \end{array}$ $173_8$

**k.** $\begin{array}{r} 10 \\ 12\overline{)123} \\ \underline{120} \\ 3 \end{array}$ $\begin{array}{r} 3 \\ 1\overline{)3} \\ \underline{3} \\ 0 \end{array}$ $A3_{12}$

**l.** $\begin{array}{r} 7 \\ 16\overline{)123} \\ \underline{112} \\ 11 \end{array}$ $\begin{array}{r} 11 \\ 1\overline{)11} \\ \underline{11} \\ 1 \end{array}$ $7B_{16}$

**2. a.** $\begin{array}{r} 4 \\ 25\overline{)123} \\ \underline{100} \\ 23 \end{array}$ $\begin{array}{r} 4 \\ 5\overline{)23} \\ \underline{20} \\ 3 \end{array}$ $\begin{array}{r} 3 \\ 1\overline{)3} \\ \underline{3} \\ 0 \end{array}$ ↓↓↓↓←←←←→→→

**b.** $\begin{array}{r} 2 \\ 125\overline{)366} \\ \underline{250} \\ 116 \end{array}$ $\begin{array}{r} 4 \\ 25\overline{)116} \\ \underline{100} \\ 16 \end{array}$ $\begin{array}{r} 3 \\ 5\overline{)16} \\ \underline{15} \\ 1 \end{array}$ $\begin{array}{r} 1 \\ 1\overline{)1} \\ \underline{1} \\ 0 \end{array}$ ↑↑↓↓↓↓←←←→

**c.** $\begin{array}{r} 1 \\ 3125\overline{)5280} \\ \underline{3125} \\ 2155 \end{array}$ $\begin{array}{r} 3 \\ 625\overline{)2155} \\ \underline{1875} \\ 280 \end{array}$ $\begin{array}{r} 2 \\ 125\overline{)280} \\ \underline{250} \\ 30 \end{array}$ $\begin{array}{r} 1 \\ 25\overline{)30} \\ \underline{25} \\ 5 \end{array}$ $\begin{array}{r} 1 \\ 5\overline{)5} \\ \underline{5} \\ 0 \end{array}$ $\begin{array}{r} 0 \\ 1\overline{)0} \\ \underline{1} \\ 0 \end{array}$ ↕↔↔↔↑↑↓←

**d.** $\begin{array}{r} 1 \\ 625\overline{)1000} \\ \underline{625} \\ 375 \end{array}$ $\begin{array}{r} 3 \\ 125\overline{)375} \\ \underline{375} \\ 0 \end{array}$ $\begin{array}{r} 0 \\ 25\overline{)0} \end{array}$ $\begin{array}{r} 0 \\ 5\overline{)0} \end{array}$ $\begin{array}{r} 0 \\ 1\overline{)0} \end{array}$ ↔↑↑↑

**3. a.** 54; 614; 8945; 3012; 71,750; 10,400

**b.** $\ddot{a}\dot{b}c$, $\ddot{c}\dot{f}f$, $\dddot{e}\,\ddot{b}\dot{h}$, $\dddot{a}$

**4.** Answers will vary depending on what is selected.

**5.**
$$\begin{array}{r}1\\64\overline{)123}\\\underline{64}\\59\end{array}\quad\begin{array}{r}3\\16\overline{)59}\\\underline{48}\\11\end{array}\quad\begin{array}{r}2\\4\overline{)11}\\\underline{8}\\3\end{array}\quad\begin{array}{r}3\\1\overline{)3}\\\underline{3}\\0\end{array}\quad 123 = 1323_4 = \text{/ ▼×▼}$$

$$\begin{array}{r}1\\256\overline{)366}\\\underline{256}\\110\end{array}\quad\begin{array}{r}1\\64\overline{)110}\\\underline{64}\\46\end{array}\quad\begin{array}{r}2\\16\overline{)46}\\\underline{32}\\14\end{array}\quad\begin{array}{r}3\\4\overline{)14}\\\underline{12}\\2\end{array}\quad\begin{array}{r}2\\1\overline{)2}\\\underline{2}\\0\end{array}\quad 366 = 11232_4 = \text{// ×▼×}$$

$$\begin{array}{r}1\\4096\overline{)5280}\\\underline{4096}\\1184\end{array}\quad\begin{array}{r}1\\1024\overline{)1184}\\\underline{1024}\\160\end{array}\quad\begin{array}{r}0\\256\overline{)160}\end{array}\quad\begin{array}{r}2\\64\overline{)160}\\\underline{128}\\32\end{array}\quad\begin{array}{r}2\\16\overline{)32}\\\underline{32}\\0\end{array}\quad\begin{array}{r}0\\4\overline{)0}\end{array}\quad\begin{array}{r}0\\1\overline{)0}\end{array}$$

$5280 = 1102200_4 =$ // *××**

$$\begin{array}{r}3\\256\overline{)1000}\\\underline{768}\\232\end{array}\quad\begin{array}{r}3\\64\overline{)232}\\\underline{192}\\40\end{array}\quad\begin{array}{r}2\\16\overline{)40}\\\underline{32}\\8\end{array}\quad\begin{array}{r}2\\4\overline{)8}\\\underline{8}\\0\end{array}\quad\begin{array}{r}0\\1\overline{)0}\end{array}\quad 1000 = 33220_4 = \text{▼▼××*}$$

**6.**
$$\begin{array}{r}2\\49\overline{)123}\\\underline{98}\\25\end{array}\quad\begin{array}{r}3\\7\overline{)25}\\\underline{21}\\4\end{array}\quad\begin{array}{r}4\\1\overline{)4}\\\underline{4}\\0\end{array}\quad 234_7$$

$$\begin{array}{r}1\\343\overline{)366}\\\underline{343}\\23\end{array}\quad\begin{array}{r}0\\49\overline{)23}\\\underline{0}\\23\end{array}\quad\begin{array}{r}3\\7\overline{)23}\\\underline{21}\\2\end{array}\quad\begin{array}{r}2\\1\overline{)2}\\\underline{2}\\0\end{array}\quad 1032_7$$

$$\begin{array}{r}2\\2401\overline{)5280}\\\underline{4802}\\478\end{array}\quad\begin{array}{r}1\\343\overline{)478}\\\underline{343}\\135\end{array}\quad\begin{array}{r}2\\49\overline{)135}\\\underline{98}\\37\end{array}\quad\begin{array}{r}5\\7\overline{)37}\\\underline{35}\\2\end{array}\quad\begin{array}{r}2\\1\overline{)2}\\\underline{2}\\0\end{array}\quad 21252_7$$

$$\begin{array}{r}2\\343\overline{)1000}\\\underline{686}\\314\end{array}\quad\begin{array}{r}6\\49\overline{)314}\\\underline{294}\\20\end{array}\quad\begin{array}{r}2\\7\overline{)20}\\\underline{14}\\6\end{array}\quad\begin{array}{r}6\\1\overline{)6}\\\underline{6}\\0\end{array}\quad 2626_7$$

**7.** $123_8 = 1 \times 8^2 + 2 \times 8 + 3 = 64 + 16 + 3 = 83$

**8.** 41, 43, 47, 53, 61, 71, 83, 97, 113
If $n$ is a multiple of 41, then all three terms of the polynomial will have 41 as a factor. Thus, the polynomial will have a factor of 41 and will not result in a prime number. If $n = 82$, the formula gives $6683 = 41 \times 163$.

**9.** 400—composite, abundant; 461—prime, deficient;
496—composite, perfect; 512—composite, deficient

**10.**

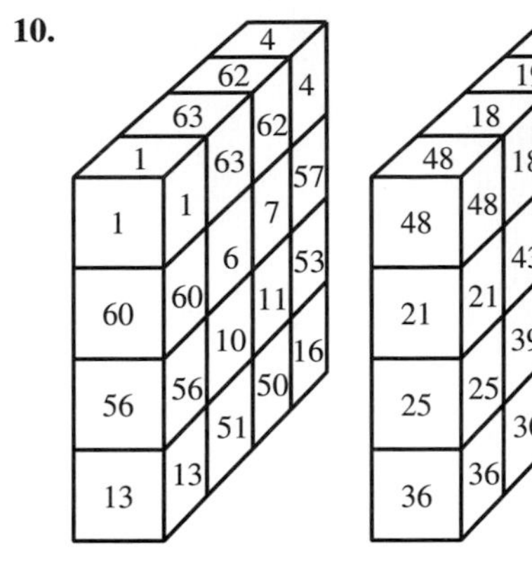


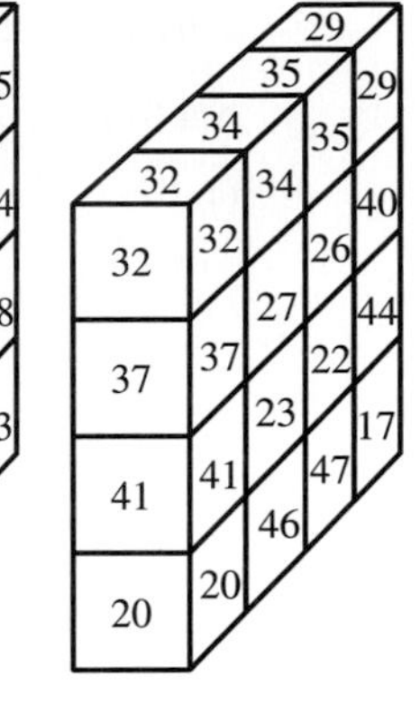


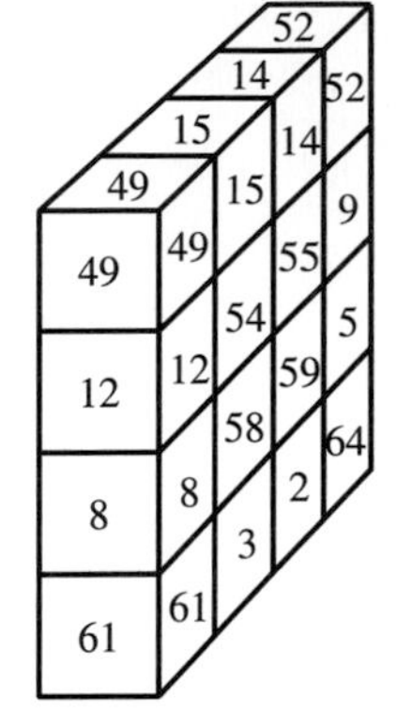


**11. b.**

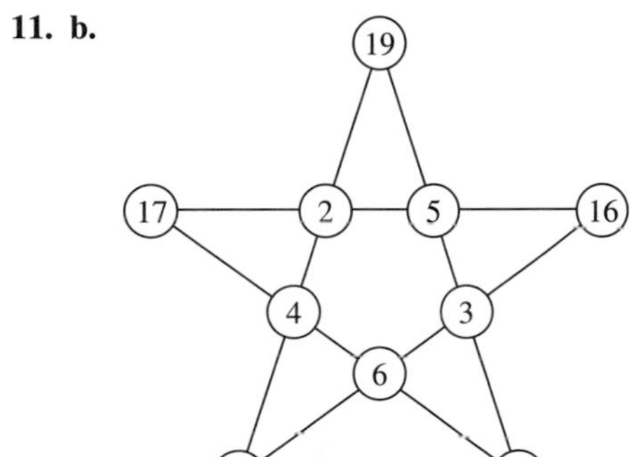


**c.**

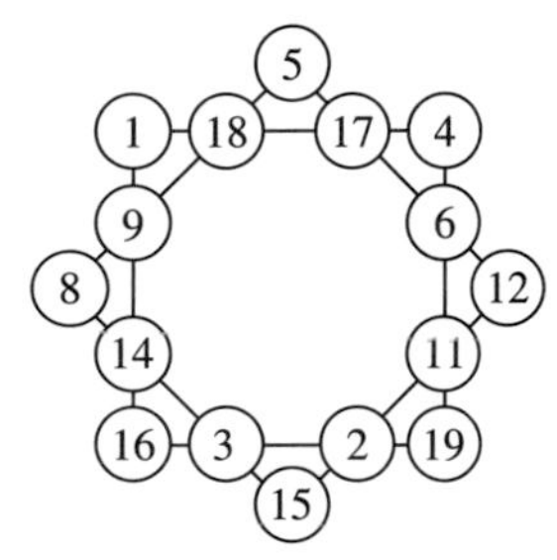


**d.**

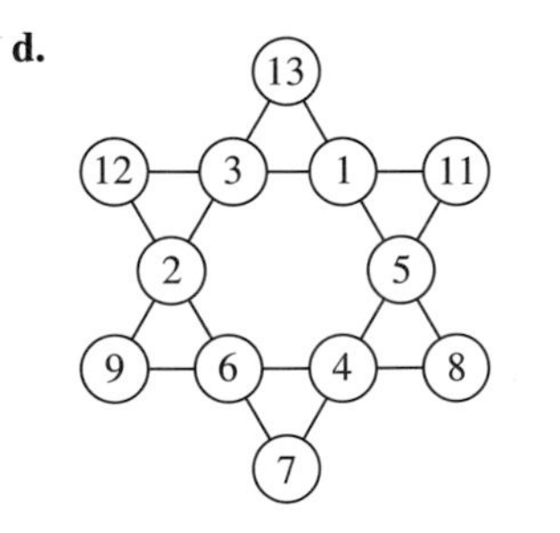


**12. a.** $\sqrt{-400}$

**b.** $-5$, $\sqrt[3]{-27}$, 0, 13

**c.** $-\sqrt{40}$, $\sqrt[4]{19}$, $\pi$

**d.** 13

**e.** All numbers except those listed in c and $\sqrt{-400}$

**f.** All numbers except $\sqrt{-400}$

**g.** 0, 13

# Chapter 2

## *Section 2.0*

**11.**

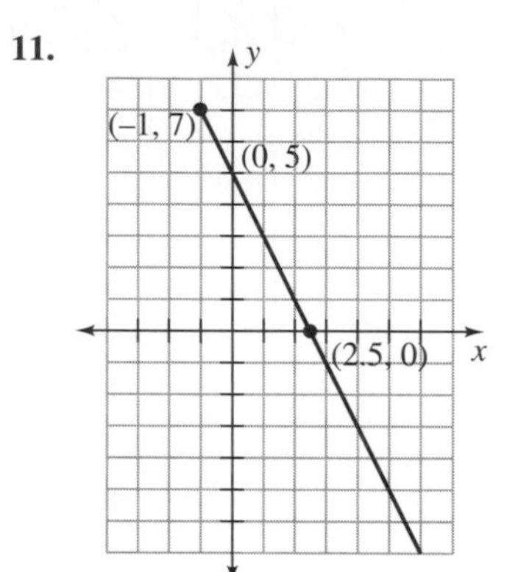


**13.**

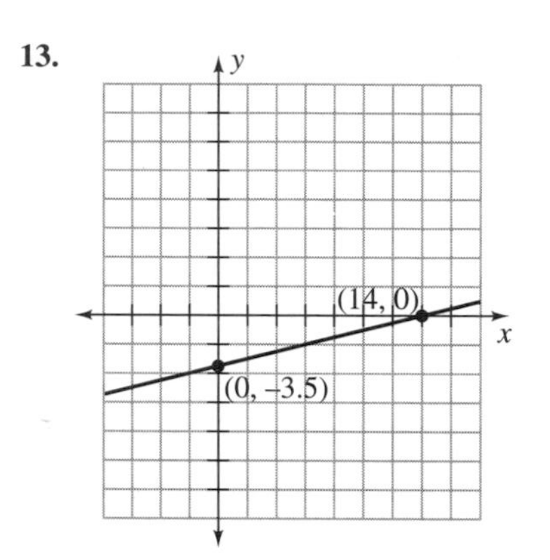


**15.**

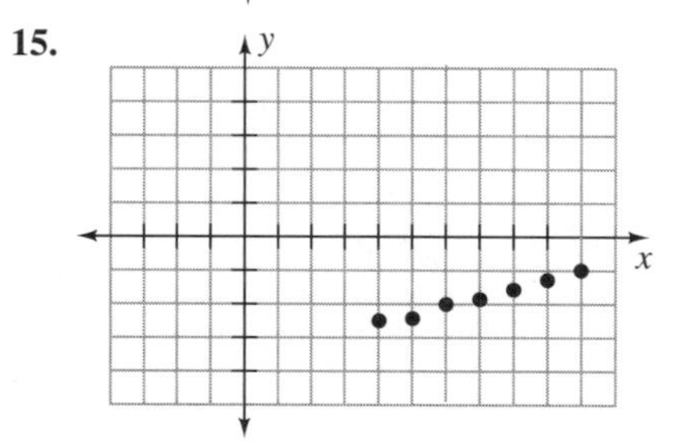


**17.**

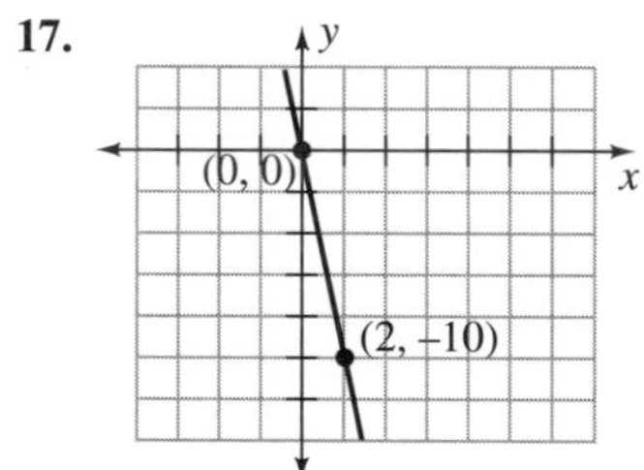


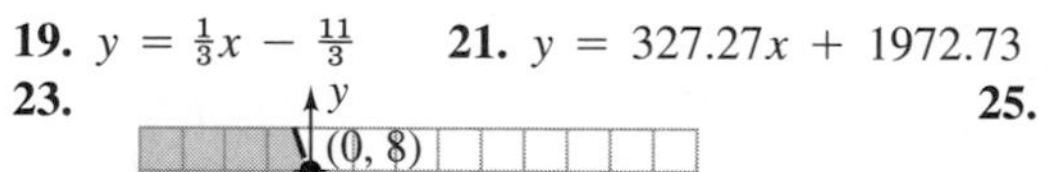

**19.** $y = \frac{1}{3}x - \frac{11}{3}$ **21.** $y = 327.27x + 1972.73$

**23.**

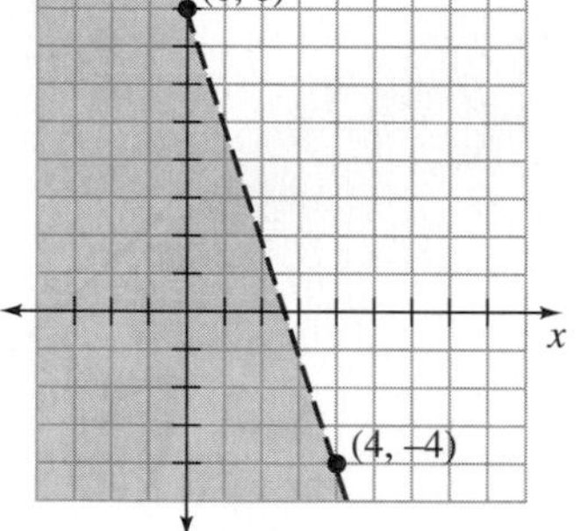
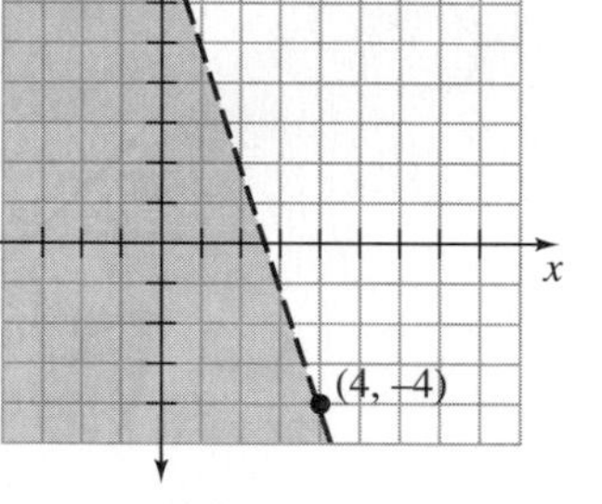


**25.**

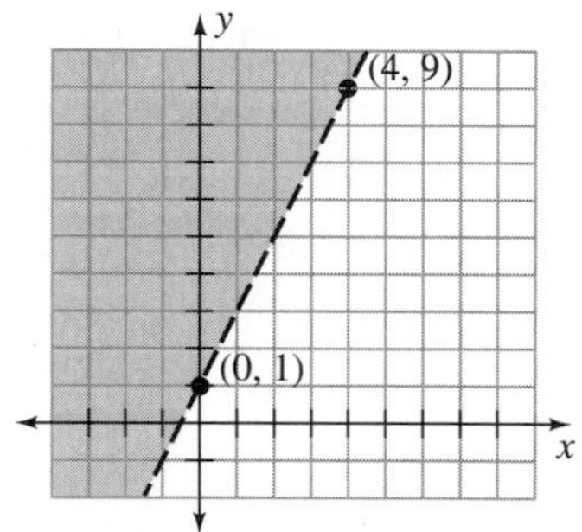


**27.**

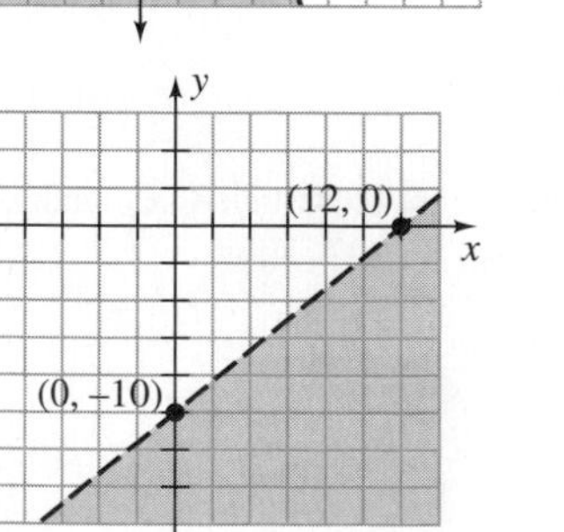


**29.**

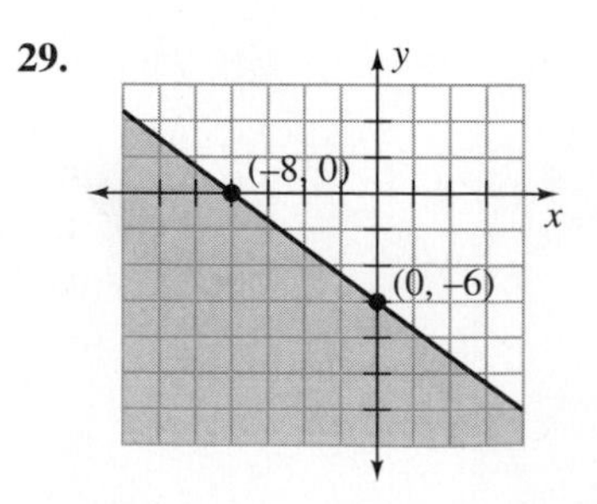

**31.**

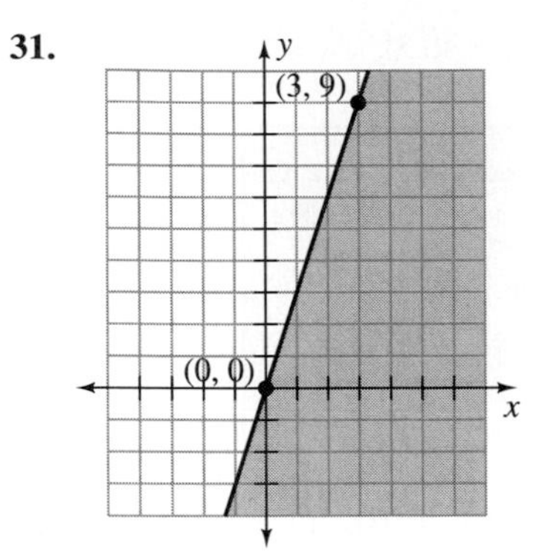


**33.** (0, 0) **35.** $(-2, 0)$ **37.** (0, 0) **39.** $(-3, -2)$

**41.**

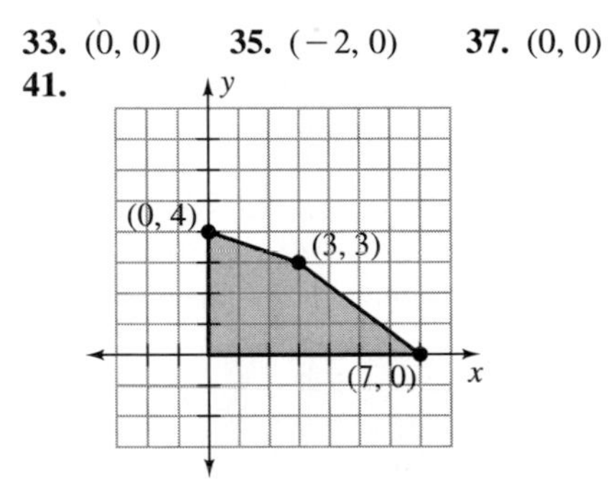


**43.**

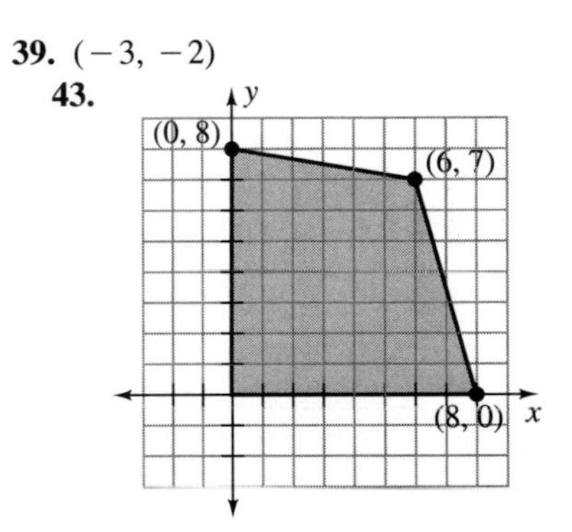


**45.** The vertex is (3, 4).

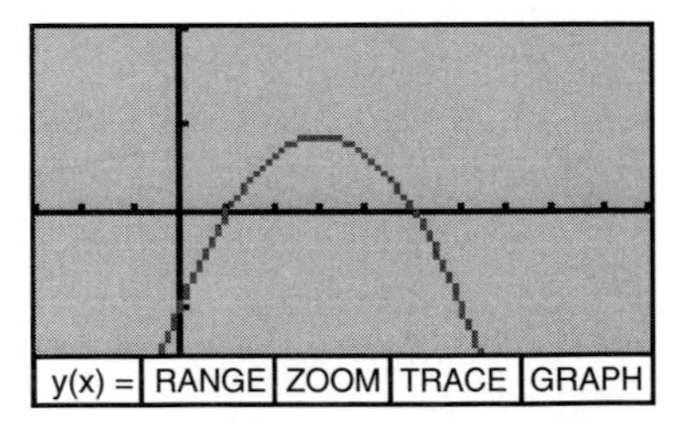


| $x$ | $y$ |
|---|---|
| 0 | $-5$ |
| 3 | 4 |
| 6 | $-5$ |

**47.** The vertex is $(0, -5.6)$.

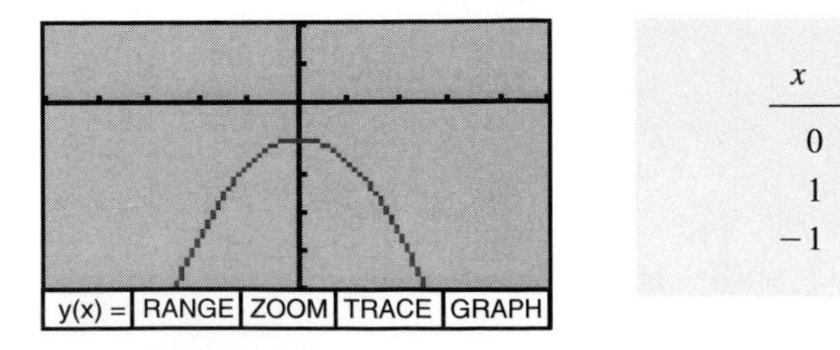


| $x$ | $y$ |
|---|---|
| 0 | $-5.6$ |
| 1 | 8.8 |
| $-1$ | 8.8 |

**49.** The vertex is $(-9, -40.5)$.

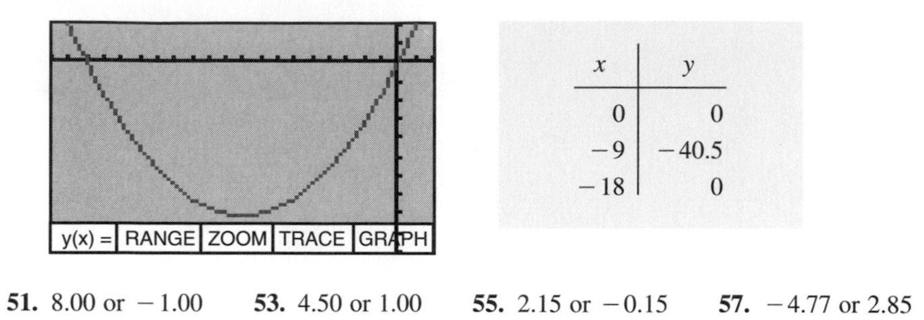


| $x$ | $y$ |
|---|---|
| 0 | 0 |
| −9 | −40.5 |
| −18 | 0 |

**51.** 8.00 or −1.00 **53.** 4.50 or 1.00 **55.** 2.15 or −0.15 **57.** −4.77 or 2.85

## *Section 2.1*

**17.** In Problem 13, the business is currently making a profit and expects the profit to increase. In Problem 14, the business is currently operating at a loss but expects to make a profit soon and expects the profit to increase. In Problem 15, the business is currently making a profit but, because of decreasing profits, expects to soon be operating at a loss. In Problem 16, the business is currently operating at a loss and expects that the losses will continue to increase.

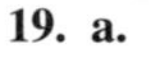
**19. a.**

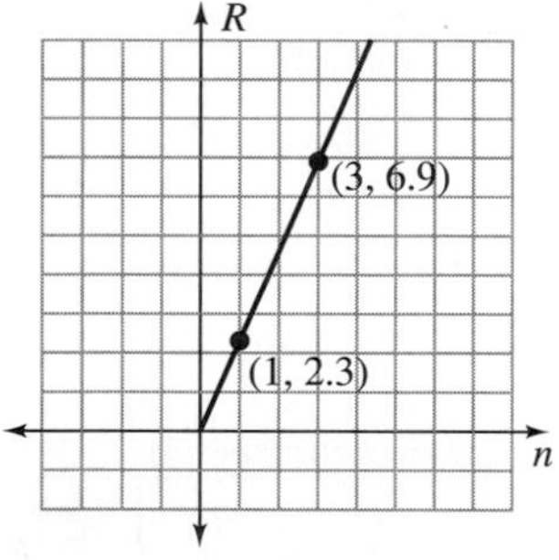


**b.** $m = 2.3$; for every additional customer, the revenue increases by \$2.30.
**c.** \$3450

**21. a.**

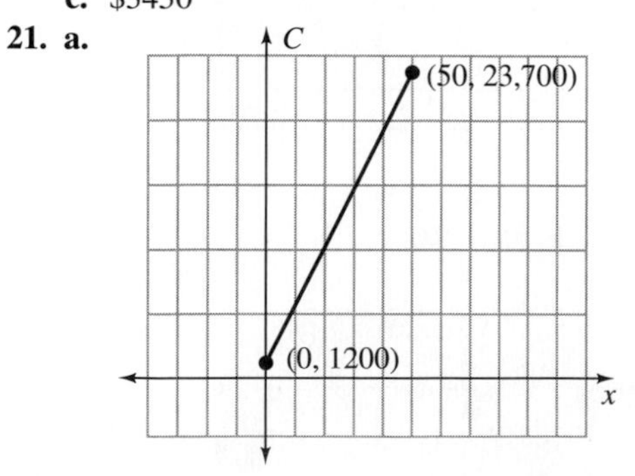


**b.** Since only a whole number of barrels may be produced, the line is actually a series of points at whole number values of $x$.

**c.** $m = 450$. This gives the cost of manufacturing one more wine barrel.
**d.** \$46,200 **e.** 219 barrels

**23. a.** $t = -0.0035a + 59$ **b.** $-25°F$

**25. a.** If $B$ = number of layers of bricks and $H$ = height of the wall in inches, $H = 8B + 10$.

**b.**

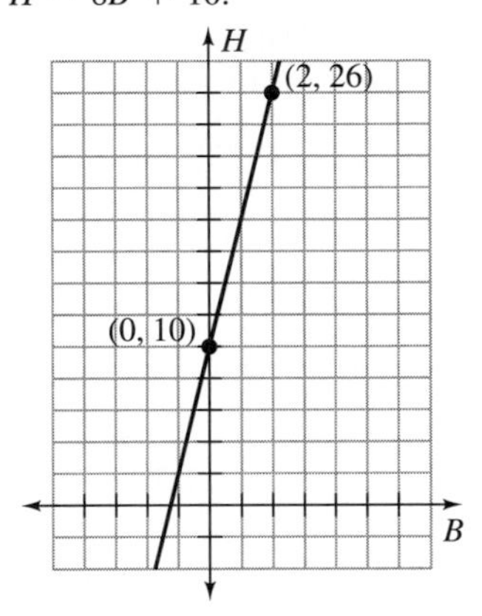


**c.** In fact, $H = \frac{2}{3}B + \frac{5}{6}$

**27. a.** Using $T$ for temperature and $a$ for altitude, $T = -0.002a + 212$ or $a = -500T + 106{,}000$.

**b.** 183.78°F

**29. a.** $c = 0.13737k + 37.0$
$c$ = cost in dollars
$k$ = number of kilowatt hours above baseline amounts

**b.** \$73.09

**c.** $\approx 551$ kwh

**31. a.** For each increase in the height, there is a constant increase in the weight.

**b.** Using $H$ = height in inches and $W$ = weight in pounds, for women we have $W = 4.5H - 148.5$ and for men $W = 4H - 128$.

**33. a.** The graph is a line. **b.** $P = 3000t - 3000$ **c.** \$27,000

## Section 2.2

**7.** Maximum $P = 75$ at (10, 9).
**9.** Minimum $P = 27$ at (0, 9).
**11.** Maximum $P = 288$ at (14, 12).
**13.** Maximum $P = 116$ at (9, 14); minimum $P = 25$ at (2, 3).
**15.** Maximum $P = 40$ at (0, 8).
**17.** Maximum $P = 88$ at (0, 8).
**19.** Minimum $C = 50$ at (6, 4).
**21.** Minimum $C = 95$ at (5, 5).
**23.** Minimum $P = 16$ at (8, 0); maximum $P = 56$ at (4, 12).
**25.** 60 gm of mix A and 100 gm of mix B gives 4.8 gm of fat.
**27.** Use 400 lb of each mix. Minimum cost is \$6000.
**29.** 18 mountain bikes and 2 touring bikes. Maximum $P = \$1460$.

## Section 2.3

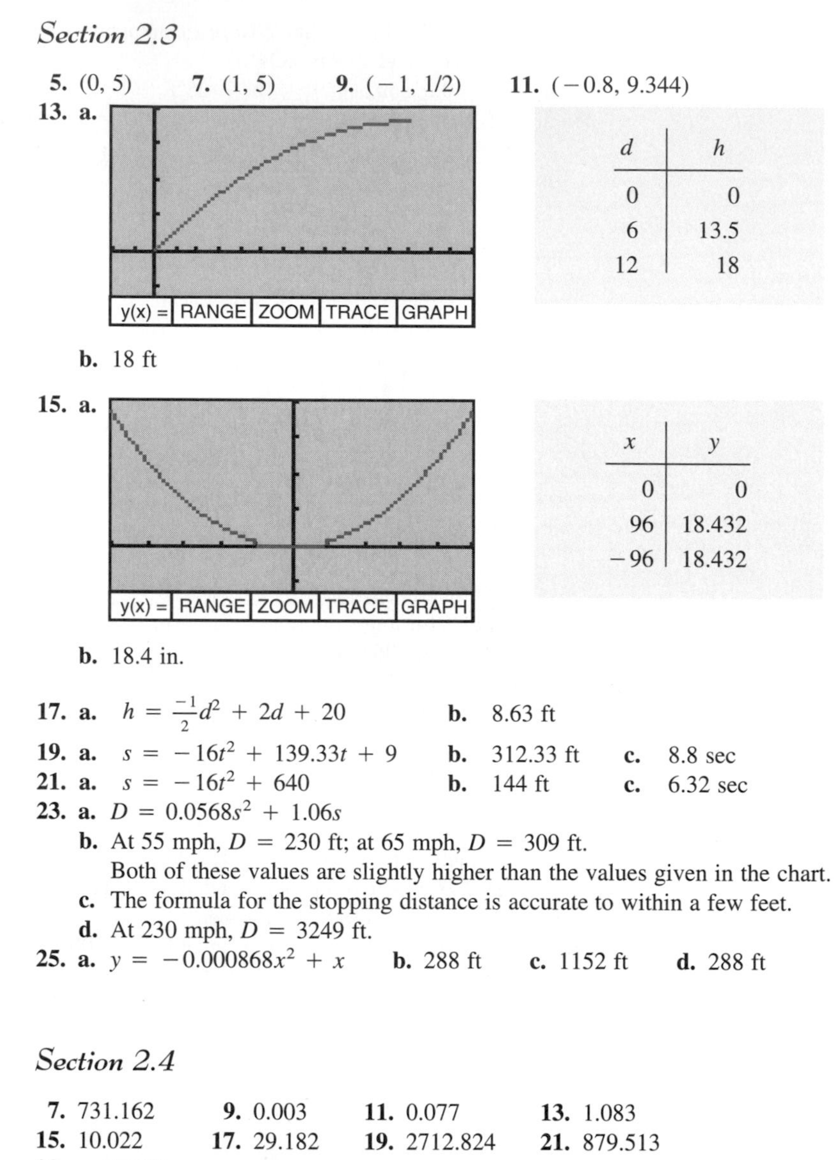


**5.** (0, 5) **7.** (1, 5) **9.** (−1, 1/2) **11.** (−0.8, 9.344)

**13. a.**

| $d$ | $h$ |
|---|---|
| 0 | 0 |
| 6 | 13.5 |
| 12 | 18 |

**b.** 18 ft

**15. a.**

| $x$ | $y$ |
|---|---|
| 0 | 0 |
| 96 | 18.432 |
| −96 | 18.432 |

**b.** 18.4 in.

**17. a.** $h = \frac{-1}{2}d^2 + 2d + 20$ **b.** 8.63 ft

**19. a.** $s = -16t^2 + 139.33t + 9$ **b.** 312.33 ft **c.** 8.8 sec

**21. a.** $s = -16t^2 + 640$ **b.** 144 ft **c.** 6.32 sec

**23. a.** $D = 0.0568s^2 + 1.06s$

**b.** At 55 mph, $D = 230$ ft; at 65 mph, $D = 309$ ft. Both of these values are slightly higher than the values given in the chart.

**c.** The formula for the stopping distance is accurate to within a few feet.

**d.** At 230 mph, $D = 3249$ ft.

**25. a.** $y = -0.000868x^2 + x$ **b.** 288 ft **c.** 1152 ft **d.** 288 ft

## Section 2.4

**7.** 731.162 **9.** 0.003 **11.** 0.077 **13.** 1.083

**15.** 10.022 **17.** 29.182 **19.** 2712.824 **21.** 879.513

**23.** −12.937

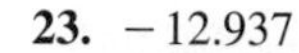

**25.**

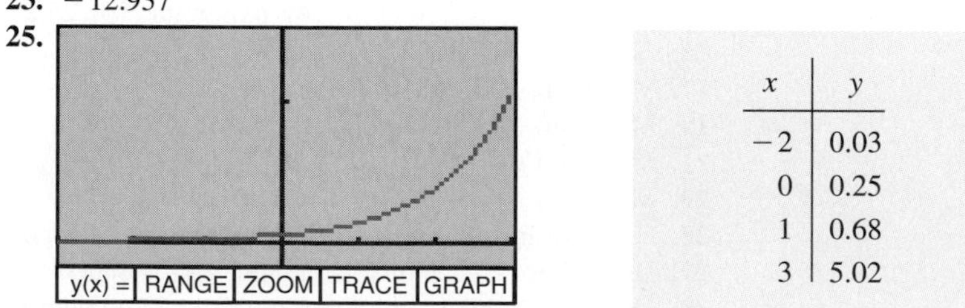


| $x$ | $y$ |
|---|---|
| −2 | 0.03 |
| 0 | 0.25 |
| 1 | 0.68 |
| 3 | 5.02 |

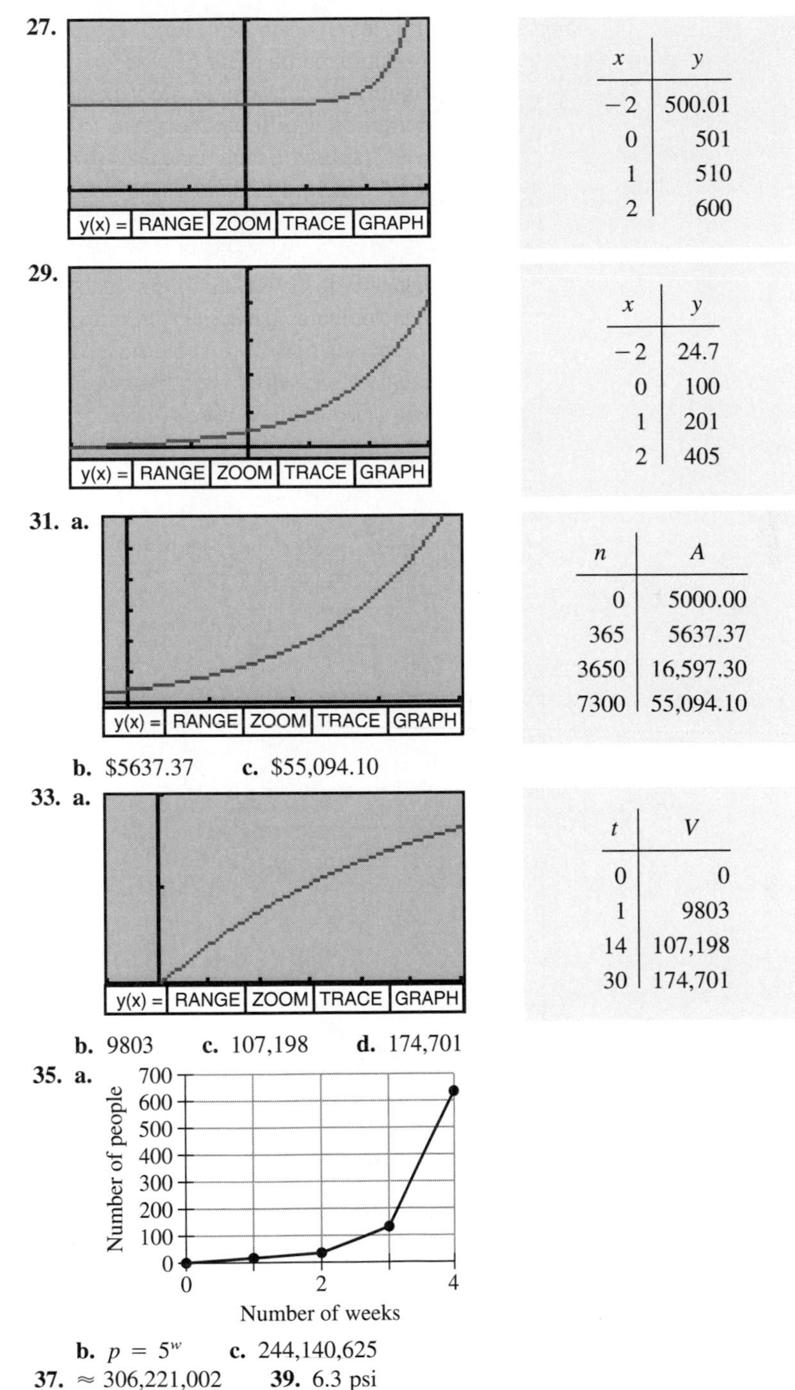


**27.**

| $x$ | $y$ |
|---|---|
| $-2$ | 500.01 |
| 0 | 501 |
| 1 | 510 |
| 2 | 600 |

**29.**

| $x$ | $y$ |
|---|---|
| $-2$ | 24.7 |
| 0 | 100 |
| 1 | 201 |
| 2 | 405 |

**31. a.**

| $n$ | $A$ |
|---|---|
| 0 | 5000.00 |
| 365 | 5637.37 |
| 3650 | 16,597.30 |
| 7300 | 55,094.10 |

**b.** \$5637.37 **c.** \$55,094.10

**33. a.**

| $t$ | $V$ |
|---|---|
| 0 | 0 |
| 1 | 9803 |
| 14 | 107,198 |
| 30 | 174,701 |

**b.** 9803 **c.** 107,198 **d.** 174,701

**35. a.**

**b.** $p = 5^w$ **c.** 244,140,625

**37.** $\approx$ 306,221,002 **39.** 6.3 psi

**41.** $3.689 \times 10^{19}$ kernels of corn, 10,541,000,000,000,000 lb

## *Section 2.5*

**7.** 22.291 **9.** 16.741 **11.** 25,212.463 **13.** 4.956

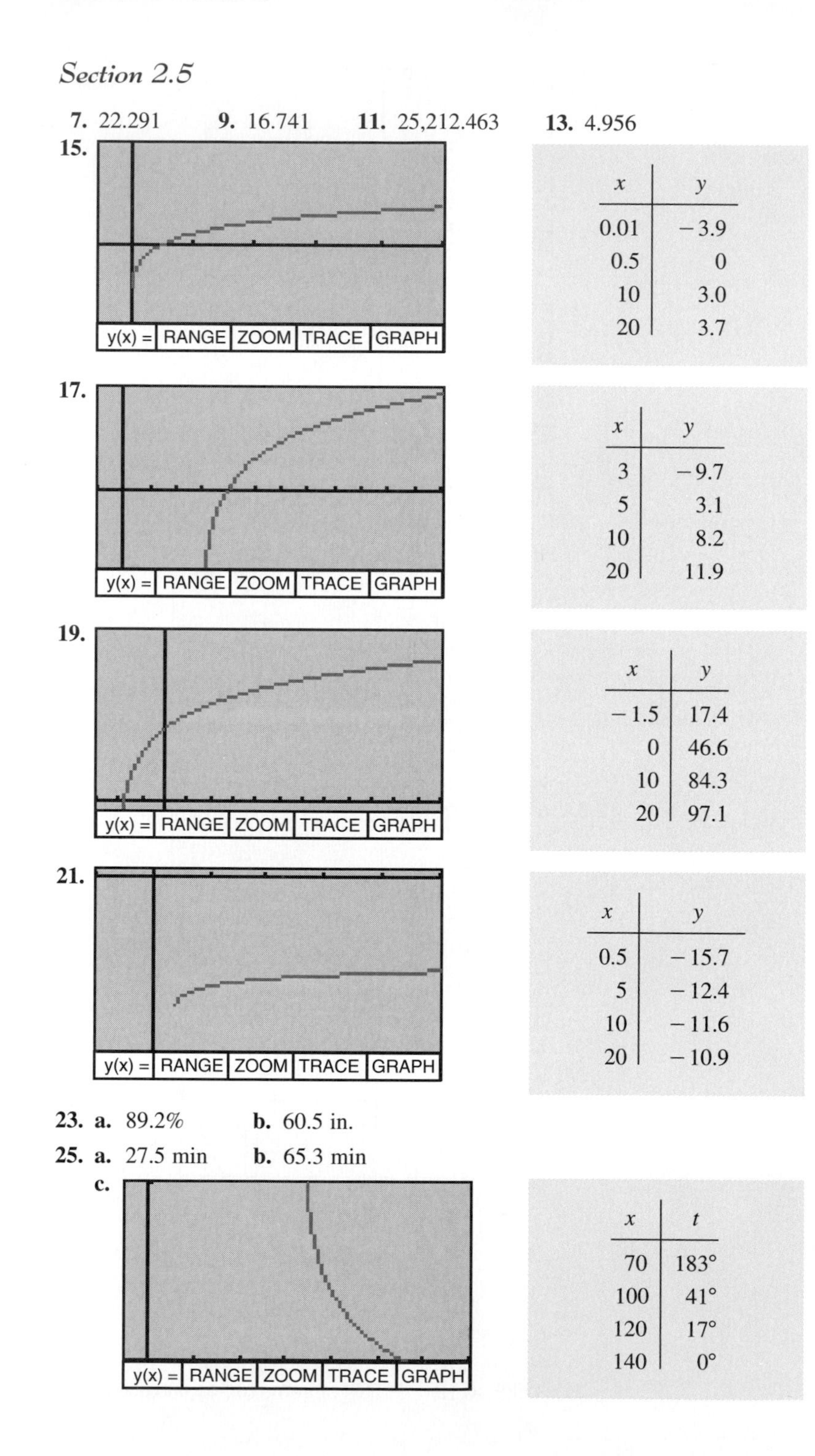


**15.**

| $x$ | $y$ |
|---|---|
| 0.01 | −3.9 |
| 0.5 | 0 |
| 10 | 3.0 |
| 20 | 3.7 |

**17.**

| $x$ | $y$ |
|---|---|
| 3 | −9.7 |
| 5 | 3.1 |
| 10 | 8.2 |
| 20 | 11.9 |

**19.**

| $x$ | $y$ |
|---|---|
| −1.5 | 17.4 |
| 0 | 46.6 |
| 10 | 84.3 |
| 20 | 97.1 |

**21.**

| $x$ | $y$ |
|---|---|
| 0.5 | −15.7 |
| 5 | −12.4 |
| 10 | −11.6 |
| 20 | −10.9 |

**23. a.** 89.2% **b.** 60.5 in.

**25. a.** 27.5 min **b.** 65.3 min

**c.**

| $x$ | $t$ |
|---|---|
| 70 | 183° |
| 100 | 41° |
| 120 | 17° |
| 140 | 0° |

**27. a.**

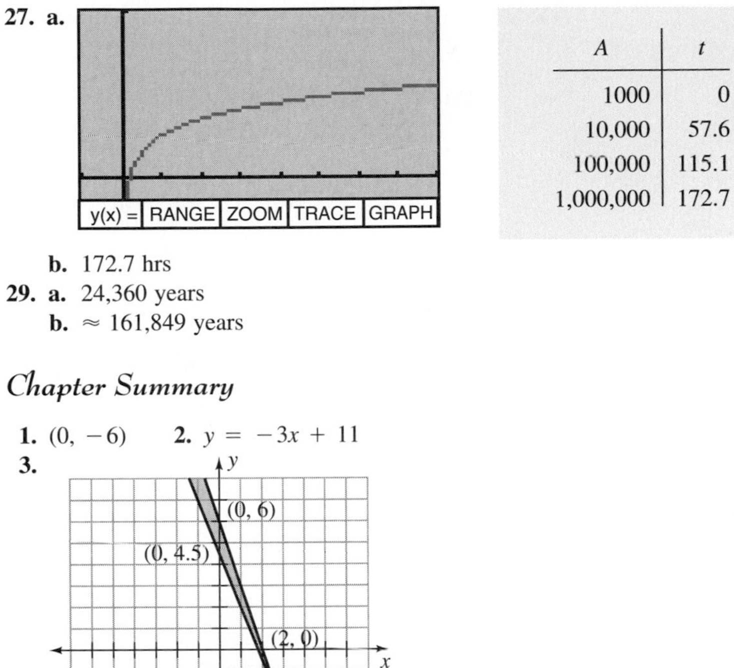


| $A$ | $t$ |
|---|---|
| 1000 | 0 |
| 10,000 | 57.6 |
| 100,000 | 115.1 |
| 1,000,000 | 172.7 |

**b.** 172.7 hrs

**29. a.** 24,360 years

**b.** $\approx$ 161,849 years

## *Chapter Summary*

**1.** $(0, -6)$ **2.** $y = -3x + 11$

**3.**

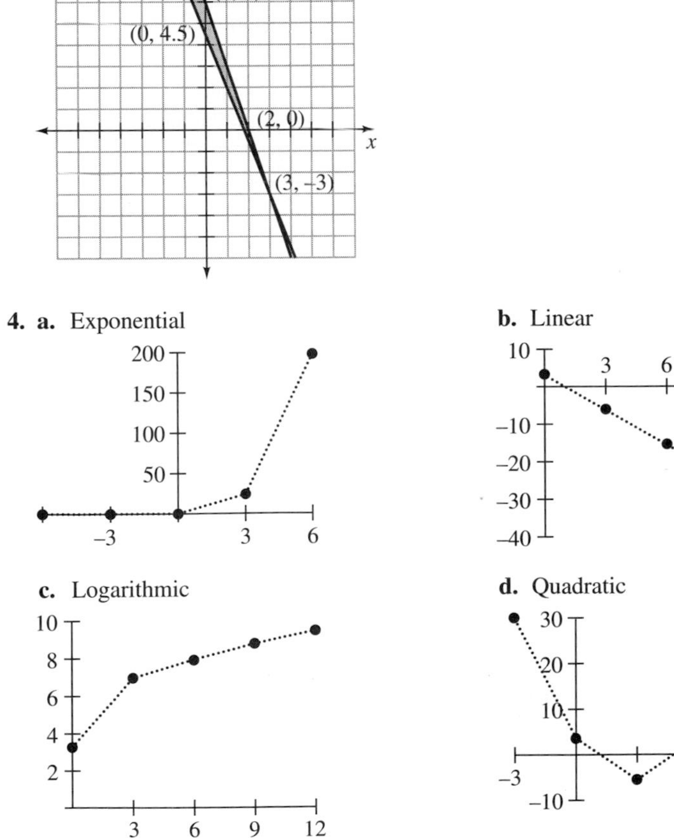


**4. a.** Exponential

**b.** Linear

**c.** Logarithmic

**d.** Quadratic

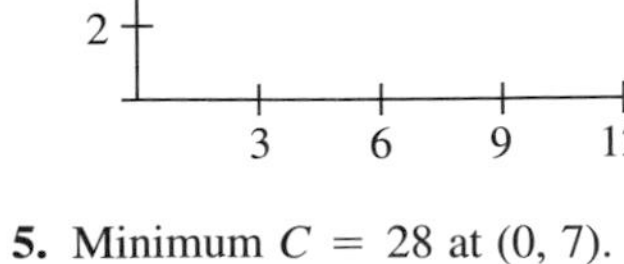

**5.** Minimum $C = 28$ at $(0, 7)$.

**6.** Maximum $P = 40$ at $(8, 0)$.

**7. a.**

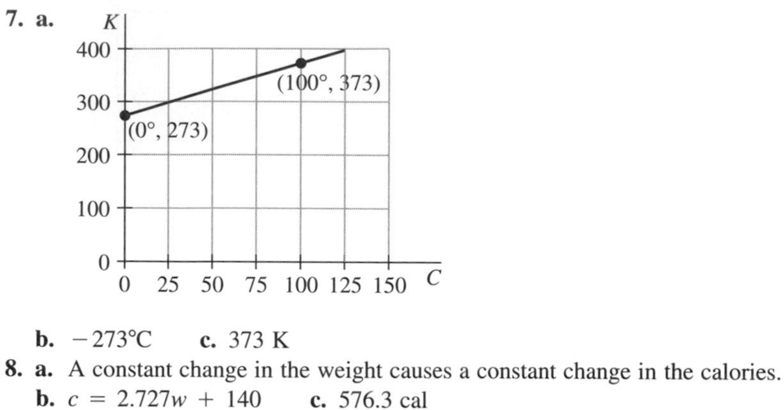


**b.** $-273$°C **c.** 373 K

**8. a.** A constant change in the weight causes a constant change in the calories.

**b.** $c = 2.727w + 140$ **c.** 576.3 cal

**9. a.** The number of logs in each row changes by a constant amount.

**b.** $L = -2r + 249$ where $L$ is the number of logs in a row and $r$, the row number, is an integer with $1 \leq r \leq 124$.

**c.** This line represents the dots where $r$ is an integer.

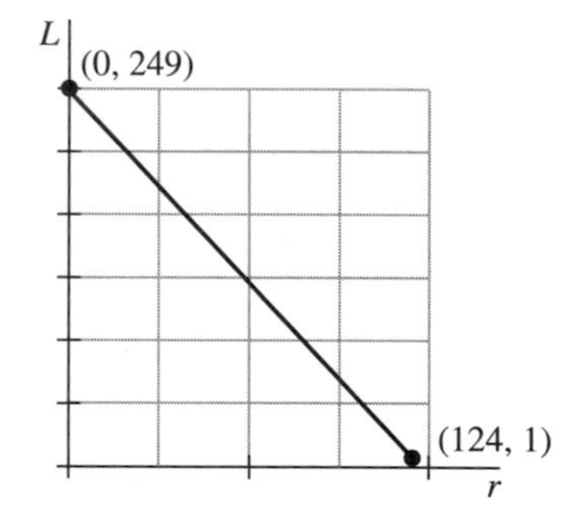


**d.** 149 **e.** 124 rows

**10. a.** $s = 293a - 2535$ **b.** 2739 stamps **c.** 43

**11. a.**

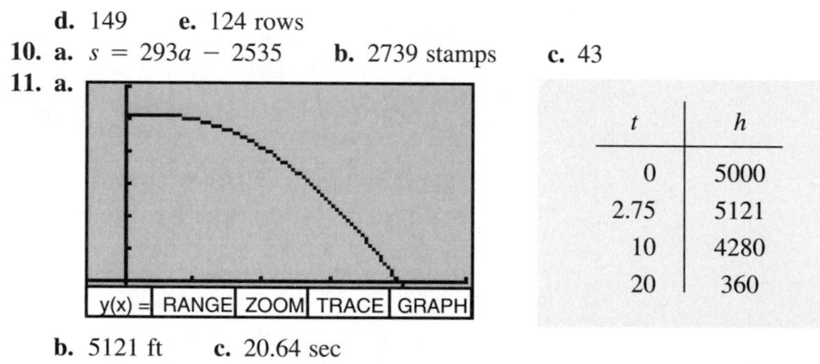


| $t$ | $h$ |
|---|---|
| 0 | 5000 |
| 2.75 | 5121 |
| 10 | 4280 |
| 20 | 360 |

**b.** 5121 ft **c.** 20.64 sec

**12. a.** $h = -0.001758x^2 + 0.616x$ **b.** 4 ft 10.5 in. and 24 ft 3.5 in.

**13. a.** $P = \frac{3}{2}t^2 - \frac{1}{2}t$

**b.**

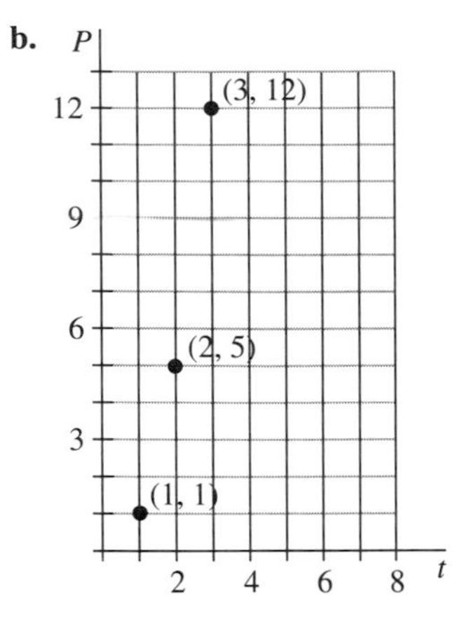


**c.** 14,950; a pentagon with one hundred dots on a side and all the smaller pentagons contained inside.

**14. a.**

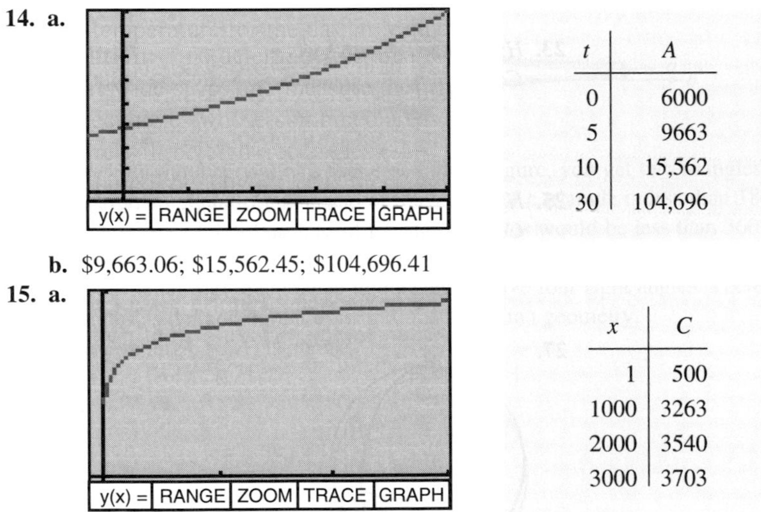


| $t$ | $A$ |
|---|---|
| 0 | 6000 |
| 5 | 9663 |
| 10 | 15,562 |
| 30 | 104,696 |

**b.** $9,663.06; $15,562.45; $104,696.41

**15. a.**

| $x$ | $C$ |
|---|---|
| 1 | 500 |
| 1000 | 3263 |
| 2000 | 3540 |
| 3000 | 3703 |

**b.** $3,263.10; $3,540.36

**c.** $3.26; $1.77. As the number of trucks produced increases, the cost per truck decreases.

**16. a.** $A = 3^s$, where $s$ = the square on the Monopoly board and $A$ = amount of money.

**b.** $19,683; no

**c.** No, more than $282,000,000,000 is needed.

**d.** $12,157,665,460,000,000,000/$40,000,000,000 $\approx$ 303,941,637 yr

## *Section 3.2*

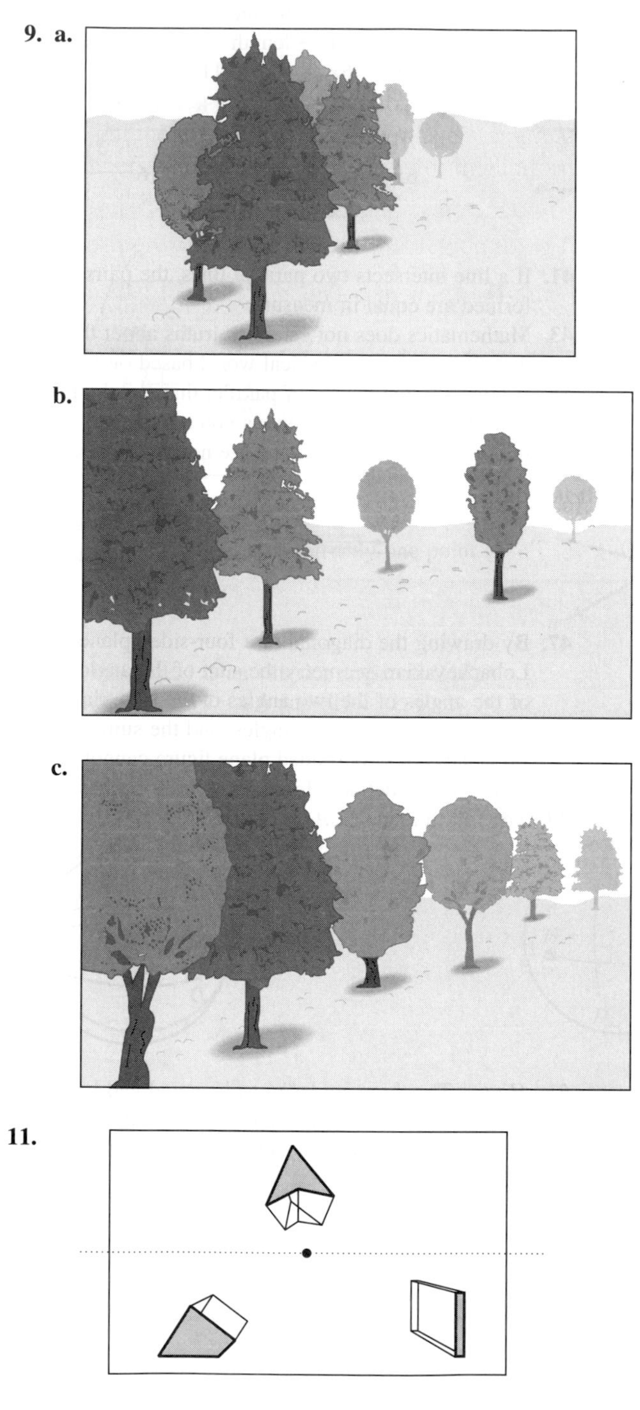

**13.**

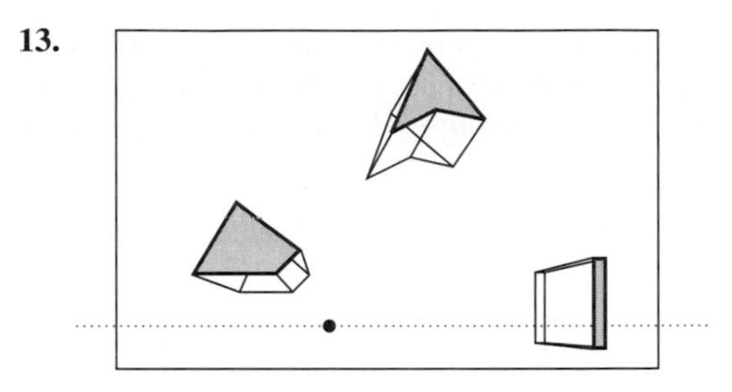

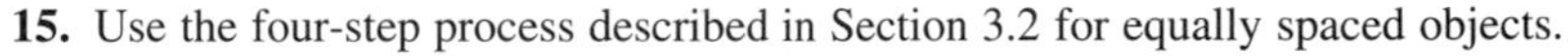

**15.** Use the four-step process described in Section 3.2 for equally spaced objects.

**17.**

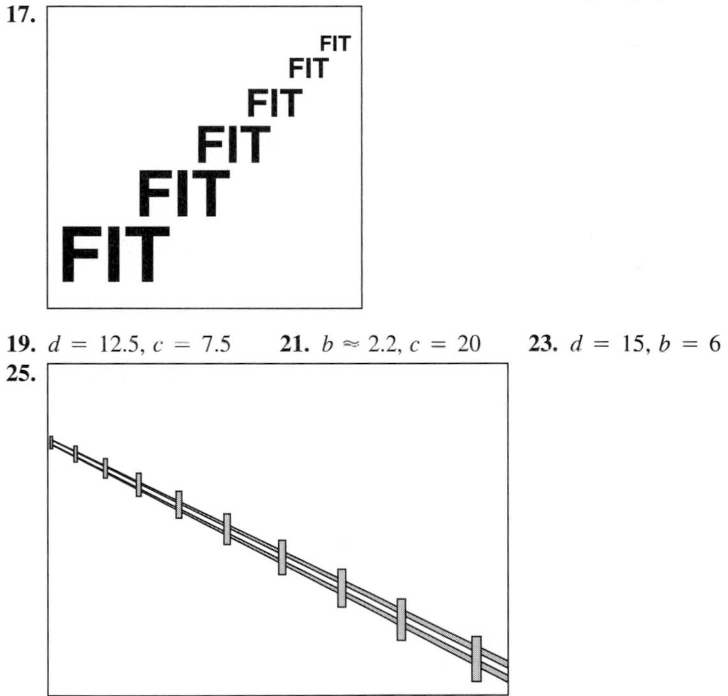


**19.** $d = 12.5$, $c = 7.5$ **21.** $b \approx 2.2$, $c = 20$ **23.** $d = 15$, $b = 6$

**25.**

**27.** Overlapping shapes **29.** One-point perspective

**31.** None of the methods are used. This is an example of a ''flat painting.''

**33.** $13\frac{1}{3}$ in.

**35.** The tracks seem to get closer together. This suggests a one-point perspective.

**37.** Answer depends on what art is examined.

## *Section 3.3*

**9.** The point is 2.6 cm from one end of the segment.

**11.** The point is 6.3 cm from one end of the segment.

**13.** No **15.** No **17.** 37.3 ft, 14.2 ft **19.** 92.0 m, 35.1 m

**21.** 5.6 cm, 14.6 cm **23.** 22.2 in., 58.3 in.

**25.** $w = 9.9$ cm, $b = 9.9$ cm, $t = 6.1$ cm **29.** $w = 8.1$ in., $l = 13.1$ in.

**31.** $w = 5.6$ cm, $l = 14.6$ cm **33.** $w = 14.6$ cm, $l = 23.7$ cm

**37.** The ratios of consecutive Fibonacci numbers are approximately the Golden Ratio. Thus, three consecutive Fibonacci numbers give approximate Golden Boxes.

**43.** $\dfrac{180(1 + \sqrt{5})}{3 + \sqrt{5}}$

## *Section 3.4*

**21.** $S = 540°$

**23.** $S = 1980°$

**25.** $S = 1440°$

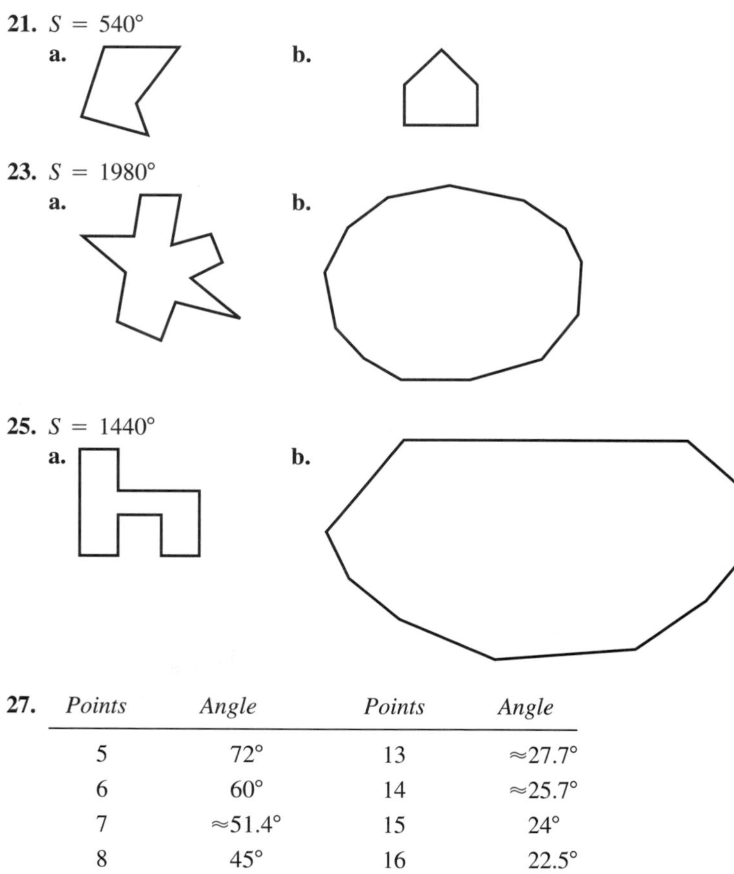


**27.**

| *Points* | *Angle* | *Points* | *Angle* |
|---|---|---|---|
| 5 | 72° | 13 | ≈27.7° |
| 6 | 60° | 14 | ≈25.7° |
| 7 | ≈51.4° | 15 | 24° |
| 8 | 45° | 16 | 22.5° |
| 9 | 40° | 18 | 20° |
| 10 | 36° | 20 | 18° |
| 11 | ≈32.7° | 30 | 12° |
| 12 | 30° | 36 | 10° |

**29.** $A = 135°$

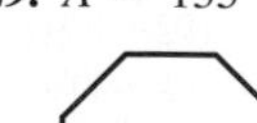

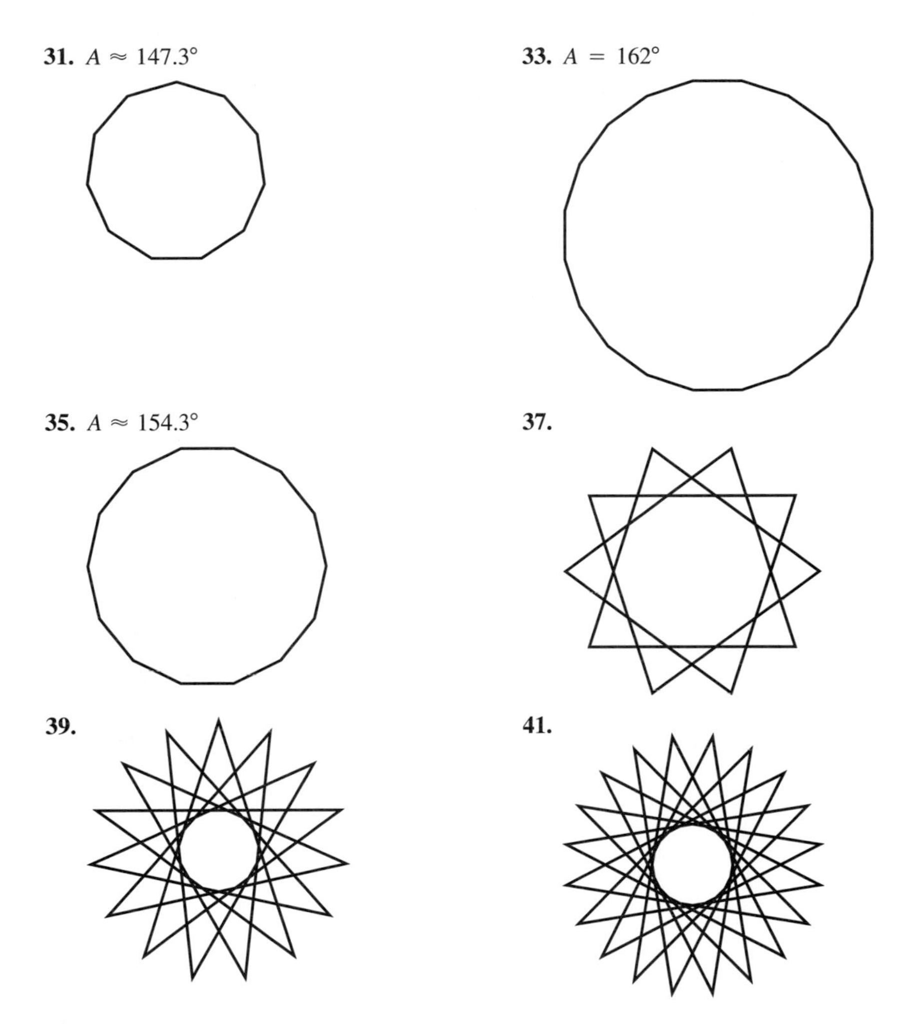


## *Section 3.5*

11. A regular heptagon has angles of $128\frac{4}{7}°$. Since 360° is not evenly divisible by $128\frac{4}{7}°$, regular heptagons will not tessellate the plane.
13. A regular decagon has angles of 144°. Since 360° is not evenly divisible by 144°, a regular decagon will not tessellate the plane.

In Answers 15–25, each diagram is approximately 40% of actual size and shows four to seven tiles.

15.

17.

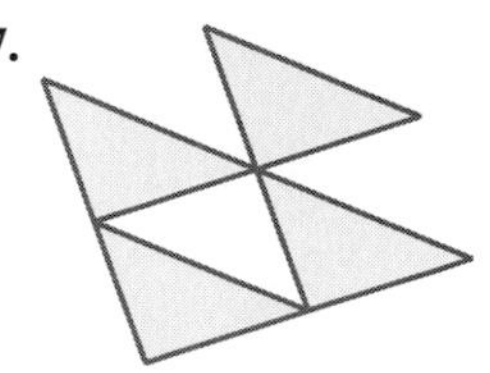

**19.** **21.** **23.** **25.**

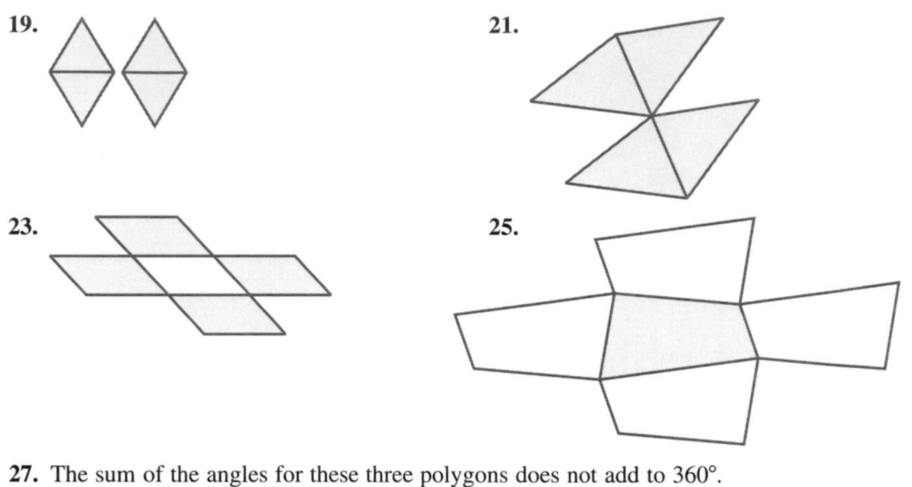

**27.** The sum of the angles for these three polygons does not add to 360°.

**29.**

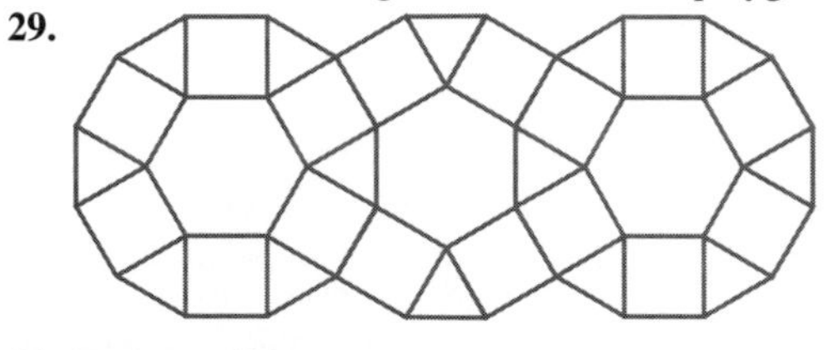

**31.** Two possible answers are to use two octagons and a square or two pentagons and a decagon.

**35.**

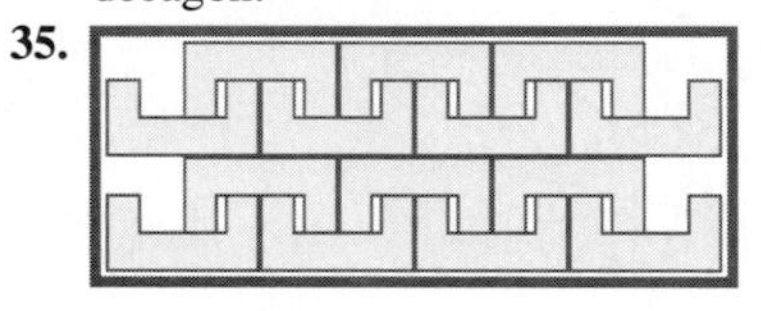

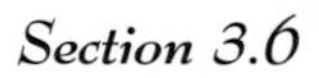

## *Section 3.6*

**9.**

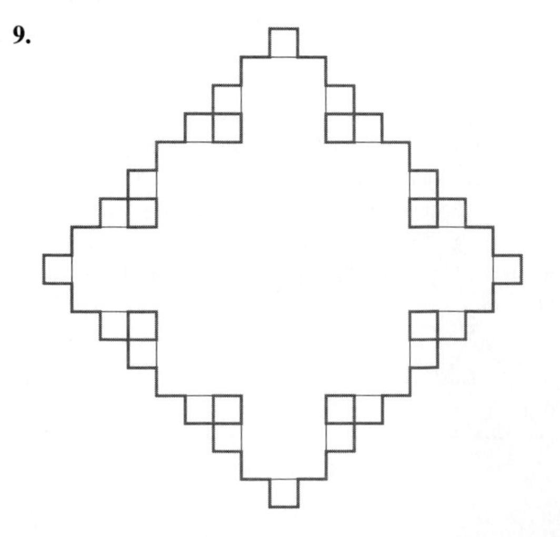

**11.** **13.**

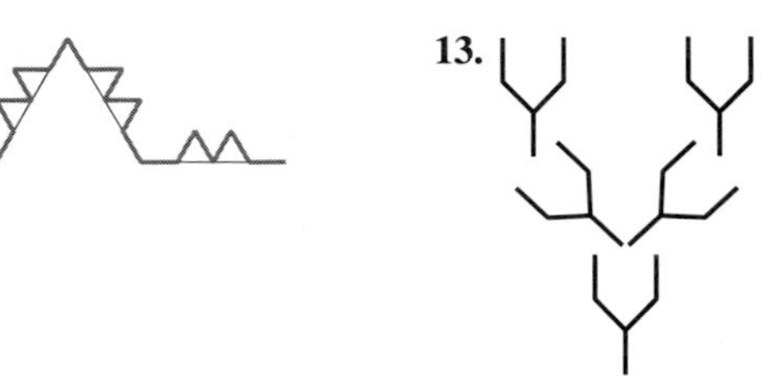

**17.** 1.46 **19.** 1.29 **21.** 1.46 **23.** 0.68

## *Chapter 3 Summary*

**1.** See Section 3.1. **2.** See Section 3.1.

**3.**

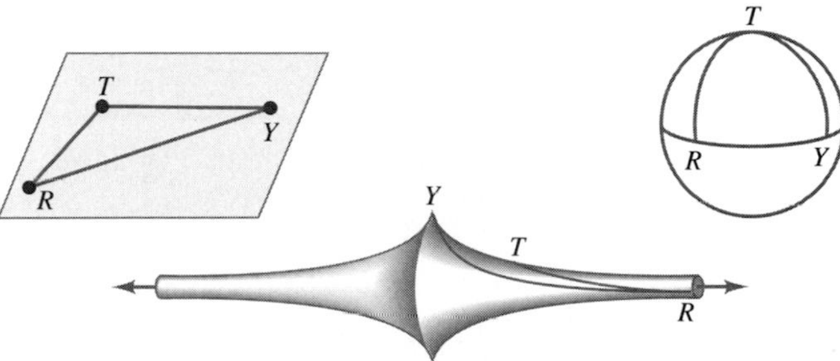


**4.**

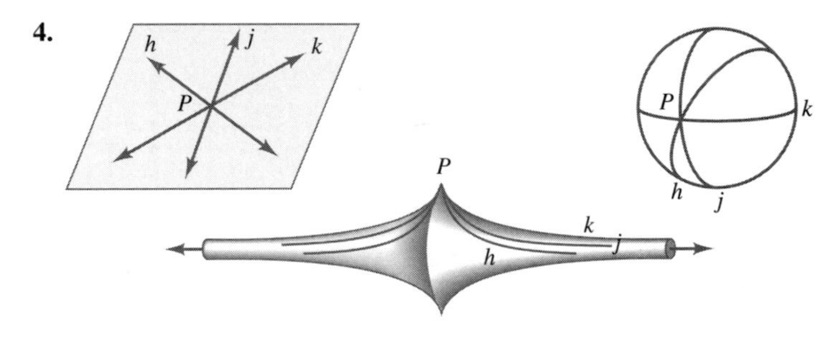


**5.** See Section 3.1.

**6. a.**

**b.**, **c.**

**7.**

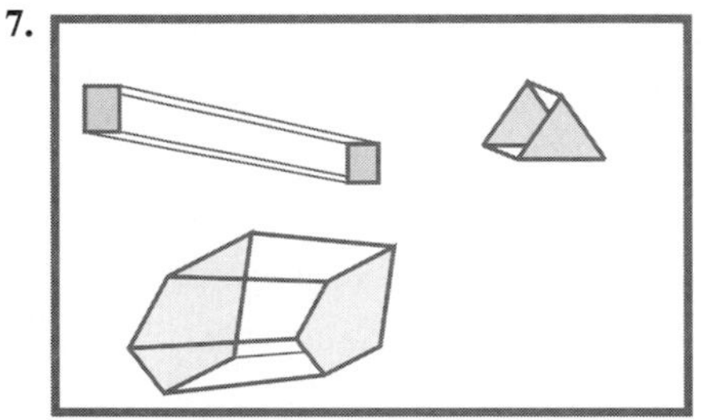

**8.** Pannini uses a one-point perspective with a vanishing point below and to the right of the center of the painting. He also uses overlapping shapes and diminishing sizes in the people in the painting.

**9.** Since the line segment is $\approx$ 1.75 in. long, the segment should be divided $\approx$1.1 in. from one end.

**10.** 13.8 in

**11.** Either the ratio of the diagonals or the ratio of the sides should be approximately 1.62.

**12.** **a.** **b.**

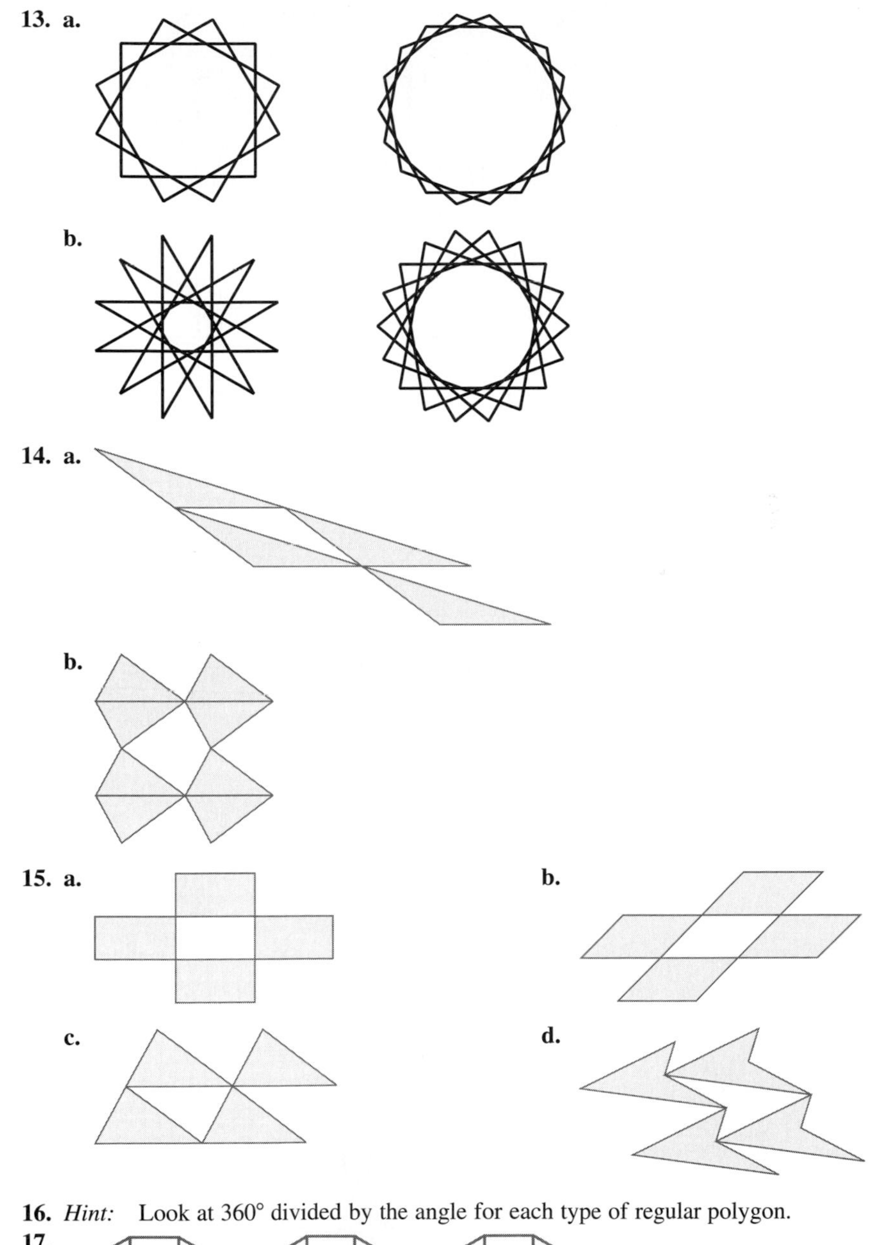


**16.** *Hint:* Look at 360° divided by the angle for each type of regular polygon.

**17.**

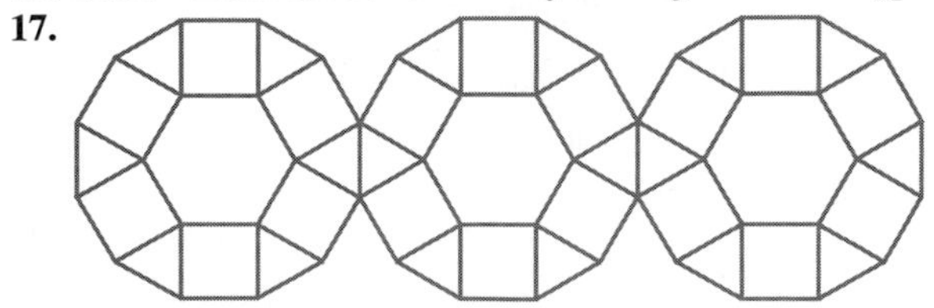

**18.**

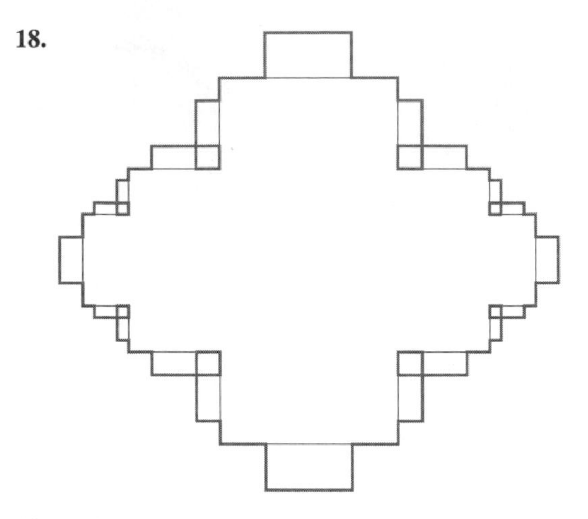

**19.**  1.29

## Chapter 4

### *Section 4.1*

**11.** **a.** $\{x|x \text{ is an even whole number} \leq 10\}$
**b.** $\{0, 2, 4, 6, 8, 10\}$

**13.** **a.** {apples, bananas, peaches, tomatoes, beans, peas, sprouts}
**b.** {tomatoes}

**15.** The symbol is being used correctly and the statement is true.

**17.** The symbol is being used correctly but the statement is false. One possible true statement is $\mathcal{T} \not\subset \mathcal{F}$.

**19.** An incorrect symbol is being used. A true correct statement is $5 \in I$.

**21.** The symbol is being used correctly but the statement is false. The statement is $5 \in I$.

**23.** The symbol is being used correctly and the statement is true.

**25.** The symbol is being used incorrectly since both $\mathcal{F}$ and $I$ are sets. The proper symbol to use between two sets is the subset symbol. One possible correct statement is $\mathcal{F} \subset I$.

**27.** 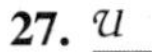

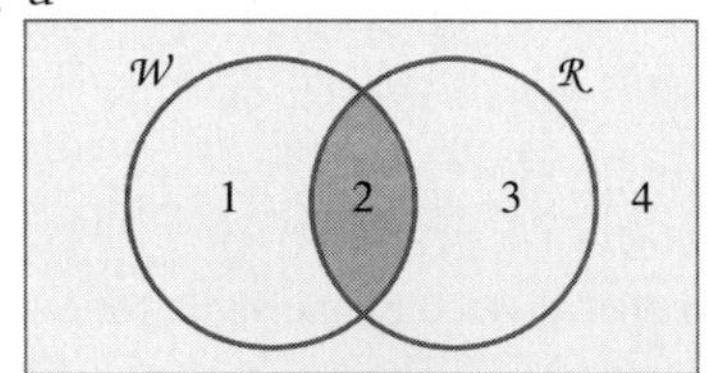


1 = {all women without red hair}
2 = {all women with red hair}
3 = {all people who have red hair but are not women}
4 = {all people who do not have red hair and are not women}

**29.** $\mathcal{U}$

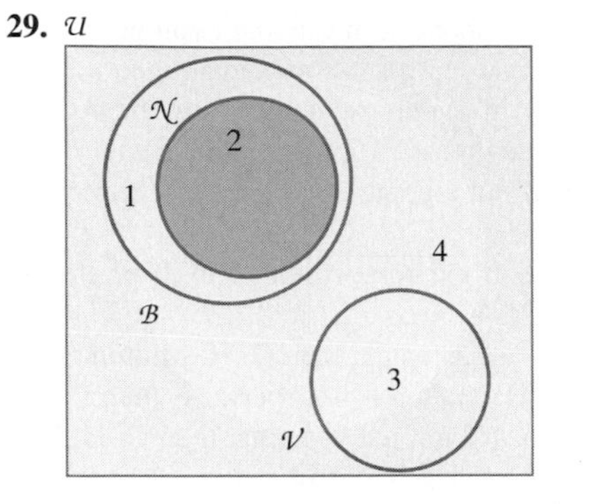


1 = {all books in the library that are not novels}
2 = {all books in the library that are novels}
3 = {all videos in the library}
4 = {all items in the library that are not books and not videos}

**31.** **a.** There are four subsets. **b.** There are eight subsets.
**c.** There are 16 subsets. **d.** There are $2^{20} = 1{,}048{,}576$ subsets.

**33.** 68

**35.** $\mathcal{A} \cap \mathcal{B}$ is the empty set.

**37.**

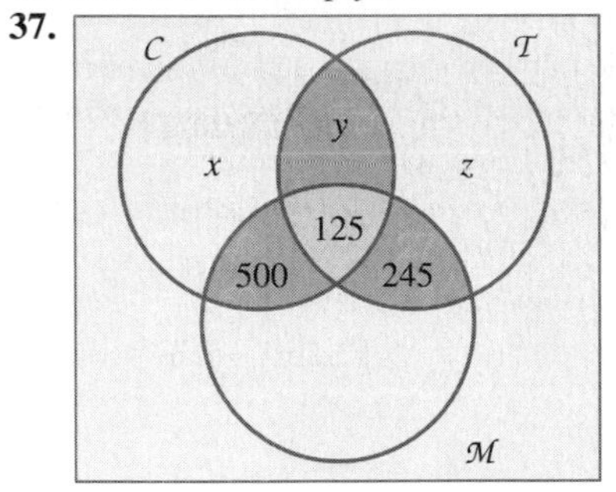


$x + y + z = 130$

**39.** **a.** 120 **b.** 40 **c.** 80 **d.** 130

## *Section 4.2*

**19.** It is not a statement because it is neither true nor false.
**21.** It is not a statement because it is neither true nor false.
**23.** A statement
**25.** It is not a statement because it is neither true or false.
**27.** My car is not in the shop.
**29.** I like sitting around doing nothing.
**31.** That is not an example of an exponential equation.
**33.** Some fish cannot live under water.
**35.** All numbers are prime numbers.
**37.** No trees are always green.
**39.** Some of the numbers are positive.

**41.** *Converse:* If the phone is in use, then you get a busy signal.
*Inverse:* If you do not get a busy signal, then the phone is not in use.
*Contrapositive:* If the phone is not in use, then you do not get a busy signal.

**43.** *Converse:* If the point is 16 in. from the center of the circle, then it is on the circle.
*Inverse:* If it is not a point on the circle, then it will not be 16 in. from the center of the circle.
*Contrapositive:* If the point is not 16 in. from the center of the circle, then it is not on the circle.

**45.** *Converse:* If I am listening, then G. H. Mutton is speaking.
*Inverse:* If G. H. Mutton is not speaking, then I am not listening.
*Contrapositive:* If I am not listening, then G. H. Mutton is not speaking.

**47.** *Converse:* If the figure is not a hexagon, then it has five sides.
*Inverse:* If the figure does not have five sides, then it is a hexagon.
*Contrapositive:* If a figure is a hexagon, then it does not have five sides.

In Problems 49–60, there may be other correct answers.

**49.** If $x$ is an integer, then $x$ is a real number.
**51.** If the product of two real numbers is zero, then at least one of them is zero.
**53.** Not possible
**55.** If $x$ is a number, then $x$ is a house.
**57.** If $x$ is a whole number, then $x$ is a prime number.
**59.** If $x$ is a prime number, then $x$ is a composite number.
**61.** Logic is the science of correct reasoning.
**63.** Deduction is the process of reasoning in which conclusions are based on accepted statements.
**65.** A paradox is a statement or group of statements that is in contradiction with itself.
**67.** Answers vary.
**69.** $\mathcal{B} \rightarrow \mathcal{A}$

## *Section 4.3*

**9.** It is not correct. It contains a converse error. This would be correct:
When it is midnight, I am asleep.
It is midnight.
Therefore, I am asleep.

**11.** It is not correct. It contains an inverse error. This would be correct:
If you are a farmer in Polt County, then you grow corn.
Farmer Ron does not grow corn.
Therefore, Farmer Ron is not a farmer in Polt County.

**13.** It is a correct syllogism.

**15.** If a whole number greater than 2 is even, then it is divisible by 2.
If a whole number is divisible by 2, then it is not a prime number.
Therefore, if a whole number greater than 2 is even, it is not a prime number.

**17.** Anyone that treats you with kindness is a nice person.
My teacher, Mrs. Santos, is always very kind to me when I am sick.
Therefore, Mrs. Santos is a nice person.

**19.** If you are serious about school, you will have less time to watch TV.

**21.** One of many possible answers:

$A$: $x$ is a whole number.
$B$: $x$ is an integer.
$C$: $x$ is a rational number.
$D$: $x$ is an irrational number.
$E$: $x$ equals $\sqrt{2}$.

**23.** In a triangle, the longest side is opposite the largest angle and the shortest side is opposite the smallest angle.

**25.** The new quadrilaterals are parallelograms.

**27.** The angle formed by the line segments connecting the endpoints of the diameter to a point on a semicircle is a right angle.

**29.** A possible inductive argument:

The plane ride is long and cramped.
The hotels in Honolulu are big and impersonal.
The beaches in Honolulu are noisy and crowded.
The restaurants in Honolulu are often expensive.
Therefore, you should not go to Honolulu, Hawaii.

**31.** **a.** This is an inductive argument. **b.** This is a deductive argument.

**33.** Some numbers are rational.

**35.** If the car is full of gasoline, we can see Vernal Falls.

**37.** By using the contrapositive of the third statement, we get $P \rightarrow \sim S$.

**39.** One possible example is

$P$ = The object is a square.
$Q$ = The object is a rectangle with four congruent sides.
$R$ = The object is a triangle.
$S$ = The object is a circle.
The conclusion is that if the object is a square, the object is not a circle.

## *Section 4.4*

**7.** $M$ = it is midnight, $A$ = asleep

$M \rightarrow A$
$A$
$\therefore M$

**9.** $F$ = farmer in Polt County, $C$ = grows corn

$F \rightarrow C$
$\sim C$
$\therefore \sim F$

**11.** $S$ = scalene, $W$ = two equal sides, $H$ = three equal sides

$\sim S \rightarrow (W \vee H)$
$\sim H$
$\therefore W$

**13.** $W$ = winning golfer, $G$ = good hand-eye coordination, $P$ = positive attitude

$W \rightarrow (G \wedge P)$
$\therefore (G \wedge P) \rightarrow W$

**15.** $B$ = voted for Bush, $P$ = voted for Perot, $C$ = voted for Clinton

$(B \vee P) \rightarrow \sim C$

**17.**

| $A$ | $C$ | $A \to C$ | $\sim C$ | $(A \to C) \vee \sim C$ |
|---|---|---|---|---|
| T | T | T | F | T |
| T | F | F | T | T |
| F | T | T | F | T |
| F | F | T | T | T |

Statement is always true.

**19.**

| $A$ | $\sim A$ | $C$ | $\sim A \to C$ | $\sim C$ | $(\sim A \to C) \to \sim C$ |
|---|---|---|---|---|---|
| T | F | T | T | F | F |
| T | F | F | T | T | T |
| F | T | T | T | F | F |
| F | T | F | F | T | T |

Statement is true whenever $q$ is false.

**21.**

| $A$ | $B$ | $\sim B$ | $A \wedge \sim B$ | $\boldsymbol{\sim(A \wedge \sim B)}$ | $\sim A$ | $\boldsymbol{\sim A \vee B}$ |
|---|---|---|---|---|---|---|
| T | T | F | F | **T** | F | **T** |
| T | F | T | T | **F** | F | **F** |
| F | T | F | F | **T** | T | **T** |
| F | F | T | F | **T** | T | **T** |

Since both statements have the same truth values, the statements are equivalent.

**23.**

| $\boldsymbol{A}$ | $B$ | $A \vee B$ | $\sim B$ | $A \vee \sim B$ | $\boldsymbol{(A \vee B) \wedge (A \vee \sim B)}$ |
|---|---|---|---|---|---|
| **T** | T | T | F | T | **T** |
| **T** | F | T | T | T | **T** |
| **F** | T | T | F | F | **F** |
| **F** | F | F | T | T | **F** |

Since both statements have the same truth values, the statements are equivalent.

**25.**

| $P$ | $Q$ | $P \to Q$ | $(P \to Q) \wedge P$ | $((P \to Q) \wedge P) \to Q$ |
|---|---|---|---|---|
| T | T | T | T | T |
| T | F | F | F | T |
| F | T | T | F | T |
| F | F | T | F | T |

Since the last column is always true, the argument is correct.

**27.**

| $A$ | $\sim A$ | $Q$ | $A \to Q$ | $\sim A \wedge (A \to Q)$ | $(\sim A \wedge (A \to Q)) \to A$ |
|---|---|---|---|---|---|
| T | F | T | T | F | T |
| T | F | F | F | F | T |
| F | T | T | T | T | F |
| F | T | F | T | T | F |

Since the last column is not always true, the argument is not correct.

**29.** $E$ = elections become TV popularity contests; $G$ = good looking, smooth-talking candidates get elected.

$$E \to G$$
$$\therefore \sim E \to \sim G$$

| $E$ | $G$ | $E \to G$ | $\sim E$ | $\sim G$ | $\sim E \to \sim G$ | $(E \to G) \to (\sim E \to \sim G)$ |
|---|---|---|---|---|---|---|
| T | T | T | F | F | T | T |
| T | F | F | F | T | T | T |
| F | T | T | T | F | F | F |
| F | F | T | T | T | T | T |

Since the last column is not always true, the argument is not correct.

**31.** $M$ = High school grads have poor math skills.
$W$ = High school grads have poor writing skills.
$J$ = less able to get a job in the computer industry

$$M \to J$$
$$W \to J$$
$$\therefore M \to W$$

| $M$ | $W$ | $J$ | $M \to J$ | $W \to J$ | $(M \to J) \wedge (W \to J)$ | $M \to W$ | $((M \to J) \wedge (W \to J)) \to (M \to W)$ |
|---|---|---|---|---|---|---|---|
| T | T | T | T | T | T | T | T |
| T | T | F | F | F | F | T | T |
| T | F | T | T | T | T | F | F |
| T | F | F | F | T | F | F | T |
| F | T | T | T | T | T | T | T |
| F | T | F | T | F | F | T | T |
| F | F | T | T | T | T | T | T |
| F | F | F | T | T | T | T | T |

Since the last column is not always true, the argument is not correct.

**33.** Constructing a truth table shows that the argument is not correct.

## *Section 4.5*

**9.** This flow chart describes the process that helps decide why a lawnmower will not start. The choices are limited to the lawnmower being out of gas, the spark plug wire not being connected, the blade being engaged, or some unknown reason.

**11.** **a.** Stand.
**b.** Stand if dealer shows a card $2 \leq x < 7$, otherwise draw another card and then reevaluate your hand.
**c.** Draw another card and then reevaluate your hand.
**d.** Draw another card and then reevaluate your hand.
**e.** Stand.

**f.** To show how to win the game, add the following to the existing flowchart. Existing decision block:

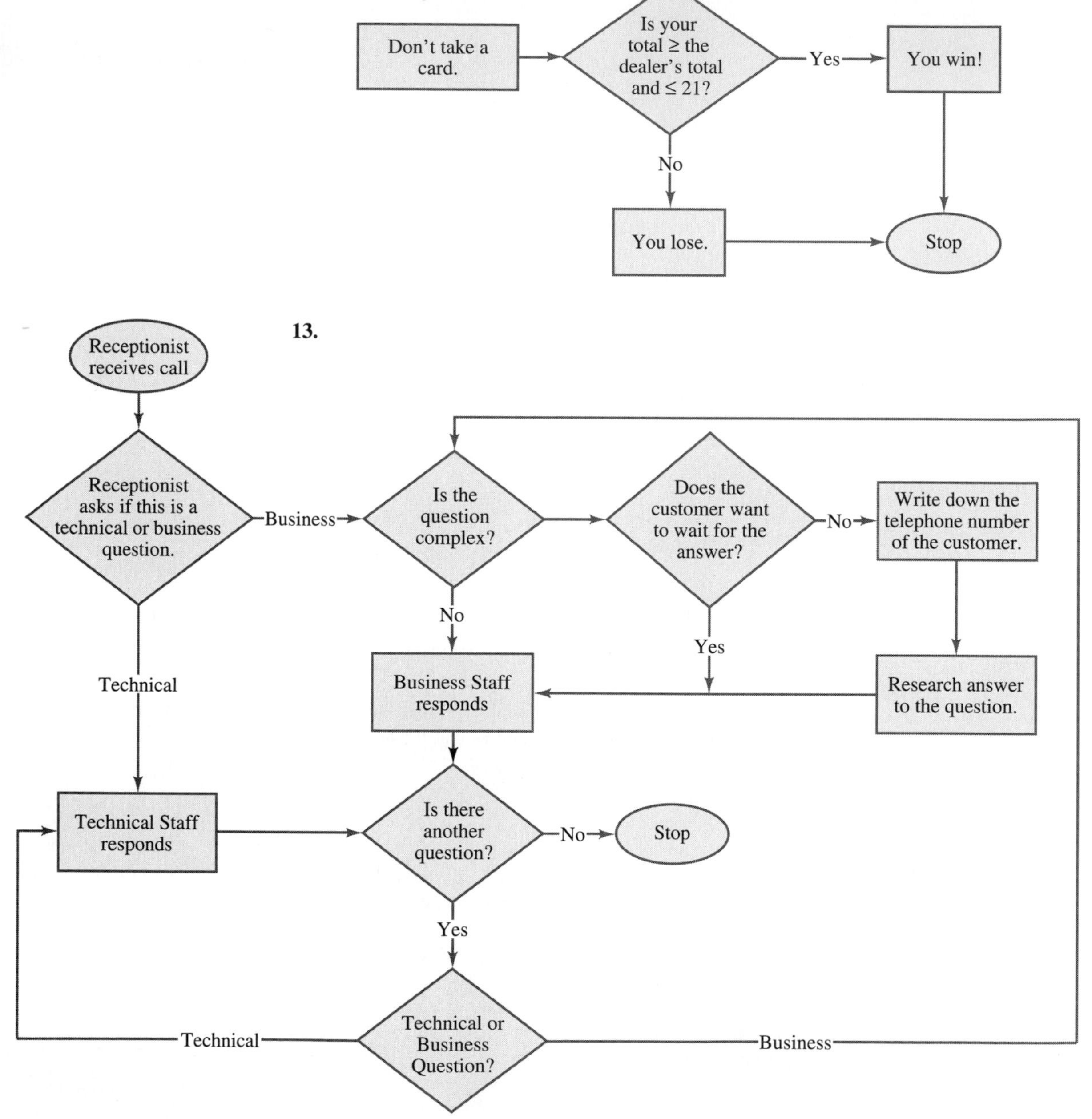

**15.**

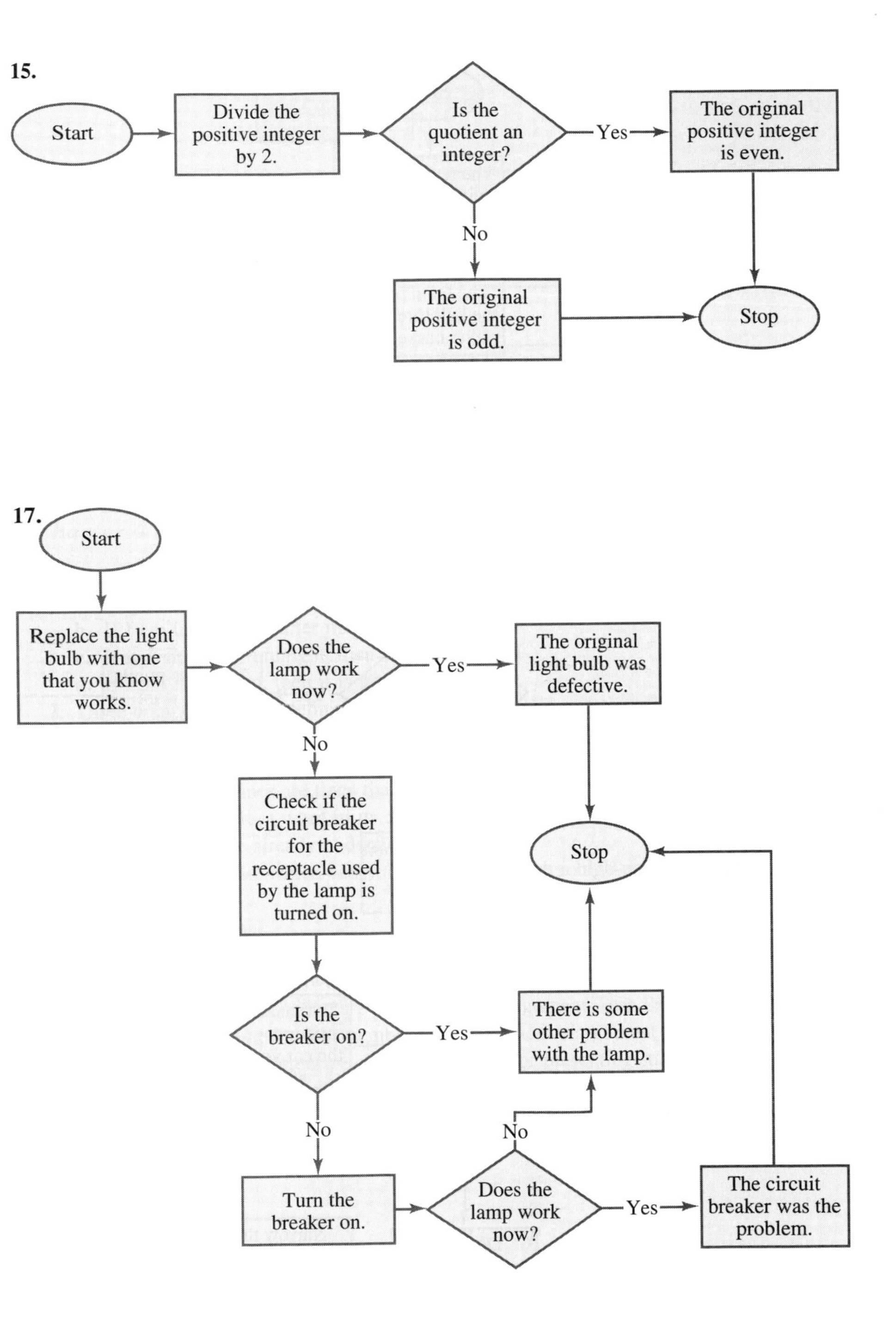
Start
Divide the positive integer by 2.
Is the quotient an integer?
Yes
The original positive integer is even.
No
The original positive integer is odd.
Stop
17.
Start
Replace the light bulb with one that you know works.
Does the lamp work now?
Yes
The original light bulb was defective.
No
Check if the circuit breaker for the receptacle used by the lamp is turned on.
Stop
Is the breaker on?
Yes
There is some other problem with the lamp.
No
No
Turn the breaker on.
Does the lamp work now?
Yes
The circuit breaker was the problem.

Thus the argument is

If the geometry is Riemannian, then there are no parallel lines.
If there are no parallel lines, then the geometry is non-Euclidean.
If the geometry is non-Euclidean, then at least one of Euclid's postulates is changed.
∴ If the geometry is Riemannian, then at least one of Euclid's postulates is changed.

**6.** 170

**7. a.** Let $H$ = headache, $G$ = grumpy, $S$ = silent; $H \rightarrow (G \vee S)$.

**b.** Let $G$ = good weather, $B$ = play baseball, $P$ = have a picnic, $\sim G \rightarrow (\sim B \wedge \sim P)$.

**c.** Let $M$ = study math, $G$ = good job, $H$ = harder to advance, $(\sim M \wedge G) \rightarrow H$.

**8.**

| $\boldsymbol{A}$ | $B$ | $A \vee B$ | $\sim B$ | $A \vee \sim B$ | $\boldsymbol{(A \vee B) \wedge (A \vee \sim B)}$ |
|---|---|---|---|---|---|
| **T** | T | T | F | T | **T** |
| **T** | F | T | T | T | **T** |
| **F** | T | T | F | F | **F** |
| **F** | F | F | T | T | **F** |

Since the two statements have the same truth values, the statements are equivalent.

**9.** Let $P$ = parrot, $W$ = cracks walnuts, $A$ = my animal

The argument is

$$(P \rightarrow W) \wedge (A \rightarrow \sim W)$$
$$\therefore A \rightarrow \sim P$$

| $P$ | $W$ | $A$ | $P \rightarrow W$ | $\sim W$ | $A \rightarrow \sim W$ | $(P \rightarrow W) \wedge (A \rightarrow \sim W)$ | $\sim P$ | $A \rightarrow \sim P$ | $((P \rightarrow W) \wedge (A \rightarrow \sim W)) \rightarrow (A \rightarrow \sim P)$ |
|---|---|---|---|---|---|---|---|---|---|
| T | T | T | T | F | F | F | F | F | T |
| T | T | F | T | F | T | T | F | T | T |
| T | F | T | F | T | T | F | F | F | T |
| T | F | F | F | T | T | F | F | T | T |
| F | T | T | T | F | F | F | T | T | T |
| F | T | F | T | F | T | T | T | T | T |
| F | F | T | T | T | T | T | T | T | T |
| F | F | F | T | T | T | T | T | T | T |

Since the last column is always true, the argument is correct.

**10.**

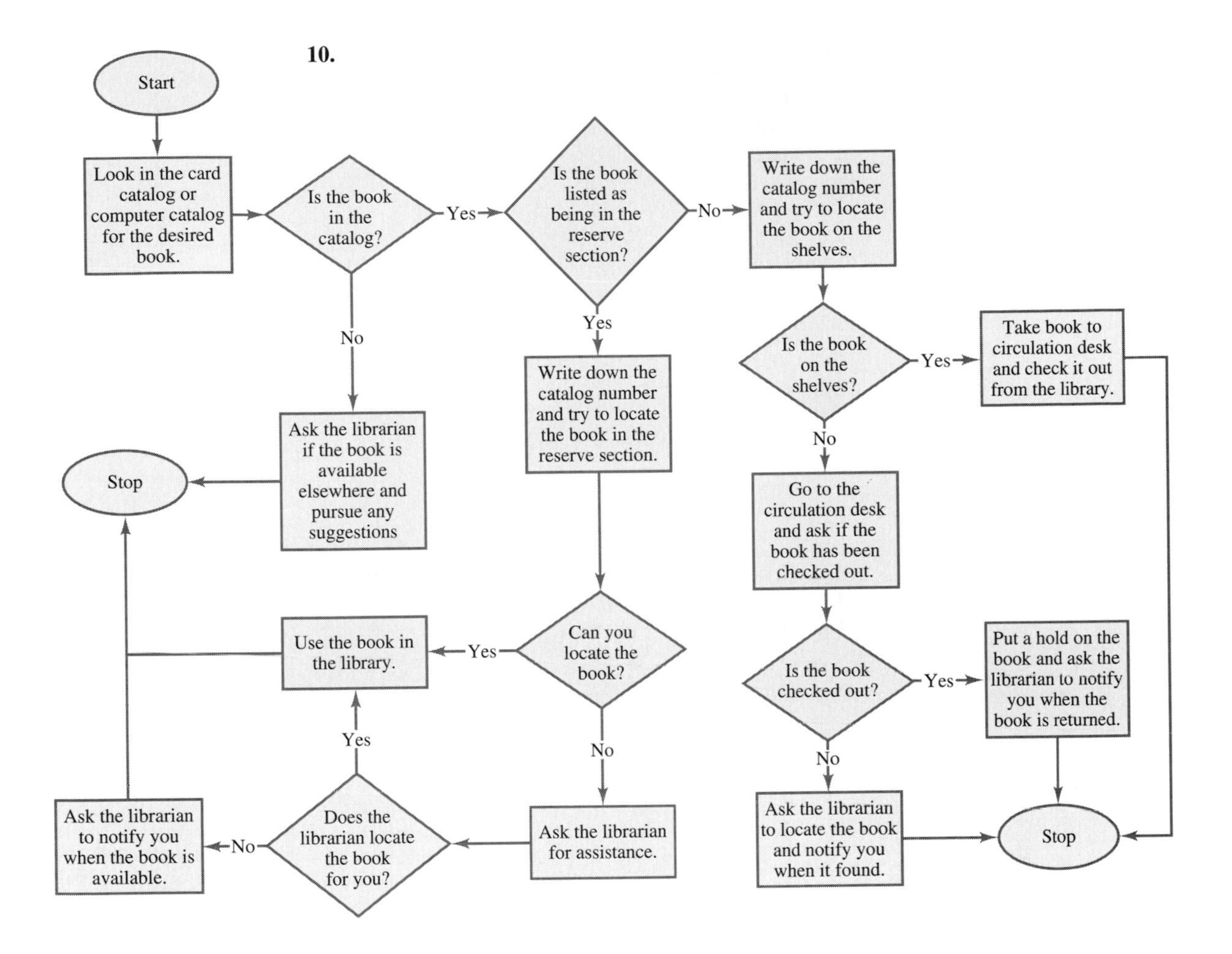


## Chapter 5

*(Round-off rules for chapter are used.)*

### Section 5.0

**9.** Legs: $k$, $d$; hypotenuse: $y$
**11.** 8) Opposite: $y$; adjacent: $d$ 9) Opposite: $d$; adjacent: $k$ 10) Opposite: $k$; adjacent: $y$
**13.** 32.6 **15.** 40 **17.** 53.4 **19.** 6.7
**21.** Yes **23.** No **25.** No **27.** Yes
**29.** 7.8 **31.** 84.9 ft **33.** 22.6 in.
**35.** Actually draw the next four right triangles on the root spiral **37.** Yes
**39.** Use $x$, $x + 2$, and $x + 4$ as the sides of the right triangle.

## Section 5.1

**13.** $\sin A = 3/5$ $\sin B = 4/5$
$\cos A = 4/5$ $\cos B = 3/5$
$\tan A = 3/4$ $\tan B = 4/3$

**15.** $\sin A = 4/5$ $\sin B = 3/5$
$\cos A = 3/5$ $\cos B = 4/5$
$\tan A = 4/3$ $\tan B = 3/4$

**17.** 0.2181 **19.** 0.1392 **21.** 0.7208
**23.** 4.1022 **25.** 0.9999 **27.** 82.9°
**29.** 42.1° **31.** 87.0° **33.** 68.7°
**35.** $\angle B = 67°$, $a = 7.0$, $b = 16.6$
**37.** $\angle A = 9.7°$, $\angle B = 80.3°$, $b = 64.1$
**39.** $\angle X = 56.4°$, $\angle Y = 33.6°$, $y = 12.9$
**41.** $\angle Y = 21.5°$, $x = 5965.8$, $z = 6412.0$
**43.** $\angle A = 61.9°$, $\angle B = 28.1°$, $\angle C = 90°$
**45.** $\angle A = 53.1°$, $\angle B = 36.9°$, $\angle C = 90°$
**47.** $\angle A = 53.1°$, $\angle B = 36.9°$, $\angle C = 90°$
**49.** $h = 32.8$
**51.** $h = 92.4$
**53.** $x \approx 6.7$

## Section 5.2

**7.** 2.6 mi **9.** 231.8 m **11.** pole: 50.0 ft, wire: 51.4 ft
**13.** 29.5 ft **15.** 114.9 ft **17.** 2192.9 ft
**19.** 65.3 sec **21.** 14.1 ft, 12.1° **23.** 29.5 ft

## Section 5.3

**7.** $n = 26.1$ **9.** $u = 407.5$ **11.** $u = 41.0$ **13.** $n = 272.3$
**15.** $d = 32.6$ **17.** $g = 4.1$ **19.** $\angle O = 60.9°$
**21.** $\angle C = 78.6°$, $t = 23.0$, $a = 27.4$
**23.** $\angle C = 66.2°$, $\angle U = 41.7°$, $\angle P = 72.1°$
**25.** $\angle G = 30.9°$, $\angle T = 81.6°$, $e = 42.3$
**27.** $\angle P = 44.9°$, $\angle E = 54.2°$, $\angle T = 80.9°$
**29.** Using the Law of Sines or right triangle trigonometry, $AD = 42.9$ and $RD = 47.3$.
**31.** It is impossible to have a triangle in which the sum of the lengths of two sides is less than the length of the third side.

## Section 5.4

**7.** 460.0 ft **9.** 27.2 ft **11.** 33.0 ft **13.** 1603.5 m
**15.** 438.6 mi **17.** 12.0 ft **19.** 49.6°, 44.5°, 85.9° **21.** Methods may vary.
**23.** 1494.9 ft

## Chapter 5 Summary

**1.** $\angle A = 28.1°$, $\angle C = 61.9°$
**2.** $\angle N = 66°$, $n = 8.8$, $f = 3.9$
**3.** $\angle N = 84°$, $r = 20.3$, $n = 29.0$
**4.** $\angle B = 68.7°$, $o = 141.5$, $r = 57.6$
**5.** $\angle M = 73.1°$, $\angle Y = 44.4°$, $a = 38.0$
**6.** $\angle D = 60.3°$, $\angle E = 88.2°$, $\angle B = 31.5°$ **7.** 41.8° **8.** 117.8 m
**9.** 4979.0 ft **10. a.** 17.0 ft, 176.2 ft **b.** 48.0 ft, 178.9 ft **11.** 213.0 ft
**12.** 22.2 mi, 13.7 mi, 12.1 mi **13.** 851.8 mi **14.** 3.7 mi
**15.** N 11.2° E, N 48.2° W **16.** 2950.8 yd, 2987.6 yd

# Chapter 6

*(Answers may vary slightly depending on calculator used and rounding procedures.)*

## Section 6.0

**7.** 12.513503 **9.** 0.386883 **11.** 35.949641 **13.** 0.270489
**15.** 0.301386 **17.** 54.097832 **19.** 32.290749 **21.** 1.055751
**23.** 1.181352 **25.** 1.597138 **27.** 0.661722 **29.** 2.033424
**31.** 5.467435 **33.** 2.384929 **35.** 1.865152 **37.** −1.861654
**39.** 1.547411 **41.** −1.666667 **43.** 0.687699 **45.** 0.187321
**47.** 4.095786 **49.** 120.007483

## Section 6.1

**7.** $400.00 **9.** $312.50 **11.** $56.25 **13.** $120.09
**15.** $3720.00 **17.** $35,466.67 **19.** $3605.00 **21.** $1090.16
**23.** $2500.00 **25.** 2.5 yr **27.** 25% **29.** $3,542.80
**31.** 1.16 yr **33.** 3.98% **35.** 8.11% **37.** 6.00%
**39.** 15 yr **41. a.** $1100.00 **b.** $1210.00 **c.** $1331.00

## Section 6.2

**9.** $2540.98 **11.** $6970.93 **13.** $6074.43 **15.** $5323.11
**17.** $5568.53 **19.** $1311.17 **21.** 3.98 yr **23.** 14.22 yr
**25.** 11.90 yr **27.** 19.52% **29.** 7.78% **31.** 7.35%
**33.** 7.8% daily **35.** $20,808.41
**37. a.** $562.72 **b.** $594.39 **c.** $627.83
**39.** $9264.84 **41.** 5.78 yr **43.** $11,628,024,100 **45.** 1.0937125
**47.** 1.2676 **49.** $398.50 **51.** $389.19

## Section 6.3

**7.** $13,954.01 **9.** $34,320.46 **11.** $409.11 **13.** $432.86
**15.** 10.22 yr **17.** 17.08 yr **19.** $192,646,74
**21. a.** $973,151.26 **b.** $57,600.00 **c.** $915,551.26
**d.** $778,181.07 **e.** $91,200.00 **f.** $686,981.07
**23. a.** $77,405.71 **b.** $1,166,691.60
**25.** 12.62 yr **27.** $307.52
**29. a.** $139.74 **b.** $47,507.66

## Section 6.4

**9.** $10,345.11 **11.** $4087.86 **13.** $609.11 **15.** $3019.28
**17.** 15.46 yr **19.** 16.97 yr **21.** $1204.88
**23. a.** $215.41 **b.** $10,339.68 **c.** $2339.68
**25. a.** $135.38 **b.** 23 months **c.** $748.24
**27.** $160,625.11

**29.**

| Payment | Interest | Principal | Balance |
|---|---|---|---|
| | | | \$2000.00 |
| \$520.79 | \$33.00 | \$487.79 | \$1512.21 |
| \$520.79 | \$24.95 | \$495.84 | \$1016.37 |
| \$520.79 | \$16.77 | \$504.02 | \$ 512.35 |
| \$520.80 | \$ 8.45 | \$512.35 | \$ 0.00 |

**31.**

| Payment | Interest | Principal | Balance |
|---|---|---|---|
| | | | \$2000.00 |
| \$675.35 | \$13.00 | \$662.35 | \$1337.65 |
| \$675.35 | \$ 8.69 | \$666.66 | \$ 670.99 |
| \$675.35 | \$ 4.36 | \$670.99 | \$ 0.00 |

**33. a.** \$1367.60 \$1048.82
**b.** \$246,168.00 \$377,575.20
**c.** The first loan because it has lower total payments.

## *Chapter Summary*

**1.** \$12,400.00 **2.** 7.21%
**3. a.** \$1100.00 **b.** \$1103.81 **c.** 1104.71 **d.** \$1105.17
**4.** \$5846.10 **5.** 14.32% **6.** \$304.28
**7. a.** \$24,160.79 **b.** \$8160.79 **c.** \$175,036.37
**8.** 291 deposits
**9. a.** \$395.14 **b.** \$17,225.04
**10.** 139 mo or 11 yr, 7 mo **11.** \$169,425.52
**12.**

| Payment | Interest | Principal | Balance |
|---|---|---|---|
| | | | \$3000.00 |
| \$530.17 | \$51.00 | \$479.17 | \$2520.83 |
| \$530.17 | \$42.85 | \$487.32 | \$2033.51 |
| \$530.17 | \$34.57 | \$495.60 | \$1537.91 |
| \$530.17 | \$26.14 | \$504.03 | \$1033.88 |
| \$530.17 | \$17.58 | \$512.59 | \$ 521.29 |
| \$530.14 | \$ 8.86 | \$521.29 | \$ 0.00 |

# Chapter 7

## *Section 7.0*

**9.** 362,880 **11.** 479,001,600 **13.** 56 **15.** 50,116
**17.** 1 **19.** 1 **21.** 336 **23.** 300,696
**25.** 40,320 **27.** 1 **29.** $2/9 \approx 0.2222$ **31.** 2520
**33. a.** 3 **b.** 3 **c.** 1
**35. a.** 3 **b.** 6 **c.** 6
**37.** 10 **39.** $C_{56,3} = 27{,}720$

## Section 7.1

**15.** **a.** 1/52 **b.** 1/13 **c.** 10/13 **d.** 1/4 **e.** 1/2 **f.** 2/13
**17.** 195
**19.** **a.** 0.85 = 17/20 **b.** 3:17 **c.** 17:3
**21.** **a.** 2/5 = 0.4 **b.** 3/5 = 0.6 **c.** $15
**23.** **a.** 1/28,561; 1:28,560 **b.** $57,120
**25.** **a.** 144 pairs: (1, 1), (1, 2), (1, 3), . . . , (12, 12)
**b.** 1/144 **c.** 0 **d.** 1/18 **e.** 7/36 **f.** 1
**27.** **a.** $32.00
**b.** Chosen Journey in the 9th was highly favored by the bettors while Miz Interco was not.
**c.** $71.00
**29.** **a.** 8/25, **b.** 1/75, **c.** 4/75, **d.** 12/37 **e.** 24/73, **f.** 1/55

## Section 7.2

**9.** Permutations or Basic Counting Law
**11.** Basic Counting Law
**13.** Combinations
**15.** Basic Counting Law
**17.** Permutations or Basic Counting Law
**19.** Permutations or Basic Counting Law
**21.** Combinations
**23.** 16; (on, on, on, on); (on, on, on, off); (on, on, off, on); (on, off, on, on); (off, on, on on); (on, on, off, off); (on, off, on, off); (off, on, off, on); (off, off, on, on); (on, off, off, on); (off, on, on off); (off, off, off, on); (off, off, on, off); (off, on off, off); (on, off, off, off); (off, off, off, off)
**25.** $10^9$ = 1,000,000,000
**27.** 3; (BBG), (BGB); (GBB)
**29.** $P_{7,3} = 210$
**31.** 1024
**33.** $C_{52,5} = 2{,}598{,}960$
**35.** **a.** 1024 **b.** 10,240
**37.** $C_{32,13} \times C_{20,0} = 347{,}373{,}600$
**39.** $C_{20,2} = 190$
**41.** $C_{4,2} \times C_{5,1} = 30$
**43.** 175,760,000
**45.** $C_{20,2} - 10 = 180$
**47.** $C_{20,6} = 38{,}760$

## Section 7.3

**7.** **a.** 1/216, 1:215 **b.** 1/72, 1:71 **c.** 1/36, 1:35
**9.** **a.** $(C_{4,1} \times C_{4,4})/C_{52,5} = 1/1649{,}740 \approx 0.000001539$
**b.** $(C_{4,1} \times 12 \times C_{4,4})/C_{52,5} = 1/54{,}145 \approx 0.00001847$
**11.** **a.** $(C_{8,3} \times C_{96,2})/C_{104,5} \approx 0.002777 \approx 1/360$
**b.** $(13 \times C_{8,3} \times C_{96,2})/C_{104,5} \approx 0.03610 \approx 1/27.7$

**13.** $1/P_{8,2} = 1/56$
**15. a.** $(C_{20,0} \times C_{60,5})/C_{80,5} \approx 0.2272 \approx 1/4.4$
**b.** $(C_{20,5} \times C_{60,0})/C_{80,5} \approx 0.0006449 \approx 1/1551$
**c.** $(C_{20,4} \times C_{60,1})/C_{80,5} \approx 0.01209 \approx 1/82.7$
**17. a.** $(C_{6,6} \times C_{43,0})/C_{49,6} = 1/13{,}983{,}816 \approx 0.00000007151$
**b.** $(C_{6,5} \times C_{43,1})/C_{49,6} = \approx 0.00001845 \approx 1/54{,}201$
**c.** $(C_{6,4} \times C_{43,2})/C_{49,6} = \approx 0.0009686 \approx 1/1032$
**19.** 8/125; = 8:117 **21.** $1/190 \approx 0.005263$
**23.** 5/14 **25.** 1/10
**27. a.** $(C_{1,1} \times C_{29,2})/C_{30,3} = 1/10$
**b.** $(C_{1,1} \times C_{14,2})/C_{15,3} = 1/5$
**c.** 1
**29.** 1/907,200, 1:907,199
**31.** $(C_{4,4}/C_{75,4}) = 1/1{,}215{,}450 \approx 0.0000008827$
**33. a.** 3/2000, 3:1997
**b.** 3/2000, 3:1997
**c.** 3/2000, 3:1997
**d.** 9/2000, 9 to 1991
**e.** ≈ $55.31

## Section 7.4

**9.** E.V. ≈ −0.3333; lose: $33.33 **11.** $20.00
**13. a.** E.V. ≈ 0.225 **b.** No **c.** 22.5¢
**15.** ≈11 **17.** ≈ −0.1407 **19.** E.V. ≈ −0.0526
**21.** $1.00 Ticket: E.V. ≈ −0.2664; $2.00 Ticket: E.V. ≈ −0.5199; $3.00 Ticket: E.V. ≈ −0.7735.
It does not matter which amount you wager. The casino expects about 26.6¢ profit per dollar wagered.
**23.** The second procedure
**25. a.** E.V. ≈ −0.28 **b.** $50.00

## Chapter 7 Summary

**1. a.** 35 **b.** 840 **c.** 1,088,430 **d.** 26,122.320
**2.** The 0.5 is the theoretical probability. This does not mean that exactly 1/2 of any number of tosses will be heads. A large number of tosses will produce a probability that approaches 0.5.
**3.** 1/100; Every dollar bet will give $99 in winnings.
**4. a.** 274,625 **b.** 262,080
**5.** 7/144 **6.** 72
**7. a.** 1056 **b.** 12,672 **c.** 164,736
**8.** 1/720 ≈ 0.001389, 1:719
**9. a.** $(C_{1,1} \times C_{13,1})/C_{14,2} = 1/7$ **b.** 1/7
**10.** E.V. ≈ 0.78
**11. a.** E.V. ≈ 1.4375 **b.** ≈ 14 points
**12. a.** 4845/1,581,580 ≈ 1/326; 1:325 **b.** 325/326; 325:1 **c.** E.V. ≈ −0.23
**d.** Over a long period of time you can expect to lose about 23¢ on every dollar bet.

# Chapter 8

## *Section 8.0*

**9.** 181 **11.** 139 **13.** 3375 **15.** 55 **17.** 67/21

**19.** $3x_1 + 3x_2 + 3x_3 + 3x_4 + 3x_5 - 4 = 3(x_1 + x_2 + x_3 + x_4 + x_5) - 4$

**21.** $\sum_{i=0}^{2} (3 + x_i)^2 = \sum_{i=0}^{2} (9 + 6x_i + x_i^2) = 27 + 6x_0 + 6x_1 + 6x_2 + x_0^2 + x_1^2 + x_2^2$

**23.** $5 \sum_{i=1}^{n} x_i + 3 \sum_{i=1}^{n} y_i$ **25.** $\dfrac{\sum_{i=1}^{n} x_i}{4}$

**27.** $\sum_{i=1}^{5} i$ **29.** $\sum_{i=1}^{5} 2i$ **31.** $\sum_{i=1}^{6} i^2$

**33. a.** 9

**b.** $\dfrac{\sum_{i=1}^{n} x_i}{n}$

## *Section 8.1*

**9. a.**

| Class | Frequency |
|---|---|
| 1–3 | 3 |
| 4–6 | 4 |
| 7–9 | 3 |

**b.**

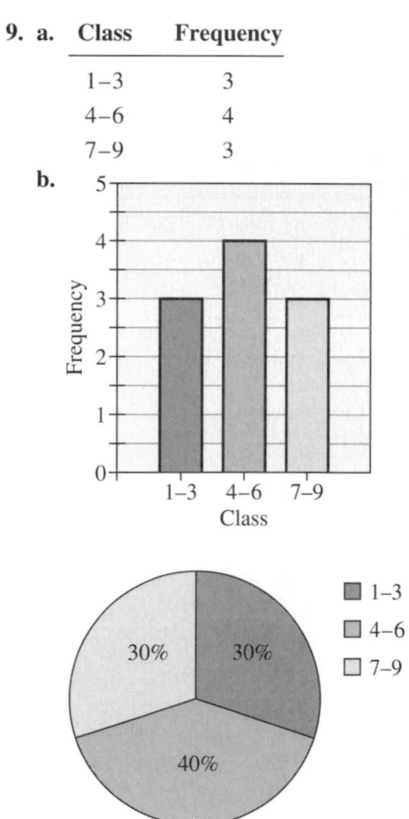


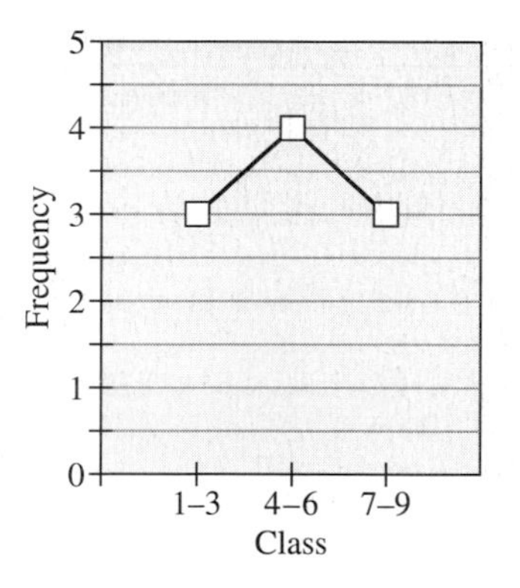

**11. a.**

| Class | Frequency |
|---|---|
| 10–11 | 1 |
| 12–13 | 4 |
| 14–15 | 2 |
| 16–17 | 4 |
| 18–19 | 1 |

**b.**

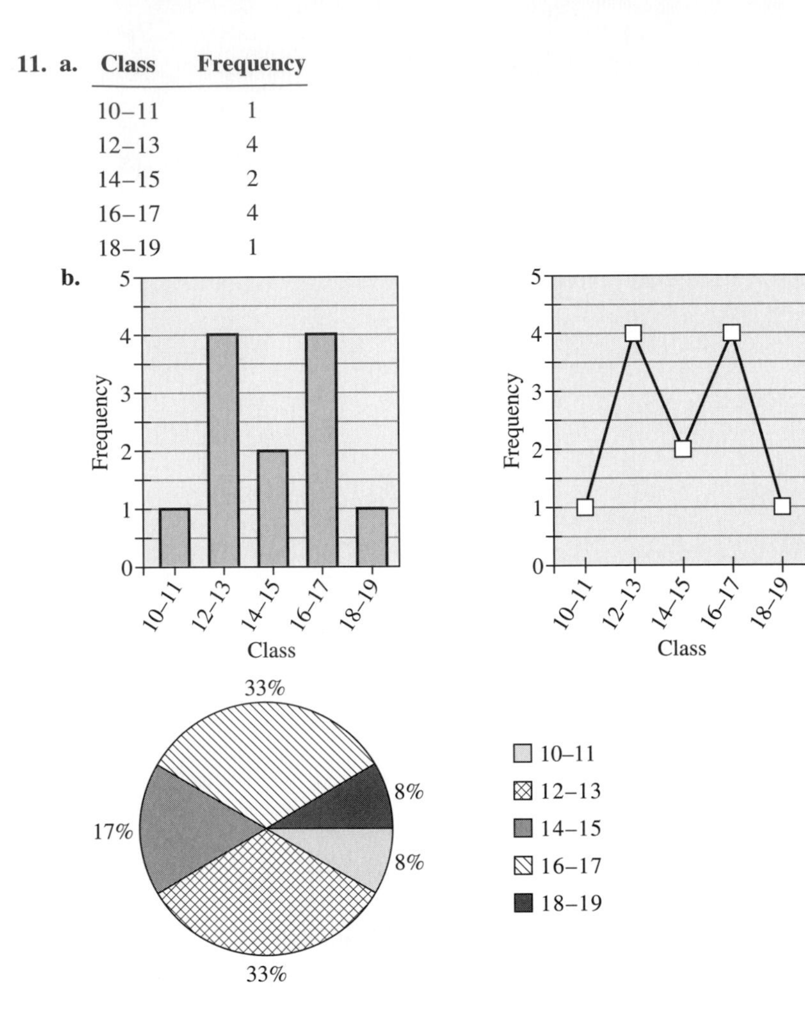


**13. a.**

| Class | Frequency |
|---|---|
| 40–49 | 1 |
| 50–59 | 0 |
| 60–69 | 3 |
| 70–79 | 6 |
| 80–89 | 6 |
| 90–99 | 4 |

**b.**

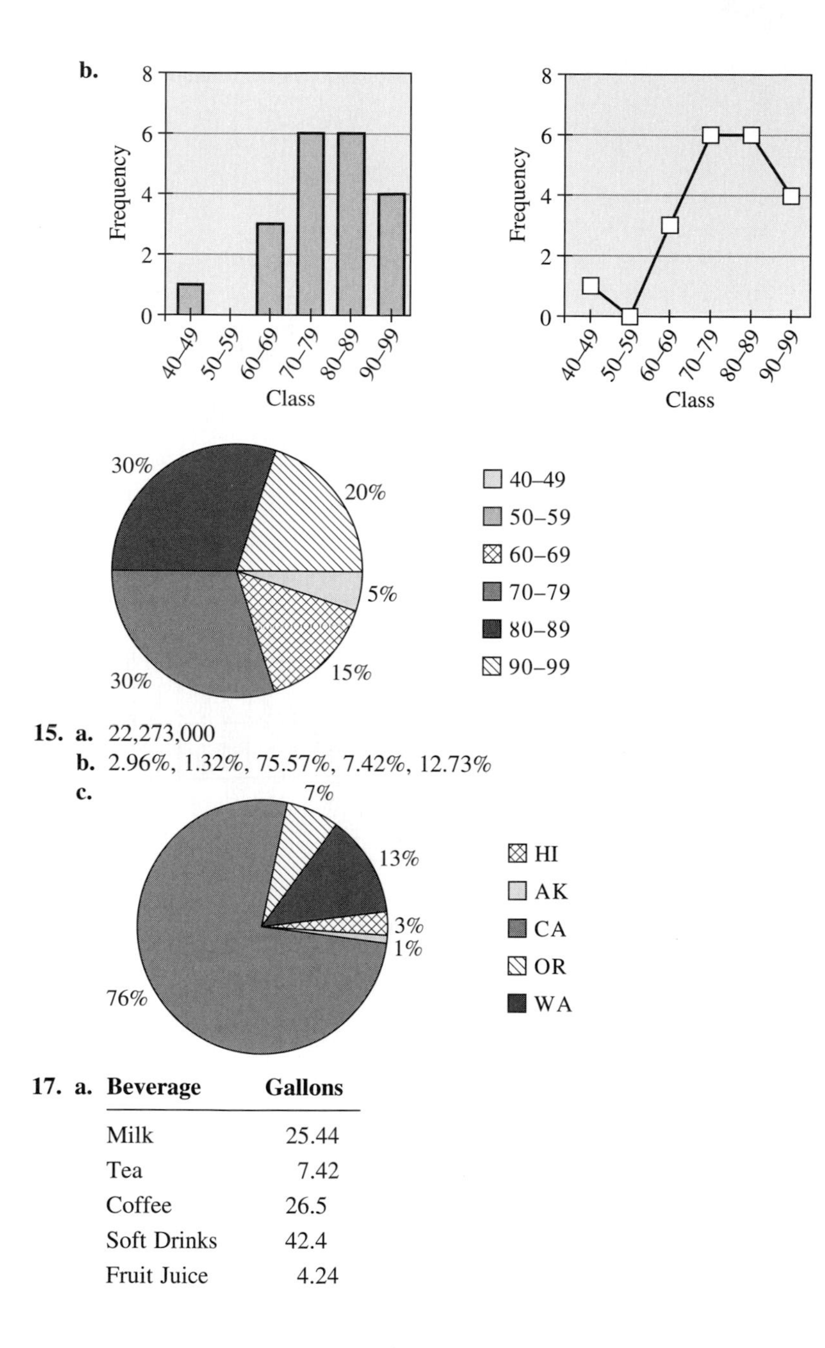


**15. a.** 22,273,000

**b.** 2.96%, 1.32%, 75.57%, 7.42%, 12.73%

**c.**

**17. a.**

| Beverage | Gallons |
|---|---|
| Milk | 25.44 |
| Tea | 7.42 |
| Coffee | 26.5 |
| Soft Drinks | 42.4 |
| Fruit Juice | 4.24 |

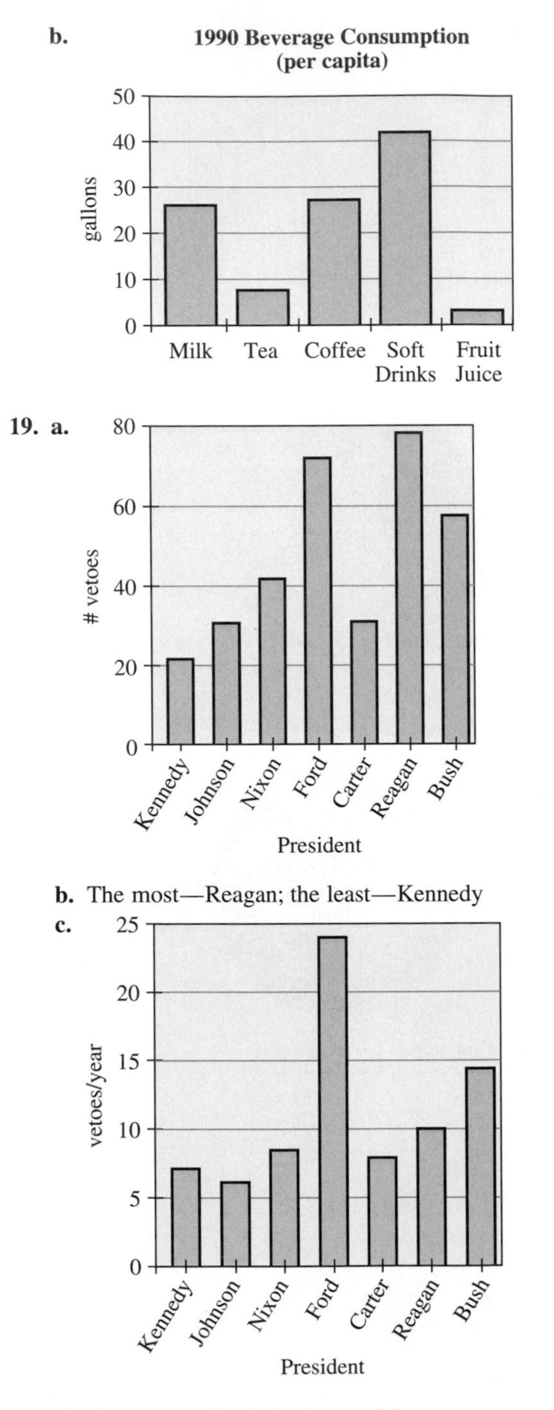


**b.** The most—Reagan; the least—Kennedy

**d.** The most—Ford; the least—Johnson

**f.**

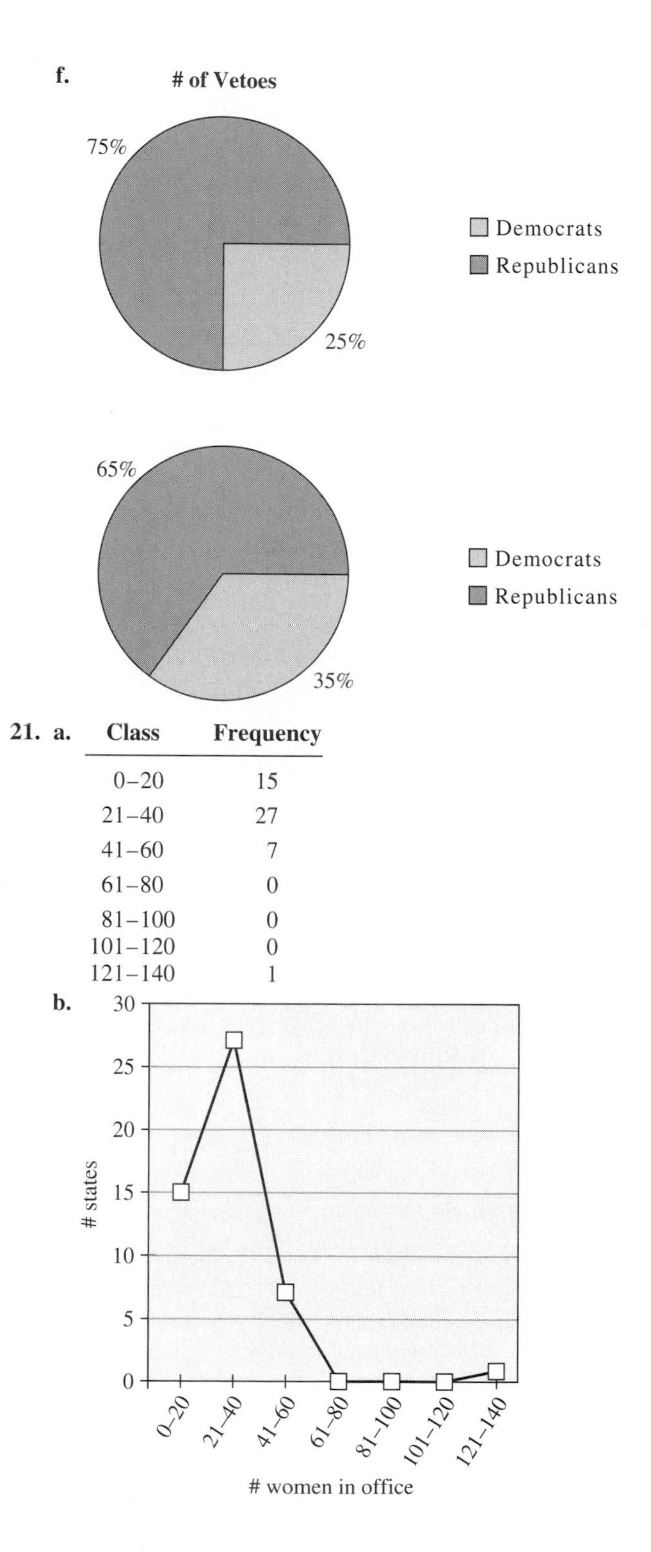


**21. a.**

| Class | Frequency |
|---|---|
| 0–20 | 15 |
| 21–40 | 27 |
| 41–60 | 7 |
| 61–80 | 0 |
| 81–100 | 0 |
| 101–120 | 0 |
| 121–140 | 1 |

**b.**

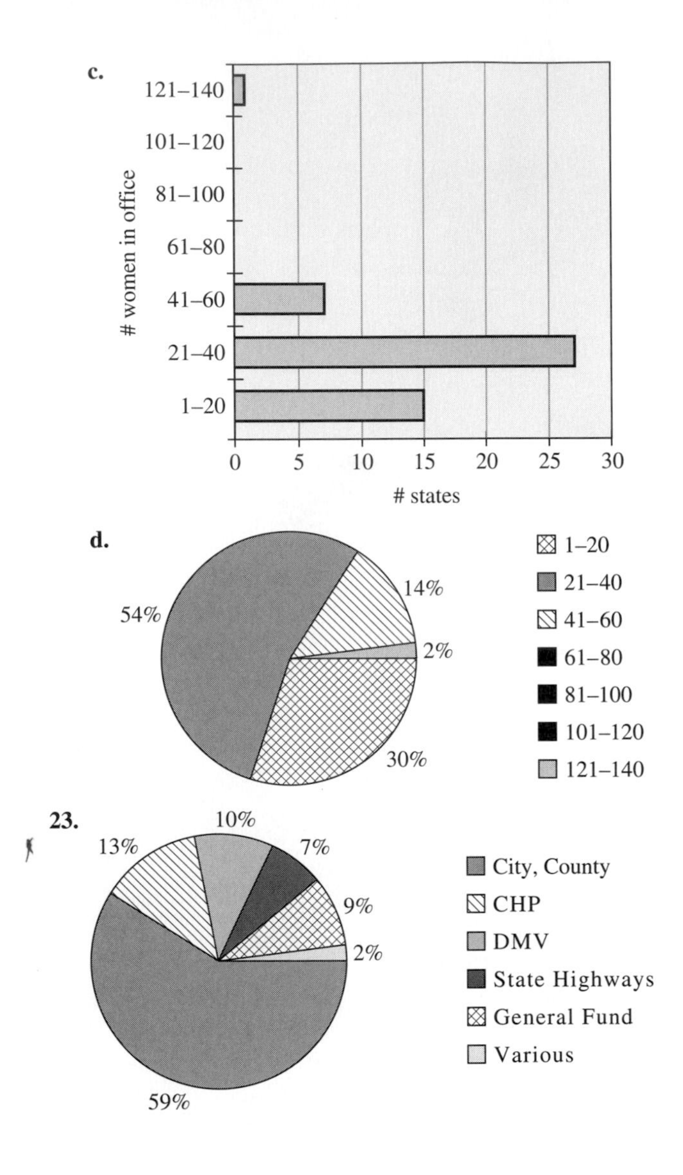


## *Section 8.2*

**11. a.** 5.8 **b.** 6.5 **c.** Bimodal with modes 2 and 9

**13. a.** 24.65 **b.** 25 **c.** 25

**15. a.** Sodium: mean = 797.3 mg, median = 747.5 mg
Cholesterol: mean = 34.8 mg, median = 32.5 mg

**b.** Tacos: sodium—mean = 430.25 mg, median = 415 mg
cholesterol—mean = 37 mg, median = 32 mg
Burritos: sodium—mean = 1131.2 mg, median = 1148 mg
cholesterol—mean = 36.8 mg, median = 33 mg

**17.** Mean = 81, median = 90, mode = 94

**19. a.**

| | Golf | Tennis | Bowling |
|---|---|---|---|
| Mean | $462,473.70 | $917,457.50 | $59,191.30 |
| Median | $416,925.00 | $623,851.50 | $55,666.00 |

**b.** Mean = $479,707.50, median = $408,301.00

**21. a.** 50.9 yr **b.** Median class = 45–64; modal class = 45–64

**23. a.** 25.5

**b.**

| Class | Frequency |
|---|---|
| 1–20 | 15 |
| 21–40 | 27 |
| 41–60 | 7 |
| 61–80 | 0 |
| 81–100 | 0 |
| 101–120 | 0 |
| 121–140 | 1 |

**c.** Modal group = 21–40 **d.** Mean = 29.3

## *Section 8.3*

**11. a.** 12 **b.** 4.38 **c.** 4.38

**d.** Since the new set of data is less spread out, the standard deviation is smaller.

**13. a.** 8 **b.** 2.45 **c.** 2.45

**15. a.** 56 **b.** 12.74 **c.** 12.74

**17.** Sodium—365 mg, cholesterol—14.6 mg

**19. a.** $4,038,000 **b.** $2,603,000

**c.** $3,500,000; $2,500,000, bimodal

**d.**

| Salary | No. of Players |
|---|---|
| $1–$1,000,000 | 2 |
| $1,000,001–$2,000,000 | 9 |
| $2,000,001–$3,000,000 | 14 |
| $3,000,001–$4,000,000 | 12 |
| $4,000,001–$5,000,000 | 3 |

**e.**

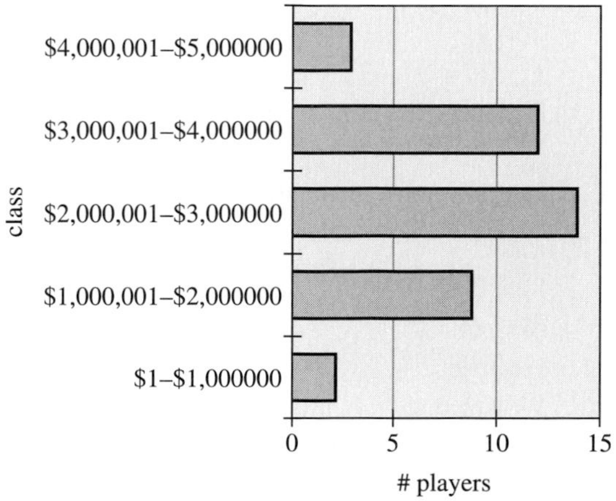

**f.** \$2,625,001 **g.** \$1,004,675

## Section 8.4

**11.** 0.3413 **13.** 0.7692 **15.** 0.1156 **17.** 0.1335
**19.** $c \approx -1.68$ **21.** 0.85 **23.** $c = -2.575$.
**25. a.** 0.1325 or 13.25% **b.** 0.5832 or 58.32%
**27. a.** 0.4090 **b.** 0.2964 or 29.64% **c.** 0.1158 or 11.58%
**29.** Approximately 48 cigarettes per day

## Section 8.5

**9.** $z = 1.28$ **11.** $z = 1.44$ **13.** 59%–67% **15.** 45.2%–50%
**17.** 4.2% **19.** 2.3% **21.** 99.97% **23.** 95.22%
**25.** 3385 **27.** 456
**29. a.** 94.52% **b.** 1037 **c.** 4145
**31. a.** 95.66% **b.** 77.8%–84.2%
**33. a.** 96.68% **b.** 461 **c.** 709
**35.** 98.18%
**37. a.** 3.68% **b.** No

## Section 8.6

**9.**

*Number of Seats Using Hamilton's Method*

| County | Population | Quotas | Initial No. of Seats | No. of Seats |
|---|---|---|---|---|
| Bristol | 48,859 | 4.87 | 4 | 5 |
| Kent | 161,135 | 16.06 | 16 | 16 |
| Newport | 87,194 | 8.69 | 8 | 9 |
| Providence | 596,270 | 59.42 | 59 | 59 |
| Washington | 110,006 | 10.96 | 10 | 11 |
| Totals | 1,003,464 | 100.00 | 97 | 100 |

**11.**

*Number of Seats Using Jefferson's Method*

| | | Divisor = 10,035 | | Divisor = 9774 | |
|---|---|---|---|---|---|
| County | Population | Quota | No. of Seats | Pop/Divisor | No. of Seats |
| Bristol | 48,859 | 4.87 | 4 | 4.999 | 4 |
| Kent | 161,135 | 16.06 | 16 | 16.49 | 16 |
| Newport | 87,194 | 8.69 | 8 | 8.92 | 8 |
| Providence | 596,270 | 59.42 | 59 | 61.01 | 61 |
| Washington | 110,006 | 10.96 | 10 | 11.25 | 11 |
| Totals | 1,003,464 | 100.00 | 97 | | 100 |

**13.**

*Number of Seats Using Webster's Method*

| County | Population | Divisor = 16,248<br>Quota | No. of Seats |
|---|---|---|---|
| Kent | 110,993 | 6.83 | 7 |
| New Castle | 441,946 | 27.20 | 27 |
| Sussex | 113,229 | 6.97 | 7 |
| Totals | 666,168 | 41.00 | 41 |

**15.**

*Number of Courses Using Hamilton's Method*

| Course | No. of Students | Quotas | Initial No. of Courses | No. of Courses |
|---|---|---|---|---|
| College Algebra | 120 | 4.00 | 4 | 4 |
| Calculus | 80 | 2.67 | 2 | 3 |
| Math for Business | 235 | 7.83 | 7 | 8 |
| Liberal Arts Math | 165 | 5.50 | 5 | 5 |
| Totals | 600 | 20.00 | 18 | 20 |

**17.**

*Number of Courses Using Jefferson's Method*

| | | Divisor = 30 | | Divisor = 27 | |
|---|---|---|---|---|---|
| Course | No. of Students | Quota | No. of Courses | New quota | No. of Seats |
| College Algebra | 120 | 4.00 | 4 | 4.44 | 4 |
| Calculus | 80 | 2.67 | 2 | 2.96 | 2 |
| Math for Business | 235 | 7.83 | 7 | 8.70 | 8 |
| Liberal Arts Math | 165 | 5.50 | 5 | 6.11 | 6 |
| Totals | 600 | 20.00 | 18 | | 20 |

**19.**

*Number of Trucks Using Webster's Method*

| District | Population | Divisor = 544<br>Quota | No. of Trucks |
|---|---|---|---|
| Bayview | 25,456 | 46.80 | 47 |
| Downtown | 32,723 | 60.16 | 60 |
| East End | 27,568 | 50.69 | 51 |
| Greenhaven | 16,475 | 30.29 | 30 |
| Ingleside | 8,696 | 15.99 | 16 |
| Riverview | 11,458 | 21.07 | 21 |
| Totals | 122,376 | 225.00 | 225 |

*Number of Sites Using Hamilton's Method*

| District | Population | Quotas | Inital No. of Sites | No. of Sites |
|---|---|---|---|---|
| 1 | 90,000 | 90.91 | 90 | 91 |
| 2 | 166,000 | 167.68 | 167 | 168 |
| 3 | 102,000 | 103.03 | 103 | 103 |
| 4 | 131,000 | 132.32 | 132 | 132 |
| 5 | 147,000 | 148.48 | 148 | 148 |
| 6 | 156,000 | 157.58 | 157 | 158 |
| Totals | 792,000 | 800.00 | 797 | 800 |

**23.**

*Number of Sites Using Jefferson's Method*

| District | Population | Pop./Divisor | No. of Sites |
|---|---|---|---|
| 1 | 90,000 | 91.19 | 91 |
| 2 | 166,000 | 168.19 | 168 |
| 3 | 102,000 | 103.34 | 103 |
| 4 | 131,000 | 132.73 | 132 |
| 5 | 147,000 | 148.94 | 148 |
| 6 | 156,000 | 158.05 | 158 |
| Totals | 792,000 | | 800 |
| Divisor | 987 | | |

**25.**

*Number of Days Using Webster's Method*

| District | Voters | Voters/Divisor | No. of Days |
|---|---|---|---|
| 1 | 167,000 | 26.02 | 26 |
| 2 | 116,000 | 18.07 | 18 |
| 3 | 237,000 | 36.93 | 37 |
| 4 | 231,000 | 35.99 | 36 |
| 5 | 215,000 | 33.50 | 33 |
| Totals | 966,000 | | 140 |
| Divisor | 6,418 | | |

## Chapter 8 Summary

**1. a.** 4.9 **b.** 5 **c.** 3 and 6 **d.** 8 **e.** 2.55

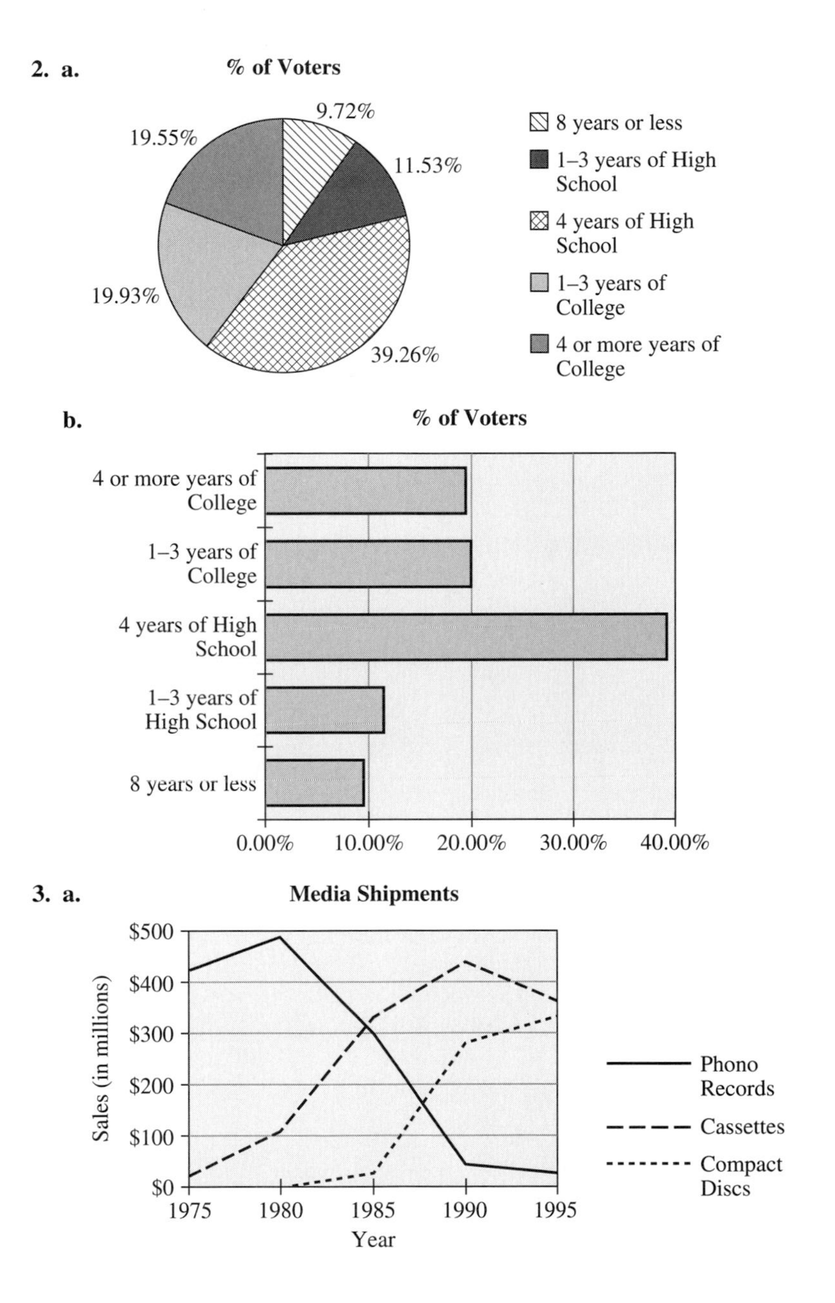
2. a.
% of Voters
9.72%
11.53%
39.26%
19.93%
19.55%
8 years or less
1–3 years of High School
4 years of High School
1–3 years of College
4 or more years of College
b.
% of Voters
4 or more years of College
1–3 years of College
4 years of High School
1–3 years of High School
8 years or less
0.00%
10.00%
20.00%
30.00%
40.00%
3. a.
Media Shipments
$500
$400
$300
$200
$100
$0
Sales (in millions)
1975
1980
1985
1990
1995
Year
Phono Records
Cassettes
Compact Discs

**4. a.** Range = 144,000, median = 109,500, mode = 98,000

| Voters in District | No. of Districts |
|---|---|
| 30,001–60,000 | 7 |
| 60,001–90,000 | 5 |
| 90,001–120,000 | 6 |
| 120,001–150,000 | 10 |
| 150,001–180,000 | 5 |
| 180,001–210,000 | 1 |

**b.**

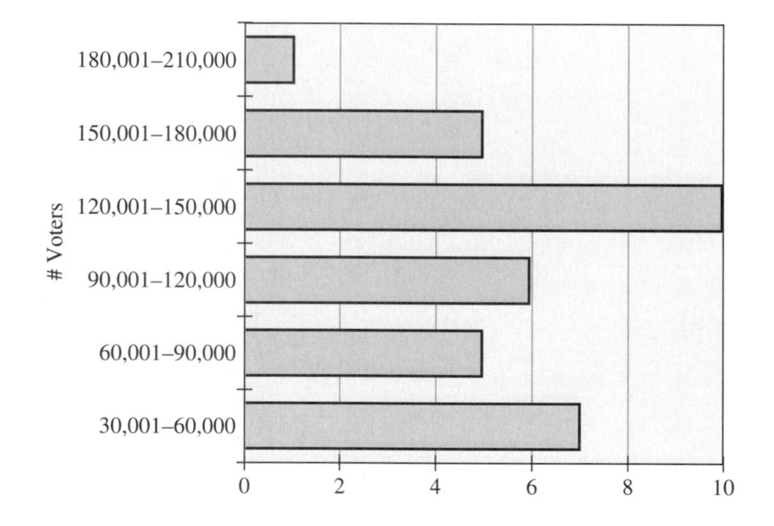


**c.** 108,530 **d.** 43,513

**5.** Company A because of less variation in the product

**6. a.** 36.24% **b.** 11.22% **c.** 0.6217 **d.** 0.2483

**7.** *A*'s: 91.2–100
*B*'s: 80.4–91.19
*C*'s: 65.6–80.39
*D*'s: 54.8–65.59
*F*'s: 0–54.79

**8. a.** 47%–55% **b.** 95.96% **c.** 1037 **d.** 4145

**9. a.**

*Number of Seats Using Hamilton's Method*

| County | Population | Quotas | Initial No. of Seats | No. of Seats |
|---|---|---|---|---|
| Kent | 110,993 | 3.50 | 3 | 3 |
| New Castle | 441,946 | 13.93 | 13 | 14 |
| Sussex | 113,229 | 3.57 | 3 | 4 |
| Totals | 666,168 | 21.00 | 19 | 21 |

**b.**

*Number of Seats Using Webster's Method*

| County | Population | Divisor = 31,722<br>Quota | No. of Seats |
|---|---|---|---|
| Kent | 110,993 | 3.50 | 3 |
| New Castle | 441,946 | 13.93 | 14 |
| Sussex | 113,229 | 3.57 | 4 |
| Totals | 666,168 | 21.00 | 21 |

**c.**

*Number of Seats Using Jefferson's Method*

| | | Divisor = 31,722 | | Divisor = 29,463 | |
|---|---|---|---|---|---|
| County | Population | Quota | No. of Seats | Pop./Divisor | No. of Seats |
| Kent | 110,993 | 3.50 | 3 | 3.77 | 3 |
| New Castle | 441,946 | 13.93 | 13 | 15.00 | 15 |
| Sussex | 113,229 | 3.57 | 3 | 3.84 | 3 |
| Totals | 666,168 | 21.00 | 19 | | 21 |

## Chapter 9

### Section 9.1

**11.**

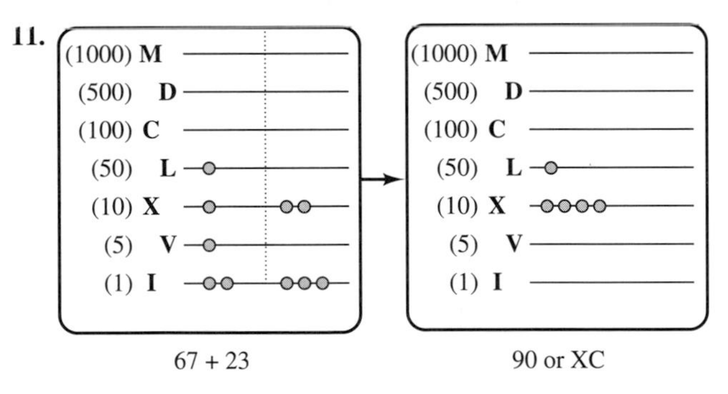


67 + 23 90 or XC

**13.**

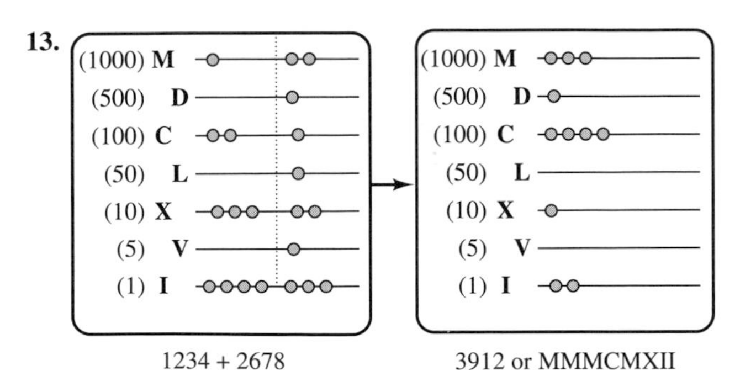


1234 + 2678 3912 or MMMCMXII

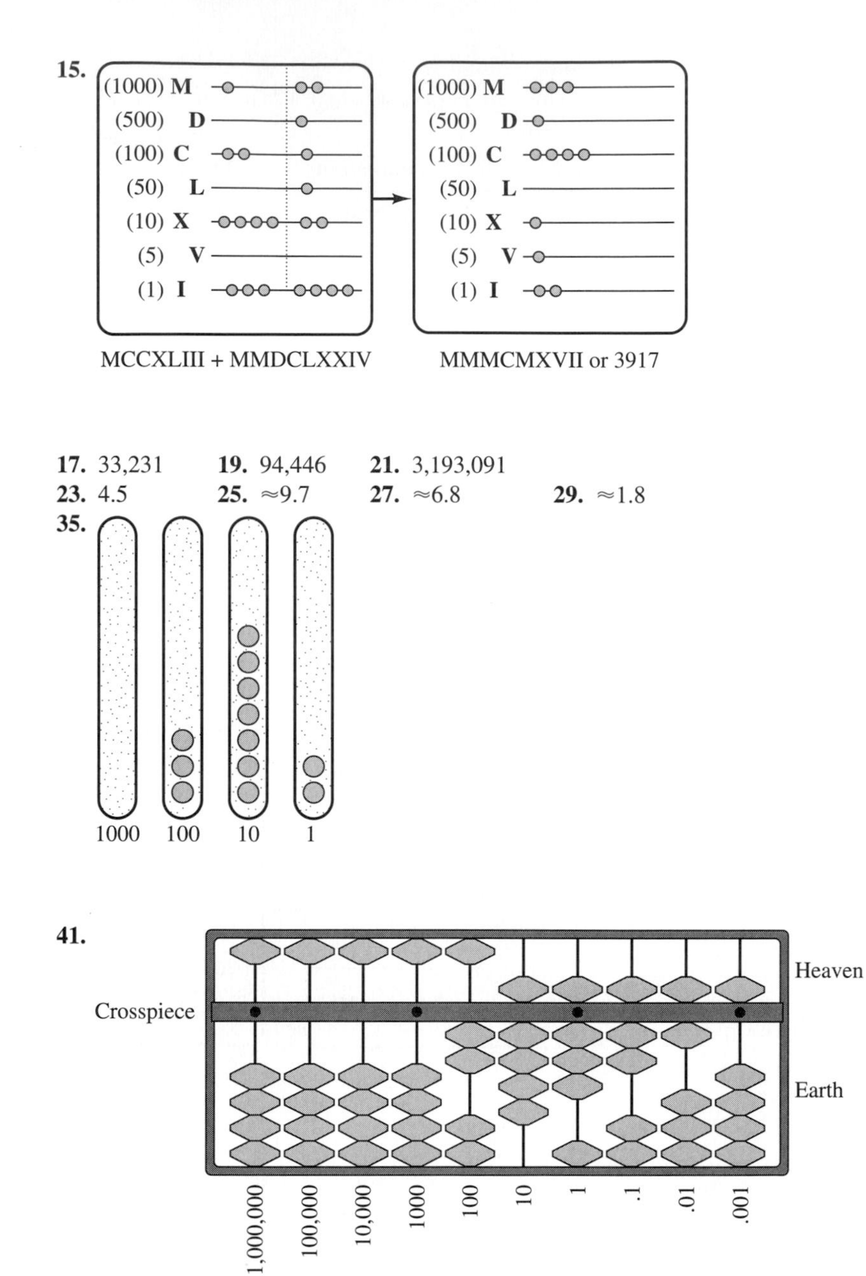


**17.** 33,231 **19.** 94,446 **21.** 3,193,091

**23.** 4.5 **25.** ≈9.7 **27.** ≈6.8 **29.** ≈1.8

**43.**

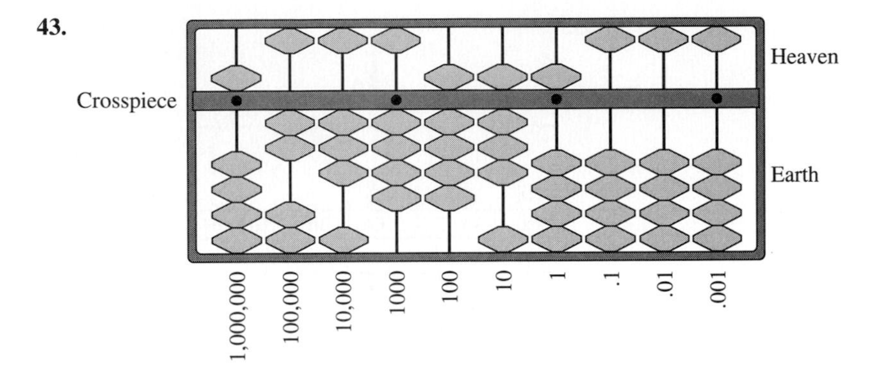


**49.** The ten-thousand-mile wheel has made three-tenths of a rotation.
The thousand-mile wheel has made three rotations.
The hundred-mile wheel has made thirty rotations.
The tens of miles wheel has made three hundred rotations.
The mile wheel has made three thousand rotations.
The tenths of a mile wheel has made thirty thousand rotations.

## Section 9.2

**19.** **a.** $25_8$ **b.** $15_{16}$
**21.** **a.** $144_8$ **b.** $64_{16}$
**23.** **a.** $5656_8$ **b.** $BAE_{16}$
**25.** $1\ 100\ 101_2$ **27.** $100\ 101_2$ **29.** $101\ 011_2$ **31.** $1\ 001\ 001_2$
**33.** $1\ 100\ 011\ 110_2$ **35.** $100\ 000\ 100_2$ **37.** $110\ 001\ 110\ 001_2$ **39.** To be or not?
**41.** 0100 1000, 0100 1111, 0101 0000, 0100 0101
**43.** 0100 1101, 0110 1111, 0110 1101

## Section 9.3

**9.** 0.2 **11.** 0.037037 **13.** 48 **15.** 51.709
**17.** 12.513503 **19.** 0.386883 **21.** 1.055751 **23.** 1.181352
**25.** $10^{13}$ **27.** $4.7 \times 10^{-9}$ **29.** $-0.033333$ **31.** 0.580952
**33.** 238,112.87; $a^m \times a^n = a^{m+n}$
**35.** 306.01968; $\frac{a^m}{a^n} = a^{m-n}$
**37.** 3.3233702; $\log(a + b) = \log a + \log b$
**39.** An error message such as **-E-** is displayed. Division by zero is not defined.
**41.** An error message such as **-E-** is displayed. It is not possible to find the common logarithm of zero.

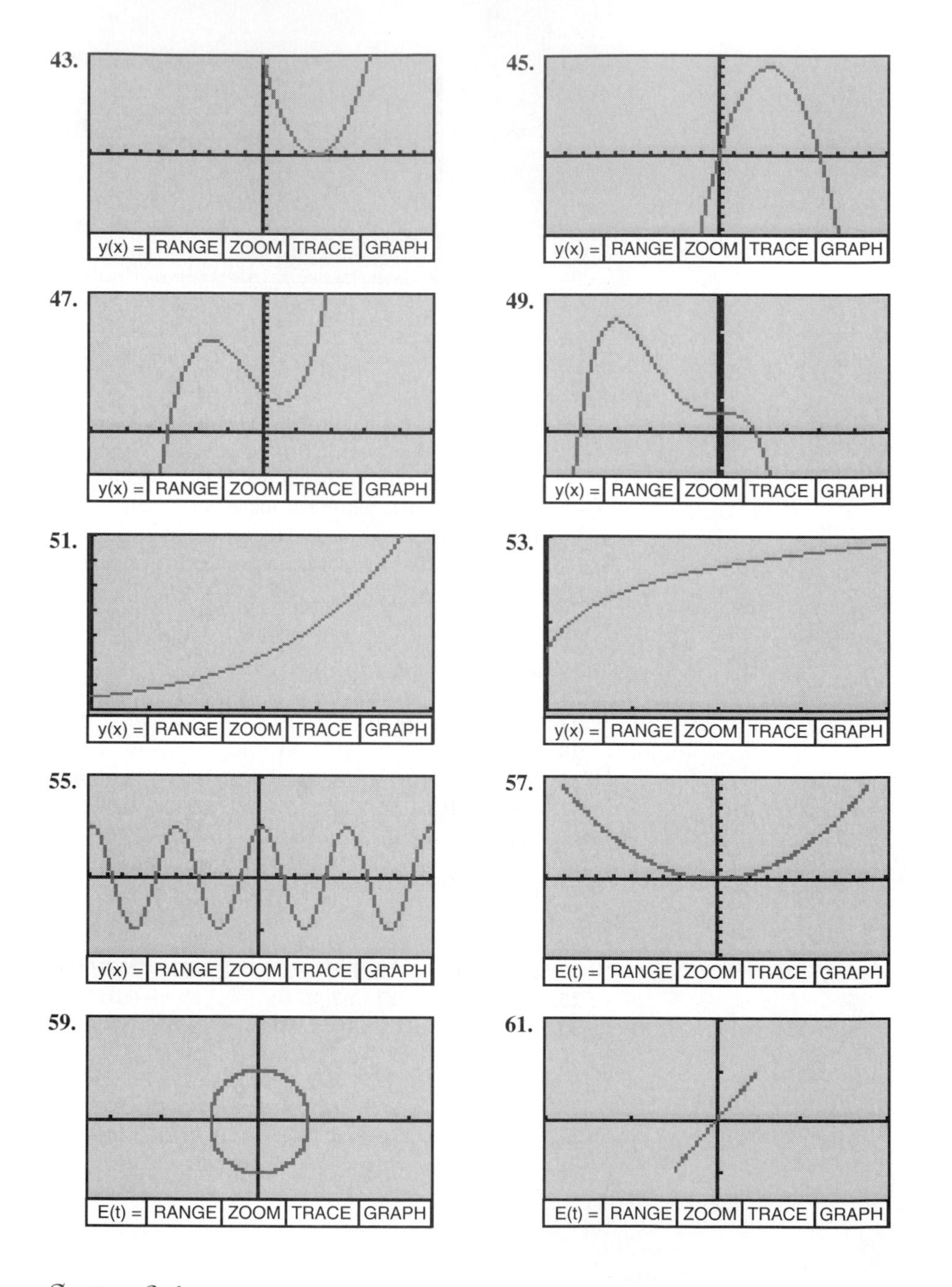


## *Section 9.4*

**9.** Mortgage payments, income taxes, gradesheets, car service records

**11.** ''There'' is correctly spelled but incorrectly used.

**13. a.** Multiply the amount in B2 by the amount in C2.
**b.** B2*C2 (note: * is used for multiplication)
**c.** Add the amounts in D2 through D7.
**d.** D2 + D3 + D4 + D5 + D6 + D7

**15.** Because the writer makes extensive use of word processing, he/she should purchase a good word processing package. If this package does not contain communications or graphics features, then integrated software with those two features or separate software packages are in order.

**17.** An integrated package with word processing, graphics, and database features would allow you to create the text and the drawings for the newsletter, and the database feature would allow you to create a mailing list.

**19.** A database program or an integrated package that includes a database would be the best program for this use. It would allow you to search for recipes of a certain type.

**21.** Research required. **23.** Research required.

**25–31.** Access to a computer and certain software is required.

## *Chapter 9 Summary*

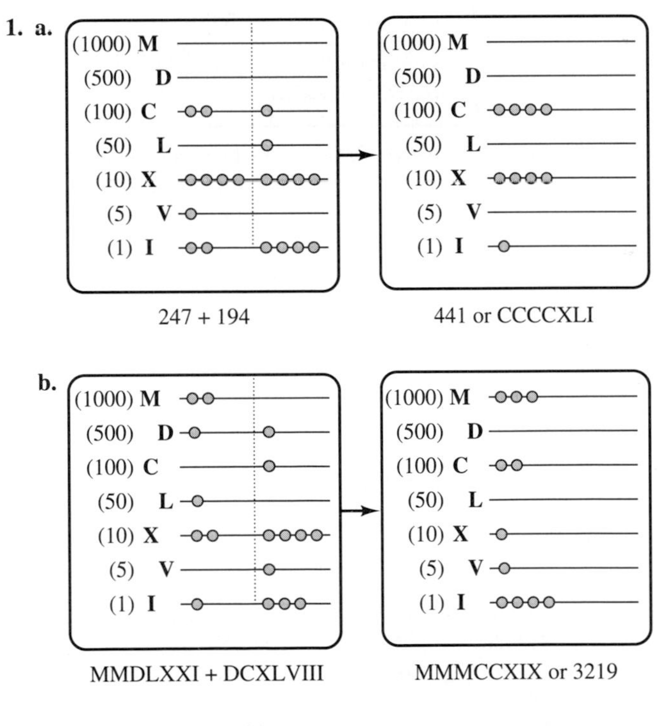


**2. a.** 1862 **b.** 27,783

**3. a.** ≈6.8 **b.** ≈8.9 **c.** ≈1.9

**4.** Transistors, integrated circuits, microprocessors. See text of Section 9.2 for the history.

**5.** Binary numbers have two values, 1 and 0. The *on* and *off* of electronic circuitry can be easily represented by these values.

**6.** 0100 1010, 0111 0101, 0111 0011, 0111 0100, 0010 0000, 0110 0100, 0110 1111, 0010 0000, 0110 1001, 0111 0100, 0010 0001

**7.** $563_8$, $173_{16}$

**8.** **a.** $111\ 011\ 010_2$ **b.** $11\ 100\ 111_2$

**9.** **a.** 0.1528 **b.** 2443.6261 **c.** 1.3662
**d.** 76,120,000,000 or $7.612 \times 10^{10}$

**10.** **a.**

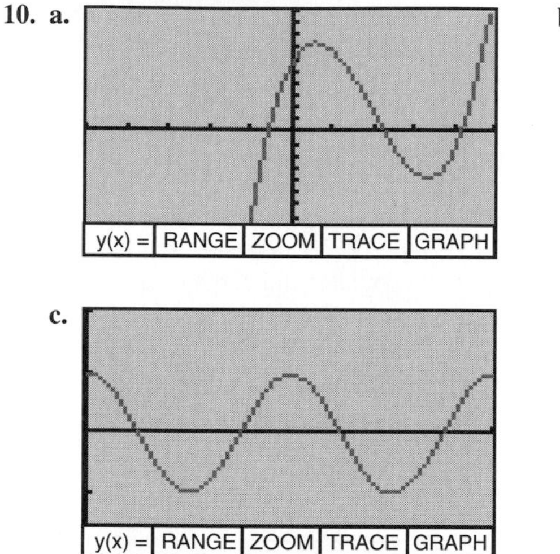


**b.**

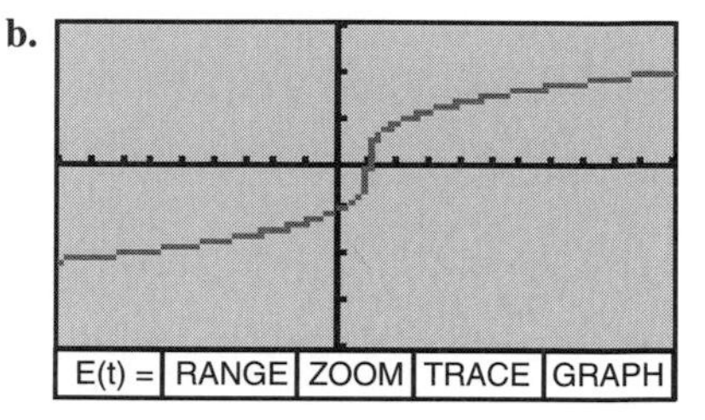


**c.**

**11.** **a.** All the words are spelled correctly but some are incorrectly used. A correct version would be "I have four pairs of sandals for you."
**b.** As shown in part (a), "for pears" should be changed to "four pairs." Thus, the grammar checker did not correct all the errors.

**12.** **a.** Word processor **b.** Spreadsheet **c.** Database
**d.** Communications software **e.** Graphics software **f.** Database
**g.** Wordprocessor **h.** Database **i.** Educational software

# Chapter 10

## *Section 10.0*

**13.** **a.** 8 **b.** $-15$ **c.** $-12$
**15.** **a.** 8 **b.** $3M - 6$ **c.** $M^2 + 7M$
**17.** **a.** 8 **b.** $-3 + 3h$ **c.** $8 + 9h + h^2$
**19.** **a.** 8 **b.** $3x - 3h - 6$ **c.** $x^2 - 2xh + h^2 + 7x - 7h$
**21.** Yes **23.** Yes **25.** Yes **27.** No
**29.** No **31.** Yes **33.** No
**35.** **a.** 2 **b.** $-1/3$
**37.** **a.** No **b.** Yes

**39.** $m = \dfrac{f(x + h) - f(x)}{h}$

**41.** **a.** 12 **b.** $12 + 12h + 3h^2$ **c.** $12h + 3h^2$ **d.** $12 + 3h$
**43.** **a.** $x^2 + 2xh + h^2$ **b.** $2xh + h^2$ **c.** $2x + h$

## Section 10.1

**11.** **a.** B, D, F **b.** C, G **c.** A, E
**13.** At $x = 3$, the derivative has a value between $-3$ and $-1$.
**15.** **a.** 5 **b.** 5
**17.** **a.** $-4/3$ **b.** $-4/3$
**19.** 0 **21.** $-2$ **23.** 6 **25.** $4x$ **27.** $2x$
**29.** $-4x + 3$ **31.** 1 **33.** 2 **35.** $-6$
**37.** **a.** $2x$ **b.** $3x^2$ **c.** $4x^3$ **d.** $23x^{22}$

## Section 10.2

**9.** 2 **11.** $-2x$ **13.** $9x^2 - 5$ **15.** $4x^3 + 3x^2 + 2x + 1$
**17.** $20x^4$ **19.** $20x^4 - 10x$ **21.** $5x^4 - 6x + 7$
**23.** **a.**

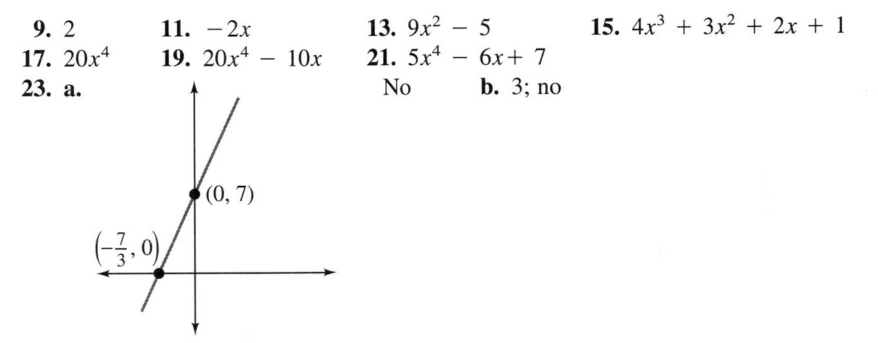


No **b.** 3; no

**25.** **a.**

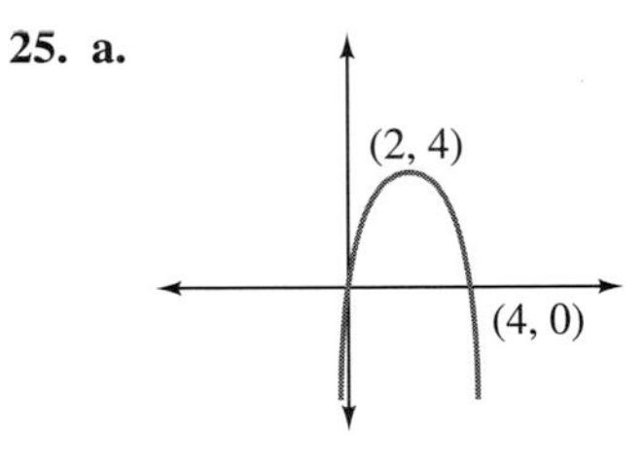


Yes **b.** $-2x + 4$; $x = 2$ **c.** Maximum

**27.** **a.** (−1, 4) (1, −4) **b.** $6x^2 - 6$ **c.** $x = -1$; $x = 1$

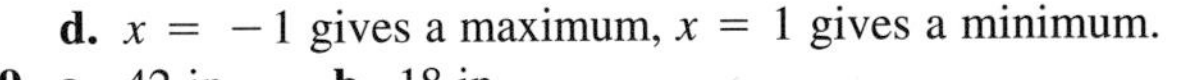

**d.** $x = -1$ gives a maximum, $x = 1$ gives a minimum.
**29.** **a.** 42 in. **b.** 18 in.
**31.** 150 ft by 300 ft; area = 45,000 sq ft
**33.** 2 ft; 4 ft; 2 2/3 ft, $V = 64/3$ cu ft
**35.** **a.** 20,000 **b.** \$16.20

## Section 10.3

**5.** 144 **7.** 96 **9.** 41.25 **11.** 292
**13.** 9750 sq ft **15. b.** 62 **17. b.** $\approx 1.35$

## Chapter 10 Summary

**1. a.** 22 **b.** $2m^2 - 3m + 2$ **c.** $2h^2 + 9h + 11$
**d.** $2x^2 + 4xh + 2h^2 - 3x - 3h + 2$
**2. a.** 1 **b.** $3w^2 + 5w - 1$ **c.** $3h^2 + 35h + 99$
**d.** $3x^2 + 6xh + 3h^2 + 5x + 5h - 1$
**3.** 3 **4.** $3x^2 + 6$ **5.** 23 **6.** $-11$
**7.** $(-2, 18)$ **8.** $(3, -27)$ **9.** 300 sq ft **10.** 375,000 sq ft
**11.** 176 **12.** 114 **13.** 7 lbs

# Index